44

Springer Series in Chemical Physics

Edited by Robert Gomer

Springer Series in Chemical Physics

Editors: V. I. Goldanskii R. Gomer F. P. Schäfer J. P. Toennies

Volumes 1–39 are listed on the back inside cover

Secondary Ion Mass Spectrometry

SIMS V

Proceedings of the
Fifth International Conference, Washington, DC,
September 30 – October 4, 1985

Editors:
A. Benninghoven, R. J. Colton, D. S. Simons,
and H. W. Werner

With 388 Figures

Springer-Verlag
Berlin Heidelberg New York Tokyo

Professor Dr. Alfred Benninghoven
Physikalisches Institut,
Universität Münster, Domagkstraße 75,
D-4400 Münster, Fed. Rep. of Germany

Dr. Richard J. Colton
Naval Research Laboratory,
Chemistry Division,
Washington, DC 20375, USA

Dr. David S. Simons
National Bureau of Standards,
Center for Analytical Chemistry,
Gaithersburg, MD 20899, USA

Dr. Helmut W. Werner
Philips Research Laboratories,
NL-Eindhoven, The Netherlands

International Organizing Committee: A. Benninghoven *(Chairman)*, V.T. Cherepin, C.A. Evans, Jr., K.F.J. Heinrich, J. Okano, G. Slodzian, H.W. Werner

Honorary Chairman: R.F.K. Herzog

Local Organizing Committee: R.J. Colton *(Chairman)*, D.S. Simons *(Vice-Chairman)*, W.M. Bone, A.M. Boone, B.I. Dunlap, R.B. Freas, L.L. Hollingsworth, D.A. Kidwell, M.M. Ross, E.G. Shafrin, J.A. Watson, J.R. Wyatt

Program Committee: D.S. Simons *(Chairman)*, J.E. Campana, W.H. Christie, J.C. Huneke, R.W. Linton, C.W. Magee, N. Winograd, M.L. Yu

Sponsoring Companies: Atomika; Cambridge Mass Spectrometry Ltd.; CAMECA Instruments, Inc.; Charles Evans & Associates; Extranuclear; FEI Company; General Ionex Corp.; Geo-Centers, Inc.; Instruments SA, Inc., Riber Division; Kratos Analytical; Leybold-Heraeus; Perkin-Elmer, Physical Electronics Division; Physicon Corp.; VG Instruments, Inc.; Walter C. McCrone Associates, Inc.

Sponsoring Societies and Organizations: American Society for Mass Spectrometry, Microbeam Analysis Society, Office of Naval Research

---

ISBN-13:978-3-642-82726-6 e-ISBN-13:978-3-642-82724-2
DOI: 10.1007/978-3-642-82724-2

Softcover reprint of the hardcover 1st edition 1986

2153/3150-543210

Preface

This volume contains the proceedings of the Fifth International Conference on Secondary Ion Mass Spectrometry (SIMS V), held at the Capitol Holiday Inn, Washington, DC, USA, from September 30 to October 4, 1985. The conference was the fifth in a series of conferences held biennially. Previous conferences were held in Münster (1977), Stanford (1979), Budapest (1981), and Osaka (1983).

SIMS V was organized by Dr. *R.J. Colton* of the Naval Research Laboratory and Dr. *D.S. Simons* of the National Bureau of Standards under the auspices of the International Organizing Committee chaired by Prof. *A. Benninghoven* of the Universität Münster.

Dr. Richard F.K. Herzog served as the honorary chairman of SIMS V. While Dr. Herzog is best known to the mass spectrometry community for his theoretical development of a mass spectrometer design, known as the Mattauch-Herzog geometry, he also made several early and important contributions to SIMS. In 1949, Herzog and Viehbock published a description of the first instrument designed to study secondary ions produced by bombardment from a beam of ions generated in a source that was separated from the sample by a narrow tube. Later at the GCA Corporation, he brought together a team of researchers including H.J. Liebl, F.G. Rüdenauer, W.P. Poschenrieder and F.G. Satkiewicz, who designed and built, and carried out applied research with the first commercial ion microprobe.

We also acknowledge the valuable assistance of a number of colleagues and organizations, listed on the previous page, who contributed much of their time or financial support to SIMS V. The success of this conference was due largely to the efforts of the individuals who served on local organizing and program committees and to the organizations that sponsored SIMS V. We owe them a great deal of thanks.

The conference was attended by about 300 participants from 12 countries. There were 163 papers presented in 21 sessions that included 9 invited papers, 2 invited symposia featuring 13 papers, and 141 contributed oral and poster papers. The proceedings contains 149 separate papers and focuses on 13 main topics of the conference: (1) Fundamentals of sputtering and ion formation; (2) detection of sputtered neutrals; (3) detection limits and quantification methods; (4) recent progress in instrumentation; (5) techniques closely related to conventional SIMS such as accelerator-based SIMS and laser microprobe mass spectrometry; (6) SIMS combined with other techniques such as X-ray photoelectron spectroscopy, Auger electron spectroscopy, or ion scattering spectrometry and applications to surface studies; (7) ion microscopy and image analysis; (8) depth profiling and semiconductor applications; (9) metallurgical applications; (10) biological applications; (11) geological applications; (12) particle-induced emission from organics; and (13) organic applications, including fast atom bombardment mass spectometry. In addition, Dr. R. Honig of RCA Laboratories presents a personal view of the history of secondary ion mass spectrometry.

To encourage care in poster preparation, a competition was held on each of the two poster days. Judging was based on a combination of visual appearance, organization, and technical content. The first-, second-, and third-place prize winners were:

1a. M. Ross, J. Campana, R. Colton, and D. Kidwell: "Characteristics of Ion Emission in Desorption Ionization Mass Spectrometry" (Pt. 14, Paper 4).
1b. M. Dowsett, R. King, H. Fox, and E. Parker: "Computer Aided Design of Primary and Secondary Ion Optics for a Quadrupole SIMS Instrument" (Pt. 5, Paper 15).
2a. T. Krishnamurthy, L. Szafraniec, E. Sarver, D. Hunt, S. Missler and W. Carmichael: "Amino-Acid Sequencing of Norwegian Fresh Water Blue-Green Algal (*Microcystis aeruginosa*) Peptide by FAB-MS/MS Technique" (Pt. 14, Paper 13).
2b. J. Gavrilovic: "Depth Resolution in Profiling of Thin GaAs-GaAlAs Layers" (Pt. 9, Paper 26).
3a. S. Tomita, K. Okuno, F. Soeda, and A. Ishitani: "Applications of SIMS Technique to Industrially Used Organic Materials" (Pt. 14, Paper 17).
3b. R. Day, M. Waters, and J. Rasile: "Incorporation of Chromium in Sputtered Copper Films and Its Removal During Wet Chemical Etching" (Pt. 10, Paper 5).

SIMS V also featured two special half-day symposia that discussed recent progress in two very active research areas. The symposium on the "Detection of Sputtered Neutrals" (Pt. 3) discussed the various post-ionization techniques such as hot electron gases and glow discharges, electron beams, and resonant and nonresonant multi-photon ionization used to detect sputtered particles. The efficiencies of the various post-ionization techniques are compared to SIMS in the review by Dr. W. Reuter of IBM (Pt. 3, Paper 6).

The symposium on "Particle-Induced Emission from Organics" (Pt. 13) compared the ion formation processes of several "soft" ionization techniques that use different excitation sources such as fission fragments or ion, neutral, or photon beams to desorb molecular species from organic materials. These techniques are used extensively to determine the structure and molecular weight of biomolecules and other high-mass species.

Finally, on behalf of the conference, we would like to thank all of the participants who helped to make SIMS V a successful and memorable conference.

Washington, DC
January 1986

R.J. Colton
(Chairman of Organization for SIMS V)

Contents

* Invited Paper

Part III **Symposium: Detection of Sputtered Neutrals**

Part VIII Ion Microscopy and Image Analysis

Part IX Depth Profiling and Semiconductor Applications

Part XIV Organic Applications Including Fast Atom Bombardment Mass Spectrometry

Part I

Retrospective

The Growth of Secondary Ion Mass Spectrometry (SIMS): A Personal View of Its Development

R.E. Honig

RCA Laboratories, Princeton, NJ 08540, USA

1. Introduction

"Those who do not REMEMBER the past
are condemned to RELIVE it"

Santayana

On September 1, 1985, Secondary Ion Mass Spectrometry (SIMS) celebrated its 75th birthday, a milestone that provides us with a good reason to take a close look at the high points in its history. Furthermore, the workers who are engaged today in this active field constitute already the fourth generation of scientists, thus it is appropriate for all of us to pay heed to Santayana's wisdom quoted above, in order to avoid reinventing the wheel.

This historical review is written from a personal, subjective standpoint, but makes every effort to bring out the highlights in an even-handed manner. Reviewing this field is no easy task because at this time there exists no simple descriptor to define it adequately in its entirety. A computer search quickly shows that of some five hundred papers published in 1984 only about half can be retrieved by using the key words "Secondary Ion Mass Spectrometry (SIMS)", while most of the remainder may respond to the non-descriptive label "Fast Atom Bombardment (FAB)". Yet neither term includes, for example, "Secondary Neutral Mass Spectrometry (SNMS)", a recently developed important sub-group. Many attempts have been made to resolve this quandary, and over the years some 30 different, inadequate acronyms have been coined by as many workers [1]. Nevertheless, the subject matter of this review must be clearly defined at this point. It is the mass spectrometry of secondary particles, either charged or neutral, which are emitted when solids or liquids absorb energy from primary particles, including photons. Since it is impractical to rename the SIMS V Conference or the present review in this all-encompassing cumbersome fashion, we will continue to use the term "SIMS" <u>in the broadest sense</u>.

2. Ancient History

Not only did Sir Joseph John THOMSON (1856-1940) discover the electron in 1897, but he was also the first to observe and identify correctly in 1910 [2] positive secondary ions produced when primary ions bombard a metal surface in a discharge tube (Fig. 1). The following decade was dominated by World War I which severely curtailed and delayed all research in this area. Another decade passed before Dempster's laboratory in Chicago became interested in "reflected" ions [3], which were later correctly identified as negative secondary ions by WOODCOCK [4] and THOMPSON [4a]. In 1931, they used one of Dempster's 180° mass spectrometers to record the first known negative secondary ion spectra (Fig. 2), obtained when NaF and CaF_2 targets

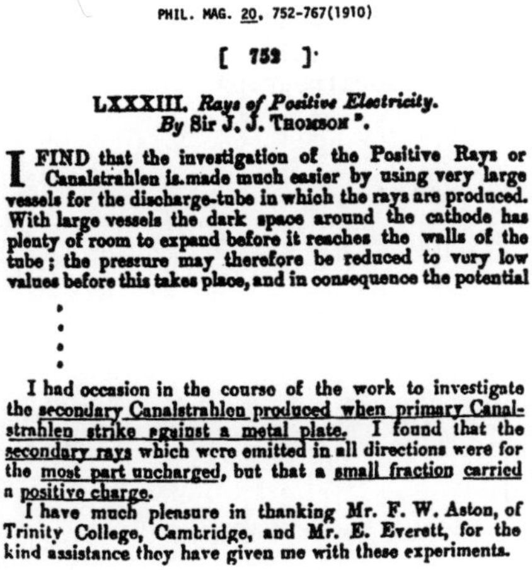
PHIL. MAG. 20, 752-767(1910)

[752]

LXXXIII. *Rays of Positive Electricity.*
By Sir J. J. Thomson *.

I FIND that the investigation of the Positive Rays or Canalstrahlen is made much easier by using very large vessels for the discharge-tube in which the rays are produced. With large vessels the dark space around the cathode has plenty of room to expand before it reaches the walls of the tube; the pressure may therefore be reduced to very low values before this takes place, and in consequence the potential

•
•
•
•

I had occasion in the course of the work to investigate the secondary Canalstrahlen produced when primary Canalstrahlen strike against a metal plate. I found that the secondary rays which were emitted in all directions were for the most part uncharged, but that a small fraction carried a positive charge.

I have much pleasure in thanking Mr. F. W. Aston, of Trinity College, Cambridge, and Mr. E. Everett, for the kind assistance they have given me with these experiments.

Fig.1. Sir J.J. Thomson's Discovery of Secondary Ions in 1910

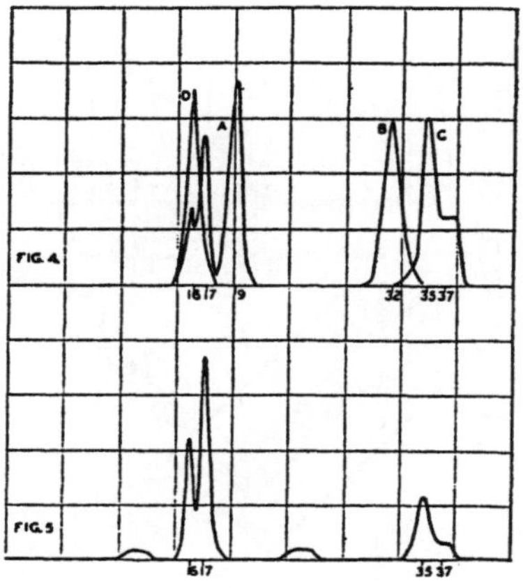

Fig.2. Woodcock's Secondary Negative Ion Spectra (1931)

were bombarded with 500 eV primary Li^+ ions. This work was apparently not known in 1936 to ARNOT and MILLIGAN [5] who believed that they could convert Hg^+ into Hg^- ions on a Ni surface. This erroneous process of "ion formation by double electron conversion" was discounted in 1938 by SLOANE and PRESS [6] who found no trace of Hg^- in a baked vacuum system, but instead observed negative secondary ions of adsorbed light elements that have large electron affinities. The study of secondary ions suffered a second setback during World War II, and it was not until 1949 that a brief letter was published by HERZOG and VIEHBÖCK [7] which outlined their novel concept of two separate electric fields to accelerate the primary and secondary ions. The letter mentioned that "metals and salts" had been studied, but presented no specific results. It was only through the kind assistance of Dr. R. F. Herzog that access was recently gained to a copy of VIEHBÖCK'S doctoral dissertation [8] on which their letter was based. The thesis shows that Viehböck had constructed a discharge source to produce the primary ions used to bombard various targets, including Fe, Ni, Cu, Al, $CaWO_4$, $Ba(OH)_2$, and H_2TeO_3(?). The resulting positive secondary ions were accelerated and then analyzed in a Thomson parabola apparatus (Fig. 3). The equipment was based on a design patented by HERZOG in 1942 [9]. It is likely that the secondary ion yields were enhanced by the fact that Viehböck had to use air to produce the primary ions, since in 1948 rare gases were unavailable in Vienna.

In 1950, a project was started at RCA Laboratories to establish whether secondary ions emitted during primary ion bombardment could be used to characterize the surface composition of oxide-coated cathodes. My early concept of a double spectrometer for primary and secondary mass analysis was presented [10] in 1951 by my colleague R. H. PLUMLEE at the First International Conference on Mass Spectrometry at NBS, Washington, D.C. Figure 4 shows the instrument actually constructed [11], in which the primary ion selector was eliminated in order to maximize its overall sensitivity. The bulk of the early results was obtained in 1954/55, but publication of this work was delayed until 1958.

At about the same time, VEKSLER and coworkers in Russia studied [12] positive ion emission from Ta and Ni surfaces under Cs^+ bombardment, attempting to reconcile secondary ion coefficients with the Langmuir-Saha equation.

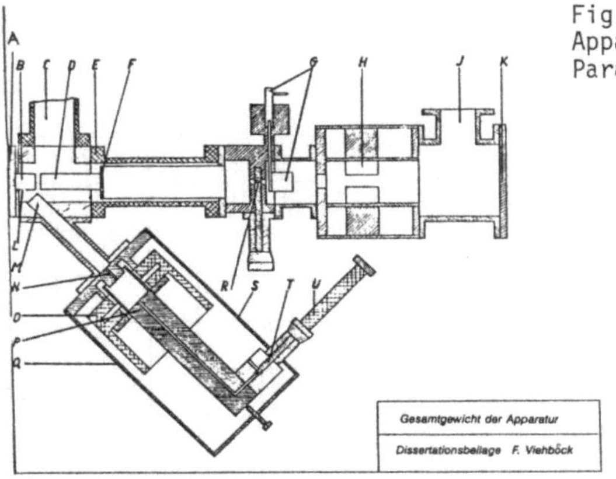

Fig.3. Herzog and Viehböck's 1948 Apparatus: Positive Ion Source and Parabola Spectrograph

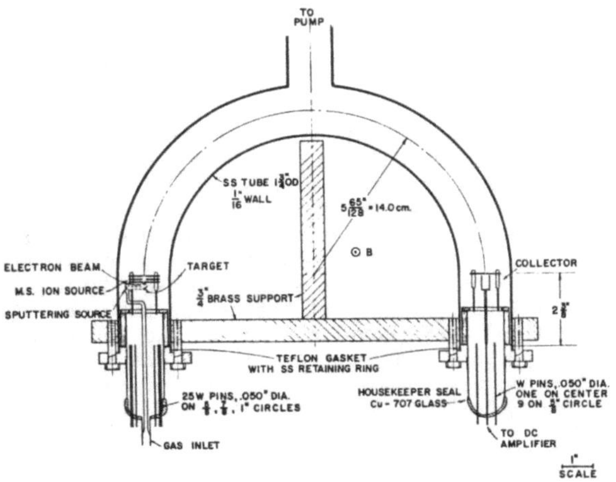

Fig.4. Secondary Ion Mass Spectrometer Constructed at RCA Laboratories in 1954

3. Modern Times

Beginning in about 1960, SIMS activities intensified, first slowly, then at an ever-increasing pace, as seen in Fig. 5 which shows annual SIMS publications as a function of time. The steep rise since 1970 suggests that by 1990 we can expect some 1000 papers per year! Such a deluge demands of the reviewer that he limit himself to major developments. For this reason, we will in the remainder of this paper focus our attention on selected topics in the inorganic and organic fields.

During the past two decades, significant advances made in ion optical design (e.g., LIEBL and HERZOG [13]) resulted in the development of three basic types of instruments: mass analyzers with moderate lateral resolution (Fig. 6); and the ion microprobe and the ion microscope (Fig. 7). The mass analyzers utilize high primary ion current densities (> 10^{-6}A cm^{-2}) to perform trace analyses (concentration levels < 1 ppm) and depth profiling, as demonstrated by HERZOG and collaborators [14], WERNER and DEGREFTE [14a], and EVANS and PEMSLER [14b]. This mode of operation is known as "DYNAMIC SIMS". As an example of a mass analyzer used successfully for accurate

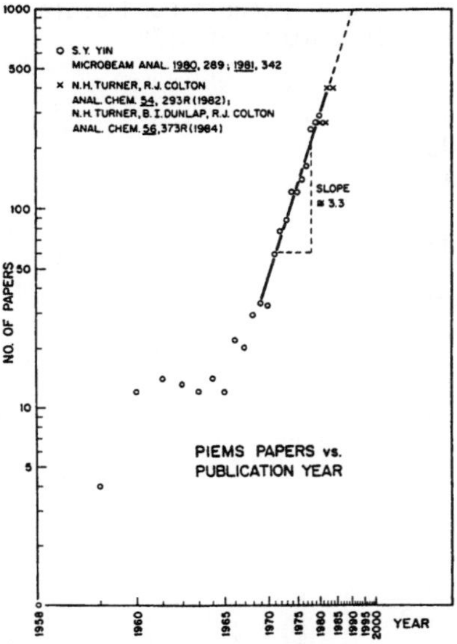

Fig.5. The Exponential Rise of Annual SIMS Publications Since 1969

SIMS DEVELOPMENT I

MODERATE LATERAL RESOLUTION (< 10 μM)

HERZOG/VIEHBÖCK	1948/49
VEKSLER	1956
HONIG	1958
LIEBL/HERZOG	1963

"DYNAMIC SIMS"
INORGANIC DEPTH
PROFILES, TRACES:

HERZOG ET AL.	1966
WERNER	1967
EVANS	1970
WITTMAACK	1973
MAGEE ET AL.	1978

"STATIC SIMS"
SURFACE DESORPTION:
BENNINGHOVEN 1970
"ORGANIC SIMS":
BENNINGHOVEN ET AL. 1976

FIG. 6

SIMS DEVELOPMENT II

HIGH LATERAL RESOLUTION (0.02-10 μM)

ION MICROSCOPE		ION MICROPROBE	
CASTAING/SLODZIAN	1960	LIEBL ARL "IMMA"	1967
ROUBEROL ET AL.	1969	"UMPA"	1971
"CAMECA IMS 300"		"COALA"	1972
ROUBEROL ET AL.	1980	WITH LIQUID METAL ION SOURCE:	
"CAMECA IMS 3F"		KROHN, RINGO	1975
		RÜDENAUER ET AL.	1982
		BAYLY, WAUGH, & ANDERSON	1983
		LEVI-SETTI ET AL.	1983

FIG. 7

depth profiling, Fig. 8 shows the quadrupole analyzer designed by MAGEE et al. [15]. It employs primary beam rastering combined with electronic gating, to insure that the secondary ions to be analyzed come exclusively from the center of the flat-bottomed crater. The idea of beam rastering, taken over from the electron-probe microanalyzer, had been applied to SIMS by ROUBEROL and DAGNOT [16] in 1970, while the combination rastering/electronic gating was first proposed by WHATLEY et al. [17] in 1972.

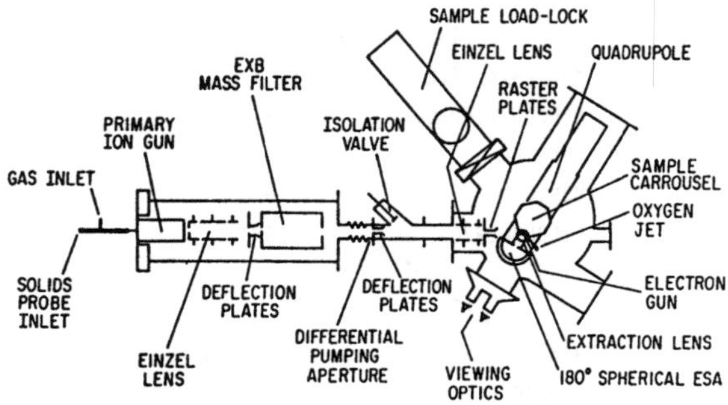

Fig.8. Magee's Quadrupole Instrument of 1978

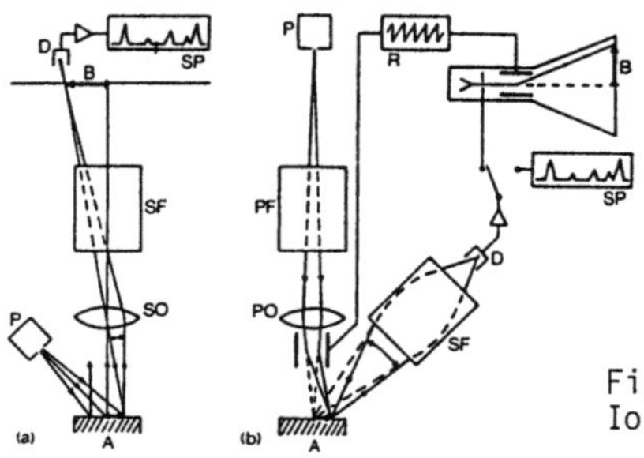

Fig.9. Schematics of: (a) Ion Microscope (b) Ion Microprobe (After Liebl)

As reviewed by LIEBL [18], SIMS instrumentation with high lateral resolution (Fig. 7) developed along two parallel paths:

1) In the ion MICROSCOPE, shown schematically in Fig. 9a, the sample is irradiated with a BROAD beam of primary ions. High lateral resolution, on the order of one μm, is achieved by forming on the screen a magnified image of the object with a NARROW beam of secondary particles which have passed through appropriate optics and a filter for mass analysis. The prototype of the ion microscope, developed in 1960 by CASTAING and SLODZIAN [19], led to two generations of sophisticated commercial instrumentation. The "CAMECA IMS 300", developed by ROUBEROL et al. [20] in 1969, offered many interesting features, including a prism-mirror-prism combination for mass- and energy-analysis, and a multiple detection and recording system. It has been applied by many groups to a wide variety of practical problems, e.g. semiconductor studies at Philips Eindhoven [21]. The second-generation "CAMECA IMS 3f", developed in 1980 by ROUBEROL et al. [22], incorporates many improvements, including a triple-focusing mass spectrometer capable of high mass resolution.

2) The ion MICROPROBE (Fig. 9b) uses a focused primary ion beam of micrometer dimensions that is rastered across the sample. A selected portion of the secondary ions is analyzed in a suitable mass filter. A successful commercial instrument, named "IMMA", was developed by LIEBL [23] in 1967 and produced by ARL. It includes a duoplasmatron ion source; a primary ion mass filter and electrostatic focusing and rastering system; a double-focusing secondary mass spectrometer; and a multiple ion detection system. This design achieved a lateral resolution of about one μm.

Substantial further improvements in lateral resolution have been made recently by incorporating into a microprobe the liquid metal ion source developed by KROHN and RINGO [24] in 1975 which is capable of producing a primary ion beam less than 0.1 µm in diameter. The potential application of this source to SIMS studies has been explored in some detail by RÜDENAUER et al. [25]. A recent commercial instrument, the "VG Scientific MIG 100" ion microprobe, is based on the liquid gallium source, and its designers, WAUGH et al. [26], have demonstrated that lateral resolutions of about 0.2 and 0.1 µm are attainable with Ga^+ ions of 10 and 27 keV energy, respectively. A further improvement in lateral resolution (~ 20 nm) has been reported by LEVI-SETTI et al. [27] for their Ga^+ scanning ion microprobe which operates at 47 keV primary energy. However, this optimum lateral resolution is obtained only at the expense of depth resolution which decreases with increasing primary energy.

4. Practical Considerations

What are the practical problems that this highly developed instrumentation has been applied to? Initially they were, almost exclusively, inorganic in nature and mainly of interest to the electronics industry, two of the key areas being depth profiling and analysis for trace impurities. A knowledge of concentration profiles, as a function of depth, of major components and dopants is of prime importance in semiconductor research. It is not surprising that many such papers have been published by many groups in many countries. To name just a few: Werner and collaborators in Eindhoven; Wittmaack in Germany; Huber and colleagues in France; Evans and coworkers, and Magee in USA. This literature has been thoroughly reviewed by ZINNER in two recent papers [28,29]. Another field of interest is quantitation, which is difficult to achieve because the secondary ion yields range over many orders of magnitude from element to element, and are subject to severe matrix effects, in particular due to oxygen coverage of the sample surface. This problem is reduced, but not eliminated, by saturating the sample surface with oxygen, first proposed by SLODZIAN and HENNEQUIN [30] in 1966 and BENNINGHOVEN [31] in 1967, and subsequently applied by ANDERSEN [32] to practical problems.

In this context, it should be mentioned, if only briefly, that the theoretical aspects of secondary ion emission have been studied over many years by many workers, among them SROUBEK [33], WITTMAACK [34], BLAISE and NOURTIER [35], WILLIAMS [36,37], YU [37a], GNASER [37b], and VASILE [37c].

Matrix effects caused by ion yield variations can be avoided by utilizing post-ionized secondary NEUTRALS instead of secondary IONS. In early experiments, HONIG [38] had employed a low-voltage (~100 eV) plasma to produce the primary ions used for sputtering, and at the same time post-ionize the secondary neutrals sputtered from the sample. This approach was investigated in detail by COBURN et al. [39,40] and by OECHSNER and GERHARD [41]. The latter group perfected this method in a series of subsequent studies reviewed in [42], gave it [43] a name and acronym of its very own ("Secondary Neutral Mass Spectrometry - SNMS"), and has recently demonstrated [44] its outstanding depth resolution of about one nm. A commercial version of this instrument has been described by MÜLLER et al. [45] and is being built by Leybold-Heraeus.

However, the plasma used for post-ionization of neutrals in SNMS requires a special geometry, i.e., plane sample surfaces of mm dimensions, which make microanalysis difficult. For this reason, alternatives have been explored that use different methods of post-ionization in conjunction with a separate, focused primary ion beam. GNASER et al. [46] use a raster/gated primary ion beam, post-ionize the secondary neutrals by electron impact, suppress the ionized residual gas molecules in an energy filter, and mass-analyze in a quadrupole. An equivalent system has been described by LIPINSKY et al. [47]. Unfortunately, the low efficiency of electron impact ionization severely limits at this time the detection sensitivity of this method.

An interesting alternative method of post-ionization has recently been developed by BLAISE and CASTAING [48,49] which is based on the thermalization and ionization of sputtered particles in a Ta cell kept at 3000 K. This approach appears to be completely free of matrix effects and strictly quantitative, but cannot detect some major elements, e.g., B, C, and O. A different and very promising approach to post-ionizing neutrals has recently been taken by BECKER and GILLEN [50,51,52] who employ NONRESONANT multiphoton ionization at 248 nm, in conjunction with a "reflectron"-type time-of-flight mass spectrometer. Preliminary results obtained for an NBS copper standard demonstrate that this is a semi-quantitative survey method, capable of detecting impurities at the ppm level.

Over the years, much attention has been given to making depth profiles more quantitative, in particular by using implanted standards, a concept proposed by GRIES [53] in 1970. This method has been used by many workers, e.g. GITTINS et al. [54] and CROSET and DIEUMEGARD [55], and has more recently been discussed in detail by LETA and MORRISON [56].

5. Organic Applications

In recent years, interest in organic systems, in particular large involatile, thermally labile molecules, has increased greatly and produced a flood of organic SIMS publications. While there exist some early exploratory studies, e.g. on ethyl monolayers adsorbed on Ge by HONIG [57] in 1962, and on polymers by DILLON et al. [58] in 1968, it was largely BENNINGHOVEN'S "STATIC SIMS" approach [59] which first made possible the analysis of organic systems by ion bombardment without destroying the desorbed molecules in the process. To insure that most of the secondary ions are emitted from areas not previously damaged, the total dose of primary ions should not exceed 10^{12} - 10^{13} cm^{-2}, corresponding to an aggregate bombarded area amounting to 1-10% of the total area. Typical bombardment conditions are: primary current densities of 10^{-9} A cm^{-2} or less for about 1000 seconds. To increase the sensitivity of the method, it is necessary to work with large sample areas. Initially, static SIMS was applied by BENNINGHOVEN to monomolecular layers adsorbed on metals [59], and only several years later, in 1976, to the desorption and analysis of organic molecules [60]. Since 1976, a large volume of work has been published by Benninghoven and collaborators, as well as by many other groups, including Colton et al., Cooks et al., Hercules et al., and Kambara et al. Further details concerning these studies will be found in recent reviews by BENNINGHOVEN [61], COLTON et al. [62,63], and BURLINGAME et al. [64,65].

An alternative to static SIMS, "FAST ATOM BOMBARDMENT (FAB)", was first published by BARBER et al. [66,67,68] in 1981/2, and has instantly become a highly successful method for the identification and structural determination of organic samples soluble in a liquid matrix, such as glycerol. The two initial publications [66,67] emphasized only the fact that the primary particles used were neutrals rather than ions. This minimizes the charging-up of organic samples which frequently are insulators, and presumably also reduces radiation damage, especially in polymers, as recently demonstrated by BROWN et al. [69]. However, the use of neutrals is not novel, since they had been employed in prior work by LEHRLE et al. [70], DILLON et al. [58], DEVIENNE, ROUSTAN et al. (reviewed in [71]), and BORCHARDT et al. [72] to investigate inorganic and organic insulators.

It is only BARBER'S third publication [68] that discloses the truly novel feature of this method: the use of a liquid matrix, such as glycerol, in which the organic sample is dissolved, extending its life from minutes to hours. To emphasize this aspect, ABERTH et al. [73] suggest that the term "LIQUID SIMS" be used, rather than the non-specific "FAB". Initially, the success of the liquid matrix had been ascribed to diffusion which allows fresh sample molecules to circulate freely to the surface. More recently, BAROFSKY et al. [74] and WONG and RÖLLGEN [75] have suggested that it is the abnormally high sputter yield of glycerol which continually exposes fresh solute molecules to the primary beam. The long sample life in the "LIQUID MATRIX SIMS" or "FAB" source is clearly its major advantage over the older static SIMS approach. It has made the new method popular with many laboratories, and has sparked the design of several small neutral particle sources. The large volume of recent studies has been reviewed by FENSELAU [76], COLTON et al. [62,63], and BURLINGAME et al. [64,65].

In recent years, several alternative methods for the desorption and subsequent identification of large involatile molecules have been developed. In each case, the primary energy supplied can be degraded to a level that matches desorption energies, allowing a fraction of the molecules to be removed as intact ions, frequently as $(M + 1)^+$ and $(M - 1)^-$. It is surprising to find that widely differing desorption methods produce mass spectra that are remarkably similar.

The first method to be discussed here is laser desorption ("LDMS"). The use of laser sources in mass spectrometry is not new, dating back to 1963 when the first exploratory experiments, made by HONIG and WOOLSTON [77,78] with a ruby laser, produced positive ions from metals, semiconductors, and insulators. Subsequently, VASTOLA and PIRONE at Penn State studied [79] laser-induced vaporization of solids, including organic compounds. However, it was only in the seventies that KAUFMANN, HILLENKAMP and coworkers [80,80a] in Germany coupled nanosecond-pulse lasers to a sophisticated time-of-flight mass spectrometer to study in detail the desorption of large organic molecules. The TOFMS is well suited for this work, since its mass range is unlimited and it records complete spectra simultaneously and instantaneously. A different approach was taken by KISTEMAKER and his group [81] in Holland, who used a conventional magnetic spectrometer equipped with a channel-plate array that recorded simultaneously a limited mass range. Since then, LDMS has been widely applied to many problems, especially to microparticulates. Two commercial versions, the "LAMMA 1000" (Leybold-Heraeus) and "LIMA" (Cambridge Mass Spectrometry Ltd.) utilize the TOF "Reflectron" principle [82] capable of improved mass resolution. The field of LDMS has been reviewed in detail by CONZEMIUS and CAPELLEN [83], complemented by a bibliography covering the period 1963-1982 [84].

Recently, LINDNER and SEYDEL [85] have demonstrated that in transmission-type LDMS a combination of thick sample ($\sim 20\,\mu m$) and high laser-power density ($\sim 10^{11}$ W cm^{-2}) will maximize the molecules desorbed intact from the solid or liquid surface. They explain the desorption process in terms of a shock-wave-driven process. An interesting instrumental development is MARSHALL'S combination of laser desorption with Fourier transform mass spectrometry ("FTMS") [86,87] which detects ions simultaneously over a wide mass range, is capable of very high resolution, but requires UHV conditions. FREISER and coworkers [88,89] and McIVER et al. [90] have applied this combination to metal ions, while GROSS et al. [91] have used it to study nonvolatile, thermally labile organic molecules. Other FTMS combinations have been investigated recently. They include Cs^+ Desorption/FTMS by CASTRO and RUSSELL [92], GC/MPI/FTMS by GROSS and coworkers [93], and Quadrupole/FTMS by McIVER et al. [94] and HUNT et al. [95,96]. The features of FTMS and its various applications have been reviewed by GROSS and REMPEL [97].

The characterization of large involatile organic molecules was given a considerable boost by the "^{252}Cf PLASMA DESORPTION (PDMS)" method developed by MACFARLANE and coworkers [98,99,100]. It utilizes the energy deposited in the sample by fission fragments from a radioactive ^{252}Cf source to desorb the organic ions, and identifies them in a time-of-flight mass spectrometer (TOFMS). This technique has since been applied in several laboratories to large molecules of interest to biochemists, e.g. to proteins with molecular weights up to 25,000 u (SUNDQVIST et al. [101]). The method was recently reviewed by SUNDQVIST and MACFARLANE [102] who discuss in detail various models of the desorption mechanism, techniques and instrumentation, and the TOF spectra obtained. A special TOFMS of the "Reflectron" type has been designed and built by DELLA NEGRA and LE BEYEC [103] to improve the mass resolution of their system. This design also permits coincidence measurements to be made of metastable ions that have decomposed in flight [104].

A technique similar to plasma desorption, but more flexible in several aspects, is based on the bombardment by heavy ions with MeV/u energies produced in a tandem accelerator. The "HEAVY ION INDUCED DESORPTION (HIID)" method was developed independently by DÜCK et al. in Germany [105] and SUNDQVIST and his Group in Uppsala [106,107]. The spectra obtained by HIID are similar to PD spectra, but afford further insight into the desorption process because energy, charge state, angle of incidence, and mass of the primary ions are adjustable parameters. Desorption yields have been measured as a function of these variables, allowing some tentative conclusions to be drawn concerning the role of electronic energy loss and the desorption mechanism [107,108,109,110]. Various parameters of plasma desorption (both ^{252}Cf PDMS and HIID) have been reviewed and discussed in detail by SUNDQVIST and coworkers [111,112].

A simple alternative desorption technique has been developed by STANDING and coworkers [113], who use a pulsed primary Cs^+ ion beam with 3 - 30 keV energy to desorb organic ions, in conjunction with a TOF analyzer. It has been applied to large organic molecules, and contributes to an understanding of alkali halide clusters, a subject that has been reviewed recently by COLTON et al. [114]. Another basic approach to produce negative ions from organic molecules has been taken recently by DELLA NEGRA et al. [115]. It is called "SPONTANEOUS DESORPTION TOFMS (SDMS)" and is based on BECKEY'S FIELD DESORPTION [116] method. It is attractive because of its very simple TOF instrumentation, and it has produced negative ion spectra comparable to PDMS.

The different methods to desorb large involatile organic molecules from solid or liquid substrates, as outlined above, produce spectra that are remarkably similar in nature. For various molecules, direct comparisons have been made between laser desorption and ^{252}CfPD [117,118], between ^{252}CfPD and SIMS [119,120], and HIID and SIMS [121]. The similarity of both positive and negative secondary ion spectra obtained by these methods is surprising because the modes of primary energy input differ widely: LSS theory shows that in the keV/u region (SIMS and FAB) primary energy is deposited through NUCLEAR interactions, while in the MeV/u region (^{252}CfPD and HIID) ELECTRONIC interactions play a major role. Various hypotheses have been proposed to explain how primary energies at the keV and MeV levels are transformed into bond-breaking vibrational energies, leading to the desorption of intact molecular ions [122-127,109]. It is clear that further work will have to be done before these issues are fully resolved.

6. Conclusion

Secondary Ion Mass Spectrometry--no matter under what label--is a fast moving field, whose vibrancy belies its biblical age. We can be certain that its next 75 years will bring many exciting developments that will contribute to a better understanding of surface science.

Acknowledgement

It is a pleasure to acknowledge the kind assistance of the following colleagues who made prepublication information available to the author: P. J. Gale (RCA), O. Ganschow (Münster), W. O. Hofer (Jülich), D. F. Hunt (Virginia), Y. Le Beyec (Orsay), R. D. Macfarlane (Texas A&M), A. G. Marshall (Ohio State), F. W. Röllgen (Bonn), K. G. Standing (Manitoba), and B. Sundqvist (Uppsala). Special thanks go to C. A. Evans, Jr. (San Mateo), R. K. Lewis (IBM), and B. L. Bentz, and C. W. Magee (RCA) for fruitful discussions.

References

1. R. E. Honig: Int. J. Mass Spectrom. Ion Processes 66, 31 (1985)
2. J. J. Thomson: Phil. Mag. 20, 252 (1910)
3. R. B. Sawyer: Phys. Rev. 35, 1090 (1930)
4. K. S. Woodcock: Phys. Rev. 38, 1696 (1931)

4a. J. S. Thompson: Phys. Rev. 38, 1389L (1931)

5. F. L. Arnot, J. C. Milligan: Proc. Roy. Soc. A 156, 538 (1936)
6. R. H. Sloane, R. Press: Proc. Roy. Soc. A 168, 284 (1938)
7. R. F. K. Herzog, F. P. Viehböck: Phys. Rev. 76, 855L (1949)
8. F. Viehböck: "Über eine neue Ionenquelle", Doctoral Dissertation (Unpublished), University of Vienna (1948)
9. R. F. K. Herzog: "Massenspektrometer", German Patent DRP H172192 IXa/42h (1942)
10. R. H. Plumlee: "Mass Spectrometric Studies in Solids", in Mass Spectroscopy in Physics Research, N.B.S. Circular 522, p. 232 (Washington, D.C. 1953)
11. R. E. Honig: J. Appl. Phys. 29, 549 (1958)
12. V. I. Veksler, M. B. Ben'iaminovich: Zhur. Tekh. Fiz. 26, 1671 (1956); Transl.: Sov. Phys. Tech. Phys. 1, 1626 (1957)
13. H. J. Liebl, R. F. K. Herzog: J. Appl. Phys. 34, 2893 (1963)
14. A. E. Barrington, R. F. K. Herzog, W. P. Poschenrieder: J. Vac. Sci. Techn. 3, 239 (1966)

14a. H. W. Werner, H. A. M. de Grefte: Vakuum-Tech. 37, 17 (1967)

14b. C. A. Evans, J. P. Pemsler: Anal. Chem. 42, 1040 (1970)

15. C. W. Magee, W. L. Harrington, R. E. Honig: Rev. Sci. Instr. 49, 477 (1978)
16. J.-M. Rouberol, J.-P. Dagnot: in Proc. Septième Congrès Internat. de Microscopie Electronique, Grenoble (1970), p. 267
17. T. A. Whatley, C. B. Slack, E. Davidson: in Proc. Sixth Int. Conf. on X-Ray Optics and Microanalysis (Univ. Tokyo Press 1972), p. 417
18. H. J. Liebl: Anal. Chem. 46, 22A (1974)
19. R. Castaing, G. Slodzian: in Proc. Eur. Conf. Electron Microscopy (Delft 1960) Vol. I, p. 169; J. Microscopie 1, 395 (1962)

20. J.-M. Rouberol, J. Guernet, P. Deschamps, J.-P. Dagnot, J.-M. Guyon de la Berge: in Proc. Fifth Int. Conf. on X-Ray Optics and Microanalysis (Springer, Berlin 1969), p. 311
21. W. K. Hofker, H. W. Werner, D. P. Oosthoek, H. A. M. de Grefte: Appl. Phys. 2, 265 (1973)
22. J.-M. Rouberol, M. Lepareur, B. Autier, J. M. Gourgout: in Proc. Eighth Int. Congress X-Ray Optics and Microanalysis, D. R. Beaman, R. E. Ogilvie, D. B. Wittry, Eds. (Pendell Pub. Co., Midland, MI 1980), p. 322
23. H. J. Liebl: J. Appl. Phys. 38, 5277 (1967)
24. V. E. Krohn, G. R. Ringo: Appl. Phys. Letters 27, 479 (1975)
25. F. G. Rüdenauer, P. Pollinger, H. Studnicka, H. Gnaser, W. Steiger, M. J. Higatsberger: in Secondary Ion Mass Spectrometry SIMS III, A. Benninghoven, J. Giber, J. László, M. Riedel, H. W. Werner, Eds. (Springer, Berlin 1982), p. 43
26. A. R. Waugh, A. R. Bayly, K. Anderson: in Secondary Ion Mass Spectrometry SIMS IV, A. Benninghoven, J. Okano, R. Shimizu, H. W. Werner, Eds. (Springer, Berlin 1984) p. 138
27. R. Levi-Setti, Y. L. Wang, G. Crow: J. Phys. (Paris) 45, C9-147 (1984)
28. E. Zinner: Scanning 3, 57 (1980)
29. E. Zinner: J. Electrochem. Soc. 130, 199C (1983)
30. G. Slodzian, J.-F. Hennequin: C. R. Acad. Sc. (Paris) 263, 1246 (1966)
31. A. Benninghoven: Z. Naturforsch. 22a, 841 (1967)
32. C. A. Andersen, Int. J. Mass Spectrom. Ion Phys. 2, 61 (1969)
33. Z. Sroubek: Surf. Sci. 44, 47 (1974)
34. K. Wittmaack: in Inelastic Ion-Surface Collisions, N. H. Tolk, J. C. Tully, W. Heiland, C. W. White, Eds. (Academic Press, New York 1977) p. 153
35. G. Blaise, A. Nourtier: Surf. Sci. 90, 495 (1979)
36. P. Williams: Surf. Sci. 90, 588 (1979)
37. P. Williams: Appl. Surf. Sci. 13, 241 (1982)
37a. M. L. Yu: Phys. Rev. B 29, 2311 (1984)
37b. H. Gnaser: Appl. Surf. Sci. 18, 389 (1984)
37c. M. J. Vasile: Phys. Rev. B 29, 3785 (1984)
38. R. E. Honig: in Advances in Mass Spectrometry, J. D. Waldron, Ed. (Pergamon Press, London 1959) p. 162
39. J. W. Coburn, E. Kay: Appl. Phys. Letters 18, 435 (1971)
40. J. W. Coburn, E. Taglauer, E. Kay: J. Appl. Phys. 45, 1779 (1974)
41. H. Oechsner, W. Gerhard: Phys. Letters 40A, 211 (1972)
42. H. Oechsner: in Thin Film and Depth Profile Analysis, H. Oechsner, Ed. (Springer, Berlin 1984) p. 63
43. H. Oechsner, E. Stumpe: Appl. Phys. 14, 43 (1977)
44. H. Oechsner, G. Bachmann, P. Beckmann, M. Kopnarski, D. A. Reed, S. M. Baumann, S. D. Wilson, C. A. Evans, Jr.: in this volume
45. K. H. Müller, K. Seifert, M. Wilmers: J. Vac. Sci. Technol. A 3, 1367 (1985)
46. H. Gnaser, J. Fleischhauer, W. O. Hofer: Appl. Phys. A 37, 211 (1985)
47. D. Lipinsky, R. Jede, O. Ganschow, A. Benninghoven: J. Vac. Sci. Technol. A3, 2035 (1985)
48. G. Blaise, R. Castaing: C. R. Acad. Sci. Paris B 284, 449 (1977)
49. G. Blaise, Scanning Electron Micros. 1985/I, p. 31
50. C. H. Becker, K. T. Gillen: Anal. Chem. 56, 1671 (1984)
51. C. H. Becker, K. T. Gillen: Appl. Phys. Letters 45, 1063 (1984)
52. C. H. Becker, K. T. Gillen: J. Vac. Sci. Technol. A 3, 1347 (1985)
53. W. H. Gries: Int. J. Mass Spectrom. Ion Phys. 30, 97 (1979): see Ref. 1
54. R. P. Gittins, D. V. Morgan, G. Dearnaley: J. Phys. D: Appl. Phys. 5, 1654 (1972)

55. M. Croset, D. Dieumegard: Corrosion Sci. 16, 703 (1976)
56. D. P. Leta, G. H. Morrison: Anal. Chem. 52, 277 (1980)
57. R. E. Honig: in Advances in Mass Spectrometry, Vol. 2, R. M. Elliott, Ed. (Pergamon Press, New York 1962), p. 25
58. A. F. Dillon, R. S. Lehrle, J. C. Robb, D. W. Thomas: in Advances in Mass Spectrometry, Vol. 4, E.E. Kendrick, Ed. (Institute of Petroleum, London 1968) p. 477
59. A. Benninghoven: Z. Physik 230, 403 (1970)
60. A. Benninghoven, D. Jaspers, W. Sichtermann: Appl. Phys. 11, 35 (1976)
61. A. Benninghoven: in Ion Formation from Organic Solids, A. Benninghoven, Ed. (Springer, Berlin 1983) p. 64
62. N. H. Turner, R. J. Colton: Anal. Chem. 54, 293R (1982)
63. N. H. Turner, B. I. Dunlap, R. J. Colton: Anal. Chem. 56, 373R (1984)
64. A. L. Burlingame, A. Dell, D. H. Russell: Anal. Chem. 54, 363R (1982)
65. A. L. Burlingame, J. O. Whitney, D. H. Russell: Anal. Chem. 56, 417R (1984)
66. M. Barber, R. J. Bordoli, R. D. Sedgwick, A. N. Tyler: J.C.S. Chem. Comm. 1981, 325 (1981)
67. M. Barber, R. J. Bordoli, R. D. Sedgwick, A. N. Tyler: Nature 293, 270 (1981)
68. M. Barber, R. J. Bordoli, G. J. Elliott, R. D. Sedgwick, A. N. Tyler: Anal. Chem. 54, 645A (1982)
69. A. Brown, J. A. van den Berg, J. C. Vickerman: J. Chem. Soc. Chem. Comm. 1984, 1684 (1984)
70. R. S. Lehrle, J. C. Robb, D. W. Thomas: J. Sci. Instr. 39, 458 (1962)
71. F.M. Devienne, J.-C. Roustan: Org. Mass Spectrom. 17, 173 (1982)
72. G. Borchardt, H. Scherrer, S. Weber, S. Scherrer: Int. J. Mass Spectrom. Ion Phys. 34, 361 (1980)
73. W. Aberth, K. M. Straub, A. L. Burlingame: Anal. Chem. 54, 2029 (1982)
74. D. F. Barofsky, A. M. Ilias, E. Barofsky, J. H. Murphy: presented at the 32nd Annual ASMS Conference, San Antonio (1984), p. 182
75. S. S. Wong, F. W. Röllgen: Biomed. Mass Spectrom. (submitted)
76. C. Fenselau: in Ion Formation from Organic Solids, A. Benninghoven, Ed. (Springer, Berlin 1983) p. 90
77. R. E. Honig, J. R. Woolston: Appl. Phys. Letters 2, 138 (1963)
78. R. E. Honig: Appl. Phys. Letters 3, 8 (1963)
79. F. J. Vastola, A. J. Pirone: in Advances in Mass Spectrometry, Vol. 4, E. Kendrick, Ed. (Inst. Petroleum, London 1968) p. 107
80. R. Kaufmann, F. Hillenkamp, E. Remy: Microsc. Acta 73, 1 (1972)
80a. F. Hillenkamp, E. Unsöld, R. Kaufmann, R. Nitsche: Appl. Phys. 8, 341 (1975)
81. M. A. Posthumus, P. G. Kistemaker, H. L. C. Meuzelaar, M. C. Ten Noever de Brauw: Anal. Chem. 50, 985 (1978)
82. B. A. Mamyrin, V. I. Karataev, D. V. Shmikk, V. A. Zagulin: Sov. Phys.-JETP 37, 45 (1973)
83. R. J. Conzemius, J. M. Capellen: Int. J. Mass Spectrom. Ion Phys. 34, 197 (1980)
84. R. J. Conzemius, D. S. Simons, Z. Shankai, G. D. Byrd: in Proc. 18th Ann. Conf. Microbeam Anal. Soc. 1983, p. 301
85. B. Lindner, U. Seydel: Anal. Chem. 57, 895 (1985)
86. M. B. Comisarow, A. G. Marshall: Chem. Phys. Letters 25, 282 (1974); J. Chem. Phys. 64, 110 (1976)
87. A. G. Marshall: Acc. Chem. Res., submitted (1985)
88. R. B. Cody, R. C. Burnier, W. D. Reents, Jr., T. J. Carlin, D. A. McCrery, R. K. Lengel, B. S. Freiser: Int. J. Mass Spectrom. Ion Phys. 33, 37 (1980)
89. R. B. Cody, R. C. Burnier, B. J. Freiser: Anal. Chem. 54, 96 (1982)
90. M. G. Sherman, J. R. Kingsley, D. A. Dahlgren, J. C. Hemminger, R. T. McIver, Jr.: Surf. Sci. 148, L25 (1985)

91. D. A. McCrery, E. B. Ledford, Jr., M. L. Gross: Anal. Chem. 54, 1435 (1982)
92. M. E. Castro, D. H. Russell: Anal. Chem. 56, 581 (1984)
93. T. M. Sack, D. A. McCrery, M. L. Gross: Anal. Chem. 57, 1290 (1985)
94. R. T. McIver, Jr., R. L. Hunter, W. D. Bowers: Int. J. Mass Spectrom. Ion Proc. 64, 67 (1985)
95. D. F. Hunt, J. Shabanowitz, R. T. McIver, Jr., R. L. Hunter, J. E. P. Syka: Anal. Chem. 57, 765 (1985)
96. D. F. Hunt, J. Shabanowitz, J. R. Yates III, R. T. McIver, Jr., R. L. Hunter, J. E. P. Syka, J. Amy: Anal. Chem. 57, 2728 (1985)
97. M. L. Gross, D. L. Rempel: Science 226, 261 (1984)
98. D. F. Torgerson, R. P. Skowronski, R. D. Macfarlane: Biochem. Biophys. Res. Comm. 60, 616 (1974)
99. R. D. Macfarlane, D. F. Torgerson: Int. J. Mass Spectrom. Ion Phys. 21, 81 (1976)
100. R. D. Macfarlane, D. F. Torgerson: Science 191, 920 (1976)
101. B. Sundqvist, P. Roepstorff, J. Fohlman, A. Hedin, P. Håkansson, I. Kamensky, M. Lindberg, M. Salehpour, G. Säwe: Science 226, 696 (1984)
102. B. Sundqvist, R. D. Macfarlane: Mass Spectrom. Rev., in press (1985)
103. S. Della Negra, Y. Le Beyec: Int. J. Mass Spectrom. Ion Proc. 61, 21 (1984)
104. S. Della Negra, Y. Le Beyec: Orsay Report IPNO-DRE-85-01 (1985)
105. P. Dück, W. Treu, W. Galster, H. Fröhlich, H. Voit: Nucl. Instr. Methods 168, 601 (1980)
106. P. Håkansson, A. Johansson, I. Kamensky, B. Sundqvist: IEEE Trans. Nucl. Sci. NS-28, 1776 (1981)
107. P. Håkansson, I. Kamensky, B. Sundqvist: Nucl. Instr. Methods 198, 43 (1982)
108. H. Voit, H. Fröhlich, P. Dück, B. Nees, E. Nieschler, W. Bischof, W. Tiereth: IEEE Trans. Nucl. Sci. NS-30, 1759 (1983)
109. R. D. Macfarlane: in Ion Formation from Organic Solids, A. Benninghoven, Ed. (Springer, Berlin 1983) p. 32
110. E. Nieschler, B. Nees, N. Bischof, H. Fröhlich, W. Tiereth, H. Voit: Radiat. Eff. 83, 121 (1984)
111. B. Sundqvist: Nucl. Instr. Methods 218, 267 (1983)
112. B. Sundqvist, A. Hedin, P. Håkansson, I. Kamensky, M. Salehpour, G. Säwe: Int. J. Mass Spectrom. Ion Proc. 65, 69 (1985)
113. K. G. Standing, R. Beavis, W. Ens, B. Schueler: Int. J. Mass Spectrom. Ion Phys. 53, 125 (1983)
114. R. J. Colton, J. E. Campana, D. A. Kidwell, M. M. Ross, J. R. Wyatt: Appl. Surf. Sci. 21, 168 (1985)
115. S. Della Negra, Y. Le Beyec, P. Håkansson: Orsay Report IPNO-DRE-84-127 (1984)
116. H. D. Beckey: Int. J. Mass Spectrom. Ion Phys. 2, 500 (1969)
117. B. Schueler, F. R. Krueger: Org. Mass Spectrom. 14, 439 (1979)
118. B. Schueler, F. R. Krueger: Org. Mass Spectrom. 15, 295 (1980)
119. E. Ens, K. G. Standing, B. T. Chait, F. H. Field: Anal. Chem. 53, 1241 (1981)
120. P. J. Gale, B. L. Bentz, B. T. Chait, F. H. Field, R. J. Cotter: Anal. Chem., submitted (1985)
121. I. Kamensky, P. Håkansson, B. Sundqvist, C. J. McNeal, R. Macfarlane: Nucl. Instr. Methods 198, 65 (1982)
122. A. Benninghoven: in Secondary Ion Mass Spectrometry SIMS III, A. Benninghoven, J. Giber, J. Laszlo, M. Riedel, H. W. Werner, Eds. (Springer, Berlin 1982) p. 438
123. B. J. Garrison: J. Am. Chem. Soc. 104, 6211 (1982)
124. C. W. Magee: Int. J. Mass Spectrom. Ion Phys. 49, 211 (1983)

125. R. D. Macfarlane: Acc. Chem. Res. 15, 268 (1982)
126. B. J. Garrison: J. Am. Chem. Soc. 105, 373 (1983)
127. A. Hedin, P. Håkansson, B. Sundqvist, R. E. Johnson: Phys. Rev. B 31, 1780 (1985)

Part II

Fundamentals

Density-Functional Studies of the Atom-Surface Interaction and the Ionization Probability of Sputtered Atoms

N.D. Lang

IBM Thomas J. Watson Research Center, Yorktown Heights, NY 10598, US

Abstract

A brief description of studies of the atom-surface interaction using density-functional methods is given. The ionization probability of atoms sputtered from metal surfaces is then discussed using results of these studies in conjunction with a resonant tunnelling approach. Good agreement of this theory with experimental data of Yu on the sputtering of adsorbed alkali atoms is obtained.

We begin our discussion with a brief outline of the calculation of the ground-state properties of an atom interacting with a metal surface a fixed distance away. [1] We simplify the description of the metal substrate by using the jellium model to represent it: that is, the semi-infinite lattice of positive ions of the metal is smeared out into a semi-infinite homogeneous positive background. This model eliminates much of the detail of the substrate (which is now specified only by the value of its average electron density), but it is able to represent many of the properties of the metal that will interest us here. [1]

Our study of the metal-atom system is based on the density-functional theory of inhomogeneous electron systems due to HOHENBERG, KOHN and SHAM. [2] The equations that must be solved self-consistently to obtain the electron density distribution have a Hartree-like form, with a potential that includes not only electrostatic but exchange and correlation effects as well. For the latter we make the commonly used local-density approximation: the relevant exchange and correlation properties in any volume element are taken to be those of a uniform infinite electron gas with a density equal to that in the volume element. [2]

When an atom interacts with a system whose electronic states form a continuum, the discrete levels of the atom which are degenerate with the continuum broaden into resonances. The resonances formed in this way when Li, Si and Cl atoms interact with a high-electron-density metal are shown in Fig. 1 (at their calculated equilibrium separations). [3] (Explicitly, what is shown is the differ-

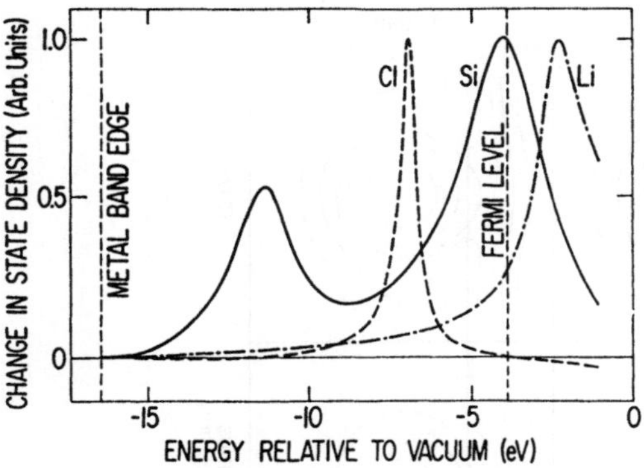

Fig. 1. Change in eigenstate density due to atomic chemisorption. The curves correspond to metal-atom separations which minimize the total energy. The lower Si resonance corresponds to the 3s level; the corresponding level for Cl is a discrete state below the bottom of the metal band. The metal electron density is 0.03 electrons/bohr^3. (From Ref. 3.)

ence in density of eigenstates between the metal-atom system and the bare metal.) The states constituting the Cl 3p resonance are below the Fermi level, and are therefore occupied; those constituting the Li 2s resonance are essentially empty. This implies that charge transfer has taken place, toward the Cl and away from the Li, as would be expected from the electronegativities of these atoms. The prohibitively large energy required either to fill or empty the Si 3p level forces this resonance to straddle the Fermi level, resulting in the formation of a covalent, rather than ionic, bond.

The electron densities [3] associated with the three fundamental bond types are exhibited for comparison in Fig. 2. In the top row of the figure are presented contours of constant total electron density for the three prototype systems. Note the way in which the contours rapidly regain their bare-metal form away from the immediate region of the atom. This is just a manifestation of the short range of metallic screening. Note also the way in which the metal contours bend toward the positively charge Li and away from the negatively charged Cl.

The detailed charge rearrangements associated with the atom-surface interaction are displayed in the second row of Fig. 2, which gives contour maps of the difference between the electron density in the metal-atom system and the superposition of bare-metal and free-atom densities. The solid contours indicate regions of charge accumulation; broken contours indicate regions of charge depletion.

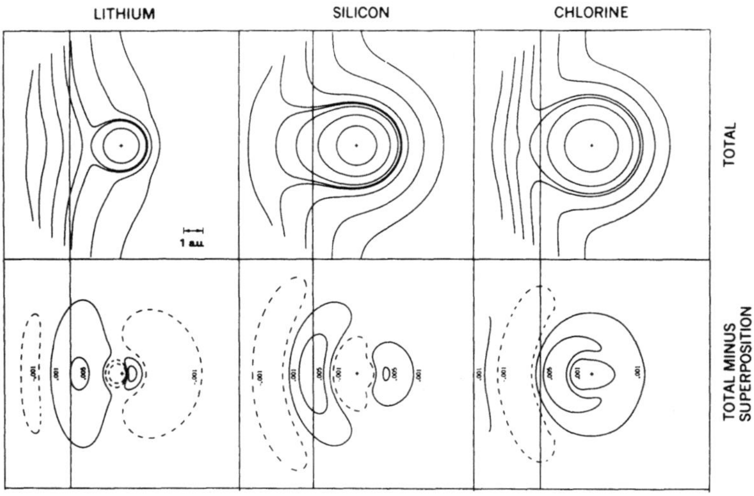

Fig. 2. Upper row: contours of constant electron density in a plane normal to the metal surface containing the atom nucleus (indicated by +): the metal is to the left-hand side; the solid vertical line indicates the positive-background edge. Atoms are at equilibrium distance from the metal. Contours are not shown outside inscribed circle of each square; contour values are selected to be visually informative. Second row: total electron density minus superposition of atomic and bare-metal electron densities (electrons/bohr^3). (From Ref. 3.)

The density-difference contours in the case of Li reveal the complexity underlying the notion of charge transfer. Taken alone, the density-difference plot indicates only that electrons have been displaced from the vacuum side to the metal side of the atom. The plot does not distinguish between the two possible interpretations of the displacement: polarization of the atom, or ionization (followed by metallic screening). The state-density plot in Fig. 1, however resolves the ambiguity; the fact that virtually none of the states in the resonance associated with the Li 2s level are occupied clearly indicates that the ionization/screening interpretation is the correct one.

We now consider the ionization probability of secondary atoms emitted from layers chemisorbed on a metal surface. We use the results of our density-functional calculations to determine the values of the parameters (such as resonance width at equilibrium) that appear in our analysis; this analysis follows the approach discussed by BLANDIN, NOURTIER and HONE [4], and later by NØRSKOV and LUNDQVIST [5], BRAKO and NEWNS [6], and LANG. [7] The metal is treated as a non-interacting Fermi gas of work function Φ. We consider the interaction of the metal with a single atom, which moves along a classical trajectory $\vec{r}(t)$. The atom has a non-degenerate valence state $|a\rangle$ of energy ε_a, lying in the metal conduction band. This level has a finite width (as in Fig. 1) because of its interaction with the states $|k\rangle$ of the metal, via matrix elements $V_{ak} = \langle a|V|k\rangle$, with V the perturbation due to the atom-metal interaction. Charge transfer between atom and metal is taken to involve the transfer of an electron between the broadened level $|a\rangle$ and metal states of the same energy. (We thus assume that Auger processes can be neglected relative to this resonant transfer process. [8])

Use of the trajectory approximation means that we can write ε_a and V_{ak} in the electronic Hamiltonian as $\varepsilon_a(\vec{r}(t))$ and $V_{ak}(\vec{r}(t))$, with $\vec{r}(t)$ the trajectory. Spin effects are neglected. The Hamiltonian is then written [9]

$$H(t) = \sum_k \varepsilon_k n_k + \varepsilon_a(\vec{r}(t))n_a + \sum_k [V_{ak}(\vec{r}(t))c_a^\dagger c_k + \text{h.c.}], \tag{1}$$

where $c_a^\dagger$ and $c_k^\dagger$ are electron creation operators for $|a\rangle$ and $|k\rangle$ respectively, and $n = c^\dagger c$. We do not take account of interactions of the form $Un_{a\uparrow}n_{a\downarrow}$ in H, which would represent a major complication in the theory. We wish instead to discuss the simplest possible model that gives a reasonable picture of the experimental data.

Following Ref. 4 the simplifying assumptions are made that the $\vec{k}$ and $\vec{r}$ dependence of $V_{ak}(\vec{r}(t))$ are separable:

$$V_{ak}(\vec{r}(t)) = V_k u(\vec{r}(t)) \tag{2}$$

and that the energy dependence of the level-width function

$$\Delta(\varepsilon) = \pi \sum_k |V_{ak}|^2 \delta(\varepsilon - \varepsilon_k) \tag{3}$$

can be neglected. The instantaneous resonance width is then given by

$$\Delta(\vec{r}(t)) = \Delta_0 |u(\vec{r}(t))|^2 \tag{4}$$

where we take $|u|^2 = 1$ for $t = 0$. Note that Δ is the half width at half maximum of the adsorbate resonance. With these assumptions, an expression for the occupation probability $\langle n_a(t)\rangle$ has been obtained in Refs. 4-7.

We neglect any dependence parallel to the surface, and take Δ and ε_a to depend only on z, the coordinate along the surface normal. We set $\hbar = 1$. We also use a simple exponential form for Δ:

$$\Delta(z) = \Delta_0 e^{-\gamma z} \tag{5}$$

($z = 0$ is the atom position at $t = 0$).

We will give the occupation probability in the limit $t \to \infty$ for a simple case in which ε_a varies linearly with distance and in which the velocity is constant (see Ref. 7). In particular, we take

$$\varepsilon_a(z) = \varepsilon_F + b(z - z_c) \tag{6}$$

with z_c the distance at which the level crosses ε_F, and we take $z = v_\perp t$. Neglecting terms of $O(e^{-\frac{\Delta_0}{\gamma v_\perp}})$, which is generally a good approximation [7],

$$<n_a(\infty)> = \frac{1}{2} + \frac{\zeta}{\pi}\mathrm{Re}\int_0^\infty \frac{dx}{\cosh x}\left(\frac{2\Delta_0}{\gamma v_\perp}\cosh x\right)^{i\zeta x} \times e^{-i\zeta\gamma z_c x}\,\Gamma(-i\zeta x) \tag{7}$$

where

$$\zeta = \frac{2b}{\gamma^2 v_\perp} . \tag{8}$$

In the limit in which $\zeta \to -\infty$ (this sign of ζ corresponds to $b<0$, which as we shall see is appropriate for the case of positive-ion formation discussed here), the probability that the sputtered atom is a positive ion far from the surface is* [7]

$$P^+ \equiv 1 - <n_a(\infty)> = \exp\left(-\frac{2\Delta(z_c)}{\gamma v_\perp}\right). \tag{9}$$

Comparisons with evaluations of (7) show this limiting form to be quite accurate for many cases of interest.

We now consider the case of a sputtered adsorbed Cs atom, which was studied experimentally by YU. [11] At small coverages, Cs atoms are chemisorbed as Cs^+ on metal surfaces, with the empty 6s level lying above the metal Fermi level just as the empty Li 2s level lies above the Fermi level in Fig. 1.

---

* This result has a simple semi-classical derivation (see e.g. Ref. 10). It was also obtained earlier by BRAKO and NEWNS (Ref. 6), although with a different criterion of validity.

Hence if the work function Φ of the surface is larger than the ionization potential I of Cs (3.9 eV), as indeed it is in Yu's experiment, the 6s level of the sputtered Cs atom will always face empty states of the metal as the Cs atom escapes and little neutralization by electron tunneling can occur. By using an adatom dipole layer to adjust the relative positions of the Fermi level and the vacuum level (this is done in the experiment by using adsorbed Li to decrease Φ, beyond the amount which the Cs itself decreases it), the 6s level can be deliberately forced to "cross" the Fermi level (when $I > \Phi$) as the Cs atom escapes, making electron tunneling energetically possible.

When the sputtered alkali atom is far from the surface, the valence level is an energy I below the vacuum level: $\varepsilon_a(\infty)-\varepsilon_F = \Phi-I$. For atom positions closer to the surface, the level $\varepsilon_a(z)$ is higher because of the image effect, i.e. it is easier to remove an electron from the level because the final-state energy of the system is lowered by the image interaction between the alkali and the metal. We thus write (in atomic units)

$$\varepsilon_a(z)-\varepsilon_F = \Phi-I+\frac{1}{4(z-z_{im})}, \tag{10}$$

where z_{im} is the position of the image plane. [12]

We will describe this behavior of ε_a by a linear form (6) in the vicinity of z_c, with a negative slope b. We employ the convention that the case in which there is no crossing corresponds to $z_c \to \infty$; and we take the velocity to be constant, which is appropriate for the sputtered-atom kinetic energy in the experiment (cf. Ref. 7). We then use expression (9) to determine the ionization probability. Note the central role in this expression of $\Delta(z_c)$, the resonance half-width at the Fermi-level crossing.

We see from (9) that P^+ does not begin to drop significantly as soon as Φ is decreased below I, i.e. as soon as there is a crossing of the curve $\varepsilon_a(z)$ by the Fermi level at some finite distance z_c. The reason is simply that Φ must continue to decrease until z_c comes in close enough to the metal to give an appreciable width $\Delta(z_c)$; before this happens, $P^+ \doteq 1$. As z_c then decreases further with Φ, P^+ decreases rapidly because of the rapid increase in $\Delta(z_c)$.

In order to use (9), (5), and (10), which can be combined as

$$P^+ = \begin{cases} 1 & \Phi \geq I \\ \exp\left(-A\ \exp\left[-\dfrac{\gamma}{4(I-\Phi)}\right]\right) & \Phi < I \end{cases}, \tag{11}$$

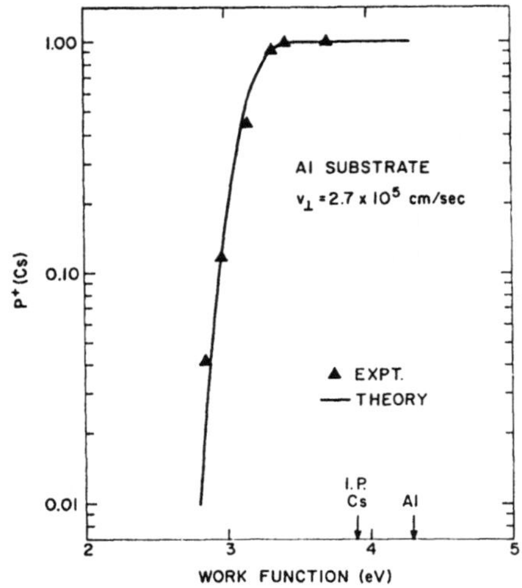

Fig. 3. Comparison between theoretical P^+ for Cs on an Al substrate and the corresponding experimental data. Given the approximate nature of the determination of such quantities as Δ_0 and γ, the extremely close agreement between theory and experiment should be regarded as somewhat coincidental; however it is clear that a reasonable account of the data is obtained from a completely *a priori* calculation. (From Ref. 11.)

where $A = 2\Delta_0(\gamma v_\perp)^{-1} \exp(-\gamma z_{im})$, to determine P^+ for the Cs/Al case, we must know the values of the quantities Δ_0, γ and z_{im}. Density-functional calculations for Cs of the type described above [3] as well as for the bare surface [12] give $\Delta_0 \sim 0.5$ eV and $z_{im} \sim -3$ bohr (recall that z is measured relative to the atom equilibrium position, with $z \to \infty$ in the vacuum) (see Ref. 11 for additional details). We furthermore use a value $\gamma = 0.7$ bohr^{-1}, determined as described in Ref. 11. Figure 3 compares the computed curve of P^+ vs. Φ with that determined experimentally by YU [11] for the Cs/Al case. The results are seen to be quite close.

References

1. For a general survey, see N. D. Lang: *Theory of the Inhomogeneous Electron Gas)*, ed. S. Lundqvist and N. H. March (Plenum, New York, 1983), pp. 309-389.
2. P. Hohenberg and W. Kohn: Phys. Rev. **136**, B864 (1964); W. Kohn and L. J. Sham: Phys. Rev. **140**, A1133 (1965).
3. N. D. Lang and A. R. Williams: Phys. Rev. Lett. **37**, 212 (1976); Phys. Rev. B **18**, 616 (1978).
4. A. Blandin, A. Nourtier, and D. W. Hone: J. Phys. (Paris) **37**, 369 (1976).
5. J. K. Nørskov and B. I. Lundqvist: Phys. Rev. B **19**, 5661 (1979).
6. R. Brako and D. M. Newns: Surf. Sci. **108**, 253 (1981).

7. N. D. Lang: Phys. Rev. B **27**, 2019 (1983).
8. J. K. Nørskov, D. M. Newns, and B. I. Lundqvist: Surf. Sci. **80**, 179 (1979).
9. P. W. Anderson: Phys. Rev. **124**, 41 (1961); D. M. Newns: Phys. Rev. **178**, 1123 (1969).
10. N. D. Lang and J. K. Nørskov: Physica Scripta **T6**, 15 (1983).
11. M. L. Yu and N. D. Lang: Phys. Rev. Lett. **50**, 127 (1983).
12. N. D. Lang and W. Kohn: Phys. Rev. B **7**, 3541 (1973).

Origin of the Chemical Enhancement of Positive Secondary Ion Yield in SIMS

K. Mann and M.L. Yu

IBM Thomas J. Watson Research Center, Yorktown Heights, NY 10598, USA

1. Introduction

It is well known that the presence of electronegative atoms like oxygen or nitrogen on a sample surface can enhance the probability p^+ of positive secondary ion emission in the sputtering process under noble gas ion bombardment by several orders of magnitude. There is, however, still some uncertainty about the physical origin of this large chemical enhancement. The strong increase of p^+ could be either due to local bonding configurations upon chemisorption of electronegative gases ("bond breaking model" [1]), or, on the other hand, caused by changes in the (global) electronic band structure, as it has been shown for the enhancement of the negative ionization probability p^- on alkali adsorption ("resonant tunneling mechanism" [2]). In this paper, we present energy and angle-resolved static mode SIMS data from oxygenated and nitrogenated Si(100) surfaces obtained under well defined experimental conditions, which strongly favor the local bond breaking model.

2. Experimental

The experiments are performed in a UHV-chamber ($p < 1x10^{-10}$ Torr) with LEED, X-ray and Ultraviolet photoemission (XPS and UPS) for in situ sample analysis. Standard cleaning procedures yield a sharp Si(100) p (2 x 1) reconstruction pattern in LEED as well as the characteristic surface state peaks in the UPS spectra. The crystals are bombarded with a scanned Ar^+ beam (1nA / 500 eV, $4x4mm^2$, incident at ~ 30° off normal), providing proper static mode SIMS conditions. After having passed an electrostatic energy filter of about 0.5 eV resolution, the sputtered substrate ions are detected with a Quadrupole mass spectrometer, as a function of the oxygen (nitrogen) coverages on the surface as monitored by the XPS-O1s (N1s) signals (area under peak). Simultaneously, the oxide (nitride) growth is studied by examining the chemical shift of the Si 2p peak. All the XPS measurements are performed under grazing emission angle (80° off normal) in order to enhance surface sensitivity and have XPS and SIMS sampling depths of the same order (~ 3Å).

3. Results and Discussion

Figure 1a shows the Si^+ secondary ion yield as a function of nitrogen coverage. N_2 was cumulatively adsorbed on Si(100) at 1000°C. Except for the case of very small exposures we observe a strong linear increase of the Si^+ signal with increasing coverage (up to ≃ 1 monolayer), suggesting that the ionization probability p^+ is di-

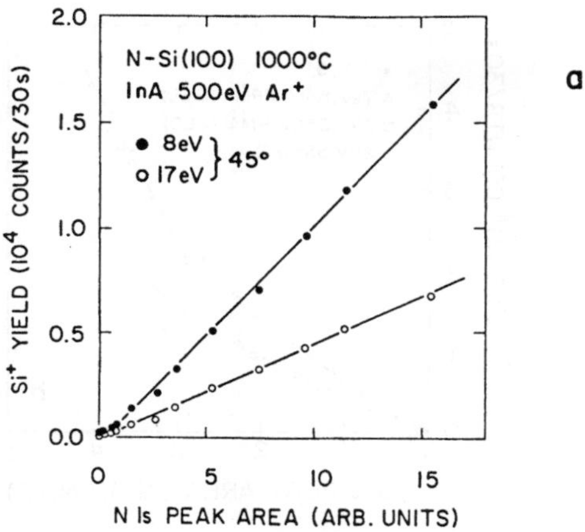

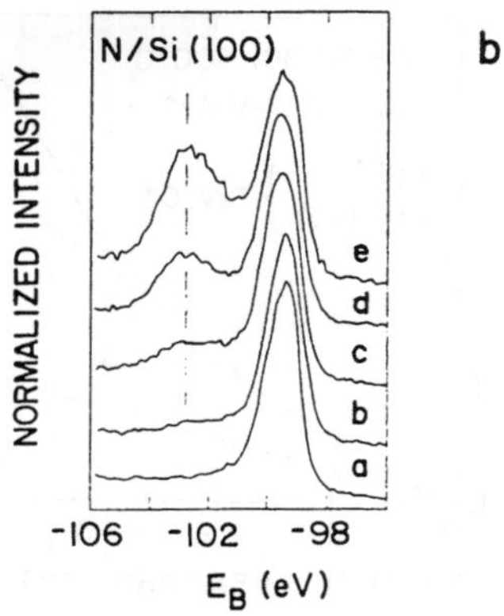

Fig. 1 (a) Sputtered Si^+ yield as a function of nitrogen coverage as monitored by the XPS N1s signal, for two different emission energies 8 eV and 17 eV and emission angle $\Theta = 45°$

(b) XPS Si2p spectra for various stages of nitrogen adsorption, from clean surface up to ~ 1 monolayer coverage (a - e)

rectly proportional to the number of Si-N bonds formed during nitridation. This result, which is well compatible with the bond breaking model, is supported by the corresponding XPS Si 2p spectra of Fig. 1b: starting from a clean Si(100) surface, nitrogen adsorption leads to a shoulder and finally to a broad second peak with ~ 3.15 ± 0.15eV shift, which agrees well with that of Si_3N_4 layers on Si. Apparently at 1000°C the mobility of the nitrogen atoms is sufficiently high to form Si_3N_4 only. Therefore, the average nitrogen coordination number, which determines p^+ (Si) following the bond breaking picture, remains unchanged for all exposures and a linear coverage dependence of the Si^+ yield is expected from this theory.

In the case of oxygen adsorption on Si (100) at T = 600° (Fig. 2) we find two regions of oxygen coverage, both exhibiting a linear increase of the Si^+ yield, but with very different slopes. The chemical shift of the Si 2p peak indicates that only suboxides are formed in the low coverage region (up to about 1 monolayer). The SiO_2 growth at higher exposures, however, leads to a stronger increase of p^+ per unit oxygen coverage. This is expected from the bond breaking model due to the higher oxygen coordination number. On the other hand, the O^-ion signal is found to be directly proportional to the oxygen coverage over the whole range (Fig. 2a), indicating that the chemical state of the oxygen atoms does not change much in these two regions.

The sensitivity of the Si^+ enhancement to the oxide chemistry decreases with increasing emission velocity (energy). As seen in Fig. 2b, the ratio of the slopes between the SiO_2 and the suboxide regions decreases from 2.6 to 1.7 when the emission velocity goes up by $\sqrt{3}$ (7eV to 21 eV). Qualitatively, both the bond breaking model

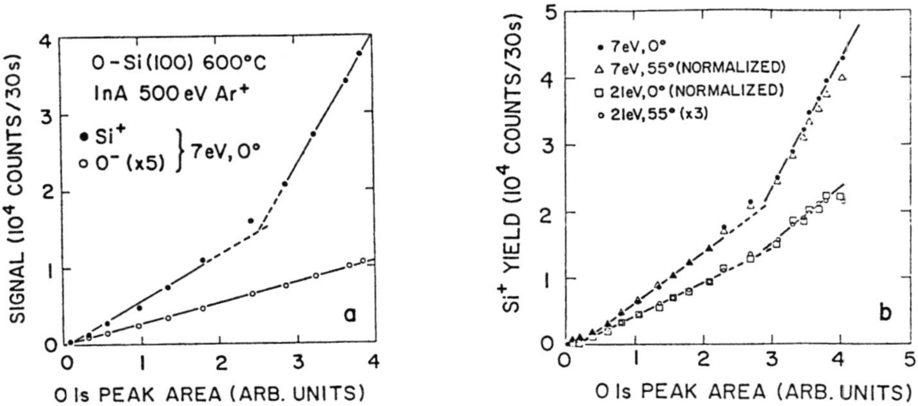

Fig. 2 (a) Sputtered Si^+ and O^- yields as a function of oxygen coverage ($\leq$ 2 monolayers)

(b) Si^+ yield as in (a), but for different emission energies and angles (E = 7 eV, 21 eV; $\Theta = 0°, 55°$). For both energies, the 0° and 55° data are normalized to the same value at 2.1 units of oxygen coverage

and the tunneling model give velocity dependences. But in the tunneling model, it is the vertical component of the emission velocity $v_z = (2E/m)^{1/2} \cos \Theta$ that is important [3] . However, by normalizing the low oxygen coverage Si^+ yields at two emission angles ($\Theta = 0°$ and 55°), the higher coverage portions show no obvious angular dependence.

The linear dependences of the Si^+ yield on the oxygen (nitrogen) coverage and the lack of correlation with v_z suggest that the local bond breaking is a more suitable description of the oxygen (nitrogen) matrix effect. We also do not see any obvious relationship between the Si^+ yield and the large changes in the valence band during oxidation and nitridation, as monitored by UPS. This is distinctly different from the Cs matrix effect where the electron tunneling model offers a good theoretical explanation.

References

1. P. Williams: Surface Sci. 90, 588 (1979) and G. Slodzian: Physica Scripta T6, 54 (1982)
2. P. J. Martin, A. R. Bayly, R. J. MacDonald, N. H. Tolk, G. J. Clark and J. C. Kelly: Surface Sci. 60, 349 (1976)
3. M. L. Yu: Phys. Rev. Lett. 47, 1325 (1981)

Ion Pair Production as Main Process of SIMS for Inorganic Solids

C. Plog, G. Roth, W. Gerhard, and W. Kerfin

Dornier System GmbH, Department NTFG, Postfach 1360,
D-7990 Friedrichhafen, F. R. G.

1. Introduction

A model of secondary ion emission was proposed [1-3] stating that neutrally emitted molecules dissociate into ion pairs above the surface. A schematic view of this process is shown in Fig. 1.

Positive monometal secondary ions of "first type" are produced by the dissociation of monometal ion molecules $MeO_n^o \rightarrow MeO_{n-1}^+ + O^-$. In Fig. 1 (left) the MeO^o dissociation is shown as an example.

Negative monometal secondary ions are produced (together with positive monometal secondary ions of "second type") by the dissociation of dimetal ion molecules $Me_2O_m^o \rightarrow MeO_{m-k}^- + MeO_k^+$ (with k<m-k). Fig. 1 (right) shows the $Me_2O_3^o$ dissociation as an example. (For silicon the yield of second type ions is in the percent range of the yield of the first type ions only.)

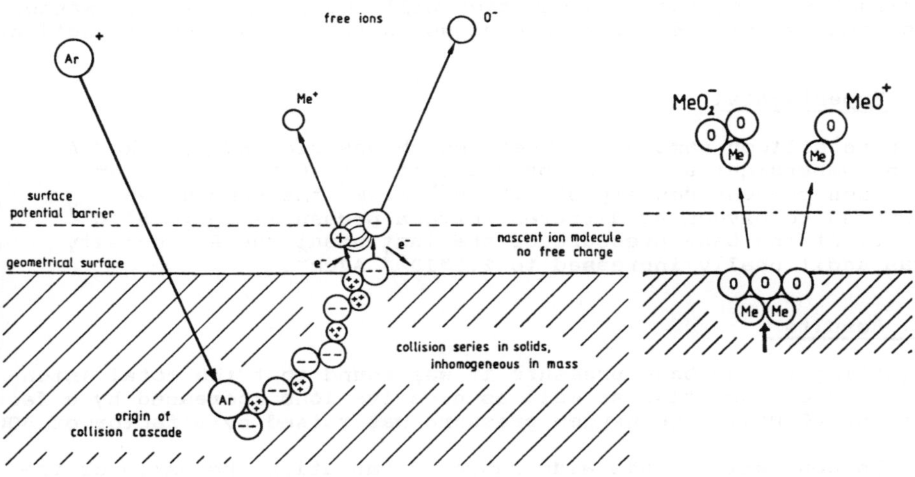

Fig. 1 According to [1-3] secondary ions are produced at the end of collision series as oppositely charged pairs.
(left): Production of positive monometal secondary ions by dissociation of (e.g.) $MeO^o \rightarrow Me^+$ and O^-
(right): Production of negative monometal secondary ions by dissociation of (e.g.) $Me_2O_3^o \rightarrow MeO_2^- + MeO^+$

Multimetal secondary ions are not produced by the "direct" but by the "statistical" emission process [4], in which a neutral molecule couples to a part of the dissociating ion molecule [1,2].

This principle of generating secondary ions from ionic solids by dissociating ion molecules entails a law of charge conservation: the number of positive secondary ions equals the number of negative ones.

Me^+ and O^- often dominate the spectra of positive and negative secondary ions. Then the law of charge conservation means a synchronized behaviour of both ions, if the entire ion energy distribution is considered and not only a small part of it [5]. Such a synchronization of Me^+ and O^- was already found [3,6,7,8]. "One might conclude that they originate from the same molecule" [6].

2. "Clean" Surfaces

Although SIMS experiments at "fully" oxidized metals were the experimental base of the model, its basic assumption - the dissociation of ionic bonds - is not restricted to that case. The polar character of a bond is an individual property of a molecule and therefore present wherever ionic bonded atoms exist at the surface. In this sense a "partly" oxidized metal means a metal which is only partially covered with ionic bonds.

It is an important question, down to which coverage of a metal with ionic bonds the model [1-3] describes the production of secondary ions. We have investigated this question experimentally using the total intensities of all positive and negative ions as well as the Me^+ and O^- intensities alone as monitor for the charge conservation. If the oxygen coverage is reduced towards a "clean" surface, these intensities will change in the same manner as long as the dissociation process dominates the ion production.

3. Experimental

A pure silicon sample was used, which was cleaned by 3 keV Ar^+ ion bombardment at a base pressure of $1.5*10^{-9}$ Torr. At an Ar^+ ion current density of $2.5*10^{-6}$ A/cm^2 the sample was exposed to oxygen at different oxygen pressures up to $8*10^{-7}$ Torr. At the base pressure of the instrument the Ar^+ density was additionally increased to $2.5*10^{-4}$ A/cm^2.

4. Results

Starting at the base pressure, it was found that the total intensity of all positive as well as negative ions increased by a factor of 40 while the oxygen pressure was raised by a factor of 500.

In contrast to this wide range of variation the ratio of the total intensities $Me_mO_n^+$ and $Me_mO_n^-$ as well as the ratio of Si^+ and O^- were found to be constant within a factor of 1.5 . The ratios even hardly changed, when the primary ion current density was raised by a factor of 100 at the base pressure. The dependence of Si^+/O^- on the oxygen pressure is shown in Fig. 2.

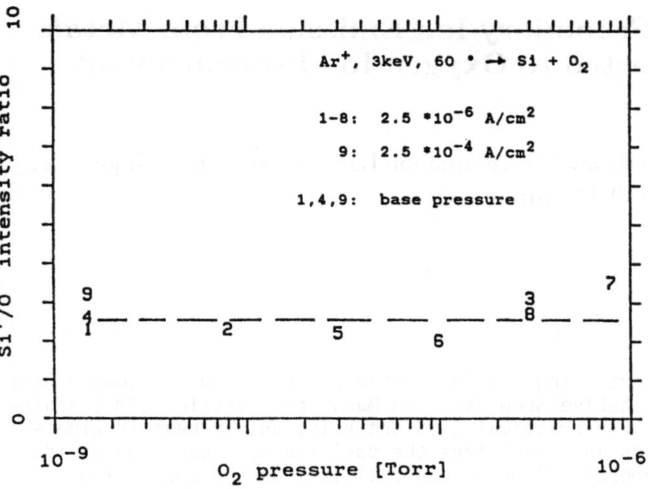

Fig. 2 Dependence of the Si^+/O^- intensity ratio (indicated by the numbers 1 to 9) on the oxygen pressure. The sequence of the numbers indicate the sequence of measuring.

5. Conclusion

The model of ion pair production elementary contains a constancy of these intensitiy ratios. As the "bond breaking" model independently produces Me^+ and O^- ions at different places at the surface, it can only explain such a constancy by an additional assumption of the geometrical arrangement.

As the constancy was found even at surfaces sometimes called "clean" (e.g. [9]), the ion pair production seems to be an extremely effective source of secondary ions. The ions produced by other ionization processes (like positive ions by Auger deexcitation) can only be observed under experimental conditions leading to essentially cleaner surfaces.

6. References

1. W. Gerhard and C. Plog Z. Phys. B 54 (1983) 59
2. C. Plog and W. Gerhard Z. Phys. B 54 (1983) 71
3. C. Plog and W. Gerhard Surf. Sci. 152/153 (1985) 127
4. W. Gerhard Z. Phys. B 22 (1975) 31
   Thesis Würzburg, Fed. Rep. Germany 1973
5. K. Mann and M.L. Yu SIMS-V conference Washington (1985)
6. H.W. Werner Develop. Appl. Spectr. 7A (1969) 239
7. K. Wittmaack Surf. Sci. 112 (1981) 168
8. H. Gnaser Int. J. Mass Spec. Ion Proc. 61 (1984) 81
9. H. Oechsner, W. Rühe and E. Stumpe Surf.Sci. 85 (1979) 289

Sputtering and Secondary Ion Emission From Metals and Alloys Subjected to Oxygen Ion Bombardment

Y. Taga

Toyota Central Research and Development Laboratories Inc., Nagakute, Aichi-gun, Aichi-ken 480-11, Japan

1. Introduction

The understanding of sputtering and ionization phenomena by ion bombardment is necessary for quantitative Secondary Ion Mass Spectometric (SIMS) analysis of solid surfaces. In practical SIMS analysis, an O_2^+ beam is widely used because it enhances and stabilizes the positive secondary ion yields [1]. Although many reports [2] have been published on the sputtering yields of metals by noble gas ion bombardment, there has been little systematic study for O_2^+ bombardment.

On the other hand, the ionization phenomena of sputtered atoms or molecules are directly responsible for the quantitative interpretation of the SIMS results. Although many theoretical models [3] have been proposed to describe the ionization phenomena of sputtered particles, a unified explanation cannot be given even at present. In addition, it is also well recognized that the ionization process is very much sensitive to the states of the surface, the matrix and the effect induced by the primary ion beam, especially under oxygen ion bombardment.

Precise measurements of sputtering and secondary ion yields may therefore result in the reliable data on the degree of ionization of sputtered particles. One of the most reliable and suitable methods to investigate the secondary ion emission mechanism will be to measure the secondary ion energy spectra under O_2^+ bombardment in ultra high vacuum (UHV) conditions.

The present authors have established [4] the methods of simultaneous determination of sputtering and secondary ion yields for thin film targets with known thickness and composition. In addition, a UHV-SIMS system was also constructed for the measurement of the precise secondary ion energy spectra [5],[6]. Systematic study of sputtering and secondary ion emission under O_2^+ bombardment has been thus extensively carried out in our laboratory in combination with thin film preparation technology [7]-[9].

This paper is mainly devoted to a synthetic presentation of the work done by "Toyota Group", whose bibliography can be seen more extensively in the cited references.

2. Sputtering yields

The most popular parameter characterizing sputtering process is the so-called sputtering yield, which is defined as the mean number of atoms removed from the surface of a solid per incident particle. The classical method for determining the amount of removed material is to weigh the target before and after irradiation and deduct the change in mass of the target. On the other hand, geometrical thickness or crater depth change before and after irradiation is also measured with a mechanical microstylus.

Provided the target consists of a thin film of known thickness and composition, either unsupported or supported by a substrate of different materials, and provided the irradiation is laterally homogeneous, depth profiling technique using surface analytical tool such as SIMS and AES may be used to reveal when the target is sputtered away. Sputtering-through of thin films has also been used as a yield-measuring technique by BENNINGHOVEN [10], who utilized SIMS. Preferential sputtering in connection with the unavoidable atomic mixing at the interface may give rise to an apparent shift of the interface. This effect, however, seems to be negligible if the film is enough thicker than atomic mixing layer.

Another important parameter in this method is the density of the film to be sputtered away. It is generally considered that the density of thin film with a thickness larger than 1000 Å is identical with that of bulk metals [11]-[13]. Recent progress in thin film technology also guaranteed the high density metallic film by the so-called magnetron sputtering technique.

In this study, the multiple-target magnetron sputtering [7],[8] technique was used to deposit the thin film samples with known thickness and composition. Multi-layer films with a known thickness and alloy films with known thickness and composition were deposited onto the mirror-polished Si wafer substrate at 35°C. Quantitative determination of the thickness and composition of the films were performed by electron probe microanalysis (EPMA) [14],[15], chemical analysis (CA) and mechanical stylus technique (MST). Experimental conditions for thin film preparation and characterization were described elsewhere [16],[17].

We have already reported [4] that the sputtering yield of metals under O_2^+ bombardment is inversely proportional to the energy-transfer factor, i.e., the sputtering yield increases with decrease in the efficiency of the energy transfer from the incident ions to the target atoms. Recently, OKAJIMA [18] measured the sputtering yield of several metals under Ar^+ bombardment and showed a linear plot on a log-log scale between the sputtering yield and the cohesive energy of the target materials. By taking OKAJIMA's data into account together with ours, it is concluded that the sputtering process of pure metals under oxygen ion bombardment can be simply explained in terms of the classical atomic collision model, irrespective of the chemical nature of the primary ions.

On the other hand, the total sputtering yields Y of Ti-Al [16] and Cu-Al [17] alloy films as a function of Al concentration are shown in figure 1. The broken lines in the figure indicate the straight lines connecting sputtering yields of Al and Ti and Al and Cu. It was found that the total sputtering yields of single phase Ti-Al and Cu-Al alloy films did not change monotonically with alloy composition. These results indicated that the alloying with Al reduced the partial sputtering yields of Cu, Ti and Al, since the total sputtering varies in accordance with the straight lines as shown in fig. 1 if each partial sputtering yield in alloy films is equal to that of pure element.

No sputtering yield data of Ti-Al and Cu-Al alloys have been reported. However, BETZ [19] and MEYER and WEHNER [20] reported that the yield of single phase alloy changed monotonically with the alloy composition. On

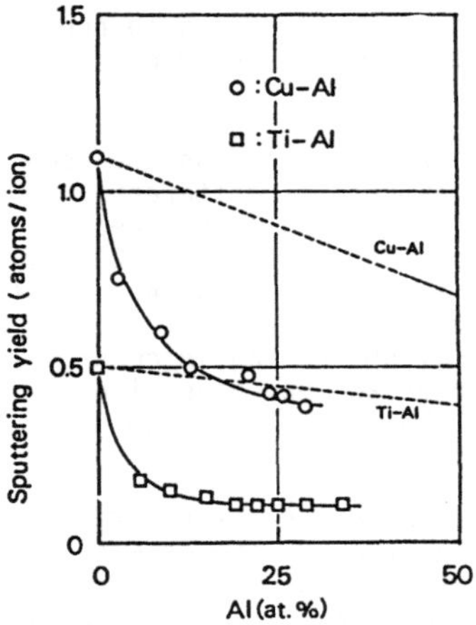

Fig. 1 Changes in total sputtering yields of Ti-Al and Cu-Al alloy films as a function of Al content.

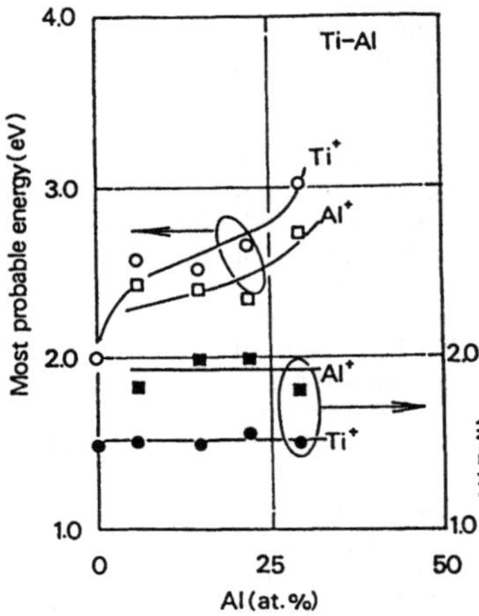

Fig. 2 Changes in most probable energy and the tail factor of Ti-Al alloy film as a function of Al content.

the other hand, GNASER el al. [21] observed the existence of a minimum yield in the Au-Pd alloy system. By taking the published data into consideration together with ours, the sputtering behaviour of alloys is not simply attributed to the solubility.

A detailed study of secondary ion energy spectra can help to identify the dominant mechanism in the ionization and sputtering processes. Attempts have been made to measure the energy distribution of sputtered ions of Ti-Al alloy films by using a UHV-SIMS apparatus [22]. The tail factor of the energy distribution of secondary ion species can be expressed as a function of E^{-N} at higher energies, where E is the energy of secondary ions and N is a fitting parameter. Changes in the most probable energy and tail factor N of energy distribution of sputtered ions of Ti-Al alloy films were shown in figure 2 as a function of Al content. Although the most probable energies for Al and Ti ions increase with increase in Al content, the tail factor N values remains unchanged,as shown in the figure.

According to Sigmund's theory [23], the sputtering yield Y is inversely proportional to the surface binding energy Us of the sample, which is usually expressed by the sublimation energy. However, Us in alloy under oxygen ion bombardment cannot be simply obtained with the present knowledge. On the other hand, the energy distribution of sputtered atoms is derived from the theoretical model by THOMPSON [24] as

$$N(E) = CE / (E + E_b)^3 \qquad \text{(i)}$$

where C is a material constant and E_b is the surface binding energy. The most probable energy of sputtered atoms is obtained to be $E_b/2$ from the eq. (i). Secondary ion energy distribution $N^+(E)$ is expressed by using the ionization probability $R^+(E)$ as

$$N^+(E) = N(E) \cdot R^+(E) \ . \qquad \text{(ii)}$$

Although several mechanisms have been proposed for the formation of sputtered ions, it is much more difficult to specify the form of $R^+(E)$ appropriately. On the other hand, the shift in most probable energy can also result from an increased work function. However, it is quite difficult to measure the changes in work function of Ti-Al alloys during O_2 ion bombardment. Therefore the shift in most probable energy of sputtered ions shown in fig. 2 can be primarily explained in terms of the increase in surface binding energy of sputtered atoms. In this case, the ionization probability $R^+(E)$ of Al and Ti becomes constant with Al content in Ti-Al alloy films. Based on the above discussion, it can be concluded that the decrease in the total sputtering yields of Ti-Al alloy films with Al content is explained in terms of the increase in surface binding energy as a result of alloying.

3. Secondary ion yields and degree of ionization

Measurement of the positive secondary ion current was performed with the ARL-IMMA. The mass filtered O_2 ions struck the target at 18.5 keV at normal incidence and were rastered to provide uniform sputtering over an area of 70 µm × 70 µm. The primary ion beam current and diameter were 10 nA and 10 µm, respectively. Electronic gating was used to limit the data acquisition on positive secondary ion current to the central part (5 %) of the crater, to avoid crater edge effects. The chamber pressure was maintained at 10^{-5} Pa during ion bombardment. The positive ion yield was finally determined after isotope correction of the measured secondary ion current.

It is well known that in SIMS study, small changes in the experimental conditions are very much influential on secondary ion emission. Therefore, reproducibility was certified three times, and each measurement shown in the following figures denotes the average data.

We have shown in the previous report [4] that a linear relationship was found between the ionization potential and a modified degree of ionization which can be expressed by the the degree of ionization normalized by the secondary O_2^+ current. This modification was also applied for Ti-Al and Cu-Al alloy films and the results are shown in figs. 3 and 4. It was found that the modified degree of ionization of Al was constant, and that of Cu

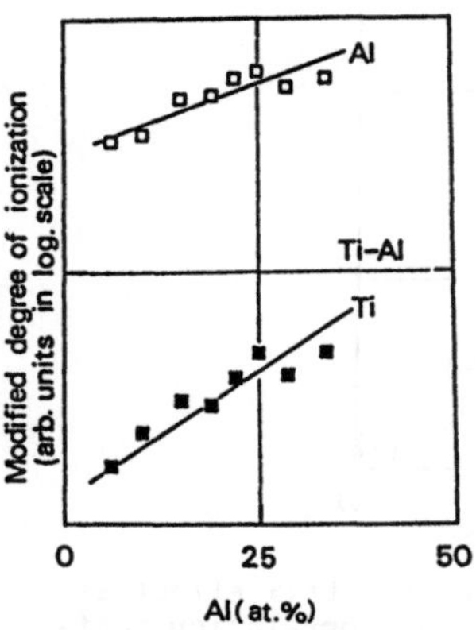

Fig. 3 Modified degree of ionization in Ti-Al alloy film as a function of Al content.

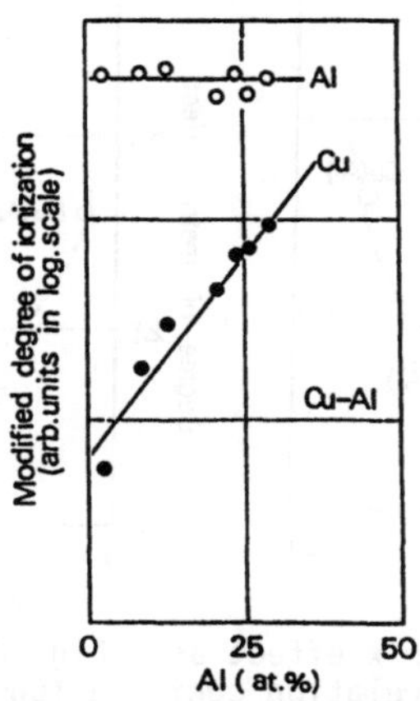

Fig. 4 Modified degree of ionization probabilities in Cu-Al alloy film as a function of Al content.

varied exponentially with Al content. The gradient of the linear relationship between the logarithm of the modified degree of ionization and the Al content indicates a degree of matrix effect due to Al content. From fig. 4, the gradients of Al and Cu were evaluated to be zero and 4.0, respectively. From fig. 3, in Ti-Al alloy the gradients of Al and Ti were evaluated to be 1.0 and 2.0, respectively. The comparison of the gradients between Ti and Cu revealed that the matrix effect due to Al content affected Cu more strongly than Ti. It was also shown that the degree of matrix effect on Al changed due to the second component; namely, Ti and Cu. This result suggests that the degree of matrix effect depends not on a characteristic of individual components, but rather on a correlation between the characteristics of the components. We tried to devise a quantitative representation of the correlation coefficient between the characteristics for the purpose of understanding the factors governing the matrix effect. As candidates for the characteristic concerning secondary ion emission under oxygen ion bombardment, the first ionization potential E_i, and the heat of formation of oxide ΔG, have been proposed by the present authors [4] and by YU and REUTER [25], respectively. On the other hand, as candidates for a representation of the correlation coefficient, we propose the quantity $X/(X+X')$, where X and X' are the characteristic values of alloy components.

The degrees of matrix effect are shown in figs. 5 and 6 as a function of the correlation coefficients of E_i and ΔG. It was found that the degree of matrix effect steadily increased and decreased with increasing the correlation coefficient of E_i and that of ΔG, respectively. These results indicate that the matrix effect under oxygen ion bombardment more strongly affects the element with larger first ionization potential E_i and smaller heat formation of oxide ΔG than those of the second component. Using these trends, it is possible to predict when matrix effects will be a problem. Then, using the appropriate calibration lines, it is possible to correct for these matrix effects. Finally, the relationship observed can be used to improve the quality of both qualitative and quantitative SIMS analyses.

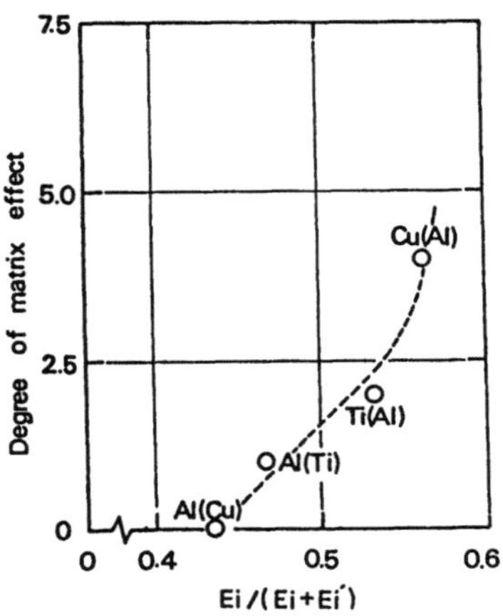

Fig. 5 Degree of matrix effect as a function of the correlation coefficient of first ionization potentials E_i of alloy components.

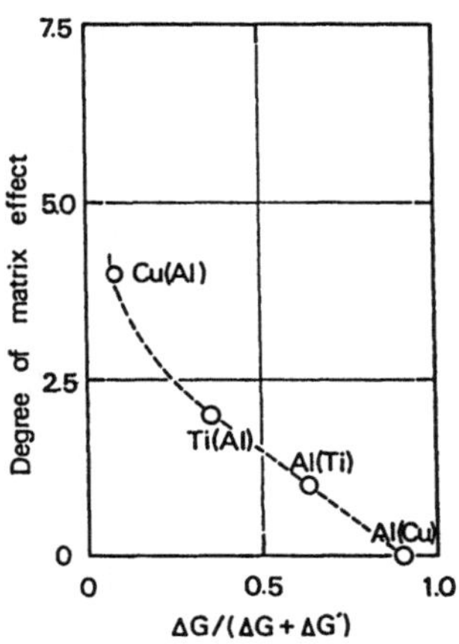

Fig. 6 Degree of matrix effect as a function of the correlation coefficient of heats of formation of oxide ΔG of alloy components.

Recently, GALUSKA and MORRISON [26] also investigated the influence of matrix effects on ion yields in Group III-V compound and found that elements that have a small ionization probability are enhanced to a greater extent by oxygen ion bombardment or by the presence of oxygen than those that have a large ionization probability.

4. Conclusion

The sputtering yield and the ionization probability under oxygen ion bombardment were determined for Cu-Al and Ti-Al alloys. The results revealed that:
(1) The alloying with Al reduced partial sputtering yields;
(2) the modified ionization probability, which was expressed by the ratio between ionization probability and secondary oxygen ion intensity, varied linearly with Al content;
(3) the gradient of the exponential relationship between the modified degree of ionization and Al content, which indicated the degree of matrix effect due to Al content, steadily increased and decreased with increasing the correlation coefficient of the first ionization potential and that of the heat of formation of oxide, respectively.

5. Acknowledgements

The author wishes to express his thanks to Director Dr. H. Morimoto of Toyota Central Research and Development Laboratories for his continuous encouragement and guidance in this research project. To Dr. K. Satta, I am grateful for his helpful discussion and comments during the course of this research. Thanks are also due to my colleagues, Messrs. K. Inoue and T. Ohwaki for their kind cooperation and discussion throughout the present work.

1. C.A. Andersen and J.R. Hinthorne: Anal. Chem. 48, 832 (1973)
2. H.H. Andersen and H.L. Bay: in Sputtering by Particle Bombardment, Vol.1, Ed. R. Behrisch (Springer, New YOrk, 1981) p.145
3. P. Williams: Surface Sci. 90, 588 (1979)
4. Y. Taga, K. Inoue and K. Satta: Surface Sci. 119, L363 (1982)
5. T. Ohwaki and Y. Taga: Jap. J. Appl. Phys. 23, 1466 (1984)
6. T. Ohwaki and Y. Taga: Surface Sci. 157, L308 (1985)
7. Y. Sawada and Y. Taga: Thin Solid Films 110, L129 (1983)
8. Y. Sawada and Y. Taga: Thin Solid Films 116, L55 (1984)
9. T. Motohiro and Y. Taga: Thin Solid Films 112, 161 (1984)
10. A. Benninghoven: Z. Angrew. Phys. 27, 51 (1969)
11. A.R. Wolter, J. Appl. Phys. 36, 2377 (1965)
12. T.E. Hartman: J. Vacuum Sci. Technol. 2, 239 (1965)
13. J. Edgecume: J. Vacuum Sci. Technol. 3, 28 (1966)
14. D.F Kyser and K. Murata: IBM J. Res. Develop. 352 (1974)
15. Y. Oda and K. Nakajima: Trans. Japan Inst. Metals 16, 697 (1975)
16. K. Inoue and Y. Taga: Surface Sci. 140, 491 (1984)
17. K. Inoue and Y. Taga: Surface Sci. 147, 427 (1984)
18. Y. Okajima: J. Appl. Phys. 51, 715 (1980)
19. G. Betz: Surface Sci. 92 283 (1980)
20. W.K. Meyer and G.K. Wehner: J. Vacuum Sci. Technol. 16, 808 (1979)
21. H. Gnaser, J. Marton, F.G. Rudenauer and W. Seiger: Spectrochim. Acta 37B, 797 (1982)
22. T. Ohwaki, K. Inoue and Y. Taga: to be published.
23. P. Sigmund: Phys. Rev. 84, 383 (1869)
24. M.W. Thompson: Phil. Mag. 18, 377 (1968)
25. M.L. Yu and W. Reuter: J. Appl. Phys. 52, 1478 (1981)
26. A.A. Galuska and G.H. Morrison: Int. J. Mass Spec. Ion Phys. 61, 59 (1984)

Measurements of the SIMS Isotope Effect

S.A. Schwarz

Bell Communications Research, 600 Mountain Ave.,
Murray Hill, NJ 07974, USA

1. Introduction

Several models of ion emission predict that the heavier isotopes of a given element will be under-represented in the SIMS mass spectrum. Isotope effect measurements by the Orsay group [1,2] and by SHIMIZU and HART [3] have demonstrated this behavior. The present study supplements this work with additional velocity-dependent measurements of the isotope effect in various materials. Negative ion emission and dimer formation are also investigated. Several interesting experimental observations are reported here.

2. Experiment

A Cameca IMS-3f magnetic sector instrument with a primary beam mass filter was employed for these studies. The energy slit was narrowed to obtain ~5 eV resolution while restricting the count rate to <500,000 cps, well within the linear region of the detector. An 8 keV O_2+ beam and a 14.5 keV Cs+ beam were used for positive and negative ion detection respectively, with sufficient presputtering to reach steady state. High purity foils of Mo, Pd, Cd, Sn, and W as well as a Ge ingot were examined. Each of these elements offers several isotopes for examination. Data was obtained by biasing the sample to the appropriate voltage, then recording and adding several mass spectra together to avoid beam instability effects. Individual mass peaks were integrated and a linear correction applied to correct for the constant mass resolution and parabolic magnetic field dependence encountered in the mass spectra mode. The measured isotopic intensities were divided by the standard isotopic ratios [4] and plotted as in Fig. 1. (The standard dimer isotopic ratios were computed as random combinations of the elemental ratios.) A least squares linear fit weighted by the isotopic abundances was then applied to these plotted deviations. The resultant slope of this linear fit yields the average % deviation per amu from the true isotopic ratios. This factor, multiplied by -M/100 (where M is the mass of the common isotope), gives the factor α as defined in [1]. The experimental procedure described here provides good signal-to-noise capability, is resistant to mass interferences and magnet or voltage drifts, and is preferable to the peak switching mode of operation [1-3] when several strong isotopes are available for analysis.

3. Results

Figure 1 illustrates the measured isotopic deviations (%) observed for Ge_2+ biased at 0V. The circle diameters are proportional to the isotopic abundance and the error bars indicate the counting statistics only. The quality of the fit supports the random combination assumption.

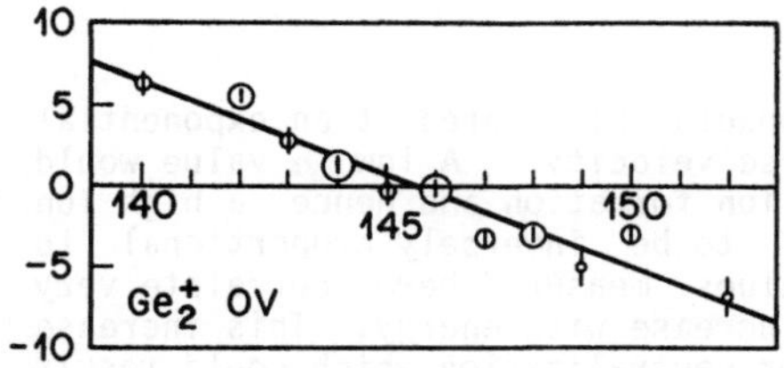

Fig. 1 Measured isotope abundance deviation (%) for Ge_2+ at OV bias with weighted least squares fit shown

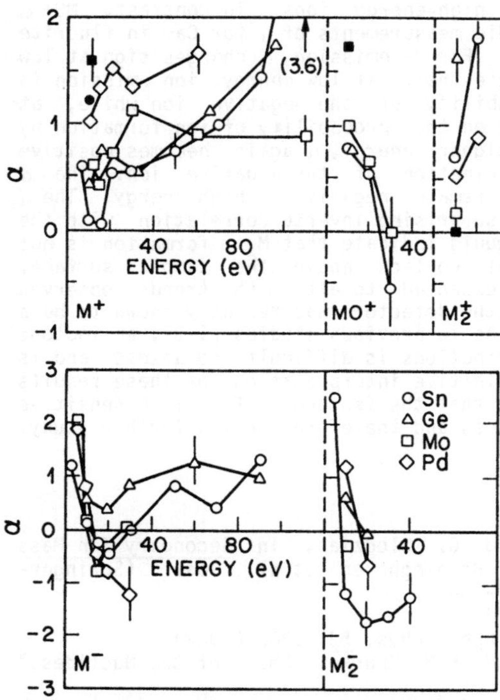

Fig. 2 Measured α values vs. secondary ion energy for M+, MO+, and M_2+ emission under O_2+ bombardment and for M- and M_2- emission under Cs+ bombardment in high-purity Mo, Pd, Sn, Cd, W, and Ge samples

Figure 2 shows the measured α values for M+, MO+, M_2+, M-, and M_2- species. Each point represents one experimental run. Several duplicate runs suggest an experimental reproducibility of ~20%. Only a few error bars are shown owing to the complexity of the figure. The larger error bars indicate runs with low count rates for which the error decreases as the square root of the total number of counts. In several of the MO+ runs it was necessary to correct for hydride interference by subtracting from each mass peak a percentage (<2%) of the previous peak. The magnitude of the hydride interference is easily ascertained in the mass spectra. A OV bias corresponds roughly to the maximum secondary ion intensity. The actual ion energies plotted in the figure are assumed to be 10 eV greater than the applied bias. The results for Mo+ are in good agreement with [3], exhibiting a maximum α of ~1 at 30 eV and a relatively level plateau at higher energies.

4. Discussion and Conclusions

The neutralization and bond-breaking models [1,2] predict an exponential dependence of the ion yield on inverse velocity. A low α value would correspond with little inhibition of ion formation and hence a high ion yield. The parameter α is expected to be inversely proportional to velocity. For M+ emission, the α values measured here correlate very roughly with ion yields and tend to increase with energy. This increase suggests a longer interaction time for neutralization which could result from subsurface emission of the high-energy ions. In contrast, MO+ α values decrease with energy, as did measurements of α for Ca+ in fluorite [1], suggesting surface emission. For M- emission, α changes sign at low energy (~10 eV). This may indicate that, at low energy, ion emission is dominated by the survival probability of the negative ion while, at higher energies, emission depends on the probability of ion formation by electron attachment. At still higher energy, α again becomes positive and may indicate subsurface origination of the negative ion. The α values for M_2- emission appear to remain negative at high energy. The α values for M_2+ emission show a surprising inverse correlation with the corresponding M+ values. This would indicate that M_2+ formation is not simply a result of ion-neutral contact above the sample surface. Experimental artifacts are not expected to alter the trends observed here. Isotopic fractionation in the detector was recently shown to be a comparatively small effect [5]. As in previous studies [1-3], an isotope effect in emitted angular distributions is difficult to assess and is assumed to be small. While a definitive interpretation of these results is not yet possible, it is clear that the isotope effect is a sensitive indicator of the ionization process and therefore merits further study.

References

1. J.C. Lorin, A. Havette, and G. Slodzian, in Secondary Ion Mass Spectrometry - SIMS III, A. Benninghoven et al., eds. (Springer-Verlag, Berlin 1982), pp. 140-150
2. G. Slodzian, ibid., pp.115-123
3. N. Shimizu and S.R. Hart, J. Appl. Phys. 53, 1303 (1982)
4. F.W. Walker, G.J. Kirouac, and F.M. Rourke, "Chart of the Nuclides," (General Electric Co. 1977)
5. A.J. Fahey, K.D. McKeegan, and E.K. Zinner, these proceedings

SIMS Study of Self-Diffusion in Liquid Tin and of Associated Isotope Effects

U. Södervall, H. Odelius, and A. Lodding

Physics Department, Chalmers University of Technology, S-41296 Gothenburg, Sweden

G. Frohberg, K.H. Kraatz, and H. Wever

Institute of Metal Research, Technical University Berlin, D-1000 Berlin 15

1. Spacelab Investigation of Self- and Isotope Diffusion in Liquid Metal

A Spacelab collaboration project is directed at diffusivities in liquid metals under gravity-free, g(0), conditions. Terrestrial measurements yield effective diffusion coefficients, D(eff) = D + D(g) + D(w) , where the last two terms are convection effects due to, respectively, gravitation and the presence of walls in the anneal vessel. Determinations of D have therefore suffered in accuracy and reproducibility, and this has impeded the testing of atomic models of liquid diffusion. In the Spacelab project D(g) is eliminated and D(w) may be assessed by measurements of D(eff) in capillaries of different diameters. The aim is an accuracy sufficient for quantitative study not only of the temperature dependence of D, but also of the isotope effect, defined as $E_D = 2\, d\ln D/d\ln M$ (where M is the atomic mass).

Self-diffusion in tin was the first objective of the g(0) study. The anneals took place in temperature-stabilized, 5 cm long, 1 or 3 mm diam., graphite capillaries. A drop of isotopically enriched metal, nearly pure 124-Sn or a 50-50 mixture of 112-Sn and 124-Sn, had been molten under vacuum on to the butt end of the tin cylinder, providing thin-film type diffusion geometry /1/. Consequently after an anneal time t, at a distance x from the butt end, the atomic fraction c_i of isotope i was to be represented by a straight line in a ln c* versus x^2 diagram (see fig.1), referring to a normalized concentration, $c^* = (c-c_0)_i/\{1-(c-c_0)_i\}$, with c_{0i} denoting the natural abundance in non-enriched metal. The diffusion coefficient could then be obtained /1,2/ as

$$D_i = -(1/4t)\, d\ln c^*/d(x^2) \qquad (1)$$

and the mass effect of two isotopes i and j (here 124-Sn and 112-Sn) as

$$E_D = 2\,\frac{d\ln(c_i^*/c_j^*)}{d\ln c_{mean}}\;\frac{M_{mean}}{M_i-M_j}\quad . \qquad (2)$$

The use of SIMS was a necessary base for the investigation; a tracer sensitive isotope analysis was to be combined with the possibility to study the variations in $c-c_0$ both along and across the diffusion column, with a resolution of the order of 5 µm. The crucial requirements were set by the need to measure the ratio $(c-c_0)_i/(c-c_0)_j$ within less than ca 3% at the low tracer end of the column, i.e. where $c-c_0 \cong 10^{-3}$. The c_0-s being ca 0.05 and 0.1, each of the 4 c-terms in the ratio must be determined within ca 0.1 %. To achieve such accuracy, at least 10^6 pulses have to be counted. This was effected by sputtering for ca 30 to 60 minutes at each spot, recording in 20 to 40 cycles the ion counts for all 10 Sn-isotopes. All concentrations at each cycle were obtained by on-line computer after correction for counting deadtime /3/ and for hydride pick-up, equating the sum of all isotopes to 100 %.

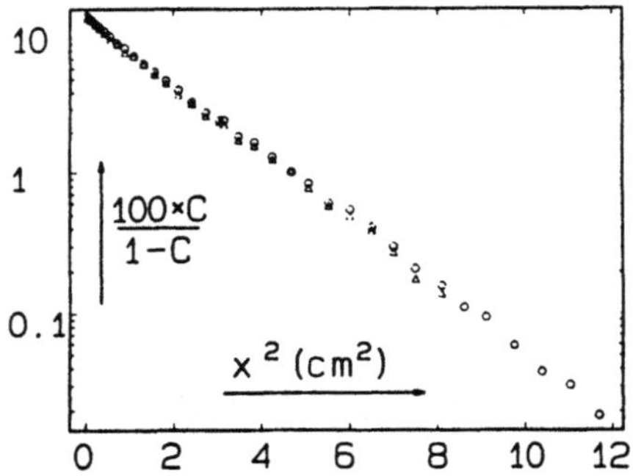

Fig.1 Gaussian diffusion profiles of two Sn isotopes in liquid tin. Normalized concentrations c_i^* & c_j^* versus square of penetration. Rings: 112-Sn. Triangles: 124-Sn

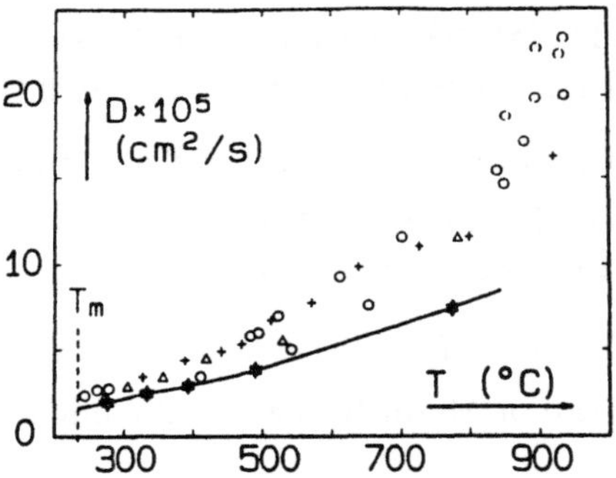

Fig.2 Self-diffusion in liquid tin. Whole-drawn and stars: present results, calculated for mean isotope mass. Triangles: results from ref. /5/. Rings: ref. /6/. Crosses: ref. /7/

A Cameca IMS-3F instrument was used. The diffusion profile was recorded by step-scan with 1 to 3 mm intervals along the axis of the flat semicylinder surface obtained by cutting the column into two halves. O_2^+ primary ions, ca 1 μA, were scanned over ca 60x60 μm². Each analyzed spot was 10 μm in diameter. It was important to prevent a possible drift in isotope fractionization due to the SIMS process itself (see section 2 below); relatively high offset in sample voltage, 100 eV, was therefore employed.

Table 1. Diffusion coefficients in liquid tin ($10^5 cm^2/s$), 112-Sn and 124-Sn in 1 mm capillaries, 124-Sn in 3 mm capillaries. Isotope effect of self-diffusion. (Error margins of E_D: ≈ 25%)

T (°C)	D_{j1}	D_{i1}	D_{i3}	E_D
270	2.14	2.10	2.16	0.44
328	2.47	2.41	2.41	0.41
388	2.97	2.88	2.87	0.52
491	3.99	3.90	3.95	0.48
775	7.61	7.42	7.70	0.56

A typical in-depth profile plot is shown in Fig.1. The results are listed in Table 1. Excellent correlation between the D values is seen at each temperature: a) the capillary diameter effect is found negligible, in contrast to many earlier measurements of liquid diffusion /4,5/; b) the isotope effect, possibly the first ever reproducibly measured in capillary-type diffusion anneals of liquid metals, lies within a reasonable range (E ≅ 0.5 suggesting cooperative position adjustments in diffusion /1,2/).

Figure 2 shows the temperature dependence of D in liquid tin in relation to results obtained in earlier measurements. The present points lie lowest, implying freedom from convection, and show by far the least scatter. The technique of annealing at g(0), using stable isotope tracers and evaluating the concentration profiles by SIMS might therefore be considered as a possible standard procedure for future investigations of liquid diffusion.

2. Isotope Fractionization Effects in SIMS

In isotope measurements by SIMS, mass fractionization due to the sputtering and ionization processes is of obvious relevance. The ionic emission R_i of an isotope i is dependent on the mass M_i via $R_i \sim M_i^{\alpha}$, where α may vary with the emitted species, the matrix and the exit kinetic energy of the ion /8,9/. If the natural abundancies of 2 isotopes are measured, their relative ion yield will not normally equal the actual abundancy ratio $(R_i/R_j)_o$. By definition,

$$\frac{d\ln(R_i/R_j)}{d\ln(M_i/M_j)} = \alpha \cong \frac{(R_i/R_j)-(R_i/R_j)_o}{(R_i/R_j)_{mean}} \Big/ \frac{M_i-M_j}{M_{mean}} \quad . \tag{3}$$

In connection with the measurements of isotope diffusion, a study of α has been made in several isotopic matrices. Fig.3 shows the ratios of measured and true abundancies, R_i/R_{io} /10/ for Sn isotopes under similar conditions as in the diffusion study. The slope of the straight line, according to eq.3, to a first approximation yields $\alpha \cong -1.5$. This applies at ion exit energies, E_{kin}, above ca 70 eV. In Fig.4, the measured α values are plotted versus E_{kin} for elemental Sn, Ge and Cu matrices. It is seen for all three cases that $-\alpha$ rises steeply at low E_{kin}, but tends to stabilize at higher energies. In the diffusion study a relatively high offset in sample voltage has been chosen, analyzing the Sn ions in the 75 to 125 kV range of E_{kin}.

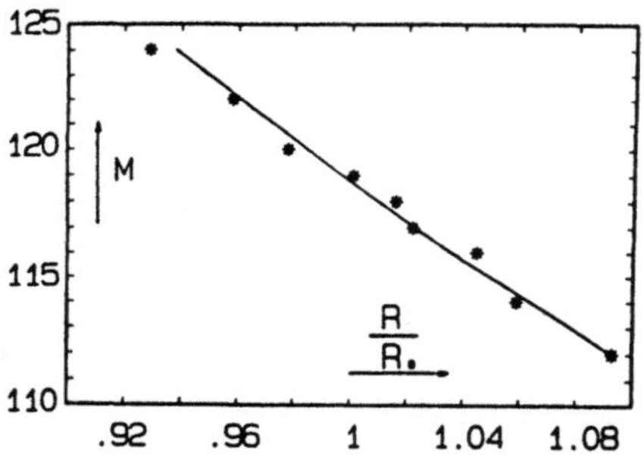

Fig.3 Specific SIMS yields of Sn^+ isotope ions, emerging with energies $\geq$ 70 eV from tin. Slope of line yields fractionization factor α

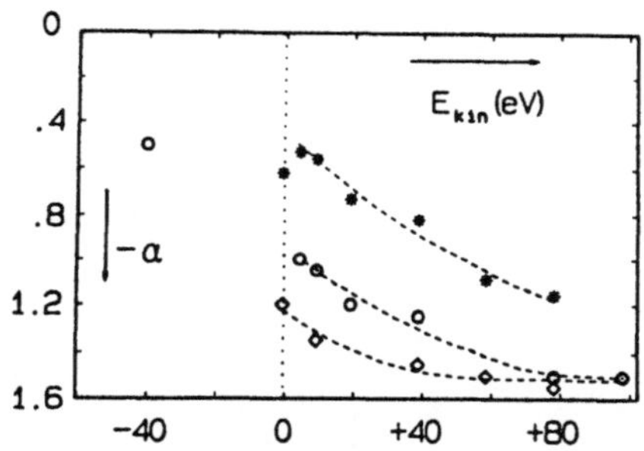

Fig.4 Isotope fractionization factor versus secondary ion exit kinetic energy for elemental tin (rings), germanium (asterisks) and copper (diamonds)

For all 3 isotope systems in Fig.4, α is seen to behave similarly, i.e., is of the order of -1/2 (as expected e.g. for thermionic emission) at low E_{kin}, but approaches ca -3/2 at high offsets. The latter tendency (but not the decrease of $-\alpha$ at low E_{kin}) has also been reported from isotope studies on minerals /8,11/. The present investigations in pure Sn are planned to be followed by diffusion studies in binary systems of Sn and other isotope metals. It is then to be considered, that in a given matrix α may vary from one species to another. Thus, as an offspin of a recent SIMS study /12/, the following preliminary results can be quoted for a mixed borosilicate glass at high E_{kin}: $\alpha_{Li} = -0.38 \pm 0.10$; $\alpha_{Mg} = -0.95 \pm 0.20$; $\alpha_{Si} = -1.15 \pm 0.35$; and $\alpha_{Zr} = -0.65 \pm 0.15$.

References

1. Y. Adda and J. Philibert: La Diffusion dans les Solides (PUF, Paris 1966)
2. Th. Hehenkamp, A. Lodding, H. Odelius and V. Schlett: Acta Metall. 27, 827 (1979)
3. H. Odelius and U. Södervall: in "SIMS IV" (Benninghoven & al., eds.; Springer Ser. Chem. Phys. 36, Berlin-NY), pp 311-316 (1984)
4. D.A. Rigney: in "Liquid Metals" (Evans and Greenwood, eds.; Inst. Phys. Conf. Ser. 30, Bristol-London), pp 619-635 (1976)
5. G. Careri, A. Paoletti and M. Vicentini: Nuovo Cim. 10, 1088 (1958)
6. C.H. Ma and R.A. Swalin: Trans. AIME 230, 426 (1964)
7. S.N. Krykov, E.A. Soldatov and V.I. Irkov: Vestn. Mosk. Univ., Khim. 12(3), 332 (1971)

8. G. Slodzian, J.C. Lorin and A. Havette: J. Physique (Paris) 41, L-555 (1980)
9. A. Lodding: in "Reviews on Analyt. Chem. (Niinistö, ed.; Akademiai Kiadò, Budapest), Ch.6 (1982)
10. H.E. Duckworth: Mass Spectroscopy (Cambridge U. P.), pp 177-182 (1958)
11. J.C. Lorin, A. Havette and G. Slodzian: in "SIMS III" (Benninghoven & al., eds.; Springer Ser. Chem. Phys. 19, Berlin-NY), pp 140-149 (1982)
12. H. Odelius, A. Lodding, L.O. Werme and D.E. Clark: Scanning Elec. Microsc. 85-III, 927 (1985)

Silicon-Induced Enhancement of Secondary Ion Emission in Silicates: A Study of the Matrix Effects

N. Shimizu

Department of Earth, Atmospheric & Planetary Sciences,
Massachusetts Institute of Technology, Cambridge, MA 02139, USA

One of the elusive goals in application of SIMS techniques to geochemistry and cosmochemistry is to devise a procedure by which secondary ion intensities can be quantitatively converted to concentrations. Although an empirical "working curve" approach has been successful for trace elements (see, for instance, RAY and HART[1]), quantitative analysis of the major elements has been difficult due mainly to the complex relationships between secondary ion intensity and concentration known as matrix effects. The matrix effects occur among different mineral groups (qualitatively similar to the "SIMS matrix effect" of DELINE et al. [2]) and even within a given mineral group as chemical composition of the solid solution varies (e.g., SHIMIZU et al. [3]). A better understanding of the matrix effects is imperative in establishing a quantitation procedure for geologic materials as most minerals form complex and extensive solid solutions, and is also essential, in more general terms, in understanding the physics of the secondary ion emission process.

This paper is intended to describe the observations of secondary ion emission of Mg, Ca, Al and Si in relation to the chemical composition and the secondary ion energy in two binary silicate glass systems, $CaAl_2O_4$-SiO_2 (CAS glasses) and $MgAl_2O_4$-SiO_2 (MAS glasses). A Cameca IMS 3f ion microprobe was used with a beam of negatively charged oxygen ions with the net energy ranging from 12.62 to 12.70 keV and with a current typically around 0.2 nA, and the beam was focused to a spot of 5–8 μm in diameter. The positively charged secondary ions were accelerated to 4.42–4.50 keV, analyzed with combined electrostatic and magnetic sectors, and detected by an electron multiplier in pulse-counting mode (with a dead time approximately 30 ns). The intensity data were acquired under steady-state sputtering conditions and were corrected for background and dead time.

The kinetic energy distribution was determined with a resolution of ±2 eV for $^{40}Ca^+$, $^{27}Al^+$ and $^{28}Si^+$ in nine CAS glasses ranging in composition (in atomic %) from 8.98%Si, 60.9%Al, 30.1%Ca (CA9S1W) to 82.4%Si, 11.8%Al, 5.85%Ca (CA2S14). It was observed that the width of the energy distribution as measured by the energy spread at the half maximum intensity (Full Width at Half Maximum or FWHM) for these ions varied systematically with chemical composition. For instance, FWHM's for $^{40}Ca^+$ decreased from 15 eV in CA9S1W (lowest in Si) to 7.5 eV in high-Si glasses such as CA2S8 with 72.7%Si, and so did FWHM's for $^{27}Al^+$ from 14 eV to 9 eV. In other words, the energy distributions of these ions were increasingly more sharpened with increasing Si abundance. At the same time, FWHM's for $^{28}Si^+$ decreased from 17 eV in CA9S1W to 12 eV in CA2S8, i.e., the energy distribution of $^{28}Si^+$ was increasingly more broadened with increasing Ca and Al abundances.

It was also observed that the intensities of Ca^+, Al^+ and Si^+ measured at the peaks of the energy distribution curves varied systematically with composition. Figure 1 illustrates the relationship between intensity and cationic

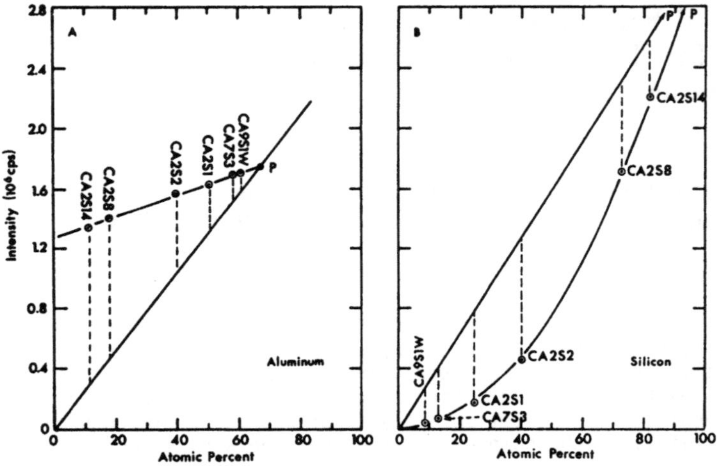

Fig. 1: Secondary ion intensities of Al^+(A) and Si^+(B) in CAS glasses measure at the peak of energy distribution curves as a function of cationic abundances of Al and Si.

abundance for Al (Fig. 1A) and Si (Fig. 1B). The point P is on the extension of the trend defined by the data points and represents the estimated intensity in an endmember composition (Al in $CaAl_2O_4$ for Fig. 1A and Si in SiO_2 for Fig. 1B). The line connecting the origin and P is the expected intensity-abundance relationship if secondary ion yield is proportional to atomic abundance. Note that all data points in Fig. 1A are above the line, showing that Al ionization is enhanced. Note further that the degree of enhancement (as measured by the length of the vertical dashed line in Fig. 1A) increases with decreasing Al, i.e., with increasing Si. Enhancement of Ca ionization occurs concurrently with a similar magnitude to Al. Figure 1B shows that Si ionization is increasingly more suppressed with increasing abundances of Ca and Al.

The fact that the intensities of Ca and Al ions emitted with initial energies of 80±10 eV do not show such enhancement with Si, coupled with the increasing sharpness of the energy distribution curves of Ca and Al with increasing Si (see above), suggests that the enhanced ionization observed here is essentially due to addition of low-energy Ca and Al ions as a function of Si abundance.

In their studies of secondary ion emission in binary alloys in the presence of oxygen, YU and REUTER [4,5] found two general rules. Rule I states that in a binary alloy A-B, in which A forms a stronger oxide bond that B, the presence of A enhances ionization of B, while ionization of A is suppressed by the presence of B. Rule II states that for the same binary alloy A-B, the presence of A sharpens the energy distribution of B^+, while the presence of B broadens the energy distribution of A^+. Since Si forms a stronger oxide bond than Ca, Mg, and Al, judging from the energy of chemical bond rupture (or energy of dissociation of gaseous oxides) compiled by SAMSONOV [6], the results of the present study obtained in silicate glasses are perfectly consistent with the above rules found in alloys, showing striking similarities in ion emission between the materials with grossly different physical properties, and confirming the notion that near-surface metal-oxygen interaction is important in secondary ion emission in oxygen-bearing environments [4,5].

The observations of SLODZIAN et al. [8] that FWHM's of $^{40}Ca^+$ is greater in calcite ($CaCO_3$) than in a feldspar (calcium alumino-silicate) are consistent with the present results in that the presence of Si in feldspar sharpened the energy distribution of $^{40}Ca^+$.

An additional observation was made on isotopic fractionation associated with the enhanced ionization. It was found that the magnitude of maximum isotopic fractionation of Si decreased with increasing Si from 6.38%/amu in CA9S1W to 4.72%/amu in CA2S8. Together with the data in pure Si metal (3.73 %/amu: SHIMIZU and HART [7]), it is concluded that the magnitude of isotopic fractionation increased with increasing addition of Ca and Al, or with increasing degree of suppression of Si ionization. Isotopic fractionation of Ca in CAS glasses was, however, found to be analytically insignificant (around 0.1-0.2%/amu).

As a tentative interpretation, a mechanism for both enhanced ionization and sharpening of ion energy distribution with increasing Si is visualized as follows: There are a number of collisions between neutral atoms, molecules and ions when they leave the surface, and these collisions may result in exchange of charge and oxygen atom between collisional partners. Such a collisional reaction between metal oxide (MO) and Si^+ could result in a reaction:

$$MO + Si^+ \longrightarrow SiO + M^+$$

If the conditions are suitable for this reaction to proceed, more M^+ are produced if more Si^+ are present (enhanced ionization due to Si). If the collision is elastic (kinetic energies of colliding partners are more or less maintained), the energy distribution of M^+ has a "signature" of the sputtered MO, which has a sharp low-energy peak. Thus, as enhanced ionization proceeds, the energy distribution of M^+ would become more sharpened.

1. G. Ray and S. R. Hart: Intern. J. Mass Spec. Ion Phys. 44, 231 (1982)
2. V. R. Deline, C. A. Evans, Jr. and P. Williams: Appl. Phys. Lett. 33, 578 (1978)
3. N. Shimizu, M. Semet and C. J. Allegre: Geochim. Cosmochim Acta, 42, 1321 (1978)
4. M. L. Yu and W. Reuter: J. Appl. Phys. 52, 1478 (1981)
5. M. L. Yu and W. Reuter: J. Appl. Phys. 52, 1489 (1981)
6. G. V. Samsonov: "The Oxide Handbook", Plenum Press (1972)
7. N. Shimizu and S. R. Hart: J. Appl. Phys. 53, 1303 (1982)
8. G. Slodzian, J. C. Lorin and A. Havette: J. Physique 23, 555 (1980)

Ionization Probabilities of Polycrystalline Metal Surfaces

M.M. Brudny and K.D. Klöppel

Department of Physical Chemistry, University of Siegen, P.O.B. 101240, D-5900 Siegen, F. R. G.

1. Introduction

Quantification of secondary ion (SI) yields presupposes knowledge of a large variety of apparative and sample specific parameters. Although, at least for metals, the main portion of in vacuum sputtered material is in neutral form, still enough ions are emitted to establish qualitative analysis of most metallic components. When oxygen or oxygen containing gases are admitted, SI yields of the most metals increase by orders of magnitude, thus enabling material analysis down to monolayer range. This "oxygen effect" is often ascribed to the high electronegativity of oxygen, which may be the cause of a local polarisation of the metal bond at the surface, favouring heterolytic detachment.

Despite the analytical importance of secondary ionization, no satisfying theoretical interpretation of fundamental processes has been published up to now. As a possible reason for this lack of a reliable theory, one can point out that experimental sputtering and SI yields are not known accurately.

2. Experimental

Sputtering and positive SI yields were measured for polycrystalline metals of transition groups as a function of oxygen pressure in the range of 0.02 to 5 mPa. The lower pressure limit is controlled by the residual gas pressure with primary ion source in operation, while the upper limit is the maximum oxygen pressure for which diffusion of reactant gas into primary ion source could be prevented. Primary argon ions of 1.8 keV were applied keeping their current constant at 1.9 μA corresponding to a current density of 95 $\mu A/cm^2$. SI currents were measured with a double focussing MAT 311 A mass spectrometer [1]. Sputtering rates were obtained by difference weighing under atmospheric pressure using a microbalance Mettler ME 22 [2].

3. Results

The sputtering yields of the investigated metals at low and high oxygen pressures are listed in table 1. The highest sputtering yield was obtained for silver at residual gas atmosphere and the lowest one for vanadium under the high oxygen pressure. A rough estimation shows that reductions of sputtering yields by oxygen admission (ΔY) represents the oxygen affinity of the metal surfaces. The sputtering yield of silver is almost independent of oxygen pressure, while electronegative metals such as chromium,

Table 1 Sputtering yields of investigated metals at low and high oxygen pressures

p_{O_2} (mPa)	Ti	V	Cr	Mn	Rh	Ag
< 0.02	2.31	2.57	4.12	5.01	4.22	6.95
> 0.20	1.16	0.91	1.70	1.44	3.75	6.70
ΔY	1.15	1.66	2.42	3.57	0.47	0.25

vanadium and manganese show sharp transitions between two plateau values. The first values characterize the clean surfaces, the second ones stand for the oxidized surfaces. The step-like transitions between these two plateau values occur at nearly the same oxygen pressure of about 0.06 mPa, e.g. the ratio of hitting oxygen atoms to sputtering primary ions is close to one (see Fig.1).

According to linear cascade theory, the sputtering yield is inversely proportional to the energy U_0 with which the target atoms are bound to the surface. For clean metal surfaces the heats of sublimation represent U_0. In the case of oxidized metal surfaces the binding energies are more nearly approximated by the enthalpy changes for the transition of most stable metal oxides into gaseous metal atoms and oxygen molecules, i.e. by the difference of sublimation energies of the metals and the heats of formation of the oxides [2]. Plotting the calculated sputtering yields - according to [3] with respect to changes of surface binding energies due to oxidation - versus the measured sputtering yields, the agreement is satisfying (see fig.2).

A comparison of sputtering and SI yields as a function of oxygen pressure is shown in fig.3 for a chromium target. Measured

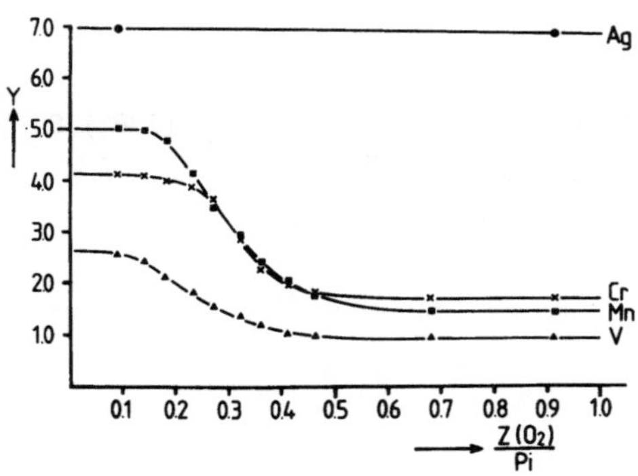

Fig.1 Sputtering yields of metals as a function of the ratio of hitting oxygen molecules to sputtering primary argon ions

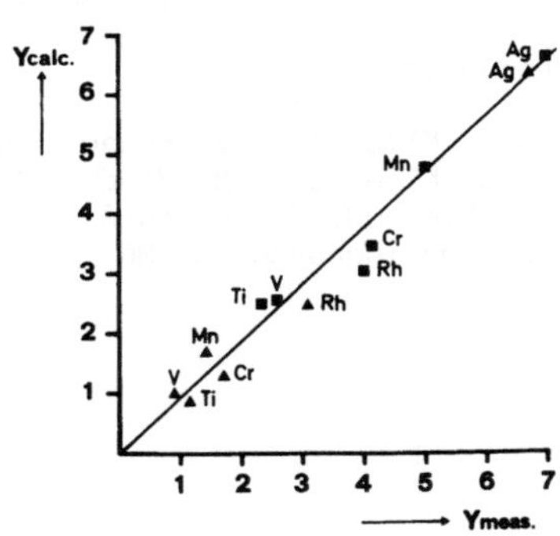

Fig.2 Proportionality between measured and calculated sputtering yields

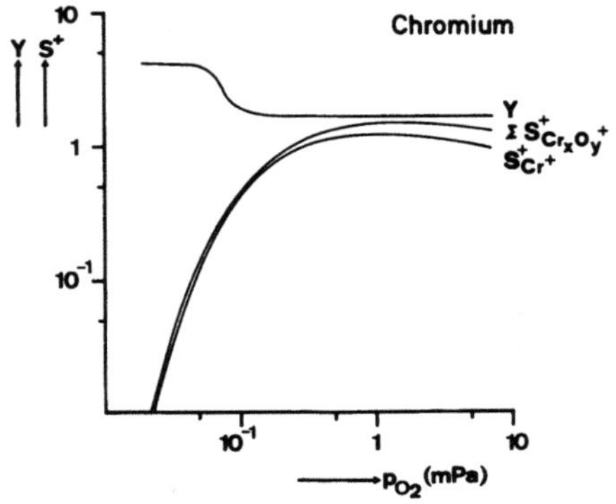

Fig.3 Positive SI yields S^+ and sputtering yields Y from chromium as a function of oxygen pressure

Me	S^+	Y	β^+
Ti	0.47	1.16	0.405
V	0.91	0.91	1.000
Cr	0.68	1.70	0.400
Mn	0.12	1.44	0.083
Rh	0.11	3.75	0.029

Table 2 Normalized positive SI yields, sputtering yields and ionization probabilities of metals under oxygen pressure of 2 mPa

SI yields were related to the SI yield of Cr^+ ions emitted from an oxidized chromium surface under static conditions [4]. The reduction of sputtering yield by admission of oxygen is higher than the SI yield enhancement, taking into account the logarithmic scale of intensity. Evaluating the vanadium data in this way, the SI yield of all emitted cluster ions would enhance the measured sputtering yield. A reasonable explanation for this discrepancy might be that positive SI yields under dynamic conditions are lower than those under static conditions. Assuming that at high oxygen pressures the main fraction of sputtered vanadium clusters is positivly charged, i.e. $\beta^+ = 1$ for oxidized vanadium surfaces, the corresponding positive SI yields and ionization probabilities are given in table 2. Deliberating that these data are limiting values, preliminary measurements of SI currents indicate that ionization probabilities under dynamic conditions are still somewhat lower.

References

1. K.D. Klöppel, G.v. Bünau, Int. J. Mass Spectrom. Ion Phys. 39 (1981) 85
2. K.D. Klöppel, M.M. Brudny, G.v. Bünau, ibid., in press
3. P. Sigmund, Topics in Appl. Phys. 47 (1981) 9
4. A. Benninghoven, A. Müller, Phys. Letters 40A (1972) 169

Reactivity and Structure of Sputtered Species

R.B. Freas and J.E. Campana
Naval Research Laboratory, Chemistry Division,
Washington, DC 20375, USA

1. Introduction

A high-pressure, fast-atom bombardment (FAB) ion source has been constructed [1] to study ion/molecule reactions of sputtered ions and neutrals. Pressures in this ion source are typically on the order of 0.1 to 0.5 Torr. Studies of metal cluster ion chemistry have been performed using this chemical ionization/fast-atom bombardment (CI/FAB) source with the mass-analyzed ion kinetic energy spectrometry (MIKES) technique. Our experiments were performed using a reverse-geometry, double-focusing mass spectrometer. Xenon gas was used as the fast-atom beam.

Post-desorption ionization (PDI) of sputtered neutrals is accomplished by ion/molecule reactions through the generation of reactant ions in a conventional manner. The PDI of sputtered neutrals has been observed to increase the abundance of an $[M+H]^+$ species by almost three orders of magnitude. Other processes occurring within the ion source include collisional stabilization and proton transfer, charge exchange, association, and condensation reactions [2].

2. Formation and Reactions of Metal Cluster Ions

The fast-atom bombardment of metal foils produces sputtered metal cluster ions M_x^+ (Reaction 1). When the desorption occurs within the high-pressure ion source, the bare metal cluster ions react with gas-phase species to form unique metal cluster product ions (Reaction 2). The neutral reactant species can be any volatile species such as organic molecules (alkanes, alcohols, or alkyl halides), inorganic molecules (NH_3, O_2, or CO), or organometallic compounds. The reactions of the bare metal cluster ions (Reaction 2) may be analogous to the sorption of the reactant species on metal surfaces.

$$M_{(s)} \quad + \quad Xe\ (keV) \quad \text{-------->} \quad M_x^+ \qquad (1)$$

$$M_x^+ \quad + \quad yR \quad \text{-------->} \quad [M_xR_y]^{+*} \quad \text{-------->} \quad \text{Products} \qquad (2)$$

When oxygen is admitted to the ion source, iron and cobalt cluster ions react to produce abundant metal/oxygen cluster ions $[M_xO_{x-1}]^+$, $[M_xO_x]^+$, and $[M_xO_{x+1}]^+$. The $[M_xO_y]^+$ cluster ions were not detected by sputtering of the target foil after O_2 was removed from the ion source. The oxide cluster ions were observed by sputtering cobalt oxide. These results suggest that oxygen is not extensively chemisorbed on the metal surface, and it is an indication that the oxide cluster ions were formed in the gas-phase.

3. Structure and Bonding of Metal Cluster Product Ions

In an experiment to probe structure and bonding using tandem mass spectrometry, a mass-selected product ion is dissociated (unimolecular or bimolecular dissociation, Reaction 3) and the fragment ions are energy analyzed.

$$[M_xR_y]^+ \xrightarrow{\text{Dissociation}} [M_iA_j]^+ + [M_kB_l]^+ + \;.\;.\;. \qquad (3)$$

The collision-induced dissociation (CID) spectra of metal/oxygen cluster product ions show major fragmentations that correspond to losses of M-O units or formation of $[M_iO_i]^+$ fragment ions. These pathways are outlined in Fig. 1 and they indicate the formation of stable structures of stoichiometry $[Co_4O_4]^+$, $[Co_3O_3]^+$ and $[Co_2O]^+$.

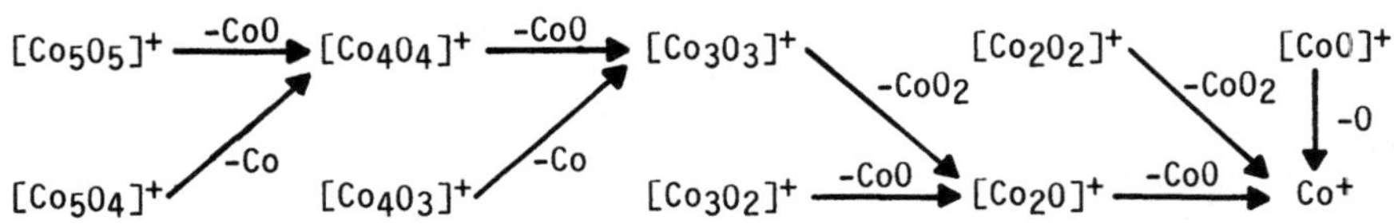

Figure 1. Major CID pathways of $[Co_xO_x]^+$ and $[Co_xO_{x-1}]^+$

The different fragmentation pathways of the $[Co_xO_x]^+$ and $[Co_xO_{x-1}]^+$ species in Tables 1 and 2 are indicative of structural differences in the cluster ions related to their stoichiometry. We are inferring structures for the cluster ions based upon the observed collision-induced fragmentations (Tables 1 and 2) and an ionic model (consisting of Coulomb and Born-Mayer repulsive interactions [3]). The fragmentations of the ionic models by cleavage of the fewest bonds yield the most stable structures for the product ions and correspond to observed CID pathways [4].

Isobutane and Fe, Co, Cu, and Ag cluster ions react to form metal/isobutane product ions, $[M_x(C_4H_{10})_y]^+$ (x=1-5; y=1-3). The CID fragment ions of cobalt/isobutane cluster ions in Table 3 indicate the bonding interactions

Table 1. Percent abundances of fragments from the CID of $[Co_xO_x]^+$

	Neutral fragment							
Parent Ion	Co	CoO	CoO_2	Co_2O_2	Co_2O_3	Co_3O_3	Co_3O_4	Co_4O_5
$[Co_5O_5]^+$	-	100	8	64	8	78	31	18
$[Co_4O_4]^+$	-	100	12	25	33	4	15	
$[Co_3O_3]^+$	18	85	100	12	59			
$[Co_2O_2]^+$	56	15	100					

Table 2. Percent abundances of fragments from the CID of $[Co_xO_{x-1}]^+$

	Neutral Fragment					
Parent Ion	Co	CoO	Co_2O	Co_2O_2	Co_3O_3	Co_4O_4
$[Co_5O_4]^+$	100	-	79	-	79	45
$[Co_4O_3]^+$	100	29	-	94	71	
$[Co_3O_2]^+$	48	100	4	52		
$[Co_2O]^+$	11	100				

Table 3. Percent abundances of fragment ions from the CID of $[Co_xC_4H_{10}]^+$

$[CoC_4H_{10}]^+$ Fragment	%	$[Co_2C_4H_{10}]^+$ Fragment	%	$[Co_3C_4H_{10}]^+$ Fragment	%
$[CoC_4H_9]^+$	2	Co_2^+	100	$[Co_3C_4H_8]^+$	10
$[CoC_4H_8]^+$	33	$[CoC_4H_8]^+$	2	Co_3^+	90
$[CoC_4H_7]^+$	2	$[CoC_3H_6]^+$	2	$[Co_2C_4H_8]^+$	21
$[CoC_3H_6]^+$	100	$[CoCH_3]^+$	1	$[Co_2H]^+$	86
$[CoCH_3]^+$	8	Co^+	14	Co_2^+	100
Co^+	55			Co^+	21

between the metal atom(s) and the isobutane C-C or C-H bonds. The cobalt ion Co^+ activates C-C and C-H bonds (exothermic bimolecular reactions [5]), and this is also observed in the CID of $[CoC_4H_{10}]^+$. The dicobalt ion Co_2^+ does not react exothermically [5], and the CID of the $[Co_2C_4H_{10}]^+$ species shows no activation of hydrocarbon bonds. In contrast to Co_2^+, $[Co_3C_4H_{10}]^+$ interacts with C-H bonds as observed from the CID loss of H_2.

The $[Co_xO_{x-1}]^+$ clusters react with isobutane in the CI/FAB ion source to form $[Co_xO_{x-1}C_4H_8]^+$ as well as $[Co_xO_{x-1}C_4H_{10}]^+$ products. The CID fragmentations show the abundant loss of H_2 from $[Co_2OC_4H_{10}]^+$ (Table 4). The $[Co_2O_2C_4H_{10}]^+$ cluster does not lose H_2 by CID (Table 4) to give $[Co_2O_2C_4H_8]^+$, and $[Co_2O_2C_4H_8]^+$ was not observed in the mass spectra. Thus, the oxygen-deficient cluster ions dehydrogenate isobutane.

Table 4. Percent abundances of fragments from the CID of $[Co_xO_yC_4H_{10}]^+$

	Neutral Fragment					
Parent Ion	H	H_2	O	H_2O	C_4H_8	C_4H_{10}
$[CoC_4H_{10}]^+$	2	33			-	55
$[CoC_4H_8]^+$	3	15			100	
$[Co_2OC_4H_{10}]^+$	-	90	10	-	-	100
$[Co_2OC_4H_8]^+$	3	3	12	11	100	
$[Co_2O_2C_4H_{10}]^+$	5	2	-	2	-	100

Thus, CI/FAB mass spectrometry provides a tool for the study of ion/molecule reactions of sputtered metal cluster ions. Tandem mass spectrometry provides a probe of the structure and bonding of metal cluster product ions. In addition, the cluster reaction product ions are inherently ions; they do not need to be photoionized as in molecular beam studies, and hence are not susceptible to perturbations upon ionization.

4. References

1. J.E. Campana and R.B. Freas: J. Chem. Soc. Chem. Commun. 1414 (1984)
2. R.B. Freas and J.E. Campana: J. Am. Chem. Soc. in press
3. T.P. Martin: J. Chem. Phys. 72, 3506 (1980)
4. R.B. Freas, B.I. Dunlap, and J.E. Campana: unpublished results
5. R.B. Freas and D.P. Ridge: J. Am. Chem. Soc. 102, 7129 (1980)

Local Thermal Equilibrium Model of Molecular Secondary Ion Emission

P.J. Todd

Analytical Chemistry Division, Oak Ridge National Laboratory, Oak Ridge, TN 37831, USA

1. Introduction

Secondary emission of intact ions characteristic of involatile analytes dissolved in glycerol may be viewed as the result of two not necessarily independent processes. First, all the intermolecular bonds of the involatile solute must be broken without breaking a single intramolecular bond. Second, the secondary ion precursor must become charged either before, during, or after it is emitted into the gas phase. It is not yet clear how the physical and chemical properties of a sample affect either of these two processes, but it is clear that chemical effects dominate secondary ion emission. Because chemical effects are the result of processes involving less than a few eV/molecule, it follows that secondary ion emission involves a third process, in which the keV energy of the primary particle is dissipated rapidly to a collection of sample molecules, such that the kinetic energy/molecule of sample is insufficient to cause pyrolysis, but sufficient to cause desorption. When viewed this way, secondary ion emission can be described as the result of the behavior of a critical fluid. Such a fluid is best described by its thermodynamic properties, and postulating that secondary ion emission proceeds via a critical fluid state implies the existence of a local thermal equilibrium within the sputtering region.

2. Temperature Limitation

If the energy ΔE of a primary particle is quantitatively transferred to the η moles of sample within the sputtering volume, the temperature rise of the sputtering volume is implicit in Equation (1):

$$\frac{\Delta E}{\eta} = \int_{T_0}^{T} C_v dT \qquad (1)$$

where $C_V = (\partial E/\partial T)_V$. However, the value of C_V follows the behavior [1]:

$$\lim_{T \to T_c} C_V = \infty \qquad (2)$$

where T_c is the material's critical temperature. Consequently, the thermodynamic properties of the matrix, its physical properties, cause a thermal clamp to be placed upon the temperature of the sputtering volume. The temperature of the sputtering volume is limited to the critical temperature of the matrix, which for glycerol is 450°C.

The existence of the "thermal clamp" makes it possible to rationalize the detection of high molecular weight ions from thermally labile compounds. Equations (1) and (2) imply that all things being equal. the kinetic energy of the primary particle is linearly related to the sputtering volume, but the internal energy distribution of sputtered molecules is independent of the primary ion kinetic energy. It is well known that organic SIMS spectra are essentially independent of primary beam energy.

3. Intermolecular Bond Breaking

Many compounds such as glycerol (and analytes as well) are involatile because of extensive intermolecular dipole-dipole and hydrogen bonds. For thermally labile compounds, the average energy necessary to break these bonds is greater than the critical energy for pyrolysis. However, dipole-dipole and hydrogen bonds are directional bonds. Were molecules in such a system allowed to rotate freely, the net potential energy of these bonds would become zero and directional bonds would be broken. At and above T_c, free rotation is known to occur [2]. Thus, a second property of critical fluids—the absence of directional intermolecular bonds, satisfies the condition that all the intermolecular bonds be broken for emission into the gas phase. The "thermal clamp" provided by the transition from liquid to critical fluid limits the rate of pyrolysis. That is, the rate of intermolecular bond breaking is more rapid than the rate of pyrolysis.

4. Ionization

The generation of a charged species from such a system is easily rationalized. The initial impact of the Ar^+ ion with surface molecules is sufficient to eject electrons and/or protons, leaving the thermally excited region charged. Even though impure glycerol is a conductor, conduction occurs by ionic motion, which is slow. Consequently, the heated region remains charged during the sputtering event. In such a case, as the excited region vents its molecules into the gas phase the species which has the highest affinity for the charge will become charged if equilibrium is permitted. Indeed, positive organic secondary ions $(M+H)^+$ generated from glycerol solution are all organic bases.

In addition to evidence concerning basicity, the foregoing mechanism requires that the analyte be dissolved in glycerol before the impact event. Hence, solubility is a requirement for enhanced secondary ion emission. Indeed, we find that when aniline is introduced via gas phase onto an irradiated glycerol surface, no aniline secondary ion emission is observed; aniline is insoluble in glycerol. Alkylamines, however, when thus introduced do indeed show intense secondary $(M+H)^+$ ion currents. Alkyl phosphonates [3], which are powerful surfactants show similar behavior: they are insoluble in glycerol, but wet a glycerol surface; no secondary ions could be detected when dimethylmethylphosphonate was admitted onto a glycerol surface under primary irradiation. However, when admitted onto a polyphosphoric acid surface, intense secondary ions were detected.

Research sponsored by the U. S. Department of Energy, Office of Basic Energy Sciences, under Contract DE-AC05-84OR21400 with Martin Marietta Energy Systems, Inc.

5. References

1. R. S. Berry, S. A. Rice, J. Ross: Physical Chemistry (John Wiley and Sons, New York 1980).

2. T. L. Hill: Lectures on Matter & Equilibrium (Benjamin, New York 1966).

3. G. S. Groenewold and P. J. Todd: Anal. Chem. 57, 886 (1985).

Time-of-Flight Investigations of Secondary Ion Emission from Metal Halides

B. Schueler, R. Beavis, G. Bolbach[1]*, W. Ens, D.E. Main, and K.G. Standing*

Physics Department, University of Manitoba, Winnipeg, Manitoba R3T 2N2, Canada

1. Introduction

The bombardment of metal halides (MX) by keV - primary particles causes the emission of secondary atomic (M^+, X^-) and cluster ($(MX)_nM^+$; $(MX)_nX^-$; n = 1,2, ...) ions. Once the secondary cluster ions leave the primary interaction region, they can undergo metastable decays due to the vibrational energy stored in the cluster bond during the initial production process. A large fraction of secondary cluster ions from metal halides, as well as molecular ions from organic materials, has indeed been observed to be metastable. We have studied the metastable decays of some CsI clusters and determined the corresponding rate constants.

2. Experimental

A schematic diagram of a detector assembly recently installed in the Manitoba time-of-flight mass spectrometer [1] is shown in Fig. 1. The pulsed primary ion beam (usually Cs^+) of ~3ns f.w.h.m. hits the target at an angle of approx. 30° relative to the spectrometer axis. Secondary ions ejected are electrostatically accelerated between target (HV) and grid (ground potential) and enter the drift region. As indicated in Fig. 1, charged parent and fragment species are deflected in an electrostatic ion mirror by 90° into one detector. Neutral particles penetrate the mirror and continue along the spectrometer axis into a second detector. Both detectors and the mirror are mounted on a linear motion feedthrough so that their distance to the target (drift region) can be continuously varied from 0.45 m to 1.45 m. Both detector outputs are processed simultaneously by the data-acquisition system [2].

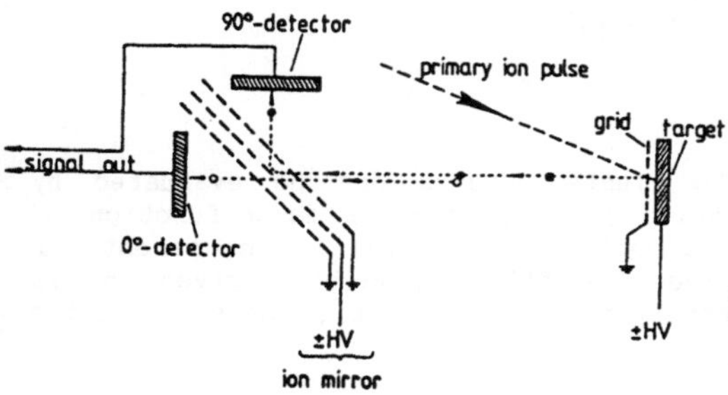

Fig. 1 Schematic diagram of the time-of-flight spectrometer with the ion mirror and two detectors.

[1] present address: Institut Curie, 11 rue Pierre et Marie Curie, 75231 Paris, Cedex 05, France

For a given accelerating voltage, the time-of-flight of an ion is proportional to the drift length. In a linear time-of-flight system (no mirror) an ion will be detected at its original mass if it survives the acceleration time (~100ns). Subsequent decays in the drift region merely broaden the mass lines. The total flight time is the sum of accelerating and drift time. With an ion mirror placed in the drift region, the total flight time is the sum of accelerating time, drift time to the mirror, deflection time in the mirror and the drift time from the mirror to the detector. Parent ions can only be recorded in the 90° detector if they survive the deflection period in the mirror. In this case the ion lifetimes have to be of the order of several microseconds.

If a parent ion of mass m_1 undergoes a fragmentation (e.g. $m_1^+ \rightarrow m_2^+ + m_3^0$) in the drift region (between grid and mirror) the velocity of the centre of mass is unchanged. The kinetic energy of the parent ion is shared among the fragments according to their masses during break-up. Each individual fragment has therefore a lower kinetic energy than its parent ion. Due to the reduced kinetic energy, the fragment will penetrate the mirror less deeply than a parent ion, and thus also have a shorter flight time to the detector. This allows the approximate determination of the fragment ion mass. A part of the charged particle mass spectrum of CsI clusters in the 90°-detector is shown in Fig. 2. The fragment peaks indicated by -CsI and -2CsI correspond to the fragments formed by the loss of one or two neutral CsI molecules from the parent ion. The experiments do not determine whether or not the loss of 2CsI occurs in one step.

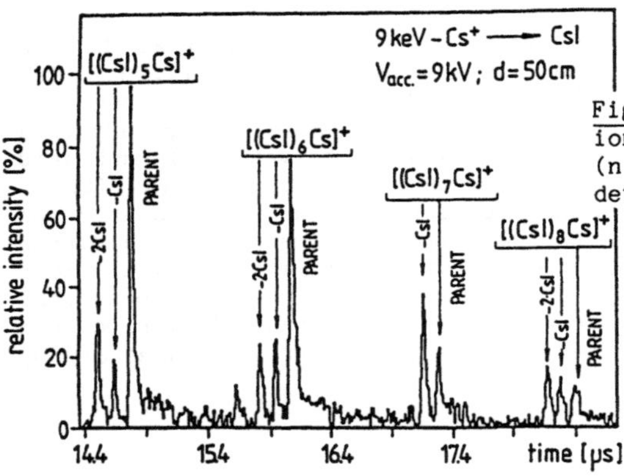

Fig. 2 Charged parent and fragment ion spectrum of $(CsI)_nCs^+$ clusters (n = 5 to 8) observed in the 90° detector.

3. Results and Discussion

The rate constants of metastable cluster decays can be evaluated by experimentally determining the ratios $y_p(d)/y_p(d = 0)$ as a function of the distance target-mirror (drift length), where $y_p(d)$ is the parent ion yield at a certain distance. Since the flight time of a given ion is proportional to d, this ratio corresponds to $\exp(-k_1 t)$, where k_1 is the first order rate constant.

The most direct way of determining the slope of $\exp(-k_1 t)$ would be to plot $y_p(d)/y_{Cs}(d)$ as a function of the distance, where $y_{Cs}(d)$ is the Cs^+ ion yield at distance d. This procedure, however, leads to unsatisfactory results (positive slopes for n = 1,2) due to the different radial velocity distributions of atomic and cluster ions, and therefore different detection probabilities. Because of this problem we are going to install a variable diameter iris in front of the ion mirror to ensure equal acceptance angles for different distances. For the present, the problem can be somewhat circumvented by evaluating the ratios $y_p(d)/y_1(d)$ at each distance, where $y_1(d)$ is the yield of the first cluster $CsICs^+$, which did not show any noticeable decay lines and was thus assumed to be stable.

Figure 3 shows the experimental yield ratios $y_p(d)/y_p(d = 0)$ for $[(CsI)_nCs]^+$ cluster ions (n = 2 - 7) as a function of the distance between target and mirror. The slopes were determined as described above. The data points were shifted so that the fitting lines pass the points $y_p(46cm)/y_t(46cm)$, where $y_t(46cm)$ is the sum of parent and fragment ion intensities of the respective cluster at d = 46 cm. With this normalization, the fitting lines were found to pass the 100% point at d = 0 within 5%, which may serve as a confirmation of the evaluation procedure. The corresponding rate constants and half-lives are listed in Table 1. The evaluation errors are ≈20%.

Although small, the rate constants given in Table 1 increase rapidly with increasing cluster size. In agreement with other measurements (e.g.[3]) n = 6 seems to be a relative stability point in the cluster distribution. The small values of the rate constants e.g. $k(n = 1) \sim 10^3 s^{-1}$ indicate that the emitted cluster ions are fairly "cold" and the vibrational temperatures involved are estimated to be much smaller than ~2000 K.

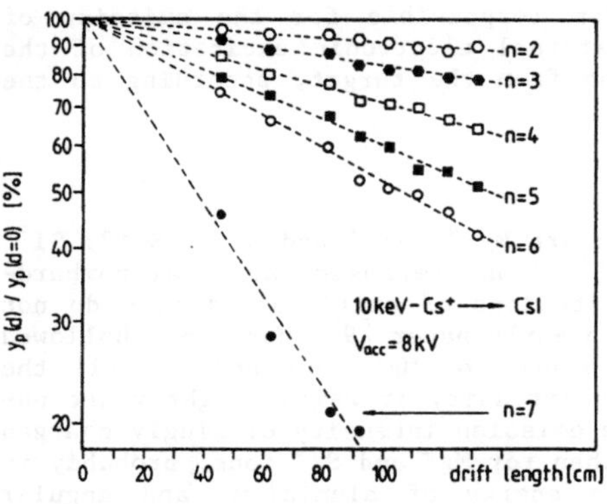

Fig. 3 Dependence of the ratios $y_p(d)/y_p(d=0)$ on the distance between target and mirror

Table 1 Rate constants and half lives for clusters $(CsI)_nCs^+$

n =	$k[s^{-1}]$	$t_{1/2}[\mu s]$
1	$\leq 1.0 \cdot 10^3$	≥700
2	$4.9 \cdot 10^3$	140
3	$8.1 \cdot 10^3$	85
4	$1.2 \cdot 10^4$	58
5	$1.7 \cdot 10^4$	41
6	$2.0 \cdot 10^4$	35
7	$3.3 \cdot 10^4$	21

4. Acknowledgement

This work was supported by grants from the U.S. National Institutes of Health, and from the Natural Sciences and Engineering Research Council of Canada. G. Bolbach received support from CNRS (France).

1. K.G. Standing et.al., Int. J. Mass Spectrom. Ion Phys. 53, 125 (1983)
2. W. Ens et.al., submitted to Nucl. Instr. and Methods
3. M.A. Baldwin et.al., Int. J. Mass Spectrom. Ion Proc. 54, 97 (1983)

Multicharged Secondary Ions from Light Metals (Mg, Al, Si)

J.-F. Hennequin, R.-L. Inglebert, and P. Viaris de Lesegno

C.N.R.S., Laboratoire P.M.T.M., Université Paris-Nord, F-93430 Villetaneuse, France

1. Introduction

Only light metals (Mg, Al, Si) bombarded with noble gas ions in the some keV energy range show a non-negligible emission of true multicharged secondary ions [1]. The few multicharged ions observed from heavy metals at very high bombardment density [2] are in fact mostly created by direct ionization of sputtered neutral atoms by primary ions in the vacuum at some distance from the target [3,4].

The aim of this paper is to collect our recent experimental results on light-metal ions with those published over some twenty years and to discuss them according to the model of kinetic emission [5], where ionization appears as the result of Auger decay, outside the target, of inner 2p-shell excited sputtered particles. The emission of Auger electrons under ion bombardment of light metals [6,7] proves the efficiency of such a collisional process, though the main mechanism responsible for the emission of singly charged ions seems to be an external electronic excitation of the outgoing particle during its separation from the target, according to the so-called surface excitation model [8].

2. Experimental Results

The observed multicharged ionic species are Mg^{2+}, Al^{2+} and Al^{3+}, Si^{2+}, Si^{3+} and perhaps Si^{4+} ; but Mg^{3+}, Al^{4+} or Si^{5+} ions are absent at usual bombardment energies [1,2]. The angular distributions of multicharged ions do not seem to have been studied again since an early paper [9] where very hollowed distributions (under-cosine law) were found for Mg^{+} ions and for all the doubly charged ions, the total emission intensity of which slightly decreases from Mg^{2+} to Si^{2+}. By contrast, the emission intensity of singly charged monatomic ions is higher for Al^{+} ions than for Mg^{+} and Si^{+} ions, probably in connection with the weaker ionization energy of aluminium, and angular distributions of Al^{+} and Si^{+} ions are much nearer to the cosine law.

The energy spectra of monatomic ions from light metals are shown in Fig. 1 for 10 keV-Ar^{+} ion bombardment at 35° incidence in ultrahigh vacuum and for secondary ions emitted with a mean take-off angle of 45° within an angular cone of nearly 25° to the mean axis [10]. As the used energy bandwidth was ca. 3 eV, the low energy maxima are somewhat distorted. Nevertheless, normalization of all spectra at the same height allows easier comparisons. As previously observed [11], the energy spectra are spread over several hundred eV, but only Al^{+} and Si^{+} spectra can be fitted by a power law ε^{-n} for initial energies 10 eV $\lesssim \varepsilon \lesssim$ 100 eV (with $n \simeq 1.2$ for Al^{+} and $n \simeq 1.0$ for Si^{+}). All other spectra show a shoulder (Mg^{+}, Al^{2+}) or a flat maximum (Mg^{2+}, Al^{3+} and Si^{2+}) between 100 to 500 eV. A steep decrease is lastly observed in the energy spectra of Mg^{+} ions and of doubly charged

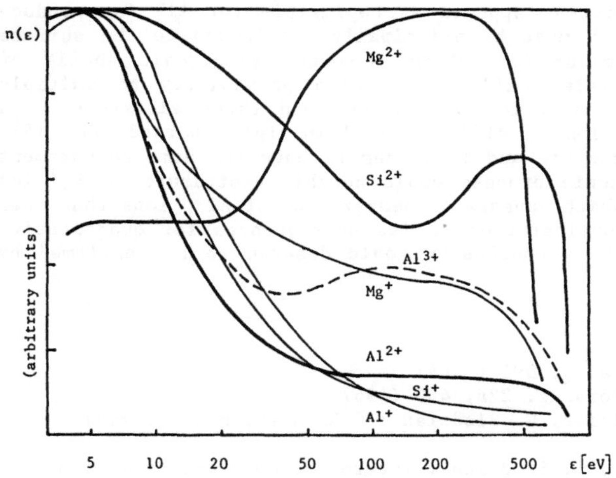

Fig. 1 Energy spectra of monatomic ions from light metals (10 keV-Ar^+) roughly normalized at their maximum : Mg^+(x1.5), Mg^{2+}(x25) ; Al^+(x1), Al^{2+}(x10), Al^{3+}(x2000) ; Si^+(x10), Si^{2+}(x40).

ions (and not in those of singly charged Al^+ and Si^+ ions) beyond 500 eV for Mg^+ and Mg^{2+}, 700 eV for Al^{2+} and 800 eV for Si^{2+} ions.

Recent experiments with Fe-Al and Cu-Al alloys brought some new informations on multicharged ions [10,12]. By contrast with singly charged ions, the energy spectra of which do not depend on Al concentration, more high-energy multicharged ions are proportionally emitted from dilute alloys than from pure aluminium [13]. Moreover, the emission intensities of Al^{2+} and Al^{3+} ions at a given initial energy, normalized to those of pure aluminium, are nearly equal to the square of the Al atomic concentration, whereas the normalized intensities of Al^+, Fe^+ or Cu^+ ions are equal to the concentration of the corresponding element [12-14]. The ratio $I(Al^{2+})/I(Al^{3+})$ does not depend on Al concentration and decreases with initial energy ε as $\varepsilon^{-0.9}$ until 100 eV at least [12] ; it would become constant at higher initial energies [15] and does not depend on the primary ion energy [14]. Similarly, the ratio $I(Si^{2+})/I(Si^{3+})$ is independent of the tilt angle of the target, as well as the mass and the energy of the incident ions [16].

3. Discussion and Conclusion

Such experimental results on light metals quite agree with the kinetic emission model, which specially predicts a broad energy spectrum with a break at high energy [17]. A multicharged ion is generally emitted at the end of a very short cascade of collisions inside the target. After the first collision of the primary ion against a metal atom, the recoil atom cannot be directly ejected, and several symmetric collisions occur between target atoms. If the collision energy is high enough, an inner electronic vacancy can be created in the 2p level of one or of both colliding atoms. The subsequent decay gives rise to an Auger electron. But the life-time of excited atoms is long enough to allow some of them to escape out of the target [18], as neutral but strongly excited particles. Thereby, external de-excitation by Auger effect leads to kinetic secondary ions, associated with Auger electrons [7,12,19].

However, the kinetic process appears as negligible for Al^+ ion production, the emission of which must be principally attributed to the surface excitation process. A similar excitation of outer electronic shells of 2p-excited sputtered neutrals could give a high probability to multiple Auger de-excitations, with ejection of two or even three electrons by a shake-off effect and creation of Al^{2+} and Al^{3+} multiply charged ions [5]. Silicon and phosphorus (from GaP and InP) seem to have the same comportment as aluminium, but the kinetic process could notably contribute to Mg^+ ion emission from magnesium, which appears as nearer that of Al^{2+} ions than that of Al^+ or Si^+ ions. In consideration of its consequences for quantitative analysis of magnesium, such a conclusion would deserve to be confirmed by further investigations.

References

1. G. Slodzian : Ann. Phys. 9, 591 (1964)
2. H.E. Beske : Z. Naturforschg. 22a, 459 (1967)
3. J.-F. Hennequin, G. Blaise, G. Slodzian : C.R. Acad. Sci. (Paris) B 268, 1507 (1969)
4. M. Bernheim, G. Blaise, G. Slodzian : Intern. J. Mass Spectrom. Ion Phys. 10, 293 (1972/73)
5. P. Joyes : J. Physique 30, 243 and 365 (1969) ; Rad. Eff. 19, 235 (1973)
6. J.-F. Hennequin, P. Viaris de Lesegno : Surf. Sci. 42, 50 (1974)
7. K. Wittmaack : Nucl. Instrum. Meth. 170, 565 (1980)
8. G. Blaise, A. Nourtier : Surf. Sci. 90, 495 (1979)
9. J.-F. Hennequin : J. Physique 29, 957 (1968)
10. R.-L. Inglebert : Thesis, Université de Paris-Nord (1985)
11. J.-F. Hennequin : J. Physique 29, 655 (1968)
12. J.-F. Hennequin, R.-L. Inglebert, P. Viaris de Lesegno : Surf. Sci. 140, 197 (1984)
13. R.-L. Inglebert, J.-F. Hennequin : In Secondary Ion Mass Spectrometry SIMS IV, Springer Series in Chemical Physics 36, 49 (1984)
14. D. Brochard, G. Slodzian : J. Physique 32, 185 (1971)
15. M. Bernheim : Thesis, Université de Paris-Sud (1973)
16. K. Wittmaack : Nucl. Instrum. Meth. B 2, 674 (1984)
17. V.I. Veksler : Sov. Phys. Solid State 24, 997 (1982)
18. T.D. Andreadis, J. Fine, J.A.D. Matthew : Nucl. Instrum. Meth. 209/210, 495 (1983)
19. C. Benazeth, N. Benazeth, M. Hou : Surf. Sci. 151, L 137 (1985).

Population of Ion Clusters Sputtered from Metallic Elements with 10 keV Ar^+

F.G. Satkiewicz

Applied Physics Laboratory, The Johns Hopkins University,
Johns Hopkins Road, Laurel, MD 20707, USA

1. Introduction

Theories on ion production from the sputtering of solids have largely ignored the question of ion clusters which are known to be formed. Perhaps one of the reasons is the absence of systematic data on the extent of clustering.

2. Experimental

The sputter-ion source mass spectrometer used in this work was the GCA IMS 101B prototype, a double-focusing instrument in which momentum separation is followed by charge separation[1]. The "energy window" or that portion of the initial secondary ion energy distribution (SIED) selected for analysis is set by changing the secondary acceleration voltage while the deflection voltage on the electrostatic sector is kept constant. The energy bandwidths employed for the results reported here were ~ 5 eV and ~ 50 eV. Pure samples were sputtered with a focused beam (15 ma/cm^2) at 45° angle of incidence against a background pressure of 10^{-7} Torr. Under these conditions the influence of residual oxygen on ion intensities was found to be negligible.

3. Results and Discussion

Ion intensities from polyatomic spectra taken for 21 elements were used to calculate the percentage of ions in the form of clusters, C, from:

$$C = \frac{100\{[(Im)_c]_{X_2^+}+\ldots+[(Im)_c]_{X_n^+}\}}{\{[(Im)_c]_{X^+} + [(Im)_c]_{X_2^+}+\ldots+[(Im)_c]_{X_n^+}\}} \qquad (1)$$

where $(Im)_c$ is the maximum intensity of the species X_n^+ in the energy range, R. Corrections for isotope abundancy and mixing were made. The results are shown in Table 1. $(E)_{Im}$ is the energy at maximum intensity. Columns 2, 3, 6, and 7 were obtained directly from (1); columns 4 and 8 (for nine elements) were obtained by numerical integration of the area under the energy distributions and multiplying by the corresponding $(Im)_c$. This accounts for differences in the SIED's. It was found that reducing the acceleration voltage to extend the mass range for the higher Z elements did not significantly alter the relative ion intensities. No further corrections were made. The intensity ratio of X_2^+/X^+ ranged almost 2 orders of magnitude, leading to the question of the property of the element which varies accordingly. Since bonds have to be broken, it appeared reasonable to investigate the relationship of clustering to properties such as melting point, surface tension, heat of fusion, and heat of sublimation. A linear regression analysis was carried with various properties (taken from [3]) and is shown in Table 2.

The correlation coefficient, r^2, is a measure of the curve fit, (a value of 1.0 corresponding to a perfect fit). The heat of sublimation (ΔH_s) alone

Table 1 Extent of Clustering from Sputtering Metals with 10 keV Ar^+

X \ R,eV	Ion Cluster %, C $(E)_{Im}\pm25$	$(E)_{Im}\pm2.5$	0-500	X \ R,eV	Ion Cluster %, C $(E)_{Im}\pm25$	$(E)_{Im}\pm2.5$	0-500
Li	4.9			Zn	2.8		
Be	31			Ga	5.8		
Mg	1.5	0.96	0.22	Zr	36		
Al	6.8	8.9	4.9	Nb	46		
Ti	22			Mo	47	42.6	25.8
Cr	17			Pd	24		
Mn	17	14.1	5.5	Ag	24	18.8	4.4
Fe	29	22.6	8.7	Ta	44		
Ni	20.5	23.8	14.9	W	51		
Co	37	29.1	13.4	Pb	6.7		
Cu	27	21.0	3.4				

Table 2 Correlation Coefficient in Plots of Percentage of Ion Clusters, C, and Various Parameters for 21 Metals[3]

Atomic #, Z	0.18	Surface tension, γ	0.795
Melting point, T_m	0.811	Vickers hardness	0.46
Heat of fusion, $\Delta H_f°$	0.804		
Heat of sublimation, $\Delta H_s°$	0.825	$\Delta H_s°(V_{MS})^{-2/3}f^{-1}N^{-1/3}$	0.863

gives a good fit with C and a better one with ΔH_s per molar area (given by $f(V_{MS})^{2/3}N^{1/3}$, where f is the surface packing factor [4], V_{MS} is the molar volume, and N is Avogadro's number). The results of the latter fit are shown in Fig. 1(A) and Fig. 1(B). The results for the 21 elements were taken using a wide energy window. The narrower energy window was used for 9 elements in Fig. 1(B). The correlation coefficient for the same 9 elements for the data set used in Fig. 1(B) is 0.749 vs 0.863. This difference may be due to the

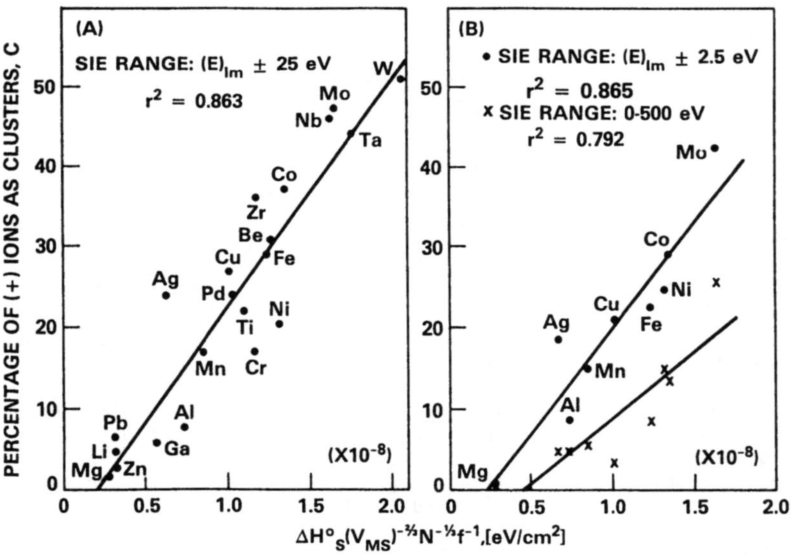

Fig. 1 Correlation between (+) ion cluster population and heat of sublimation

weighting of X^+ in two cases. This is evidenced by the lowering of r^2 when the full distribution of X^+ is used for the same elements (Fig. 1(B)). The lowering is due to the high energy tails in the X^+ SIED.

This correlation would appear to be the opposite of what might be expected at first glance, but several points can be made: 1. The correlation does not compare the total number of ions formed upon sputtering, but merely the percentage of those formed detected as clusters (n>1) for a given element. 2. Thermodynamically, less energy is involved with the ejection of clusters than the total as single atoms (you don't need to supply the energy of dissociation for bonds in the clusters). The sputtering process involves direct ejection of clusters whose stability once formed is greater, the higher the heat of dissociation [5]. 3. The correlation with heat of sublimation for atomic species exists since the ΔH_s ratio for dimer/monomer is approximately constant for at least 9 of the elements studied [6]. 4. The probability of ionization of the cluster is an unknown factor in this correlation, but apparently doesn't vary much between clusters. 5. The correlation with melting point, heat of fusion, and surface tension suggest that the key bonding issue is the force that holds the atoms of a fluid together; once again this may be due to an indirect relation between these properties and the heat of sublimation.

4. Conclusion

The extent of ion clustering from sputtering 21 metallic elements was found to be dependent on the heat of sublimation. The degree of correlation diminished when the high energy tails of the X^+ SIED were included in the calculations, implying that ion clustering is mainly associated with low energy sputtered atoms and ions.

Acknowledgment

This work was supported in part by Naval Sea Systems Command under Contract N00024-85C-5301.

References

1. R. F. K. Herzog, W. P. Poschenrieder, F. G. Satkiewicz: Rad Effects 18, 199 (1973).
2. F. G. Satkiewicz: Proc. 27th Conf. on Mass Spectrometry and Allied Topics, Seattle, WA, p. 303, 1979.
3. G. V. Samsonov: Handbook of Physicochemical Properties of the Elements, IFI Plenum, N.Y., 1968.
4. A. S. Skapski: Acta Met 4, 576 (1956).
5. P. Joyes: J. Phys. Chem. Solids 32, 1269 (1971).
6. B. deB. Darwart, Bond Dissociation Energies in Simple Molecules NSRDS-NBS31, Jan. 1970.

Ejection and Ionization Efficiencies in Electron and Ion-Stimulated Desorption from Covalently Bound Surface Structures

T.R. Hayes and J.F. Evans

Department of Chemistry, University of Minnesota, Minneapolis, MN 55455, USA

1. Introduction

In this study we have examined the effects of substrate and modifier structure on the ion and electron-induced ejection of ions from surface species which are covalently attached to a metal oxide. The primary goal was to compare the ejection of parent and fragment ions induced by ion-stimulated desorption (1 KeV Ar^+) to those ejected by electron-stimulated desorption (1 KeV) to ascertain which excitation method yielded a more reliable or sensitive mode for analysis of such surface structures. In the process of examining the analytical aspects of ESD vs SIMS we have observed marked similarities in the secondary ion energy spectra and total dissociation cross-sections for the two techniques, yet the ionization efficiencies are drastically different. Some indirect insight into the role played by the electronic component of the collision cascade in SIMS of these and related systems is thus provided.

2. Experimental

In all of the experiments reported here organosilanes were reacted with surface hydroxyl groups on metal oxides through gas-solid reactions in a dedicated reactor appended to the UHV system in which the desorption experiments were carried out [1]. The metal oxides were those formed by the thermal oxidation of silicon, titanium (supported as a thin film on SiO_2/Si) and zirconium. The attached silanes had structures M-O-Si($R_1R_2R_3$), where M represents the metal of interest. We restrict our discussion here to the trimethylsilyl (TMS) modified surfaces (R_1=R_2=R_3=-CH_3) of SiO_2, TiO_2 and ZrO_2. Details of the experimental aspects of sample preparation and instrumentation, as well as a discussion of the effects of modifier size on SIMS and ESD positive ion spectra, can be found elsewhere [1-4].

3. Results and Discussion.

The most probable dissociative pathway for bound TMS groups under 1.0 KeV Ar ion bombardment is cleavage of the Si-C bond, resulting in methyl group ejection. This process results in the detection of secondary ions of the formula $(CH_3)_nSi_m^+$, where n=3, 2, 1 and 0, and m=1 or 0. The perdeuterated group behaves in an analogous fashion. Under conditions of static SIMS, CH_3Si^+ is the most abundant ion derived from TMS, and $(CH_3)_2Si^+$ the least abundant. Methyl groups which are created *via* ion impact-induced dissociative processes are ejected primarily as neutral radical species.

The relative yield of the parent ion (TMS^+) in SIMS is found to be only slightly substrate dependent. The chemical nature of the substrate is also found to modulate the total dissociation cross-sections of bound TMS and d_9-TMS groups under ion bombardment in a way which is more complicated than consideration of the simple, binary events in the collision cascade would predict. (E.g., the total dissociation cross-section for ZrO_2 is highest, yet the mismatch in Zr-O masses is the largest.)

TMS^+, $CH_3{}^+$, $H_2{}^+$ and H^+ are the most abundant ionic desorbates which arise from TMS groups bound to SiO_2 under 1.0 KeV electron bombardment. The detection of intact parent ions of TMS groups bound to SiO_2 indicates electron bombardment stimulates transitions which lead to the dissociation of the Si-O bond (the Si being that of the silane). Although the nature of the excitation is unknown, it appears to be highly localized state, because it selectively ruptures the Si-O bond and leads to the desorption of an intact parent ion.

The total dissociation cross-sections for ESD from TMS/SiO_2 and TMS/TiO_2 are 2.3 (+/-0.5) x 10^{-15} cm^2 and 6.3 (+/-1.6) x 10^{-16} cm^2, respectively, whereas for SIMS the corresponding values are 1.1 (+/-0.1) x 10^{-14} cm^2 and 8.8 (+/-0.7) x 10^{-15} cm^2. Furthermore, in neither case is a discernable isotope effect found when perdeuterated TMS is bound to these surfaces.

In general, ESD ion yields are much lower than those from SIMS. Specifically, considering TMS modified SiO_2, the SIMS parent ion yield and the ESD parent ion yield differ by more than two orders of magnitude. No parent ion signal is detected in the case of TMS modified TiO_2, and $CH_3{}^+$ intensities are at least 10 times lower than those in the case of TMS/SiO_2. The lower ESD ion yields for groups bound to TiO_2 are rationalized by a much higher probability for neutralization of near-surface ions (<u>via</u> electron transfer from the surface) in the case of TiO_2 than SiO_2. The electronic structure of TiO_2 is such that resonance neutralization of $(CH_3)_3Si^+$ and $CD_3{}^+$ is expected to be energetically favorable, while unfavorable at the SiO_2 surface. This is because the SiO_2 valence band edge is at least 1 eV and 2.5 eV lower than the energy of the ion ground states of $CH_3{}^{\cdot}$ and $(CH_3)_3Si^{\cdot}$, respectively. Abundant TMS^+ desorption from ZrO_2, whose valence band edge is 0.4-1.7 eV lower than the TMS^+ gas phase ground state, is also consistent with this argument.

Examination of the energy distributions for ion vs electron stimulated desorption for TMS^+ from SiO_2 shows that these are almost identical. Hence arguments which rely upon higher velocities for secondary ions in SIMS and, therefore, lower neutralization probability are not valid. It appears that under ion bombardment the electron density of states in the local plasma is very similar for the collision cascade at the various substrates. It is suggested that the electron attachment phenomena which occur as the plasma decays will be of primary importance in affecting neutralization for this model. Such is not the case in ESD, where the density of states which is important is that of the bound electrons of the valence band in the substrate. An alternative explanation for the enormous differences in ion yield between ESD and SIMS, yet one which is consistent with the similarity in secondary ion energy spectra, is based on the assumption that a significant contribution to the SIMS signals for both substrates arises from ionization events which occur some distance away from the surface. That is, a large fraction of the ion signal in SIMS might result from dissociative relaxation of neutrals which are ejected as electronically excited species. These species dissociate to yield a neutral, a positive ion and an electron. The similarity in secondary ion energy spectra (ESD vs SIMS) is thus rationalized by the assumption that these sputtered, electronically excited neutrals are in states which are similar to the electronically excited states which are precursors to ESD of the same ions (TMS^+). Of course, a third explanation is that the similarity of the secondary ion energy spectra is purely coincidental. Only further experimentation in which the ejected neutrals are also examined will resolve this point.

References

1. T. R. Hayes, J. F. Evans and T. W. Rusch, J. Vac. Sci. Technol., A4, 1768 (1985).
2. T. R. Hayes and J. F. Evans, J. Phys. Chem., 88, 1963 (1984).
3. T. R. Hayes and J. F. Evans, Surface Sci., in press.
4. T. R. Hayes, PhD Dissertation, University of Minnesota, 1984.
5. J. A. Kelber and M. L. Knotek, Surface Sci., 121, L499 (1982).

Part III

Symposium: Detection of Sputtered Neutrals

Electron Gas SNMS

H. Oechsner

Fachbereich Physik der Universität Kaiserslautern and
Sonderforschungsbereich 91, D-6750 Kaiserslautern, F.R.G.

1. Why SNMS?

Though the secondary ions ejected from an ion bombarded solid surface are immediately available for mass spectrometric detection, surface analysis by sputtered secondary neutral particles offers several essential advantages compared to SIMS. This justifies the additional experimental complications connected with appropriate postionization techniques that are prerequisite for Secondary Neutral Mass Spectrometry SNMS.

The main advantages of SNMS arise from the following points:

- Since in most cases the sputtered particle flux consists almost exclusively of neutrals, the SNMS-signal of a species X remains stable even when the corresponding SIMS-signal varies by orders of magnitude due to the sometimes strong influence of varying surface composition on the ionization probability in secondary ion formation.
- While in SIMS the formation and the ionization process of an analyzed particle occur in the same event, both processes are decoupled in SNMS [1]. Hence, the postionization probability for a species X is a constant of the respective SNMS equipment.

From both reasons, SNMS is not expected to be influenced by matrix effects as is SIMS. This has been confirmed experimentally [2,3].

The SNMS-signal $I(X^\circ)$ of an ejected species X is directly proportional to the formation probability of X or its partial sputtering yield Y_X, i.e. $I(X^\circ) = D_X Y_X$ with D_X being a constant detection factor for X. For atomic sputtering under stationary conditions the partial sputtering yield Y_X is directly proportional to the atomic concentration c_X of X in the sample. Consequently, we obtain $I(X^\circ) \sim c_X$ which is the basis for the accurate and straightforward quantification of SNMS [1,2]. Because of the proportionality $I(X^\circ) \sim Y_X$ SNMS is also very useful for fundamental sputtering studies, in particular for investigations of the formation of sputtered molecules [4]. The formalism for the formation of molecules like MeO(Me: metal) derived from such studies can again be employed for the quantitative evaluation of molecular SNMS-signals from an analyzed sample [1].

2. Electron Gas SNMS

Postionization of sputtered neutrals is in principle possible by electron impact ionization with an electron beam crossing the flux of sputtered neutral particles (see e.g. [5] and corresponding contributions to this conference). However, only moderate postionization factors α° being at best on the order of 10^{-5} are obtained with such arrangements [6]. This is mainly due to the short dwelling times of the sputtered neutrals (mean energy on the order of 10 - 15 eV [7]) in the electron beam volume. On the other hand, the ionizatio probabilities for secondary ion emission from an oxidized metal surface are sometimes as high as several 10^{-2} [8]. Hence, to combine a detection sensitivity equal to that in oxygen-assisted SIMS with the above mentioned advantages of SNMS, postionization factors $\alpha°$ around 10^{-2} have to be established.

To achieve such conditions, a) the density n_e of the postionizing electrons has to be considerably increased above the n_e-values in appropriate electron beams ($n_e \cong 10^7$ cm^{-3} in an electron beam of 100 eV and 1 mA cm^{-2}), and b) the ionizing volume has to be expanded spatially. Both requirements are met by employing a dense and energetic Maxwellian electron gas for postionization. In the corresponding experimental technique, for which the acronym SNMS has been introduced at first [7], such an electron gas is produced by ionizing noble gas atoms at low pressures by an electrodynamic resonance effect, the so-called electron cyclotron wave resonance [9]. For Argon at pressures of a few 10^{-4} mbar electron densities n_e of 10^{10} - 10^{11} cm^{-3} are achieved at electron temperatures T_e corresponding to 15 eV. Such high n_e-values become possible since the electron space charge is compensated by the positive noble gas ions in the generated low-pressure high-density plasma. For a travelling path of about 5 cm $\alpha°$-values close to 10^{-1} are achieved for noble metal atoms, while $\alpha°$ is of the order of 10^{-2} for other elements. $\alpha°$ is a convolution of the Maxwellian energy distribution of the "plasma" electrons, the ionization function of the particles X to be ionized and - via the dwelling time - of the energy distribution of the sputtered particles X [1]. Due to the low working pressure, interfering interactions between sputtered particles and the working gas become negligible.

Corresponding SNMS-arrangements are schematically shown in Fig. 1. The postionizing plasma maintained in highly purified noble gas is produced in an internal chamber of an UHV-system. No plasma excitation electrodes are necessary inside the ionization volume. Contrary to electron beam arrangements, no impurities from hot filaments are introduced into the analytical system. In consequence of the quasi-neutrality of the postionizing plasma, no trapping of postionized sputtered particles by space charge potential wells occurs.

Besides a conventional ion gun (Fig. 1a) the positive ions from the resonant low-pressure plasma can be used for the ion bombardment of the sample (direct bombardment mode of SNMS, Fig. 1b). Plasma ions are extracted by means of a simple ion optics in front of the sample surface at bombarding energies from a few keV down to only 20 eV. Under appropriate operation conditions extremely high lateral homogeneity of the bombarding ion current with a density around

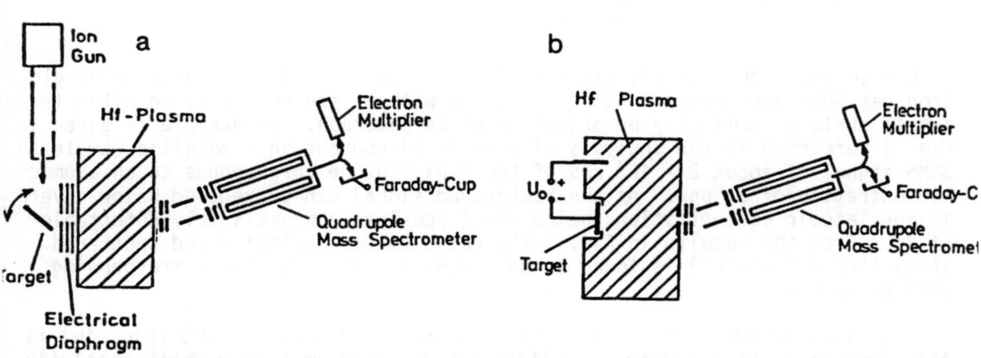

Fig. 1 Schematic diagrams of arrangements for electron gas SNMS. (a) Sample bombardment with an external ion gun. (b) Direct bombardment mode

10^{-3} Acm^{-2} is achieved [1,10]. Low bombarding energies around 10^2 eV under homogeneous ion bombardment have been shown to be a necessary condition for e› tremely high depth resolution[1,11]. At such low bombarding energies bombard- ment-induced atomic mixing is confined to the uppermost 2 atomic layers, which corresponds to the ejection depth of the analyzed particles [1].

The external bombardment mode of SNMS (Fig. 1a) can be readily applied for insulating samples. Then the electrical diaphragm between the sample and the ionizing volume is opened for plasma electrons to such an extent that sample charging by primary ions is compensated [1]. Both the direct bombardment mode and a separate ion gun can be applied without any change of the sample positic in another apparatus for electron gas SNMS [12]. High lateral resolution and even surface imaging by SNMS [13] are possible when properly focused ion beams are used in the external bombardment mode.

3. Surface and Depth Profile Analysis by Electron Gas SNMS

Both operation modes of SNMS (see Fig. 1) have been applied for quantitative surface and depth profile analysis of metals [1,2], semiconductors [11] and ir sulators [12]. The accurate quantification of SNMS is demonstrated in Fig. 2 where the concentrations c_x obtained by SNMS for the constituents in a metal sample are compared with the results from conventional wet chemistry analysis of the same sample.

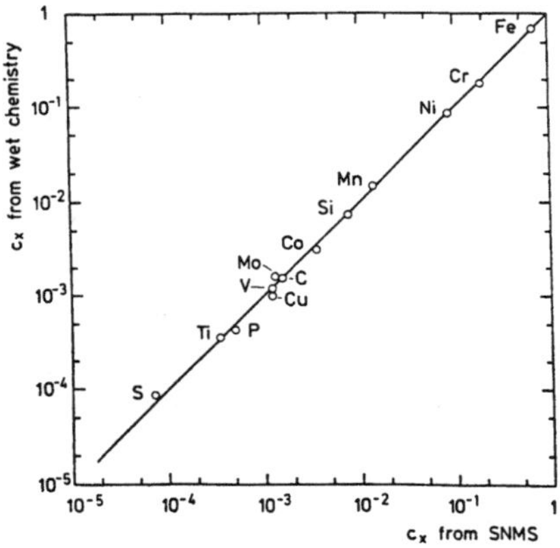

Fig. 2 Correlation plot for the atomic concentrations of various elements in a metallic sample determined by SNMS and by conventional wet chemistry [1]

Due to the high postionization factors α° the detection sensitivity of elec tron gas SNMS approaches that of SIMS, but without the necessity to alter the sample surface chemically by oxidation or Cs-coverage. An example is given in Fig. 3 referring to the analysis of a small Ni-content in a metallic sample. SNMS-signal of about 2 · 10^4 cps of the ^{62}Ni-isotope corresponds to an atomic concentration of 34 ppm. As the background signal can be reduced below severa 10 cps, atomic concentrations below the 1 ppm-level become safely detectable [1,2]. From the results in Fig. 3 the ratio between detected and sputtered atoms ("useful yield") is found to be around 3 · 10^{-8} for the corresponding SNMS equipment.

A particular advantage of the direct bombardment mode of SNMS (Fig. 1b) is the combination of accurate quantification and extremely high depth resolutio A corresponding example is shown in Fig. 4. Structural details of the unanne depth profile of P implanted into Si become visible as well as the presence o

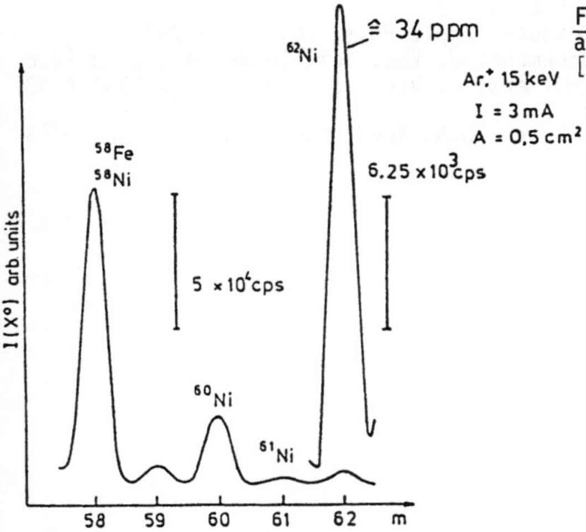

Fig. 3 Trace analysis by SNMS of a small Ni content in a metal sample [2]

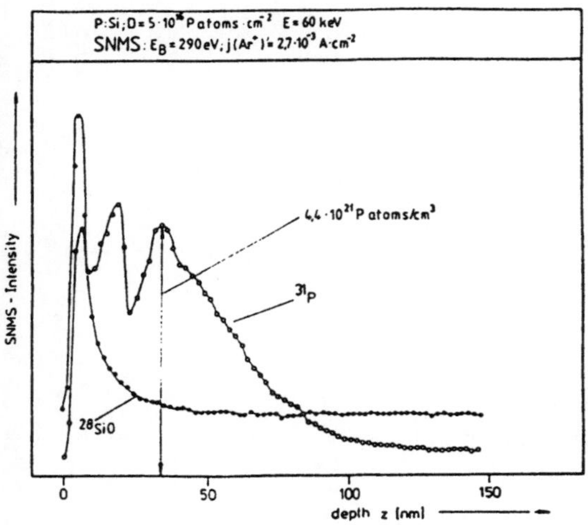

Fig. 4 High resolution depth analysis of a structured P implantation profile in oxide-covered Si by low-energy SNMS [11]

a thin residual surface oxide [11]. Additional examples for high resolution dep profile analysis of multilayer structures with a periodicity down to a few atomi distances are reported in another paper in the present volume [14].

References

1 H. Oechsner: Thin Film and Depth Profile Analysis, ed. H. Oechsner, Springer 1984, p. 63
2 K.H. Müller and H. Oechsner: Microchim. Acta Supp. 10, 51 (1983)
3 H. Oechsner, W. Rühe and E. Stumpe: Surface Sci. 85, 289 (1979)
4 H. Oechsner: SIMS III, ed. A. Benninghoven et al. Springer 1982, p. 106
5 R.E. Honig: J. Appl. Phys. 29, 549 (1958)
6 H. Oechsner: Nucl. Instr. Meth. Phys. Res. B, in print
7 H. Oechsner: Appl. Phys. 8, 185 (1975)
8 A. Benninghoven: Surface Sci. 53, 596 (1975)

9 H. Oechsner: Plasma Physics 16, 835 (1974)
10 E. Stumpe, H. Oechsner and H. Schoof: Appl. Phys. 20, 55 (1979)
11 H. Oechsner, H. Paulus and P. Beckmann: J. Vac. Sci. Techn. A3, 1403 (198
12 K.H. Müller, K. Seifert and M. Wilmers: J. Vac. Sci. Techn. A3, 1367 (198
13 K.H. Müller: unpublished results
14 H. Oechsner, G. Bachmann, P. Beckmann, D.A. Reed, S.M. Baumann, S.D. Wils
C.A. Evans Jr.: this conference

Glow Discharge Mass Spectrometry

W.W. Harrison, K.R. Hess, R.K. Marcus, and F.L. King

Department of Chemistry, University of Virginia,
Charlottesville, VA 22901, USA

The glow discharge is a simple, yet effective device which has found application in atomic absorption, atomic emission, and elemental mass spectrometry. It takes the form of a low-pressure gas discharge wherein rare gas ions are attracted to a sample cathode and efficiently sputter atoms from the surface. Figure 1 shows the essential components of the source. The discharge has been powered by dc, pulsed dc, and rf supplies, usually operating in a low power regime, with optimum currents and voltages determined by such factors as source configuration, discharge gas, and pressure. Nearly all the discharge voltage is dropped across the cathode dark space, serving to accelerate the argon ions onto the cathode, which is formed directly from the sample or compacted from a sample/matrix mixture. The glow discharge approach is simple, inexpensive, and shows relatively few matrix effects. Sensitivity in the ppb range is seen for most elements. A disadvantage of the technique is that the gas discharge produces certain background species which can create interferences for some elements, if alternative isotopes are not available.

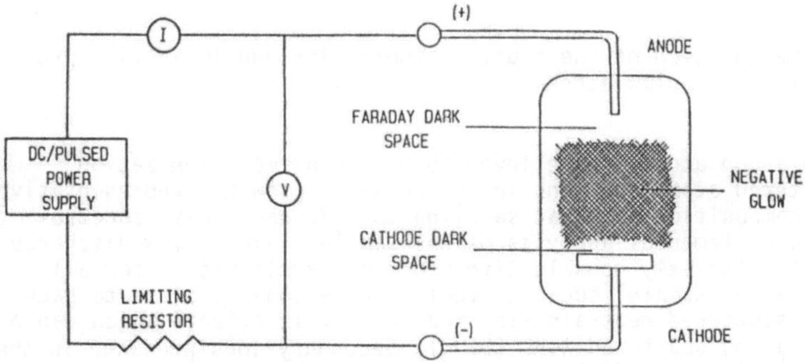

Figure 1. Schematic diagram of the glow discharge source

1. Glow Discharge Processes.

The glow discharge serves both to atomize and ionize the sample. As shown in Fig. 2, accelerated argon ions bombard the cathode (sample) surface and transfer their momentum to the sample atoms through a series of collisions. Atoms at the surface may receive sufficient energy to overcome binding forces and thus are released (sputtered) into the discharge plasma. Sputtered atoms diffuse into the negative glow and are subjected to energetic collisions, leading to excitation and ionization. Electrons

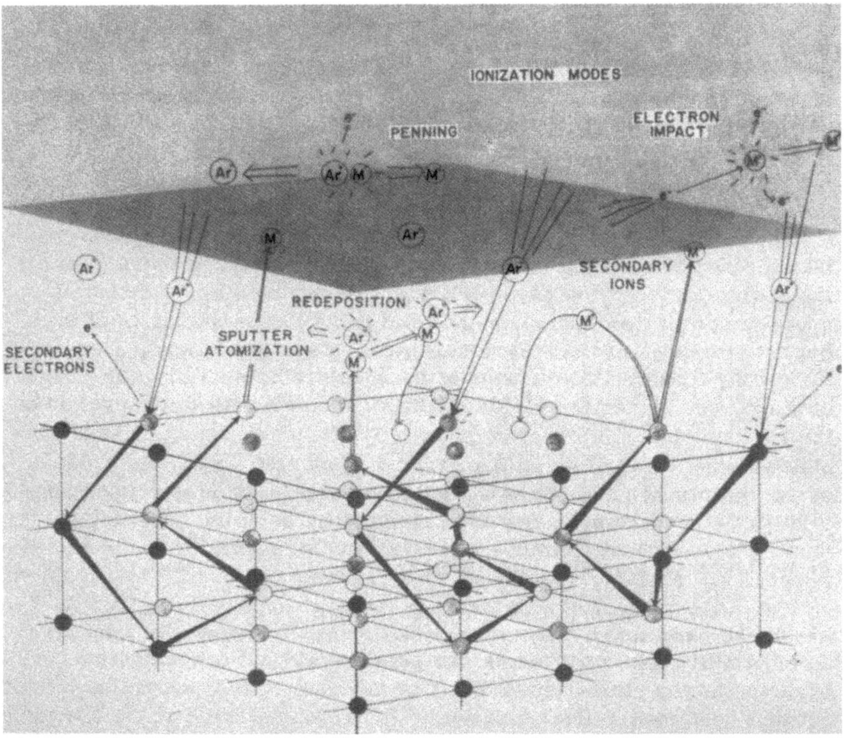

Figure 2. Representation of the sputter atomization and ionization processes in a glow discharge.

and metastable argon atoms are believed to be the primary ionization species. Sputtered atoms and ions in the negative glow are representative of the sample composition, so that sampling the discharge mass spectrometrically allows elemental analysis of the sample. Since glow discharge sputtering is a relatively nonselective process, sensitivities for all elements in an alloy sample (such as steel) are reasonably close to each other. Use of sputtered neutrals also reduces matrix effects which can be a problem when using sputtered ions (SIMS). Secondary ions produced in the glow discharge would normally be returned to the surface by the cathode fall potential. Many sputtered atoms are also knocked back onto the sample by collisions near the cathode due to the short mean-free-path at 1 torr pressure. Collisions in the cathode dark space produce a distribution of energies for the sputtering argon ions, ranging from full discharge potential to near thermal. Sputter yield is affected by discharge current, pressure, and the mass of the discharge gas, most often argon.

2. Experimental

Ion sampling of glow discharges has been done by both quadrupole and magnetic sector instruments. While the only commercial instrument on the market uses the latter approach, it is the quadrupole which has been used by most practitioners. Representative of glow discharge mass spectro-

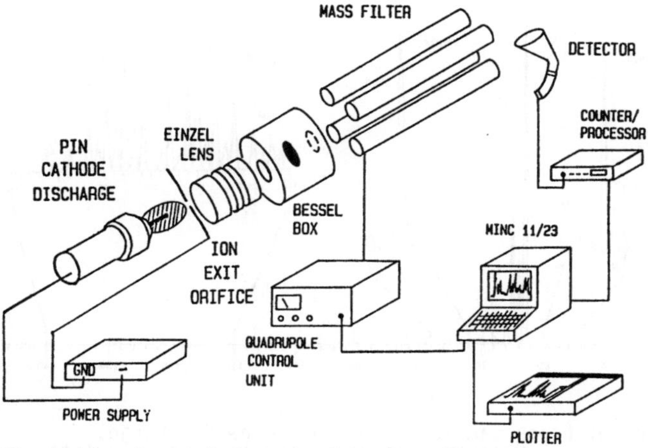

Figure 3. General layout of a glow discharge mass spectrometer

meters is that shown in Fig. 3. The only real experimental concern, compared to other ion sources, is the differential pumping to permit interfacing of the relatively high pressure (0.1 - 1.0 torr) source to the low pressure demands of the analyzer. An energy analyzer is useful in removing high energy ions which cause spectral tailing and degradation of resolution. Instruments should have the capability of fast repetitive scans for qualitative analysis plus peak-switching integration for quantitative measurements.

Glow discharge mass spectra consist primarily of singly charged ions of the sputtered atoms. The ion energy band-pass can be set to enhance these species. A few molecular ions (M^+, MO^+, MAr^+, M_2^+) do appear at high sensitivity and can cause interference at times.

Several other discharge configurations, in addition to that shown in Fig. 3, have been used. Samples in the form of discs work quite well. Also, hollow cathode discharges (another form of glow discharge) offer particular advantages for high sputter rates. Pulsed discharge operation can also yield higher sputter rates for glow discharges due to the higher operating voltage of the pulse. Pulsed discharges permit acquisition of time-resolved spectra, whereby enhanced sample-to-background signals may be obtained.

3. Applications

The major use of the technique to date has been for elemental analysis of bulk solids. It is competitive with spark source mass spectrometry for this purpose while being more stable, less complex, and less expensive. It may also find interest among the solids optical emission community because (a) mass spectra are much simpler than optical spectra and (b) mass spectrometry surveys the entire periodic table in a single scan, giving isotopic information. Figure 4 shows a typical glow discharge mass spectrum for a brass sample (low gain), showing the relative distribution of ion peaks. At higher gains, the minor and trace constituents appear.

Glow discharge mass spectrometry can also be used for compacted samples. Metal powders or filings can be pressed in a die and run

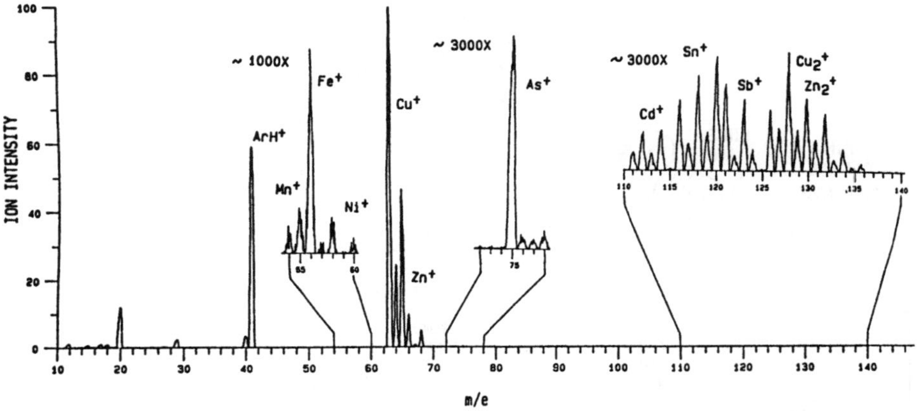

Figure 4. Mass spectrum of brass sample in an argon glow discharge. Concentrations, ppm: Mn(55), Fe(370), Ni(130), As(90), Cd(55), Sn(160), Sb(120).

directly. Nonconducting materials, such as ores, soils, and glasses, can be mixed in high purity graphite or silver, compacted into a pin or disc, and analyzed. This will open up the technique to many additional applications beyond metals and alloys. With some care, thin metal films may also be profiled. Since the glow discharge is a surface technique, it can sputter its way through the layers while sequential mass spectra identify the transitions. Typical sputter rates are 10-100 μg min^{-1} in argon, but this can be reduced by adjustment of discharge conditions and by use of other discharge gases.

General References

1. B. Chapman, Glow Discharge Processes, John Wiley & Sons, New York, 1980.
2. J.W. Coburn and W.W. Harrison, Appl. Spec. Rev. 17, 95 (1981).
3. W.W. Harrison, "Glow Discharge Mass Spectrometry" in Inorganic Mass Spectrometry, Chemical Analysis Series, John Wiley & Sons, New York, in press.
4. W.D. Westwood, Prog. Surf. Sci. 7, 71 (1976).

Electron Beam Postionization of Sputtered Neutrals

O. Ganschow +

Physikalisches Institut der Universität Münster, Domagkstr. 75, D-4400 Münster, F.R.G.

1. Introduction

Historically, electron beam postionization of sputtered neutrals, using the ionizer of a gas phase mass spectrometer /1/, was the first way to sputtered neutral mass spectrometry (SNMS). It was revived and further developed in recent years by several groups /2-5/ after Oechsner and his coworkers demonstrated the prospects /6-9/ of SNMS as a tool of quantitative mass spectrometric thin film and bulk analysis of solids.

With respect to SIMS, the advantage of SNMS is clearly the limited role of matrix dependence of secondary ion yields: A variation of the ionization probability of a sputtered species, e.g. from 10^{-5} to 0.5, introduces a variation of SIMS sensitivities of nearly five decades, but a factor of two only in SNMS calibration by complementarity.

With respect to Auger and photoelectron spectroscopy it offers the advantage of overcoming preferential sputtering as an artefact disabling quantitative calibration of concentrations in sputter depth profiling. In addition even the worst implementations of SNMS exceed the sensitivity of electron spectroscopy. To make this advantage real and to control the possible trade-offs of SNMS implementations one has to look at the physical problem of sputtered particle postionization. This is illustrated in Fig. 1, /5/ showing schematically the energy distribution of various radiations resulting from ion bombardment of solid surfaces and in addition that of residual gas: In a SIMS instrument the task is to collect secondary ions as efficiently as possible and at the same time to suppress all other radiations to result in a good dynamic range.

In SNMS the principal problems are the same, but in addition one needs efficient postionization along with suppression of the secondary ion signal, which would destroy quantitation, and suppression of the residual gas signal, which would limit the dynamic range by mass interference. Both suppression ratios must be 10^{-6} or smaller for "pure" SNMS spectra.

+ now with Leybold-Heraeus, Bonner Str. 498, D-5000 Köln 51, Fed. Rep. of Germany

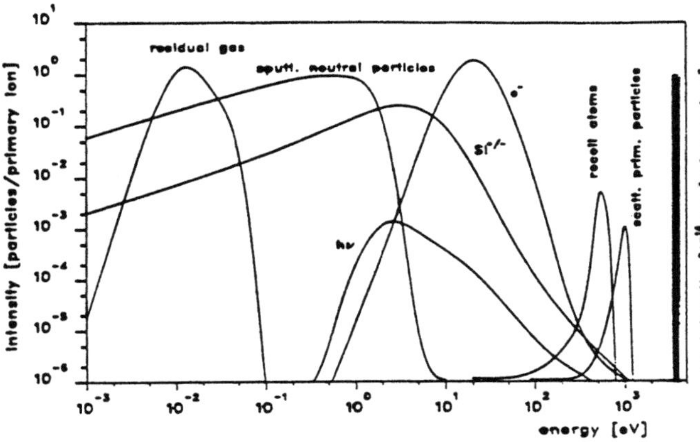

Fig. 1 Schematic energy distributions of various types of radiation created by ion bombardment of solids and postionization

2. Experimental

Figure 2 shows a schematic of our most recent electron beam impa SNMS instrument /11/. It is an improvement of our first desi /5,10/. With respect to the secondary ion and residual gas sign suppression schemes it differs from other approaches /2-4/ electron beam impact postionization.

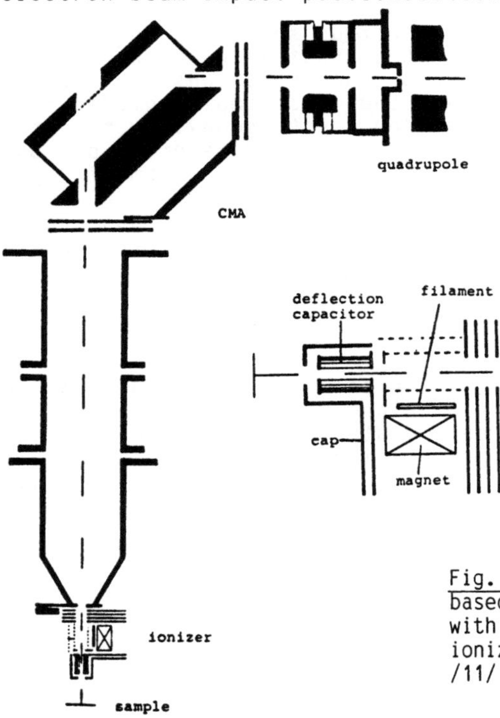

Fig. 2 Schematic of a quadrupole-based SIMS/SNMS/RGA instrument with electron beam impact postionization and energy filtering /11/

The beam path starts with an electrically insulated cap, serving as extraction electrode in the SIMS mode and as a repeller for secondary ions in the SNMS mode. However, usually secondary ions are discriminated by the subsequent deflection capacitor, which, by the semitubular shape of its electrodes, can be used as an electrostatic lens for focusing secondary ions in the SIMS mode. This discrimination is much more efficient than retarding fields /2,4/ for high energy secondary ions. The ionizer uses a magnetically collimated electron beam perpendicular to the sputtered neutral beam axis. The magnetic collimation increases the electron current density available and in addition keeps the electron space charge potential well inside the ionizer more uniform. As is well known, this reduces the spread of the formation potential for both sputtered neutrals and residual gas. Suppression for the latter signal is achieved by the electrostatic diaphragm following the ionizer. It imposes a weak retarding field, just sufficient to reflect the residual gas ions, having a thermal energy distribution, and transmitting the postionized neutrals with their kinetic energy of some eV.

In principle one could also try to use the dispersion of the subsequent energy filter, like in Ref. /4/ to separate sputtered neutrals and residual gas. However, in order to keep up with a suitable transmission of the energy filter, pass energies typically must be in the 80 - 100 eV range at a base resolution E/E of 1 %. Hence, the energy difference between postionized neutrals and residual gas ions is too small to achieve the required signal suppression ratio of 10^{-6}.

The most critical figures of any SIMS or SNMS instrument are the useful yield and the signal to background ratio. In an electron beam impact postionizer derived from mass spectrometer ionizers, saturation of the extractable ion current with respect to the electron emission current occurs as a result of the electron space charge and limits the useful yield. The data shown in Fig. 3

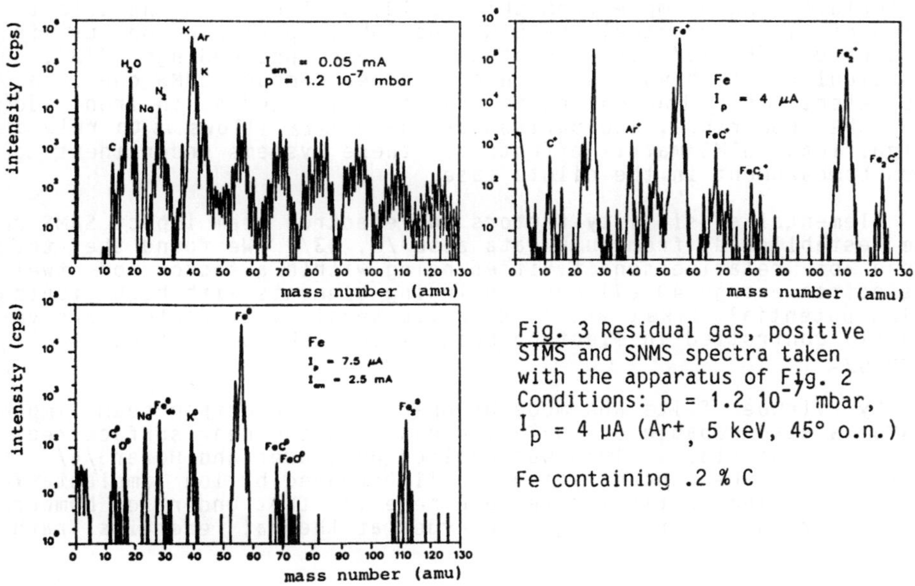

Fig. 3 Residual gas, positive SIMS and SNMS spectra taken with the apparatus of Fig. 2
Conditions: p = 1.2 10^{-7} mbar, I_p = 4 μA (Ar^+, 5 keV, 45° o.n.)

Fe containing .2 % C

give a residual gas spectrum, positive SIMS and SNMS of an ir sample containing 0.2 % of carbon. The SNMS data in the giv sputtering situation correspond to a useful yield of the order 10^{-9}, whereas in the SIMS mode 10^{-4} of the secondary ion flux detected. It is seen, that the signal to background ratio of th instrument in the SNMS mode is around 10^{-5}. This is improved by o decade with respect to our previous design /5,10/. Mostly this due to the reduction of the structureless background former observed.

The alkaline signals present in both the residual and the SN spectra are due to insufficient outgassing.

3. Applications

Electron beam impact SNMS, allows switching between residual g analysis, positive SIMS, negative SIMS and SNMS by changing ele trode potentials only. Furthermore it is performed in a true U environment. Thus it allows one to inspect all channels of sputt emission in one experiment.

In a metal-oxygen system this feature allows one to establi secondary ion yields and ionization probabilities as a function oxygen concentration, which is one of the key issues in quantit tive evaluation of the SIMS matrix effect and secondary ion form tion mechanisms. We did such experiments for silicon /5/ and ir /12/. In the case of silicon this revealed that under oxyg saturation as much as 45 % of the atomically sputtered Si ionized and that as long as no SiO_2 is formed the ionizati probability increases exponentially with oxygen concentration. the case of iron the maximum ionization probability was 17 Both figures can be compared against literature data about usef yields obtained in existing SIMS machines.

In another type of application we looked for bulk analysis materials. An example from this field, dealing with characteriz tion of ultrapure steels by SIMS and SNMS, performed on our fir apparatus /5, 10/ is presented in these proceedings /13/. particular it shows that by electron beam impact SNMS one is ab to detect fairly low concentrations of C /14/ and Al in iron, do to the ppm range. Comparison of these data allows us to rule o significant SIMS matrix effects for these systems under inert g ion bombardment in the dilute case.

Elemental sensitivity factors for electron beam impact SNMS c be established from bulk data also /5, 13/. We found the tot span of relative sensitivities being within a factor of twel (electron energy 40 eV) even including elements with high ioniz tion potentials like C and N. Relative sensitivity factors are ve easily controlled in electron beam impact SNMS by setting electr energy.

By virtue of its RGA mode of operation, electron beam impa SNMS is also capable of analyzing gases in the near surface regi of solid materials. This was applied by Gnaser and Hofer /4/ detect the He concentration profile created by ion implantati in metals. The particular feature here is that under ion bombar ment for depth profiling He emerges at thermal energies rath

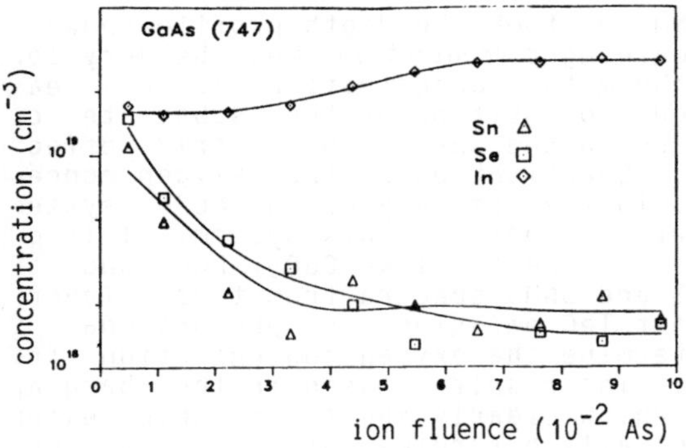

Fig. 4 Dopant redistribution in GaAs. Total eroded depth: approx. 150 nm

than being sputtered. It was detected by tuning the transmission energy window to the RGA mode. Thus they were able to detect the onset of gas agglomeration as a function of the implanted He fluence.

SNMS spectra in general exhibit a much smaller variety of mass lines than SIMS spectra do, in particular if compounds are considered. This reduces mass interference, which often causes problems in SIMS analysis of semiconductors at small concentrations using quadrupole based intruments. Figure 4 shows an example of Sn and Se surface enrichment in a GaAs MBE layer, ranging over approximately 150 nm. At the depth resolution indicated by the distance of the data points 10^{-18} atoms/cm represents the detection limit of our first instrument /5/. In addition to the Sn and the Se in survey spectra, In was detected as a contaminant.

As a final example we consider the quantitative analysis of the elemental composition of an anodic oxide layer on $Cd_{0.3}Hg_{0.7}Te$ /15/. More precisely this layer is supposed to consist of a mixture of TeO_2, $CdTeO_3$ and $HgTeO_3$. Two questions regarding these layers cannot be answered by SIMS, XPS and AES: First the apparent Hg concentration shown by electron spectroscopy seems to be very low, which might be due to preferential sputtering or reflect a genuine Hg depletion. Second, as a result of preferential sputtering, the oxygen concentration cannot be determined accurately from electron spectroscopy or SIMS.

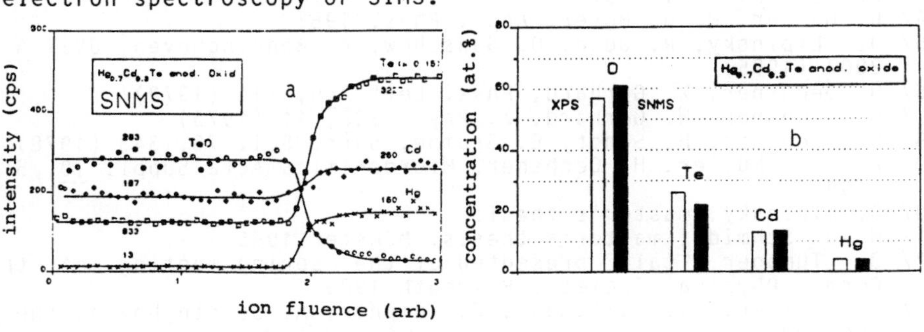

Fig. 5 a) SNMS depth profile of an anodic oxide layer on $Cd_{0.3}Hg_{0.7}Te$. Layer thickness 90 nm. /15/

b) Comparison of SNMS analysis and XPS analysis

Using our electron beam SNMS we took the depth profile shown : Fig. 5a. It is seen that again Hg concentration must be very lo\ Quantitative evaluation of the metal concentrations is very ea: from these data, as the known composition of the substrate c serve as an internal standard in the absence of matrix effec also applicable to the oxide. Quantitation of the oxygen conce tration can be carried out in the same way as in Me/O syste /5, 12/ if oxygen interacts with Te only in this system. This c be inferred from the absence of species like CdO, HgO and t presence of TeO in both SIMS and SNMS spectra from these layer The basic idea then is, that TeO emission is proportional c_{Te} c_O. This allows us to determine the oxygen concentration fr the Te and TeO signals only. The result is seen in the bargra representation of Fig. 5 b, which clearly shows to which exte oxygen is sputtered preferentially but rules out that Hg is spu tered preferentially.

4. Conclusion

Electron beam postionization allows state-of-the-art SNMS. With useful yield around 10^{-9} and a signal-to-background ratio of 1 it allows elemental analysis down to the ppm range at yet reaso able sample consumption. Its advantage is that it is fully U compatible and allows fast switching to SIMS and residual g analysis. In quantitation it allows one to separately track all spu tered fluxes but, different from electron gas SNMS in the dire bombardment mode on conductive samples, secondary ion emission not automatically eliminated from quantitation. Electron be impact postionization can easily be combined with other methods surface analysis.

5. Acknowledgements

Thanks are due to Prof. Benninghoven for his strong support this work. The cooperation of R. Jede, U. Kaiser, D. Lipinsk H.J. Schmidt, P. Sander, L. Wiedmann, E. Manske, J. Tümpne R. Wilsch, J. Coellen and W. Mai is gratefully acknowledged.

/1/ R. Honig, J. Appl Phys, 29, 549 (1958)
/2/ T. R. R. Lundquist, JVST, 15, 4 (1978)
/3/ G. P. Können, J. Groner, A. Haring, A E. de Vries, Ra Effects 21, 171 (1974)
/4/ H. Gnaser, W. O. Hofer, Appl. Phys. 1985
/5/ D. Lipinsky, R. Jede, O. Ganschow, A. Benninghoven, JVST A 2007 (1985)
/6/ H. Oechsner, W. Gerhard, Phys. Lett A40, 211 (1972)
/7/ H. Oechsner, W. Gerhard, Z. Phys. B22, 41 (1975)
/8/ H. Oechsner, H. Schof, E. Stumpe, Surf. Sci. 76, 343 (1978)
/9/ K. H. Müller, H. Oechsner, Microchimica Acta Suppl. 10, 5 (1983)
/10/ D. Lipinsky, master's thesis
/11/ H. J. Schmidt, master's thesis, Münster 1985
/12/ J. Tümpner, talk presented at the spring meeting of t German Physical Society, Bayreuth 1985
/13/ E. Laiber, J. Dittmann, J. Tümpner, A. Benninghoven, the proceedings
/14/ D. Lipinsky, R. Jede, J. Tümpner, O. Ganschow, A. Bennin hoven, JVST A3, 2035 (1985)
/15/ U. Kaiser, thesis, Münster 1985

General Postionization of Sputtered and Desorbed Species by Intense Untuned Radiation

C.H. Becker and K.T. Gillen

Chemical Physics Laboratory, SRI International,
Menlo Park, CA 94025, USA

1. Introduction

Since HONIG [1] first used electron bombardment ionization to examine the neutral material sputtered from a sample, there has been a general realization that detection of sputtered neutrals could and should be performed. Although the convenience of detecting sputtered or desorbed secondary ions led to a much more intense development of secondary ion detection techniques, the motivation for examining the neutrals is more than just the fundamental interest in the emission process and its ion-to-neutral branching ratios. A prime practical motivation is the difficulty in quantifying the secondary ion yields in SIMS and laser desorption mass spectrometry. Since the neutrals are generally the major component of the sputtered or desorbed material, their intensities are much less dependent on matrix effects that can vary the ionization fraction over a wide range.

With the motivation clear, the question is how best to accomplish neutral detection in a way that is efficient, clean, widely applicable, and quantifiable. Our approach to postionization is to use the intense radiation from a focused, pulsed, untuned, powerful UV laser to cause ionization, principally by nonresonant multiphoton ionization (MPI), of material desorbed or sputtered by a separate surface probe beam. These photoions are then both separated from secondary ions and mass analyzed by a reflecting time-of-flight mass spectrometer. We call the method surface analysis by laser ionization (SALI) [2,3].

Our method needs to be contrasted with the other emerging laser-based approach to postionization of sputtered or desorbed neutral material, selective resonant ionization. In the resonant approach, first reported by WINOGRAD et al. [4] and now being investigated in several laboratories[5,6], laser radiation is tuned to a specific transition of a desired atom (or small molecule, if the spectroscopy is well characterized) to excite that atom to an intermediate excited state that is then ionized by a subsequent photon. The goal is generally to achieve highly selective, extremely sensitive detection of the specific element (or isotope) being examined. A mass spectrometer is generally employed between the ionization zone and the detector both for isotope analysis, and for increased selectivity and background suppression.

Our approach is quite distinct. We use an untuned laser of much higher power in an attempt to achieve efficient (even saturated) ionization for all the neutrals within the focused laser zone; we attempt to <u>minimize</u> selectivity in the ionization step and to thereby generate ions that are characteristic of the composition of the surface being probed. Time-of-flight mass spectrometry is used as an efficient method to disperse and collect all of the ions generated from each laser pulse.

Resonantly-enhanced MPI techniques will generally require lower laser powers than SALI for efficient ionization, implying that ionization can be made to occur over a larger laser focal volume. This fact and the suppression of isobaric interferences would suggest a more efficient detection of a single desired species and a lower ultimate detection limit (higher sensitivity) for the selective resonant MPI technique. However, it will not always be possible to extract and detect efficiently the ions produced over a larger photoionization volume, especially in situations requiring mass analysis. More importantly, resonant techniques can only examine one quantum state at a time; atoms sputtered over a large range of quantum states or in molecules or aggregates will have correspondingly lower sensitivities (and quantitation will require laser or multilaser probing of all important atomic quantum states along with estimation of the molecular fractions). Hence, the nonresonant technique can be expected to sometimes compete favorably with the resonant approach for ultimate sensitivity limits on a single species. When several species are being examined, the ease with which SALI simultaneously detects all masses is clearly attractive, and quantitation of one species relative to another is more straightforward.

With its emphasis on general postionization of sputtered particles, SALI is in some ways more closely related to electron bombardment postionization techniques, including plasma ionization [7]. By separating the sputter (or desorption) process from the ionization step, these techniques all attempt to remove the most troublesome matrix effects (variations in ionization probabilities) from SIMS (or other ion desorption techniques) and replace the ionization step by a more easily characterized process. Although electron bombardment ionization cross-sections and branching ratios are known for a much larger variety of systems than are MPI nonresonant cross-sections, this situation will change as high-power ultraviolet lasers achieve more widespread use. Ionization efficiency has been shown to be considerably greater using intense laser radiation; and the very high sensitivity of SALI has already been demonstrated [2,8,9].

2. Laser Power Dependencies of Ionization Yields

Important to the success of SALI in analytical applications is the achievement of efficient, quantifiable ionization for all the sputtered or desorbed components of a material. At sufficiently low laser power, an N-photon nonresonant ionization process will produce ions at a rate that grows as the Nth power of the laser power density. As the laser power density is increased, the ionization probability becomes large and the growth in the total ion signal decreases from the N-power law, eventually leveling off. Although most species can be ionized with pulsed ultraviolet lasers, power density requirements (10^7 to 10^{11} Watt/cm^2) imply that the laser generally must be focused to a small spot above the surface being examined. Even when all species are ionized in the region of tightest laser focus, differences in sensitivity will occur because the volume of efficient, saturated ionization will extend further into the wings of the laser beam spatial profile for species that have larger ionization cross-sections. However, at some point at high powers the size of the mass spectrometer extraction zone can limit the collected signal.

It should be noted that under typical focus conditions, about 10^{-2} of the sputtered fraction for a 1 microsecond ion beam pulse can be intercepted by the laser. Also note that the short pulse photoionization ($\lesssim 10^{-8}$ s) acts as a density detector, and there is a velocity influence on the density (and detection probability); velocity distributions can be directly measured however, and generally they play a weak role in detection probability.

In this section, we describe experiments examining the laser power dependence of ionization by an ArF excimer laser (193nm, 6.43 eV) for a range of neutrals sputtered from samples of known composition by a 3 keV Ar^+ beam or introduced to the ionization zone as a permanent gas. We will show typical data for one and two-photon ionization of atoms for laser power densities up to approximately 10^9 watt/cm^2. The experimental setup has been described elsewhere [2,9]. Each data point presented here is an average at 100-300 laser shots.

Figure 1 shows the power dependence of Xe two-photon ionization for a gaseous sample at low pressure (about 10^{-8} torr). The slope is within experimental error of the expected value of two and is consistent with the expectation based on the measured Xe cross-section [10] that deviations from the power law should occur at powers of about 10^{10} W/cm^2. Figure 2 again shows the Xe data and contrasts it with power dependencies for Al and Na sputtered from a glass sample and for various charge states of W sputtered from a W mesh. W is very easy to ionize (two photon) at these power levels, and W^{++} is unusually easily formed for a doubly charged ion; even the triple ionization occurs with a slope smaller than that of Xe. Al is ionized by one photon, which is resonant with an allowed transition to an autoionizing state within the continuum; hence the ionization rate is very large and the ionization volume has grown large enough that there is little increase in signal with laser power. Na (one-photon ionization) has an intermediate slope (0.6) indicating again a saturated ionization zone that is growing with laser power.

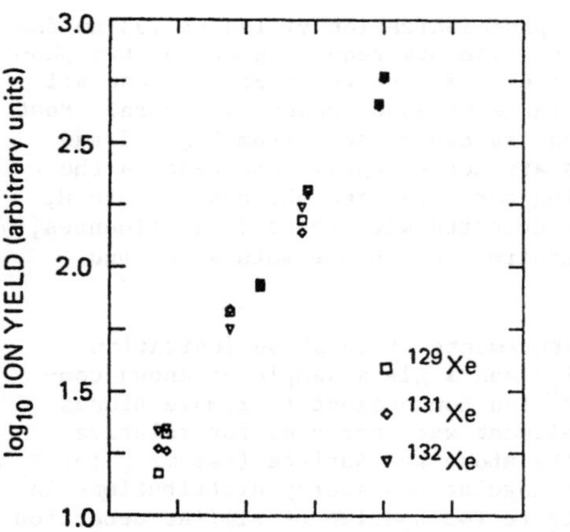

Fig. 1. Relative Xe ionization yields at 193 nm s a function of laser power (isotopes scaled by the known ratios)

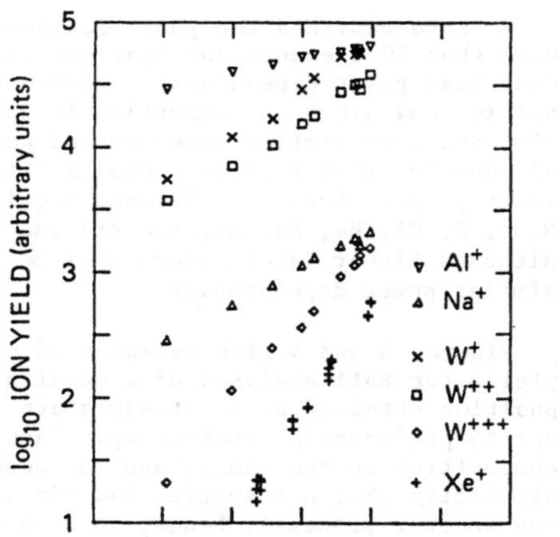

Fig. 2. Laser power dependence of ionization yields at 193 nm for Al and Na sputtered from glass, W sputtered from the metal, and gaseous Xe (W charge state ratios as measured; other atoms arbitrarily normalized)

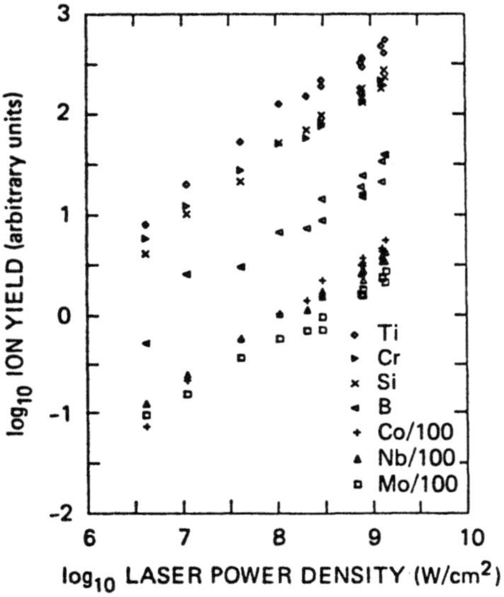

Fig. 3. Relative SALI detection yields at 193 nm for different elements sputtered from a multicomponent alloy

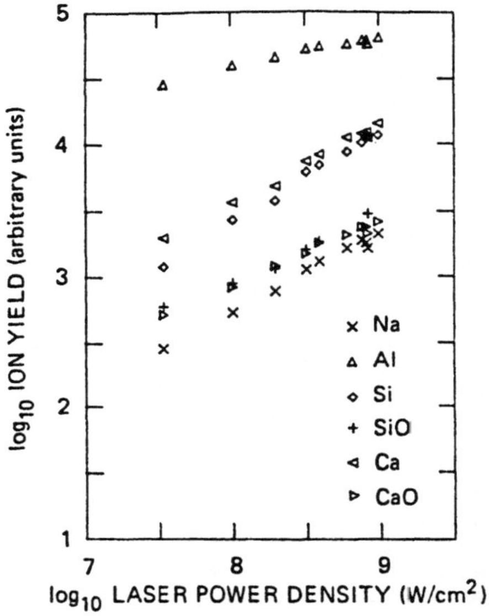

Fig. 4. Relative SALI detection yields for Al, Si, Ca, and Na sputtered from NBS glass 610

We have examined the power-dependent photoionization yields at 193 nm for more than 30 elements and many molecules; elements requiring one or two photons have power dependencies between the extremes shown on Fig. 2, and all but Xe show signs of saturation in the range of laser power used here. Most elements have similar power dependencies (as can be seen from Figs. 3 and 4), meaning that relative sensitivities are not strongly dependent on the laser power. Even the elements requiring more than two photons (atomic H, N, O, F, Cl, He, Ne, Ar, and Kr) can be detected with these laser fluences, although higher laser powers will be required to achieve saturation and similar power dependencies.

Figures 3 and 4 give examples of measurements of relative ionization yields for SALI analyses of a metal alloy and a glass sample of known composition obtained under steady-state Ar^+ ion bombardment to remove biases due to preferential sputtering. Each element was corrected for relative composition in the sample and for density above the surface (assuming for simplicity that all species had similar angular and energy distributions in the sputter process); hence, in each figure two species of similar detection sensitivity would overlap. In Figure 4 the CaO and Ca signals should be added to give the total Ca contribution; likewise for Si and SiO.

Variations in sensitivity are seen to be small, all within an order of magnitude except for B in Fig. 3 and Na in Fig. 4 (possibly sputtered predominantly as Na^+, which could be detected by time-of-flight SIMS in this apparatus [3]). Using the Al^+ signal in Fig. 4 as an upper limit, it is clear that the variations in sensitivities will decrease as laser power density increases. High and quite uniform ionization efficiencies appear to be achievable for most species in the near future, with intense and spatially well defined untuned UV radiation.

Acknowledgement

We thank the San Francisco Laser Center for loan of an excimer laser, and we acknowledge SRI Internal Research and Development Funds for financial support.

References

1. R. E. Honig, J. Appl. Phys. 29, 549 (1958).
2. C. H. Becker and K. T. Gillen, Anal. Chem. 56, 1671 (1984).
3. C. H. Becker and K. T. Gillen, J. Vac. Sci. Technol. A3, 1347 (1985).
4. N. Winograd, J. P. Baxter, and F. M. Kimock, Chem. Phys. Lett. 88, 581 (1982)
5. M. J. Pellin, C. E. Young, W. Calaway, and D. M. Gruen, Surface Sci. 144, 619 (1984).
6. D. L. Donohue, W. H. Christie, D. E. Goeringer, and H. S. McKown, Anal. Chem. 57, 1193 (1985).
7. H. Oechsner and W. Gerhard, Surf. Sci. 44, 480 (1974).
8. C. H. Becker and K. T. Gillen, Appl. Phys. Lett. 45, 1063 (1984).
9. C. H. Becker and K. T. Gillen, J. Opt. Soc. Am. B 2, 1438 (1985).
10. A. W. McCown, M. N. Ediger, and J. G. Eden, in Excimer Lasers-1983, C. K. Rhodes, H. Egger, and H. Pummer eds., American Institute of Physics Conference Proceedings No. 100, New York (1983), pp. 128-141.

Multi-Photon Resonance Ionization of Emitted Particles

G.A. Schick, J.P. Baxter, J. Subbiah-Singh, P.H. Kobrin, and N. Winograd

The Pennsylvania State University, Department of Chemistry,
152 Davey Laboratory, University Park, PA 16802, USA

1. Introduction

It is clear from the scope of this symposium that major advances have been achieved in our ability to detect and characterize the uncharged species that desorb from ion bombarded surfaces. The use of plasma sources and lasers have been particularly useful in enhancing the sensitivity of detection over the traditional weight-loss methods. In this paper, we describe the use of multi-photon resonance excitation as a post-ionization technique for detecting desorbed neutrals with high sensitivity and selectivity. The discussion will focus on where this approach is most likely to be successful and where it should probably not be utilized. Examples of present applications will be cited, with a special emphasis on the use of laser techniques to determine the energy and angle distributions of particles emitted from single crystals.

2. Experimental Strategy

The basic experimental configurations for using multi-photon resonance ionization (MPRI) in ion bombardment experiments have been described.[1,2] Briefly, a pulsed ion beam of typically 0.2–3.0 μs serves to generate a burst of desorbed particles localized in space moving away from the target surface. At an appropriate time, a pulsed laser (6 ns) tuned to an atomic resonance of the emitted particle is directed a few mm above and parallel to the target. As discussed by others [3], with sufficient laser power and excited state lifetimes, all neutrals with excitations in resonance with the photon field and with appropriate ionization potentials are converted quantitatively into ions. Due to the pulsed nature of the experiment and the selective nature of the ionization processes itself, the resulting ions may be efficiently counted by accelerating them with an external potential and measuring their time-of-flight.

We have utilized two different experimental geometries to characterize the desorbed neutrals. To maximize sensitivity [1], the size of the laser beam is made as large as possible, and the laser is fired at a time when the maximum number of neutrals should be present in the photon field. Typically, for this experiment, the 3 μs ion beam pulse causes neutrals to be emitted into a 6 mm diameter laser beam 1 mm above the target fired 0.5 μs after the end of the ion beam pulse. With this scheme, it is possible to count more than 20% of the atoms originally present on the sample surface [1]. We have designed a second experimental geometry to provide spatial and temporal information about neutral trajectories in order to obtain basic data relative to the ion/solid collision event [2]. In this configuration, the laser beam is focused in the shape of a ribbon 0.1 mm thick and 6 mm wide and directed 1 cm above and parallel to the target.

Focusing allows ionization of those neutrals which are in a specific region of space. Hence, by varying the time between a 200 ns ion pulse and the 6 ns laser pulse, the velocity distribution of the neutral may be determined. Furthermore, the resulting ion may be imaged to a microchannel plate which allows the position of the neutral in the laser to be determined with respect to the position of the incident ion beam on the target. With knowledge of the nature of the extraction fields and instrumental factors associated with time-of-flight measurements and position-sensitive detectors, it is feasible to calculate the take-off angle of the neutral atom.[2] This energy and angle-resolved neutral (EARN) detector is a powerful device for measuring neutral trajectories with enormous sensitivity. Note, however, that by utilizing a focused laser beam a large distance above the target for velocity selection purposes, the overall efficiency is dramatically reduced when compared to the first experimental configuration.

3. Comparison of MPRI Methods to Other Post-Ionization Schemes

In general, the post-ionization methods may be regarded as inherently selective or non-selective in nature. Clearly, since the MPRI method required tuning a dye laser wavelength to an atomic resonance, only those atoms or molecules with appropriate electronic structure will be ionized. In essence, the laser wavelength serves as the mass selector. The laser-induced-fluorescence methods [4] exhibit the same basic property. Non-resonant laser ionization as demonstrated using the SALI method [5] , as well as plasma ionization [6] and electron impact ionization are, of course, non-selective methods which ionize all types of species. The advantage of non-selective methods is that multi-element detection is easily accomplished, but usually at the cost of generating interferences. The selective methods find their place when interferences are not tolerable.

The question of ultimate sensitivity is always a difficult one when attempting to compare various methods. The major advantage of the MPRI method is that it possesses the highest overall sensitivity when considering the efficiency of transporting the desired neutral from the solid to the detector for counting. Using an unfocused laser, more than 70% of the emitted neutrals may be directed into the photon field and ionized with nearly 100% efficiency and independent of velocity up to 50 eV [1]. There are two serious complications in practice which limit the desired sensitivity. First, the lasers are pulsed at a rate of 30 Hz. With a 3 μs ion pulse of peak current 100 μA, the average current is only 10 nA. With a sputtering yield of 5 atoms/incident ion, only about 10^{11} atoms/s will find themselves in the photon field. For the cw methods such as plasma ionization and electron impact ionization, this number may be 4 orders of magnitude higher, yielding a larger ultimate signal for low level impurities even though the ionization efficiency for a given atom may be much lower. Another consequence of this problem is that it will be difficult to couple MPRI methods to experiments using focused ion beams for microscopy purposes. For example, a 200 Å ion beam may only have a peak current of 10 pA which would inject only 10000 atoms/s into the laser. This would obtain plenty of signal for bulk components, but would clearly restrict analysis of trace components since one would have to count for inordinate times to generate sufficient signal. A second problem with achieving the desired sensitivity is that the MPRI methods are not only sensitive to atom type, but also to the electronic state of the atom. Thus, if a significant fraction of the ground state neutrals are desorbed in excited states, or as molecules, they may not be counted. The SALI method and other non-selective ionization methods do not suffer from this problem. In summary, then, we believe that the MPRI method possesses the highest ultimate sensitivity of any known method, but it may not generate the biggest signals.

4. Energy and Angle Distributions from Rh{111}

We have obtained preliminary data from the EARN detector utilizing a Rh{111} single crystal target. The details of the experimental set-up are given elsewhere [7]. The Rh{111} crystal is cleaned by cycles of heating in oxygen to 950K and ion sputtering such that no carbon, oxygen or sulfur intensity could be measured using Auger spectroscopy, and LEED measurements produced sharp {111} diffraction spots characteristic of a well-ordered surface [8]. A well-formed p(2x1) LEED pattern could be generated by exposure of the clean surface to 10 L oxygen at room temperature. Experimental results for both surfaces are shown in Figure 1 and Figure 2. The total dose of incident 3 keV Ar^+ ions was less than $10^{12}/cm^2$ for each case, and no effects of beam damage could be discerned by Auger spectroscopy or LEED.

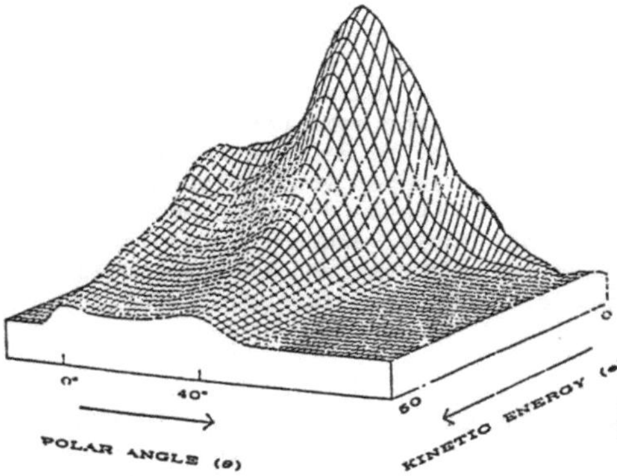

Fig 1. Kinetic Energy and Polar Angle (θ) distribution for Rh atoms desorbed from Rh{111}. The ion beam is incident normal (θ=0°) to the surface.

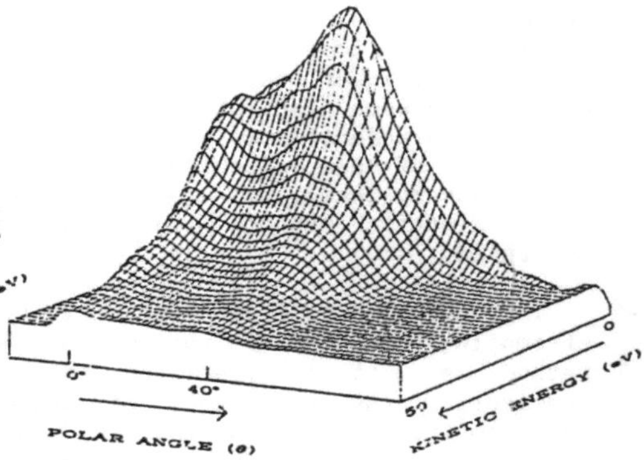

Fig 2. Kinetic Energy and Polar Angle distributions for Rh atoms desorbed from p(2x1) O on Rh{111}.

Several features of these results are of interest. Notice that from Fig 1 the intensity of the peak at θ=0° is somewhat more important than the one at θ=35°, at higher kinetic energies. In addition, the effect of surface channeling becomes apparent at high kinetic energies since the width of the angular distribution narrows considerably and the magnitude of the angular anisotropy is much greater. When the surface is covered with a fraction of a monolayer of oxygen as shown in Fig 2, the shape of the distributions change a great deal. The θ=0° peak is much more intense at low kinetic energy, the θ=35° peak shifts in by about 5° relative to the clean surface and is almost completely wiped out at high kinetic energy. We presume these effects are due to a blocking of the take-off trajectory of the Rh atoms by the chemisorbed oxygen atoms. Classical dynamics calculations are in progress to attempt to find a quantitative analysis of these effects. In any case, the results point out how sensitive these angular distributions are to surface contamination.

5. Conclusions

The MPRI method for detecting neutrals desorbed from ion-bombarded surfaces has many areas of unique application. Specifically, it is a useful trace analysis tool for detection of specific atomic species where interferences may be a problem. The spatial and temporal aspects of the experiment make it possible to determine energy and angle distributions. It will be interesting to see new applications evolve. For example, molecular species and excited states have been studied only briefly. Atoms with rather complicated ionization schemes such as oxygen and carbon have yet to be studied. The advent of more powerful lasers with higher repetition rates is also in the cards.

6. Acknowledgments

We wish to acknowledge the financial support of the National Science Foundation, the Air Force Office of Scientific Research, the Office of Naval Research, and the IBM Corporation.

7. References

1. F. M. Kimock, J. P. Baxter, D. L. Pappas, P. H. Kobrin and N. Winograd: *Anal. Chem.*, **56**, 2782 (1984).
2. P. H. Kobrin, J. P. Baxter and N. Winograd: *J. Vac. Sci. Tech. A* 3(3), 1596 (1985).
3. G. S. Hurst, M. G. Payne, S. D. Kramer, J. P. Young: *Rev. Mod. Phys.*, **51**, 767 (1979).
4. W. Husinsky, G. Betz, I. Girgis, F. Viehbock and H. Bay: *J. Nucl. Mater.* **128**, 577 (1984).
5. C. H. Becker and K. T. Gillen: *Anal. Chem.*, **56**, 1671 (1984).
6. O. Oechsner and W. Gerhard: *Surf. Sci.*, **44**, 480 (1974).
7. P. H. Kobrin, G. A. Schick, J. P. Baxter and N. Winograd: *Rev. Sci. Instru.*, submitted.
8. L. A. DeLouise and N. Winograd: *Surf. Sci.* **138**, 417 (1984).

Post-Ionization of Sputtered Particles: A Brief Review

W. Reuter

IBM Thomas J. Watson Research Center, Yorktown Heights, NY 10598, USA

Introduction

Definition of terms:

α^+ : The fraction of sputtered particles of element Me in the post-ionizing volume in the Me^+ state.

T: The overall transmission including estimates on losses due to the fact that only a fraction of the sputtered particles is in the post-ionizing volume.

$\alpha \times T$: This is the total useful yield, i.e., the fraction of ions detected for a monoisotopic element per sputtered atom. Not included are electronic gating losses reducing the useful yield by a factor of 10 to 100 in SIMS, e-beam and thermal post ionization.

By now secondary ion mass spectrometry (SIMS) has been developed into a highly sensitive technique for the detection of impurities at surfaces and interfaces and for the determination of implant and diffusion profiles in thin films and semiconductors. In favorable cases relative detection sensitivities (RDS) extend into the low ppb range and absolute detection sensitivities (ADS) as high as 10^{-2} (1 atom detected in 100 atoms sputtered) can be obtained if high transmission ($T \sim 10^{-1}$) machines (magnetic mass spectrometers) are used and the particular analytical conditions yield high ionization fractions ($\alpha^{+,-} > 0.1$) .

The outstanding sensitivity of SIMS in respect to other analytical techniques [1] is however counter-balanced by our lack of a fundamental understanding of the secondary ion emission process. Only recently for the relatively simple case of Cs^+ emission from a metal surface under Ne^+ bombardment has a quantitative understanding of the ionization process been reported [2]. In spite of many valuable contributions towards a better understanding of SIMS or more generally of ion solid interactions, the case of concern to the practicing analyst, i.e., secondary ion emission mechanisms under O_2^+ or Cs^+ bombardment, is at best qualitatively understood presently. The ionization probabilities for elements in a given matrix and analytical condition vary over about four orders of magnitude and are also matrix dependent, i.e., standards are required similar in composition to the unknown sample. Such standards are rarely available, except for the special case of low dose implants, where $\alpha^{+,-}$ is constant and can be calibrated if the implant dose is known.

In view of the complexity of the secondary ionization process particularly under reactive ion bombardment we have seen recently various efforts based on very different principles to decouple the sputter and ionization process. The objective is obvious: At steady state bombardment conditions the elemental composition of the sputtered material is identical to the bulk (not the sputter induced altered surface) composition of the target. If the sputter process produces within the analytical accuracy discussed here ($\pm$ 10%) only one species in the ground state (Me_o) regardless of the matrix composition and if the post-ionization process is independent of any other elements present in the post-ionizing region then, and only then, can the technique be easily calibrated and yield quantitative information on the bulk composition of the target. These conditions usually prevail for metallic targets. It has been shown [3, 4] that for most metallic targets, under inert gas ion bombardment, more than 95% of all sputtered species are emitted as metal atoms with the notable exception of some elements (Cu, Ag, Au,) with one outer shell electron for which Me_2/Me can be equal or larger than 0.1. If all species, regardless of their charge state can be measured, and either the sputter process or post ionization suppress molecular ion formation, quantification should be possible even for samples with strong ionic bonds.

Post-ionization techniques should either employ fast removal rates [3] or sputter under UHV conditions to retain the simplicity in the quantitative data reduction in the analysis of metallic targets under inert gas ion bombardment. Since residual gas constituents are also subject to Post-ionization, UHV conditions will reduce such interference problems. Some post-ionization techniques (e-Beam and Hot Electron Gas) allow electronic discrimination based on the differences in the ion energy distribution of post-ionized residual gas constituents and sputtered particles.

Although inert gas ion bombardment produces predominantly metal atoms from metal targets, we will find only a fraction of these in the ground state. Multiphoton resonance ionization (MPRI) has recently been used by Pellin et al. [5] to probe the abundance of the ground and the excited states of iron sputtered with Ar^+ from iron and iron implanted silicon and found some matrix dependence in the distribution of the Fe states probed, although the ratio of the ground state to the sum of all states was fairly matrix independent ($.64_{Fe}$ vs $.54_{Fe\ in\ Si}$) . This issue which is obviously of importance in the quantification of MPRI, should however be irrelevant in non resonant photon ionization if saturation can be achieved over the entire volume of post-ionization accepted by the time of flight mass spectrometer. In the post-ionization by a hot electron gas or e-beams at best second order corrections may be required, since only low lying excited states within a few hundred millivolts above the ground state are significantly populated which should have ionizaton cross sections nearly equivalent to those in the ground state. This is supported by the experimental results [6] , that elemental sensitivity factors relative to a common standard element (Fe) have been found in many cases to be independent of the matrix.

Provided the sample is metallic and the target is bombarded with inert gas ions, the reader should conclude from the discussion above that any of the post-ionizing techniques presently pursued can be calibrated with pure element or composite standards without the requirement of matrix matching to yield compositional information under steady state sputtering conditions with about ± 10% or better accuracy, provided the sample is metallic and the target is bombarded with inert gas ions.

Far more complicated is the quantification scheme if the sample to be analyzed has strong ionic bonds. Direct secondary ion emission of positively or negatively charged particles is not negligible and may become the predominant species in the sputter process. Clever schemes have been incorporated in existing post-ionizing systems to reject direct secondary ion emission [5, 7, 25] in the detection of post ionized particles, which does not eliminate the problem in the quantification process (see, however, thermal post-ionization) unless matrix matched standards are available (back to SIMS). Furthermore it has been shown by Oechsner et al. [8] that for samples with strong ionic bonding (Ta_2O_5, Nb_2O_5, WO_3) about half of the sputtered particles are sputtered into the MeO channel. The same authors [8] demonstrated that in simple metal oxide targets quantitative information on the metal oxygen ratio can be obtained with post ionization. Another group [7] has shown that if all constituents of significant abundance in the sputter process (Me^+, Me^-, Me_0, MeO^+) are measured, quantification even in ionic targets is possible. It appears uncertain at this time whether a generally applicable quantification scheme can be developed for ionic targets of complex composition (eg. ceramics, glasses, minerals) even if all species of significant abundance in the sputtered material can be measured by a particular technique.

In what follows the author will discuss the strength and problems of SIMS and various post-ionizing methods. The reader should bear in mind that post-ionization of sputtered particles is a rapidly evolving field and that the figures given in the text and the table will certainly require revisions as these techniques are further perfected.

<u>SIMS:</u> [9, 10, 11]

The ionized fraction $\alpha^{+,-}$ is of the order of 0.1 in favorable cases. In such a case ADS are as high as one atom detected in one hundred atoms sputtered when high transmission (T ~ 0.1) magnetic mass spectrometers are used. If $\alpha^{+,-}$ is about 0.1, RDS are in the low ppb range regardless of the transmission of the system. With tight electronic gating (~1% of the crater area is accepted) up to six orders of magnitude in concentration range (dynamic range) can be covered in depth profiling, however at the expense of the useful yield. Excellent SIMS systems are available.

$\alpha^{+,-}$ varies by about four orders of magnitude and is strongly matrix dependent requiring standards similar in composition to the sample, the only, but very important disadvantage of SIMS.

<u>e^- Beam Post-Ionization</u> [7, 12]

$\alpha^+ \sim 10^{-4}$ based on the measured useful yield of $\sim 10^{-9}$ and an estimated transmission of 10^{-5} [7] and is fairly constant within a factor of four at 60 eV electron energy (5 to 10 times I_p). The technique is easily calibrated with available standards. Small hardware investment is required and the post-ionizer can be incorporated into quadrupole based SIMS systems. The design should allow easy removal of the post-ionizer for SIMS operations to maintain the medium transmission (10^{-4}) typical of quadrupole base SIMS systems. RDS is in the low ppm range if relatively large Ar^+ currents (~ 10μA) are used. Noteworthy is the recently demonstrated [13] capability of this technique to detect helium inspite of its high ionization potential (24 eV) with a sensitivity of $5x10^{18}$ atoms/cm^3 .

Present systems have relatively low useful yields. Using typical performance data of residual gas analyzers and estimating the particle density in the post ionizer at 1μA Ar^+ sputter current a count rate of $6x10^4$ cps/1μA has been calculated by Lipinsky et al. [7] for pure element standards and a filament current of 1 mA in good agreement with their experimental results, i.e., α x T ~ 10^{-9} . The dynamic range is probably about 10^4 due to the low useful yield of 10^{-9} further reduced to 10^{-11} with electronic gating.

Hot Electron Gas

The ionization fraction is large [14] (~ 10^{-2}) and within a factor of four the same for all elements [15] . For metallic samples the technique can be easily calibrated using available standards. In constrast to all other techniques discussed here, focussing of the primary beam is not required for depth profiling studies and the sputter beam energy can be reduced to 100 - 300 eV without loss in the high beam current (a few mA/cm^2). Under such conditions near monolayer depth resolution has been demonstrated [16] . High depth resolution, however, requires close matching of the beam energy and the sample-plasma distance, with the latter only 0.6 mm at 100 eV [17] (1.3 mm at 400 eV). With the reduced linear sputter rates at low beam energies (300 eV) and the low transmission (~ 10^{-7}) the RDS is about 100 ppm, reducing to about 1 ppm if higher beam energies are used. Further improvement in the equipment should reduce the background signal and reduce the RDS levels (see table). This post-ionization technique is the only one presently commercially available with the data in Table 1 quoted for the commercially available system. The dynamic range has not been quoted yet, but is probably not better than 10^4 due to the low useful yield and the inability to employ electronic gating in this technique.

MPRI - Multi Photon Resonance Ionization [5, 18, 19]

Since resonance photon excitations have very high cross sections, saturation can be easily achieved with tunable dye lasers, i.e., the ionization probability is unity for the state probed, except for a few elements (C, N, O, As, P) difficult to saturate with presently available laser powers. This technique is element and state selective, i.e, mass spectroscopy is not required and isobaric interferences($^{56}Fe \rightarrow {}^{28}Si_2$) are eliminated. However a general elemental survey analysis, while not impossible, is difficult to do. This argument may become irrelevant in the future if instruments become available combining resonance and non resonance photon ionization. The technique can be calibrated fairly easily with pure element standards if the sample is metallic. The relative sensitivity factors, sputter yield corrected, should be the same for all elements within a factor of about two, reflecting the differences in the abundance of the neutral ground state atoms generally probed by this technique. While not enough data are available yet, it appears that the ground state abundance of a particular element in different metallic matrices is fairly invariant. (~ ± 10%). Sputter yields of sample and standard must be known for the quantitative data reduction. In all other post-ionization techniques discussed here, all elements in the sample and standard are detected and the quantification can be reduced to the determination of the relative concentration of two constituents, eliminating the need to know the sputter rates. For samples of strongly ionic character the quantification is subject to the same problems already discussed in the introduction. Useful yields as high as 10^{-3} have been reported in the literature [5] . This has now been improved to $\sim 10^{-2}$ [20] and may ultimately reach the limit of about 0.1, set by the abundance of ground state atoms sputtered ($\sim$ 0.5) and the fraction of sputtered atoms in the post ionizing volume ($\sim$ 0.2). The same group recently demonstrated [20] the depth profiling capability by using a focussed raster scanned 3 keV Ar^+ beam pulsed for 4 μs on the center of the crater during data acquisition cycles, detecting 30 ppb of Fe in silicon using 10,000 laser shots or 1000 sec at the 10 Hz rep. rate. The same group has now reduced the measuring time to 50 sec with a 200 Hz laser with a detection limit of 1 ppb ($T \sim 10^{-2}$) . It is important to note here that in MPRI and MPI crater edge discrimination required to cover a large dynamic range in the depth profiling mode is easily accomplished in the pulsed mode of operation with the sputtering beam on center of the crater during data acquisition cycles. In SIMS and the e^- beam technique electronic gating reduces the total useful yield αT quoted in table 1 by one to two orders of magnitude. No data are yet available on the dynamic range, but the range should be comparable to SIMS.

MPI- Non-Resonant Multi Photon Ionization

Although non-resonant multi photon laser-ionization is a low cross section process, it has been shown [21, 22] recently that near saturation conditions ($\alpha \sim 1$) can be reached with large laser power densities of about 10^{10} W/cm^2 , except for some elements, (21) requiring more than two photon processes. This is presently achieved by focusing the laser beam to 0.2 mm (MPRI nonfocussed $\sim$ 6 mm), resulting in a reduction of the fraction of sputtered particles in the post-ionization volume. ($\sim 10^{-2}$ MPI vs $\sim 2x10^{-1}$ MPRI). In contrast to MPRI ground and excited sputtered atoms are probed under saturation conditions. In fact laser intensities may approach levels sufficiently large to achieve high degrees of ionization for all chemical species within the laser beam. At such high laser power densities the formation of doubly ionized species becomes non-negligible [21, 23] . Since in the reflection TOF mode both post-ionized species and direct secondary ion emission can be measured, this technique may ultimately be capable to analyze at least semiquantitively ($\pm$ 30%) the composition of samples with strong ionic bonds. The total transmission ($\sim 10^{-4}$) is presently fairly low [24] , but will increase with increasing laser power (defocussed operation) and improved TOF ion optics.

Values for ADS and RDS reported so far in the literature [22] are very encouraging. About 10^{-16} g are presently detectable. With an area of 0.1 cm^2 sputtered, less than 0.1 monolayers are removed to achieve 1 ppm detectabilities. The RDS values for the trace constituents of known concentration fluctuated only within a factor of three, indicating that saturation of ionization can be achieved for many elements. It is noteworthy [25] that here as well as in other post-ionizing techniques Ga and As have nearly identical sensitivities, although their first ionization potentials differ by a factor of two. It appears that the technique can be calibrated within about $\pm$ 10% accuracy using non-matrix matched standards. The data accumulation time for the 1 ppm detection level was one hour at the 10 Hz laser operation. Improvements of up to two orders of magnitude in the RDS and ADS and/or the time required for the analysis are expected in the near future as higher laser powers (defocussed operation), higher repetition rates (200 Hz), better TOF ion optics and better noise reduction techniques are incorporated in the system. It is reasonable to expect that MPI will be able to do a general survey analysis at the 100 ppb detection level within about 10 min, which is about comparable to the time required for a SIMS survey analysis. This would mean that at comparable useful yields of about 10^{-3} (only reached in favorable cases in SIMS) the duty cycle problem in the MPI mode is approximately counterbalanced by the advantage of the simultaneous recording of all masses. The major advantage over SIMS is however the relative invariance of the useful yield for many elements and that MPI may be easily calibrated to give quantitative information within ± 10% for metallic targets. The latter point has yet to be demonstrated. Depth profiling as already demonstrated by Pellin [20] should also be possible with good crater edge discrimination in the MPI technique. It should however be emphasized here, that in the predominant mode of operation in SIMS only one particular dopant element is profiled and duty cycle limitations in photon ionization even at 200 Hz operation will increase the analysis time of the latter by about two orders of magnitude. The fact that the gain in SIMS is only 2 orders of magnitude although the gain in on-time is 10^4 (200 Hz, 0.5μs −MPI) is offset by the typical 10^{-2} loss in the gating process in SIMS. This statement, however, is correct only if $\alpha_{SIMS} > 0.1$. In the more general case of $\alpha_{SIMS} < 10^{-3}$ the analysis time should be equal or smaller in the MPI than in SIMS. The dynamic range in the MPI should be comparable to SIMS but this has not yet been demonstrated.

Thermal Post-Ionization (26) (27)

While beautifully simple in its concept, thermal post-ionization produces only very low ionized fractions $\left(\alpha = 10^{-7} \right.$ at an ionization potential of 8 eV$\left. \right)$ at achievable furnace temperatures of 3000°K . In addition the overall transmission is low (10^{-4}), although magnetic mass spectrometry is used ($T \sim 10^{-1}$), due to the small fraction of sputtered particles (10^{-3}) entering the post-ionizing furnace. This translates into a count rate of a few hundred counts per second for a monoisotopic pure element (I_p = 8eV) at a few μA of primary Ar^+ beam current. Since multiple collisions occur with the furnace wall, thermal equilibrium requiring only one collision is established and the population of the Me^+ state can be accurately calculated for a given temperature with the ionization potential of the element and the work function of the furnace material as the input parameters. Direct secondary ion emission from the target entering the furnace is subject to the same equilibrium conditions. Most molecules are nearly completely dissociated at 3000°K, except for those with very high binding energies (30% dissociation at 10 eV). The temperature can be accurately calibrated and maintained within 5°K via the Ta^+ emission of the Ta furnace. Quantitative data on the sample composition can be easily obtained via available standards. In at least two cases it has been demonstrated [26] that samples with fairly strong ionic bonding ($Al_2O_3 - Cr_2O_3$ powder and olivine $SiO_4Mg_xFe_yCa_{0.02}$) can be analyzed quantitatively with metal alloy standards. In neither case were molecular species such as Me_2 and MeO detected. Although the presence of MeO species may escape detection due to their high ionization potentials, most metal oxide bonds have bonding energies below 8 eV and should be dissociated at 3000°K. The capability to analyze such samples quantitatively has yet to be demonstrated by other post-ionizing techniques. Thermal post-ionization has been recently combined [26] with electron impact ionization by adding a mounted filament in the furnace biased at 50 volt negative in respect to the furnace wall. Count rates for pure elements are of the order 10^5 counts/sec for pure elements, i.e., the useful yield is about 10^{-8}.

<u>Glow discharge Mass Spectrometry (GDMS)</u> [28,29]

In this commercially available technique sputtered particles are fairly efficiently ionized ($\alpha \sim 10^{-2}$) [30] in a multi-collisional process at relatively high discharge pressures of about 1 torr. For the limited cases studied so far, α is matrix independent and constant within a factor of four with some systematic dependence on the ionization potential [31]. With an ionization potential dependent correction term, elemental sensitivity factors should be constant within a factor of two. GDMS has primarily been applied for the analysis of bulk impurities with a detection sensitivity

<u>Table 1.</u> α, T, and αxT (see text) for the respective techniques. All data given are experimental data. cps/μA are counts per second per μA Ar^+(~ 3 keV) and for 1 mA filament emission current for the e^-beam technique, except for the hot electron gas, and GDMS where the numbers quoted * are cps/mA more applicable for these two techniques. Arrows indicate expected improvements by a factor of 10 or larger with presently available technologies.

	α^+	T	αxT	cps/μA	cps/Bkgd
SIMS	$10^{-1} - 10^{-5}$	$10^{-1}/10^{-4}$	$10^{-2} - 10^{-9}$	$10^{11} - 10^4$	~1
e-BEAM	10^{-4}	10^{-5}	10^{-9}	10^5	~1
HOT e^- GAS	10^{-2}	10^{-7}	10^{-9}	10^{7*} 1 keV 10^{5*} .3 keV	50^* ↓
MPRI	1 (Me_0)	10^{-2}	10^{-2}	10^8 200Hz 4μs	50 ↓ 200Hz 4μs
MPI	1 (Me)	10^{-4} ↑	10^{-4} ↑	$5x10^4$ ↑ 10Hz 3μs	1 ↓ 10Hz 3μs
THERMAL 3000°K	10^{-7} I_p = 8eV	10^{-4}	10^{-11} I_p = 8eV	10^2 I_p = 8eV	?
GDMS	10^{-2}	10^{-5}	10^{-7}	10^{9*}	~ 1^*

of one ppb. While depth profiling is in principal feasible [32] the depth resolution and dynamic range will be limited by the lack of discrimination of edge effects and back sputtering of particles from the high pressure plasma onto the sample surface.

Summary:

With some hesitation the author summarized in Table 1 performance characteristics of the various techniques discussed. Undoubtedly in the rapidly evolving field of post-ionization the numbers will have to be revised as improvements are made in existing systems. In some cases arrows indicate expected improvements by more than a factor of 10 but not yet demonstrated, based on convincing arguments by the workers in the respective fields. No data are given on achievable depth resolutions. One would not expect significant differences here, since most post-ionizing techniques will employ Ar^+ ion beams of about 1 μA at a few keV as a reasonable compromise, dictated by primary ion beam optics considerations for focused rastered beam operation and the sputter rates. Hot electron gas post-ionization is the notable exception, allowing reduction of the bombardment energy to about 100 eV, however at the expense of much lower sputter rates.

Many contributions to this conference deal with SIMS artifacts introduced by bombardment with reactive ions (O_2^+), a necessary requirement for high sensitivity. Inert gas ion bombardment as employed by all postionization techniques will reduce such problems. Finally it should be noted that all techniques based on a sputtering process are subject to the same quantification problems if non steady state conditions prevail, such as in the analysis of interfaces.

References

1. W. Reuter, Nucl. Instrum. Meth. 218, 391 (1983).
2. M. L. Yu and N. D. Lang, Phys. Rev. Lett. 50, 127 (1983).
3. H. Oechsner and W. Gerhard, Surf. Sci., 44, 480 (1974).
4. H. Oechsner and E. Stumpe, Appl. Phys. 14, 43 (1977).
5. M. J. Pellin, C. E. Young, W. F. Calaway and D. M. Gruen, Surf. Sci. 144, 619 (1984).
6. K. H. Müller and H. Oechsner, Mikrochimica Acta 10, 51 (1983).
7. D. Lipinsky, R. Jede, O. Ganschow, and A. Benninghoven, J. Vac. Sci. Technol. A3, 2007 (1985).
8. H. Oechsner, H. Schoof and E. Stumpe, Surf. Sci. 76, 343 (1978).
9. A. Benninghoven, F. G. Rüdenauer and H. W. Werner, Seconday Ion Mass Spectrometry, Publ. John Wiley, NY, 1986.
10. K. Wittmaack, Nucl. Instrum. Methods 168, 343 (1980).
11. K. Wittmaack, Surf. Sci., 89, 668 (1979)
12. H. Gnaser, J. Fleischhauer and W. O. Hofer, Appl. Phys. A 37, 211 (1985).
13. H. Gnaser, H. L. Bay and W. O. Hofer, to be published Nucl. Instr. Meth. B15 (1986).
14. H. Oechsner, W. Rühe and E. Stumpe, Surf. Sci. 85, 289 (1979).
15. K. A. Müller, K. Seifert, and M. Wilmers, J. Vac. Sci. Technol. A3, 1367 (1985).

16. P. Beckmann, H. Oechsner, D. A. Reed, S. M. Baumann, S. D. Wilson and C. A. Evans. In this Proceeding.
17. E. Stumpe, H. Oechsner and H. Schoof, Appl. Phys. 20, 55 (1979).
18. D. L. Donahue, W. H. Christie, D. E. Goeringer and H. S. McKown, Anal. Chem. 57, 1193 (1985).
19. F. M. Kimock, J. P. Baxter, D. L. Pappas, P. H. Kobrin and N. Winograd, Anal. Chem. 56, 2782 (1984).
20. M. J. Pellin, C. E. Young, W. F. Calaway and D. M. Gruen, Nucl. Instrum. Meth. accepted.
21. C. H. Becker and K. T. Gillen, In this Proceeding.
22. C. H. Becker and K. T. Gillen, Anal. Chem., 56, 1671 (1984).
23. C. H. Becker and K. T. Gillen, J. Opt. Soc. America, B2, 1438 (1985).
24. C. H. Becker and K. J. Gillen, Appl. Phys. Lett., 45, 1063 (1984).
25. C. H. Becker and K. T. Gillen, J. Vac. Sci. Technol. A3, 1347 (1985).
26. G. Blaise, Scanning Electr. Microscopy, Vol. I, 31 (1985).
27. G. Blaise, and R. Castaing, C. R. Acad. Sci., Paris Series B 284, 449 (1977).
28. J. W. Coburn and W. W. Harrison, Appl. Spectros. Rev. 17, 95 (1981).
29. W. W. Harrison, K. R. Hess, R. C. Marcus and F. L. King, In this Proceeding.
30. W. D. Westwood, Prog. Surf. Sci. 7, 71 (1976).
31. J. C. Huneke, Charles Evans Associates. Private communication.
32. J. W. Coburn, E. W. Eskstein and E. Kay, J. Appl. Phys. 46, 2828 (1975).

Quantitatve Analysis Using Sputtered Neutrals in an Ion Microanalyser

P. Williams

Department of Chemistry, Arizona State University, Tempe, AZ 85287, USA

Recently, a number of new approaches have been made to the problem of thin film analysis using the sputtered neutral flux. Following the early studies using electron impact ionization [1,2], and the plasma approach [3], refined by OESCHSNER [4], workers have returned to the electron impact approach [5] and have investigated the new technique of multiphoton ionization [6,7]. Common to all these approaches is the desire to optimize the efficiency of ionization and collection of the sputtered neutrals, and, to varying degrees, capabilities for concurrent sputtered ion analysis have received less emphasis. We have chosen instead to take an analyser optimized for sputtered ion detection--the Cameca IMS 3f--and to investigate the efficiency with which sputtered neutrals could be detected in this system without compromising the sputtered ion performance.

The high extraction field (1 V/μm) existing at the sample surface in the IMS 3f precludes the introduction of a low-energy electron beam for ionization. However, it has long been known that the primary ion beam can cause ionization of the sputtered species by impact in the gas phase [8] and we have investigated the efficiency and quantitative analytical capabilities of gas-phase sputtered neutral ionization by primary ion impact.

Ions produced in the gas phase are distinguishable in the IMS 3f from ions directly sputtered from the sample surface because the former are accelerated through less than the full sample potential. By adjusting the energy- resolving slit to accept ions with energies in the range 4400-4500 eV and then switching the sample potential to 4600 V, ions formed 100 - 200 μm above the surface could pass through the slit. Switching the sample potential to ~4450 V allowed directly-sputtered secondary ions to be sampled with high efficiency. For a primary current of 1 μA of Ar^+, focussed into a spot ~100 μm diameter, at 11.5 keV impact energy, the secondary ion signals of Ga^+ and As^+ from gallium arsenide were ~5 x 10^9 c/s and ~x 10^7 c/s respectively. The energy-deficient ion signals for both were ~5 x 10^5 c/s. However, it was noted that for species with high secondary ion yields, e.g. Ga, Al, and Si, the energy-deficient signal appeared not to arise solely from gas-phase ionization. As BERNHEIM et al. [9] had noted earlier, and WITTMAACK [10] has also found, these signals did not show the simple second-order dependence on primary current density expected for a gas-phase ionization process, but rather a dependence between first and second order, varying with the ion species and primary current. To explain their results, BERNHEIM et al. [9] suggested such processes as ejection of atoms in very long-lived autoionizing states, or in clusters which dissociated in the gas phase. We have considered another possibility, namely that the signal derives in some way from the intense flux of secondary ions. Evidence for this comes from the observation that increasing the sputtered ion yield by a factor ~30 by sputtering with oxygen ions increased the energy-deficient signal by the same proportion. Such a signal could arise by charge-exchange between sputtered ions and atoms above the sample, or by scattering of the intense directly-sputtered ion flux into the energy-deficient channel. The latter explanation is supported by the fact that the energy-deficient signal is never less than about 10^{-4} of the directly-sputtered flux, a factor characteristic of scattering effects in single deflection analysers.

Regardless of the origin of the interfering signal, it was necessary to eliminate it. This we did by using doubly-charged ions, which for most species are formed with low efficiency in direct sputtering processes, and also would be unlikely to form in the processes suggested by BERNHEIM et al. [9]. Under the bombarding conditions noted above, the signals for Ga^{++} and As^{++} directly sputtered were ~10^4 c/s, and for the energy-deficient signal ~10^3 c/s. Even if 10^{-4} of the directly-sputtered signal were scattered into the energy-deficient trajectory, this would now represent a negligible component. In fact, for a doubly-charged ion to scatter without losing a charge is improbable, and the probability of double charge exchange in the sputtered flux is similarly low.

To evaluate the quantitative analytical capabilities of the approach, two sample sets were evaluated: a) $Al_xGa_{1-x}As$ epitaxial layers (0<x<0.4), b) $TaSi_x$ co-sputtered layers (0<x<2). The stoichiometry of these samples had been determined by electron microprobe and Rutherford backscattering (RBS) analyses respectively. The Al^{++}/As^{++} and Ta^{++}/Si^{++} gas-phase signal ratios were found to vary linearly with the respective analytical stoichiometries, to within better than 5%.

These findings confirm the suitability of sputtered neutral analysis for determination of thin film and bulk stoichiometries in samples where sputtered ion yields are low. The Cameca IMS 3f can be readily used for such analyses with no modification. Sensitivities are low, due to the necessity to use doubly-charged ions and the relatively low sputtering currents employed. However, the collection efficiency of the IMS 3f is high due to the high field in the ionization region. Normalized to the same sputtering current, signals are an order of magnitude lower than those obtained in SNMS [4]. If the rejection of signals arising from the sputtered ion flux can be accomplished, singly-charged gas-phase ions can be used, which should offer an improvement in sensitivity of about two orders of magnitude.

A full account of this work has been submitted to Nuclear Instruments and Methods B. The work was supported by the National Science Foundation Grant DMR 8206028. I thank Rick Hervig and Lori Streit for their assistance, and S. S. Lau for providing the TaSi samples and analyses.

References

1. R. E. Honig: J. Appl. Phys. 29, 549 (1958)
2. R. C. Bradley: J. Appl. Phys. 30, 1 (1959)
3. J. W. Coburn, E. Kay: Appl. Phys. Lett. 19, 350 (1971)
4. H. Oechsner, W. Gerhard: Surf. Sci. 44, 480 (1974)
5. J. Schou, W. Hofer: Appl. Surf. Sci. 10, 383 (1982)
6. N. Winograd, J. P. Baxter, F. M. Kimock: Chem. Phys. Lett. 88, 581 (1982)
7. C. H. Becker, K. T. Gillen: Anal. Chem. 56, 1671 (1984)
8. J. F. Hennequin, G. Blaise, G. Slodzian: C. R. Acad. Sci. Paris 268, B1507 (1969)
9. M. Bernheim, G. Blaise, G. Slodzian: Int. J. Mass Spectrom. Ion Phys. 10, 293 (1972-73)
10. K. Wittmaack: Nucl. Instrum. Methods 132, 381 (1976)

Quantitative Analysis of Iron and Steel by Mass Spectrometry

J. Dittmann, F. Leiber, J. Tümpner, and A. Benninghoven

Thyssen Stahl AG, D-4100 Duisburg, F. R. G.
Universität Münster, Physikalisches Institut, Domagkstr. 75,
D-4400 Münster, F. R. G.

1. Introduction

Today's iron and steel production asks for a fast quantitative analysis of bulk material over a large scale of elements and concentration. Usually different optical spectroscopies are used to scan the whole element and concentration range.
The outstanding sensitivity, wide mass range and additional chemical information capability led us to another attempt of quantitative SIMS analysis (1) by a calibration curve approach. The additional requirement of a fast analysis resulted in the development of a target introduction system and a mechanical target treatment inside the apparatus. Problems, possibly occuring because of peak interferences, matrice effects and adsorption of residual gas components are proved by additional analysis of emitted secondary neutrals, where these effects are expected to be of minor importance (2).

2. Instrumentation

The instrument (Fig.1) consists of five vacuum chambers evacuated by turbomolecular pumps. They are used for independent target introduction, UHV-milling,

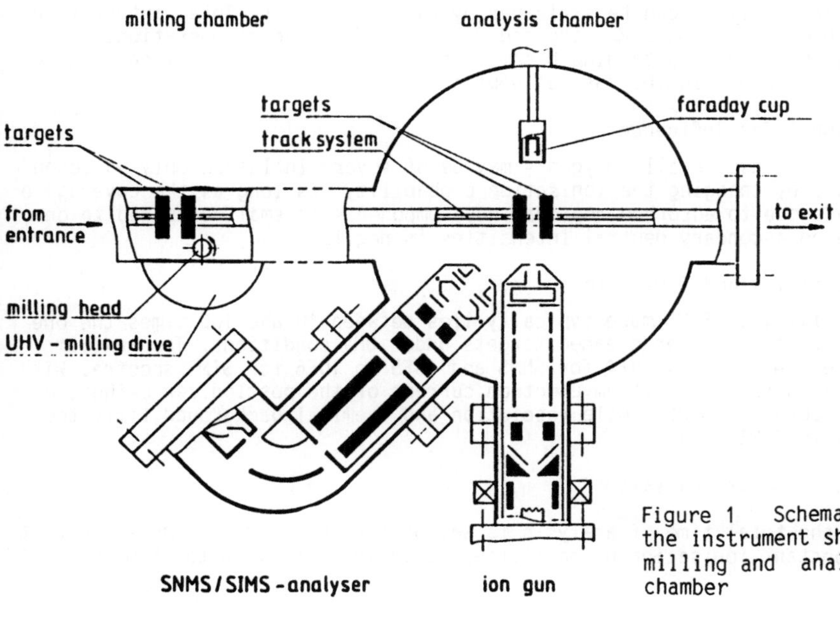

Figure 1 Schematic of the instrument showing milling and analysis chamber

analysis and target exit. The different chambers may be separated by lock valves. Up to eight targets - a pair of two on each targetholder - may stay inside the apparatus at once. Two fed into, two under mechanical pretreatment, two being analysed and another two leaving the instrument. The targets are brought through the instrument on a track system by independent pushing rod drives.
The milling chamber is equipped with an especially developed UHV-compatible milling motor. The geometry of the milling cutter itself is optimized to give smooth surfaces, while milling is performed under UHV conditions. To avoid shreds on top of the targets, milling and analysis is done upside down.

The analysis chamber is equipped with a fine focusing raster scan ion gun. It is able to supply primary ion currents up to 15 µA Ar to the target as long as the chosen energy is higher than 3 keV, while the beam diameter is kep below 1 mm. Energy may be varied between 1 and 10 keV. Scanning area is 10 mm to 20 mm. Beam current, energy, focus, scanning area and position are under co puter control. Two quadrupole mass filters - one for on-line residual gas analy sis, the other equipped with special ion optics for combined secondary neutral and secondary ion spectroscopy (3) - are mounted in the analysis chamber.

All instrument functions are controlled by a PDP 11/24 computer. Data storage and presentation is done by an adapted measurement control and graphic software package.

3. Results

The performance of the instrument has been tested by applying it to a selecti of well defined binary, ternary and multicomponent standards.

3.1 Calibration Curves

Under well defined experimental conditions there is a simple linear behaviour of relative intensity with concentration of species for emitted secondary neutrals as well as secondary ions. So it is for the binary system Fe-C in th concentration range from 8 to 10000 ppm (see fig. 2.1-2.2). This is also true for vanadium in a multicomponent environment (see fig 2.3-2.4). Here deviatio from linear behaviour can be explained by rate statistics. Total integration time has been 0.25 s in SIMS mode and 2.5 s in SNMS mode of operation. Analysing the target containing 8 ppm of carbon we chose 2.5 s, when looking for ions, and 250 s in the case of neutrals.

3.2 Residual Gas Influence

Residual gas - escpecially oxygen - may be of severe influence only on second ion signals by changing the ionisation probability. As long as the coverage o the surface due to adsorbed residual gas components is small compared to one, influence on secondary neutral intensities is neglible.

3.3 Sensitivity and S/N-ratio

The sensitivity in SIMS mode typically lies between 10 and 100 times the one in SNMS mode for sputter cleaned targets under UHV conditions. The maximum attainable S/N ratio is 10E4 for SNMS and 10E5 to 10E6 for SIMS spectra. With statistical noise and maximum electron current of the postionizer being the limiting factor in SNMS, while resolution and chemical background limit the S/N-ratio in SIMS mode.

3.4 Total Time of Analysis

For SIMS the total time of analysis is set by the necessary presputter dose t obtain constant ionisation probabilities. Usually we took 10 to 15 minutes of

sputter time, while bombarding the target with 20 µA/cm² of Argon for targets not milled. Integration times of 1/4 s have been sufficient for the detection of components down to a few ppm.
In SNMS mode presputter time may be shortened because the influence on the analysis is only fixed by the actual surface concentration of the adsorbed component as long as the ionisation probability is small compared to one. Presputter times of 2 to 5 minutes have been proved to be sufficient in most cases. Spectra were run repetitively and averaged adding to a total integration time of 2 to 250 (8 ppm C) seconds.

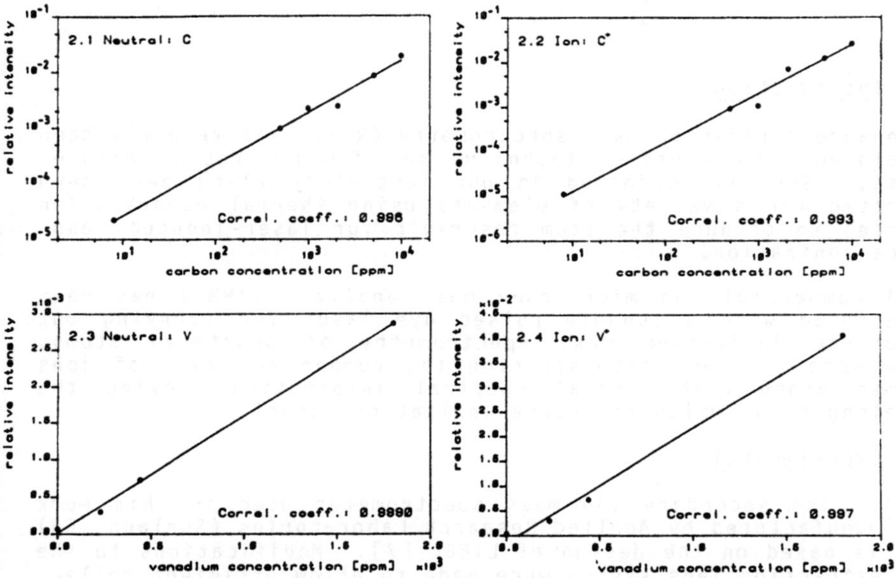

Figure 2 - Double log. plot of carbon to iron intensity ratio versus concentration for emitted neutral (2.1) and ion (2.2)
- Linear plot of vanadium to iron intensity ratio versus concentration for emitted neutral (2.3) and ion (2.4)

References

1 K.Tsunoyama, T.Suzuki, Y.Ohashi and K.Kishidaka, "Proc. 2nd Int. Conf. SIMS", Stanford, August 1979, 146
2 H.Oechsner and E.Stumpe, Applied Physics, 14 ,43-47, (1977)
3 D.Lipinsky, R.Jede, O.Ganschow and A.Benninghoven, J. Vac. Sci.Technol., in press

Ion Microprobe Mass Spectrometry Using Sputtering Atomization and Resonance Ionization

D.L. Donohue, W.H. Christie, and D.E. Goeringer

Analytical Chemistry Division, Oak Ridge National Laboratory, Oak Ridge, TN 37831, USA

1. Introduction

Resonance ionization mass spectrometry (RIMS) has recently been developed into a useful technique for isotope ratio measurements. Studies performed in our laboratory [1-6] have been reported for a variety of elements using thermal vaporization sources to produce the atom reservoir for laser-induced resonance ionization.

A commercial ion microprobe mass analyzer (IMMA) has been interfaced with a tunable pulsed dye laser for carrying out resonance ionization mass spectrometry of sputtered atoms. Data were obtained demonstrating the number and type of ions formed, along with optical spectral information showing the wavelengths at which resonance ionization occurs.

2. Experimental

IMMA. The secondary ion mass spectrometer used in this work was manufactured by Applied Research Laboratories (Sunland, CA) and is based on the design of LIEBL [7]. Modifications to the ion extraction lens system were made to allow efficient collection of resonance ions. For this work the secondary mass analyzer was operated at mass resolutions ($M/\Delta M$) on the order of 200. Primary ions were generated by using a mixed gas consisting of approximately 2 atom % Ar, 34 atom % N_2 and 64 atom % O_2 in the duoplasmatron ion source. This allowed the sputter targets to be bombarded with operator-selected mass analyzed ion beams of Ar^+, N_2^+ or O_2^+.

3. Laser Source and Light Optics

The laser source used in this study was a Chromatix CMX-4 flashlamp-pumped dye laser with a bandwidth of approximately 3 cm^{-1} and an optical pulse width of about 1 μsec. A useful wavelength range of approximately 580-610 nm was generated with the Rhodamine-6G dye used in all the experiments. The pulse energy measured directly at the laser output was about 5 mJ with the available energy above the sample surface estimated to be from 10-50 times less. The laser beam was spatially filtered and focused to approximately a 0.01 cm diameter spot directly above the sample. The horizontal and vertical positions relative to the sample surface were varied by an adjustable mirror mount. The elevation of the laser beam was lowered to a plane parallel and just above the sample surface by a periscope assembly mounted in the sample chamber.

Signal Processing. Photomultiplier anode pulses produced by bundles of ions reaching the Daly detector were processed by a preamplifier and pulse-shaping amplifier. The output pulse was sampled by a gated integrator triggered by the sync pulse from the laser; the gate delay was adjusted to coincide with the output pulse which occurs approximately 25 μsec after the laser sync pulse.

4. Samples

Samples consisting of natural U and Sm metal embedded in low vapor pressure epoxy were polished until a smooth metallographic finish was obtained. The cross-sectional area of each metal sample was approximately 0.1 cm^2.

5. Results and Discussion

Sensitivity. The ion bundle produced in a single laser pulse (nominally 1 μsec long) was found to be 3 ± 1 μsec wide at the detector following a delay of 23 μsec (for Sm) or 30 μsec (for U) due to time-of-flight through the secondary mass spectrometer. The number of ions in each bundle could be counted using a fast oscilloscope to monitor the output of the ion multiplier preamplifier. The largest signals obtained for ^{238}U using Ar^+ bombardment were found to have > 100 ions per pulse while those for ^{152}Sm were found to have 40 ions per pulse.

An estimate of the sampling efficiency can be obtained from the conditions which gave > 100 U^+ ions per laser pulse. The current of Ar^+ ions incident on the surface was 21 nA, or 1.3 x 10^{11} ions/sec. However, due to the pulsed nature of the laser, only a small fraction of this current was effective in producing U atoms for the resonance ionization process. This fraction is determined by the average transit time of atoms through the laser volume (for short duration laser pulses) or by the actual length of the laser pulse (for pulse lengths of ≥ 100 nsec). The above argument assumes a 0.1-0.01 cm laser beam diameter and an average kinetic energy of 10 eV for the U atoms. In the case reported here, the nominal laser pulse duration was 1 μsec, which means that 1.3 x 10^5 Ar^+ ions incident on the surface resulted in > 100 U^+ ions being detected. Therefore, the gross efficiency of the sputtering-resonance ionization process was 1 ion detected for every 1300 Ar^+ ions striking the sample.

Mass Spectra. It was of interest to know what ionic species are produced by the laser beam. Therefore, mass spectra were scanned over the mass range 0-300 amu near the resonant wavelengths for U and Sm. At the resonant wavelength of 591.4 nm, the U^+ peak is observed. At higher mass, the UO^+ peak appears with almost equal intensity, in spite of the fact that Ar^+ was the bombarding species. In fact, the UO^+ peak occurs at all wavelengths produced by the dye laser, indicating a nonresonant ionization process. The effect of detuning the laser 0.2 nm resulted in the disappearance of the U^+ signal with no change in the UO^+ intensity. A similar effect was observed for Sm. Both Sm^+ and SmO^+ signals were observed with the SmO^+ signal occurring at all wavelengths produced by the laser, whereas the resonant Sm^+ signal was sharply tunable.

These results suggest that residual oxygen is interacting with the sample to produce metal oxide species which are sputtered and ionized by the laser in a non-resonant process. The SIMS spectra of these metals under Ar^+ bombardment also show strong MO^+ and MO_2^+ peaks. This phenomenon will be the subject of further study.

6. Future Studies

Future work will involve refinements of the ion optical system to obtain higher extraction and transmission efficiency of the laser generated ions. The use of pulsed ion beam sputtering is being pursued in order to maximize the efficiency of atom production. In addition, a more powerful laser system is being acquired to produce saturated resonance ionization over a larger volume. Investigations of different sample types will involve use of resin beads and small particles of Sm or U oxides, to test the capabilty of the technique for measuring real-life samples.

7. References

1. D. L. Donohue, J. P. Young, D. H. Smith: Int. J. Mass Spectrom. Ion Phys. 43, 293-307 (1982).
2. J. P. Young, D. L. Donohue: Anal. Chem. 55, 88-91 (1983).
3. D. L. Donohue, J. P. Young: Anal. Chem. 55, 378-379 (1983).
4. D. L. Donohue, D. H. Smith, J. P. Young, H. S. McKown, C. A. Pritchard: Anal. Chem. 56, 379-381 (1984).
5. J. P. Young, D. L. Donohue, D. H. Smith: Int. J. Mass Spectrom. Ion Processes 56, 307-319 (1984).
6. D. L. Donohue, W. H. Christie, D. E. Goeringer, H. S. McKown: Anal. Chem. 57, 1193-1197 (1985).
7. H. Liebl: J. Appl. Phys. 38, 5277-5283 (1967).

This research was sponsored by the U. S. Department of Energy, Office of Basic Energy Sciences, under Contract No. DE-AC05-84OR21400 with Martin Marietta Energy Systems, Inc.

Part IV

Detection Limits and Quantification

Memory Effects in Quadrupole SIMS

J.B. Clegg

Philips Research Laboratories, Redhill, Surrey, RH1 5HA, England

1. Introduction

Semiconductor assessment studies often require the measurement of dopant concentrations at the $10^{14}cm^{-3}$ level and to achieve this detection sensitivity with SIMS it is necessary to reduce sources of background interferences [1]. One source of background is the cross contamination or memory which occurs when samples of different atom composition are analysed. It should be remembered that most of the sample sputtered by the primary beam is deposited on adjacent mechanical surfaces and some of this material is sputtered back onto the target surface by reflected primary ions [2] and energetic secondary ions [3]. Thus when the sample matrix is changed, for example GaAs to Si, a sensitive SIMS system will show a memory due to Ga and As and both elements will apparently be present in the Si sample. As the magnitude of the memory effect depends on the experimental conditions [3] and on the system configuration in the target region, it is important to characterise particular types of SIMS instruments with respect to this effect. Such measurements have been reported for the As and Si cross-contamination occurring in an ion microscope system [3]. This paper gives details of a quantitative assessment of the memory present in a high sensitivity quadrupole system.

2. Experimental

The investigations were performed using the DIDA ion microprobe described elsewhere [4]. The system utilises secondary ion optics which are not closely coupled to the sample surface (separation distance ~3 cm) in order to reduce the memory effect. Our system has been modified in the target region by the addition of a tantalum cover plate (5 cm x 5 cm) to the einzel lens assembly to mask stainless steel and aluminium components. For this study two single crystal Si and GaAs samples were used as primary targets; the GaAs was undoped with a Si concentration of $<1x10^{15}$ cm^{-3}. Both samples were mounted on opposite faces of the eight-sided sample holder and were raster scanned with a focused primary beam of 12 keV O_2^+ (1.8 μA) at normal incidence. Prior to recording the secondary ion intensities, the sample surfaces were pre-sputtered for several minutes to achieve equilibrium analysis conditions. Electronic gating (25% linear) was used to suppress crater wall and edge effects. Positive ions were detected and molecular ion interferences in the mass specrum were reduced by applying a negative bias to the sample holder [1]. Element calibration factors, derived from the analysis of standard samples, were used to quantify the cross contamination levels.

3. Results and Discussion

The results shown in Fig. 1 were obtained after the system had been heavily contaminated with GaAs ($1.6x10^{18}$ sputtered atoms). With the rastered area of

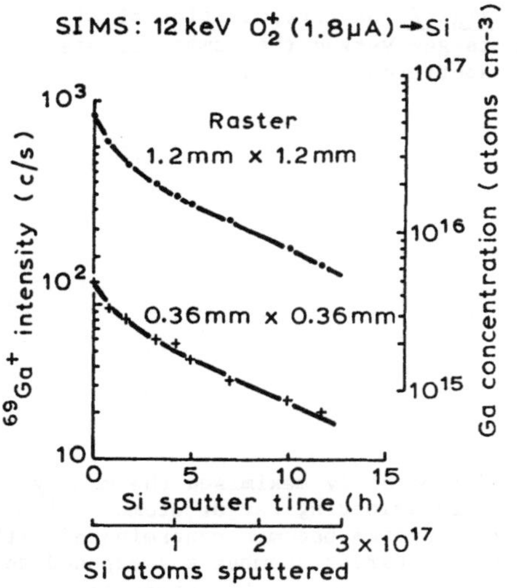

Fig. 1 Reduction in the level of Ga contamination as a function of Si sputter time

1.2x1.2 mm^2, the initial Ga contamination of the Si is 5×10^{16} cm^{-3}. Continued sputtering of the Si target leads to a reduction in the memory as the GaAs layer previously deposited on adjacent surfaces, is masked by the new Si layer. The decay curve represents the superimposition of two exponential decay processes with half-lives ($t_{\frac{1}{2}}$) of 0.5 h and 5h. The slower process controls the rate at which the memory is reduced and it is necessary to sputter 1×10^{17} Si atoms for a two-fold reduction. After prolonged sputtering the lowest Ga concentration ($5 \times 10^{13} cm^{-3}$) is not determined by memory but by system noise. The decay curve obtained with the smaller rastered area shows that the Ga intensity scales almost directly with the area, using a fixed duty cycle of the electronic gate. Additional measurements show that the intensity is also directly proportional to the primary ion current. Thus to optimise the signal to background ratio it is advantageous to reduce the rastered area [5]. The initial contamination levels of As in Si were found to be 1×10^{18} cm^{-3} and < 3×10^{17} cm^{-3} with a $t_{\frac{1}{2}}$ value of 1.5h. Experiments with the ion beam switched off for 10 minute periods showed that the As contamination continued to build up during this period. This indicates that the transport of As back to the Si surface involves not only direct line of sight back sputtering but the adsorption of a volatile species present in the vacuum ambient. The latter process is absent with the Ga memory. Clearly many processes control the memory effect, and the contamination levels of Ga and As do not reflect the original atom composition of the target.

Analysis of the samples in the reverse order with the Si sputtered for 18h followed by the analysis of the undoped GaAs sample, gave initial Si contamination levels of 2×10^{17} cm^{-3} and 5×10^{16} cm^{-3} with the two raster sizes. In contrast to the Ga memory, a direct proportionality between the background signal and the rastered area was not observed with scan widths less than 0.5 mm. The $t_{\frac{1}{2}}$ value for the Si decay was 4h (corresponding to 2.6×10^{17} sputtered Ga and As atoms) and the lowest Si concentration measured with the small rastered area was 5×10^{14} cm^{-3}. In this case the detection limit is not set by Si memory but by the gas phase adsorption of CO (28 amu) onto the sample surface.

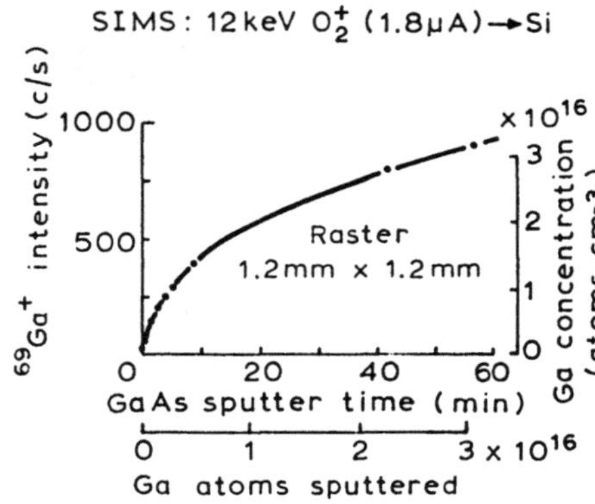

Fig. 2 Ga contamination of Si target versus the number of pre-sputtered Ga atoms

The experiments described so far have deliberately maximised the memory effect by sputtering large numbers of the contaminating target atoms. Also of interest is the rate at which the "clean" system becomes contaminated with a new matrix element. To investigate this, the target chamber was exposed to known numbers of Ga atoms followed by the analysis of the Si target to determine the corresponding Ga contamination. This process was repeated, so that a total of 3×10^{16} Ga atoms were sputtered. The results of these measurements are shown in Fig. 2 where it can be seen that with less than 3×10^{15} Ga atoms sputtered the contamination follows an approximate linear increase. With further sputtering, the gradient of the build-up curve becomes less steep and ultimately reaches a saturation level in the mid 10^{16} cm^{-3} region (see Fig. 1). This build-up curve may be used to estimate the reduction in depth profile dynamic range when eroding through regions of high and low concentration. For example, the analysis of a layered Si sample containing Ga at 10^{21} cm^{-3} in a region 0.1 μm thick, will lead to 1.4×10^{14} Ga atoms being sputtered or a memory of 4×10^{14} cm^{-3}. The expected dynamic range (peak to background ratio) is 2.5×10^6. Smaller rastered areas give a reduction in both the number of atoms sputtered and the memory effect, and dynamic ranges in excess of 10^8 are estimated. As a practical dynamic range of 7×10^5 is observed with our system, these estimates show that memory is not the main factor determining the dynamic range.

Finally, attention is drawn to the fact that beam contact with the target holder or with other materials can lead to cross contamination. For example, a 1 min exposure of the beam (1.8 μA) to the Ti sample holder or the KBr phosphor (used for beam focusing) produces a Ti^+, K^+ and Br^- background equivalent to a concentration in the low 10^{15} cm^{-3} region in Si, with a scan width of 1.2 mm.

In conclusion, this work has shown that by taking suitable precautions, the memory effect can be minimised so that it is not a significant factor in determining element detection limits. These precautions are (i) the batch-wise analysis of samples with the same matrix composition (ii) minimising beam contact with other materials (iii) pre-sputtering the new target material and (iv) minimising the rastered area.

1. J.B. Clegg, Surf. Inter. Anal. 2, 91 (1980)
2. K. Wittmaack, private communication
3. V.R. Deline, Nucl. Instr. and Meth. Phys. Res. 218, 316 (1983)
4. K. Wittmaack, M.G. Dowsett and J.B. Clegg, Inter. J. Mass Spectrom. Ion Phys. 43, 31 (1982)
5. J.J. Le Goux and H.N. Migeon, SIMS III, Springer Series in Chemical Physics, Vol. 19, pp. 52-56

Correction for Residual Gas Background in SIMS Analysis

*G.J. Scilla**

Xerox Corporation, Rochester, NY 14644, USA

Introduction

SIMS analyses for elements which are also constituents of the residual vacuum gases are often limited by the background signal introduced when these gases strike and stick to the sample surface and are subsequently ionized during sputtering. If the sensitivity of the analysis is sufficient, it is this background which will define the analytical detection limit. The factors which determine the severity of the problem include the partial pressure of molecules containing the analyte in the vicinity of the sample, the sticking coefficient of these molecules on the modified (by the primary ion beam) sample surface, and the primary ion current employed in the analysis (eg. - sputtering rate versus residual gas arrival rate at the surface).[1]

Several methods can be employed to reduce the severity of the problem. Perhaps the best approach to minimizing the background is employing UHV instrumentation. A second approach utilizes a cold plate in the vicinity of the sample to reduce the level of species such as H_2O. In addition, standard background subtraction procedures can be employed; however, these generally improve the accuracy of the analysis but do not significantly improve analytical detection limits. Finally, the use of the highest sputtering rates compatible with the other constraints of the analysis (principally depth resolution) results in reducing the relative contribution of the analyte present in the residual gas background to that present in the sample. The methodology described in this paper is an extension of this last approach to minimizing the effects of background on detection limits and accuracy. The method was necessitated by the need to quantitate the levels of H in a – Si:H thin films, utilizing a SIMS instrument with a base pressure of 10^{-6} torr. The concentration range of interest in this study was from < 1 atom % H to 30 atom %.

Experimental

An Applied Research Laboratories IMMA was employed in this study. In all cases, the data were collected with a background pressure of 1 – 2×10^{-6} Torr. An O_2^+ primary beam at either 15 kV or 20 kV was employed and the $^1H^+$ and $^{30}Si^+$ secondary ions were monitored. Electronic gating of approximately 40% was also utilized with a rastered area of $100\mu m \times 100\mu m$. The H levels in a – S:H films were quantitated employing infrared spectroscopy and solid state proton NMR. The IR data were

*Present address: IBM T. J. Watson Research Center, Yorktown Heights, NY 10598

quantitated using the Si-H stretch at 2000cm^{-1}. These films were then employed as standards to calculate the SIMS relative sensitivity factors (RSF), after correction of the SIMS data for H background employing the method described below.

Results and Discussion

Standard background correction procedures for this type of analysis would call for the measurement of the background H^+ intensity on an analytical blank such as Float Zone (FZ) Si and subtraction of that value from the intensity measured on the a – Si:H samples. This approach is not valid in this case since 1) the background on a – Si:H is not the same as that on FZ Si, and 2) it is experimentally observed that the background signal varies from one a – Si:H sample to another, most likely due to differences in sticking coefficient related to structure and H concentration. The background correction procedure employed in this study is based on the principle that under dynamic SIMS conditions, residual gas background is a constant independent of primary beam current at constant analysis area. (Note that background will vary linearly with raster area in an instrument employing electronic gating and will be constant in an instrument with mechanical gating). In addition, the signal originating from the analyte present in the sample and that of the matrix ion will increase linearly with increasing beam current (again, constant analysis area). Given the above behavior, one would expect that a plot of the intensity of an analyte (or matrix) ion without background interference will be linear with an intercept of zero, while a plot of intensity of a secondary ion with a residual gas background signal will be linear with an intercept equal to the value of the background for that particular sample. This behavior was experimentally confirmed by measuring the $^1H^+$ and $^{30}Si^+$ secondary ion intensities with varying primary beam current (i_p) on a sample of a – Si:H and F.Z. Si. The data are illustrated in Fig. 1. (Note that the Si signal displayed in Figure 2 is that obtained from the a – Si:H sample; the Si signal from the FZ sample exhibits the same behavior and was omitted for clarity). These data demonstrate the different backgrounds obtained on a – Si:H and c-Si. The intercept background value obtained from the a – Si:H plot can be subtracted from the measured $^1H^+$ intensity at any given i_p and a background corrected secondary ion intensity ratio obtained for use in a relative sensitivity factor (RSF) quantification scheme. In addition since the slope of each line is related only to the signal originating from the sample and is not dependent on background, the ratio of the slopes yields the background corrected secondary ion yield:

$$i_H^+ = m_H \, i_p + i_{H,BK}^+ \qquad (1)$$

$$i_{Si}^+ = m_{Si} \, i_p \qquad (2)$$

$$\frac{m_H}{m_{Si}} = \frac{\left(i_H^+ - i_{H,BK}^+\right)/i_p}{i_{Si}^+/i_p} = \frac{i_{H,\,corrected}^+}{i_{Si}^+} \qquad (3)$$

where m's are the slopes, i_x the secondary ion intensities and i_p the primary ion intensity. An alternate graphical method for obtaining the background corrected sec-

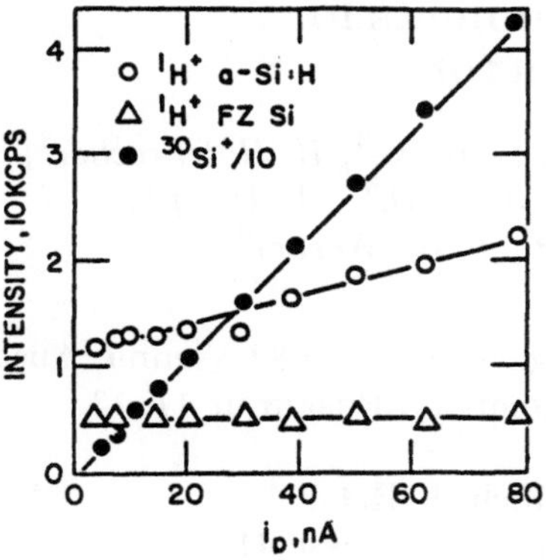

Fig. 1 Secondary Ion Intensity vs Primary Ion Current

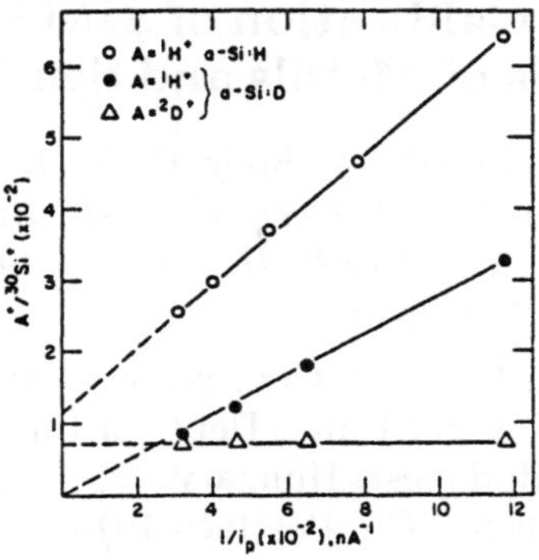

Fig. 2 Secondary Ion Intensity Ratio vs $1/i_p$

ondary ion intensity ratio comes from taking the ratio of equations 1 and 2 and substituting with equation 3 yielding:

$$\frac{i_H^+}{i_{Si}^+} = \frac{i_{H,\,corrected}^+}{i_{Si}^+} + \frac{k}{i_p} \tag{4}$$

where k is a constant containing the background intensity and the slope of Si line. As seen from equation 4, a plot of secondary ion intensity ratio against $1/i_p$ should be linear with an intercept equal to the background corrected ratio, and have a slope of k. Figure 2 displays the data from an a – Si:D sample prepared by RF plasma deposition of SiD_4 and a representative plot from a – Si:H. The plot of D^+/Si^+ secondary ion intensity ratio is invariant with i_p^{-1}. The H^+/Si^+ ratio from the a – Si:D sample varies linearly with i_p^{-1} and has an intercept near zero while that from the a – Si:H sample is linear with i_p^{-1} and has a non-zero intercept. Comparison of the H data by IR with the SIMS data after background correction using the i_p^{-1} method has demonstrated good day-to-day reproducability of the method as well as it's accuracy (assuming good standards) in the concentration range of interest.

Conclusions

The precedure described above can be employed to effectively correct for residual gas background. It is particularly useful in situations where a suitable analytical blank is not available, and when background values may vary sample-to-sample.

References

1. C. Magee and E. Botnick, J. Vac. Sci. Tech, 19, 47 (1981).

Cross - Calibration of SIMS Instruments for Analysis of Metals and Semiconductors

F.G. Rüdenauer[1], *W. Steiger*[1], *M. Riedel*[2], *H.E. Beske*[3], *H. Holzbrecher*[3], *H. Düsterhöft*[4], *M. Gericke*[5], *C.-E. Richter*[5], *M. Rieth*[5], *M. Trapp*[5], *J. Giber*[6], *A. Solyom*[6], *H. Mai*[7], *G. Stingeder*[8], *H.W. Werner*[9], *and P.R. Boudewijn*[9]

[1] Austrian Research Ctr., Seibersdorf, Lenaugasse 10, A-1082 Vienna, Aust
[2] Eötvös Lorand Univ., Dept. of Physical Chemistry, Puskin ut 11–13, H-1088 Budapest, Hungary
[3] KFA Jülich, ZCh, Postfach 1913, D-5170 Jülich, F.R.G.
[4] Humboldt Univ. Berlin, Physics Section, Invalidenstrasse 41, DDR-1040 Berlin, GDR
[5] VEB Werk f. Fernsehelektronik, Ostendstrasse 1-5, DDR-1040 Berlin, GD
[6] Technical Univ. Budapest, Physics Institute, Budafoki ut 8, H-1111 Budapest, Hungary
[7] Academy of Sciences, ZFW, Helmholtzstrasse 20, DDR-8027 Dresden, GD
[8] Technical Univ. Vienna, Inst. for Analytical Chemistry, Getreidemarkt 9, A-1060 Vienna, Austria
[9] Philips Research Lab., Eindhoven, The Netherlands

1. Introduction

It is known from previous round robin experiments [1] that relative sensitivity factors, obtained on identical samples by different SIMS instruments, frequently disagreee by a factor of up to 50 and that element concentrations, derived from the raw peak height data by some quantification algorithm, still disagree by a factor of ∿5. Two methods, both based on the relative sensitivity factor (RSF) quantification scheme, have been suggested to obtain interlaboratory standardization of quantitative SIMS analyses: (a) *"Standard - Transfer"*; for every element/matrix system to be analyzed, a well characterized external standard sample is distributed to each laboratory. Previous to analysis of the unknown, RSFs are determined from the appropriate standard and used for the quantification of the unknown. The method rests on the avail ability of identical standards in each laboratory and not on the agreement in raw ion intensity data. Each laboratory has to determine its own RSFs previous to each individual analysis. the accuracy is of the order of 20% [2] , determined mainly by the accuracy of the standard composition. (b) *"Cross-Calibration"*; here, the aim is to tune different instruments to give identical RSFs from identical samples. If this can be realized, RSFs can be transferred between instruments and laboratories so that the workload of determining RSFs can be split between laboratories. Standard samples still are required for each unknown/matrix system,but essentially only in that laboratory determining the particular RSF. The accuracy of the method is determined by the quality of the standard and of the instrument tuning.

The following heuristic "cross - calibration hypothesis" forms the base of quantification using transferred RSFs: *if two instruments, under identical bombarding and environmental conditions, produce identical RSFs from a single calibration sample, their windows in secondary ion energy and directional distributions are identical; they therefore should produce identical RSFs from samples other than the calibration sample.* The validity of this hypothesis can be reasoned, but not rigorously proven by its logical reversal: "if any two SIMS instruments are looking at the same windows in secondary energy and directional distribution, they certainly will produce identical RSFs from all samples analyzed under standardized bombarding and environmental conditions.

This suggests the following cross-calibration strategy: (a) a multielement "primary calibration standard" (PCS) is distributed to all participating laboratories. It is not necessary to specify the elemental concentrations in the PCS with great precision; its composition should however be homogeneous, stable and reproducible. (b) a set of "standard operating conditions" (SOC) is adopted, specifying bombardment and environmental parameters, e.g. primary ion, energy, current density, bombardment angle, oxygen backfill pressure, precleaning procedure. (c) Each individual instrument is tuned to obtain identical peak height ratios for the PCS - elements; this tuning generally involves setting of secondary ion energy bandpass and is a highly empirical procedure [3] . Thereby, a "virtual operating point" (VOP) (= set of peak height ratios or RSFs) is established which all participating instruments can approach with minimal deviation.

2. Experimental Results

Seven laboratories were participating in the present cross-calibration experiment. Preliminary results have been reported [3] . The SIMS instruments used were two IMS 3fs, two IMMAs, one BALZERS quadrupole SIMS, 2 homemade quadrupole SIMS and one quadrupole ion microprobe. A metallic glass of the composition $B_{15}Fe_{75}W_{10}$ was selected as PCS, enabling proper instrument tuning over a wide mass range. The following RSFs for B and W (with respect to Fe) define a "virtual operating point" (VOP) which all 7 instruments can approach with an average "tuning error factor" $\overline{TUE}$ = 1.39 (defined as RSF ratio between instrument i and VOP, averaged over B, W and all instruments i):

$$RSF(B) = 0.250 \qquad RSF(W) = 0.0335 \qquad \text{..."VOP - tuning"} \qquad (1)$$

If cross-calibration is not adopted and each instrument is individually tuned to maximum sensitivity for any of the PCS elements (as has been done previously), the average tuning error increases dramatically: $\overline{TUE}(B)=6.4$, $\overline{TUE}(Fe)=6.1$, $\overline{TUE}(W)=3.52$.

Using the same VOP tuning as in (1), RSFs were determined for the elements X = V, Cr, Ni, Zr, Nb, Ta from metallic glasses of nominal composition $B_{15}Fe_{80}X_{5}$ 3 .In this case the RSFs agreed within an average "transfer error factor" $\overline{TRE}$ = 1.68 . A detailed analysis showed, that the error factor TRE_{su}, incurred in transferring a RSF between a "source" instrument s and a "user" instrument u, decreased with decreasing individual tuning error factors TUE_s , TUE_u (of source and user instruments respectively); for small detuning the following approximation can be derived from the data:

$$TRE_{su} < 4.25\ [(TUE_s-1)^2+ (TUE_u-1)^2]^{1/2} \qquad (2)$$

Further measurements are being performed on the following semiconductor samples: GaAs, $GaAs_{.62}P_{.38}$, $GaAs_{.31}P_{.69}$, GaP, InSb, InAs, InP, $Si(B,Ag)_{impl}$. So far, results are available from two instruments only: an IMS 3f and a quadrupole ion microprobe. Figure 1 shows the correlation shows the correlation between the 10 RSFs derivable from these samples. An average transfer error fac-

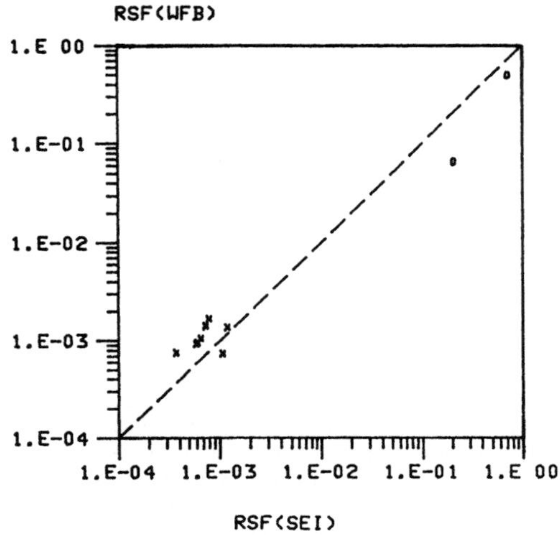

Fig. 1. Correlation between RSFs obtained from 8 samples by cross-calibrated IMS 3f (WIE) and quadrupole ion microprobe (SEI)

tor $\overline{TRE}_{su} = 1.61$ is obtained. The value predicted from (2), using the tuning errors obtained on the two instruments previous to sample analysis is 1.69 .

3. Conclusions

(a) SIMS instruments of widely varying designs can be tuned to yield, from a particular calibration sample, RSFs agreeing to within better than 40%. (b) Instruments, cross-calibrated on a particular "primary calibration standard (PCS)" also produce similar RSFs (within 60%) for other elements in matrices similar to that of the PCS. The agreement seems to be of the same order even if the matrices of the PCS and the unknown are widely different (less rigorously tested). (c) RSFs can be transferred between any two cross-calibrated instruments. (d) The error incurred in the use of transferred RSFs depends on the tuning accuracy (to the PCS) of both the "source" and the "user" instruments and generally is of the order of 60%. (e) The analytical quantification accuracy (using transferred RSFs) depends on the accuracy of the external standard used for determination of that particular RSF and the RSF-transfer error. Since the latter generally is dominating, analytical accuracy also is of the order of 60%. (f) According to the cross-calibration hypothesis transfer of RSFs should not be limited to and from the present instruments. It should also be possible to accomodate "newcomer" instruments provided they are tuned to the VOP (1).

1. D.E. Newbury in "Proc. of the 1st Japan-US Seminar on SIMS" (Hawaii (1975), Univ. of Illinois Internal Report, undated.
2. Y. Homma, S. Kurosawa, Y. Yoshioka, M. Shibata and K. Nomura, Anal. Chem., in print.
3. F. Rüdenauer, W. Steiger, M. Riedel, H.E. Beske, H. Holzbrecher, H. Düsterhöft, M. Gericke, C.-E. Richter, M. Rieth, M. Trapp, J. Giber, A. Solyom, H. Mai and G. Stingeder, Anal. Chem. 57, 1636 (1985).

On-Line Ion Implantation: The SIMS Primary Ion Beam for Creation of Empirical Quantification Standards

H.E. Smith, M.T. Bernius, and G.H. Morrison

Baker Laboratory of Chemistry, Cornell University, Ithaca, NY 14853, USA

1. Introduction

The variation in ionization probability and sputter yield with changes in matrix composition in SIMS demands a quantification standard with a matrix that matches that of the sample of interest in order to obtain the most accurate quantitative analyses. To this end, ion implantation is commonly used to create external standards, and, in the case of analytes that are homogeneously distributed with depth, superimposed internal standards [1]. Publications that suggested the development of an in-situ ion implantation gun for SIMS instruments [2] and the use of the concentration profile of the continuously implanted primary beam for LTE quantification [3] motivated the use of the CAMECA IMS 3f primary ion beam in this laboratory to obtain a two-step, on-line ion implantation/depth profile quantification of carbon concentration in silicon [4], the result of hydrocarbon contamination in the chemical vapor deposition process. The empirical determination is performed by superimposing a 20 keV C^+ implant at 40 nA upon residual levels of carbon in the Si matrix, followed by a depth profile of this implanted area with an O_2^+ beam. Both ion beams are produced from CO_2 and were resolved with the primary beam mass analyzer, so that the entire experiment was performed in vacuo. This pilot experiment illustrates that such on-line ion implantation creates quantification standards of excellent accuracy and precision in comparison to conventional high-energy ion implant standards. The ability to quantify upon demand and to supervise implant purity and dose accuracy are attractive benefits of such an analytical scheme.

2. Experimental Method

The salient experimental features of on-line ion implantation are listed in Table [1].

Table 1: Experimental Features of On-Line Ion Implants vs. Conventional High Energy Implants

	On-Line	Conventional
Implant Species	Z = 1 to 18 at present e.g., H,B,C,N,O,F,P,S	all elements
Acceleration Energy	6 to 22 keV	30 to 300 keV
Projected Range	0 to 0.3 μm	0 to 3.0 μm
Range Spread	0 to 0.08 μm	0 to 0.15 μm
Implantation Area	600 x 600 μm^2 limited by primary beam raster	several cm^2 determined by choice of aperture
Dose Accuracy	area-limited: ~10%	better than 5%

3. Results and Data Treatment

In order to obtain an accurate measurement of implanted fluence (in atoms cm^{-2}), the absolute ion current must be known [4], as well as the area of bombardment. Regarding the area, the implantation beam is focused to a small spot (< 50μm diameter) and rastered ($500^2 \mu m^2$). In this way, the total area is known to better than 10% accuracy (limited by the uncertainty in beam diameter), with a 450^2 μm^2 area of uniform bombardment. A 60 μm diameter area is analyzed in the depth profile. In general, reflection of the implant beam and the extent of sputtering it causes must be considered (5,6). For 20 keV C^+ onto Si at 3 x 10^{15} atoms cm^{-2}, the error due to reflection (~5%) and sputtering (< 1%) were not corrected for.

The shallow distribution of these on-line, low energy ion implants demands a different approach to data reduction after depth profiling, in order to obtain the most accurate results. Since the surface concentration of implanted depth-profiling primary ion species reaches an equilibrium only after a significant portion of the previously implanted analyte distribution has been sputtered away, the collected signal originating from the ion implant may be higher or lower than that collected for the same implant under steady-state conditions, depending on the effect of the primary beam species upon the analyte ion yield and the matrix sputter yield. Further error can be introduced by surface native oxides and by adsorbed contaminants containing analyte atoms or interfering species. For the cases where the peak of the analyte ion implant is deeper than the sputtered depth required to reach a surface equilibrium of depth profiling ion concentration, the problem can be circumvented by doubling the measure of the deeper half-integral of the implanted distribution, assuming a symmetrical peak shape. Table [2] illustrates the improvement of quantitative accuracy obtained by using this method.

The sputtered depth required to achieve an equilibrium concentration of implanted depth profiling species is approximately twice that of the projected range of that species [6]. Figure [1], obtained from LSS range calculations [7], illustrates that this method of data reduction is appropriate for quantification of symmetrical 20 keV atomic ion implants of mass 1 to approximately 30 when an 8.5 keV O_2^+ beam is used to depth profile, and for ions of mass 1 to 60 when a 6 keV O_2^+ beam is used. For heavier ions implanted at 20 keV, it is necessary to force the surface concentration of primary beam species to an early equilibrium, either by flooding the vacuum chamber or by presputtering the matrix prior to analyte implantation with the depth-profiling ion beam.

Table 2: Residual Carbon in Silicon Determined with On-Line Ion Implants and Conventional Implants

	x 10^{19} atoms cm^{-3}		Accuracy†††	Precision
1. Conventional External Standard Ion Implant†	3.4 (±0.6)	mean = 3.7 x 10^{19}		
	4.1 (±0.6)			
2. On-Line Ion Implants††				
a) Integrated Signal	1.8 (±0.4)		-51%	22%
b) Doubled Half-Integral	3.4 (±0.4)		- 8%	12%

† Thirteen replicate analyses in each of two separate determinations
†† Nine replicate analyses, two methods of data reduction
††† Percent relative difference compared to the mean external standard determination

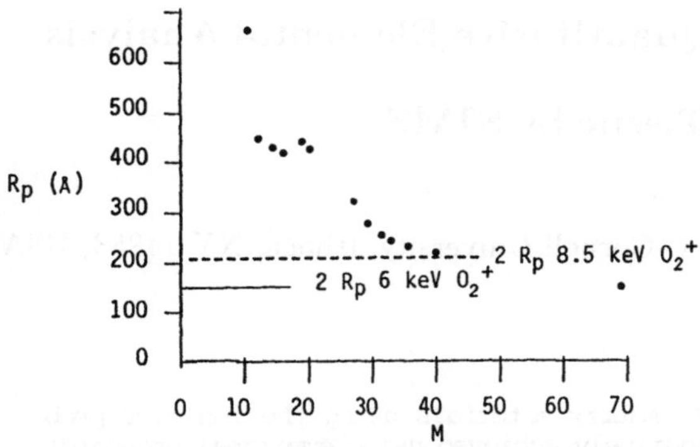

Fig. 1: Projected Range vs. Mass of Various Atomic Ions in ^{28}Si at 20 keV (normal incidence)

4. Acknowledgments

The authors thank J. Tom Brenna for valuable input, the National Science Foundation and the Office of Naval Research for support of this research, and the National Research and Resource Facility for Submicron Structures at Cornell University for use of the ion implanter.

References

1. Leta, D.P. and Morrison, G.H., Anal. Chem. 52 514 (1980)
2. Rudenauer, F.G.. "Instrumental Aspects of Spatially 3-Dimensional SIMS Analysis", in Secondary Ion Mass Spectrometry SIMS III, ed. by A. Benninghoven, J. Giber, J. Laszlo, M. Riedel, H.W. Werner, Springer Series in Chemical Physics, Vol. 19 (Springer-Verlag: Berlin, Heidelberg, New York 1981) p. 2.
3. Steiger, W., Rudenauer, F.G., Gnaser, H., Pollinger, P., Studnicka, H., Mikrochim. Acta Suppl. 10, 111 (1983)
4. Smith, H.E. and Morrison, G.H., Anal. Chem. (in press)
5. Bottiger, J., Davies, J.A., Sigmund, P., Winterbon, K.B., Radiat. Eff. 11 69 (1971).
6. Gries, W.H., Int. J. Mass Spec. Ion Phys. 30 97 (1979)
7. Gibbons, J.F., Johnson, W.S., Mylroie, S.W., "Projected Range Statistics", 2nd ed., Dowden, Hutchinson, & Ross, Inc.: Stroudsberg, PA, 1975

Matrix Effects in the Quantitative Elemental Analysis of Plastic-Embedded and Ashed Biological Tissue by SIMS

J. T. Brenna and G.H. Morrison*

Baker Laboratory of Chemistry, Cornell University, Ithaca, NY 14853, USA

Quantitative analysis of single matrix materials using the ion microprobe mode of SIMS analysis is now routinely achieved using empirical standards. Quantification in multilayer/multimatrix specimens has also been realized in recent years (1). However, ion imaging based on SIMS has been chiefly qualitative in nature due to artifacts in the SIMS measurement. Among these, the inability to precisely calibrate ionization probability variations due to matrix composition variation over an ion image field is probably the most difficult variable to account for. This is true because precise matrix composition from pixel is seldom known with sufficient accuracy to allow fabrication of external standards.

Matrix effects in biological materials have been examined in a single microprobe study (2) of dried, homogenized Standard Reference Materials. However, no studies of matrix effects has been reported to date for ion imaging experiments. In order to assess the importance of matrix-induced variability on ionization probability, practical ion yield (τ) maps have been constructed, using beryllium as a reference element. The specific samples used for this study were intact plastic-embedded mouse intestine prepared using aqueous fixation methods, and ashed mouse intestine (3).

EXPERIMENTAL

Several factors are important in the choice of Be as the reference element: 1) absence of the element in the sample, 2) ability to generate an ion beam containing the element in an ion implanter, 3) low mass so as to limit specimen damage during implantation, 4) absence of mass spectral interferences during analysis, and 5) sensitivity to matrix effects. Beryllium is one of the most toxic elements and so is virtually absent from biological materials. It is routinely generated in conventional ion implanters as a $^{9}Be^{+}$ beam, and can be produced with sufficient intensity at low energy. As a low-mass ion, it causes negligible sputter damage to the sample during implantation. In these specimens, the most prominant molecular interferences are oxide and organic cluster ions, all of which occur at higher masses than Be analysis. Experimentally it was found than the background at mass 9 is negligible in the absence of intentionally introduced Be. Lastly, Be, with an ionization potential of 9.3 eV, tends to be more sensitive to matrix effects than physiologically interesting elements with lower ionization potentials in controlled studies of well characterized materials (4,5).

A $^{9}Be^{+}$ beam was generated in an Accelerators, Inc. 300R ion implanter by PF_5 sputtering of solid BeO. The implantation energy was 20 keV in order

*present address: IBM, Systems Technology Division, Endicott, New York

to keep the entire implant within the half micron thin section. Ion current was 0.5 µA and total fluence was 1×10^{15} ions/cm^2. A depth profile of an intact section established the mean implant range to be 0.2 µm and the bulk of the implant to reside in the top half µm of the section.

τ maps were constructed by imaging Be throughout the entire depth of the implant in an image depth profile (6). This was accomplished with the aid of a video recording system (7) interfaced to the MIDAS ion-imaging system (8). The entire image series was then summed to remove the effects of implant range variation and differential sputtering. The resulting calibrated image is the τ map.

The τ map for intact plastic sections showed little variation within tissue, but an overall enhancement of ion yield in tissue relative to regions of pure plastic adjacent to tissue. Quantitative analysis reveals the τ (tissue)/ τ (plastic) ratio to be 1.37. The τ maps for ashed material revealed significant variation within tissue. In particular, ion yields for nuclei and muscle fibers were greater than in cytoplasmic regions. The maximum variability in these images was τ (nuclei)/ τ (plastic) = 1.14.

DISCUSSION

These results have specific ramifications for the quantitative analysis of each specimen type. For intact plastic specimens, intratissue matrix effects appear to be small, thus the most serious sample-dependent artifact in ion images is differential sputtering. The first order correction for differential sputtering proposed by Patkin et al. (9) is an effective method for calibrating for this phenomenon. Application of this method along with appropriate empirical calibration should remove the bulk of artifactual contrast from ion images.

For ashed specimens, ionization probability variations are of the order of a few percent. Ignoring this effect, the most significant artifact in analysis of these samples is image evolution (3). The solution to this problem is the integration of total analyte signal through the depth of the section, in a procedure analogous to that used to generate the τ maps. Since the ash matrix should be attainable from any initial state of the sample, e.g. freeze-dried or frozen-hydrated, more rigorous methods of sample preparation such as these, which preserve elemental integrity, may be used and the results of this study applied.

The method of τ map construction demonstrated here may be applied to any sample type for the purposes of assessing the effects of matrix composition variations on ionization probabilities. The critical factor in any such analysis is indepth homogeneity within the range of the implant. The information derived from such studies may be of theoretical, as well as analytical significance.

ACKNOWLEDGMENTS

The authors would like to thank the National Science Foundation, the National Institutes of Health, and the Office of Naval Research for financial support of this work.

REFERENCES

1) Galuska, AA, Ph.D. Thesis, Cornell University, 1984.

2) Ramseyer, GO; Morrison, GH, Anal. Chem. 1983, 55, 1963.

3) Brenna, JT; Morrison, GH, Anal. Chem. 1984, 56, 2671.

4) Leta, DP: Morrison, GH, Anal. Chem. 1980, 52, 514.

5) Meyer, C; Maier, M; Bimberg, DJ, J. Appl. Phys. 1983, 54(5), 2672.

6) Patkin, AJ; Morrison, GH, Anal. Chem. 1982, 54, 2.

7) Brenna, JT; Moran, MG; Morrison, GH, Anal. Chem. 1985, in press.

8) Furman, BK; Morrison, GH, Anal. Chem. 1980, 52, 2305.

9) Patkin, AJ; Chandra, S; Morrison, GH, Anal. Chem. 1983, 55,2507.

Reproducible Quantitative SIMS Analysis of Semiconductors in the Cameca IMS 3F

G.D.T. Spiller and T. Ambridge

British Telecom Research Laboratories, Ipswich IP5 7RE, U.K.

1. Introduction

The use of ion-implanted standards for the quantification of SIMS depth profiles is well established [1]. The absolute accuracy obtainable depends on the accuracy with which the implanted dose is known (either from accurate dosimetry on the implanter or from independent calibration of the specimens), but the experimental errors involved in making SIMS measurements also need to be assessed. Quantification is usually carried out using the relative sensitivity factor (RSF), measured from the dose or calculated peak concentration of the implant standard, which is defined as:

Concentration = (RSF) x (element counts/second)/(matrix counts/second).

Particular care is needed in SIMS analyses when pre-determined RSFs are used, so that the standard is NOT re-profiled each time a calibration is required. This paper describes some of the procedures which have been adopted at British Telecom for quantitative analysis on the Cameca IMS 3F, and assesses the experimental errors that may be introduced.

2. Effect of Sample Holder

The effect of the tantalum foil specimen holder on the measurement of RSFs was assessed using the 18-hole holder supplied by Cameca. This holder comprises a number of 3 mm square holes, and 1.5 mm x 3 mm slots. During analysis, the sample is at ±4500V (for the analysis of +ve and -ve secondary ions respectively) and perturbations in the equipotentials between the sample and the immersion lens (5mm away), caused by the edges of the foil, may result in variations in the collected ion current.

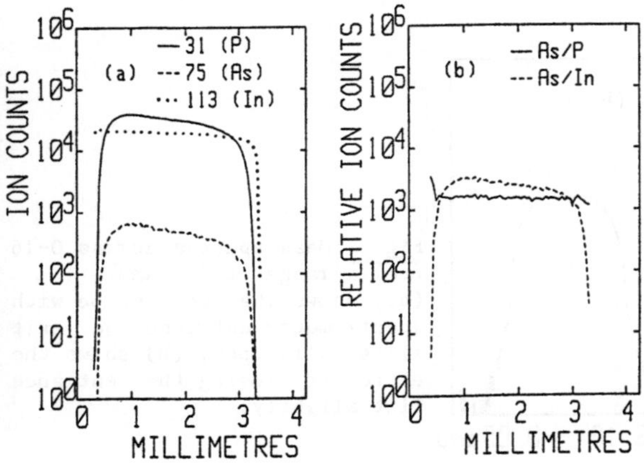

Fig. 1 (a) Line scans across a central 3 mm square hole, from an As-doped InP sample. Indium is recorded with -120V offset; no offset is used on the other elements.
(b) Ratio of As ion counts to P and In (an arbitrary multiplier of 1 E5 is included)

Line scans were recorded across selected holes from a uniformly As-doped InP epilayer, using oxygen primary ions and recording As and matrix signals P-31 (without voltage offset) and In-113 (with -120 V offset, to reduce intensity), using a ±25 eV window. Figure 1 shows a line scan recorded for a central 3mm x 3mm hole. The results can be summarised as follows:

(i) The ion signals fall off steeply at the edge of the holder, and there is also a variation in the absolute signal levels across the hole, particularly for As and P. Similar line scans on different specimens have been reported before [2]. The effect was more severe on the outer holes and the narrow slots. These variations in absolute intensity were increased if the foil was distorted with use. The apparently constant level for In is fortuitous, and arises from a shift in the energy distribution across the hole - tending to increase the signal measured with offset - and a decrease in the total signal level.

(ii) The (As/P) ratio is constant (to within the statistical error on the signals) over 2.5 mm of the 3 mm aperture.

(iii) The (As/In) ratio varies by over a factor of 2 over 2 mm.

From these measurements, it is recommended that:

- Analysis craters should be at least 0.5 mm from the foil edge.
- Only the central two holes should be used for quantitative analyses.
- Matrix signals should be recorded without voltage offset, because of absolute shifts in the energy distribution across the holes.

To confirm the reliability of these guidelines, several measurements were made on a Mg implanted GaAs standard, using As-75 without offset as reference, and without making any changes to the alignment or the magnet settings. The standard deviation on the RSF (calculated from the peak concentration, to eliminate possible errors introduced by the crater measurement) was ±3% over the nine measurements.

3. Reproducibility of Magnet Settings

When setting up the instrument for a depth profile, the magnet is tuned manually or automatically to each mass to be profiled. The 'autocalibration' routine supplied by Cameca scans across the peak and finds the centre of gravity. In manual tuning, the maximum of the secondary ion current for each mass is identified from the current meter.

A mass spectrum across O-16 over a range of ± 0.05 amu is shown in Fig.2a, with the spectrometer entrance and exit slits fully open (mass resolution at the minimum of ~300); the peak is rounded, and slightly asymmetric. Although the peak shape may be improved with alignment, it is evident that the centre

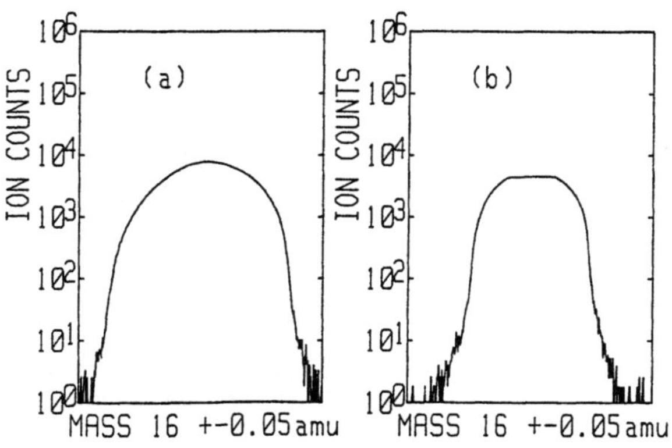

Fig. 2 Mass spectra across O-16 over a range of 0.1 amu. (a) shows the peak shape with spectrometer entrance and exit slits fully open, (b) shows the effect of closing the entrance slit slightly.

of gravity selected in the automatic mode may not correspond precisely to the maximum counts. Closing the spectrometer entrance slits slightly produces flat-topped peaks (Fig. 2b), which can be set with greater reliability: the loss in peak intensity is about a factor of 2. The use of the 'autocalibration' routine is preferred, as manual tuning may not be reliable - even with flat-topped peaks - for low masses (narrower peaks) and low signal levels. (It should be noted that the autocalibration routine cannot be used under high mass resolution conditions).

4. Overall Assessment

This paper has highlighted some of the areas which need to be considered in making quantitative measurements, and suggestions have been made for overcoming possible problems. Further complicating factors, such as non-linear dependence of secondary ion current on primary beam current in certain samples [2], have not been considered here. Experience has shown that satisfactory measurements on Si and III-V semiconductor materials can be made with the procedures described: repeated measurements on selected standard samples over periods of several months gave a variation in RSF of between ± 10-15%. If greater accuracy is required, concurrent profiling of standard and unknown sample is essential. A measurement error of less than ±5% should be possible. Absolute accuracy will be limited by the accuracy and uniformity of the standard.

The use of voltage offset has been shown to give rise to possible errors. Although the use of alternative matrix species is usually possible, caution is needed when offset is used to eliminate mass interferences in the detection of trace impurities [3]. In such cases, occasional reprofiling of the appropriate standard is recommended.

Acknowledgement is made to the Director of Research of British Telecom for permission to publish this paper. Part of this work has been carried out with the support of Procurement Executive, UK Ministry of Defence, sponsored by DCVD.

1. Newbury D E and Simons D, SIMS IV: p101 (1983)
2. Holland R and Blackmore G W, Surface and Interface Analysis 4: p174 (1982)
3. Blattner R J and Evans C A; SEM Conference: p55 (1980)

Part V

Instrumentation

Imaging SIMS at 20 nm Lateral Resolution: Exploratory Research Applications

R. Levi-Setti, G. Crow, and Y.L. Wang

The Enrico Fermi Institute and Department of Physics,
The University of Chicago, Chicago, IL 60637, USA

1. Introduction

Notwithstanding the high brightness of liquid metal ion sources (LMIS), ion probes based on their use still encounter fundamental current limitations, as their size is reduced. Due to the intrinsically large energy spread (5-10 eV) of the ions extracted from LMIS, in fact conventional electrostatic round-lens focusing columns are predominantly chromatic-aberration limited, yielding probes of size-independent current density. The probe current then decreases with the square of the probe size [1]. Thus, the recently developed University of Chicago (UC)-Hughes Research Laboratories (HRL) scanning ion microprobe (SIM) yields currents of 40 keV Ga^+ or In^+ ions typically in the range 1.5 - 30 pA, for probe sizes in the range 20 - 90 nm [2],[3]. Consequently, in addition to a number of intrinsic and instrumental limitations [4], the SIMS sensitivity, proportional to the amount of target material which can be sputtered and analyzed within acceptable scan times, also deteriorates rapidly as the probe size is reduced. Even with the use of a high-transmission SIMS system, such as that implemented in the UC-HRL SIM, useful applications of probes of a few pA, at the limits of spatial resolution presently attained, become restricted to selected favorable problems. These will involve primarily the mapping of segregated components of elements of low ionization potential or high electron affinity. Nevertheless, such minuscule probe currents can be exploited to unique advantage when it is of the essence to limit sputter etching of the sample to a few surface atomic monolayers. Several examples of high spatial resolution SIMS imaging, involving the above concepts, will be shown here. These originate from current exploratory investigations of interdisciplinary applications in cosmochemistry, materials science, and biomedical research. Our observations broaden the scope of LMIS SIMS applications, already explored by several authors [5],[6]. Several previous reports detail the schematics and performance of the UC-HRL SIM [2],[3],[7].

2. High-Resolution SIM and SIMS Imaging

Topographic SIMS images. As repeatedly stressed [3],[8], a small SIM probe size does not necessarily imply that comparable size resolution can be attained in either topographic or SIMS imaging. In fact, several parameters, function of the probe-target interaction and of the methods of image acquisition, must be carefully optimized to achieve image resolution approaching the probe size. For ion-induced secondary electron (ISE) or ion (ISI) imaging of the surface topography, the signals collected with channel electron multiplier detectors are generally so copious that no limitations due to picture element (pixel) statistics are encountered even with probes of a few pA. The precaution is then that of optimizing the pixel sampling in relation to raster size and size of the scanned area [2],[9]. The concept of optical magnification is involved here, which for a digital raster, is the one that maps contiguous sample elements of size equal to the probe size dp, onto contiguous pixel elements on the CRT. Thus with raster size 1024 x 1024 pixels, as commonly used in the UC-HRL SIM, the optimal scan area has side dimension of 1024 x dp, or 20 x 20 μm at e.g. dp = 20 nm.

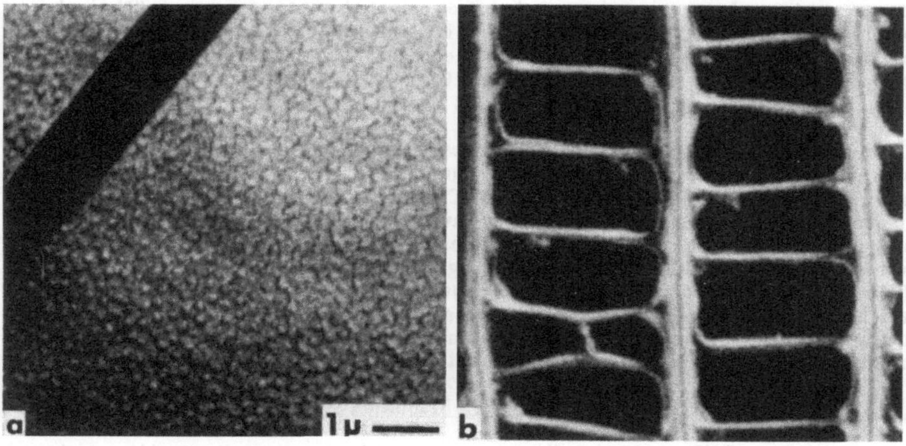

Fig. 1. (a), ISI topographic image of lens in the compound eye of the fruit fly. 1.6 pA, 40 keV In^+ probe. 256 sec. exposure. (b), ISE topographic image of butterfly wing scale. 20 pA, 40 keV Ga^+ probe, 32 sec. exposure.

Two examples of high-resolution ISE imaging are shown in Fig. 1. Fig. 1a shows the structure of the surface of a lens element in the compound eye of *Drosophila melanogaster*, in a scan with a 40 keV, 1.5 pA In^+ probe. The network of small protuberances represents the corneal nipple array, an antireflection device common to insect eye lenses [10]. Visible are crevices separating individual nipples, ~ 50 nm wide. The sharp edge of the 1 μm-thick bristle (seta) in the upper left corner has a fall-off width of ~20 nm. Image recording with the In^+ probe at this magnification (10 x 10 μm^2 scan) is feasible only at probe currents of a few pA, as given by a 5 μm optical aperture. At 10 pA, the entire lens surface is seen to collapse during visual focusing. Figure 1b shows a portion of a 10 x 10 μm^2 scan of a scale from the wing of a butterfly, obtained with a 20 pA Ga^+ probe defined by a 25 μm aperture. Feature separation of structures still below the 100 nm level are demonstrated in this micrograph. It should be emphasized that image resolution with highly destructive probes such as these, depends to a great extent on the damage to the object during image acquisition.

SIMS elemental maps. Much more critical are the conditions for high-resolution SIMS imaging where, in addition to optimally sample the object surface as above, one wishes to:
a) maximize SIMS sensitivity to gain pixel statistics and depth resolution,
b) allow the dwell-time/pixel to exceed the flight time of the collected ISI through the SIMS system to preserve synchronous image display or bandwidth, a mass-dependent condition, and in favour of pixel statistics,
c) limit the dwell-time/pixel to preserve surface O2 coverage and ensuing ion fraction enhancement (in absence of O2 flooding), and in favour of depth resolution.

Optimization of SIMS imaging requires a delicate balancing of conflicting requirements. In the UC-HRL SIM, the back-sputtered ISI's are collected, energy analyzed and transported to the entrance of an RF quadrupole mass filter, by an efficiently designed optical system, 2 cm thick. The overall SIMS detection efficiency is ~ 1/250 sputtered atoms for e.g. $^{40}Ca^+$ in fluoroapatite [3]. This leads to mass resolved counting rates, in this case, of ~$2x10^4$ cps/pA for a 40 keV Ga^+ probe, somewhat higher for In^+, at residual pressures of ~10^{-8} Torr. To maximize SIMS sensitivity, the accepted ISI energy window (<~10 eV for ΔM <~1 amu) can be positioned anywhere on the ISI energy spectrum, by varying the sample offset potential over the range ± 200 eV. A compromise satisfying requirements b) and c) is generally found, so that several high resolution maps of the same area may be obtained, prior to signal fall-off due to surface O2 depletion. This is facilitated in cases where O2 multilayers may be present, or with O-rich matrices.

As reported in detail elsewhere [2],[3],[8], we have shown that high quality, gray level tone SIMS maps can be obtained at resolutions approaching the probe size, with Ga^+ or In^+ probes at currents as low as 1 pA. Clearly, such feasibility is restricted to a limited number of ISI species. These probes have proven equally effective in terms of ion yields, for both negative and positive ions, approaching the performance of O^- probes. Statistical limitations will generally degrade resolution, for sparse distributions of low-ion-fraction species. For such species however, high-resolution imaging can still be achieved when these segregate in small areas such as nucleation centers, islands or at grain boundaries. A critical matching of instrument performance to the research problem is then called for to achieve and exploit high-resolution SIMS imaging.

3. Exploratory Research Applications of High Resolution SIMS Imaging

Cosmochemistry. Differentiated bulk structures in silicate minerals provide a rich supply of elements suitable for high-resolution SIMS imaging. The study of meteorite material, chondrites in particular, is most rewarding from this standpoint. These are a class of primitive stone meteorites consisting of aggregates of spheroidal bodies (chondrules), composed of crystalline silicate minerals (e.g., olivine, pyroxene), cemented together by a glassy matrix. In turn, the chondrules are cemented in a matrix made up of a variety of fine-grained differentiated minerals and metal inclusions. The chondrules, which range in size from a few µm to several mm, represent quenched melt droplets thought to have originated in the early solar nebula. Several unresolved questions regarding both the chondrules and their matrix can be investigated by high-resolution SIMS.

We have already reported on our exploratory study of a chondrule rim [11] and on observations of the interior of chondrules, in the UC-HRL SIMS [2],[8]. We show in Fig. 2a a ^{24}Mg map of a section of a chondrule in the Mezo-Madaras chondrite. Magnesium is

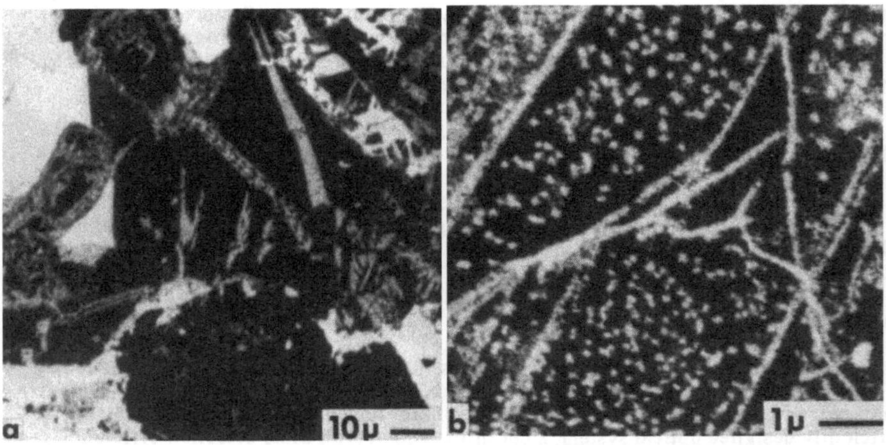

Fig. 2. (a), $^{24}Mg^+$ map of interior of chondrule in the Mezo-Madaras meteorite. 8 pA, 40 keV Ga^+ probe, 512 sec. scan. From Ref. [2], courtesy of J. de Physique. (b), $^{35}Cl^-$ map of freshly cleaved, stage-4 $SbCl_5$-intercalated graphite. 1.6 pA, 40 keV Ga^+ probe, 512 sec. scan.

associated here with Fe in olivine ($(Mg\ Fe)_2\ SiO_4$). Complementary to this map is a Na map [2], indicating that the olivine crystals are cemented by glass. The fine radiating structures of the olivine crystals suggest a rapid quench. The polished section was lightly coated with Au-Pd, to prevent charging. The coating was rapidly sputtered away from the scan area, prior to mapping. Maps of this and other areas of the same sample have been obtained for e.g. Na^+, Mg^+, Al^+, Si^+, Ca^+, Fe^+, and O^-. Typical counting rates are $\sim 2 \times 10^3$ cps/pA for Na in the glassy areas, comparable to those for Mg in olivine. Fe in olivine yields $\sim 10^2$ cps/pA, $O^- \sim 4 \times 10^2$ cps/pA. Since most mineral phases are segregated in areas with sharp boundaries, high quality maps can be obtained over the entire range of counting rates.

Materials science. SIMS imaging of graphite intercalation compounds (GIC) has given a rare opportunity to take advantage of the ion current limitations in highly focused probes. GIC's consist of alternating layered structures obtained by inserting foreign elements and compounds between graphite lattice planes. The stage of a GIC refers to the order number of lattice planes separating successive intercalant layers. When a GIC is cleaved, the intercalant distribution may become exposed. For high intercalation stages, and in view of the low sputtering yield of C, it becomes feasible to map by SIMS the intercalant distribution of a single monolayer, provided the eroded depth does not exceed the distance between intercalant layers. In a study [12] of $SbCl_5$-intercalated, highly oriented pyrolithic graphite (HOPG), we have shown that single-monolayer mapping of Cl^- can be obtained for stage-4 GIC's, using a 1.6 pA 40 keV Ga^+ probe. An example of such mapping is shown in Fig. 2b. The remarkable feature observed here, and in innumerable other maps, is the coagulation of the intercalant into well-defined domains 100-200 nm wide, which may coalesce into lines at lattice steps. This interplanar intercalate topology has been interpreted as demonstrating the intercalation model proposed by Daumas and Herold [13]. For stage-2 $SbCl_5$ GIC's, the Cl^- maps suggest that a second intercalant monolayer is detected. The domains fade into a blur when an 8 pA probe is used, clearly due to the increased sputter rate and consequent overlap of several intercalant monolayers. High-resolution SIMS imaging in this application has clearly become a method of surface, rather than bulk analysis.

The crystallization of amorphous metal-metalloid alloys (metallic glasses) leads to elemental segregation on a scale of dimensions such that high-resolution SIMS imaging is most informative. We have examined the shiny surface of $Fe_{75}\,B_{15}\,Si_{10}$ ribbons after crystallization carried out in a quartz tube at a residual Ar pressure of $\sim 10^{-6}$ Torr. Two

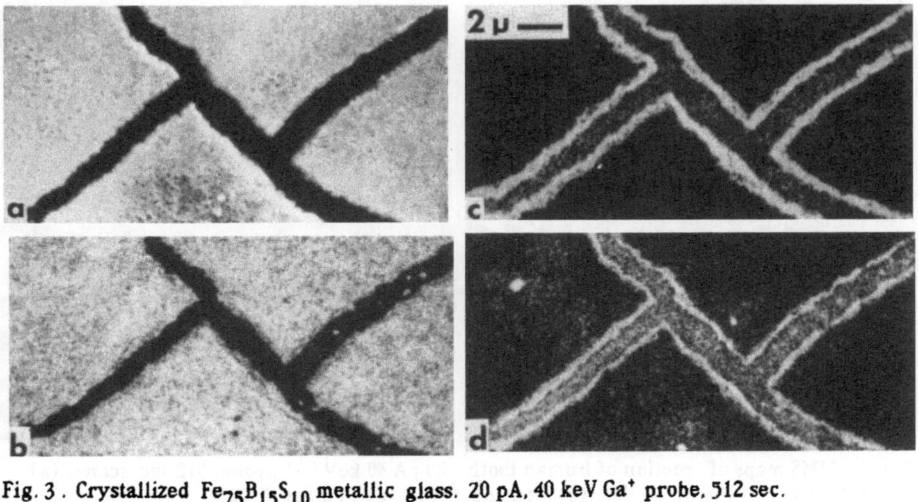

Fig. 3 . Crystallized $Fe_{75}B_{15}S_{10}$ metallic glass. 20 pA, 40 keV Ga^+ probe, 512 sec. exposures. (a) $^{16}O^-$; (b) $^{28}Si^+$; (c) $^{43}(BO_2)^+$; (d) $^{56}Fe^+$.

crystallization phase transitions for this alloy are known to occur at 846 and 860 K. The sample was heated at ~ 900 K for ~4 hours. Figure 3 shows maps for O^-, Si^+, BO_2^+ and Fe^+ respectively, of a typical structure. A boron map, not shown here, is similar to that of BO_2 in Fig. 3c. The surface of the ribbon exhibits a network of fine shallow crevices. These seem to represent grain boundaries where Fe and B segregate, with identical distributions, presumably in a Fe_2B eutectic crystallization phase [14], which appears oxydized at the surface. Similarly, the O^- and Si^+ maps suggest a SiO_2 layer covering most of the sample surface. It is quite evident that high-resolution SIMS imaging will add a new dimension to this area of investigation, by providing an insight which seems needed for a more detailed interpretation of the results obtained by other approaches, e.g. Mossbauer spectroscopy [15].

Biomedical applications. Since many of the elements of low ionization potential or high electron affinity are also, for the same reasons, biologically important, it is generally feasible to exploit high-resolution SIMS imaging in several areas of biomedical research. Biomineralizations, such as e.g. in bone and teeth, have been an all-time favorite for SIMS studies [16],[17]. We have already shown high-resolution SIMS maps for Ca in kidney stones and for Na, K and Ca in the skull bone of neonatal mouse [18]. Labelling of the same kind of bone cultured *in vitro* in a ^{44}Ca medium, has clearly outlined the active Ca fixation sites [19].

Of renewed interest may be the details of structures we observe in polished sections of the human tooth. Figure 4a is a $^{40}Ca^+$ map of a region of the enamel, close to the front surface of the tooth (located beyond the left border in Fig. 4a). A lesion in progress is clearly outlined, and seen to proceed via dendritic, fractal-like processes. Figure 4b shows an O^- map of the dentine. Here the cross-sections of the dentinal tubules contrast with the matrix by their higher oxygen content. There is a one to one correspondence between O^- and Ca^+ maps, both components of fluoroapatite. Figures 4c and 4d are Al^+ and Ca^+ maps of the same dentine area. Here the Al is seen in the matrix, and abundant in the core of the dentinal tubules, while absent in the Ca-rich peritubular dentine.

Soft biological tissue can also be mapped at high resolution. Figure 5 shows SIMS maps of human erythrocytes, fixed with glutaraldehide, and mounted on Ta (5 a,b) or stainless steel (5 c,d) supports with polylysine. We understand that this preparation procedure may be objectionable, and that freeze-dried samples are preferable for SIMS,

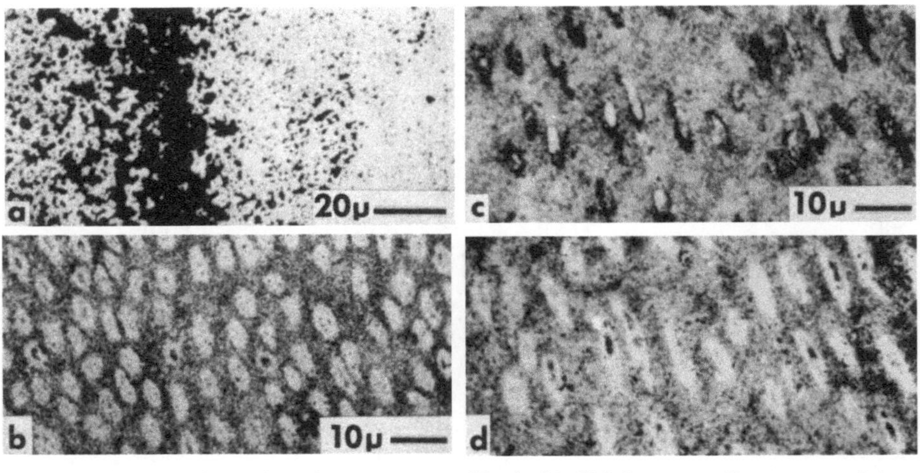

Fig. 4. SIMS maps of section of human tooth. 20 pA 40 keV Ga^+ probe. 512 sec. scans. (a) Enamel, $^{40}Ca^+$: (b) Dentine, $^{16}O^-$; (c) Dentine, $^{27}Al^+$; (d) Dentine, $^{40}Ca^+$.

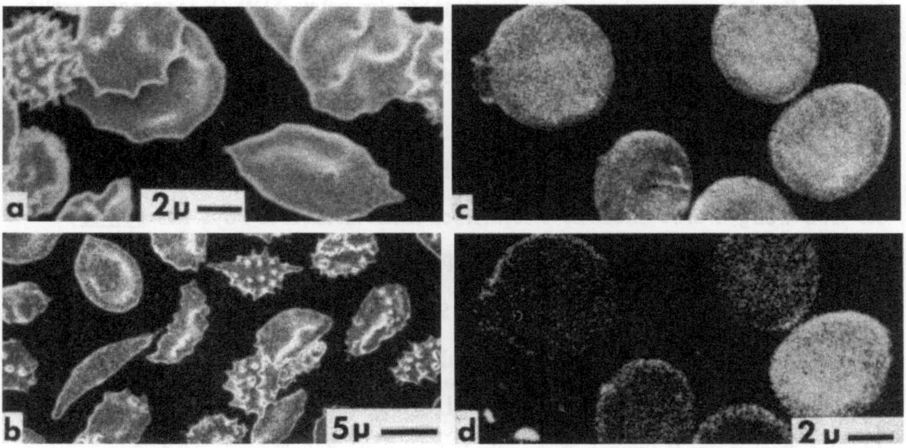

Fig. 5. SIMS maps of human erythrocytes. 20 pA, 40 keV Ga^+ probe, 512 sec. scans. (a),(b) $^{26}(CN)^-$ maps from sickle-cell sample. (c),(d) Na^+ and $^{40}Ca^+$ maps of normal red blood cells.

due to the possibility of ion loss or exchange [20]. SIMS maps for CN^-, possibly originating from peptide bonds in proteins, are shown in Fig. 5a and 5b for a sample of sickled erythrocytes. Spiculated bodies (echinocytes) originate from storage-induced modifications of the erythrocyte morphology. Figures 5c and 5d show maps, respectively for Na and Ca, of normal erythrocytes. The unusually Ca-rich cell in Fig. 5d may be senescent. Its apparently abnormal Ca content cannot be due to a SIMS artifact due to possible non-uniform local fields, in view of the approximately uniform Na yields exhibited by all the erythrocytes in the same field of view.

4. Conclusions

SIMS imaging at high spatial resolution can make a virtue of the limitations in probe current resulting primarily from chromatic probe aberrations. This requires the use of an efficiently designed SIMS system. At present, a wide range of applications can benefit from this extension of conventional SIMS methodology, via a careful match of problem with instrument performance. A broader utilization of ultra-fine probes for SIMS may result from the coupling of LMIS-based instruments with laser resonant or non-resonant post-ionization of the sputtered atoms [21].

5. Acknowledgements

This work was supported by the National Science Foundation (Grant No. DMR-8007978) and partially by the NSF Materials Research Laboratory at the University of Chicago. We wish to thank Dr. V. R. V. Ramanan of Allied Corporation for providing us with the samples of metallic glass and relevant information, and Dr. G. Boas for a tooth sample. Dr. T. Steck has kindly discussed with us several aspects of erythrocyte morphology and biochemistry, and carefully prepared for us many samples.

6. References

1. R. Levi-Setti and T. R. Fox, Nucl. Instr. Methods 168 (1980), 139.
2. R. Levi-Setti, Y. L. Wang and G. Crow, J. Physique 45-C9 (1984), 179.
3. R. Levi-Setti, G. Crow and Y. L. Wang, Scanning Electron Microsc. II (1985), 535.
4. P. Williams, Scanning Electron Microsc. II (1985), 553.

5. F. G. Ruedenauer, in : Secondary Ion Mass Spectrometry- SIMS IV (Springer, Berlin, (1984), 133.
6. A. R. Waugh, A. R. Bayly and K. Anderson, Vacuum 34 (1984), 103.
7. R. Levi-Setti, P. H. La Marche, K. Lam and Y. L. Wang, SPIE 471 (1984), 75.
8. R. Levi-Setti, G. Crow and Y. L. Wang, in : Microbeam Analysis - 1985 (San Francisco Press, San Francisco, 1985), 209.
9. P. D. Prewett, Vacuum 34 (1984), 936.
10. W. H. Miller A. R. Moeller and C. G. Bernhard, in: The Functional Organization of the Compound Eye (Pergamon, NY: 1985), 21.
11. Y. L. Wang, G. Crow and R. Levi-Setti, Nucl. Instr. Methods in Phys. Res. B10/11, (1985), 716.
12. R. Levi-Setti, G. Crow, Y. L. Wang, N. W. Parker, R. Mittleman and D. M. Hwang, Phys. Rev. Letters 54 (1985), 2665.
13. N. Daumas and A. Herold, C. R. Acad. Sci., Ser. C 268 (1969), 373, and Bull. Soc. Chim. Fr. 5 (1971), 1598.
14. U. Koester and U. Herold, in: Glassy Metals I (Springer, Berlin, 1981), 225.
15. H.-G. Wagner, M. Ackermann and U. Goser, J. Non-Cryst. Solids 61/62 (1984), 847.
16. A. Lodding, Scanning Electron Micros. III (1983), 1229.
17. R. Lefevre, R. M. Frank and J. C. Voegel, Calcif. Tiss. Res. 19 (1976), 251.
18. D. A. Bushinsky, G. Crow, S. Deganello, R. Levi-Setti, Y. L. Wang and F. L. Coe, presented at 17th Ann. Meet. of the Am. Soc. Nephrology, 1984.
19. D. A. Bushinsky, R. Levi-Setti and F. Coe, presented at the 7th Ann. Meet. of the Am. Soc. Bone & Min. Res., 1985.
20. M. S. Burns, in: Analysis of Organic and Biological Surfaces (New York, Wiley, 1984), 259.
21. F.M. Kimock, J.P. Baxter, D.L. Pappas, P.H. Kobrin and N. Winograd, Anal. Chem. 56 (1984), 2782.

Use of a Compact Cs Gun Together with a Liquid Metal Ion Source for High Sensitivity Submicron SIMS

T. Okutani, T. Shinomiya, M. Ohshima, and T. Noda*

Toyohashi Univ. of Technol., Tempaku, Toyohashi 440, Japan

H. Tamura and H. Watanabe

Hitachi Central Research Lab., Kokubunji, Tokyo 185, Japan

1. Introduction Ion microprobe SIMS by means of positive or negative secondary ions from the specimen surface offers a very high sensitivity not only in surface analysis but in depth profiling also, allowing vast application to solids including biological samples. Particularly for electronegative elements, it is very effective to bombard the specimen surface with Cs ions to get negative secondary ions. Conventional ion bombardment under Cs flooding is also known as an alternative.

Regarding such Cs guns to attain higher secondary ion yield in SIMS, several attempts have been reported in the past conferences [1-3]. Since one has to pay attention when atomic Cs is used because it violently reacts with water in air, we tried to employ a stable mixture with Cs compound, which produces vapor of atomic Cs through chemical reaction in a crucible. The Cs gun was then mounted on a SIMS system in parallel with an easy-to-operate Ga liquid metal ion source (LMIS) which we have developed for a submicron SIMS since 1981 [4].

2. Experimental The ion microprobe SIMS unit used in this experiment is a modified Hitachi IMA-2 which has a large specimen chamber evacuated by a 2000 liter/s turbo molecular pump. As illustrated in **Fig. 1**, the standard duoplasmatron source was

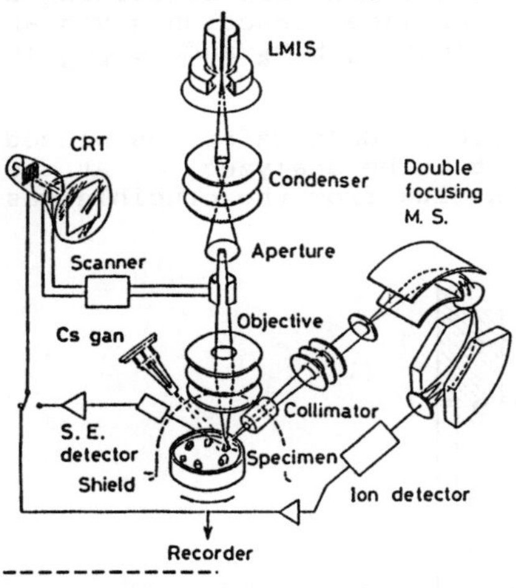

Fig.1 Microprobe SIMS with LMIS and Cs gun

***Dr. Tsuyoshi Okutani**, the first author who had devoted himself to this study, suddenly died of pneumonia during the process of severe treatment agaist acute leukaemia, as young as 34 years old, on July 13, 1985.

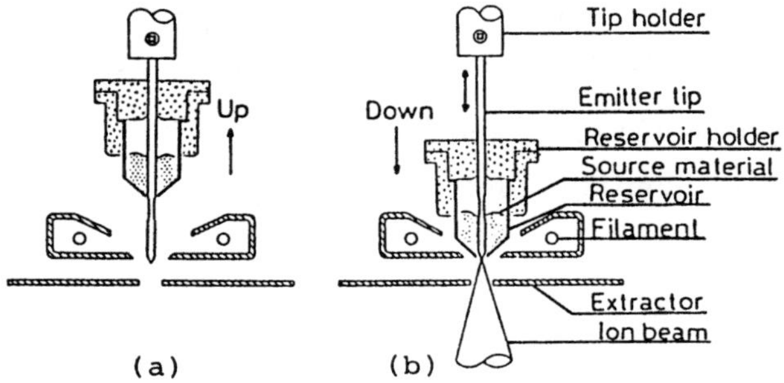

Fig.2 Mechanism of our LMIS

replaced with our LMIS, and our Cs gun was mounted on the specimen chamber in place of a blank lid.

The latest model of our LMIS employed in this experiment is newly equipped with a unique mechanism, as seen in **Fig. 2**, to adjust the positions of both the tip and the reservoir with respect to the extractor, separately from outside of vacuum, even during ion extraction. With one knob the reservoir can move up to heat the tip for cleaning and down to rub the tip for better wetting with material in the reservoir. Position of the tip is adjustable by another knob to define the objective distance for the optics. In addition, a shield cap was provided for each insulator to prevent contamination deposit due to slight evaporation from the tip even with a closed construction of the reservoir. With this LMIS, small chips of Si wafer as the source material gave a total ion current of 60 uA, composed of atomic and cluster ions only of Si. This particular study, however, employed Ga as the source metal throughout the experiment.

Major components of our Cs gun are, as shown in **Fig. 3**, a boron nitride crucible, a tungsten coil heater and a chromel-alumel thermocouple. When we heat the crucible containing a powder mixture of Cs_2CrO_4 and Si, a chemical reaction gives an atomic Cs flux through the nozzle, leaving SiO_2 and Cr_2C_3 in the crucible as solid residues.

3. Results and Discussion A piece of GaAs wafer was placed in the specimen chamber as the target to be analyzed. At the same time as the Ga ion impact, a Cs flux from the crucible was

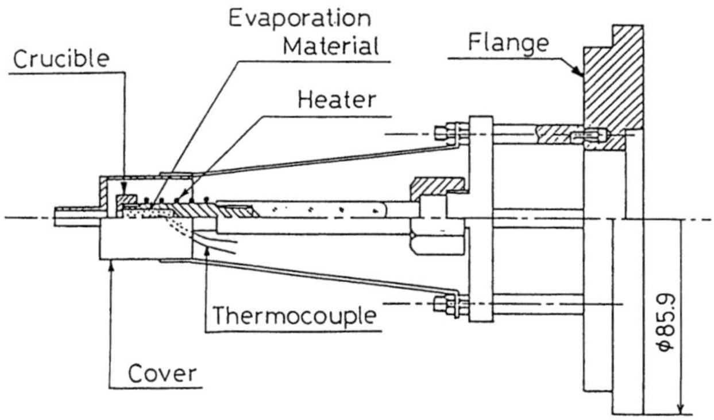

Fig.3 Construction of Cs gun

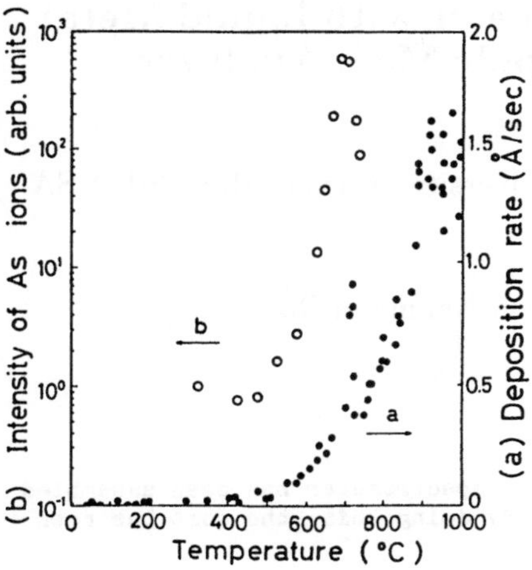

Fig.4 Cs enhanced As emission

flooded onto the specimen surface. The rate of Cs deposition was measured with a quartz oscillator placed at a point 25 mm apart, same as to the specimen, from the nozzle of Cs gun, and controlled by the crucible temperature as shown in **Fig. 4(a)**. As the temperature increased, the negative ion intensity of As, an electronegative element, exhibited a dramatic increase above 500° C due to Cs deposition. The peak as high as 300 times the intensity at 300° C appeared near 700° C, as seen in **Fig. 4(b)**, corresponding to a Cs deposition rate of 0.04 nm/s. A sharp fall beyond 700° C may result from a Cs layer forming on the specimen surface due to a higher rate of deposition of Cs than sputtering by the Ga ion bombardment.

As a smart technique in the ion microprobe SIMS, it is proposed to use a focused ion probe with an LMIS, as the primary beam system, and a Cs vapor flux onto the specimen surface for the purpose of enhancing electronegative elements, analogous to the case of detecting positive secondary ions with an O_2 flux.

Authors are greatly indebted to Professors G. Slodzian, J. Brown and R. Shimizu for their invaluable advice to improve writing of this paper.

REFERENCES

1 M. Bernheim, J. Rebiere and G. Slodzian : SIMS-II, Ed. by A.Benninghoven, C.A.Evans,Jr., R.A.Powell, R.Shimizu and H.A.Storms, Springer-Verlag, (1979) p.40
2 T. Okutani and R. Shimizu : ibid, p.186
3 M. L. Yu : SIMS-IV, Ed. by A.Benninghoven, J.Okano, R. Shimizu and H.W.Werner, Springer-Verlag, (1984) p.60
4 T. Okutani, M. Fukuda, T. Noda, H. Tamura, H. Watanabe and C. Shepherd : J. Vac. Sci. Tech. **B1**(4), 1145 (1983)

Secondary Ion Mass Spectrometer with Liquid Metal Field Ion Source and Quadrupole Mass Analyzer

J.M. Schroeer

Department of Physics, Illinois State University, Normal, IL 61761, USA

J. Puretz

Oregon Graduate Center, 19600 N.W. von Neumann Rd., Beaverton, OR 97006, USA

An ultra-high vacuum secondary ion mass spectrometer has been assembled from commercially available components. By doing this, the cost has been held to around $100,000.

Figure 1 shows a diagram of the mass spectrometer. The ion source was a fine tungsten needle coated with liquid gallium [1] coupled to electrostatic extraction, focussing, and scanning optics. [2-4] Typical beams were 0.2 nA of 10-keV Ga+ ions focussed to a spot as small as 0.5 μm. The beam could be rastered over an area up to 1.5 mm square.

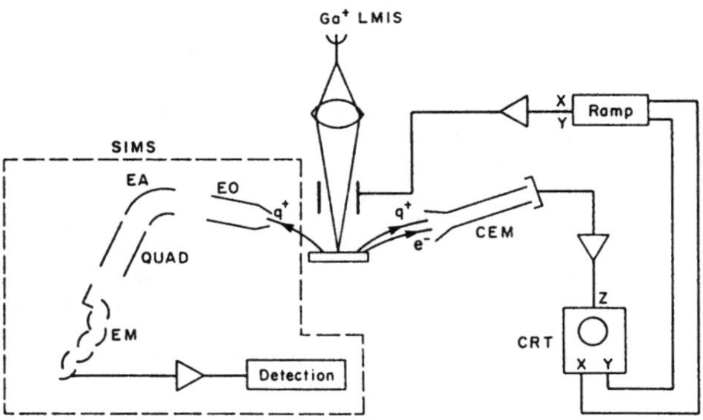

Fig. 1. Block diagram of the complete system. LMIS: liquid metal ion source and gun; CEM: continuous dynode electron multiplier for the collection of the secondary electrons or the total secondary ions, modulating the intensity of the oscilloscope; EO: SIMS extraction optics; EA: energy analyzer; QUAD: quadrupole mass analyzer; EM: 16-dynode Cu-Be electron multiplier.

Samples were mounted on a rotating target caroussel. The sputtered ions traversed extraction and focussing optics and a 90-degree spherical energy analyzer [5], then a quadrupole mass analyzer, and were finally detected by pulse counting with a Cu-Be electron multiplier. A jet of oxygen could be directed at the target to enhance the yield of secondary ions. In close proximity to the target was a continuous-dynode electron multiplier for the detection of secondary electrons or the total ion current from the target.

The instrument could be operated in several modes. In the scanning-ion-microscope mode the secondary electrons from the sample produced images similar to those from an SEM. If the total secondary ions were detected instead, the image quality was not as good as with secondary electrons. The primary beam could be used for milling structures as small as 1 μm square. We could produce standard secondary ion mass spectra. Finally, the output of the mass filter could be coupled to the z-axis of a cathode-ray tube through a commercial pulse-shaping amplifier, such that each output pulse resulted in a dot on the tube face, producing elemental distribution maps. It was very easy to switch between modes, as the same probe, the gallium ion beam, was used.

A home-made miniature gas ion source [6] was mounted on the target caroussel and could be positioned near the extraction optics so as to simulate the emission of secondary ions from a sample. This ion source aided greatly in the mechanical and electrical adjustment of the extraction optics and energy and mass filters and in the measurement of the transmission efficiency ($\sim$ 0.3%) for the secondary ions from target to detector. By using different gases and varying the ion energy, it is possible to measure the transmission and detection efficiencies for ions of different energies, masses, and chemical properties.

The gas ion source, shown in Fig. 2, typically operates at a pressure of 2×10^{-6} torr with a heater current of 2 A and an electron emission of 0.4 mA at 70 eV. The extraction grid is at ground potential, simulating the sample surface. The gas ions, simulating the secondary ions, have an adjustable energy. At 15 eV the ion current is about 1 nA. The beam energy spread is estimated to be about 1% [6], and the beam is about 2 mm in diameter 2 mm from the extraction grid.

Figures 3 and 4 show the mass spectrum and the elemental distribution maps, respectively, of a copper grid pressed into an aluminum substrate. It is interesting to note how much larger the secondary ion emission of Ga+ is from copper than from aluminum. Count rates are low as the instrument's operation had not yet been optimized. Fig. 5 shows the mass spectrum of a Ti alloy with and without oxygen flooding. Because of the high bombarding current density, only the titanium peaks are affected by the oxygen. The magnitude of the Ga+ peaks from the gallium implanted and then resputtered might be a useful reference in quantitative analysis. The instrument has

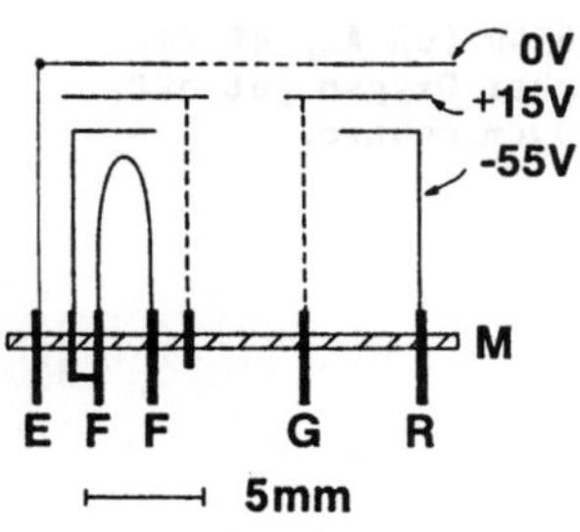

Fig. 2. Schematic diagram of the miniature gas ion source. FF: 0.127 mm W filament, R: reflector, G: grid, E: extraction grid, M: mica insulator.

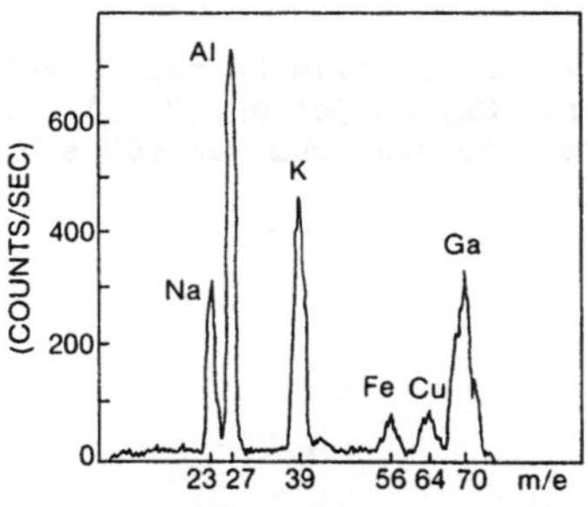

Fig. 3. Low-resolution positive SI mass spectrum of the Cu grid on an Al substrate shown in Fig. 4.

Fig. 4. Images of a Cu grid with 100 μm spacing on an Al substrate: (a) secondary electrons, (b) Al+, (c) Cu+, (d) Ga+. Exposure time was approximately one minute.

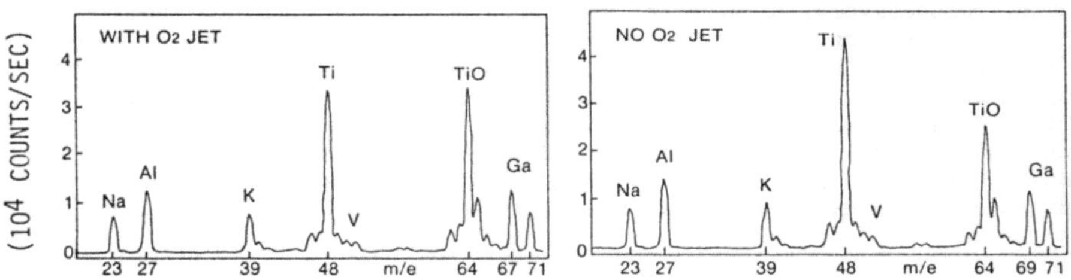

Fig. 5. Positive SI mass spectra of Ti alloy NBS 654a (6% Al, 4% V). Left: Oxygen jet on, $P(O_2) = 2.7 \times 10^{-6}$ torr. Right: Oxygen jet off. Time per spectrum was 500 sec, raster area was 200 μm square.

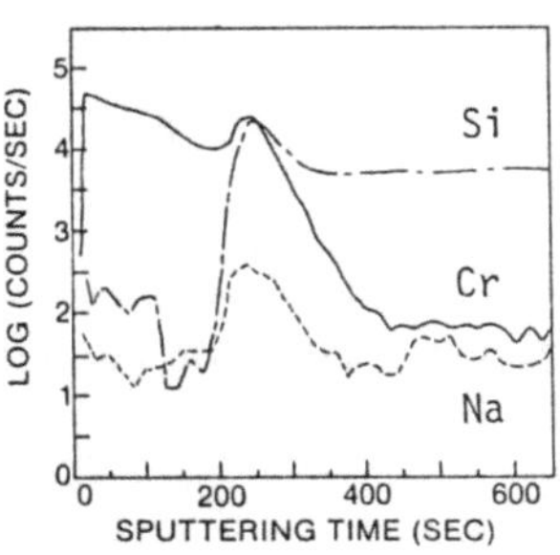

Fig. 6. Depth profile of a 70-nm Cr film on Si, using 50% gating of the rastered area to avoid crater edge effects.

recently been used in a depth-profiling mode to analyze Cr and Cu films on Si, as shown in Fig. 6. Insulating materials, such as passivation layers on integrated circuits have also been analyzed, using an electron flood gun.

<u>References</u>

1. L.W. Swanson et al., J. Vac. Sci., Technol. <u>16</u> (1979) 1864
2. Ion gun manufactured by FEI Inc., 2575 Cathryn, Hillsboro, OR 97124
3. J. Orloff and L.W. Swanson, J. Vac. Sci, Technol. <u>19</u> (1981) 1149
4. J. Puretz, J. Orloff, and L. Swanson, SPIE Vol. 471: Electron Beam, X-Ray and Ion Beam Techniques for Submicrometer Lithography III (1984) 38-76.
5. D.G. Welkie and R.L. Gerlach, in SIMS IV, (1984) p. 317
6. Non-miniaturized electron impact gas ion sources are described by K.M. Khan and J.M. Schroeer, Rev. Sci. Instr. <u>42</u> (1971), 1348, and J.-H. Wang et al., SIMS IV (1984) p. 130.

Evaluation of a New Cesium Ion Source for SIMS

D.G. Welkie

Perkin-Elmer Corporation, Physical Electronics Division, 6509 Flying Cloud Drive, Eden Prairie, MN 55344, USA

1. Introduction

Cesium is commonly used as the primary ion beam constituent in the SIMS analysis of many species to chemically enhance their negative secondary ion yields and thereby to improve their detection sensitivities. The detection sensitivity depends also on other properties of the ion beam, such as current density, beam diameter, and neutral content. These parameters are determined by the brightness of the ion source and the characteristic of the probe-forming optics. In this paper, the design and operation of a new surface ionization-type cesium ion source is described, and its performance is evaluated in conjunction with the optical column of the PHI Model 6050 Microbeam Ion Gun.

2. Description of the Ion Source

A schematic diagram of the ion source is shown in Fig. 1. The source is mounted via a standard 4 1/2" metal-seal flange. It i interchangeable with a duoplasmatron on the PHI Model 6050 optical column, which is described in detail elsewhere [1]. The principle of operation of the source is well known [2,3]. Basically, cesium metal is heated in a reservoir and the resultant vapor flows through a feed tube to a porous tungsten plug. The cesium atoms diffuse through the pores in the plug to the front surface, which is maintained at a temperature of >1100°C. by the ionizer heater. The cesium atoms are then ionized during evapor ation because the work function of tungsten (4.52 eV) is substan

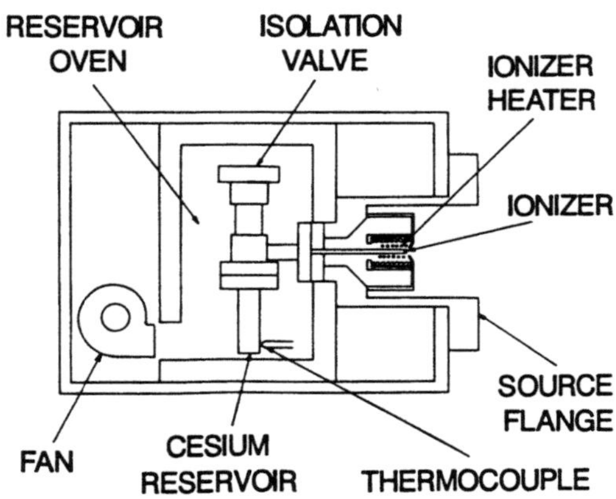

Fig. 1 Schematic diagram of the cesium ion source

tially greater than the ionization potential of cesium (3.88 eV) [3]. The ions are extracted and accelerated to an energy of up to 10 keV.

A metal-seal valve is used to isolate the cesium during bakeout and when the source is not under vacuum. The reservoir temperature is monitored with a thermocouple. The signal from this thermocouple is used in a feedback loop to regulate the reservoir oven power to maintain a constant reservoir temperature. Thus, the cesium vapor pressure remains constant, which results in a stable beam current. A fan is used to rapidly cool the cesium reservoir after operation, so that the cesium vapor condenses back in the reservoir rather than in the valve, the feed tube, or the porous plug. This ensures that the source characteristics are reproducible from one usage to the next.

3. Performance Evaluation

The most descriptive parameter for an ion source is its brightness, which is defined as the current emitted per unit solid angle per unit area. When the ion source is coupled to probe-forming optics, the final probe size can be expressed as [4]

$$d^2 = \frac{4I}{\pi^2 \beta \alpha^2} + \frac{1}{4} C_s^2 \alpha^6 + C_c^2 \left(\frac{\Delta E}{E}\right)^2 \alpha^2 \qquad (1)$$

to a good approximation, where C_s and C_c are the spherical and chromatic aberration coefficients, respectively, of the objective lens, α is the semi-angle of convergence of the beam, ΔE is the energy divergence in the beam, and E is the beam energy. The brightness can be determined, then, by measuring the probe diameter and current, provided the optical properties of the focusing column and the energy divergence in the beam are known. Plotted in Fig. 2 are the beam diameters and currents measured from the cesium ion source on the Model 6050 optical column for a range of magnification conditions. The beam diameters were determined by sweeping the beam across a knife edge and measuring the distance between beam positions that resulted in a 10% and a

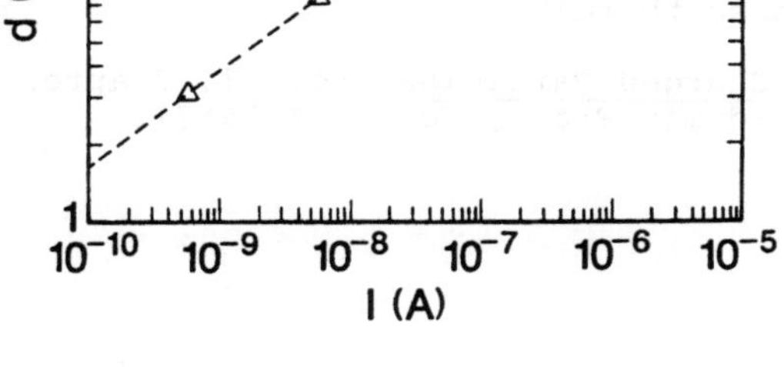

Fig. 2 Probe diameter vs. current characteristics for the cesium ion source coupled to the PHI Model 6050 Microbeam Ion Gun

- ϕ Measured data
- —— Best fit to the measured data
- - - - Theoretical d vs. I characteristic for optimum conditions
- Δ Optimum conditions for the aperture sizes available on the Model 6050 optical column

90% change in secondary electron signal. Also plotted in Fig. 2 is the fit of (1) to the data, using the brightness as the adjustable parameter. The aberration coefficients C_s and C_c were determined for the objective lens to be 93 cm and 13 cm, respectively, by computer simulation. The energy divergence for a surface ionization ion source is expected to be a few tenths of an eV, which results in negligible beam broadening within the range of beam diameters of Fig. 2. The best fit to the data, then, corresponds to a brightness of 14 $A/cm^2/str$.

At a given beam diameter, the current density is maximized only for a particular convergence angle. This optimum convergence angle varies with the beam diameter, so in order to most efficiently utilize the source brightness over a range of beam diameters, the objective aperture size must be adjustable. The Model 6050 incorporates a five-position variable objective aperture for this purpose. Assuming optimum objective apertures, and neglecting chromatic aberrations, the beam diameter is given by [4]

$$d = \left(\frac{16C_s^{2/3}I}{3\pi^2\beta}\right)^{3/8} \tag{2}$$

Equation (2) is also plotted in Fig. 2. The objective aperture used for the measurement of Fig. 2 is shown to be optimum for a beam diameter of 30 µm, although the current density achievable with this aperture is not much less than the maximum for beam diameters from about 20 µm to about 100 µm. Below 20 µm, however, smaller apertures are required in order to obtain smaller beam diameters, as well as to maximize the current density. Two other aperture sizes can be selected on the Model 6050 for this purpose. These apertures are optimum for the beam diameters indicated in Fig. 2. Hence, the ion gun optics can be optimized either for the small beam diameters required for obtaining SIMS maps or depth profiles over small areas, or for the large beam diameters useful for obtaining depth profiles over wide areas.

References

1. D.G. Welkie and R.L. Gerlach, in Secondary Ion Mass Spectrometry SIMS IV, ed. by A. Benninghoven, J. Okano, R. Shimizui and H.W. Werner (Springer, New York 1984), p. 317.

2. P. Williams, R.K. Lewis, C.A. Evans, Jr., and P.R. Hanley, Anal. Chem. 49, 1399 (1977); H.A. Storms, K.F. Brown, and J.D. Stein, Anal. Chem. 49, 2023 (1977).

3. L. Valyi, Atom and Ion Sources (John Wiley and Sons, London 1977), p. 138, and references therein.

4. T. Mulvey, in Focusing of Charged Particles, Vol. 1, Chapter 2.6, ed. by A. Septier (Academic Press, New York 1967).

Survey of Alkali Primary Ion Sources for SIMS

R.T. Lareau and P. Williams*

Department of Chemistry, Arizona State University, Tempe, AZ 85287, USA

1. Introduction

Since the first studies of negative ion yield enhancement by alkali ion bombardment [1,2], it has been natural to assume that cesium was the primary ion species of choice. Cesium is the most electropositive of the stable alkali metals and produces the lowest work function upon adsorption, implying that its effect on negative ionization efficiencies should be greater than that of the other alkalis. Further, because cesium has the lowest ionization potential of the stable alkali metals, and is the most volatile, it has appeared easiest to generate a Cs^+ ion beam. Ion sources capable of generating stable Cs^+ ion beams of several microamperes for several hundred hours have been used in SIMS since 1977 [3,4]. In 1978, DELINE et al. [5] and later CHELGREN et al. [6] observed that the ion yield for Si^- sputtered by Cs^+ from Pt_2Si was about three orders of magnitude lower than the yield of Si^- from silicon, mainly because the very high sputtering yield of Pt_2Si under Cs^+ bombardment led to a low surface concentration of implanted Cs. From the steep dependence of ionization efficiency on Cs surface concentration in a range of noble metal silicides (the Si^- yield varied with $[Cs]^{3.7}$), it appeared that sputtering with a lighter alkali species, with lower sputtering yield, might give higher ionization efficiencies due to the higher level of incorporated alkali, even though, at the same concentration, the lighter alkalis enhance negative ion yields less than does Cs [7]. The present study compares ion source performance for Na^+, K^+, Rb^+ and Cs^+ beams and yields of Si^- sputtered from a silicon target by these species.

2. Experimental

The ion source design was a modification of the Hiconex thermionic source (General lonex) reported earlier [4]. The source was mounted on the primary mass filter of a Cameca IMS 3f ion microscope and operated at an ion accelerating potential of 10 kV. An accel-decel ion extraction system gave higher ion currents (by a factor of ~5) than those obtained with a simple grounded extraction electrode. With the sample at -4.5 kV for negative ion analysis, the incident ion energy was 14.5 keV. The incident angle under these conditions was ~25° to the sample normal. Primary ion currents were measured using the primary current measurement system of the IMS 3f.

Ion yields (secondary ions detected/incident primary ion) were measured with the same instrument settings, namely with all apertures at their largest value. Measurements were made with an unrastered focussed primary

* U.S. Army, Electronics Technology and Devices Laboratory, LABCOM, Fort Monmouth, New Jersey 07703-5302

ion beam smaller than the image field selected by the largest field aperture (150 µm diameter image field). Prior to each measurement, the surface was sputtered for an extended time with the alkali beam to ensure that the measured ion yields were characteristic of a surface with a steady-state implanted level of the alkali. Sputtering yields were determined by sputtering a crater roughly 250 um square and some thousands of angstroms deep with a rastered ion beam and measuring the crater depth and size using a Dektak 11 surface profilometer.

3. Results and Discussion

Ion source performance for the four alkali metals is shown in Table 1. It may be seen that the source performance improved slightly for the lighter alkalis Rb and K, relative to Cs. This may be due to reduced space-charge broadening of the beam for these species. The lower current for Na^+ is thought to result partly from the lower volatility of sodium, and partly from the higher ionization potential. Interestingly, beam stability for all three lighter alkalis was comparable to that for Cs^+ -- 1% over a 10 min. period.

The efficient performance of the ion source, even for those metals, such as sodium, with ionization potentials higher than the often quoted thermionic work-function of tungsten (4.5 eV) results from the fact that the tungsten frit emitter is polycrystalline; the ionizing surface therefore contains low-index, high work-function planes on which ionization can proceed efficiently. Measurements using ion, rather than electron, emission show that the low index planes of tungsten have work-function values as high as 5.4 eV [8].

Relative values of the sputtering yields, normalized signals (counts/s/nA primary current) and useful yields for silicon are listed in Table I. As expected, sputtering yields were highest for Cs^+ impact, and lowest for Na^+. It may be seen that the ion signals are highest for Cs^+ bombardment, and decrease significantly with decreasing alkali mass. The variation in useful yield is smaller. Also shown in Table I is data from FRENTRUP et al. [7] who have compared the effects of implanted alkalis on Si^- yields. Their comparisons were made at low concentrations (up to ~2 atom %) and unfortunately an O_2^+ primary beam was used, the enhancing effect of which may mask the alkali effect. They found a linear dependence of enhancement on alkali concentration, in contrast to ref. 5 where, however, the matrix and the alkali concentration varied together. The fact that the variation in our normalized signals is roughly similar to the data

Table I

Relative sputtering yields and negative ion yields for silicon under alkali bombardment

	Cs	Rb	K	Na
Maximum current (µA)	0.8	1.0	1.3	0.3
Sputter Yield	4.8	3.0	2.2	1.0
Useful Si^- Yield	4.8	2.0	1.7	1.0
Normalized Signal	23.0	6.1	2.3	1.0
Alkali-enhanced Si^- Signals (From Ref. 7)	23	9	5	1

from ref. 7 suggests that a similar linear dependence holds in the present work.

4. Conclusion

We have found that a cesium ion source can also produce ion beams of Rb^+ and K^+ (with somewhat improved efficiency compared to Cs^+) and Na^+ (with lower efficiency). The improved performance with the lighter alkalis may result from lower space-charge spreading. Useful ion yields for Si^- sputtered from silicon with the various alkalis decrease monotonically with increasing alkali atomic number, as do the substrate sputtering yields. The results indicate that the effect of increased alkali surface concentration resulting from the lower sputtering yields of the light alkalis is not sufficient to compensate for the lower efficacy of the less electropositive alkalis in enhancing negative ion yields.

Acknowledgement

This work was supported by the National Science Foundation Grant DMR 8206028. Construction of the ion source was funded by a Grant-in-Aid from Arizona State University.

References

1. V. J. Krohn: Appl. Phys. 33, 3523 (1962)
2. C. A. Andersen: Int. J. Mass Spectrom. Ion. Phys. 3, 413 (1970)
3. H. A. Storms, K. F. Brown, J. D. Stein: Anal. Chem. 49, 2023 (1977)
4. P. Williams, R. K. Lewis, C. A. Evans, Jr., P. A. Hanley: Anal. Chem. 49, 1399 (1977)
5. V. R. Deline, W. Katz, C. A. Evans, Jr.: Appl. Phys. Lett. 33. 832 (1978)
6. J. E. Chelgren, W. Katz, V. R. Deline, C. A. Evans, Jr., R. J. Blattner, P. Williams: J. Vac. Sci. Technol. 16, 324 (1979)
7. W. Prentrup, J. Griepentrog, G. Kreysch, U. Muller-Jahreis: Phys. Stat. Sol. (a) 84, 269 (1984)
8. F. L. Reynolds: J. Chem. Phys. 39, 1107 (1963)

Non-Oxygen Negative Primary Ion Beams for Oxygen Isotopic Analysis in Insulators

R.L. Hervig and P. Williams

Department of Chemistry, Arizona State University, Tempe, AZ 85287, USA

1. Introduction

The use of an O^- primary ion beam to sputter insulators for secondary ion mass spectrometry (SIMS) is widespread. Negative ion sputtering delivers negative charge to the sample which can be carried away by secondary electrons. In contrast, positive ion bombardment leads to severe charging [1]. Analysis of negative secondary ions is particularly difficult with positive primary ions even using an electron gun, because the secondary electrons are accelerated away from the sample, adding to the charging problem. For oxygen analysis, the use of negative secondary ions is to be preferred, because positive sputtered ion yields for oxygen are low and the heavy isotopes of oxygen have very low abundance. However, the use of O^- primary ions significantly impedes analysis of oxygen isotope ratios because the primary species dilutes the oxygen in the sample. Thus, it seemed useful to investigate ion-beam production from other eletronegative species for oxygen analyses.

2. Experimental

The work was carried out in a Cameca IMS 3f ion microanalyser, equipped with a standard duoplasmatron ion source and a primary beam mass filter. The duoplasmatron incorporated design features [2] allowing maximum currents of 1-2 µA O^- to be obtained by eliminating the plasma expansion cup, and restricting the intermediate electrode aperture to a diameter of 1/16" (with a channel length of 1/8"). All primary ion currents reported here were measured using the primary beam Faraday cup assembly of the IMS 3f. Independent measurements using a Faraday cup at the sample position showed that the IMS 3f negative primary ion measurements were systematically high by 10-15%, presumably due to extraction of sputtered positive ions from the Faraday cup by the secondary electron suppression voltage.

3. Production of F^- and CN^-

Following REUTER [3], who obtained F^+ beams from a freon-oxygen mixture, we first investigated the production of F^- using a CF_4-O_2 mixture. Although we obtained a stable F^- beam, varying ^{18}O signals from silicates indicated that the primary beam was contaminated by $^{18}OH^-$. The same problem resulted from a CF_4-Ar mix; presumably from oxygen generated by the reaction of fluorine with oxide coatings in the duoplasmatron. We concluded that F^- is unacceptable for ^{18}O analysis.

As an alternative to fluorine, we have investigated the production of a CN^- beam. For safety reasons, HCN was ruled out as a feed gas; instead, we used a mixture of 25%CO_2:75% compressed air. This gave a maximum of 300 nA of CN^-, although 50 - 100 nA was more common. At the same time, an abundant O^- beam was obtained (maximum of 800 nA) so that conventional analyses could utilize this species without changing source conditions. Initially, we had problems with primary beam stability. By testing several mixtures we found the most stable CN^- beams were achieved with relatively high ratios of N_2/CO_2.

A more serious problem was that CN^- generation resulted in severe corrosion of the duoplasmatron. Operation for 24 hours eroded the aperture in the intermediate electrode to more than twice its initial diameter and large whiskers grew from the electrode to the anode.[1] We have solved this problem by designing a new intermediate electrode to protect against such attack. A mild steel electrode was machined with a large aperture (0.2" dia.) and a Macor machinable ceramic insert with a 1/16" aperture was pressed inside this. The magnetic electrode with an insulating, non-magnetic insert thus had a large magnetic aperture, but the optimum physical aperture. This design worked surprisingly well for routine operation with O_2^+ and Ar^+. Currents of 10 μA and 7 μA, respectively, were obtained. Currents for CN^- were lower with the new design, but degradation of the duoplasmatron was substantially reduced. We now use a mixture of CO_2 and dry N_2, giving maximum currents of 75 nA.

4. Preliminary Results on Oxygen Isotope Studies

For analyses, the CN^- current was reduced to 1-5 nA, with a 20-40 μm spot size. The beam was unrastered. Negative secondary ions were analysed from a 10 μm diameter area within the sputtered crater. The insulating samples were coated with a thin layer of gold-palladium alloy.

Analysis of oxygen in silicates with the CN^- beam gave much more stable secondary ions than were obtained under F^- bombardment. Intensities were high: $>10^7$ c/s/nA for $^{16}O^-$ and $>2 \times 10^4$ c/s/nA for $^{18}O^-$. Because our instrument has not yet been optimized for precise isotope ratio work, we have not yet attempted to determine absolute abundances, but have determined ratios of ^{18}O, measured on the electron multiplier, to ^{16}O, measured on the IMS 3f Faraday cup. The Faraday cup measurement was corrected for the offset current, but no attempt was made to obtain an absolute comparison of the multiplier and electrometer measurements. The mean of four analyses at different points on a terrestrial olivine had a standard deviation of 0.3% in the $^{18}O/^{16}O$ ratio. The standard deviation of the mean of 24 determinations of the 18/16 ratio in a single 20 minute analysis in which the integral ^{18}O signal was $>10^6$ counts was 0.1% - 0.15%.

Several problems may act to degrade reproducibility and accuracy in isotope ratio measurements:

a) Instability of secondary ion signal due to sample charging. This may not be critical, in that both isotopes should be similarly affected. With the CN^- beam, gross charging of metal-coated insulators, even for analysis of negative secondary ions, was small (20 - 30 V) for small spot sizes (20 - 40 μm) and currents (<5 nA) and stabilized in a few minutes after sputtering through the coating. Changing the sample potential to compensate for the charging maximizes the secondary current; however, differential charging across the bombarded region can produce a lateral electric field at the ion emission point which is a potential source of

[1]CAUTION: Operation with gases such as CO_2 can result in the generation of toxic $Ni(CO)_4$. Rotary pumps should be vented into a fume hood.

discrimination effects if the energy or angular distributions of the two isotopes are not identical.

b) Molecular interferance of $^{16}OH^-$ with $^{17}O^-$ and of $^{17}OH^-$ with $^{18}O^-$. The $^{17}OH^-$ contribution to the mass 18 signal becomes negligible when the primary column is baked out for ~ 18 hours prior to analysis. However, the contribution of $^{16}OH^-$ to the weak $^{17}O^-$ signal can be considerable, varying in this work from 1 to 5 times the $^{17}O^-$ signal. Precise determination of the ^{17}O abundance by SIMS appears difficult at this time.

c) Precise determination of count rates for two isotopes with grossly different abundances. For precision of 0.1% in the oxygen isotope ratio, the integrated signal at mass 18 must be > 10^6 counts. This makes the count rates at mass 16 extremely high and leads to count-loss problems. Measurement of the higher current using an electrometer raises difficulties of comparison between two distinct measurement systems. Eventually, the problem must be solved by increasing ion currents to the point where both isotope signals can be precisely measured with an electrometer.

d) Isotope fractionation in the ion emission process. Errors due to fractionation effects can be minimized by comparing $^{18}O/^{16}O$ ratios with those in a standard of closely similar chemical composition.

5. Conclusion

The minor modification of an insulating aperture in the intermediate electrode of a duoplasmatron allows stable operation with corrosive gases producing non-oxygen negative ions such as CN^-. Significantly, the presence of a large magnetic hole in the intermediate electrode gives enhanced production of positive primary ions, while the small mechanical aperture maintains good performance for negative ions. The use of CN^- primary ions to sputter insulating samples allows the detection of large O^- signals with only minor sample charging, and no contamination of the oxygen isotopic ratio by primary beam species. In preliminary measurements of the $^{18}O/^{16}O$ an internal precision of 0.1% (standard deviation) has been achieved.

Acknowledgements

This work was supported by the National Science Foundation Grant DMR 8206028.

References

1. Werner, H. W. and A. E. Morgan, J. Appl. Phys. 47 (1976) 1232
2. Williams, P. Unpublished
3. Reuter, W. in "Secondary Ion Mass Spectrometry, SIMS IV", eds. A. Benninghoven, J. Okano, R. Shimizu, H. W. Werner, Berlin: Springer 1984, p. 54

A New SIMS Instrument: The Cameca IMS 4F

H.N. Migeon, C. Le Pipec, and J.J. Le Goux
CAMECA, 103 Boulevard Saint-Denis, B.P. 6,
F-92403 Courbevoie Cedex, France

The Cameca IMS 4F is the successor to the IMS 3F. This new instrument contains all the features of its predecessor (1) plus improvements of the basic instrument as well as new accessories.

1.0 Basic instrument :

1.1 Dynamic transfer system :

The static transfer system of the IMS 3F optimizes the collection efficiency for three different analyzed areas. At constant mass resolving power the collection angle increases when the size of the analyzed area decreases (1).

In the dynamic transfer system, the deflection plates have their rotation center placed in the cross-over plane located between the immersion lens and the transfer lenses. This deflection system is synchronized with the primary beam scanning system. Thus the secondary ions emitted by a rastered primary beam will follow the trajectories corresponding to a static primary beam after this deflection system (2). This dynamic transfer system permits therefore the analysis of an area which is continuously adjustable up to 100 % of the rastered area using a transfer lens optimized on a much smaller field of view. Figure 1a shows a scanning ion image obtained with the IMS 4F using a transfer lens optimized on 65 µm without energizing the deflection plates. As shown on figure 1b the use of this deflection system increases the size of the analyzed area while maintaining the same collection efficiency.

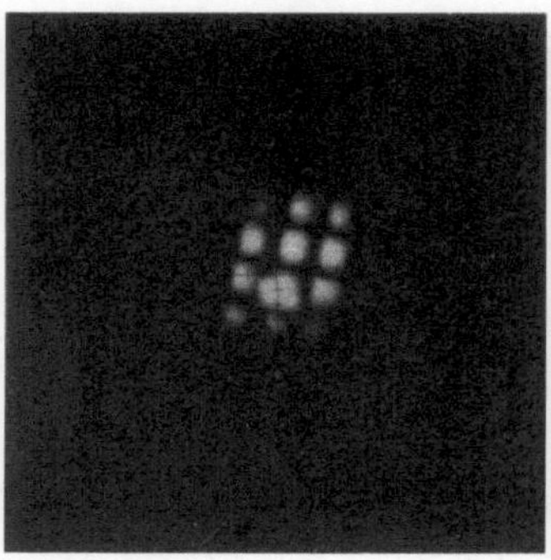

Fig. 1a

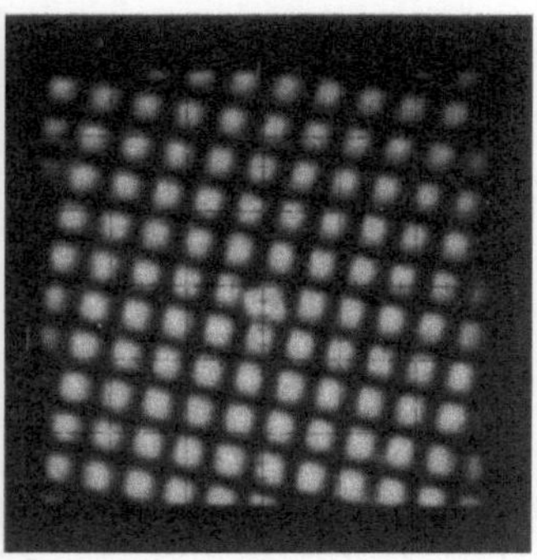

Fig. 1b

1.2 Improved detection limits of vacuum contaminants :

Compared with the IMS 3F the detection limits for H, C, N, O... are improved by about a factor of 10 thanks to an improved pumping efficiency in the primary column and the use of a liquid nitrogen cold trap surrounding the specimen.

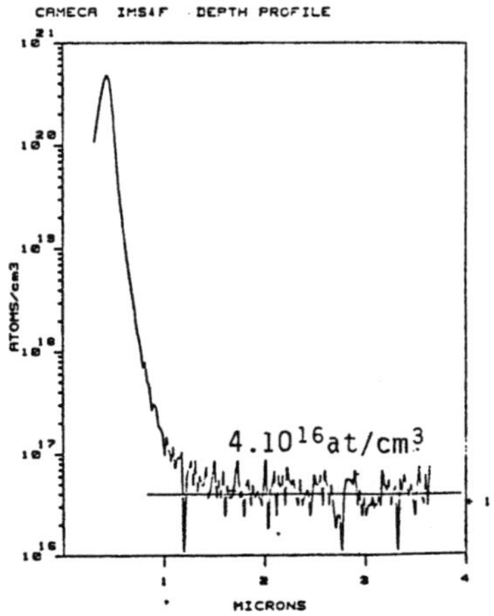

Fig. 2 is a depth profile of hydrogen in silicon showing a detection limit of 4×10^{16} at/cm^3

1.3 Elimination of instrumental background :

In the IMS 3F the detection of Cr, Fe... was limited by sputtering of stainless steel material in the instrument. A collaboration with Melle HUBER (4) allowed us to identify the sources of this contamination. In the IMS 4F, the last lens of the primary column as well as the immersion lens and entrance slit are now made of tantalum or titanium with a platinum coating. The detection limit of chromium in gallium arsenide is 5×10^{12} at/cm^3 instead of 5×10^{14}.

2.0 Accessories :

2.1 Cesium microbeam source :

This source, developed in collaboration with Slodzian, Daigne and Girard, and installed on the IMS 4F demonstrated the following preliminary results : a maximum intensity of 500 nA, a stability over 10 minutes better than 1 % and a smallest beam diameter less than 0.5 µm.
Figure 3 shows scanning ion images obtained on a gold copper sandwich, the width of the 7 gold bars decreasing progressively from 4.2 to 0.16 µm.

2.2 Normal incidence electron gun :

This new electron gun developed by Slodzian (5) allows the analyses of positive and negative secondary ions in insulating samples.

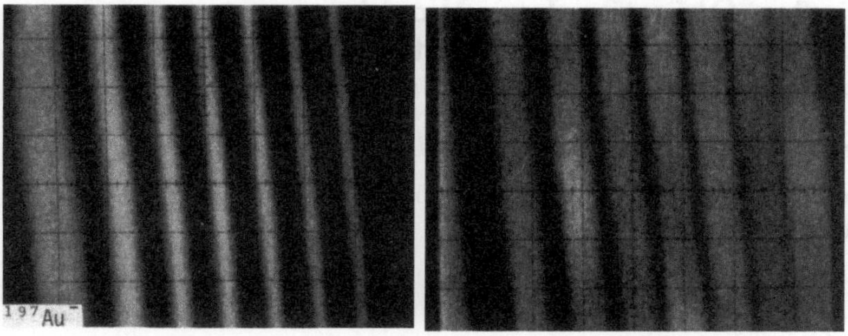

Fig. 3 Cesium microbeam source - primary beam intensity : 10 pA

2.3 Immersion lens strip :

Four interchangeable positions of the front plate of the immersion lens allow a factor of 10 reduction in the memory effect.

2.4 Scanning ion image :

Application examples of this system are shown in figures 1 and 3.

References

1. G. Slodzian, Nat. Bur. Stand. U.S. Spec. Publ. 427 (1975)
2. G. Slodzian, in Applied Charged Particule Optics, A. Septier, ed., Academic Press (1980).
3. H. Liebl, in Low-Energy Ion Beams (1977), K. G. Stephens et al., eds Conf. Ser. 38, the Inst. of Phys. (1978).
4. A. M. Huber, G. Morillot, these proceedings.
5. G. Slodzian, M. Chaintreau, R. Dennebouy, these proceedings.

The Emission Objective Lens Working as an Electron Mirror: Self Regulated Potential at the Surface of an Insulating Sample

G. Slodzian, M. Chaintreau, and R. Dennebouy

Laboratoire de Physique des Solides, Bât. 510, Université de Paris-Sud, F-91405 Orsay, France

1. Introduction

In the ion microanalyzer, when the sample is polarized with a negative voltage, -V, negative ions and secondary electrons are simultaneously collected and focused by the emission objective lens. With insulating samples, the departure of these negative particles causes a positive charge to appear on the surface. After the onset of the primary beam, this positive charging-up will rapidly increase up to the occurrence of an electrical breakdown or the deflection of the primary beam. Occasionally, an equilibrium situation can be reached (small surface conductivity, capture of secondary electrons from neighboring conducting areas,...) but even in this case the surface of the insulator is far from being equipotential, so that the nice optical focusing of the objective lens is completely disturbed, preventing image formation and efficient collection of secondary ions. One should note that because the yields of secondary electrons produced by the primary ion beam are much larger than one, bombarding with neutral or negative particles will not avoid the charging-up problems.

2. Experimental Set-up and Main Operating Features

To prevent the sample charging-up and create a nearly equipotential surface on the insulating sample, the emission lens has been made to work as an electron mirror acting on an auxiliary electron beam directed towards the surface through the objective lens. The experimental set-up is shown in figure 1.

The electron gun delivers a beam at an energy eV. The beam goes through a small magnetic prism where electrons are deviated by an angle of 90° so that their mean trajectory is made to coincide with the optical axis of the objective lens. Reflected electrons suffer a symmetrical deviation, and are presently used for alignment purpose. Since secondary ions are slightly deflected by the prism, two additional adjustable magnets are aimed at correcting this deflection. The magnetic prism provides bidirectional focusing, which allows the electron beam stemming from the gun cross-over to be focused in the cross-over plane of the objective lens located at point C on the axis.

Now, let's consider an ideal equipotential plane, -V, in place of the sample surface. The cross-over of the emission lens plays the part of the optical center of the electron mirror, which means that if the incident auxiliary beam is focused to a point on C, the electrons will spread out into a parallel beam near the surface, "land" with zero energy on the equipotential -V and then, trace their way backwards along the same trajectories. There is no room for going into a more detailed description of the optical properties. However, it should be pointed out that surface relief, local work function, insulating inclusions,... would have to be taken into account when dealing with the reflection of electrons on real surfaces. In a first step, when the goal is to produce a swarm of low-energy electrons near the sample, there is no need to go into such details.

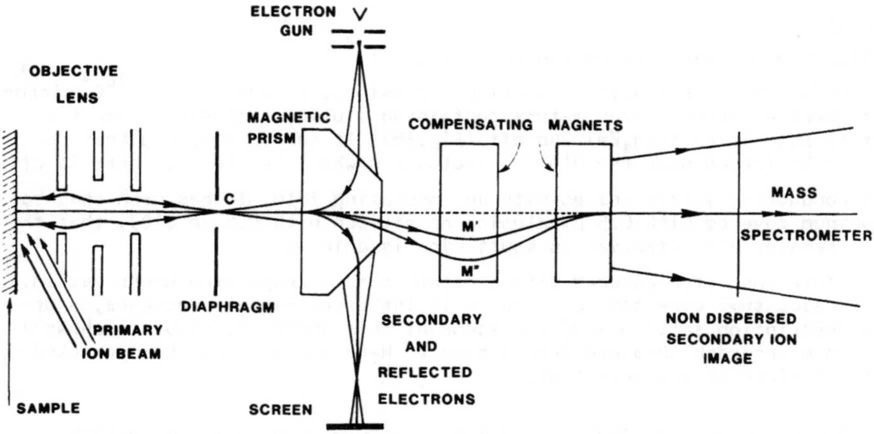

Fig. 1 Schematic diagram. M and M' represent two different secondary ion masses ; their magnetic deviations are greatly exaggerated

3. Charge Balance at the Surface

It is generally acknowledged that the yield γ of secondary electrons produced by primary electrons depends on the impact energy E_p : γ starts from zero, goes through a maximum higher than unity and then decreases. As a result, γ is equal to one for two values of E_p, E_I and E_{II} ; E_I is of the order of a few tens of electron-volts and E_{II} is usually higher than one keV [1].

Now, let's take a conducting sample covered with an insulating film of constant thickness whose free surface is at potential $-V_i$; $E_p = V-V_i$ (in electron-volts). If $E_p < E_I$, the sample becomes negatively charged (since $\gamma < 1$) and the amplitude of the negative charge increases until the surface reaches a potential $-V$; then, the electrons are reflected since they have been produced at the same potential in the electron gun. It can be seen that the negative charge distributes itself so as to produce an equipotential surface on the insulating film. When the electrons hit the surface with $E_{II} > E_p > E_I$ the sample takes a positive charge (since $\gamma > 1$) which increases with time as well as the impact energy. This increase is stopped when E_p becomes equal to E_{II}, if a breakdown has not damaged the film in the meantime. For an homogeneous film, the surface potential would be $-V_i = -V + E_{II}$. For heterogeneous samples E_{II} varies from place to place, so that the whole equilibrium situation would have to be reconsidered. This situation is not, in general, suitable for well-controlled ion collection and ion microscopy.

Once the equilibrium situation corresponding to $E_p < E_I$ has been established, the primary ion beam is set on. Negative sputtered ions and ion-induced secondary electrons leave behind them a positive charge easily cancelled by the reflecting electron beam. Whatever the local yield of secondary particles, the potential is self-regulated at each point of the surface. Of course, the surface density of available low-energy electrons sets a limit to the density of the primary ion beam that can be used. This limit mainly depends upton the yield of secondary electrons under ion impact.

4. Results

Experiments have been made on typical samples :

(i) a conducting matrix with insulating inclusions. Figure 2 shows $^{16}O^-$ micrographs obtained from an iron matrix containing alumina inclusions when the auxiliary beam is on (Fig.2a) and off (Fig.2b). It can be observed that no image can be formed near the alumina inclusions when the electron beam is off.

(ii) a conducting matrix coated with an insulating film. It has been checked on aluminum covered with 0.5 µm alumina or silicon with 0.5 µm SiO_2, that the surface remains equipotential as sputtering is going on.

(iii) a bulk insulator covered with a metallic grid evaporated under vacuum. In this case, some care has to be taken in the experimental procedure, a detailed description falls beyond the scope of this abstract. Figure 3 shows a $^{35}Cl^-$ ion micrograph obtained from a basalt. Here again it has been checked that the surface is equipotential.

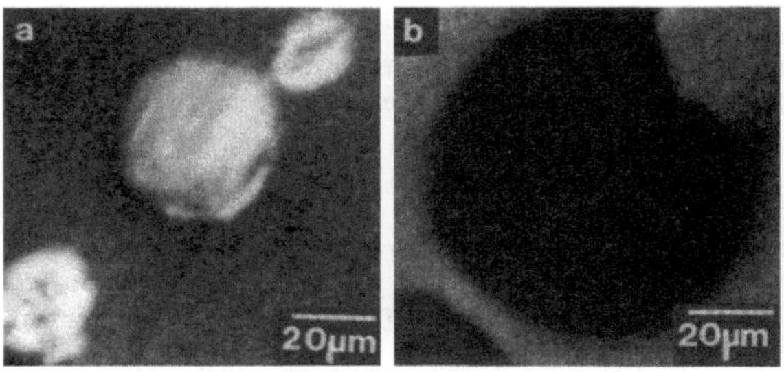

Fig. 2 $^{16}O^-$ from an iron matrix with alumina inclusions ; a) electron beam, b) beam off

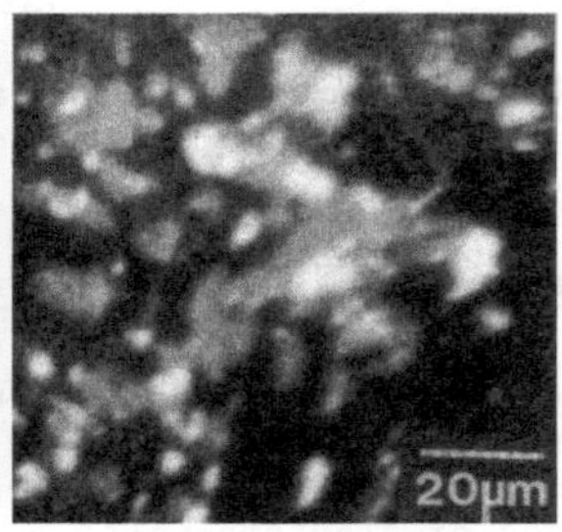

Fig. 3 $^{35}Cl^-$ from a basalt

5. Conclusions

By using an auxiliary electron beam and making the emission lens work as an electron mirror it is possible to create and maintain an equipotential surface on insulators being bombarded by a primary ion beam. This can be achieved under the experimental conditions which are the most appropriate for efficient collection of negative secondary ions and direct ion imaging.

The main advantages of secondary ion emission - ion imaging, in-depth profiling, isotopic ratio measuring,...- which, with the ion microanalyzer, have been restricted to positive ions on most insulators,become available now when using negative ions on all types of insulating samples. This is of crucial importance for those elements whose ionization yields are much higher in the negative mode,and it could concern up to about 30% of the periodic table if alkaline vapor jets or primary alkaline ions were used to enhance the ionization yields of negative species.

7. References

1. B. Gross, H. von Seggern, and J.E. West:J. Appl. Phys. 56 (8), 2333 (1984)

20 K-Cryopanel-Equipped SIMS Instrument for Analysis Under Ultrahigh Vacuum

Y. Homma and Y. Ishii

NTT Musashino Electrical Communication Laboratories, Musashino-shi, Tokyo 180, Japan

1. Introduction

The detection of trace H, C, and O by SIMS has been hampered by residual gas backgrounds in the SIMS sample chamber vacuum. Detection limits are generally of the order of 10^{18} atoms cm^{-3} for H and O in Si, and 10^{17} atoms cm^{-3} for C in Si [1,2]. To reduce the contaminating gas backgrounds, an ultrahigh vacuum (UHV) is required during analysis.

An effective method for achieving the UHV condition is to place a pumping system having a high pumping speed near the sample. We have recently developed a cryopanel pumping system placed around the sample in the SIMS chamber [3]. In this work we have investigated characteristics of this UHV system.

2. Vacuum System

Cryopanels made of copper were installed in the sample chamber of two types of instruments, a Cameca IMS 3F and an Atomika A-DIDA 3000. The cryopanel is cooled to < 20 K by a helium refrigerator (refrigeration capacity: 2 W). The cryopanel in IMS 3F is held in contact with the 12 K-refrigerator cold head by means of a copper rod 200 mm in length and 30 mm in diameter. The surface area of the panel is 383 cm^2. The area and shape of the panel is restricted because of the small space of the IMS 3F sample chamber.

A larger cryopanel is installed in the A-DIDA sample chamber. The panel is held in direct contact with the 12 K-cold head. The panel has a cylindrical shape with a surface area of about 790 cm^2. To reduce the thermal radiation from the sample chamber wall, a second panel is placed between the 20 K-panel and the chamber wall. The second panel temperature is about 80 K. A bakeable-type refrigerator is employed, and the nonbakeable components in A-DIDA have been removed or replaced: a metal sealed gate valve is used for the sample transfer system and primary ion column. Thus the system is entirely bakeable to 250°C. The total pressure in the A-DIDA chamber was measured with a Seiko I&E Modulating Ion Current Pressure Gauge [4] connected to the chamber by means of a 70 mm ConFlat flange.

3. Pump-Down Characteristics and Detection Limits

The sample chamber of the standard IMS 3F is pumped with a 680 l/s cryopump and its base pressure is $\sim 1 \times 10^{-7}$ Pa. The effect of the cryopanel on the H, C, and O backgrounds is evaluated using this instrument. Figure 1 shows the background secondary ion intensity variations for Si under Cs^+ bombardment after the refrigerator started. A step is observed in both the H^- and O^- background curves at the early stage of cooling. This step corresponds to H_2O condensation on the cryopanel. The background signals then gradually drop to 1/10 of the initial levels when the cold head is cooled to 14 K. It

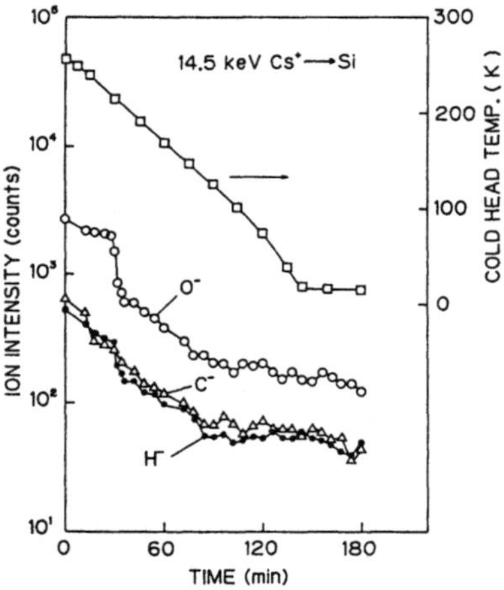

Fig.1 Variations of H, C and O backgrounds in cryopanel-equipped IMS 3F

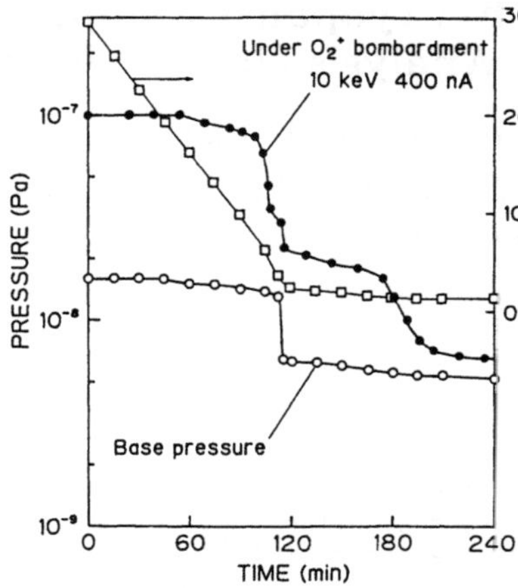

Fig.2 Pump-down curves of cryopanel in A-DIDA

should be noted that the vacuum in the IMS 3F chamber is rather poor because Viton gaskets are used,and only a moderate bake-out (<150°C) is possible. In this case the cryopanel effectively works out because of its high pumping speed. However, in the IMS 3F, it was difficult to further improve the capture efficiency for the residual gases,since the cryopanel configuration was largely restricted by the small space of the sample chamber.

In the A-DIDA, after a 250°C bake-out for 48 h, the total pressure in the sample chamber was 1.5×10^{-8} Pa by pumping down with a 220 l/s ion pump. Afterwards, the refrigerator started. Typical pump-down curves of the cryopanel are shown in Fig.2 both for the base pressure and the pressure under O_2^+ bombardment. The base pressure can be lowered to 4×10^{-9} Pa 7 h after starting the refrigerator. Since the system was entirely baked out, no pressure reduction due to condensation of H_2O occurred. A large pressure step is observed in the pump-down curve under O_2^+ bombardment at 60 K of the cold head temperature. This step corresponds to O_2 condensation. The following small step, which is also observed in the base pressure curve, may correspond to CO condensation. After the final dull step at low temperature, the pressure under O_2^+ bombardment becomes 7×10^{-9} Pa. Thus, the cryopanel is also effective for C and O background reduction in the UHV system. The main residual gas component in the A-DIDA chamber is thought to be H_2 since H_2 is a noncondensable gas at 14 K. However, it is reported that H_2 does not contribute to the H background in SIMS analysis [1].

The detection limits (background levels) of H, C, and O in Si under Cs^+ bombardment were measured using the IMS 3F. Under the existence of residual gases, background levels are determined by the removal rate of sample surface due to primary ion sputtering as well as the residual gas pressure [3]. The depth profiles of H, C, and O implanted in Si are presented in Fig.3 showing the detection limits of these elements under both high and low sputtering rate conditions. The detection limits under the high sputtering rate (~16 nm/s) using the ~7 mA/cm^2 Cs^+ beam for H, C, and O in Si are 1×10^{17}, 1×10^{16}, and 3×10^{16} atoms cm^{-3}, respectively. Even in the low sputtering rate (~0.7 nm/s) using the ~0.3 mA/cm^2 Cs^+ beam, detection limits of 2×10^{18}, 1×10^{17}, and 3×10^{17} for H, C, and O respectively can be obtained.

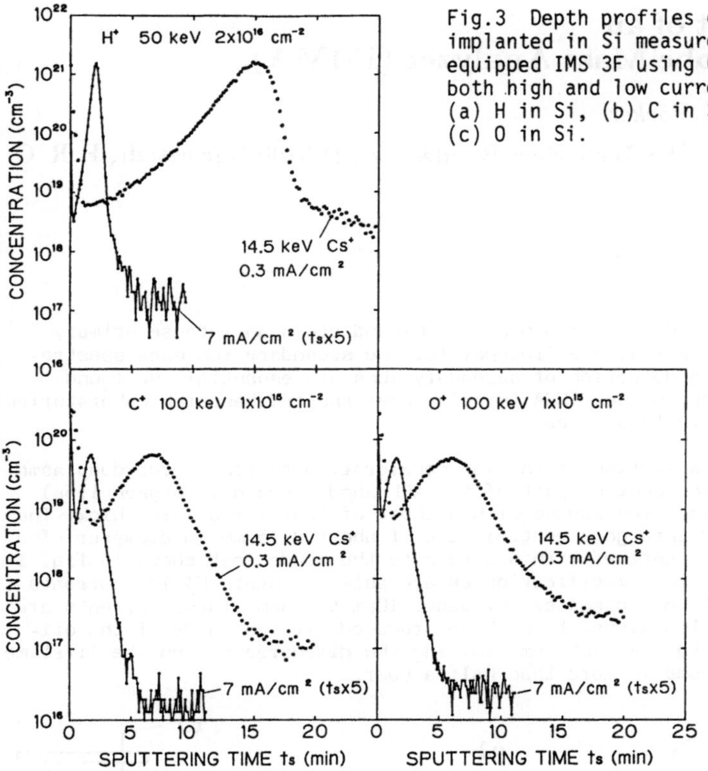

Fig.3 Depth profiles of H, C, and O implanted in Si measured by cryopanel equipped IMS 3F using Cs^+ beams of both high and low current densities; (a) H in Si, (b) C in Si, and (c) O in Si.

4. Conclusions

We have shown that the 20 K-cryopanel has a large pumping capability. An extremely low pressure during analysis of 7×10^{-9} Pa is achieved. The 20 K-cryopanel pumping system can reduce not only the H background due to H_2O condensation but also the C and O backgrounds due to condensation of low condensation temperature gases such as CO and O_2.

References

1. C. W. Magee and E. M. Botnick, J. Vac. Sci. & Technol. 19, 47 (1981).
2. K. Wittmaack, Nucl. Instrum & Methods 218, 327 (1983).
3. Y. Homma and Y. Ishii, J. Vac. Sci. & Technol. A3, 356 (1985).
4. F. Watanabe, S. Hiramatsu, and H. Ishimaru, J.Vac.Soc.Jpn. 25, 506 (1982)

Improvement of an Ion Microprobe Mass Analyzer (IMMA)

K. Miethe and A. Pöcker

Forschungsinstitut der Deutschen Bundespost, D-6100 Darmstadt, F.R.G.

For the analysis of dopant elements in semiconductors an intense primary ion beam, high transmission efficiency for the secondary ion mass spectrometer and sensitive detection of secondary ions are essential. We found that the performance of our IMMA (manufacturer Applied Research Laboratories) could be improved in these areas.

Figure 1 shows a section of the ion acceleration region at the duoplasmatron. We cut off the central part of the ARL anode (shown in broken line) and defined the discharge regime with a sheet of tantalum 0.1 mm thick. The ions are extracted through a central hole of about 0.15 mm in diameter. The sheet is fixed and centered with 4 screws to the anode (not shown in fig. 1 for clarity). With this modification we are able to obtain O_2^+ ion currents of 30nA to 80nA at beam diameters of about 10µm to 20µm. These currents are about a factor of two higher than those produced with an anode of the original layout. Due to the tantalum limiting the discharge region, the lifetime of the modified anode is more than half a year.

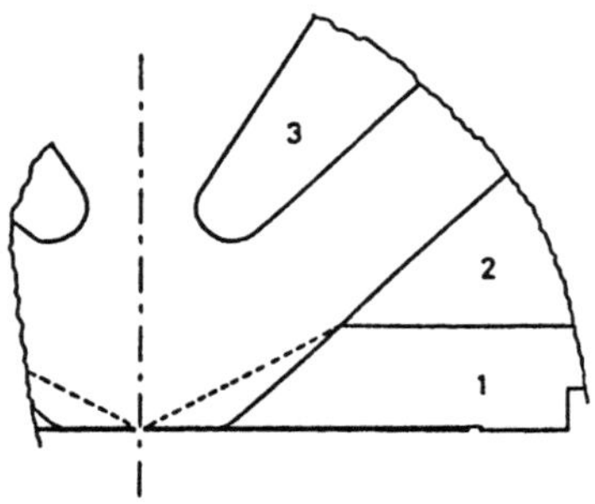

Fig. 1 Anode geometry
1 duoplasmatron anode
2 anode mounting plate
3 acceleration electrode

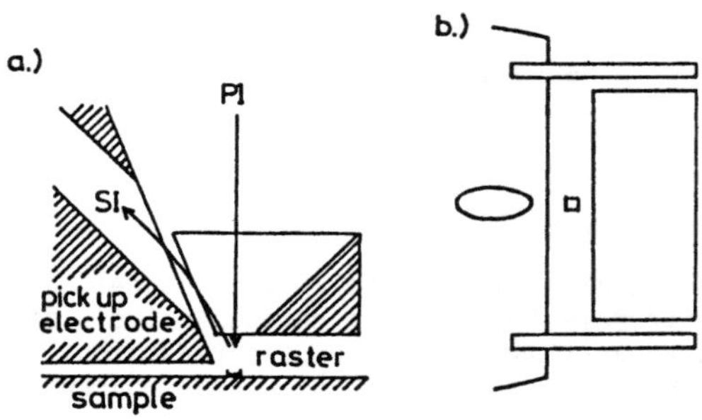

Fig. 2 Secondary ion extraction
a) section with schematic trajectories of primary and secondary ions
b) view from top

As BEDRICH et al. [1] pointed out, the secondary ion extraction of the IMMA is very sensitive to slight misalignments of the sample. There are no means to influence the trajectories of the secondary ions on their path from the sample to the entrance hole of the spectrometer. To overcome this lack we installed additional deflection plates in front of the secondary ion pick-up, as shown in fig. 2.

The voltage applied to the additional electrodes is taken from the sample voltage by resistor dividers. The increase in transmission was about 30%

compared to the best values obtained without the plates. We attribute this relatively low increase to poor focussing of secondary ions at the hole in the pick-up electrode in both cases with and without plates. However, it turned out that the distance from the sample to the pick-up electrode has lost its strong influence on the secondary ion intensities, and that slight inclinations of the sample could easily be corrected. By this way, results from different areas of a large sample or from different samples are now more comparable and reproducible.

To check the performance of the Daly-type ion detector, the high voltage of the photomultiplier and the level of the discriminator for the multiplier pulses were adjusted to yield high signal levels and a good signal-to-noise ratio. The background count rate of the ion detector is about (1-2)c/s. To check the efficiency of the detector, we measured the ion currents at the ion electron conversion target, applying a negative voltage of 100V to the scintillator to suppress the secondary electrons produced at the target. In order to assure measurement of correct ion currents at the conversion electrode, we replaced the whole ion detector with a Faraday cup. With a steel sample we obtained a signal of 2.1pA at mass 56 per nA O_2^+ primary current with fully open slits in the secondary ion mass spectrometer. With the Daly detector in position, we measured ion currents at the target electrode according to this relation at the same sample. Comparing ion currents measured with the corresponding count rates, we estimate a detection efficiency of 0.45 ± 0.05 and a dead time of 43ns for the pulse-counting mode.

As an example of the performance of our instrument, fig.3 shows the depth profile of chromium implanted into GaAs. The dynamic range of about three orders of magnitude is limited, due to a neutral beam produced in the primary lens column, which is mounted in a closed tube and therefore insufficiently pumped.

To analyze electronegative elements with better sensitivity, we installed a cesium ion source developed by LIEBL [2] in place of the duoplasmatron. The source is essentially of the same type as reported at the SIMS III

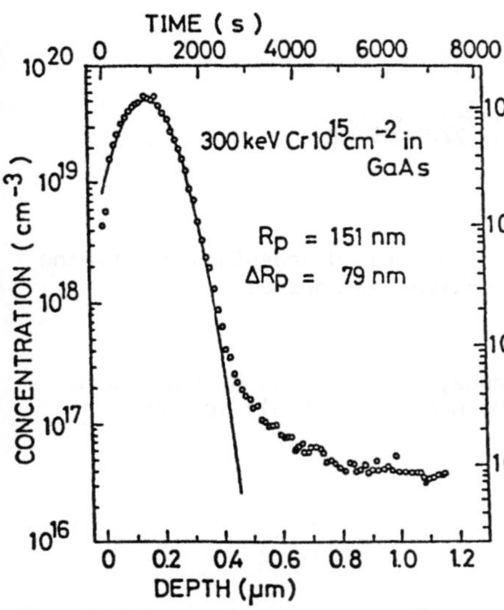

Fig. 3 Primary ions 18.5keV O_2^+

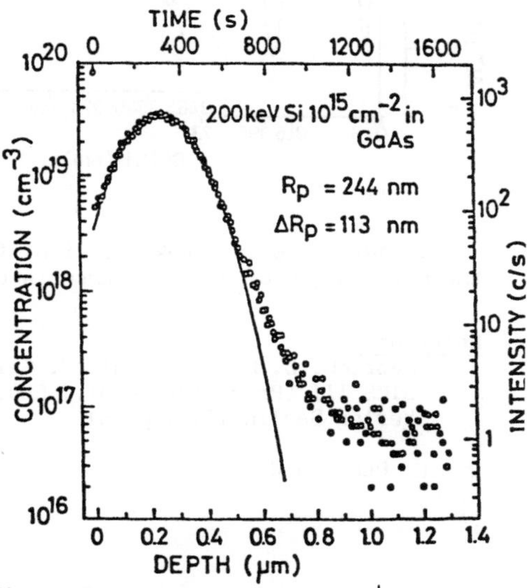

Fig. 4 Primary ions 5.5keV Cs^+

Conference, with the addition of an electrostatic lens and deflection plates mounted onto the source. The cesium ions are accelerated to 4keV and mass analyzed. They impact on the sample surface at 5.5keV for the analysis of negative secondary ions. We obtained cesium primary beam spots of (10 - 15)μm diameter with (20-30)nA current.

The low energy of (4-5.5)keV in the secondary ion extraction region results in considerable bending of the cesium primary ions away from the pick-up electrode. The distance between pick-up electrode and primary beam impact location increased to 1.5mm instead of 0.8mm in the case of (20-18.5)keV O_2^+ ions (fig. 2). With the help of the additional deflection plates, however, it was possible to deflect the secondary ions through the hole in the pick-up electrode into the spectrometer. Figure 4 shows a depth profile of silicon implanted into GaAs. The dynamic range and the detection limit in this case are limited by a mass interference at mass 28.

We checked the depth resolution of our cesium beam by taking an aluminum depth profile of a GaAlAs/GaAs superlattice with layers of 8 nm GaAs and 23.5nm GaAlAs. The aluminum signal measured up to a total depth of 0.8μm exhibits fluctuations like a damped oscillation as displayed in Fig. 5. Defining the depth resolution as distance between signal variation from 0.16 to 0.84 we estimate the depth resolution at the surface to be 6nm, increasing to 20nm at a depth of 0.8μm. We attribute this increase to a slightly inclined bottom of the crater.

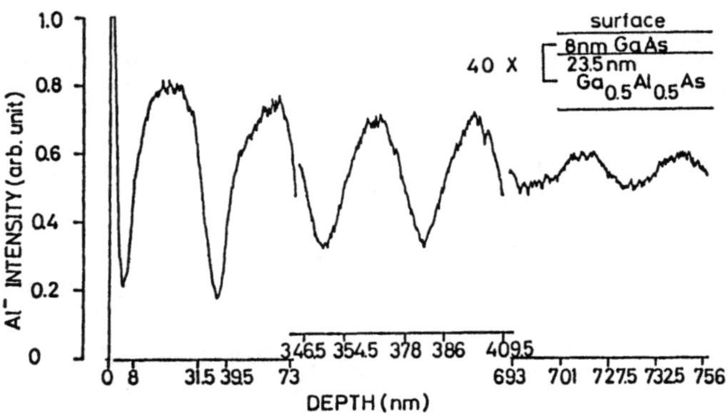

Fig. 5 Aluminum depth profile of a superlattice, primary ions 5.5keV Cs

We would like to thank Dr. Liebl for many helpful discussions concerning the performance of our instrument, and the cesium ion source.

References

1 W. Bedrich, B. Koch, H. Mai, U. Seidenkranz, H. Syhre and R. Voigtmann, SIMS III (Proc. Int. Conf., Eds. A. Benninghoven et al) Springer Verlag Berlin 1979, p. 81

2 B.L. Bentz and H. Liebl, ibid, p. 30

A High-Performance Analyzing System for SIMS

Cha Liangzhen, Xue Zuqing, Rao Ziqiang, Liu Weihua, and Tong Yuqing
Department of Radio Electronics, Tsinghua University,
Beijing, Peopl's Republic of China

Increasing the overall transmission of a secondary ion-analyzing system is the key to enhancing the sensitivity of SIMS. Secondary ion energy distribution can provide us with much useful information. In order to do high sensitivity SIMS work and to study secondary ion energy distribution, a high-performance analyzing system for SIMS has been developed.

Since it has sufficiently high energy resolution, can markedly increase the signal/noise ratio of an analyzing system and particularly, can efficiently accept the secondary ions from a large target area with a large solid angle, a retarding-accelerating energy analyzer has been applied in quadrupole-based SIMS [1]-[5].

The results of our serial studies show that by improving the matches within this kind of analyzing system the performance can be improved considerably. According to the computer calculations of the ion trajectories and the experimental studies [5], an ion optics system has been designed and set up, and its electric parameters have been adjusted carefully, so that the matches within the system have been improved greatly.

A schematic view of the secondary ion-analyzing system is shown in Fig. 1. It consists of a retarding-accelerating energy analyzer made by improved manufacturing technology, 14mm-diameter quadrupole rods and an off-axis channeltron with a 20mm-diameter entrance.

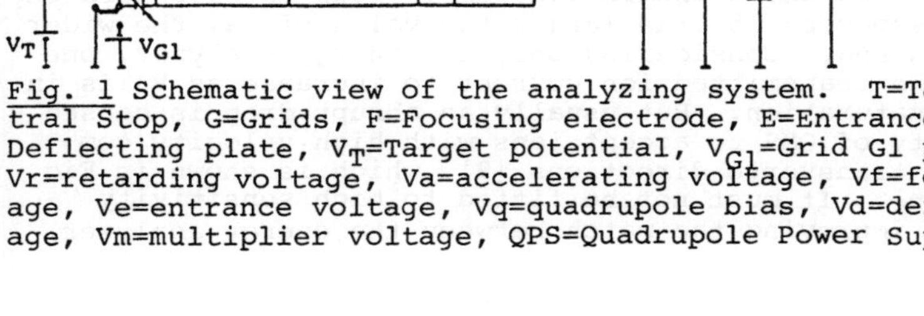

Fig. 1 Schematic view of the analyzing system. T=Target, S=Central Stop, G=Grids, F=Focusing electrode, E=Entrance of QMS, D=Deflecting plate, V_T=Target potential, V_{G1}=Grid G1 potential, Vr=retarding voltage, Va=accelerating voltage, Vf=focusing voltage, Ve=entrance voltage, Vq=quadrupole bias, Vd=deflecting voltage, Vm=multiplier voltage, QPS=Quadrupole Power Supply

Because the voltages Va, Vf, Ve and Vq are all kept fixed relative to the retarding voltage Vr in the electric connection shov in Fig. 1, all ions have the same entrance condition and axial energy in QMS when Vr is varied to measure secondary ion energy distribution [1, 4]. As a result, the calibration procedure for correcting the velocity discrimination of QMS [2] isn't needed.

Some detailed experiments with a thermal Na^+ source have been completed. The energy spread of the Na^+ source itself is less than 0.3eV. The mass resolution is kept at $\Delta M_{5\%H} \leq 1$ amu.

A group of typical transmission characteristic curves is showr in Fig. 2. It's clear that the energy passband can be adjusted conveniently by changing the electric parameters to meet the different requests for high sensitivity SIMS work or secondary ion energy spectra studies. The experimental result shows that the minimum energy passband of 0.4eV (FWHM) can be achieved. It' near the energy spread of the source and sufficiently narrow fc measuring secondary ion energy distribution in most cases. Our experience shows that the precision of the spherical grids of the energy analyzer is the key for achieving this property.

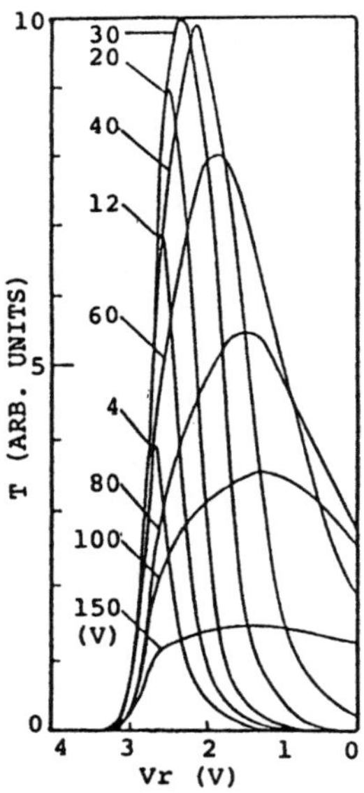

Fig. 2 Transmission versus retarding voltage Vr for various Va. The mean ion energy is about 3eV (uncorrected for contact potential differences). Vf=Ve=Vq=10V

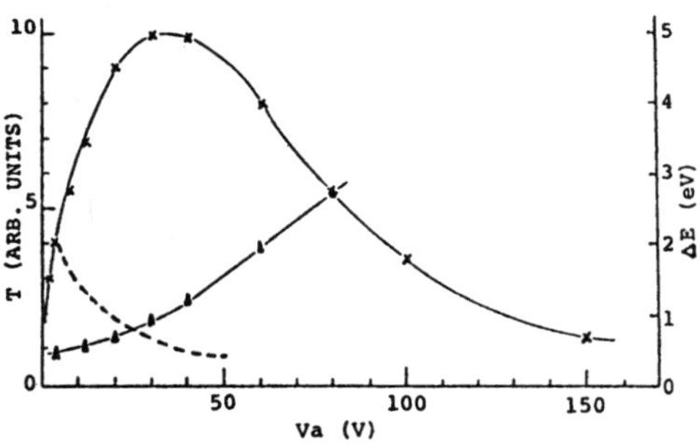

Fig. 3 Transmission (x) and energy passband ΔE (▲) as functions of Va. The solid curves are experimental results (from Fig. 2)

The solid curves in Fig. 3 are the experimental results of the energy passband and the transmission for various Va (accelerating voltage). As expected [5], the larger the value of Va, the wider the energy passband. Considering only the energy analyzer, one would expect the transmitted ion current to increase as Va is increased until saturation. But usually an abrupt drop is caused by the inability of QMS to accept ions with high-velocity components in the transverse directions [2], which is shown in Fig. as a dashed line. It must not be fitted to high sensitivity SIMS work. By improving the match between the energy analyzer and QMS in our design, however, the transmission characteristics at higher Va have been improved, just as Fig. 3 shows, which is of great significance for high sensitivity SIMS work.

For the Na^+ source, the maximum transmission of the whole system (the ratio of the detected ions to the entered ions) of 3% has been already obtained through choosing the suitable electric parameters, while $\Delta M_{5\%H} \leq$ 1 amu.

The variation of the transmission is less than 20% when the Na^+ source is moved off the axis in the transverse directions within ±2mm and 10% for the source moving along the axis within ±1.5mm. The displacements hardly influence the energy passband and the mass resolution. It makes the system particularly suitable for high sensitivity Static SIMS operations. In addition, the sample position isn't critical.

It isn't difficult to achieve a mass resolution of $\Delta M_{5\%H} \leq 0.5$ amu or $\Delta M_{50\%H} \leq$ 0.2 amu at Vq=10V.

It has been proved that this system is adapted to high sensitivity SIMS operations and secondary ion energy spectra studies as well. It is a high-performance analyzing system for SIMS and has already been used to perform analysis. Analytical results will be published elsewhere.

References:

1 A. R. Krauss and D. M. Gruen, Appl. Phys. 14, 89 (1977)
2 P. H. Dawson and P. A. Redhead, R. S. I. 48, 159 (1977)
3 R. -L. Inglebert and J. -F. Hennequin, in SIMS III, Springer, 57 (1982)
4 K. Wittmaack, Vacuum 32, 65 (1982)
5 Cha Liang-Zhen, Xue Zu-Qing, Zheng Zhao-Jia, Tong Yu-Qing, Wu Yong-Qing, in SIMS IV, Springer, 79 (1984)

Characterization of Electron Multipliers by Charge Distributions

E. Zinner, A.J. Fahey, and K.D. McKeegan

McDonnell Center for the Space Sciences, Washington University, Box 1105, St. Louis, MO 63130, USA

1. Introduction

Electron multipliers are used in the pulse counting mode in many SIMS instruments. To use the electron multiplier in this way for precise elemental and isotopic measurements, two parameters must be known: (a) the detection efficiency for ions impacting the conversion dynode and (b) the dead time of the whole pulse counting system. RUDAT and MORRISON [1,2] have reported measurements of detector yield factors for a Cu-Be ion to electron converter in a Daly-type detector installed in a CAMECA IMS-300. For electron multipliers (EM), the aforementioned parameters depend on EM type, ion energy, biasing voltage and the detected element. Furthermore, the EM gain changes with the age of the multiplier. We have found that the pulse charge distribution of the EM output is correlated with the detection efficiency. Its measurement provides a quick and reliable means to monitor the EM performance. We report measurements of the pulse charge distributions, detection efficiencies, and dead times, for various positive and negative ions.

2. Experimental

Ions having a nominal energy of 4.5 KeV are measured in a CAMECA IMS 3F with a Balzers SEV 217 electron multiplier. The EM conversion dynode is kept at ground, the total voltage applied across its 17 stages is 2740V unless stated differently. The EM output is AC coupled to the input of a fast linear amplifier (LeCroy VV101A) with a total gain of 40. The amplifier output is connected to a LeCroy 3001 multichannel analyzer, and also to a LeCroy 821 discriminator whose output pulses are counted in a LeCroy 2551 100MHz scaler. Secondary ion currents are measured in a Faraday cup (FC) by a Keithley 642 electrometer. The ion detection efficiency is obtained from the EM count rate (dead time corrected) times the electron charge divided by the FC current.

3. Results

The measured pulse charge distributions are a strong function of ion type (atomic vs. molecular), element, polarity, and energy. Fig. 1 shows the distributions for $^{16}O^-$ and $^{181}Ta^-$ (normalized to have the same area). From the distributions we calculate the mean and variance. Fig. 2 shows the variance of the pulse charge distribution for positive and negative ions as function of atomic number. There are large variations among different elements; although there is an overall decrease of the variance with increasing Z, there is more detailed structure reflecting the electronic properties of the elements. Both positive and negative ions show similar patterns, although negative ions generally yield larger pulses.

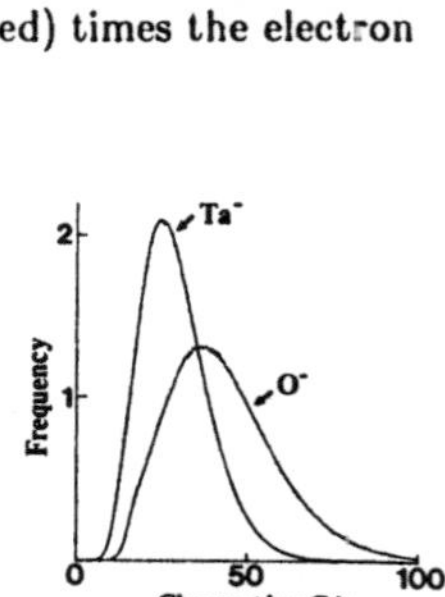

Fig. 1
Pulse charge distribution

The variance of the pulse charge distribution is related to the ion detection efficiency, as can be seen in Fig. 3. The fact that there is no single curve indicates that the probability for obtaining secondary electrons (which primarily determines the detection efficiency) is not uniquely related to the distribution of the number of secondary electrons produced by different ions, i.e. the distribution is not strictly Poisson. A similar general correlation is seen for the detection efficiency as function of the mean of the charge distribution.

Molecular Ions: Table 1 gives the mean, variance, and detection efficiency for a variety of atomic and molecular ions. There is a tendency for molecular ions to produce more secondary electrons, but the relationships are complicated.

Energy Dependence: Table 2 shows that pulse charges and detection efficiency vary with the ion energy. The effect is large for doubly charged ions which have twice the energy of singly charged ions.

Isotopic Effects: There are small but noticeable differences in the detection of isotopes of the same element. Fig. 4 shows the difference between pulse charge distributions of $^{44}Ca^+$ and $^{40}Ca^+$, and of $^{48}Ca^+$ and $^{40}Ca^+$. The detection efficiencies are .804 ± .009, .792 ± .006, and .774 ± .006 for $^{40}Ca^+$, $^{44}Ca^+$, and $^{48}Ca^+$, respectively, yielding a linear mass fractionation of -4.9‰/amu. A more precise measurement of Ti gave detection efficiences .7920 ± .0007, .7908 ± .0006, .7891 ± .0007, and .7874 ± .0007 for $^{46}Ti^+$, $^{47}Ti^+$, $^{49}Ti^+$ and $^{50}Ti^+$ resulting in a linear mass fractionation of -1.35 ‰/amu. This fractionation is much smaller than the instrumental isotopic mass fractionation [3-5] which is typically -16‰/amu for Ti^+ ions sputtered from Ti metal. However, our results show that isotopic mass fractionation in the EM detection efficiency is part of the overall instrumental mass fractionation.

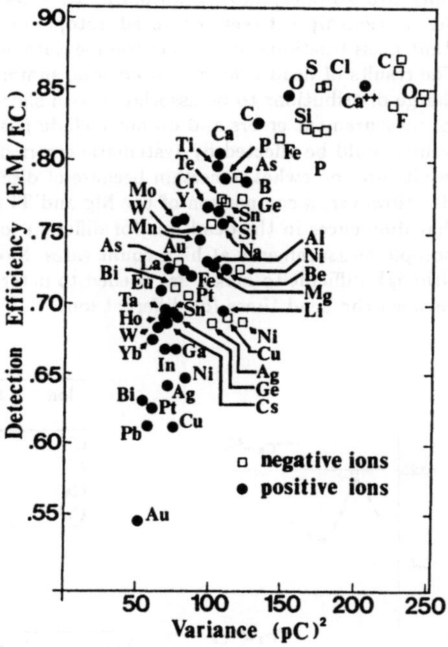

Fig. 3 Detection efficiency vs. variance

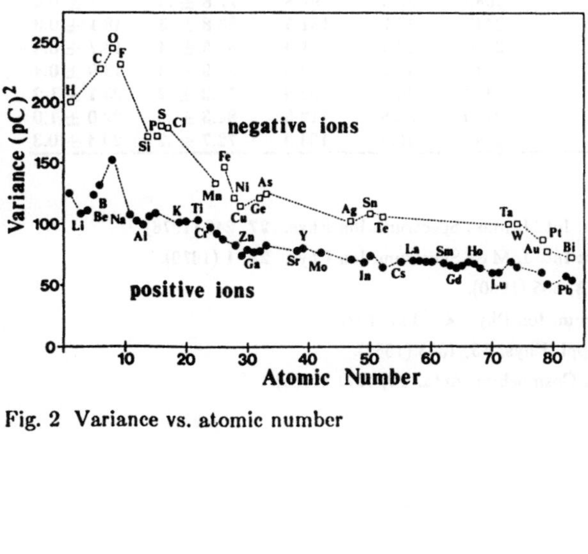

Fig. 2 Variance vs. atomic number

Table 1

Ion	Mean (pC)	Variance $(pC)^2$	Detection Efficiency (%)
C^-	37.1	228.6	86.5 ± .5
C_2^-	40.5	229.1	88.0 ± .4
C_3^-	41.0	226.0	88.5 ± .8
C_4^-	40.7	213.9	---
C_5^-	40.2	201.0	---
C_6^-	39.5	191.4	---
Si^-	33.9	176.8	82.5 ± 1.3
SiO^-	37.3	193.0	---
SiO_2^-	38.6	195.1	83.4 ± 1.0
Ti^+	32.7	131.5	80.8 ± .3
TiO^+	33.2	119.6	83.9 ± .4
Si^+	30.1	102.9	76.2 ± .2
SiO^+	32.8	112.5	84.3 ± .2

Dead time: The time resolution of our pulse counting system is limited by the pulse width at the discriminator threshold. Since different elements have different pulse charge distributions, and thus possibly different distributions of the pulse width at the threshold voltage, it cannot be assumed, *a priori*, that the dead time is the same for different elements. Dead times were determined by measuring ratios of isotopes of different abundances. For Si, Mg, and Ti the deviations from absolute ratios (after mass fractionation correction) yield the dead time. For C (only two isotopes) and Pt (absolute ratios not well known) the dead time was obtained from the relationship between measured isotopic ratios and count rate. This method assumes constant mass fractionation during the measurement and is subject to possible systematic errors. The results of dead time measurements, summarized in Table 3, show a trend for larger pulse charge distributions to be associated with shorter deadtimes. The errors given are the statistical measurement errors and do not include possible systematic errors. The Pt and C results could be affected by systematic errors due to changing mass fractionation, and the Si results are somewhat uncertain because of discrepancies for the absolute isotopic composition [6]. However, a comparison of the Mg and Ti measurements, which are of high accuracy, shows that differences in the deadtime for different ions exsist and are important for high precision isotopic measurements at high count rates. More measurements at this level of precision, although difficult to make, are needed to determine the reason for systematic differences between the dead times for different ions.

Table 2

Ion	Energy (KeV)	Mean (pC)	Variance $(pC)^2$	Det.Efficiency (%)
C^-	4.35	39.3	231.9	84.2 ± .7
C^-	4.65	42.6	270.3	85.7 ± 1.0
Ca^+	4.50	31.4	103.1	80.4 ± .9
Ca^{++}	9.00	39.9	203.1	85.4 ± .2

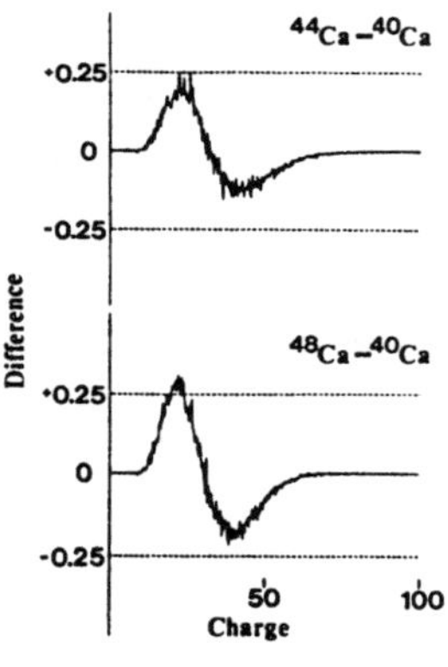

Fig. 4 Isotope effect

Table 3

Ion measured	EM voltage (V)	Mean (pC)	Variance $(pC)^2$	Detection efficiency (%)	Dead Time nsec $\pm 2\sigma$
C^-	2740	37.1	228.6	86.5 ± .5	17.5 ± 0.9
Pt^-	2740	26.3	85.6	70.6 ± .6	21.5 ± 0.6
Ti^+	2680	29.3	91.8	77.8 ± .3	18.7 ± 0.4
Ti^+	2740	32.7	131.5	80.8 ± .3	18.1 ± 0.3
TiO^+	2680	30.4	91.9	82.5 ± .4	18.7 ± 0.4
TiO^+	2740	33.2	119.6	83.9 ± .4	18.9 ± 0.4
Si^+	2740	30.1	102.9	76.2 ± .2	23.1 ± 1.2
SiO^+	2740	32.8	112.5	84.3 ± .2	22.0 ± 1.0
Mg^+	2740	29.5	103.0	72.7 ± .2	20.4 ± 0.3

References:

1. M. A. Rudat and G. H. Morrison: Int. J. Mass Spectrum. Ion Phys. **27**, 249 (1978).
2. M. A. Rudat and G. H. Morrison: Int. J. Mass Spectrum. Ion Phys. **29**, 1 (1979).
3. G. Slodzian et al.: J. Physique **23**, 555 (1980).
4. A. J. T. Jull: Int. J. Mass Spectrum. Ion Phys. **41**, 135 (1981).
5. G. Shimizu and S. R. Hart: J. Appl. Phys. **53**, 1303 (1982).
6. K. D. McKeegan et al: Geochim. Cosmochim. Acta, **49**, 1971 (1985).

Automated Collection of SIMS Data with Energy Dispersive X-Ray Analysis Hardware

D.M. Anderson and E.L. Williams

Lockheed-Georgia Company, 86 South Cobb Drive,
Marietta, GA 30063, USA

1. Introduction

Our SIMS equipment consists of a Model 160 3M Brand SIMS mounted on an Amray 1100 Scanning Electron Microscope (SEM). The original two data output devices were a storage oscilloscope and strip chart recorder. These devices were not suitable for the rapid collection and analysis of large amounts of SIMS spectral data obtained from metallic surfaces.

To improve this situation, the SIMS output signals were directed into Kevex computer-controlled energy dispersive x-ray hardware, so that many spectra could be collected and saved while the ion beam was impinging on a particular sample area. These spectra could be printed and studied later at the convenience of the experimenter.

2. Command Sequences

Command sequences were devised to allow for the automated, rapid, successive collection and storage of the spectra on floppy discs. The computer is placed in the Acquire/Select (ACQSEL) mode and the following parameters are typed into the memory as prompted by the CRT display.

CRT Prompt	Operator Response
MODE	SEQ
DWELL	29.64 millisec per channel
INC/CH	0.15
OFFSET	+ 0.3
UNITS	MU

The 29.64 millisec/ch allows for 30 sec (29.64 millisec/channel X 1012 available channels) sweeps across the desired mass range of detected ions. The 0.15 inc/ch and + 0.3 offset commands set bottom labels of the hardware.

Once the above commands have been accomplished, the operator can then program the computer to acquire the required number of identified spectra from a sample area and save the spectra.

Commands for number of successive spectra:

SETLA initiates the routine that defines the number of spectra desired to record. Up to 33 spectra on an individual spot can be collected and saved with this routine.

SETID initiates the routine that identifies the spectra to be saved. For instance, if five successive spectra are to be saved from a sample area the command sequence would be:

CRT Prompt	Operator Response
Enter ID for each Sample Pos. Analyzed	SET ID
Position 0 ID:	A1
Position 1 ID:	A2
Position 2 ID:	A3
Position 3 ID:	A4
Position 4 ID:	A5

An automatic routine was programmed with the following commands to acquire and save designated spectra:

```
ACQ
WAI
REA
SAV
1
SETPO
+
LOOP
ROF
```

Once the above routine was defined and saved, the spectra were automatically acquired and stored by typing ATO followed by the name given to the above routine; i.e., ATO, PRO. Note: There is a delay as a spectrum is being saved, so 30 seconds of sputtering between recordable spectra will be lost.

Similarly, another routine was programmed with appropriate commands to print (graph) designated spectra. Once done, the several desired spectra were printed by typing ATO, PLOT.

When it was not necessary to graph all of the obtained spectra, a recorded spectrum was recalled and printed individually with direct commands and rather than the automatic routines.

By the use of auxiliary coaxial and Kevex data cables, the CRT and keyboard of the data storage system could be physically separated from the balance of the hardware. This arrangement provided work flexibility but minimized the noise the operator had to endure during recording, saving, and printing of spectra.

3. Typical Spectrum

Figure 1 is a positive SIMS spectrum obtained employing the SIMS equipment and Kevex hardware. The mass range shown is from 0.3 to 37 mu. The specimen was a 7475 aluminum alloy section that did not bond to another section during an attempted diffusion-bonding cycle. The actual panels to be bonded were windows of the alloy surrounded by areas supposedly masked with a carbon paste. It appeared that the carbon paste had migrated during the diffusion-bonding cycle onto the clean alloy panel section, preventing any bonding. Figure 1 is the first seven 30-sec sweeps performed on the contaminated panel using argon as the ion beam. The high magnesium content at the surface is common for 7000 series aluminum alloys after heat treatment. Magnesium preferentially diffuses to the surface during any elevated-temperature operation. Very little aluminum is evident on this outermost surface layer. Sodium remained on the surface as a result of a prior cleaning operation and an insufficient rinse. Lithium was contained in the carbon paste which was later found to actually be a carbon-based grease. The lithium readily diffused onto the clean alloy section that was to be bonded. Negative SIMS analysis revealed carbon, hydrocarbon and water (OH^- ions) residue on the outer surface of the specimen.

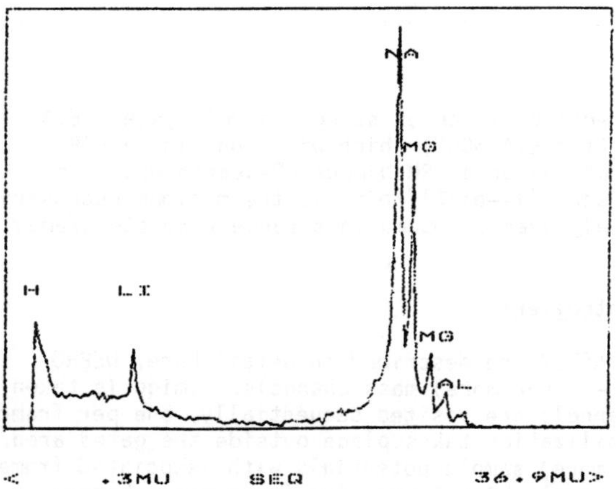

FIGURE 1. Positive SIMS spectrum for contaminated 7475 aluminum alloy surface

4. Conclusions

Computer-controlled energy dispersive x-ray hardware can be utilized with SIMS equipment to obtain rapid accumulation and storage of surface composition data from metal surfaces. Such an arrangement offers a versatile tool for failure analysis investigations.

Software and Interfaces for the Automatic Operation of a Quadrupole SIMS Instrument

M.G. Dowsett, J.W. Heal, H. Fox, and E.H.C. Parker

Solid State Research Group, Sir John Cass Faculty of Physical Sciences and Technology, 31 Jewry St., London EC3N 2EY, England

Software and interfaces developed for the automation of the EVA 2000 [1] and EVA 3000 quadrupole SIMS instruments designed and built by the authors' research group are described. The software runs on a 380Z micro-computer (Research Machines, Ltd.). It contains a number of unique features, including automatic parameter change (mass, target potential, or raster) on detection o a threshhold ion intensity, and the ability to acquire diagnostic energy and mass spectra from any plane in a depth profile. There are four sequential experimental controllers (SEC's) written in Basic: DEPRO, ESP, MASCAN, and MASTEP. These enable automatic running of multi-channel depth profiles [1], energy spectra [2], mass spectra (line or peak), and static SIMS experiments [3] respectively. Programmes run under an Interpreter (BASIMS) which has been modified to include about 40 machine code subroutines (MCS) for SIMS purposes These may be called from the keyboard for immediate control, or from the SEC's. SEC's for special experiments can be written by anyone with a knowledg of Basic, yet retain the full flexibility of the instrument by CALLing the appropriate MCS. The SEC's enable new personnel to become familiar both with the operation of the instrument, and with sequences for the main SIMS tasks. Instrumental parameters are stored on disc for automatic setting up.

1 Parameters controlled

Table 1 shows the parameters interfaced in an existing control system (EVA 2000) and one under development (for EVA 3000) which will run with a PDP 11/23+ (Digital Equipment Corporation) or an RM Nimbus (Research Machines Ltd). The precision of the mass scan (14-bit$\pm$1/2bit) is the minimum necessary to define the peak-shape adequately over a 200amu mass range ($\sim$80 steps/amu).

2 The Sequential Experiment Controllers

EVA 2000 versions of DEPRO and MASTEP are described in detail here. DEPRO controls depth profiling with 1 - 9 (or more) mass channels. Timing is taken from the raster scanner, and channels are visited sequentially, one per frame Instrumental setting-up and stabilization takes place outside the gated area. Each mass channel has 1 - 4 (or more) sample potentials with associated frame numbers (break points). A given sample bias is applied for a mass channel until a break point is reached, when the next potential in sequence is applied. Alternatively, the potential can be changed if the signal intensity for that channel crosses a threshold. Thus concentrations can be profiled ove many decades without detector saturation by using secondary ion energy selection. During the setting-up, a tuning display shows the mass and energy spectra for any channel selected, with cursors indicating the tuning points. Fine tuning of the instrument is carried out via the computer at this point, and the energy filter is manually set up.

Table 1 **Control and Interface Parameters**

Device/Parameter	EVA 2000	EVA 3000
Mass Spectrometer	14-bit DAC=0.012amu/step for 200 amu range	
Resolution	manual	14-bit DAC
Delta-M	manual	14-bit DAC
Pole bias	manual	12-bit DAC=0.07V/step ±150V range
Energy filter	manual	8 channel 12-bit DAC=0.12V/step ±500V range-bit 13 gives polarity
Sample Bias	12-bit DAC=0.07V/step for ±150V range	
Primary Ion Beam	Beam Blank	Beam Blank (onto current monitor)
Energy	On/Off	On/Off & Set in 60V steps 0-15keV
Beam Scanning	10-bit DAC X=Y	2x8-bit DAC, X & Y independent
Scan rate	manual	0.01-300sec/frame in 1/2 decades
Gate	manual	X & Y origin, X & Y width (8-bit)
Manipulator	Motor On/Off, forward/reverse, 10 preset positions	
Data Logging	RS232 serial interface	32-bit parallel binary interface
	8 decimal decades	8 decimal decades (27-bits)
	8 data points/sec	1000 data points/sec
Analogue:	3x10-bit ADC	3x12-Bit ADC
	1000 data points/sec	1000 data points/sec
Synchronisation	Frame, gate or ext.	Frame, gate, single pixel, ext.
Digital gating	No	Optional

Different raster sizes may be selected together with break points. After a break point, the next raster size is selected automatically. E.g. a large raster may be applied for several frames to pre-clean a surface, or the profile rate may be controlled.

During the profile a real-time display of the instrument parameters and active values is provided. A digital plotter displays the raw data, and the data are printed. The instrumental control loop will run at up to 3 sec/frame with printing, and faster than 1 sec/frame without. The profile may be ended after a preset number of frames, when an interface is detected, or manually from the key board. A menu of options is available once the profile is interrupted, some of which are described below.

DEPRO (and ESP) allow a rapid mass or energy spectrum to be acquired for diagnostic purposes on interruption of the profile (energy spectrum). Any desired part of the spectrum may be examined at different stages in the profile, (e.g. to identify sources of mass interference). The profile can be resumed after the operation with no loss of accuracy in the depth scale, and little loss of data. These spectra are acquired using an MCS which reads an analogue input once per line in the frame-scan whilst the electronic gate is open. Mass is incremented by 0.025 amu/line or sample potential by 0.5 V/line. Thus if the gate is open for 80 lines in a 256 line scan (a gated area ~10%), a mass scan proceeds at 2 amu/frame, and an energy spectrum at 40 eV/frame. These data are output to an analogue display (plotter or oscilloscope). This facility was used extensively in [2].

Automatic control of the sample potential (e.g. to compensate for changes in the sample resistance in multi-layer structures) is achieved in a similar way: One mass channel is designated as a reference, and tuned to the peak of the energy distribution of an intense species with a narrow energy spectrum (e.g. a matrix cluster). In this channel an MCS acquires an energy spectrum synchronised with the linescan, from the centre of the gated area, by scanning the sample potential symmetrically ±5V about the tuning point. Any shift in the energy spectrum is measured, and the voltage increment required for compensation is applied to all channels. The method only works correctly if the equilibrium surface potential is uniform over the sampled region, however Alternatives to this approach are discussed in [2].

It is not always desirable to use a large offset in the sample potential t suppress an intense species (e.g. charge compensation can be upset). DEPRO ca therefore automatically offset the mass of a particular channel when a threshhold intensity is detected. This was made extensive use of in [4].

DEPRO is available in versions which allow the running of one or more samples automatically by activating the sample manipulator between profiles. The samples may differ, and appropriate experimental parameters are set up of one or several previously written data files.

MASTEP is primarily for static SIMS applications, and was designed to assist in the minimisation of ion doses. MASTEP produces line spectra, and makes efficient use of sputtered ions by only sampling the peak maxima. It i found experimentally that the mass spectrometer (Extranuclear Laboratories Inc.) is capable of resetting to within 1/40 amu of a previous mass for a given input from the 14-bit D/A on the SIMS interface within 1hr of switching on. This performance is reproducible over several weeks, provided that the resolution, delta M, pole bias etc. are not changed. Also, the mass is proportional to the programming voltage over spans exceeding 50 amu. MASTEP therefore contains a tuning stage as follows: The instrument is tuned to 3 or more masses spread over the desired range from a test sample. A map of programming voltages corresponding to peak maxima is computed by interpolation. (In principle, several maps corresponding to different resolution and delta-M etc. could be produced.) The map is then used to step-drive the mass spectrometer and is stored as part of the data file for future setting up. MASTEP takes its sampling time from the pulse counter, an external clock, or the frame scanner, and operates in gated or ungated modes. The operator can select from single or repeated cycles where each cycle may be defined as follows: (a) stepping between single peaks (as in a depth profile), (b) stepping 1amu (or 1/2 amu to detect a doubly charged species) across any desired mass range, (c) as (b) for several mass ranges or peak groups. The cycle may be repeated immediately, after a predetermined delay (e.g. for a ga exposure), or triggered manually or by an external event. The ion beam may be on between cycles for ion dose measurements, or off to minimise the dose. Optionally, the beam may be blanked between mass peaks so that the exposure time and acquisition times are identical. For example, the data in [3] could be acquired for a dose 5×10^{6} ions/mass peak/sec. Individual sample potentials may be set up for each peak, or peak group to optimise sensitivity or suppres intense species. MASTEP has been used for ion dose experiments, static SIMS, and for sampling mass spectra at various depths in a material, where the sample depth is determined by the inter-cycle time.

References

1 M G Dowsett, E H C Parker, R M King and P J Mole, J. Appl. Phys. **54**(1983)6340-6345

2 D S McPhail, M G Dowsett, and E H C Parker, These Proceedings.

3 W F Croydon, M G Dowsett, R M King and E H C Parker, J. Vac. Sci. Technol. B3(2)(1985)604-607

4 M G Dowsett, E H C Parker and D S McPhail, These Proceedings.

Computer-Aided Design of Primary and Secondary Ion Optics for A Quadrupole SIMS Instrument

M.G. Dowsett, R.M. King, H. Fox, and E.H.C. Parker

Solid State Research Group, Sir John Cass Faculty of Physical Sciences and Technology, City of London Polytechnik, 31, Jewry St., London EC3N 2EY, England

The SLAC Electron Trajectory Program [1] is capable of tracing ion trajectories in cylindrical or linear symmetry, with or without space charge, and in both magnetic and electrostatic fields. We have applied the program in design studies of various ion-optical components for a new Quadrupole SIMS system (EVA 3000). Amongst the topics studied have been the design of the secondary ion transfer optics, the extraction of ions from a plasma source, and the design of the primary ion column final lens. Space limitations confine us to the latter topic here. Much electrostatic lens data, obtained by other means, has been published (eg [2]). Ray tracing a specific structure gives valuable extra information, eg. non-paraxial ray behaviour, space-charge effects. We have modified the SLAC program to allow computation of cardinal points and spherical aberration. Data are reported here for the symmetric and asymmetric einzel lenses shown in Fig 1. The proportions have been chosen so that their powers are similar, and so that practical implementations will fit into the same space. The lenses are intended to focus an ion beam of perveance 3×10^{-6} μperv (eg. 3μa at 10keV) into a small spot at a distance of W=40mm or less (the working distance) from the reference plane (L_A in the centre of the accelerating gap - dotted line in Fig. 1). Voltages V_1 and V_2 are measured from a point (eg. the plasma) where the ions have zero energy. V_r values >1 (where $V_r=V_2/V_1$) give better theoretical performance because trajectories remain closer to the axis. Often, however, practical focussing voltages are excessive ($\sim$-30kV for 10keV +ve ions) but there may be advantages for ion energies $\lesssim$2keV. For V_r<0.15, multiple crossovers occur inside the lenses as the mirror condition is approached. This study is confined to $0.15\leq V_r\leq 0.5$.

The positions of the focal points F1 & F2 and principal planes P1 & P2 are determined by tracing the trajectories of a set of parallel rays with radii R from left-to-right (l→r) and right-to-left (r→l) through the lenses. The trajectories of the emergent rays are extrapolated as straight lines until they cross the axis, and back till they cross the entrant ray (Fig 2a). These

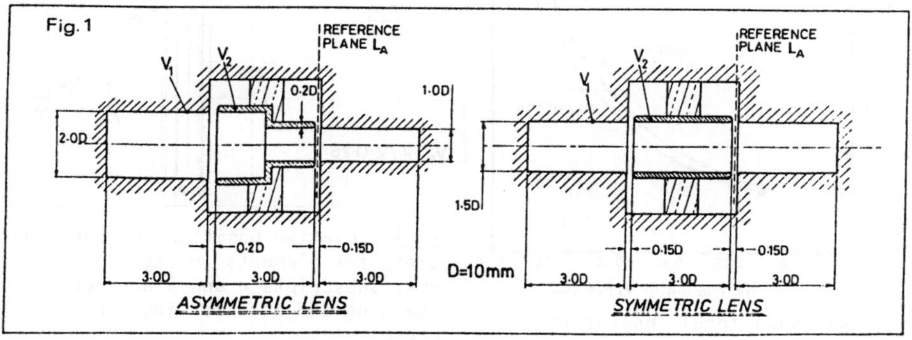

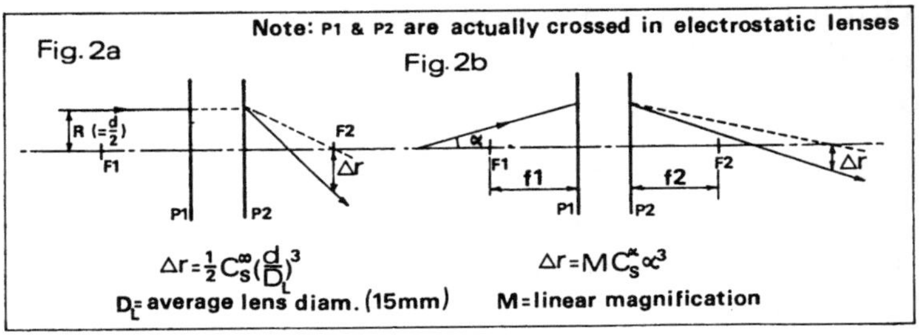

Fig 2 Definitions of geometrical lens parameter

intersections should approach the paraxial values as $R \to 0$. For the focal points, aberration theory [3] shows that the axis crossing points should fit an even ordered polynomial of the form $A + BR^2 + CR^4 + \ldots$ where A is the paraxial focus F, and B is related to C_S^{∞} (Fig 2a and [4]) the spherical aberration coefficient for an object at infinity by:

$$C_S^{\infty} = BD_L^3/4f2$$

The polynomial gave an excellent fit to ray-tracing data for a range of V_r and the paraxial values for the focal lengths are shown in Fig. 3. For the symmetric lens, $r \to l$ and $l \to r$ tracing should produce focal point and principa plane positions and shapes which are reflected in the plane of symmetry. Thi: was the first test of the accuracy of the ray tracing. Conjugate surfaces we found to be symmetrical to better than 1%. For the asymmetric lens the non-paraxial principal and focal surfaces have different shapes, and are shifted towards the small diameter gap (L_A) as expected. The curvature of the surfac showed clearly that the C_S^{∞} $r \to l > C_S^{\infty}$ $l \to r$. However, the paraxial focal lengths (f1 and f2) should be the same for both directions because the refractive indices of object and image space are equal. This is a more severe test of t method, and was again satisfied to better than 1%. In Fig. 3 the f1 and f2 a superimposed, and are indistinguishable on the scale of the plot. As a third test of the ray tracing, the cardinal points for the asymmetric lens were computed from data for accelerating immersion lenses (A1.3 in ref. 2), by treating the lens as a doublet consisting of two immersion lenses, and combining the individual parameters of each using standard optical formulae [3]. Fig. 4 shows the relationship of the individual and combined parameter for the asymmetric lens with a V_r of 0.25.

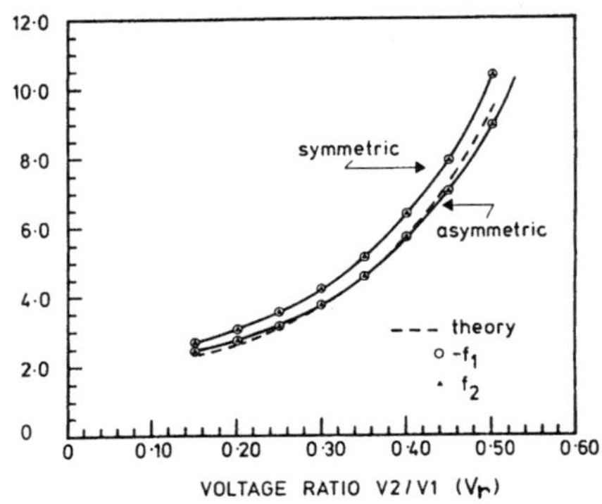

Fig 3 Paraxial focal lengths for symmetric and asymmetric lenses vs V_r

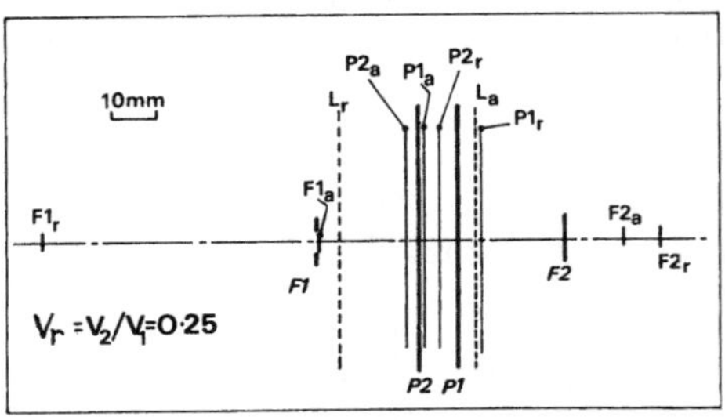

Fig 4 Relative positions of cardinal points for elements in asymmetric lens. Subscripts r and a are for retarding and accelerating gaps respectively

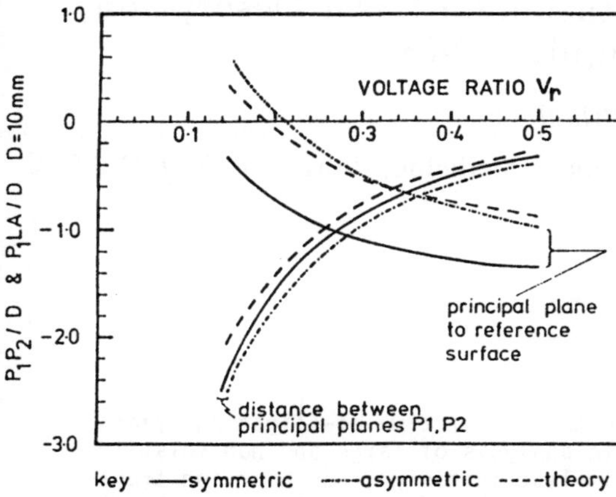

Fig 5 Principal plane separation P1P2 and displacement of P1 from L_A vs V_r

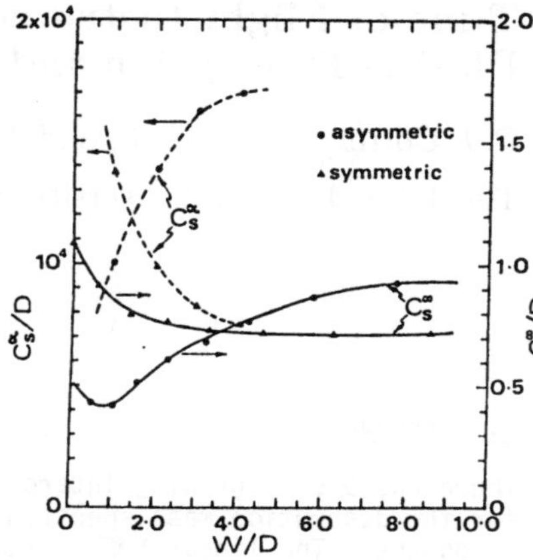

Fig 6 Spherical aberration coeffs. as a function of working distance

The dotted line in Fig. 3 shows the remarkable agreement obtained for the asymmetric lens. For the symmetric lens, this method produced focal lengths which agreed to within 10% with those from the ray tracing,but the separation and positions of the principal planes in the doublet calculation were incorrect because of perturbation of one lens field by the other. Paraxially, the lenses are fully characterised if f1, f2, the separation of P1 & P2, and the displacement of P1 from a fixed reference plane (L_A) are known. Fig 5 shows P1P2 and P1L_A as functions of V_r. The dashed lines show asymmetric data from the doublet calculation. The functional behaviour of the ray tracing and doublet data is very similar. Slight displacements are probably due to the different gaps and electrode thicknesses in the lenses compared.

In our existing SIMS instrument (EVA 2000), it is found that the maximum useful beam current is obtained for a virtual object ~9 D (ie 90mm) behind L_A. The behaviour of the lenses for this object position, and a W $\leqslant$ 40mm was investigated. In particular, the spherical aberration for finite object distances (Fig.2b and [2,4]) was investigated. For a finite object, the non-paraxial axis crossings should fit an even-ordered polynomial of the form $A'+B'\alpha^2+C'\alpha^4+...$, where A' is the conjugate image position, and B' is related to the C_s^α by:

$$C_s^\alpha = B'/M^2$$

Again, the fit of the data over a range of values of W (or V_r) was excellent, and produced results consistent with the previously calculated cardinal points. Figure 6 shows C_s^α and C_s^∞ plotted against the working distance. Both the behaviour (as a function of V_r) and magnitude of these parameters are in reasonable agreement with published data for fairly similar lenses [4]. The C_S are plotted here vs. the working distance, because this makes them directly comparable for a given image point. For W/D 2, the asymmetric lens has the lower aberration coeffs. However, the C_s^∞ are equal for a working distance of 4D and the C_s^α for the symmetrical lens is nearly a factor of two lower. The symmetrical lens is therefore a better choice over the full range of working conditions from an object 9D behind L_A to a retracted object for low-current small-spot applications.

References

1 W B Hermannsfeldt, SLAC Report-226, Stanford Linear Accelerator Centre.
2 E Harting and F H Read, Electrostatic Lenses, Elsevier, New York (1976)
3 R S Longhurst, Geometrical and Physical Optics 2nd Ed., Longmans (1968)
4 A Adams and F H Read, J. Phys. E (GB) **5** (1972)150-155

Time-of-Flight Instrumentation for Laser Desorption, Plasma Desorption and Liquid SIMS

R.J. Cotter, J. Honovich, and J. Olthoff

The Johns Hopkins University, School of Medicine, Baltimore, MD 21205, U

1. Introduction

There has been a growing interest in the use of time-of-flight analyzers for the desorption mass spectrometric analysis of large and non-volatile compounds. The linear TOF is the simplest in design, has high ion transmission, and records the products of metastable decompositions in the flight tube at the flight times of their parent species. In this paper we compare several different instrumental configurations. While spectra reflect the ionization technique, they are also dependent upon the analyzer design. Two designs are discussed: one employing a single accelerating field region, the other employing a multiple accelerating field region, in which the first (extraction) field is pulsed following the ionization event.

2. Experimental

Data were obtained on (1) a BIO-ION Nordic plasma desorption mass spectrometer with a 15 cm flight tube, 20 KV accelerating voltage and a 1 nS/channel multi-stop time-to-digital converter interfaced to an Apple IIe microcomputer and (2) a CVC 2000 time-of-flight mass spectrometer with a 1 m flight tube, 3 KV accelerating voltage, a grounded ion source, with the first accelerating region pulsed -150 V and delayed after the ionization pulse. On the latter instrument both laser desorption (Tachisto 215G carbon dioxide laser) and ion bombardment (Kratos Minibeam I) have been used. Spectra were collected using a Biomation 100 MHz waveform recorder interfaced to an Apple II+ microcomputer or using a LeCroy 3500SA signal-averaging system.

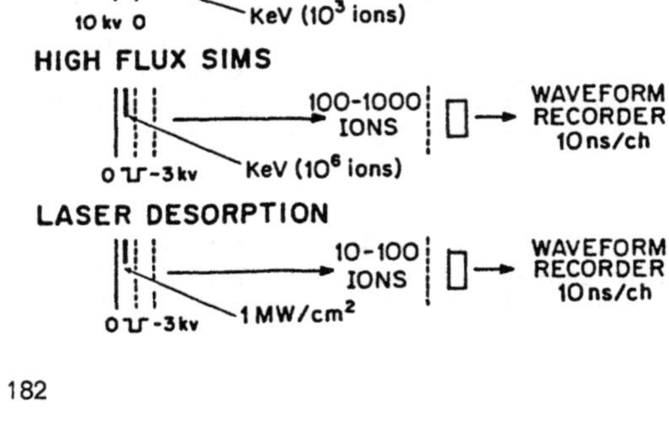

Fig. 1

3. Ion Focussing

3.1 The Static Field TOF Analyzer

The time-of-flight of ions in an instrument with a single accelerating region (s) and a drift region (D) is:

$$t = \frac{(2m)^{1/2}\, s \left[(U_0 + eV)^{1/2} \pm (U_0)^{1/2}\right]}{eV} + \frac{(2m)^{1/2}\, D}{2(U_0 + eV)^{1/2}}$$

Optimal mass resolution is achieved by keeping the source (accelerating) region small with respect to the drift region, so that the difference in arrival times of ions of the same mass is:

$$\Delta t = D(m/2)^{1/2} \left[\frac{1}{(eV)^{1/2}} - \frac{1}{(eV + U_0)^{1/2}}\right]$$

and the mass resolution reduces to:

$$\Delta M/M = 2 \left[\frac{(eV + U_0)^{1/2} - (eV)^{1/2}}{(eV)^{1/2}}\right] \cong U_0/eV$$

when the energy due to acceleration (eV) greatly exceeds the energy spread (U_0) It is important that the sample region be kept very small and parallel to the drawout grid to insure that all ions are accelerated through the same field This scheme is used for both plasma desorption [1] and SIMS [2] ,in which the start pulse for a TDC originates from a sharply defined ionization event. (Figure 1.a and b).

3.2 The Dynamic (gated) TOF Analyzer

Designed originally by Wiley and McLaren for gas phase ionization [3] , this analyzer is suitable for desorption when a less defined ionization region is used (e.g. glycerol matrix) or the ion formation pulse is wide with respect to time resolution. For SIMS experiments this permits the use of large ion bursts (in this case 1 µA for 1-10 µS, or 10^6 to 10^7 ions/pulse) as in Figure 1.c. [4] . The carbon dioxide laser also produces broad desorption times (Figure 1.d). In both cases the secondary ions from each cycle produce analog signals at the detector and are more appropriately recorded using fast analog techniques.

4. Results and Discussion

While resolution is dependent upon the initial energy spread, it is also degraded by ion decomposition in the accelerating region. In the static field analyzer, prompt as well as metastable decomposition takes place in this region, and depending upon the rate constants, will produce "tailing" [5] and a broad continuum [6] in the spectrum. In the dynamic (gated) analyzer,ion acceleration can be delayed to permit decompositions to occur before acceleration, thus eliminating these effects on the spectrum. Figures 2 and 3 show some comparative results for cyclosporin and the ganglioside GM3 respectively. In the latter, where decomposition is extensive, the molecular ion is not observed when the accelerating pulse is delayed.

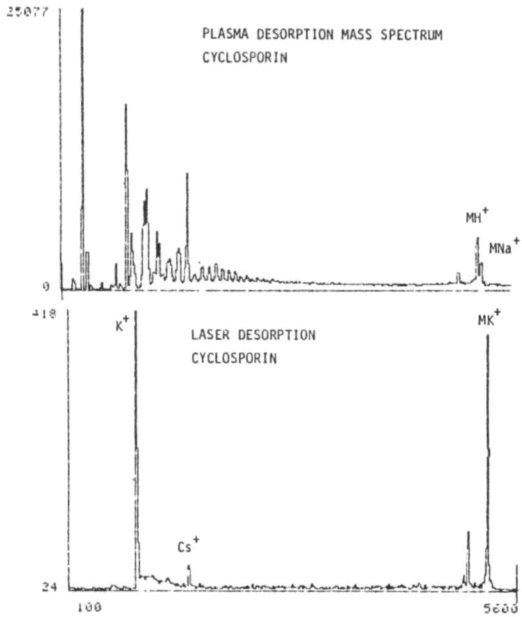

Fig. 2.

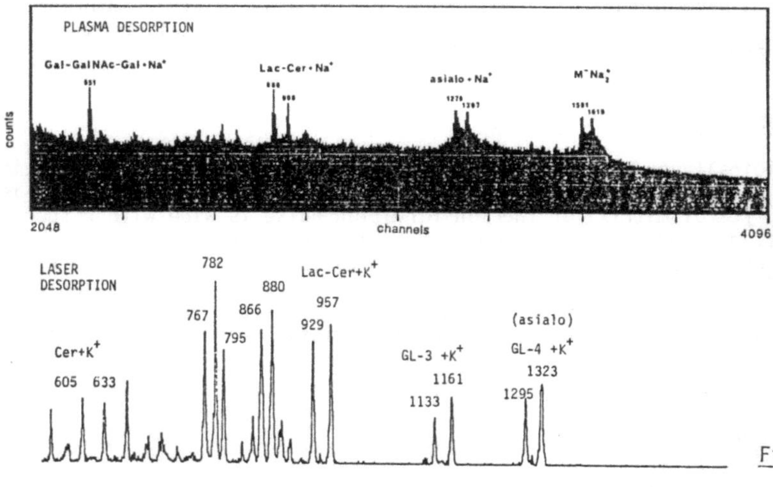

Fig. 3

1 R.D. Macfarlane and D.F. Torgerson, Science, 191, 920 (1976)
2 B.T. Chait and K.G. Standing, Int. J. Mass Spectrom. Ion Phys., 40 185 (1981)
3 W.C. Wiley and I.H. McLaren, Rev. Sci. Instrumen., 26, 1150 (1955)
4 R.J. Cotter, Anal. Chem., 56, 2594 (1984)
5 R.J. Cotter and J.-C. Tabet, Int. J. Mass Spectrom. Ion Phys., 53, 151 (1983)
6 B.T. Chait and F.H. Field, Int. J. Mass Spectrom. Ion Phys., 65, 169 (1985)

Coincidence Measurements with the Manitoba Time-of-Flight Mass Spectrometer

W. Ens, R. Beavis, G. Bolbach[1], D.E. Main, B. Schueler, and K.G. Standing
Physics Department, University of Manitoba, Winnipeg, Canada, R3T 2N2

1. Introduction

Observations of secondary ion multiplicity have been reported for high energy (~MeV/u) bombardment of large organic molecules [1,2] and CsI [2]. The average number of secondary ions detected per primary projectile was >4, and some incident particles ejected more than 32 secondary ions. Significant correlations among ions desorbed from thiamine by ^{252}Cf fission fragments have also been observed [3]. Similar measurements with ~10 keV Cs^+ ions have recently been made possible with a new data system [5] installed on the Manitoba time-of-flight mass spectrometer [4].

2. The Data System

The data system [5], illustrated in Fig. 1, is based on an 8-stop time-to-digital converter (LeCroy 4208). It is controlled by an LSI 11 computer but uses a fast front-end processor (INCAA CAPRO 68K) with 0.5 Mbyte memory to handle the high-speed tasks. After each cycle (250 µs), the CAPRO transfers the data from the TDC to a buffer before storing them in a 120 Kword $\times$ 32 bit histogram. The system is capable of data rates as high as 32000 counts/s, but is typically run at $\lesssim$ 16000 counts/s. This is ~100 times faster than our previous method (using conventional analogue techniques [4]) and ~10 times faster than without the CAPRO. In addition the advantages of digital electronics have greatly improved our resolution and accuracy and have simplified the data analysis.

Although a hard-wired system may be faster, the software control of this system allows much greater flexibility. Before the data are stored in the histogram, multiplicity is recorded and correlations may be identified. Multiplicity and correlation measurements are reported below. Recently we have installed a 45° electrostatic mirror described in more detail in an accompanying paper [6]. Neutrals pass through the mirror into one detector while charged particles are deflected through 90° into a second detector. The data system was easily modified to record both spectra (56000 channels each) simultaneously without reduction in the data rate. The TDC is configured here as two 4-stop digitizers. In addition, the CAPRO has been programmed to measure correlations between charged and neutral particles. Experiments using this method to identify pathways of metastable decay for several small peptides are reported separately [7].

3. Secondary Ion Multiplicity

Figure 2 shows the multiplicity measured for CsI and thiamine with 10 and 13 keV Cs^+ ions. In both cases events yielding more than one secondary ion

---

1 Present address: Institut Curie, Paris, France.

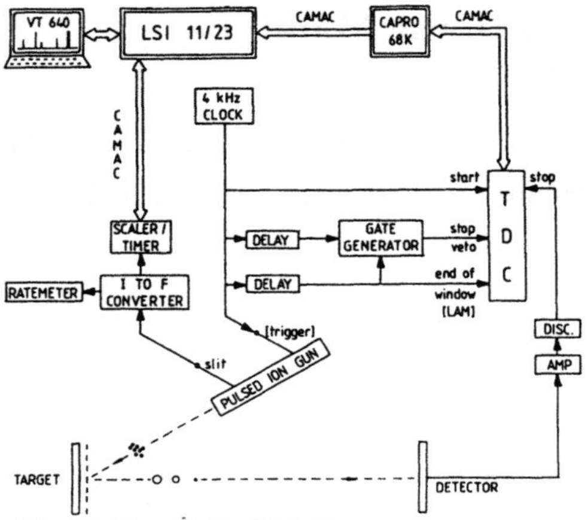

Fig. 1 The Data System

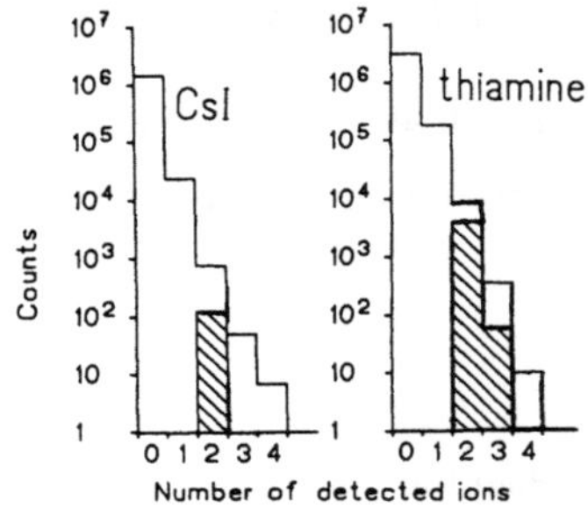

Fig. 2 Secondary ion multiplicity for CsI and thiamine for incident ~10 keV Cs^+ ions

are only a few per cent as intense as those yielding a single secondary ion. The shaded area represents the expected Poisson distribution from superposition of independent events based on the probability of detecting at least one ion per pulse. These independent events may be caused by the same primary ion or by different ions in the same primary ion pulse; ~10 ions/pulse for CsI and ~70 ions/pulse for thiamine were used. All but ~ 1% of the multiple events are accounted for by such independent events. Because of the dead time, multiple ions of the same mass are counted as one ion. The data were not corrected for this effect, but it is expected to be small.

Much of the multiplicity observed with fission fragments can also be explained by a superposition of independent events in a single track; higher multiplicity with fission fragments reflects the much higher probability of ion production. However, as shown in Fig. 3 for an organic compound, much of the tail of the measured distribution is not accounted for by independent events which would result in a Poisson distribution. Thus the correlated multiplicity also appears to be higher for fission fragments than for low-energy ions (note the log scale in Fig. 2). Neither measurement was corrected for spectrometer efficiency, but these corrections are expected to be similar.

We have also looked for correlations among specific ions from thiamine and CsI. Figure 4 shows the normal spectrum of thiamine and the spectrum taken in coincidence with the fragment ion at m/z 144; i.e. only the data from pulses which produced an ion detected at m/z 144 are recorded. The lines shown on the expanded scale are completely accounted for by independent events; the primary current was kept low to reduce this effect. Taking account of statistical error and the spectrometer efficiency, an upper limit of ~1% correlation between m/z 144 and 265 is found. Similar results are found for thiamine spectra taken in coincidence with m/z 122 and 265, and for CsI spectra taken in coincidence with the Cs^+ cation. This is in sharp contrast to the results reported for fission fragments [3]; in particular, nearly 100% correlation was found between the thiamine fragment at m/z 144 and the parent ion at m/z 265; i.e. if the fragment ion is desorbed, the parent ion is also desorbed with ~100% probability [3]. This

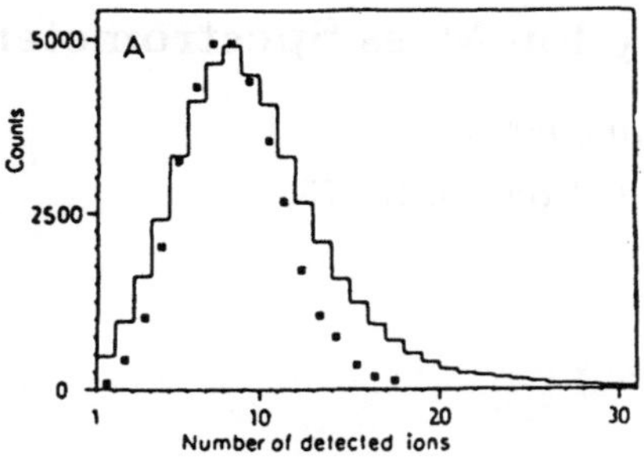

Fig. 3 Multiplicity for an organic compound with incident fission fragments. The squares represent a Poisson distribution with an average value of 8. Taken from [3]

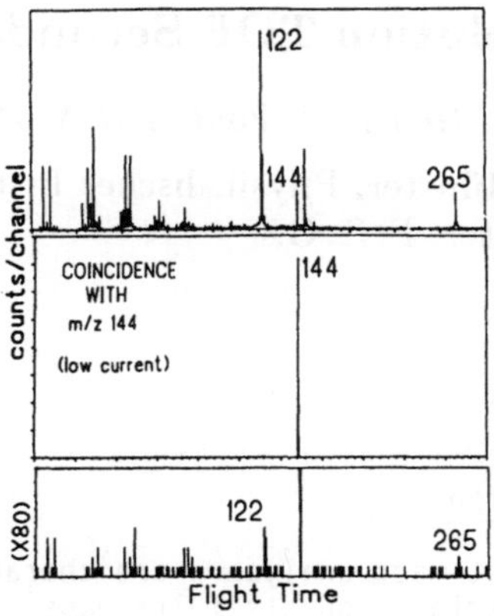

Fig. 4 Pos. ion spectra of thiamine. The lower spectra were taken in coincidence with the peak at m/z 144. The bottom trace is magnified 80 times.

is surprising, since the relative intensities of the 3 peaks (m/z 122, 144, 265) in our case are the same within 5% of those reported for fission fragments [2].

4. Discussion

We have observed very small multiplicity for keV ion bombardment compared to the results for incident MeV/u ions; the results reflect several orders of magnitude difference in yield. We have found no evidence for significant correlations among secondary ions desorbed by low-energy primary ions in contrast to the results reported for incident MeV/u ions. This appears to be consistent with recent measurements at Uppsala [9], which find no correlations at high energy, in disagreement with the earlier results [1,3].

5. Acknowledgements

This work was supported by grants from the U.S. National Institutes of Health, and from the Natural Sciences and Engineering Research Council of Canada. G. Bolbach received support from CNRS (France).

6. References

1. B.Sundqvist et al, in A. Benninghoven (Ed.), Ion Formation From Organic Solids, Springer-Verlag, Heidelberg, 1983, pp. 52-57.
2. S.Della Negra et al, Int J. Mass Spectrom. Ion Phys. 53, 215 (1983)
3. N.Fürstenau et al, Z. Naturforsch 32a, 71 (1977); N.Fürstenau, Z. Naturforsch 32a, 563 (1978)
4. B.T.Chait and K.G.Standing, Int.J. Mass Spectrom. Ion Phys. 40, 185 (1981)
5. W.Ens et al, submitted to Nucl. Instr. and Meth.
6. B.Schueler et al, these proceedings.
7. K.G. Standing et al, 3rd conf. on Ion Formation From Organic Solids, Münster, 16-18 Sept. 1985, to be published by Springer
8. I.Kamensky et al, Nucl. Instr. and Meth. 198, 65 (1982)
9. B.Sundqvist, private communication

High Resolution TOF Secondary Ion Mass Spectrometer

E. Niehuis, T. Heller, H. Feld, and A. Benninghoven

Universität Münster, Physikalisches Institut, Domagkstr. 75,
D-4400 Münster, F.R.G.

1. Introduction

Time of flight mass analyzers are characterized by a high and mass independent transmission, quasisimultaneous detection of all masses and an unlimited mass range. A new TOF-SIMS instrument is presented which combines these advantages with high mass resolution. It is equipped with a mass separated pulsed primary ion source and a reflectron type TOF analyzer.

2. Design and performance of the instrument

2.1. Primary beam system

A mass separation in the primary system is necessary in order to get a high dynamic range. Any contamination in the primary beam will distort the secondary ion spectrum due to flight time differences in the primary system. We developed a new mass separation and beam chopping method based on a pulsed 90 degree deflection. The same momentum, defined by the time integral of the HV deflection pulse, is added to primary ions of different mass but same energy. Therefore their final momenta have different directions. After the 90 degree deflection the diverging angle is drastically reduced. This provides the separation of different masses by an appropiate drift path.

An electron impact ion source is used to produce a continuous beam of noble gas ions with 10 keV energy. The deflection field is switched on for 1 µs by a transistorized pulse generator with 6 kV amplitude and 20 ns rise- and falltime. Before the ions leave the deflector the pulse is switched off. At the exit slit a short bunch of ions, perpendicular to the direction of motion, is cut out of the beam. A drift path of 25 cm with an exit slit width of 1 mm is used for the mass separation. The achieved mass resolution of 150 is in good agreement with the calculations. The beam is focused onto the target by an Einzel-lens with a spot diameter of 50 µm.

Due to axial velocity differences the ion bunch spreads resulting in a pulse width of about 20 ns at the exit of the deflector. These velocity differences can be reversed by axial bunching. A second HV pulse is switched on when the ion bunch is in the center of the bunching system. Equal flight times to the target can be achieved by adapting the amplitude of the bunching pulse. The pulse width at the target depends on the entrance slit width of the deflector and the diverging angle of the incident ion beam. A pulse width of 1.3 ns is expected for 0.5 mm and 0.5 degrees. It is increased to 1.5 ns due to the nonnormal incidence on the target.

The width of the H^+-peak in the SI spectrum reflects the primary pulse width. It has been measured with a time-to-pulse-height converter (TAC) combined with a multichannel analyzer (MCA) and a digital delay. The peak width (1.7 ns) includes the time dispersion in the analyzer. Taking this into account the resulting primary pulse width is 1.5 ns.

2.2 Mass analyzer

High resolution time of flight SIMS requires energy isochronous focusing. Several types of electrostatic TOF-MS have been proposed using mirrors and sector fields. Second order energy focusing is achieved by a two stage reflector /1/. Multiple time and space focusing sector field instruments have been designed by POSCHENRIEDER /2/ and MATSUDA /3/. As the mass resolution of the proposed stigmatic imaging sector field instruments is limited by the quadratic errors of the entrance angles we have chosen the reflectron type analyzer.

The reflection angle of the analyzer is 177 degrees; the mirror consists of a retarding gap (30 mm) and a reflecting gap (60 mm) defined by grids of 90 % transparency. The total effective flight path is 2.2 m. In order to improve the transmission an Einzel-lens is used for the beam transport. This lens provides a large acceptance angle of +/- 1.5 degrees in both directions at a magnification of 50.

The conditions for second order energy focusing are described in the literature /1/ but in a SIMS reflectron the additional time dispersion in the acceleration gap limits the mass resolution. For a restricted energy range this time dispersion can be reduced by tuning the reflecting field to a lower field strength compared to second order energy focusing conditions. A mass resolution of about 15 000 can be expected for an acceleration gap of 2 mm at an energy width of 0.4 %.

The resolution was tested with the TAC-MCA. The base peak of Crystal violet (M.W. 372.24) has a 50%-width of 2.8 ns with a total flight time of 72.50 us corresponding to a mass resolution of 13000. The 10%-resolution is about 6700.

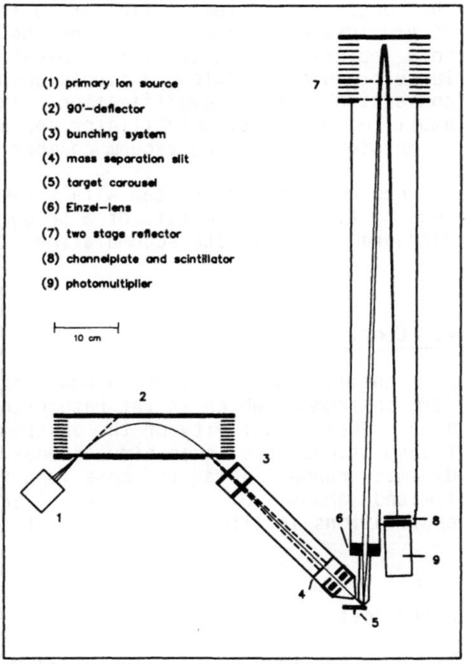

Fig.1 Schematic of the instrument

The spectrum of Cetrimoniumbromide, prepared on etched silver,(Fig. 2) demonstrates the dynamic range of the spectra. It was recorded with a 100 MHz signal averager which is limiting the mass resolution. A dynamic range of six orders of magnitude can be achieved due to the mass separation in the primary system. The quality of the primary pulse and of the energy focusing analyzer is reflected by the shape of the Ag 108.9 peak. It has a width of 10 ns at 3.5 orders of magnitude below the maximum.

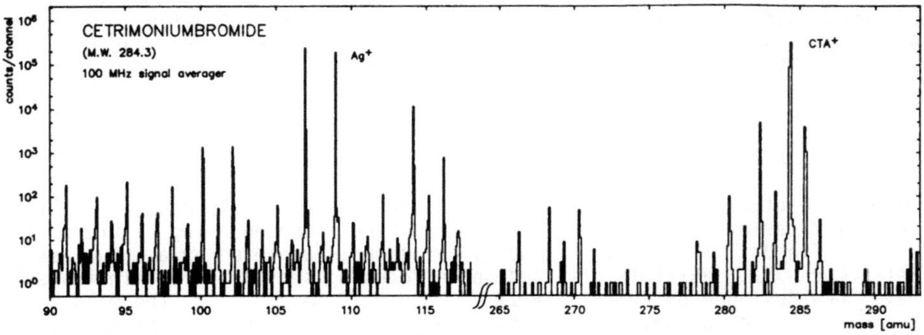

Fig.2 Positive SI spectrum of Cetrimoniumbromide prepared on etched Ag

2.3 Detection system

The detection system is similar to the one used in our first TOF SIMS /4/. It provides a postacceleration of the secondary ions up to 20 keV energy for both polarities in order to achieve a high detection efficiency even for large organic molecules. A single channel plate is used for the ion-electron conversion and preamplification. The signal is coupled to ground and amplified by a fast scintillator and a photomultiplier. The photomultiplier is mounted outside the vacuum chamber and its exit is at ground potential.

A new registration system with a time resolution of 1.25 ns is being under construction. It consists of a triggerable TDC with 256 stops and special hardware for fast data accumulation.

4. Conclusion

A transmission-resolution product of about 5000 can be attained by a TOF-SIMS instrument which is yet rather compact (overall length of the analyzer 1 m). The sensitivity of the instrument is extremely high due to the high transmission, the quasisimultaneous detection of all masses and the high dynamic range. This is most important for applications where only very limited amounts of sample material are available, for example static SIMS applications in life sciences and microanalysis.

References

1. V.I. Karataev, B.A. Mamyrin and D.V. Shmikk: Sov. Phys.-Tech. Phys., 16 (1972) 1177
2. W.P. Poschenrieder: Int. J. Mass Spec. Ion Phys.,9 (1972) 357
3. T. Sakurai, T. Matsuo and H. Matsuda: Int. J. Mass Spec. Ion Phys., 63 (1985) 273
4. P. Steffens, E. Niehuis, T. Friese, D. Greifendorf and A. Benninghoven: J. Vac. Sci. Technol. A 3 (1985) 1322

Part VI

Techniques Closely Related to SIMS

Evaluation of Accelerator-based Secondary Ion Mass Spectrometry for the Ultra-trace Elemental Characterization of Bulk Silicon

R.J. Blattner, J.C. Huneke, and M.D. Strathman

Charles Evans & Associates, 1670 South Amphlett Boulevard, Suite 120, San Mateo, CA 94402, USA

R.S. Hockett

Monsanto Electronic Materials Co., 800 N. Lindberg Boulevard, St. Louis, MO 63167, USA

W.E. Kieser, L.R. Kilius, J.C. Rucklidge, G.C. Wilson, and A.E. Litherland

Isotrace Laboratory, University of Toronto, Toronto, Canada M5S1A7

Introduction

Secondary ion mass spectrometry (SIMS), implemented using high efficiency ion extraction and transmission optics, compares favorably with other "standard" trace element analysis techniques such as spark source mass spectrometry or neutron activation analysis for bulk characterization of solid specimens. Typical SIMS detection limits range from 5 x 10^{16} atm-cm^{-3} (1 ppm) to 5 x 10^{13} atm-cm^{-3} (1 ppb). Detection limitations can arise, however, due to molecular ion interferences. Accelerator-based SIMS was developed to perform ultra-sensitive ^{14}C analyses because molecular secondary ions can be dissociated into atomic fragments in the stripper canal of a tandem ion accelerator [1]. This paper will present an evaluation of AB/SIMS analysis of B, P, As, Sb and FeSi [2].

Experimental

Samples consisted of bulk-doped silicon wafers with dopant concentrations ranging from 10^{19} - 10^{14} atm-cm^{-3}. Undoped ultra high purity float-zone (FZ) silicon (n-type with <1 x 10^{13} free carriers) was employed as a "blank." After thoroughly cleaning the sample region of the secondary ion source, a highly doped sample was sputtered to produce a pilot beam of the dopant species for instrument tuning. Subsequently, samples with lower impurity concentrations were analyzed in terms of impurity count rate (surface barrier detector) to silicon matrix ion intensity (Faraday cup at the injection end of the accelerator).

Results

Boron. Formation of a negative analyte ion is a prerequisite to AB/SIMS. For these measurements B^- was the injected species while B^{+3} was selected for detection after terminal stripping and final acceleration. The $^{11}B/^{10}B$ ratio served as an aid in qualitative identification. Data were taken (Table 1) after a high-energy ESA but before high-energy magnetic analysis.

*This work supported in part by the Defense Advanced Research Projects Agency under Contract MDA 903-83-0099

Table 1

Concentration	$^{10}B^{+3}$ (cps)	$^{11}B^{+3}$ (cps)	Si^- (namp)	$^{11}B/^{10}B$
$<1\times10^{13}$	51	2012	640	39.4
2×10^{17}	3064	8750	450	2.9

The anomalous isotopic ratio observed for the blank is attributable to a memory effect from an enriched ^{11}B specimen analyzed prior to our measurements. Additional background for both ^{10}B and ^{11}B is produced by the high doping concentration sample required for instrument tuning. The detection limit for boron was calculated to be 2×10^{15} atm-cm^{-3} based on the ^{10}B data. Linear extrapolation to a signal limited situation assuming no background and a detector "noise" of 0.1 counts per second gives a value of 1×10^{12} atm-cm^{-3} for ^{11}B.

Arsenic. Three detection protocols were investigated for As including: 1) atomic ion injection ($^{75}As^-$) with high-energy electrostatic filtering, 2) atomic ion injection with high-energy electrostatic filtering and post acceleration foil stripping followed by magnetic filtering, and 3) molecular injection of mass 103 amu (predominantly $^{75}As^{28}Si^-$) with molecular dissociation and charge stripping at the terminal followed by high-energy electrostatic filtering, foil stripping and final magnetic analysis. Foil stripping prior to final magnetic filtering was required because the high energy magnets were limited in terms of mass-energy product per unit charge. (Plus three charge state energy was typically 8 MeV.) Injection of As^- followed by terminal stripping to a charge state of +3 and foil stripping to a charge state of +10 gave an extrapolated As detection limit of 4×10^{11} atm-cm^{-3} (exclusive of any instrumental background) based on an observed signal of 320 counts per second (cps) for a known As doping concentration of 4×10^{14} atm-cm^{-3}. The ultimate detection limit could conceivably be a factor of five lower by circumventing the foil stripping process.

Antimony. Like As, intense unidentifiable beams were detected for "Sb" after the high-energy ESA using the straight through port (necessitating high energy magnetic analysis). Although it was later found to not be optimal, the analytical combination used was Sb^{+3} from the accelerator and Sb^{+6} after foil stripping. The +6 component of the charge state distribution represents about 5% of the total incident Sb^{+3} beam (effective "foil transmission"). The isotopic ratio ($^{121}Sb/^{123}Sb$) measured using the highest Sb doping concentration sample was 1.28 versus the natural abundance value of about 1.34. At an injected $^{28}Si^-$ current of 440 namp, count rates for the 121 and 123 amu isotopes were respectively 10,400 and 8,200 cps for an Sb concentration of 5×10^{18} atm-cm^{-3}. A 1×10^{14} atm-cm^{-3} doped sample gave an $^{121}Sb^{+6}$ signal of 2.2 cps at 615 namp of injected $^{28}Si^-$ while the FZ sample gave a count rate of 0.6 cps for $^{121}Sb^{+6}$ at a $^{28}Si^-$ current of 790 namp. If we extrapolate the $^{121}Sb^{+6}$ count rate for the lowest doping level sample we obtain a detection limit of 5×10^{12} atm-cm^{-3}. However, we have already noted a 20-fold loss in analyte signal resulting from foil stripping, such that additional improvement could be obtained (in the absence of background). A detection limit of 5×10^{11} atm-cm^{-3} or lower for Sb in Si should be achievable by eliminating background sources and replacing the high-energy filtering magnets.

Phosphorus. No useful data for trace levels of P in Si could be obtained due to insufficient abundance sensitivity in the low-energy mass spectrometer and/or an unresolvable interference from silicon monohydride ($^{30}Si^1H^-$) dissociation in the charge stripper canal. Results for P in Si have been reported by a group from Texas Instruments using the AB/SIMS instrument at the University of Arizona [3].

Iron. Efforts to detect Fe in a highly doped (1×10^{19} atm-cm^{-3}) silicon sample were unsuccessful. The requirement that the source produce an intense negative ion beam of the analyte species for instrument tuning precluded unambiguous detection of Fe. In future experiments it is conceivable that by introducing oxygen into the sample chamber, FeO^- or some similar molecular ion may be used as the injection species. Alternatively, a positive ion source (e.g., I^+) combined with an alkali metal charge exchange canal may allow the direct production of Fe^- [4].

IV. Conclusions

Figure 1 shows a plot of normalized signal intensity versus concentration for the doping species B, As and Sb in Si as measured by AB/SIMS. A linear relationship appears to exist at the higher concentrations, but deviations from linearity occur at lower concentrations, in particular for the FZ-Si blank. These deviations have been attributed to a memory effect in the sample region, produced by initially sputtering samples with high analyte concentrations during instrument tuning. All three dopant elements were restricted to a useful detection limit of about 5×10^{14} in these experiments because of background.

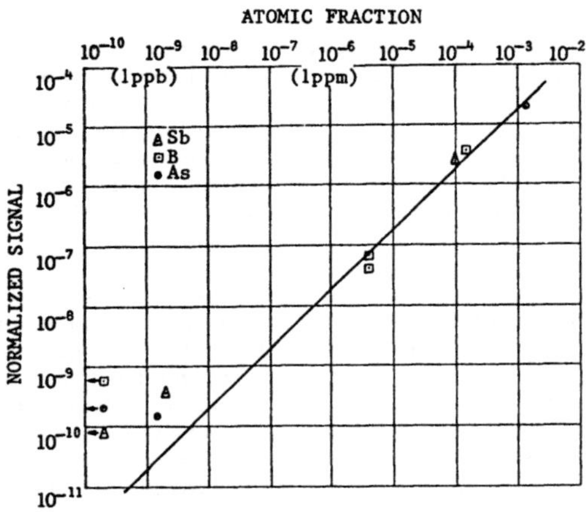

Figure 1

References

1. K. H. Purser: Revue de Physique Appliquee, v. 12, 1487-1492.

2. A. E. Litherland: "SIMS at Higher Energies" in Proc. SIMS IV, Osaka, (1983), Springer Series in Chem. Phys. 36, p.170.

3. D. J. Donahue: "Detection of Trace Amounts of ^{31}P and ^{11}B in Silicon by Accelerator Mass Spectrometry," Bulletin of the American Physical Society 28, p. 991.

4. Isotrace Laboratory 1984 Annual Report, University of Toronto, 31 December (1984).

Quantitative Analysis with Laser Microprobe Mass Spectrometry

R.W. Odom and I.C. Niemeyer

Charles Evans & Associates, 1670 South Amphlett Boulevard, Suite 120, San Mateo, CA 94402, USA

In this paper we discuss the quantitative analysis of two different materials using the laser microprobe technique. These materials include a microanalytical metallurgical sample and a series of phosphosilicate glasses (PSG) containing different P concentrations. These analyses were performed on a reflection mode laser microprobe instrument (Cambridge Mass Spectrometry Ltd., Cambridge, England, Model LIMA 2A) which has been described in detail elsewhere [1].

The metallurgical standard analyzed in this work is the CrFeNi alloy produced by the National Bureau of Standards (NBS) as a microprobe standard. This standard (NBS 479a) is comprised of 18.1% Cr, 71.0% Fe and 10.9% Ni by weight. The weight percent variation of these elements for a given specimen lie with a 99% probability within the following intervals: Cr: 17.6-18.6, Fe: 69.6-72.4, Ni: 10.7-11.1, respectively. Fifteen positive-ion laser microprobe mass spectra were obtained from this sample at different regions. These spectra were produced at relatively high laser power conditions in which the mass 56 ion signal intensity was near the saturation level of the four decade logarithmic preamplifier, which corresponds to a 10V input from the electron multiplier (EM)/line driver circuit. If one assumes a nominal 50 nsec ion signal peak width, a gain of 5×10^5 on the EM and using a 500Ω load resistor, a 10V signal corresponds to approximately 10,000 ions arriving at the EM. Several low-power density laser pulses were fired onto a given analytical region before the high-power pulse. This was done in order to "clean" the analytical surface of contamination (viz., Na, K and other environmental species). These low-power pulses did not produce visible craters in the alloy. The analytical laser pulses produced craters 2 to 5 µm in diameter and approximately 0.5 µm deep. A positive ion SIMS analysis was also performed on this sample. Oxygen primary beam bombardment was utilized and the analysis was performed on the CAMECA IMS-3f ion microanalyser. The analysis conditions included utilizing an O_2^+ beam current of 1.0 µA rastered over a 250 µm area, the number 1 field aperture and the 20 contrast diagram setting. SIMS ion imaging was employed to assure the uniform distribution of the Cr, Fe and Ni in this sample. Table 1 summarizes the results of the analyses.

Table 1 Results of a Laser Microprobe and SIMS Analysis of NBS 479a CrFeNi Microprobe Standard

Element	Atomic %	Ratio	True	Measured			
				LIMS (%RSD)	SIMS	RSF_{LIMS}	RSF_{SIMS}
Cr	17.87	52/56	0.2250	0.7273(13.1)	1.707	3.232	7.589
Ni	9.53	58/56	0.0972	0.0994(24.8)	0.0248	1.023	0.255

This table displays the measured and true atomic fraction of ^{52}Cr and ^{58}Ni to ^{56}Fe for this sample. The relative sensitivity factor (RSF) values in this table are defined by X^+/X / $^{56}Fe^+/^{56}Fe$ where X^+ and X are the ion signal intensity and atomic concentration of ^{52}Cr or ^{58}Ni, respectively.

The RSD of the laser microprobe measurements are also indicated in this table. It is interesting to note that although the SIMS RSF values for Cr and Ni differ by a factor of 30, the LIMS RSF values differ by only a factor of three. In addition, the laser microprobe Ni/Fe RSF is nearly equal to the true atomic ratio of these two elements. These relatively small differences in the measured LIMS RSF values are similar to those observed by Dingle and Griffiths (3) in the laser microprobe analysis of Ca, O, Al and Na in a silicon glass standard. These authors observed that the largest difference in the RSF values for these four elements (with respect to Si) differed by a factor of 6.

A second example of quantitative analysis by the laser microprobe was the analysis for phosphorus in a series of PSG glasses containing different P concentrations. The homogeneity of these 2 μm thick specimens was ascertained using electron probe microanalysis (EPMA) and secondary ion mass spectrometry (SIMS). In addition, the bulk P concentration in these films was determined by spectrophotometry and Rutherford backscattering spectrometry (RBS) [2]. Five PSG samples containing .48, 1.09., 1.36, 3.05 and 5.03 atomic % P were analyzed by the laser microprobe technique. Fifteen mass spectra were obtained from 15 different regions on each sample, and the laser power density in these analyses was adjusted to yield a $^{28}Si^+$ ion signal near the detector saturation level. The high-power laser pulses produced craters approximately 10 μm in diameter and 0.5 to 1.0 μm deep. The integrated peak areas for the Si isotopes, elemental phosphorus and the PO cation were evaluated from these spectra and the average value of the $^{31}P/^{28}Si$ and $(^{31}P + {}^{31}PO)/^{28}Si$ ion intensity ratios are plotted in Fig. 1. The relative standard deviation of the (P+PO)/Si ratios are also indicated in this figure. These RSD values ranged between 50 and 30% and were observed to decrease with increasing P concentration. The P/Si ratios exhibited similar RSD values. The correlation coefficients of the linear least squares fitted lines for the P/Si and (P+PO)/Si data as a function of P atomic % were .863 and .950, respectively.

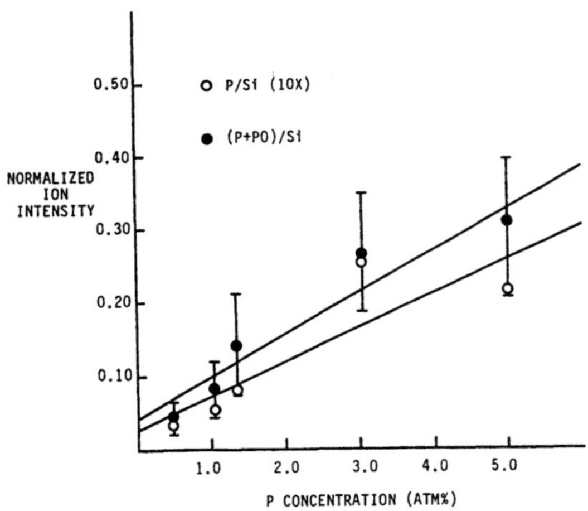

Fig. 1 P and PO cation signals normalized to ^{28}Si for different P concentrations in PSG.

Acknowledgements

Partial support for this work was provided by NIH Grant #R44 GM33123-03.

References

1. T. Dingle, B. W. Griffiths, J. C. Ruckman and C. A. Evans, Jr., Microbeam Analysis--1982, Ed. K. F. J. Heinrich, San Francisco Press, 1982.
2. P. K. Chu and S. L. Grube, Anal. Chem. 57, 1071 (1985).
3. T. Dingle and B.W. Griffiths, Microbeam Analysis-1985, Ed. J. T. Armstrong, San Francisco Press, p. 315 (1985).

Laser-Probe Microanalysis: Aspects of Quantification and Applications in Materials Science

M.J. Southon, A. Harris, V. Kohler, S.J. Mullock, and E.R. Wallach

Department of Metallurgy and Materials Science, University of Cambridge, Cambridge, CB2 3QZ, U.K.

T. Dingle and B.W. Griffiths

Cambridge Mass Spectrometry Ltd., Science Park, Cambridge, CB4 4BH, U.

Laser-probe microanalysis is a technique similar in some respects to SIMS; a primary beam, of light rather than ions, is directed onto a specimen to excite secondary ions, which are then analysed using a time-of-flight mass-spectrometer. The LIMA 2A instrument used for this work has been described elsewhere in the literature [1], as has the similar LAMMA instrument [2]. The use of the laser/time-of-flight mass-spectrometry combination produces unique advantages and disadvantages compared to other microanalytical techniques. These will be discussed with particular emphasis on possible methods for quantifying the results and illustrated by reference to specific examples.

Reproducibility and Quantification

The major obstacle to quantifying the results obtained by the LIMA instrument is the large scatter observed in both absolute and, to a lesser extent, relative peak areas in the mass-spectrum. The former was investigated by repeated analysis of pure, single-crystal silicon. These results showed that the scatter in absolute peak-areas (assumed to be directly proportional to the ion-yield) was very dependent on the laser-power used for the analysis. At a power barely high enough to ionise the silicon, the results had a scatter of about 250% (2 standard deviations for 25 results), whilst at twice this power the scatter was reduced to 20%. The shot-to-shot variation in the incident laser-power has been measured as only about 5%, and can be discounted as the primary cause of the scatter, instead it is suggested that the observed variation is due to differing degrees of absorption of the laser light by the specimen. The relative peak-areas also exhibit scatter, but this is somewhat reduced compared to the above. For a typical metallic specimen the scatter in relative areas is about 10%, and is again improved at higher laser-powers and also for elements of lower ionisation potentials.

The quantification of LIMA results obviously involves relating the measured relative ion-yields (peak area ratios) to the actual composition of the specimen. There are several possible approaches to this problem [2,3], either involving establishing a "working-curve" based on a known standard [4] or by attempting to predict the results theoretically using the Local Thermodynamic Equilibrium model [5]. Our results, derived from numerous analyses of E-glass and metallic glasses, indicate that it is possible to establish working-curves but the inherent variability of the results, as discussed above, means that large amounts of data are required to establish valid curves. Application of LTE theory to these results allows the calculation of conditions within the plasma, and the resultant values for electron temperature (in the range 10000-16000K) and electron density (10^{17} to 10^{20} cm^{-3}) fall well within those expected for a plasma in local thermodynamic equilibrium [5].

Applications

Rapid solidification of aluminium-lithium alloys provides a method for increasing the lithium content and improving the chemical homogeneity of the material. This provides a potential technique for enhancing the strength of Al-Li at elevated temperatures. Methods for characterising the chemical segregation of Li and other alloying elements in this material are therefore of importance in the understanding of its mechanical behaviour. Gas-Atomised Al-Li powder solidifies with a cellular structure, each cell about 5 μm in diameter or less. The intercellular regions have a high impurity concentration, while the intracellular regions remain relatively pure. LIMA was able to identify which elements segregate to the cell walls and to measure the partitioning of Li to these regions. Figure 1 shows significant levels of Ga, Mg, Cu, Fe and Li at the interface [6]. The minimal sample preparation required for the analysis, as well as the full elemental range capabilities of LIMA provide a ready analytical technique for a problem difficult to solve by conventional X-ray microanalysis or electron energy-loss techniques.

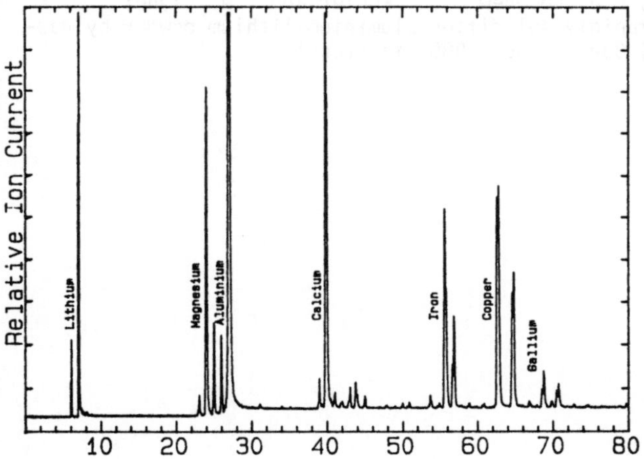

Fig. 1. Spectrum of intercellular region in Al-Li alloy.

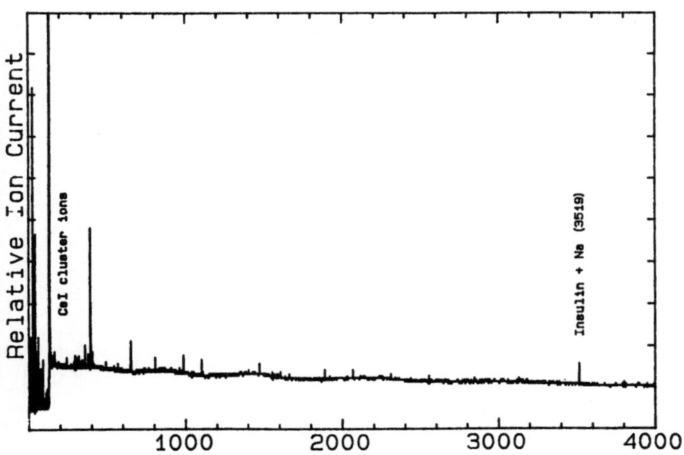

Fig. 2. Oxidised insulin molecular ion

Molecular ions with masses of several thousand amu have been detected usin LIMA. Positive and negative ion spectra have been otabined from a sample of bovine insulin (molecular mass 5780 amu) with sodium chloride as an additive to assist the production of positive molecular ions. Oxidised insulin (molecular mass 3496 amu) has also been analysed, this time combined with caesium iodide in glycerol. The caesium iodide gives a set of cluster ions of the general formula Cs_nI_{n-1} which provides a mass calibration scale up to about 2600 amu. Peaks were also observed corresponding to the addition of sodium or caesium to the insulin, at masses 3519 and 3529 respectively. Figure 2 illustrates this detection of relatively high-mass organic molecules.

References

1. M.J. Southon et al., Vacuum 34, 23 (1984).
2. R. Kaufmann, Microbeam Analysis 1982, 341.
3. M.J. Southon et al., Microbeam Analysis 1985, 310.
4. T. Dingle and B.W. Griffiths, Microbeam Analysis 1985, 315.
5. C.A. Anderson and J.R. Hinthorne, Anal. Chem. 45, 1421 (1973).
6. R.A. Ricks, P.M. Budd, P.J. Goodhew, V.L. Kohler and T.W. Clyne: "Production of fine, rapidly solidified aluminium-lithium powder by gas-atomisation", in Al-Li Conf. Proc. 1985 (in press).

Some Aspects of Laser-Ionisation Mass-Analysis (LIMA) in Semiconductor Processing

M.J. Southon, A. Harris, and V. Kohler

Department of Metallurgy and Materials Science, University of Cambridge, Cambridge, CB2 3QZ, U.K.

G.D.T. Spiller

British Telecom Research Laboratories, Ipswich IP5 7RE, U.K.

The Laser-Ionisation Mass-Analyser (LIMA) [1] is a relatively new microanalytical instrument, essentially a variant of SIMS, which exhibits capabilities that should both complement and augment more established techniques in the field of semiconductor analysis. The instrument, which has been described elsewhere in the literature [2], of a laser for sample excitation coupled with a time-of-flight mass-spectrometer for the analysis of the ions thus produced. The possibility of obtaining fully quantitative analyses from laser-probe instruments has been discussed on numerous occasions, both with regard to the LIMA [3,4] and the very similar LAMMA instrument [5,6,7]. Both theoretetical and empirical approaches have been attempted, the latter being more successful initially, although both are hindered by the relatively large spread found in the raw results. The lateral and depth resolution of the LIMA technique have not been fully assessed as yet, but, as the examples will illustrate, they appear to lie in the ranges 1-3μm and 100-300nm respectively.

The general performance and usefulness of LIMA as an analytical technique in the field of semiconductors is being assessed as part of the UK Alvey programme, with particular attention to its use for a rapid initial survey of a specimen. Some applications of LIMA to semiconductors were discussed by Evans at SIMS-IV [8], and the following two examples will illustrate our own current work.

The first example consisted of two specimens of a 1μm CVD poly-silicon layer over SiO_2, requiring a general trace impurity analysis. Figure 1 shows a LIMA spectrum from one of the specimens; the principal impurities found were carbon and oxygen, with low levels of aluminium and sulphur in addition. Peaks were occasionally found at m/e = 14 and 56 (apparently nitrogen and iron) but are more likely to be due to Si^{2+} and $(Si_2)^+$ respectively. The measured levels of the impurities showed considerable variation, with an apparently higher level of carbon at the surface, due at least in part to surface contamination. As can be seen, the levels of carbon and oxygen are comparable in concentration to the 29 and 30 isotopes of silicon (at 4.7 and 3.1% respectively), and, since previous experience has shown that relative sensitivities to these elements are about 0.5-1 compared to silicon [4], the concentrations should be in the 1-2% atomic range. SIMS analysis of the same specimen yielded comparable results, and also found slightly increased levels near the surface.

The second of these specimens contained much lower impurity levels. In this case the high background meant that SIMS was unable to definitely detect either carbon or oxygen. In contrast, LIMA could detect both these elements, and the concentrations were estimated as about two orders of magnitude lower than in the previous specimen. Once again aluminium and sulphur were also

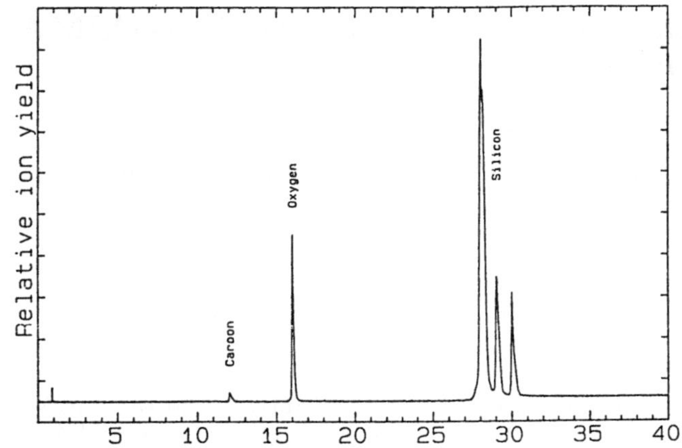

Fig. 1. Impurities in CVD poly-silicon

found. In both specimens the concentrations seemed to be about 100ppm and 500ppm respectively. In addition, both specimens yielded clear peaks at masses 133 and 181, corresponding to the elements caesium and tantalum. The presence of caesium is attributed to the prior analysis of the specimen by SIMS utilising a primary beam of caesium ions, whilst the tantalum had been produced by the sputtering of a support composed of the same material in the same SIMS instrument. In contrast, no such contamination is produced during LIMA analysis.

Evidence for both the trace sensitivity and depth resolution of LIMA was provided by a specimen which consisted of a 1μm GaInAs layer on InP. The layer was doped with zinc at 10^{20} atoms cm^{-3} to a depth of 0.4μm and with sulphur at 10^{16} atoms cm^{-3} throughout; these correspond to about 0.2%at and 0.2ppm respectively. As expected, the zinc was readily detectable, with the zinc peaks generally appearing in the spectra obtained from the first three or four shots at a given location: however, it proved difficult to detect sulphur above the noise level with any certainty. The zinc results indicate that each shot is removing about 0.1μm of the specimen, whilst the sulphur puts a lower limit on the trace sensitivity.

Figure 2 shows a LIMA spectrum derived from the GaInAs layer, showing the peaks due to these elements. Integration of the peaks yields an apparent

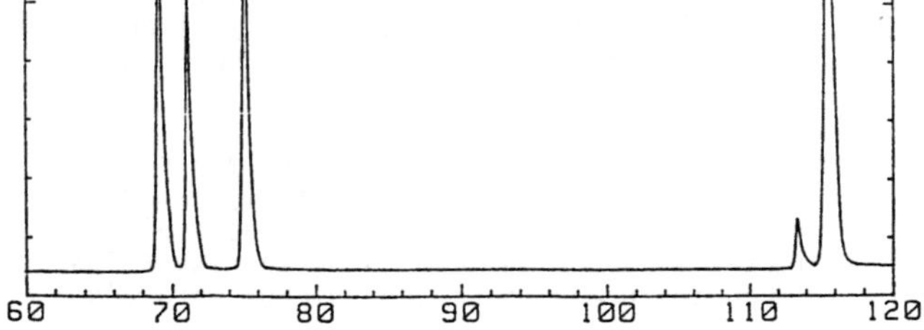

Fig.2. Apparent composition of GaInAs

stoichiometry of about $Ga_5In_5As_2$ compared with the known composition of $(Ga_{0.47}In_{0.53})As$, hence the gallium-to-indium ratio is apparently correct, but arsenic is measured as only 20% of the correct value. When the ionization energies are considered it is seen that those of gallium and indium are about the same (6.0 and 5.8eV respectively) whilst that of arsenic is distinctly higher, at 9.8eV, so arsenic would be expected to be underestimated. This is a good illustration of the use of LIMA's rapid depth profiling, where it was also found that the impurities associated with the GaInAs-InP interface; hydrogen, carbon and silicon, also occurred at reduced levels within the GaInAs layer itself. A rapid survey of the whole layer takes only ten or fifteen laser shots and would provide the basis for a more detailed analysis using other techniques.

Overall, it is suggested that LIMA will not completely supplant any established microanalytical technique, but that it does provide a valuable complement to those already available. Future work should improve the accuracy and precision of LIMA results and extend the scope of its applications.

Acknowledgement
This work has been carried out with the support of the Procurement Executive, UK Ministry of Defence, sponsored by DCVD. Acknowledgement is also made to the Director of Research of British Telecom for permission to publish this paper.

References

1. LIMA 2A, Cambridge Mass Spectrometry Ltd., Science Park, Cambridge.
2. M.J. Southon et al., Vacuum 34, 23 (1984).
3. M.J. Southon et al., Microbeam Analysis 1985, 310.
4. T. Dingle and B.W. Griffiths, Microbeam Analysis 1985, 315.
5. R. Kaufmann, Microbeam Analysis 1982, 341.
6. T. Mauney, Microbeam Analysis 1985, 299.
7. T. Mauney, Ph.D. Thesis, Colorado State University 1984.
8. C.A. Evans jr., Private communication.

Part VII

Combined Techniques and Surface Studies

The Use of SSIMS and ISS to Examine Pt/TiO_2 Surfaces

G.B. Hoflund and D.A. Asbury

Department of Chemical Engineering, University of Florida, Gainesville, FL 32611, USA

Shin-Puu Jeng and P.H. Holloway

Department of Materials Science and Engineering, University of Florida, Gainesville, FL 32611, USA

In the study of catalytic surfaces, it is very important to use techniques which are highly surface sensitive because the catalytic reactions actually occur at the outermost layer. SSIMS and ISS are two techniques which meet this criteria, and their combined use provides a particularly powerful approach for the study of complex surfaces. It is interesting to consider the similarities and differences between SSIMS and ISS because both enhance the synergism which makes the combined use much more effective than the use of either technique individually [1,2]. ISS and SSIMS are similar in that they both yield surface-sensitive, compositional information, they both employ a primary beam of inert gas ions as an excitation source and neither is well suited for determination of oxidation state or geometrical structure. ISS is essentially outermost layer sensitive while SSIMS samples the top few layers. ISS is most sensitive to heavy elements, and SSIMS is usually most sensitive to lighter species. This fact allows for a more complete mass analysis using both techniques rather than either one individually. Also, SSIMS has higher mass resolution and dynamic range than ISS.

The behavior of platinized titania surfaces has been a controversial topic since the report of strong-metal-support interaction (SMSI) phenomena by Tauster et al. [3,4]. SMSI phenomena is characterized by suppression of the amounts of H_2 and CO which chemisorb on oxide-supported metal crystallites. This suppression occurs after a high-temperature (500°C) reduction in hydrogen and is particularly pronounced when the oxide support is reducible. Annealing in oxygen followed by a low-temperature (200°C) reduction usually restores most of the chemisorption capacity. Numerous explanations of SMSI phenomena have been proposed including suggestions of an electronic interaction between the supported metal and the oxide, i.e. charge transfer [5-11], changes in metal crystallite morphology [12], encapsulation of the metal crystallites by support species which form during reduction of the support [13-24] or by impurity species, creation of special active sites at the metal-support interface, hydrogen spillover and others. Several studies demonstrate that encapsulation does occur, and that the electronic effects can be insignificant [25], small [19] or quite significant [12].

The combined use of ISS and SSIMS provides an excellent means of studying the encapsulation phenomena. Figure 1A shows

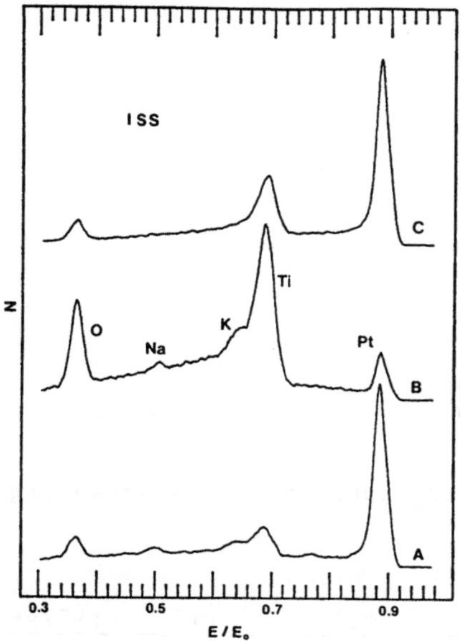

Fig. 1 ISS spectra of the Pt/TiO_2 surface A: before annealing, B: after annealing for 1 hour at 500°C and C: after Ar^+ sputtering

an ISS spectrum of a Pt/TiO_2 surface before annealing. The Pt peak is the prominent feature, and small peaks due to Ti, O, K and Na are also present. The spectrum shown in Fig. 1B is obtained after annealing the sample at 500°C. Clearly, encapsulation occurs during the anneal. This is verified by sputtering the encapsulating species off the surface, which results in the spectrum shown in Fig. 1C. This spectrum is quite similar to the spectrum shown in Fig. 1A except that the Na and K contamination has been removed by sputtering.

A SSIMS spectrum of the low-mass range is shown in Fig. 2. More peaks due to numerous species which do not appear in the ISS spectra are observed in the SSIMS spectrum due to its high sensitivity. The mass resolution is much higher, but peak identification is considerably more difficult. The tremendous variation in SIMS cross-sections makes it quite difficult to quantify the SSIMS results. However, ISS cross-sections increase with mass, so it is easier to quantify ISS results particularly when the masses of the surface species are similar. Hydrogen is observed with SSIMS whereas it cannot be seen with ISS due to its low mass.

Table 1 presents the SSIMS results regarding encapsulation. These data are difficult to obtain due to the extremely small SSIMS cross-section for Pt. Even though the Pt signals are small, the Ti^+/Pt^+ ratio increases with anneal temperature suggesting encapsulation. This is consistent with the ISS results. The larger Pt^+ signal obtained after the 500°C anneal is believed to be a real effect and due to an increase in cross-section caused by an interaction between the Pt and encapsulating species.

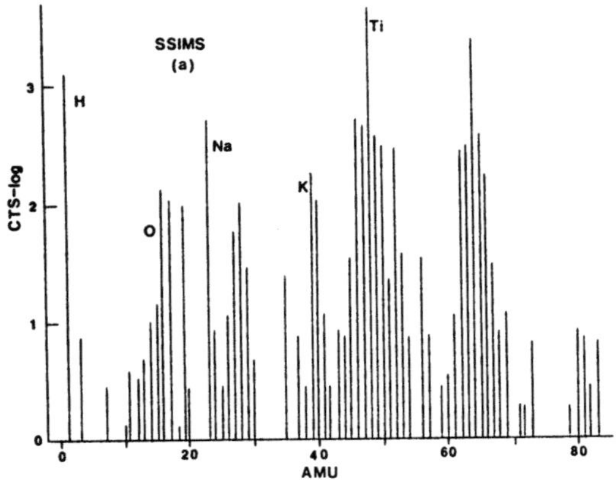

Fig. 2 SSIMS spectrum of the Pt/TiO_2 surface before annealing.

Table 1 SSIMS Encapsulation Results

SSIMS total counts for	Ti^+	Pt^+	Ti^+/Pt
Treatments			
before annealing	780	17	45
annealing at 250°C for 10 min.	2380	8	300
annealing at 500°C for 10 min.	4280	11	400

Acknowledgements

This research was supported by NSF grant # CBT-8146380.

References

1. C.A. Evans, Jr.: Anal. Chem. 47, 818A (1975)
2. R. Jede, E. Manske, L.D. An, O. Ganschow, U. Kaiser, L. Wiedmann and A. Benninghoven: "Secondary Ion Mass Spectrometry SIMS-III," eds. A. Benninghoven et al. (Springer 1979), p. 234
3. S.J. Tauster, S.C. Fung and R.L. Garten: J. Am. Chem. Soc. 100, 170 (1978)
4. S.J. Tauster and S.C. Fung: J. Catal. 55, 29 (1978)
5. C.C. Kao, S.C. Tsai and Y.W. Chung: J. Catal. 73, 136 (1982)
6. M.K. Bahl, S.C. Tsai and Y.W. Chung: Phys. Rev. B 21, 1344 (1980)
7. S.C. Fung: J. Catal. 76, 225 (1982)
8. S.C. Chien, B.N. Shelimov, D.E. Resasco, E.H. Lee and G.L. Haller: J. Catal., 77, 301 (1982)
9. B.A. Sexton, A.E. Hughes and K.J. Foyer: J. Catal. 77, 85 (1982)
10. J. Horsley: J. Am. Chem. Soc. 110, 2870 (1979)
11. S.J. Tauster, S.C. Fung, R.T.K. Baker and J.A. Horsley: Science 211, 1121 (1981)
12. M.A. Vannice, L.C. Hasselbring and B. Sen: J. Phys. Chem. 89, 2972 (1985)

13. X-Z Jiang, T.F. Hayden and J.A. Dumesic: J. Catal. 83, 168 (1983)
14. R.T.K. Baker, E.B. Prestridge and L.L. Murrell: J. Catal. 79, 348 (1983)
15. P. Meriadueau, O.H. Ellestad, N. Dufaux and C. Naccache: J. Catal. 75, 243 (1982)
16. D.E. Resasco and G.L. Haller: J. Catal. 82, 879 (1983)
17. J.A. Cairns, J.E.E. Baglin, G.J. Clark and J.R. Ziegler: J. Catal. 83, 301 (1983)
18. H.R. Sadeghi and V.E. Henrich: J. Catal. 87, 279 (1984)
19. D.N. Belton, Y-M Sun and J.M. White: J. Phys. Chem. 88, 5172 (1984)
20. S. Taketani and Y-W Chung: J. Catal. 90, 75 (1984)
21. R.A. Demmin, C.S. Ko and R.J. Gorte: J. Catal. 90, 59 (1984)
22. J. Santos, J. Phillips and J.A. Dumesic: J. Catal. 81, 147 (1983)
23. A.J. Simoens, R.T.K. Baker, D.J. Dwyer, C.R. Lund and R.J. Madon: J. Catal. 86, 359 (1984)
24. Y-W Chung, G. Xiong and C.C. Kao: J. Catal. 85, 237 (1984)
25. D.J. Dwyer, S.D. Cameron and J. Gland: Surface Sci. 159, 430 (1985)

Secondary Ion and Auger Electron Emission by Ar^+ Ion Bombardment on Al-Fe Alloys

Y. Fukuda

Tech. Res. Center, Nippon Kokan K.K., Kawasaki, 210 Japan

1. Introduction

Secondary ion emission for binary alloys has been widely studied by Secondary Ion Mass Spectrometry (SIMS) using noble gas and oxygen ions [1]. However, few studies on Auger electron emission induced by low-energy ion bombardment have been reported [2]. It was found that there exists a quadratic relationship between ion-induced Auger electron intensity and bulk aluminium concentration for Al-Fe alloys [2], which suggests that the Auger electrons are produced by symmetric collisions (Al--→ Al) [2,3]. Recently a mechanism of the secondary ion emission for aluminium has been discussed in terms of Auger deexcitation [2]. However, few studies on Auger electron emission by low-energy ion bombardment for transition metals in alloys have been reported [2]. Results on secondary ion and Auger electron emission by Ar^+ ion bombardment for aluminium and iron in Al-Fe alloys are presented.

2. Experimental

Al-Fe alloys with various compositions (Al=8-82%) were prepared by melting of aluminium (99.999%) and iron (99.99%) powder in vacuum. The bulk composition of Al and Fe was determined by EPMA. Auger and SIMS spectra were simultaneously measured during Ar^+ ion bombardment in a ultra-high vacuum chamber with base pressure $3x10^{-10}$ Torr. The alloy surfaces were bombarded with 5 keV Ar^+ ion using a differential-pumped ion gun. Ion currents on the samples were 6 μA. The electron-excited Auger spectra of the alloys were also measured.

3. Results and Discussion

Fig. 1 shows the surface concentration(obtained by electron-excited Auger spectroscopy) of aluminium for the alloys as a function of the bulk, in which the Auger sensitivity factors were assumed to be 0.05, 0.96, 0.20, and 0.31 for Al KLL(1396 eV), Al LVV(68 eV), Fe LMM(703 eV), and Fe LVV (47 eV), respectively[4]. The surface composition obtained by the low energy peaks is in good agreement with that of the bulk. However, it is less than that of the bulk in the case of using the high-energy peaks. This would be due to incorrect sensitivity factors for Al KLL and Fe LMM peaks.

Fig. 2 shows changes of the intensity ratio of Fe^+ to Al^+ ions as a function of the bulk iron composition. The intensity of secondary ions,I, is generally expressed as the following;

$$I = \gamma S I_p P C \qquad (1)$$

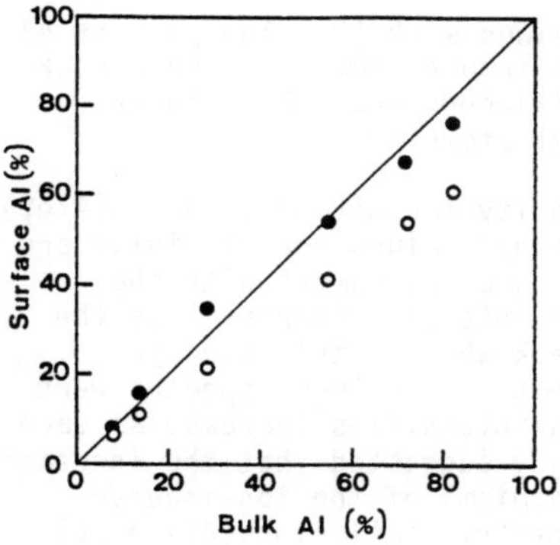

Fig.1 Al surface composition on Al-Fe alloys obtained by AES as a function of the bulk Al. ● and o were obtained by the low and the high-energy peaks in electron-excited AES, respectively.

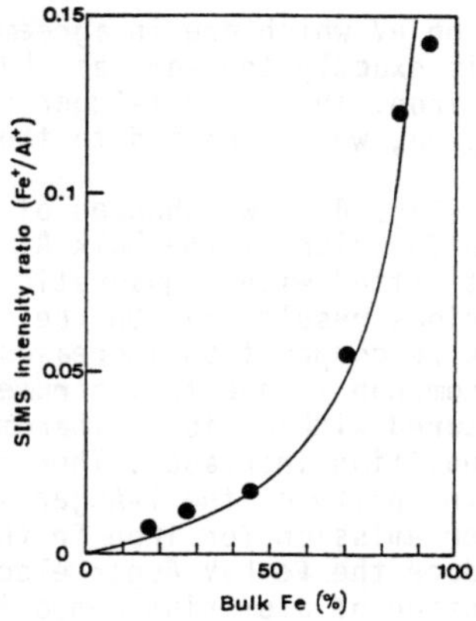

Fig.2 Changes of I_{Fe^+}/I_{Al^+} as a function of the bulk Fe composition.

in which γ , S, I_p, P, and C correspond to constant factors due to the apparatus, the sputter yield, the incident ion current, the ionization probability, and the atomic concentration of the samples, respectively. If S·P is constant in the alloys, the eq.(1) is simplified as follows since γ and I_p can be kept constant,

$$I = K\ C \tag{2}$$

Therefore, the intensity ratio of Fe^+ to Al^+ ions is,

$$\frac{I_{Fe^+}}{I_{Al^+}} = \frac{K_{Fe^+}}{K_{Al^+}} \quad \frac{C_{Fe}}{1 - C_{Fe}} \tag{3}$$

in which $C_{Fe} + C_{Al} = 1$. A best fit curve with $K_{Fe^+}/K_{Al^+} = 0.021$ is also shown in Fig. 2. The theoretical curve is in good agreement with the measured values. This implies that S·P is constant in the alloys. Since the sputter yield of Al and Fe in the alloy is constant [2], the ionization probability,P, is constant as well.

Fig. 3 shows the Ar^+ ion- and electron-induced Auger spectra of a Al (82%)-Fe alloy. Four ion-induced Auger (i-Auger) peaks appear at 47, 51, 58,

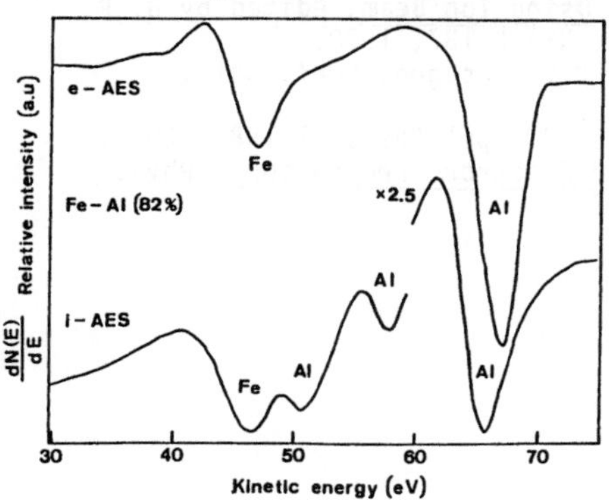

Fig.3 Ar^+ ion- and electron-induced Auger spectra of a Al(82%)-Fe alloy.

and 66 eV which are in agreement with the previous work [2]. The peak at 47 eV is exactly the same as that of the electron-induced Auger (e-Auger)peak for iron. The other i-Auger peaks,which are different from the e-Auger spectra, were ascribed to those of sputtered Al atoms [3].

Fig. 4 shows changes of the i-Auger intensity for Al(66eV) and Fe(47eV) as a function of the bulk Al composition. Obtained values for aluminium are best fitted with a quadratic curve which is in good agreement with the previous result [2]. On the other hand, the Fe intensity decreases as the bulk Fe composition increases. In order to check whether this strange phenomenon is due to instrumental effects or not, the e-Auger spectra were measured without ion bombardment (Fig.5). Both intensities increase as each composition increases. Therefore, it is strongly suggested that the increase in intensity of the i-Auger is related to mechanisms of the ion-induced Auger emission for iron in the alloys. If asymmetric collisions(Al--→ Fe) produce the Fe LVV Auger electrons, the intensity would increase as a function of aluminium composition up to 0.5 and decrease above it. Fig. 4 shows that the intensity keeps increasing, which implies that not only the asymmetric collisions but also the other mechanisms are taken into account for the Fe i-Auger emission in the alloys.

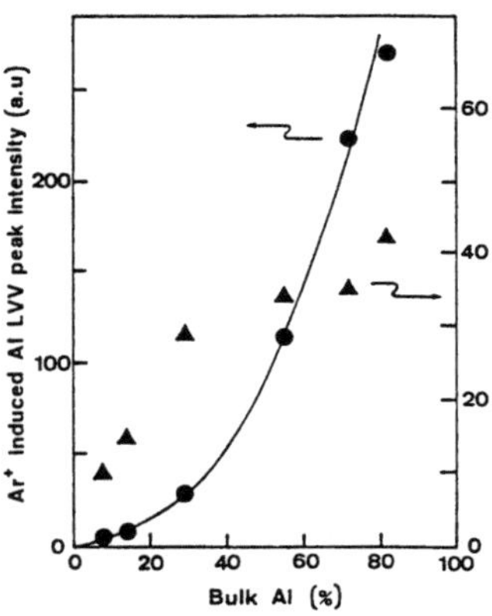

Fig.4 Changes of the i-Auger intensity for Al(66 eV) and Fe (47 eV) as a function of the bulk Al composition.

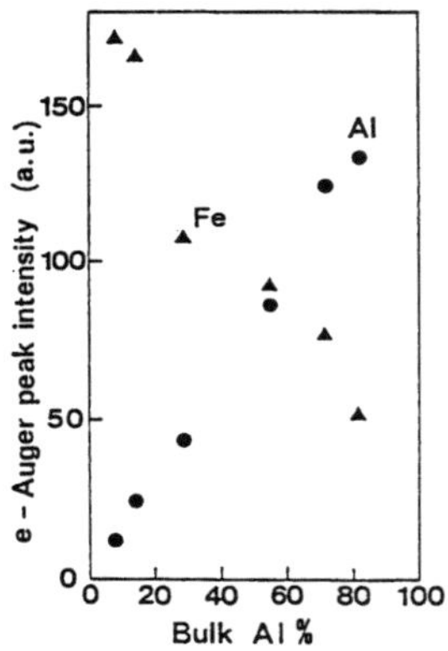

Fig.5 Changes of the e-Auger intensity for Al (67 eV) and Fe (47 eV) as a function of the bulk Al composition.

(1) G.Blaise: Material Characterization using Ion Beam, edited by J. P. Thomas and A. Cachard (Plenum Press, NY) P.137, 1978.
(2) J. F. Hennequin, R.L. Inglebert and P.V. Lesegno, Surf. Sci., 140(1984) 194.
(3) W.A. Metz, K.O. Legg and E.W. Thomas, J. Appl.Phys.,51(1980) 2888.
(4) L.E. Davis et al : Handbook of Auger Electron Spectroscopy (Phys. Electronic Ind. Inc., Minnesota) 1978.

Characterization of Planar Model Co-Mo/γ-Al_2O_3 Catalysts by SIMS, ESCA, and AES

T. Sahin and T.D. Kirkpatrick

Chemical Engineering Department, Montana State University, Bozeman, MT 59717, USA

Surfaces of Cobalt-Molybdenum/ γ-Alumina(Co-Mo/γ-Al_2O_3) petroleum hydrotreating catalysts have been studied by several techniques [1] including recent attempts by modern surface science techniques [2] (i.e. SIMS, XPS, AES). Applications of the surface science techniques are limited due to the porosity and insulating character of these catalysts. This work presents an alternative approach by replacing the porous γ-Al_2O_3 supports with planar γ-Al_2O_3 supports.

A mechanically polished 5 9's pure polycrystalline aluminum was sputter cleaned with Ar^+ until the Al(2p) XPS band indicated no presence of oxide. Growth of the γ-Al_2O_3 was accomplished in-situ by exposing the cleaned Al substrate to a mixture of O_2/H_2O at 10^{-2} Pa and 823 K for 0.5 hr. First molybdenum and then cobalt were vapor deposited on the γ-Al_2O_3 to approximately a monolayer coverage, and they were oxidized by exposing to 16 Pa O_2 and 50 Pa O_2 pressure, respectively for 0.17 hr at 823 K.

Al 2p XPS bands are shown in Figure 1. Upon crystallization to γ-Al_2O_3 the binding energy of the oxide band is 3.4 eV above that for pure Al. The results are consistent with Al 2p XPS binding energy ranges reported by Cocke [3]. The Al 2p band binding energy shifted 0.7 eV lower following Mo deposition. Mo 3d band binding energy is also broader and slightly at higher binding energy compared to the pure Mo band (see Fig. 2). This indicates surface interactions between the Mo monolayer and the γ-Al_2O_3. Probably, Mo is partially oxidized by sharing the surface oxygens of γ-Al_2O_3. Oxidation of the Mo did not result in an additional shift in Al 2p band; However, broadening of the Al 2p band occurred, coupled with an increase in intensity. This suggests additional interaction between the Mo and γ-Al_2O_3 along with migration of some Mo into the γ-Al_2O_3 lattice. The γ-Al_2O_3 crystalline structure contains octahedral sites large enough to accept Mo atoms.

Oxidation of the Mo monolayer resulted in a broader Mo 3d binding energy distribution with three distinct peaks; all occurring at higher binding energies than the unoxidized Mo3$d_{5/2}$ peak. In its elemental state the Mo 3$d_{5/2}$ binding energy is 227.? eV. Increased binding energy upon oxidation is expected due to an increased oxidation state of the metal. Handbook values of Mo 3$d_{5/2}$ binding energies for MoO_3 range from 232.0 to 233.0 eV [4]. However, this range is for bulk phase MoO_3. The oxidized Mo 3d peak shown in Figure 2 may be indicating incomplete oxidation or strong surface interactions with the γ-Al_2O_3. Following impregnation of the Co, the Mo 3d peak maintained the same shape

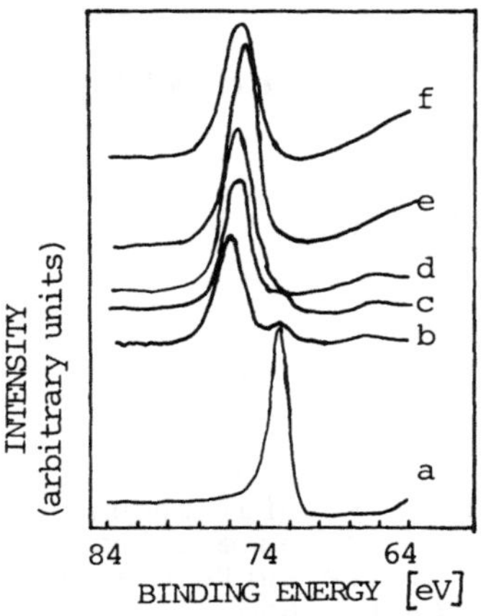

Fig. 1 Al 2p XPS spectra; (a) pure Al, (b) Al oxidized under, 10^{-2} Pa at 825 [K] for 0.5 hr, (c) after Mo deposition (d) after oxidation of Mo layer, (e) after Co deposition, (f) after oxidation of Co layer

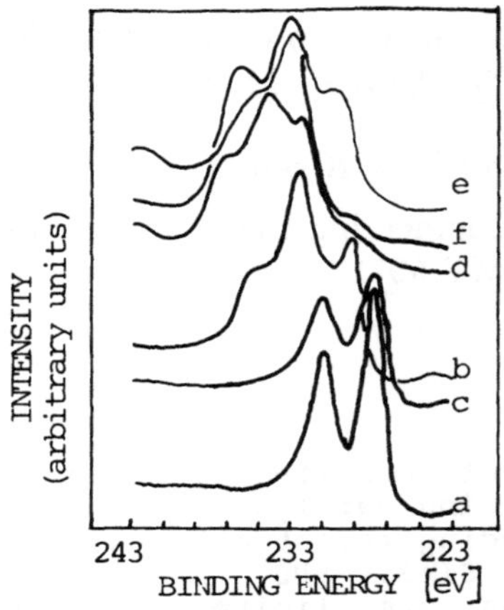

Fig. 2 Mo 3d XPS spectra; (a pure Mo metal, (b) Mo meta oxidized, (c) after Mo deposi tion, (d) after oxidation o Mo layer, (e) after Co deposi tion, (f) after oxidation o Co monolayer

but the entire peak shifted to a lower binding energy. This may b a result of a sharing of the Mo oxygens with the Co having a ne effect of lowering the oxidation state of the Mo. The Mo 3d pea returned to its more characteristic doublet form followin oxidation of the Co. This probably indicates more complet oxidation of the Mo. However, the binding energy is still abov that for pure oxidized Mo and the peak is broader than tha obtained immediately following deposition of the Mo indicatin surface interactions between the Mo and Co or γ-Al_2O_3 layers.

Subsequent deposition of Co on the Mo/γ-Al_2O_3 surface resulte in binding energy shifts in both Al 2p and Co 2p bands (see Fig. and Fig.3). Again indicating oxygen-sharing type interaction. After oxidation, Al 2p band returned to its original position, bu Co 2p band shifted to a higher binding energy (781.6 eV). Thi binding energy is higher than that obtained for pure oxidized C metal, indicative of surface interactions. Because of its smalle size, the Co atom is capable of occupying tetrahedral sites, i addition to octahedral sites, in the γ-Al_2O_3 lattice, allowin increased interaction between the Co and γ-Al_2O_3. Co usually occupies the tetrahedral sites in γ-Al_2O_3 lattice [5, 6]. It i suspected that the Mo monolayer hindered migration of the Co int the γ-Al_2O_3 lattice [5, 6].

To determine the location of the elements through the depth o the catalyst sample, SIMS depth profiling was performed. Th results are plotted in figure 4. Intensity of the Co^+ steadil declined. This behavior is expected, since Co is the last elemen deposited on the surface. Mo^+ intensity started low, and afte reaching a maximum it also declined steadily. O^- intensity sharpl declined first, reached a minimum, and then increased to a stead state level. As expected, Al^+ intensity started with a lo

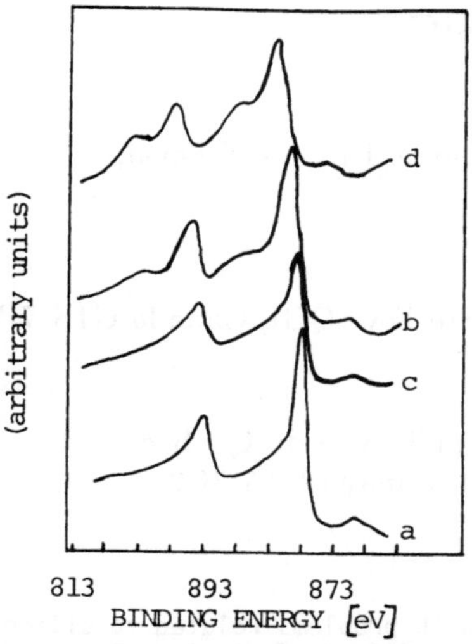

Fig. 3 Co 2p XPS spectra; (a) pure Co metal, (b) Co metal oxidized, (c) after Co deposition, (d) after oxidation of Co layer

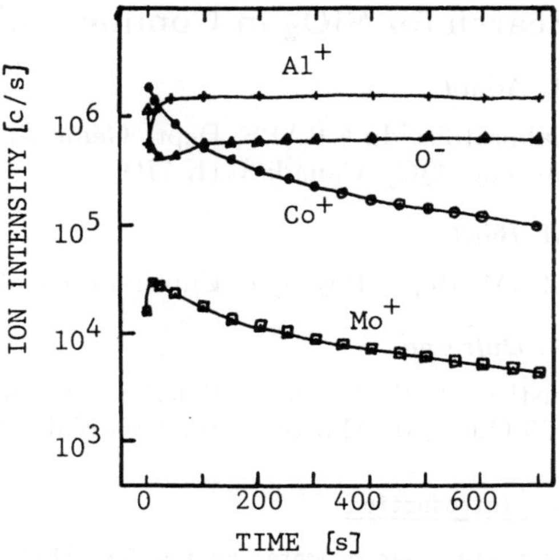

Fig. 4 SIMS depth profiling of the model Co-Mo catalyst: J_p = 20 A/cm^2, 1x1 mm crater size

intensity value and sharply increased to a steady state level since Al is present primarily in the Al_2O_3 sublayer. The maximum in the Mo^+ intensity shows that majority of Mo is present in a layer between the Co surface layer and the Al_2O_3 sublayer. This type of a distribution is expected from the sequence of metal deposition. Although we can not exclude the matrix ionization effects, the general trend of the data is consistent with the location of the elements. A survey scan taken at the center of the crater after depth profiling showed the presence of increased amount of Co and Mo. That could be due to a significant atom movement during oxidation of the Co and Mo at high temperatures, or matrix ionization effects. Overall the crater edge effects due to the geometry of our instrument, memory effects, and matrix ionization effects limited the depth resolution, and consequently the information content of the depth profiling.

The AES line analysis and surface mapping has shown that the distribution of Co and Mo was homogeneous over the sample surface.

When information gained from XPS and SIMS is combined, the presence of the majority of Co and Mo in a Co-Mo interaction complex, along with minor amounts of Co-Al_2O_3 and Mo-Al_2O_3 interactions, can be speculated at this point.

1. F.E. Massoth: in Advances in Catalysis, v. 27, p. 265
2. P. Gajardo, A. Matheieux, P. Grange, B. Delmon: Appl. Catal. 3, 347 (1982)
3. D.L. Cocke, E.D. Johnson, R.P. Merrill: Catal. Rev. Sci. Eng. 26(2), 163 (1984)
4. Physical Electronics: Handbook of ESCA.
5. K.S. Chung, F.E. Massoth: J. Catal. 64, 320 (1980)
6. K.S. Chung, F.E. Massoth: J. Catal. 64, 332 (1980)

Search for SiO_2 in Commercial SiC

A. Adnot

Laboratoire du GRAPS, Dept. Génie Chimique, Université Laval, Ste-Foy (QC), Canada G1K 7P4

M. Baril

LPAM, Dept. Physique, Université Laval, Ste-Foy (QC), Canada G1K 7P4

A. Dufresne

Institut de Recherche en Santé et Sécurité au Travail du Québec, 505 Ouest, de Maisonneuve, Montréal (QC), Canada H3A 3C2

1. Introduction

This study was prompted by the fact that health problems related to silicosis have been observed among some workers of a SiC manufacturing plant and conventional analysis methods failed to identify silica. Observation of individual particles, selected from SiC dust by polarized optical microscopy, has shown pleochroism at the surface, indicating the possible presence of patches of foreign materials. Surface analysis techniques were used in an attempt to identify the surface species on samples of SiC taken at the manufacturing plant. The ultimate aim of the study is to try to identify the structure and morphology of the SiO_2 compound, to find a possible correlation between the SiC grain size and the amount of SiO_2 on the grain, and to find if crushing of the bulk material results in fracture along planes possibly containing SiO_2. However, the preliminary study reported in this paper is intended as an evaluation of the usefulness of the surface techniques ESCA, AES and SIMS for such a problem.

2. Experimental

ESCA, AES and SIMS were performed in a commercial multi-technique apparatus. ESCA was performed on different powders and on two large SiC pieces presenting relatively flat surfaces, taken from the center of the SiC reactor. Auger results were obtained on SiC powders of different grain size, crushed from bulk material, and pressed onto indium for analysis. Important beam parameters were E = 10keV, I = 1 to 5 nanoamp, angle of incidence 45°, variable rastered area. SIMS spectra were recorded on two samples previously characterized by ESCA; the primary ion beam was Ar, energy is E = 5keV, current 5 nanoamp. A large rastered area (2.5mm x 1.3mm) was used to integrate the results over an area comparable in size to that analyzed by ESCA.

3. Results and discussion

ESCA results for the Si2p line of large sample # 1 are displayed in figure 1; spectrum A refers to the "as received" sample, and shows essentially only one component, identified as SiO_2 [1] after energy correction for charging (103.2eV). The survey spectrum shows only O, C and Si as main elements and carbon was found to be almost completely removed by a slight sputtering (removal of an equivalent SiO_2 layer of about 0.8nm). Prolonged sputtering corresponding to the removal of an equivalent SiO_2 layer of roughly 300 nm yields the spectrum B. The survey scan reveals a strong increase of the C line (compared to the slightly sputtered surface) and we identify the low

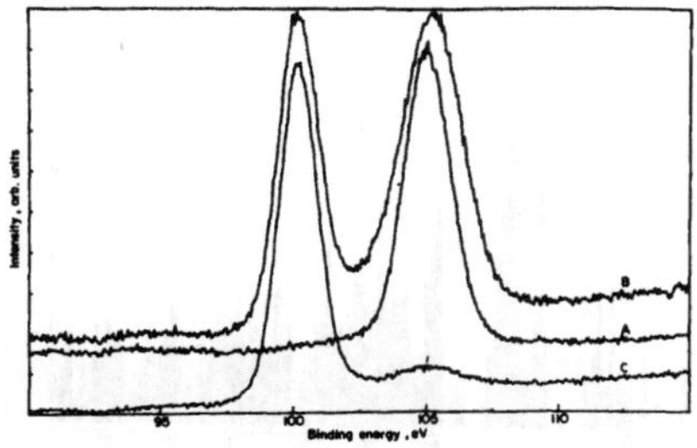

Fig. 1 Uncorrected ESCA Si 2p spectra for sample # 1; (A): as received; (B), (C): after two sputtering periods

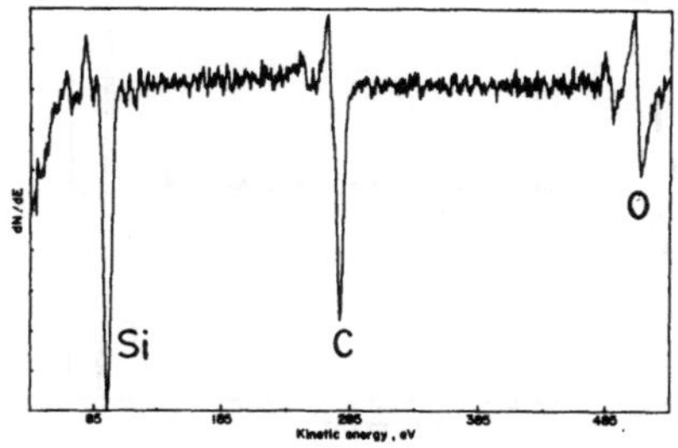

Fig. 2 Uncorrected Auger spectrum for a selected area of a commercial SiC powder

energy Si2p peak with SiC. The corrected Cls binding energy, 281.5eV, is compatible with this interpretation [2]. Further sputtering, corresponding to the removal of another equivalent SiO_2 layer of about 400 nm, yields the spectrum C, showing a strong decrease of the SiO_2 peak, accompanied by the increase of the Cls line and the decrease of the Ols line. The very long sputtering needed to change the crystal surface from SiO_2 to mainly SiC, as well as the simultaneous observation of both species, would suggest that for a flat SiC sample, there are thick patches of SiO_2 on or imbedded in the SiC surface. However, due to the roughness of the actual surface, we believe that this point needs further study. A second bulk SiC sample (# 2) was studied by ESCA after slight sputtering to remove the top carbon contamination layer; the Si2p line shows two components identified with SiO_2 and SiC, with the approximate ratio $Si(SiC)/Si(SiO_2) \approx 0.15$; of interest is the detection of Ca, with a relative concentration $Ca/Si \approx 0.02$ at. Other ESCA spectra, recorded for several powders prepared with bulk material originating from different locations inside the reactor, always reveal both SiC and SiO_2, SiO_2 being the main component.

AES was performed on several powders and on selected areas of the grains, as observed in the SEM mode. Figures 2 and 3 display the spectra recorded on two areas ($1 \times 0.4\ \mu m^2$, $2.5 \times 1\ \mu m^2$) of different grains of the same powder with grain sizes between 38 and 45 µm. According to the characteristic shapes of the Si and C signals, the spectra can be qualitatively interpreted as typical of a SiC surface with small amounts of SiO_2 (figure 2), and of a SiO_2 surface with small amounts of SiC (figure 3). Several other analyses show intermediate compositions, and almost pure graphitic carbon patches have also been detected.

SIMS has been performed on the surface characterized by the ESCA spectrum C in figure 1. Positive SIMS (figure 4) shows strong peaks associated with SiC (12, 28, 40, 48, 52, 56, 64, 68, 72, 80: C, Si, SiC, C_4, SiC_2, Si_2, SiC_3, Si_2C, C_6, Si_2C_2), and a strong Al peak; small peaks at 44, 45 may be associated with SiO_2. The main feature observed in the negative SIMS spectrum (figure 5) is the C_n series (n = 1 to 8). The SIMS results are therefore in agreement with the ESCA results. Positive SIMS from the SiO_2 rich surface (sample # 2) is displayed in figure 6; this spectrum is rather difficult to analyse and is believed to reveal surface contamination. The interesting feature is the observation of silicon oxide-related peaks (28, 44, 45, 56: Si, SiO, SiOH, Si_2) and of peaks associated with SiC on the preceeding spectrum; peak 40 is intense and confirms the presence of Ca detected by ESCA. The negative SIMS spectrum (not shown) is surprisingly weak, but reveals peaks associated with SiO_2 (60 and 76). The peak at 60 may have a

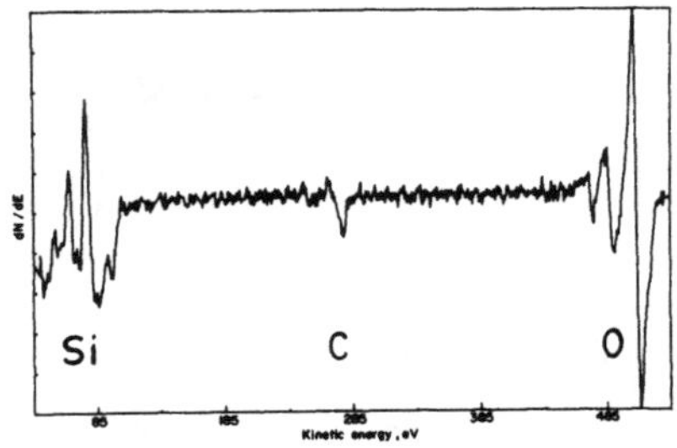

Fig. 3 Uncorrected Auger spectrum for a selected area of a commercial SiC powder

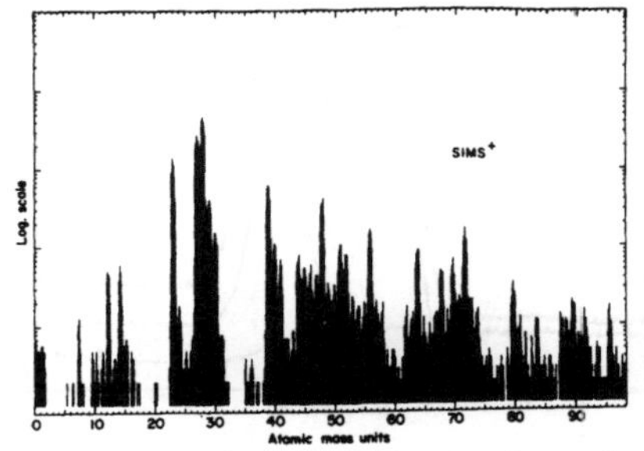

Fig. 4 Positive SIMS spectrum for sample # 1 (Figure 1C)

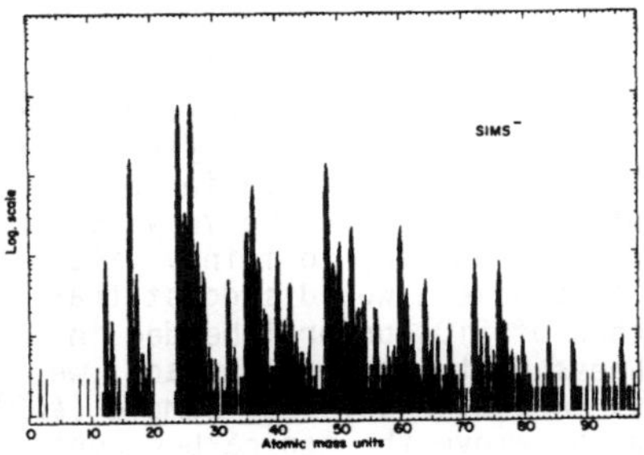

Fig. 5 Negative SIMS spectrum for sample # 1 (Figure 1C)

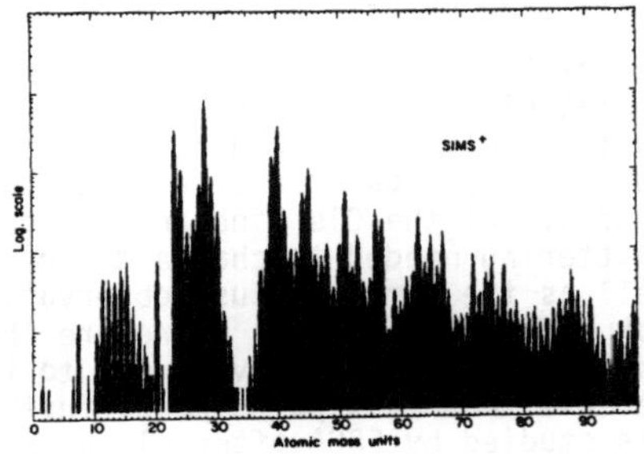

Fig. 6 Positive SIMS spectrum for sample # 2

contribution from C_5 clusters as peaks at 12, 24, 36 and 48 are indeed observed, corresponding to the C_n series. The highest peak is O^-. The SIMS results for this sample are therefore compatible with the ESCA results.

4. Conclusion

This study demonstrates that silicon oxide is indeed present on commercial SiC, as shown by surface analysis techniques. Microscopic analyses have been found possible with AES and reveal a nonhomogeneous SiO_2 distribution on the SiC powders studied. It is expected that the use of a liquid metal ion gun will permit the SIMS microscopic study as well, for detection of lower concentration impurities on selected areas of SiC material.

5. References

1. T.L. Barr, Appl. Surf. Sci., 15 1 (1983).
2. K. Miyoshi, D.H. Buckley, Appl. Surf. Sci., 10 357 (1982).

SIMS/XPS Studies of Surface Reactions on Rh(111) and Rh(331)

E. White, L.A. DeLouise, and N. Winograd

The Pennsylvania State University, Department of Chemistry, 152 Davey Laboratory, University Park, PA 16802, USA

1. Introduction

The adsorption of NO and HCN on Rh surfaces can be monitored using the Rh_2NO^+ or $RhHCN^+$ molecular cluster ions formed in static SIMS. From parallel XPS studies, for example, we find that NO adsorbs molecularly at 300K on Rh(111) but dissociates significantly on the Rh(331) surface. This dissociation is accompanied by a decrease in the Rh_2NO^+ ion yield and an increase in the Rh_2N^+ and Rh_2O^+ ion intensities. This type of chemical specificity illustrates that SIMS studies are possible on more complex systems which are difficult to monitor using XPS alone.

The crystal surfaces used in this study were the flat Rh(111) surface and the Rh(331) surface which is composed of two different (111) crystal planes. These planes produce a surface with an atomic step every three atomic rows. By comparing the reactivity of the Rh(331) crystal with that of the Rh(111) crystal, we can separate the catalytic effect of the step edges from that of the terraces which are composed of Rh(111) crystal planes.

When clean Rh(111) or Rh(331) is exposed to ethylene and heated to 700K a "carbidic" overlayer is formed. Following NO exposure, CN^- and Rh_2CN^+ ion peaks are detected by SIMS when the sample is heated, indicating a cyanide containing compound is being synthesized. We have further investigated the effect of atomic steps on the adsorption and desorption of HCN from these surfaces. The results show that adsorption is mainly molecular on Rh(111) and almost completely dissociative on Rh(331). These results are supported by an analysis of the Rh_2CN^+ and $RhHCN^+$ ion intensities.

2. Adsorption and Desorption of Nitric Oxide and Hydrogen Cyanide

Adsorption of NO at 300K results in two XPS N_{1s} peaks , a molecular NO_{ads} peak at 400.1 eV BE and a N_{ads} peak at 397.8 eV BE. SIMS results show NO^+, $RhNO^+$, Rh_2NO^+, Rh_2O^+, and Rh_2N^+ ions [1].

A comparison of a 5L exposure of NO onto the clean Rh(111) and Rh(331) surface shows approximately four times the dissociation of NO to N_{ads} and O_{ads} on the Rh(331) surface as on the Rh(111) surface. Since the Rh(111) surface is not defect-free with an estimated defect density of 5%, we propose dissociation of NO at 300K occurs on step or defect sites on the surface, with these sites then being blocked from further activity by the dissociation products. As the temperature increases to 430K, NO desorbs and dissociates as indicated by the loss of the Rh_2NO^+ ion, an increase in the Rh_2N^+ and Rh_2O^+ ion intensities, the loss of the molecular XPS peak, and an increase in the N_{ads} peak to 55 ± 5% of the total N_{1s} intensity at 300K on both surfaces. This indicates there is either an increase in the number of dissociation sites or that the sites become unblocked due to increased mobility of the dissociation products. With further heating the nitrogen desorbs from the surface presumably as N_2 [2], until at 680K no N_{1s} or Rh_2N^+ intensity remains.

Adsorption of HCN at 300K on the Rh(111) surface results in two XPS N_{1s} peaks at 396.7 ± 0.2 eV BE and 398.1 ± 0.2 eV BE and three types of CN containing SIMS ions CN^-, $RhHCN^+$, and Rh_2CN^+ (Fig 1). The presence of two N_{1s} peaks correlates with the two types of CN containing SIMS ions CN^- and $RhHCN^+$. Quantification of the XPS spectra show 50 ± 5% of the HCN dissociates upon adsorption to H_{ads} and CN_{ads}. Heating the crystal results in desorption and decomposition of the HCN_{ads} to CN_{ads} by 420K increasing the intensity of the SIMS CN^- and Rh_2CN^+ ions and reducing the XPS N_{1s} spectra to a single peak at 397.6 ± 0.2 eV BE. Continued heating results in the desorption of CN until at 680K no XPS N_{1s} or SIMS CN containing ions remains.

Adsorption of HCN at 300K onto the Rh(331) surface results in an XPS peak at 397.1 ± 0.2 eV BE with a shoulder at 398.3 ± 0.2 eV BE. SIMS results show the same nitrogen-containing ions as on the Rh(111) surface. Quantification of the N_{1s} spectra show 85 ± 5% of the HCN dissociates upon

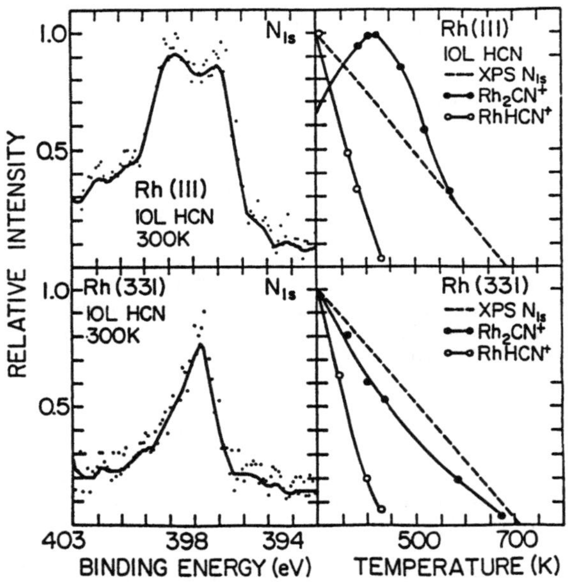

Fig 1. A. XPS N_{1s} Spectra for HCN adsorbed at 300K on the Rh(111) and Rh(331) surfaces. B. XPS N_{1s}, Rh_2CN^+, and $RhHCN^+$ intensity changes with temperature for HCN adsorbed on the Rh(111) and Rh(331) surfaces.

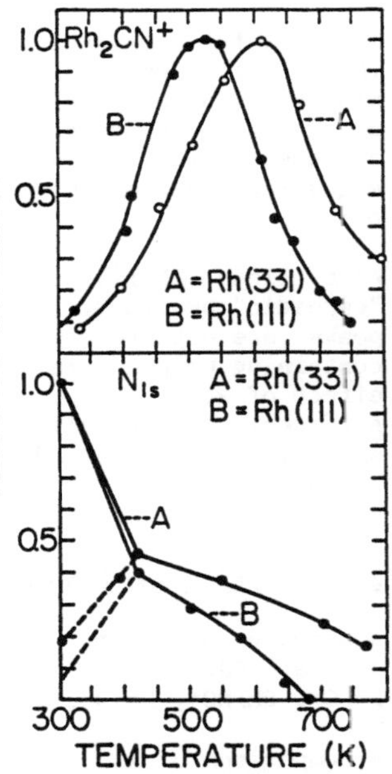

Fig 2. A. Rh_2CN^+ ion intensity changes for NO exposed on the carbon pretreated Rh(111) and Rh(331) surfaces. B. XPS N_{1s} intensity changes with temperature for NO exposure on the carbon pretreated Rh(111) and Rh(331) surfaces (—— N_{1s} total, --- dissociated N_{ads}).

adsorption. As the crystal is heated to 420K HCN desorbs and dissociates as evidenced (a) by loss of the XPS N_{1s} shoulder, (b) the shift of the XPS peak to 397.4 ± 0.2 eV BE, and (c) loss of the SIMS $RhHCN^+$ ion. There was no increase in the intensity of the SIMS CN^- or Rh_2CN^+ ions, indicating that as much CN_{ads} desorbed as HCN dissociated. Further heating decreases both the XPS N_{1s}, and SIMS CN^- and Rh_2CN^+ intensities until at 710K no intensity remains.

3. The Reaction of Nitric Oxide and Carbon on Rhodium Surfaces

Carbon pretreatment of the rhodium surfaces consists of adsorbing a submonolayer coverage of ethylene at 300K and heating the crystals to 700K to dehydrogenate the ethylene. The crystals are cooled to 300K exposed to 10L NO and sequentially heated [3].

Nitric oxide desorbs and dissociates by 430K as on the clean surface. As the temperature is increased further CN^- and Rh_2CN^+ SIMS ions increases in intensity, peaking at 510K on the Rh(111) surface and 620K on the Rh(331) surface(Fig 2). XPS results on the Rh(111) surface show N_{1s} intensity desorbs below 680K as on the clean surface. This contrasts with the SIMS CN^- and Rh_2CN^+ ion signals, which retain 30% of the total intensity at 680K. This indicates that CN is present in amounts below the XPS detection limit of a few percent of a monolayer, and that the SIMS CN^- ion species have a high ionization probability. XPS N_{1s} results from the carbon pretreated Rh(331) surface differ considerably from that of carbon pretreated Rh(111) surface and the clean Rh(331) surface. Approximately 60% of the dissociated N_{1s} intensity is present on the pretreated Rh(331) surface at 700K with a XPS N_{1s} BE of 397.4 ± 0.2 eV. Using isotopically labeled $N^{15}O$ we find that CN is the only nitrogen containing SIMS ion present above 700K. Oxygen reacted with C_{ads} forming CO or CO_2 which desorbs below 500K.

A comparison of the two surfaces shows the majority of the N_{ads} on the Rh(331) in the presence of carbon reacts to form a CN containing compound while on the Rh(111) surface very little of the N_{ads} reacts to form this compound. The CN_{ads} produced by reaction of N_{ads} and C_{ads} has a desorption temperature approximately 100K higher than the dissociated HCN indicating the CN may be part of a larger compound [4].

4. Acknowledgments

We wish to acknowledge the financial support of the National Science Foundation, the Air Force Office of Scientific Research, the Office of Naval Research, and the IBM Corporation.

5. References

1. L.A. DeLouise and N. Winograd: Surf. Sci., *159*, 199 (1985).
2. D.G. Castner and G.A. Somorjai: Surf.Sci. *83*, 60 (1979).
3. L.A. DeLouise and N. Winograd: Surf. Sci. *154*, 79 (1985).
4. F.P. Netzer: Surf. Sci. *61*, 343 (1976).

SSIMS – A Powerful Tool for the Characterisation of the Adsorbate State of CO on Metallic and Bimetallic Surfaces

A. Brown and J.C. Vickerman

Department of Chemistry, University of Manchester, Institute of Science and Technology, Manchester M60 1QD, U.K.

1. Introduction

In order to fully characterise the adsorbate state we need to know the chemical structure, the surface coverage of the various states and the bond energy. The work described in this paper will demonstrate that static SIMS can provide most of this information. However, the absence of satisfactory quantitative models of molecular ion formation has led us to adopt an empirical approach to the study of the relationship between secondary ion emission and adsorbate structure.

CO adsorption on metal surfaces has been studied very extensively by a large number of workers using a wide variety of surface analytical techniques, it is therefore possible to cross-correlate this data with that we obtain from SSIMS and this is the approach we have used.

2. Experimental

The work described has been carried out in a UHV static SIMS system comprising a VG MM12-12 quadrupole mass analyser (1-800 daltons) and a mass filtered ion/atom beam source designed and constructed at UMIST [1]. An off axis energy analyser similar to that described by Wittmaack is fitted [2]. Ar^+ beam currents of less than 0.5 nA cm^{-2} at 2 keV were used. TPD studies used an electron impact source in the energy analyser. A Leybold Hereaus ELS 22 electron energy-loss spectrometer was also incorporated. Surface cleaning of the various metals used is described elsewhere [3-6].

3. Molecular vs Dissociative Adsorption

There is a threshold heat of adsorption of CO adsorption (~250 kJ $mol.^{-1}$) below which molecular adsorption occurs and above which dissociative adsorption predominates at 300 K, [7]. Thus molecular adsorption should be found on Cu, Pd and Ni; mixed adsorption on Fe and dissociative adsorption on W. SSIMS studies of CO adsorption on these metals showed that molecular adsorption is characterised by M_nCO^+ (found on Cu, Pd, Ni and Fe) and dissociative adsorption by M_nO^+ and M_nC^+ (found on Fe and W) [3].

4. Coverage measurements of molecular adsorption

The sensitivity of the secondary ion yield to the electronic state of the substrate-adsorbate system makes quantitative data difficult to obtain. When CO adsorbs on most metals the work function increases, hence the yield of M_nCO^+ will not be a simple function of coverage. We suggest that the main influence on the M_n^+ yields will be the electronic effects [4]. Hence taking a sum of ratios $\Sigma M_nCO^+/M_n^+$ should result in a function directly related to CO coverage. There are various assumptions implicit here, eg (a)

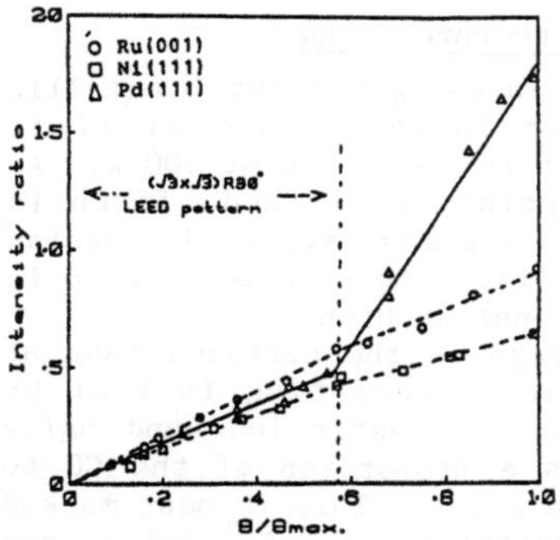

Fig 1 ($\Sigma M_nCO^+/M_n^+$) versus θ_{CO} at 300 K

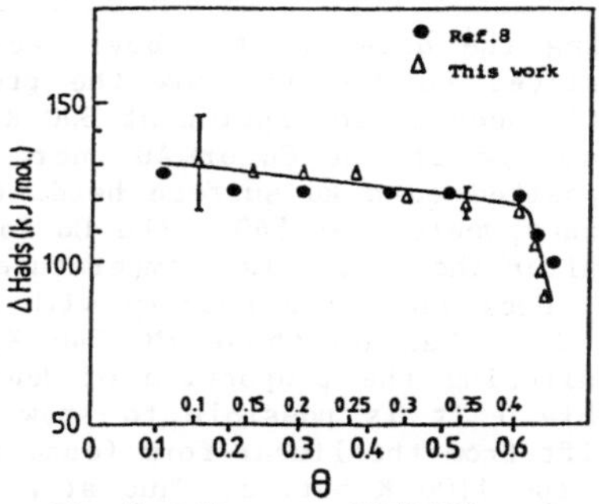

Fig 2 ΔH_{ads} on Ni(100) v θ_{CO}

that the ionisation potentials of M_nCO^+ and M_n^+ are close; (b) that the dissociation energies of the CO clusters are almost invarient with coverage.

Adsorption of CO has been studied on Cu(100), Ni(100), (111), Pd(100), (111), Pt(100) and Ru(0001) surfaces. In every case there is a linear relationship, fig.1, up to the coverage at which compression of the overlayer occurs [4-6]. The gradient varies with the metal surface and is sensitive to the strength of the adsorbate bond. Beyond compression the gradient changes, which suggests a sensitivity to the overlayer structure.

It is now possible to monitor the CO coverage as a function of temperature at constant pressure to obtain adsorption isobars and thence isosteres, and using the Clausius-Clapyron equation the isosteric heat of adsorption is obtained, fig. 2, [4]. SSIMS is one of the few methods to yield this data in a straight forward way. Agreement with the literature is excellent [8].

5. SSIMS data and the mode of adsorption.

The application of infra-red spectroscopy and high-resolution electron energy-loss spectroscopy, HREELS, to CO adsorption on metal single crystals has enabled us to compare this data with our SSIMS results for such systems. In the case of Cu(100), Ru(0001), Ni(111), (100), Pd(111), (100) and Pt(100) these comparisons showed that the relative yields of MCO^+, M_2CO^+ and M_3CO^+ are sensitive to the occupancy of the linear, bridge or triply bridge CO sites on a given surface, Table 1, [9,10]. In contrast to vibrational spectroscopy, it has been possible to quantify the proportions of each mode adsorbed on a surface as a function of coverage.

Table 1: Secondary ion cluster emission and CO adsorbate structure (θ_{CO}=0.5)

Surface	SSIMS			CO structure
	MCO^+	M_2CO^+	M_3CO^+	IR/HREELS/LEED
Cu(100)	0.9	0.1	--	linear only
Ru(0001)	0.9	0.1	--	linear only
Ni(100)	0.8	0.2	--	linear + bridge
Ni(111)	0.6	0.4	--	bridge + linear
Pd(100)	0.3	0.6	0.1	bridge only
Pd(111)	0.3	0.4	0.3	triple bridge
Pt(100)	0.65	0.35	--	linear + bridge

6. The influence of Cu and Au on CO adsorbate structure on Ru

Using the data in the above sections we have investigated two bimetallic surfaces to discover how the presence of Cu or Au on the surface of Ru influences CO adsorption at 300 K. Neither Cu or Au adsorb CO at 300 K. As coverage of the Cu or Au increases CO adsorption is blocked. If Cu is deposited on a Ru surface held at close to 1100 K almost atomic dispersion occurs, whereas at 540 K the Cu forms islands, [11]. The behaviour of Au is similar, though at high temperature dispersion is not so high.

θ_{CO} decreases more sharply with Cu or Au coverage on the surfaces made at 1100 K than on those at 540 K. What is more interesting is that by monitoring the proportion of $RuCO^+$ amongst the CO cluster ions and using Table 1 it is possible to show that Cu causes a proportion of the CO to shift from the linear form found on Ru to bridge sites. This is most marked on the 1100 K series. Thus at a θ_{CO}=0.2 the proportion of $RuCO^+$ falls from 0.9 to as low as 0.65 at a θ_{Cu}=0.4, figs 3 and 4, which indicates that up to 40% of the CO is on bridge sites. Concurrent with this behaviour, we observe the emission of new $RuCuCO^+$ ions suggesting the involvement of Cu in the adsorption sites. Au however does not have this effect. The proportion of $RuCO^+$ is 0.9 at all Au coverages and there is no evidence of a $RuAuCO^+$ ion. SSIMS shows that dispersed Cu is positively involved with Ru in the adsorption process whilst Au is not. Earlier work function studies showed that there is a significant electronic interaction between the surface Cu and the Ru, [11], so the involvement of Cu in adsorption is not surprising.

The evidence of mixed bonding sites and the quantification of the relative occupancy of linear and bridge sites is not so easily accessible from any other technique. This study demonstrates the power of SSIMS in defining the chemical structure of surface species.

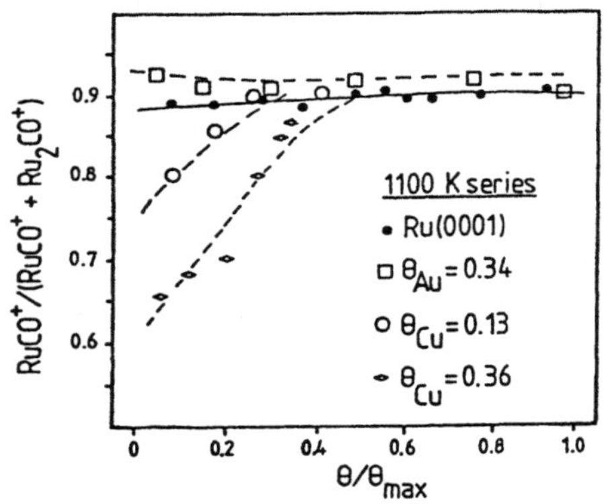

Fig 3 $RuCO^+/(\Sigma Ru_nCO^+)$ v θ_{CO} at 300 K

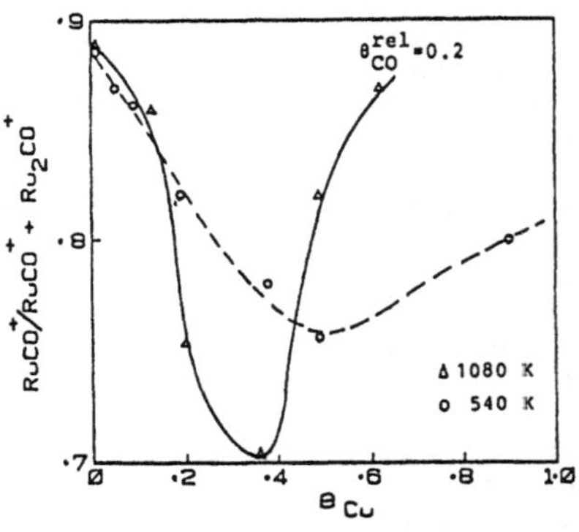

Fig 4 $RuCO^+/(\Sigma Ru_nCO^+)$ v θ_{Cu}

1 A Brown, J van den Berg, J C Vickerman, Spectrochim. Acta 40B, 871 (1985)
2 K Wittmaack, J Maul, F Schultz, Int J. Mass Spec. Ion Phys. 11, 23 (1973)
3 M Barber, J C Vickerman and J Wolstenholme, Surf. Sci. 68, 130 (1977)
4 R S Bordoli, J C Vickerman and J Wolstenholme, Surf. Sci. 85, 244 (1979)
5 A Brown and J C Vickerman, Surf. Sci. 117, 154 (1982)
6 A Brown and J C Vickerman, Vacuum 31, 429 (1981)
7 K Kishi and M W Roberts, JCS Faraday I, 71 1715 (1975)
8 J C Tracey, J Chem. Phys. 56, 2736 (1972)
9 A Brown and J C Vickerman, Surf. Sci. 124, 267 (1983)
10 A Brown and J C Vickerman, Surf. Sci. 151, 319 (1985)
11 J C Vickerman, K Christmann, G Ertl, P Heimann, F J Himpsel and D E Eastman, Surf. Sci. 134, 367 (1983)

Energy and Angle-Resolved SIMS Studies of Cl_2 Adsorption on Ag{110}; Evidence for Coverage Dependent Electronic Structure Rearrangements

D.W. Moon, R.J. Bleiler, C.C. Chang, and N. Winograd

The Pennsylvania State University, Department of Chemistry,
152 Davey Laboratory, University Park, PA 16802, USA

We present angle-resolved secondary ion mass spectrometry (SIMS) measurements for chlorine adsorbed on Ag{110} over a coverage range extending from near zero up to 0.5 monolayer, where a (2x1) LEED pattern is observed. From a detailed analysis of the ejected Cl^- kinetic energy, polar angle, and azimuthal angle distributions, as well as from the dependence of the Cl^- yield on coverage, we illustrate strong electronic and structural changes that occur over the coverage range.

The details of the angle-resolved SIMS apparatus have been described elsewhere [1]. The crystal was cleaned by cycles of heating and ion bombardment. All the SIMS measurements were performed with Ar^+ primary ions of 1 keV kinetic energy and of 2 nA current. The ion yield of Cl^- secondary ions as a function of Cl_2 exposure is shown in Fig 1. It is

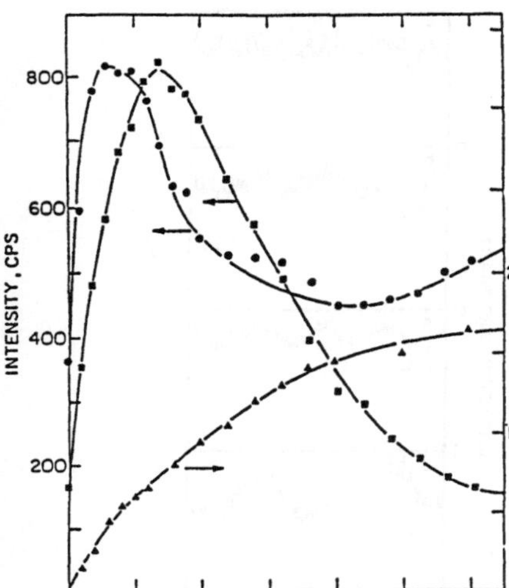

Fig. 1. Cl^- ion yield for low (●) and high (■) energy and the work function changes (▲) as a function of exposure at room temperature. The incidence angle of Ar^+ is perpendicular to the surface. The polar collection angle is 45° and the azimuthal angle is along the <001> direction.

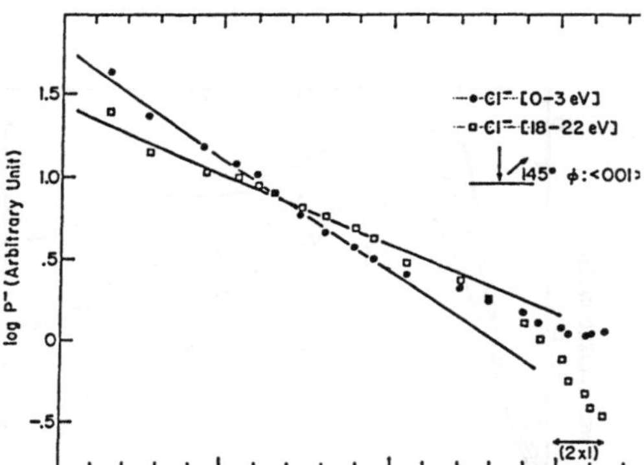

Fig. 2. Plot of log P^- vs. the change of work function, $\Delta\phi s$ for low (●) and high (■) energy Cl^- ions.

clearly not linearly dependent on the chlorine coverage. The crossover of the two Cl^- ion yield curves for low [0-3 eV] and high [18-22 eV] kinetic energies indicates a substantial change in the kinetic energy distribution of the Cl^- secondary ion. It has been shown [2] that the negative ionization probability, P^- is an exponential function of the surface work function, ϕ_s and electron affinity, A;

$$P^- \propto \exp[-(\phi_s - A)/\varepsilon_0], \quad \varepsilon_0 = \hbar\ \gamma\ v_\perp / C_1 \pi. \qquad (1)$$

Here γ and C_1 are related to the electronic configuration of the secondary-ion-substrate combination and $v_\perp$ is the normal component of the emission velocity. The observed anomalous coverage dependence of the Cl^- ion yield can be partly explained by the work function dependence of negative secondary ion emission. The plot of log P^- vs. the change of work function, $\Delta\phi_s$ in Fig 2 exhibits straight line behavior up to 1.0L. After this value, however, the curve deviates from a straight line for both energies. This observation suggests that there is a change in γ/C_1 related to the electronic configuration of the Ag - Cl bond.

The kinetic energy distribution of Cl^- secondary ions also exhibits an anomalous change in this coverage range, as shown in Fig 3. Up to 1.2L exposure, the peak position moves to a higher value before decreasing at higher exposure. The kinetic energy distribution of sputtered ions may be represented by Thomson's equation [3] of the kinetic energy distribution for sputtered neutrals multiplied by the ionization probability of Eq (1). According to the combined equation, the peak of the kinetic energy

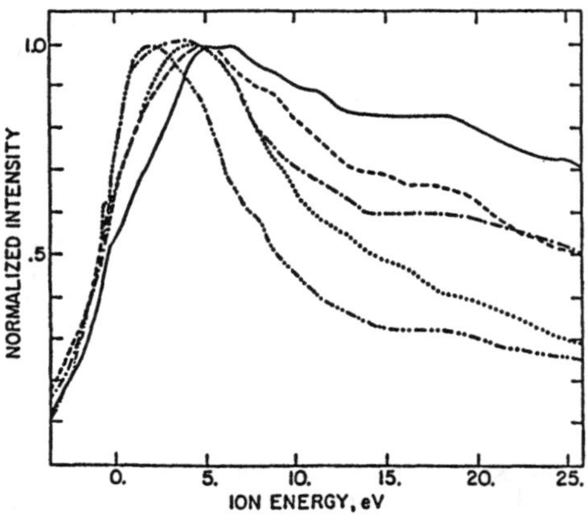

Fig. 3. Normalized energy distributions of Cl^- ions for Cl_2 various exposures; 0.1L (•••), 0.6L (---), 1.2L (---), 1.8L (-•-•), and 2.5L (-••-••).

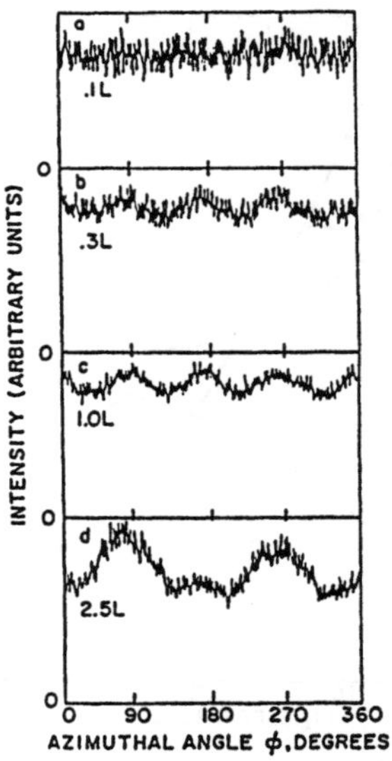

Fig. 4. Azimuthal angle distributions of high-energy (18-22eV) Cl^- ions for various exposures. ϕ=0° and ϕ=90° correspond to the <110> directions, respectively. The incidence angle is perpendicular to the surface.

distribution of secondary ions is predicted to shift to a higher value, as the surface work function increases. However, the change in the measured kinetic energy distribution of Cl^- after 1.2L could be reproduced by decreasing the γ/C_1 term in Eq (1). Therefore, the anomalous change of the Cl^- ion yield and the kinetic energy distribution is likely due to a change in the electronic structure of the Ag - Cl surface chemical bond.

A geometric structural change of the Cl adlayer is indicated by the measured azimuthal and polar angle distributions of the Cl^- secondary ions. In Fig 4, the anisotropy of the Cl^- azimuthal angle distribution is shown to increase continuously from .1L to 1.0L exposure. After 1.0L, the anisotropy increases substantially until 2.5L exposure. In Fig 5, the polar angle distribution is shown to exhibit a peak at $\theta=0°$ below exposures of 1.0L. After 1.0L, another peak emerges at $\sim\theta=15°$ and the ejection to higher polar angles is decreased. These changes suggest that the ejecting Cl^- ion experiences greater channeling by the underlying Ag atoms as its coverage is increased beyond 1.0L. Increased channeling would arise from a shortened Ag - Cl bond length, allowing the Cl atom to reside within the valley found in the <110> direction.

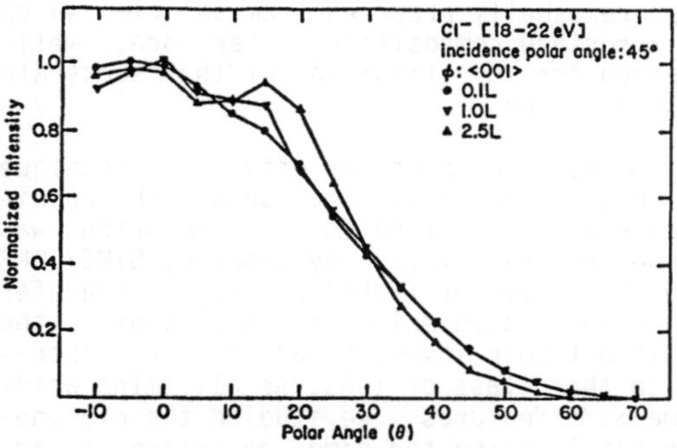

Fig. 5. Polar angle distributions of high-energy (18-22 eV) Cl^- ions for various exposures. The azimuthal orientation is along <001> and the incidence polar angle is 45°.

This electronic and geometric structural change is consistent with the depolarization of the surface dipole due to the repulsive coulombic interaction between the negatively charged chlorine adatoms, as is suggested by the work function change reported in Fig 1. With a submonolayer Cs overlayer ($\theta\sim0.05$) on Ag{110}, the anomalous changes of the Cl^- ion yield and the kinetic energy distribution were not observed. The positive charge on Cs adatoms may decrease the repulsive coulombic interaction and stabilize the negative charge, such that the Cl adatoms do not experience a structural change over this coverage range.

Acknowledgement

We wish to thank the NSF, The Air Force Office of Scientific Research, and the Office of Naval Research for Financial support.

References

1. R. A. Gibbs and N. Winograd: Rev. Sci. Instruments 52, 1148 (1981).
2. J. K. Norskov and B. I. Lundquist: Phys. Rev. B 19, 5661 (1979).
3. M. W. Thompson: Philos. Mag. 18, 377 (1968).

Molecular Secondary Ion Emission from Different Amino Acid Adsorption States on Metals

D. Holtkamp, M. Kempken, P. Klüsener, and A. Benninghoven

Universität Münster, Physikalisches Institut, Domagkstr. 75, D-4400 Münster, F. R. G.

1. Introduction

Independent of the special kind of irradiance (photons, keV-ions, fission fragments), the emission of molecular secondary ions $(M\pm H)^{\pm}$, $(M+alkali)^{+}$, $(M+Ag)^{+}$ has been observed from various biologically important compounds (amino acids, peptides, nucleotides) with high yields. However, only little understanding of the ion formation processes has been achieved so far, mainly due to the fact that samples are usually prepared from solution which obviously results in a very complex surface composition. Therefore, well-defined adsorption systems are required for investigations of the substrate influence on molecular secondary ion emission.

We have approached this problem by using a molecular effusion technique to prepare amino acid overlayers on clean metal surfaces under UHV-conditions /1/. We found that the protonated/deprotonated ion emission was strongly dependent on the kind of the substrate /2,3/; by combined SIMS/TDMS experiments we demonstrated that the formation of $(M+H)^{+}$ by proton transfer is only possible between weakly bound amino acid molecules /4/. However, the metal-containing molecular ions have not been investigated so far. Therefore, we extended our experiments to that class of ions. As all amino acids under investigation showed the same SIMS features, we studied the glycine-metal interaction for Au and Ni in detail using the same experimental arrangement as described in Ref. 4. Most results presented in this paper refer to the system glycine - Ni.

2. Results and Discussion

Figure 1 shows the substrate influence on the secondary ion emission from about two monolayer equivalents of glycine on Au and Ni. From an Au substrate the protonated/deprotonated ion yields are similar, and a large $(M-H)^{+}$ signal is observed which is not found for Ni. Furthermore, for both substrates positively and negatively charged metal-containing molecular ions are detected with comparable yields, and from Ni an $(M-2+Ni)^{-}$ emission occurs additionally. From solution-prepared amino acid samples on Ag an $(M+Ag)^{-}$ ion is generally not emitted if the amino acid is dissolved in HCl, as usual; if, however, an aqueous solution is dried on a sputtered Ag substrate, an $(M+Ag)^{-}$ signal has been observed for glycine /5/.

By using deuterated compounds of glycine the structures of the ions $(M-H)^{+}$ and $(M-2+Ni)^{-}$ were identified. It turned out that $(M-H)^{+}$ is formed by loss of one carbon-bound hydrogen, whereas the $(M-2+Ni)^{-}$ emission results from the separation of two hydrogens from the functional groups. The $(M-17)^{-}$ ion which interferes with the Ni 58^{-} signal, was found to be formed by loss of NH_3/ND_3 from the glycine molecule.

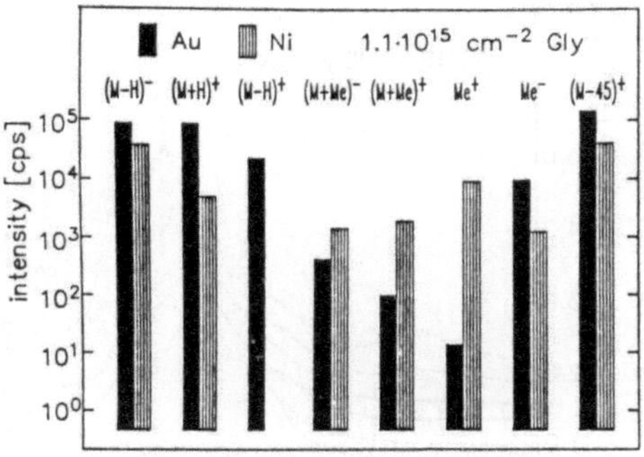

Fig.1: Characteristic secondary ion emission for glycine on Au and Ni (3.5 keV Ar^+, 6.5 nA/cm^2 on 1 cm^2). The 58^- and $(M+Ni^{58})^-$ yields are corrected for interference with $(M-17)^-$ and $(M-2+Ni^{60})^-$.

For submonolayer coverages on Ni only the negatively charged molecular ions $(M+Ni)^-$, $(M-2+Ni)^-$, and $(M-H)^-$ were detected showing a similar increase during glycine layer growth (Fig. 2), whereas on an Au substrate the same behaviour was observed for all molecular ions. The dependence of molecular secondary ion emission from a certain adsorption state is demonstrated in Figs. 3 and 4: the $(M+H)^+$ and $(M+Ni)^+$ intensities decrease at about 340 K (Fig. 3) corresponding to the small TDMS peak in Fig. 4 which has been assigned to multilayer desorption in a previous paper /4/. The appearance of this peak is correlated with the onset of the positively charged molecular ion emission (Fig. 2) showing again that the formation of both the protonated and the so-called "cationized" ion is only possible in the presence of weakly bound amino acid molecules.

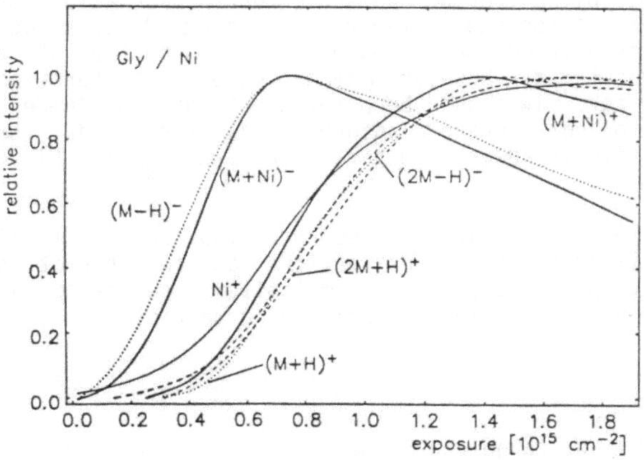

Fig.2: Molecular secondary ion emission during glycine layer growth on Ni, normalized to maximum intensity. The Ni^- and $(M-2+Ni)^-$ signals show the same increase as $(M+Ni)^-$ and $(M-H)^-$.

The $(M+Ni)^-$ emission from submonolayer coverages suggests that the amino acid is adsorbed intact which is in contrast to our previously proposed model of (M-H) - Ni complexes /4/. One could argue that this ion is formed by recombination of (M-H) and NiH clusters. We have tried to verify this mechanism by evaporating glycine onto Ni (273 K) in a D_2 atmosphere (deuterium on the surface was indicated by Ni_2D^+ /6/), but we never did observe an $(M+Ni)^-$ signal shifted by one amu. Therefore, we conclude that the amino acid is indeed adsorbed nondissociatively. Due to the rather strong interaction, however, the Ni-bound molecules decompose upon target heating, as indicated by the large H_2^+ /4/, H_2O^+, CO^+, and CO_2^+ TDMS peaks.

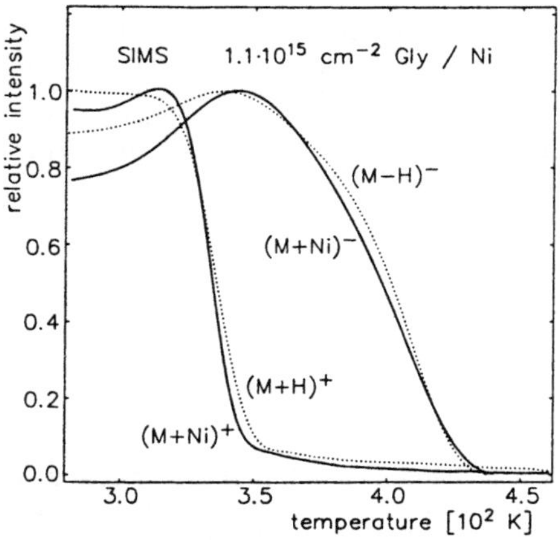

Fig.3: Molecular ion emission during temperature-programmed target heating (heating rate: 3.5 K/s)

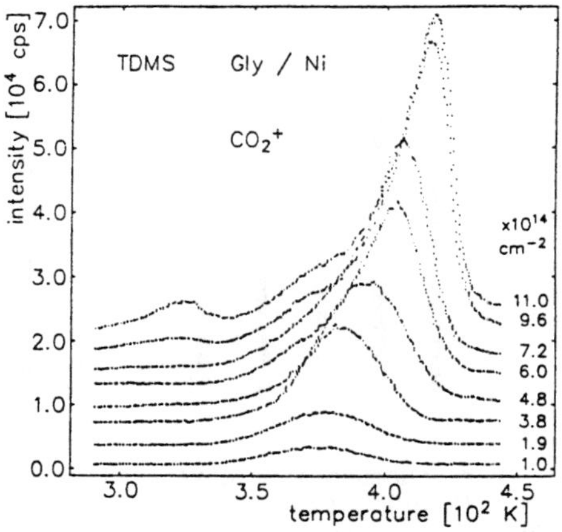

Fig.4: CO_2^+ TDMS signal for different glycine exposures to Ni (heating rate: 3.5 K/s). Similar results were obtained for H_2^+ /4/, H_2O^+, and CO^+.

One possible mechanism for $(M+Me)^\pm$ formation is the recombination of M and $Me^\pm$. While the similar behaviour of $(M+Ni)^-$ and Ni^- for monolayers would be in agreement with this mechanism, the results for Au are not: despite an Au^+/Au^- ratio of 10^{-3} the yields of $(M+Au)^\pm$ are comparable. We therefore assume that $(M+Me)^\pm$ are formed by sputtering and subsequent dissociation of Me_mM_n clusters. Concerning Ni, n=1 for submonolayers whereas $n \geq 2$ for structures from which $(M+Ni)^+$ is emitted (weakly bound molecules on top of a closed monolayer); the values for m, however, may depend on the adsorption geometry which is unknown so far. It should be pointed out that an $(M+Me)^-$ signal is a unique feature for polar organic adsorbates as amino acids. No such emission is observed for simpler adsorption systems e.g. Ni-H_2/6/, Ni-CO /7/, or Ag-CH_3OH /8/.

In conclusion, it can be stated that UHV-prepared amino acid - metal adsorption systems are well suited for investigations on ion formation mechanisms. It would be very interesting to perform similar experiments as described in this paper, for different kinds of bombarding particles.

3. References

1. D. Holtkamp, M. Jirikowsky, W. Lange and A. Benninghoven: Appl. Surface Sci. 17, 296 (1984)
2. W. Lange, M. Jirikowsky, D. Holtkamp and A. Benninghoven, in: SIMS III, Springer Series in Chemical Physics 19, p.416 (Springer, Berlin, 1983)
3. W. Lange, M. Jirikowsky and A. Benninghoven: Surface Sci. 136, 419 (1984)
4. D. Holtkamp, M. Jirikowsky, M. Kempken and A. Benninghoven: J.Vac.Sci. Technol. A 3 (3), 1394 (1985)
5. M. Junack, unpublished results
6. A. Benninghoven, P. Beckmann, D. Greifendorf, K.-H.Müller and M.Schemmer: Surface Sci. 107, 148 (1981)
7. A. Brown and J.C. Vickerman: Surface Sci. 117, 154 (1982)
8. R. Jede: Thesis, Münster 1985

Part VIII

Ion Microscopy and Image Analysis

A High-Resolution, Single Ion Sensitivity Video System for Secondary Ion Microscopy

D.P. Leta

Exxon Research and Engineering Co., Annandale, NJ 08801, USA

1. Introduction

In order to fully utilize the ion microscopy capabilities of SIMS instrument tion such as the CAMECA IMS-3F, an imaging system capable of single ion dete tion with 512 x 512 resolution and more than eight orders of magnitude dynam range is needed. Such a system has been assembled from commercially availab video equipment, which additionally features virtually unlimited optical las disk image storage, alphanumeric image labeling, and polaroid instant fi copies of stored ion images. The modular system additionally offers spee versatility and ease of use.

2. System Configuration and Capabilities

A schematic of the video system interconnections, as used in the normal r cording mode, is shown in Fig. 1. The ultra-high gain, high-resolution vid camera is manufactured by Venus Scientific Inc. and distributed by Zeiss, In (model TV-2M). It views the CAMECA's single multichannel plate/fluoresce screen through a 25mm f/0.85 video lens which is in contact with the revers 50mm camera lens provided by CAMECA as collection optics. The camera fe tures two stages of multichannel plate image amplification, which provid sufficient gain, when coupled to the instrumental image gain, to clearly vi the photon bursts arising from individual ion impacts. It is assumed that t detection efficiency of the system is limited by the fraction of ions whi enter the multichannel plate's channels. Although the gain and camera tu voltages may be manually controlled, in normal operation the camera automat cally adjusts its amplification over a range of more than 10^6 to provide viewable image. With the additional use of the CAMECA's variable gain mult channel plate to reduce the signal intensity when necessary, the collection ion images over the entire possible intensity range is readily accomplished.

Following a real-time view of the ion signals on a T.V. monitor, the seco stage of the image-collection system, a 512 x 512 x 8 video digitizer, Mic Consultant/Quantel's CRYSTAL, performs integration, recursive averaging a image subtraction at video rates. In this way ion images arising from ve low count rates which may look much like noise because of the slow ion arriv rate may be formed into viewable high-resolution images. The processor m also perform up to a 9 x 9 programmable kernal deconvolution on frozen imag to improve the apparent spacial resolution, as well as adjusting the i image's contrast and brightness after collection is completed for improv viewability. The firmware processor is connected to an I-Omega 10 Mbyte ca tridge floppy for digital image storage, and may additionally be interfaced a host computer for control and further ion-image processing.

Once the ion-image collection is complete, the image is converted back in an analog RS-170 video signal and looped through a character overlay generat

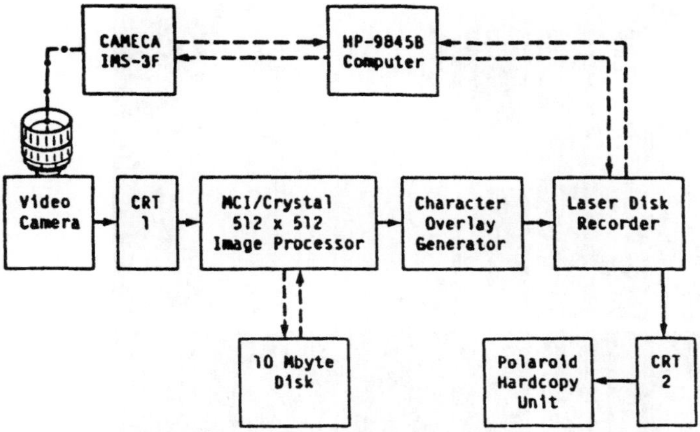

Figure 1. Schematic of video interconnections and control lines

for labeling (Knox K-50), and to a high-resolution optical laser disk recorder (Panasonic TQ2021 FBC) for permanent storage of the image with very high fidelity. The laser disk recorder, which takes a maximum of 0.5 second for image storage or retrieval, has the capacity to store 10,400 images on a disk costing <$300 in an individually addressable manner. The laser disk recorder is interfaced to the Hewlett Packard 9845B computer which controls the SIMS, for remote control operation. This allows coordination of the mass switching and the rapid data collection needed for the formation of multielement imaging depth profiles. Under software control, ion intensities may be measured by the SIMS detection circuitry during an imaging depth profile interspersed with the recording of recursively averaged ion images. In this manner the total ion dose per second represented by each ion micrograph may conveniently be recorded, providing normal depth profile data simultaneously with the collection and storage of hundreds of sequential ion images. The images may then be quantitated and corrected for autogain changes by correlation with the electron multiplier count rates.

After looping through the image recorder, the ion micrographs may be returned to the CRYSTAL for further adjustment and/or viewed on a second T.V. monitor. The loop through system design and frame time speed of the image processor allows the images to be viewed during any integration or other manipulation. When hard copies of the ion images are desired the video signal is sent to a Polaroid recording system (ADR Ultrasound PM-1) for making instant black and white prints.

3. Results and Discussion

Figures 2-5 are provided as examples of the system performance. The CAMECA IMS-3F was operated with a 5 micron contrast aperture to demonstrate the system's utility in a low signal, high lateral resolution situation, and a 150 micron field of view for the collection of all of the micrographs. Figure 2 is a one-video frame collection, 1/30 sec., of the Fe^+ image arising from a polished crosssection of a carburized 900B stainless steel pipe. The dark areas in the image are chromium and niobium carbides at the grain boundaries. The mottled appearance of the image is due to the recording of the individual ion arivals, and would not be present at higher count rates. Figure 3 shows

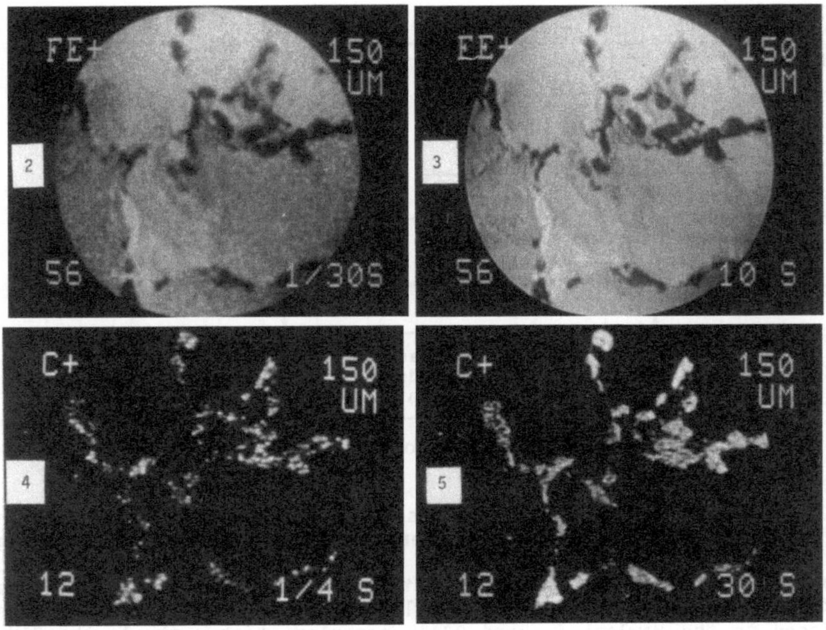

Figures 2-5. See text for descriptions.

the same ion image integrated for 10 sec. by the CRYSTAL processor. The ap pearance of noise in the image is removed and the viewability much improved b the integration. The micrograph has also been processed by a 7 x 7 kerna deconvolution, to sharpen the image. The use of such kernal deconvolutions c the image data is an approximation of fourier transforms to deconvolute th ion image data from the instrumental transfer function. Although this type c data massaging could possibly introduce errors into the image, the judiciou use of deconvolution filters generally has allowed us to bring out feature which were present in the original image but of a diffuse nature.

Figure 4 is an 8 frame (1/4 sec.) integration of the positive carbon ic signal from the same area. The ion intensity for carbon was 1500 cps and th brief exposure shows the system response to <400 ions. The image is muc improved by a longer, 30 sec. integration, as shown in Fig. 5, where th details of the carbide's distribution have become visible.

The video system enables the use of the CAMECA IMS-3F as an ultimate sensi tivity, real-time ion microscope. In comparison to 35mm camera image record ing the system is much more sensitive, faster and easy to use, and offer immediate feedback of image quality. The ability to focus and "push buttc record" secondary ion images at all intensity ranges has vastly improved th technique's use as a sensitive elemental microscope.

Dynamic Range Consideration for Digital Secondary Ion Image Depth Profiling

S.R. Bryan[1], *R.W. Linton*[1], *and D.P. Griffis*[2]

[1]Department of Chemistry, University of North Carolina, Chapel Hill, NC 27514, USA

[2]Engineering Research Services Division, North Carolina State University, Raleigh, NC 27695, USA

1. Introduction

Digital imaging systems, capable of on-line acquisition of digital image depth profiles (IDP) using ion microscopes [1-3] and ion microprobes [4], have fostered the recent development of SIMS for 3-dimensional analysis. The series of digital images which result from an IDP may represent several thousand individual small area depth profiles acquired in parallel. A local area depth profile (LADP) may be generated from any selected area within the image field by simply summing the intensities of pixels within the selected area for each image of the IDP [5-7]. The main disadvantage of image depth profiling with a direct imaging ion microscope (Cameca IMS-3F), however, has been dynamic range limitations. As a result of the high speed of data acquisition required for real-time (30 frames/s) digital imaging, many framebuffers are only 8-bits deep. Therefore, under a given acquisition time and microchannel plate gain, the intensity of any given pixel can only vary from 0 to 255 over the course of the IDP.

2. Dynamic Range Enhancement Method

In order to increase the dynamic range of the IDP technique, our original multi-element IDP acquisition program [7], has been modified, such that each mass of an IDP may be acquired with either a fixed or a variable image acquisition time and dual microchannel plate (DMCP) gain. The fixed mode allows optimum acquisition of masses having intensities which do not vary appreciably with depth. The variable IDP acquisition mode can accommodate large variations of intensity with depth, while automatically optimizing dynamic range within each image.

3. Applications

For the purpose of testing the dynamic range of the new IDP technique, a B implant (118 KeV; 5E15 cm^{-2}) in Si was depth profiled. A total of 40 B images was acquired over the 1.32 µm sputtered depth. Figure 1 is a comparison of a LADP generated using the entire 60 µm diameter image field and a conventional depth profile acquired using the electron multiplier under identical instrumental conditions. The peak B concentration was calculated using LSS theory to be 3E20 at/cm^3. The comparison illustrates that the new IDP method has similar dynamic range and detection limit to the conventional method of depth profiling. The unequal spacing of the actual data points in the IDP (Fig. 1) is a result of the variable image acquisition times used.

Figure 2 illustrates the trade-off between the size of the area used for generating the LADP and dynamic range or signal-to-noise ratio (S/N).

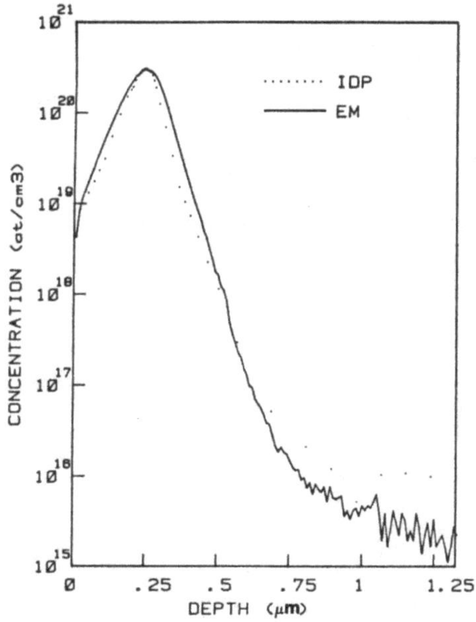

Fig. 1. A comparison of B depth profiles acquired using image depth profiling (IDP) vs. electron multiplier (EM) detection.

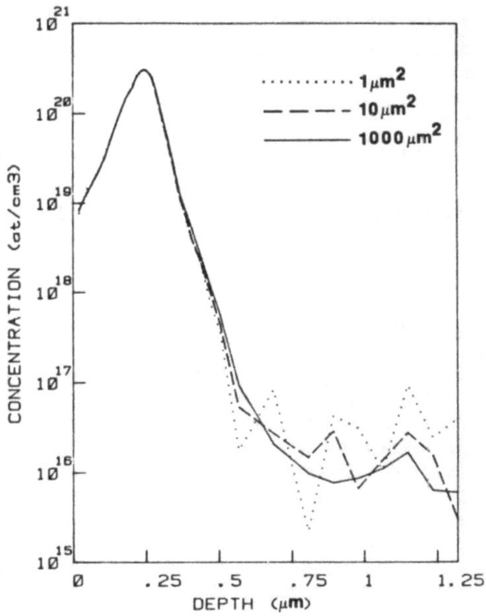

Fig. 2. A comparison of B local area depth profiles (LADP's) generated from 1 µm², 10 µm² and 1000 µm² areas

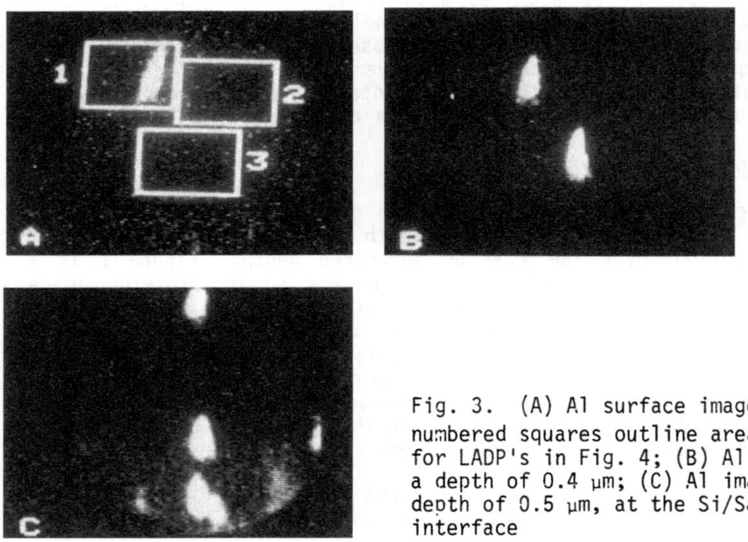

Fig. 3. (A) Al surface image; numbered squares outline areas used for LADP's in Fig. 4; (B) Al image at a depth of 0.4 µm; (C) Al image at a depth of 0.5 µm, at the Si/Sapphire interface

The three profiles in Fig. 2 illustrate that a dynamic range of about 4 orders of magnitude or better is achievable from areas as small as 1 µm. Due to statistical averaging of pixel intensities, the S/N increases and corresponding detection limit decreases roughly as $\sqrt{n}$, where n is the number of pixels in the local area.

The IDP analysis of the distribution of Al in the Si layer of a SOS (silicon on sapphire) sample, is summarized in Figs. 3 and 4. A total of 48 data sets was acquired while sputtering the 0.5 µm Si surface layer. Each data set consists of an Al and Si image. The acquisition time for Al varied from 8 s. at the high DMCP gain setting in the Si layer to 0.5 s. at the low DMCP gain setting in the region approaching the Si/sapphire interface. Figure 3 shows the Al image at 3 different depths. The surface image (Fig. 3A) contains a single Al rich particle. A second particle appears at a depth of approximately 0.4 µm (Fig. 3B). The three numbered squares in Fig. 3A outline the (60 x 60) μm^2 areas used to generate the LADP's in Fig. 4. The substantial differences in the shapes of the three profiles explains the irreproducibility of conventional depth profiles observed using a standard 60 µm diameter image field and an electron multiplier detector. In samples with 3-dimensional heterogeneity, the additional lateral spatial information obtained from an IDP may be essential for accurate interpretion of the depth distribution of the element of interest.

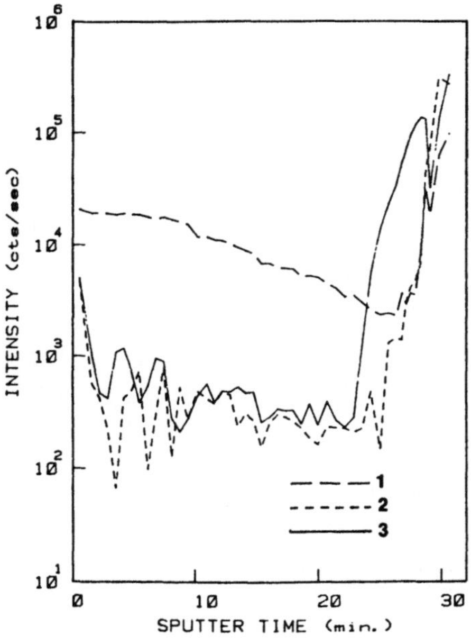

Fig. 4. Al LADP's of 3 areas in Fig. 3A. Acquisition time per image ranged from 8 sec. (8 frames at 1 sec./frame) using the high DMCP gain to 1/8 sec. (2 frames at 1/15 sec./frame) using the low DMCP gain.

References

1. B.K. Furman and G.H. Morrison: Anal. Chem. 52, 2305 (1980).
2. Y. Mishiko, K. Tsutsumi, H. Koyama, and S. Kawazu: Secondary Ion Mass Spectrometry: SIMS IV, A. Benninghoven, J. Okano, R. Shimizu, H.W. Werner, eds., Springer Verlag, New York, 1984, p. 183.
3. S.R. Bryan, W.S. Woodward, D.P. Griffis, and R.W. Linton: J. Microsc. 138(1), 15 (1985).

4. F.G. Rudenauer and W. Steiger: Mikrochim. Acta [Wein], Suppl. 2, 375 (1981).
5. A.J. Patkin and G.H. Morrison: Anal. Chem. 54, 2 (1982).
6. W. Steiger, F. Rudenauer, H. Gnaser, P. Pollinger, and H. Studnicka: Mikrochim. Acta [Wien], Suppl. 10, 111 (1983).
7. S.R. Bryan, D.P. Griffis, W.S. Woodward, and R.W. Linton: "SIMS/Digital Imaging for the Three-Dimensional Characterization of Solid-State Devices", J. Vac. Sci. Technol. A, in press, 1985.

Digital Slit Imaging for High-Resolution SIMS Depth Profiling

S.R. Bryan[1], *D.P. Griffis*[2], *R.W. Linton*[1], *and W.J. Hamilton*[3]

[1] Department of Chemistry, The University of North Carolina, Chapel Hill, NC 27514, USA

[2] Engineering Research Services Division, North Carolina State University, Raleigh, NC 27695, USA

[3] Jet Propulsion Laboratory, California Institute of Technology, Pasadena, CA 91109, USA

A critical attribute in the application of SIMS to semiconductor material and device development is the ability of SIMS to perform quantitative analyses of impurity concentrations as a function of depth. SIMS is uniquely capable of measuring impurity levels (typically parts per million) with a depth resolution of 10 nm or better for shallow depth profiles. The majority of these analyses are performed using a mass spectrometer having low mass resolution (M/delta M = 250). In several important semiconductor problems, however, spectral interferences reside within a few hundredths to a few thousanths of an atomic mass unit from the spectral line of the analyte ion. In the frequently encountered analysis of phosphorous dopant in n-type silicon, the hydride of silicon, ^{30}SiH, lies 0.00778 amu from the ^{31}P ion. Materials such as these are usually analyzed with a high resolution mass spectrometer (M/delta M values as high as 10,000) by mechanically narrowing the slits of the instrument such that only the analyte line is allowed to reach the detector. For the instrument used in this study, the Cameca IMS-3F, the adjustment of the magnetic field to align the analyte line with the exit slit is very difficult and time consuming. Not only is the adjustment of the magnetic field extremely sensitive due to the very narrow width of both the entrance slit and the exit slit, but the magnetic field adjustment is also subject to error due to hysteresis as well as drift due to temperature effects on the magnet control electronics.

In this study, we have addressed these limitations by using a custom microcomputer-based digital imaging system [1,2] to provide a synthetic digital slit mode of high resolution depth profiling using the IMS-3F. Digital ion imaging has been developed in a variety of semiconductor and materials areas with significant progress in quantifying laterally heterogeneous samples [3]. Images of slits have been acquired digitally [4].

Digital slit imaging was achieved in this investigation by projecting the mass-filtered ion image of the spectrometer entrance slit, a standard IMS-3F operating mode, onto a dual microchannel plate/phosphor screen detector. The analog light image generated was collected and digitized by a CID digital TV camera. The digital data representing the image was stored, displayed, and further processed by a microcomputer system [1,2]. Rather than narrowing the exit slit to isolate the analyte line, the exit slit was kept fully open, while successive slit images of the analyte and adjacent interferences were acquired and stored. To extend the dynamic range of the digital imaging technique, the gain of the dual microchannel plate and the image acquisition time were dynamically adjusted during the profiling. In post-acquisition processing, the pixel intensities, represen-

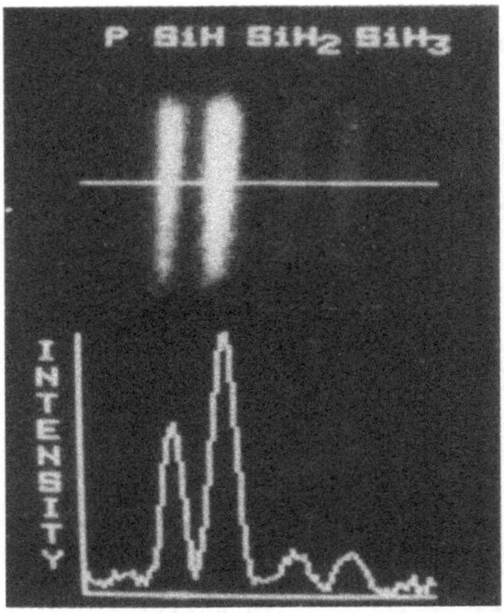

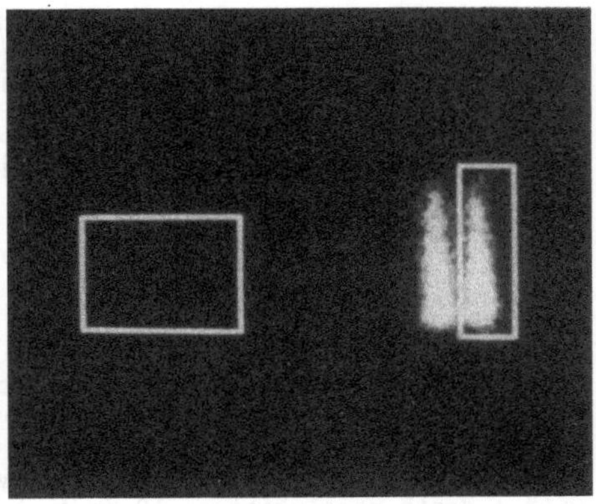

Fig. 1 Digital high resolution slit image with line intensity profile

Fig. 2 Synthetic digital slits for ^{31}P (left) and background (right)

ting only the region in the slit image containing the analyte line were summed, normalized for acquisition time and gain of the system, and background subtracted. Processing of sequential images resulted in the final high-mass resolution depth profile plot. Further details on the general instrumental approach to image depth profiling are available elsewhere [2,5].

A typical digital slit image from the silicon surface is seen in Fig. 1, where the ^{31}P and the three hydrides of the silicon isotopes appear from left to right. Fig. 1 also shows an intensity profile across the line drawn through the slit image. Such a line scan may be used dynamically to determine the position of the analyte peak, should magnetic hysteresis or instrument instabilities cause peak shifts during profiling. The areas selected for the digital slit and for the subtraction of background are seen in Fig. 2. The two higher hydride peaks seen in Fig. 1 have diminished as this image was obtained at a greater depth from the surface. The sum of the pixel intensities within this digital slit constitutes the phosphorous signal and excludes all interferences, thus providing one point of the intensity versus time profile.

A comparison of phosphorus depth profiles (implant energy 600 keV, dose $5x10^{15}$ at/cm^2) acquired using the digital slit image technique and conventional electron multiplier detection are shown in Fig. 3 and 4. The profiles were obtained with a Cs^+ primary beam at 12.5 keV and a current of 600 nA rastered over an area of (150x150) um^2. For the slit image depth profile, the image-acquisition program adjusted the dual microchannel plate gain over two values differing by a factor of 10 and the image acquisition time from 8/15 sec. at the peak of the implant to 64 sec. at the end. Roughly 5 orders of dynamic range were obtained using both detection modes. The only observable difference in the shape of the two profiles occurs where the total integrated intensity is less than 100 counts. The slit image depth profiling technique has the advantage of eliminating the difficult task of aligning the analyte line with the exit slit of the mass spectro-

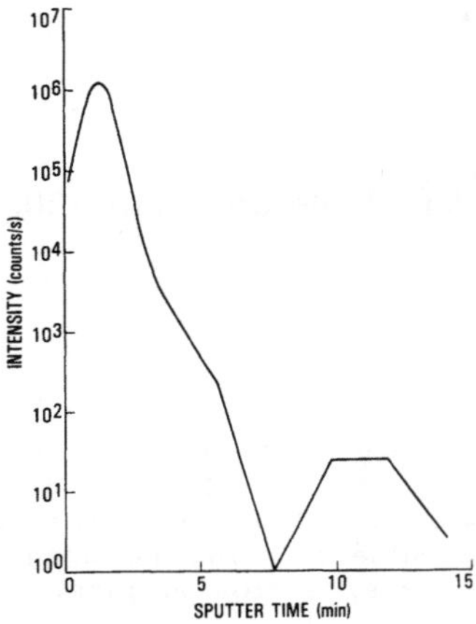

Fig. 3 High-resolution digital depth profile of ^{31}P -- raw intensity data of ion implant in Si

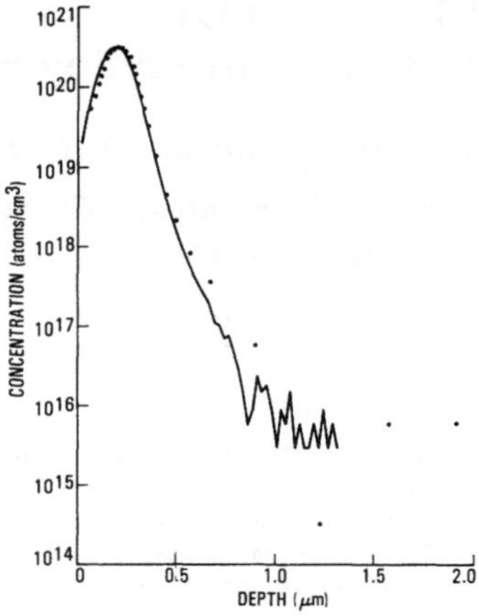

Fig. 4 Composite of digital (dotted) and mechanical (solid) high-resolution depth profiles of ^{31}P in Si

meter while potentially improving the accuracy of high-mass resolution depth profiling by compensating for instabilities of the secondary magnet.

Acknowlegement

This paper presents the results of one phase of research conducted at the Ion Microscope Facility operated by Analytical Services, Engineering Research Services Division North Carolina State Univ. through partial support for the CAMECA IMS-3F and digital imaging system by a grant from the National Science Foundation (DMR-8107499) and matching funds from the Microelectronics Center of North Carolina, North Carolina State University, and the University of North Carolina, and at the Jet Propulsion Laboratory, California Institute of Technology, for the U.S. Department of Energy, through an agreement with the National Aeronautics and Space Administration.

REFERENCES

1. S.R. Bryan, W.S. Woodward, D.P. Griffis, and R.W. Linton: J. Microsc. 138, 15 (1985).
2. S.R. Bryan, D.P. Griffis, W.S. Woodward, and R.W. Linton: "SIMS/Digital Imaging for the Three-Dimensional Characterization of Solid-State Devices," J. Vac. Sci. Technol. A, in press, 1985.
3. G.H. Morrison and M.G. Moran: "Image Processing SIMS," in SIMS IV, pp. 178-182, 1983, and references therein.
4. R.W. Odom, D.H. Wayne, and C.A. Evans, Jr.: "A Comparison of Camera Based and Quantized detectors for Image Processing on an Ion Microscope," in SIMS IV, pp. 186-188, 1983.
5. S.R. Bryan, R.W. Linton, D.P. Griffis: "Dynamic Range Considerations in Digital Image Depth Profiling," in SIMS V, These Proceedings, 1985.

Lateral Elemental Distributions on a Corroded Aluminum Alloy Surface

R.L. Crouch, D.H. Wayne, and R.H. Fleming

Charles Evans and Associates, 1670 South Amphlett Boulevard, Suite 120, San Mateo, CA 94402, USA

Stigmatic imaging SIMS has been applied as a survey tool in determinin lateral elemental distributions of selected elements in aluminum allo 2024-T3. A resistive anode encoder (RAE) was used as a detector system fo the CAMECA IMS-3f ion microscope to provide both pulse counting and imagin capability of elements associated with grain features, corrosion pits, an precipitates within the alloy.

The RAE is a device which can measure charge impact position on resistive film [1]. An ion strikes the front side of a dual microchanne plate assembly and produces a pulse of 10^6 - 10^7 electrons from the revers side. This charge pulse impacts a layer of resistive material and the charg is collected by four electrodes at the corners of the anode. The fou electrode pulses differ in magnitude depending on the impact position of th original pulse. Computation of the position is accomplished via analo electronics. Following conversion to a digital signal consisting of puls position, the information is stored and processed using IBM PC/AT base software. The end result is a 16 level pseudocolor image of the sample Image acquisition time is determined by the ion arrival rate (maximu rate = 50,000 ions/second) and the required accuracy of the pixe intensities. Resolution has been demonstrated with a sample of 1 micro lines spaced 1 micron apart. The lines can be resolved using the RAE and 150 micron image field.

Elements were selected for imaging based on mass spectra obtained usin both Cs^+ and O_2^+ bombardment and a knowledge of the alloy compositio (Table I).

The aluminum alloy samples were exposed to a chemical environment known t induce intergranular corrosion and pitting, as illustrated in Fig. 1. Thi alloy is of interest because it is commonly used in aircraft and th

Table I Composition of 2024-T3 Aluminum Alloy

Element	Weight Percent
Cu	3.8-4.9
Mg	1.2-1.8
Mn	0.3-0.9
Si	0.5
Fe	0.5
Zn	0.25
Ti	0.15
Cr	0.1
Al	balance

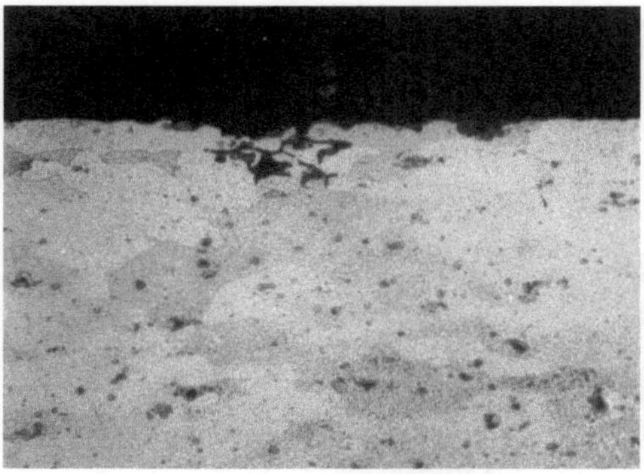

Fig. 1 Polished and etched section through pitted area in Al 2024-T3 illustrating intergranular corrosion. The depth of the pit is approximately 25 micrometers.

corrosive environment is one occasionally found in aircraft operations. The objectives of this work are 1) to determine the elements associated with corroded grain boundaries and 2) to explore the capabilities of the RAE and its associated data-processing system. Images were obtained both from the bottom of pitted areas and from pits through which a cross-section had been cut. No elements were observed at grain boundaries in the cross-sectioned samples. Direct observation of the pit bottoms yielded observable lateral distributions of several elements.

Intergranular corrosion results from exposure of this aluminum alloy to a mixture of water and approximately 0.5% vol/vol of a proprietary hydrophilic polyacrylamide thickening agent. The suspected causative agent is dissolved copper used as a catalyst and left in the polyacrylamide after synthesis. In theory, precipitation of copper occurs at grain boundaries and eventually causes the grains to separate. Because the corrosion is intergranular, the bottom of a pit consists of grain boundaries from which the overlying grains have been removed. Figure 2 shows RAE images of aluminum and copper in an uncorroded area. Aluminum is evenly distributed and copper appears in localized areas of relatively high concentration. Note that the copper does not appear to be associated with grain boundaries in the alloy. The

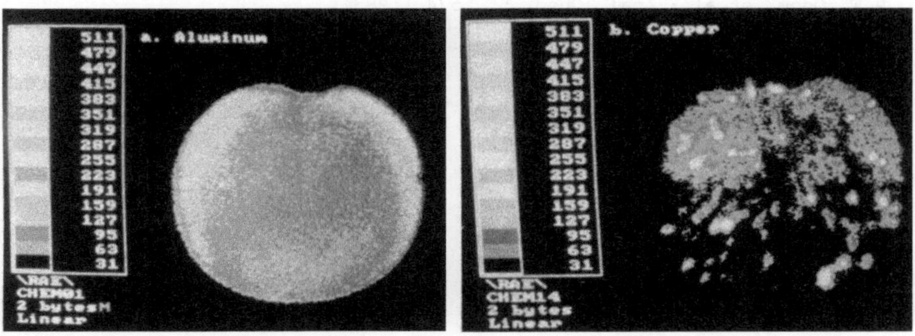

Fig. 2 RAE images obtained from the uncorroded metal surface. O_2 bombardment with positive secondary ion mass spectrometry, 400 micron imaged field. a. Aluminum. b. Copper.

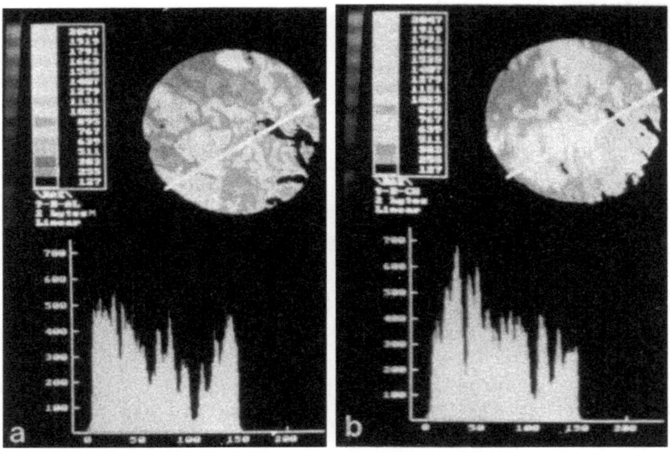

Fig. 3 RAE images and associated line scans obtained from the bottom of a pit. Cs^+ bombardment with negative secondary ion mass spectrometry, 150 micron imaged field. a. Aluminum. b. Copper.

distributions of copper and aluminum are strikingly different in the bottom of pits compared to the uncorroded area, as illustrated by the images in Fig. 3. Here the variation in aluminum ion intensity over the imaged field is most likely due to surface roughness. The more uniform distribution of copper on the pit bottom compared to the localized copper in the uncorroded area suggests that copper has plated out of solution and coated the grain.

The use of the RAE greatly facilitates the acquisition of imaging data in survey techniques such as the one described here. The integration time for a single image can be as short as a few seconds, depending on the secondary ion intensity, and the image obtained can provide quantitative information if the appropriate scale factors are applied. Thus, a large number of images can be obtained in a short time and displayed in a useful form.

References

1. R.W. Odom, et al.: Anal. Chem. 55, 574 (1983).

Improved Spatial Resolution of the CAMECA IMS-3f Ion Microscope

M.T. Bernius, Yong-Chien Ling, and G.H. Morrison

Baker Laboratory of Chemistry, Cornell University, Ithaca, NY 14853, USA

1. Introduction

The imaging mode of a stigmatic SIMS microanalyzer such as the CAMECA IMS-3f provides the unique capability of producing a magnified localized mapping of the elemental constituents on a sample surface in real time. The principal function of the instrument when used in the microscope mode is to resolve detail present in the specimen. Magnification is necessary to make the image of sufficient size so that the smallest microfeatures resolved by the instrument can be seen by the eye. Therefore it is convenient to compare instrumental parameters of the ion microscope in terms of its inherent resolution capabilities.

The lateral resolution obtained with the CAMECA instrument is generally quoted anywhere between 0.3 - 1.0 μm. A critical evaluation of such a range becomes essential before attempting to extract information from the ion image. An accurate knowledge of the spatial resolution and the factors that influence it defines the scope of microfeature recognition as is necessary prior to image-processing techniques.

2. Criterion for Image Resolution

The established criterion for resolution was proposed by Lord Rayleigh [1] in 1879 for monochromatic spectral peaks of the form $[\sin^2 \theta/ \theta^2]$. Two peaks of equal intensity were considered resolved if the principal maximum of one coincided with the first minimum of the other. The minimum between the two peaks of unit amplitude is located at $\theta=\pi/2$, so the combined amplitude here is equal to $2[\sin^2 \theta/\theta^2] = 8/\pi^2 = 0.812$. However it is experimentally possible to resolve two peak components at closer distances. By adopting a new criterion for resolvability, namely the distance between a single peak curve's inflection points (referred to here as γ) it is observed that at distances smaller than γ, peak differentiation is not possible, where at distances greater, it is.[2]

It is convenient to express the two-dimensional (spatial) spectral peaks that form an image (where the intensity of the peak is associated with contrast) in terms of Gaussians. The resolution can then be expressed in terms of Rayleigh's criterion by using the 0.812 ratio of the minimum between two peaks of equal intensity to a maximum [3], which is fulfilled for a peak separation of 3.104 times the Gaussian standard deviation, σ, or approximately 1 1/2 times γ. However, as features can be resolved at distances closer than this with the use of digital techniques, the remainder of this presentation will express resolution in terms of the smallest resolvable distance (SRD) observed associated with 1.0 γ. All error parameters presented using this method [2] represent replicate analyses from various images with many (N>5) microdensitometer image traces each.

3. Resolution Evaluation

To objectively evaluate an instrument's resolving power, it is necessary to image a specimen having adequate fine structure. Therefore standard samples were fabricated as test patterns for the ion microscope at the National Research and Resource Facility for Submicron Structures located at Cornell University. These bar patterns, made using electron lithographic techniques, allow accurate and reproducible image tuning for high spatial resolution. They also permit the study of each instrumental parameter used in imaging and their effect on image quality [2]. The micro-test patterns were used to study the resolution in the image by examining the resulting spread function which is due to the instrumental image-transfer process that is inherent in any image-generating device. The same changes in instrumental resolution as a function of the variables examined will take place regardless of the specimem used. Due to its simple composition, the micro-test patterns are much easier to work with as compared with the varied, complex structures found, for instance, in biological specimens.

Images were recorded on photographic film and were later analyzed by microphotodensitometry. Results indicate that the character of the sample (i.e. resistivity, surface texture, etc.) and the instrumental operating conditions (vacuum, contrast aperture setting, instrumental field of view, magnification factor, and microchannelplate gain) directly affect the relative resolution of the ion image. The best image resolution of 0.53 ± 0.03 µm. was achieved using the 150 µm. field of view with the highest magnification (250X as measured at the fluorescent screen), the smallest contrast aperture (20 µm.) and an operating vacuum of 10^{-7} torr.

4. Resolution Improvement

The instrumental parameter with the greatest effect on the resolution of the CAMECA IMS-3f was found to be the magnification factor, which is governed by the varifocal (zoom) dual-einzel projection lenses located before the image detector (microchannelplate/fluorescent screen). This is because of the microchannelplate blooming effect, which can be seen graphically in Figure 1. The smaller the feature being imaged, the smaller the projected

Fig. 1: **Demonstration of the microchannelplate blooming effect.** Presented is a silicon image using a low current oxygen primary beam, photographed with push-processed film (2000 ASA). The central portion is where most of the primary beam was located. The honeycomb structure is more pronounced at the periphery, where fewer channels were being used to form the image. This structure represents the "dead space" on the microchannelplate where no channels exist (because of the hexagonally-closely packed structure of the channels used in the manufacturing process. The honeycomb structure cannot be seen towards the center, demonstrating the blooming of the electrons from the channels on neighboring sides of this dead space, as all of the microchannels are being used to form this portion of the image.

feature's ion beamlet cross-sectional area (normal to the optical axis), the fewer the number of channels being used to form the image in the microchannelplate, and the greater the effect that "channel blooming" has on the formed image. To overcome this microchannelplate blooming effect, the electronic control board governing the operation of the zoom projector lenses were bypassed with a manually operated FLUKE 410B linear high voltage power supply, thus giving greater flexability to the operator with regard to the position of the ion beam crossover in this lens area. In this way, magnifications as high as 500X were achieved. With the smallest contrast aperture (20μm.)and operating vacuum of 4 x 10^{-8} torr, an instrumental magnification of 406X achieved a lateral resolution of 0.22 ± 0.02 μm. The results can be seen in Figure 2. At these enhanced magnifications, however, identification of imaged areas become difficult, as the observed field of view is diminished and detected signal intensity is decreased. Also instrument vibrational interference makes imaging under these conditions difficult at the present time.

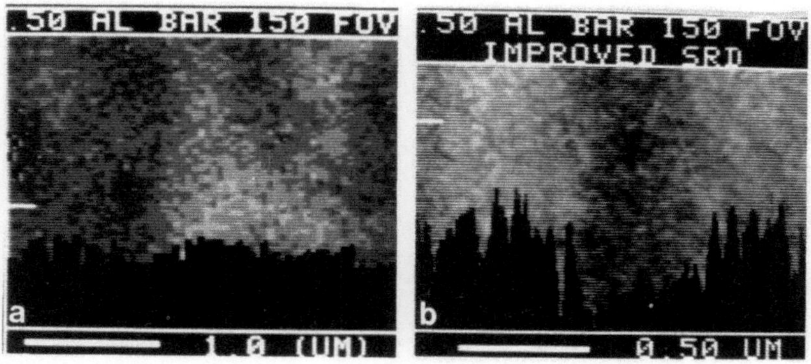

Fig. 2: **Results of instrumental resolution improvement.** Presented are digitized images of a 0.50 μm. Aluminum on Silicon square wave bar pattern with superimposed line traces, from which the instrumental spread function is derived. (Comparable images of the Aluminum bar and Silicon substrate yielded identical results.) Fig. 2a has an instrumental magnification of 250X (unmodified instrument) and a SRD of ~0.5 μm. Fig. 2b has an instrumental magnification of 406X (manual magnification adjustment) and a SRD of 0.22 ± .02 μm.

5. Summary and Conclusions

The best lateral resolution obtained with an unmodified CAMECA IMS-3f ion microscope is 0.53 μm. with an instrumental magnification of 250X using the "observable criterion" (ie:γ) for resolution proposed. With slight modification, a lateral resolution of 0.22 μm. is achieved with an instrumental magnification of 406X. Both these values were obtained using the 150 μm. instrumental field of view.

Thus, following a recent study [2], we have improved instrumentally the resolving power of the CAMECA IMS-3f ion microscope, and now have necessary information for digital image-processing techniques to further enable microfeature identification and analysis without distorting the information content in the ion image.

6. Acknowledgments

The authors would like to thank the National Science Foundation and the Office of Naval Research for their financial support, and the NRRFSS at Cornell University for the assistance in fabricating the test patterns.

References

1. Lord Rayleigh, Phil. Mag.,S.5, 8, (1879), pg. 261
2. Bernius, M.T., Ling, Y.C., Morrison, G.H., Anal. Chem. Jan. 1986 (in press)
3. Born, M., and Wolf, E., "Principle of Optics", Pergamon Press, Oxford (1964), pg. 333.
4. Lampton, M., Sci. Am., 245 (Nov. 1981) pp 62-71.

Video Tape System for Ion Imaging

M.G. Moran, J.T. Brenna†*, and G.H. Morrison*

Baker Laboratory of Chemistry, Cornell University, Ithaca, NY 14853, USA

1. Introduction

In recent years, several schemes have been proposed for direct acquisition of digital representations of stigmatic SIMS images [1-3]. These approaches suffer some common drawbacks. In order to construct a reasonable representation of an ion image, a considerable amount of raw data (ca. 100 Kbyte/image) must be manipulated. In the systems described by [1,2] data cannot be acquired from the ion microscope and written to random access memory at the same time. Consequently, some data are missed while the record of previous data is being written. This problem becomes worse as spatial quantization (i.e. more image pixels) of the image increases. Additionally, in applications such as image depth profiling [4], investigations of differential sputtering by construction of "burn through" maps [5,6] and image integration [7], many images must be recorded. The sheer volume of digital data can rapidly outstrip the capacity of contemporary random access mass storage units. Furthermore, on-line data acquisition puts considerable pressure on the operator to make decisions about data recording during an experimental run. This can be difficult in some cases, such as those involving high sputter rates and rapid image evolution. As the sampling in SIMS is destructive, it is crucial to record data in a "one pass" mode, making these decisions even more critical. Recently, improvements in experimental strategy have ameliorated these difficulties to some extent [8], but the acquisition of copious amounts of data remains problematic.

In order to alleviate some of these problems, we have integrated the so called VICAR system [9] (Video Image Continuous Acquisition Recorder) into our current MIDAS [1] system. Figure 1. shows a schematic of the system layout. This consists of a SONY VO5850 industrial quality 3/4" video tape recorder which serves as an essentially unlimited serial access memory. A LYON-LAMB VAS-IV video controller is the key to the system and serves two functions. One is the ability to uniquely label each video frame of data by writing and reading frame code information to and from the tape. Secondly, this device can control the video tape recorder, and is interfaced to our DEC PDP 11/34 computer. This brings the entire post acquisition data digitization process under computer control. Finally, a FORTEL CCDHPS time base corrector is used to correct time base errors in the video signal arising from mechanical instabilities in the video recorder. A similar approach has been briefly described by SCHUETZLE et al. [10], but has not been quantitatively evaluated. More recently, LETA has proposed an elegant system for high sensitivity ion imaging using a video disk apparatus and a 512x512 frame buffer [11].

---

† Present address: IBM, Systems Technology Division, Endicott, NY 13760

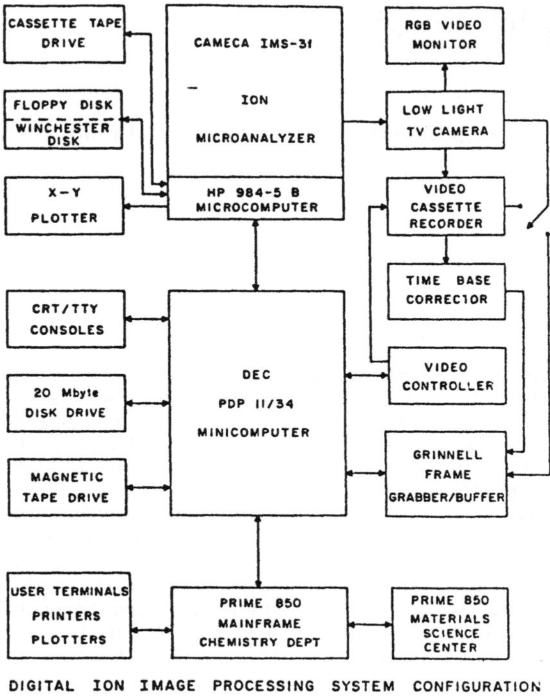

Figure 1. System configuration with video tape acquisition capability.

2. Results and Discussion

We have investigated the linearity of response, inherent noise level, and image resolution capability of the VICAR system. Basically, these comparisons were done by simultaneously recording data using the MIDAS system and the video recorder. The corresponding data were then digitized from video tape and compared to the other data set. Linearity was checked by imaging a single target of varying brightness levels. Linear regression of a plot of pixel values from corresponding areas of the images obtained by the two methods yields a slope of 1.006 with a correlation coefficient of .997 and a set of unstructured residuals.

System noise is an important consideration. In MIDAS, the dominant sources of noise are the 3f microchannelplate, the low light intensity video camera and the quantization noise of the digitizer. Using a pool of 3 data sets we have calculated pixel standard deviations for n=3600 pixel values. The confidence limits on the pixel standard deviations were found to be $27.0 \leq \sigma \leq 28.2$ for MIDAS and $26.8 \leq \sigma \leq 28.1$ for the combined MIDAS/VICAR system. The VICAR system adds virtually no noise to the images, and if anything, acts as a low pass filter. This is important as noise degrades spatial resolution and raises analytical detection limits.

Image resolution was investigated by recording a scene of a NBS 1010a standard resolution chart. We know from previous experience that the frame buffer's spatial quantization of an image into a 240x256 pixel representation is the limiting factor of image resolution in the MIDAS

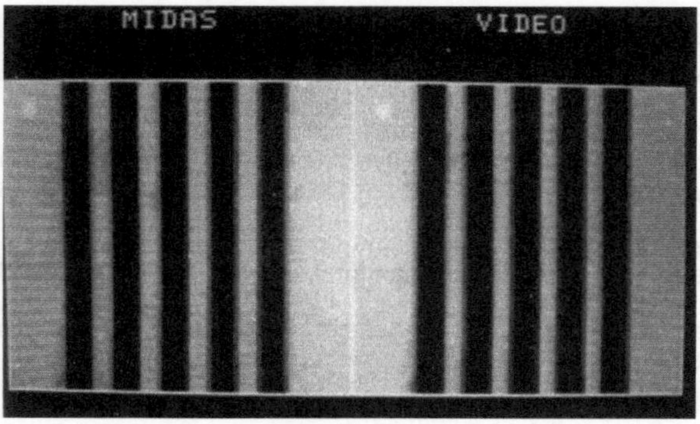

Figure 2. Resolution chart from MIDAS and VICAR respectively.

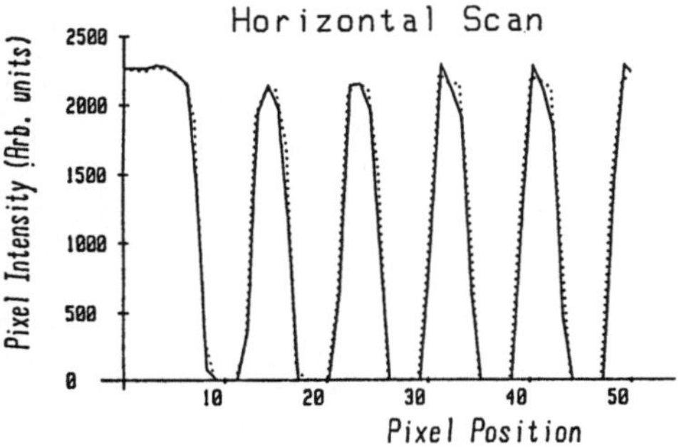

Figure 3. Horizontal line scans from image in Figure 2.

system. Figure 2. shows the side by side rendition of a set of resolution bars by MIDAS and VICAR, respectively. Figure 3. shows horizontal line scans obtained from these images. We conclude that image resolution is essentially kept intact by the VICAR system. This statement applies to the spatial quantization of 240x256 currently used by MIDAS. This point will be rechecked upon upgrading MIDAS to 480x512 image representations.

3. Conclusion

The VICAR system provides a convenient, flexible way to acquire many stigmatic SIMS images in real time. Also, the data can be readily retrieved and reduced under computer control after the experimental run, when time is not an important factor. Our evaluation of the system shows excellent linearity of response, low inherent noise, and very good preservation of image resolution.

4. Acknowledgment

The authors would like to thank the National Science Foundation and the Office of Naval Research for their financial support.

References

1) B.K. Furman, G.H. Morrison: Anal. Chem. 52, 2305 (1980).
2) S.R. Bryan, W.S. Woodward, D.P. Griffis, R.W. Linton: J. Micros. 138(1), 15 (1985).
3) R.W. Odom, B.K. Furman, C.A. Evans, C.E. Bryson, W.A. Petersen, M.A. Kelly, D.H. Wayne: Anal. Chem. 55, 574 (1983).

4) A.J. Patkin, G.H. Morrison: Anal. Chem. 54, 2 (1982).
5) A.J. Patkin, S. Chandra, G.H. Morrison: Anal. Chem. 54, 2507 (1982).
6) R.W. Linton, M.E. Farmer, P. Ingram, S.R. Walker, J.D. Shelburne: Scanning Electron Micros. III, 1191 (1982).
7) J.T. Brenna, G.H. Morrison: submitted to Anal. Chem.
8) S.R. Bryan, R.W. Linton, D.P. Griffis, "Dynamic Range Considerations for Digital Secondary Ion Image Depth Profiling" SIMS-V in press (1985)
9) J.T. Brenna, M.G. Moran, G.H. Morrison: in press Anal. Chem.
10) D. Shuetzle, T.L. Riley, T.E. deVries, T.J. Prater: Mass Spec. Rev. 3, 527 (1984).
11) D.P. Leta "A High-Resolution Single Ion Sensitivity Video System for Secondary Ion Mass Spectrometry" SIMS-V in press (1985).

Application of a Framestore Datasystem in Imaging SIMS

A.R. Bayly, D.J. Fathers, P. Vohralik, J.M. Walls, A.R. Waugh, and J. Wolstenholme

VG Scientific Ltd., The Birches Industrial Estate, Imberhorne lane, East Grinstead, Sussex, RH19 1UB, U.K.

1. Introduction

The advances made in UHV SIMS instrumentation (scanned ion probe/quadrupole analyser) have increased analytical capabilities in several areas : depth profiling using high current reactive ion (O_2, Cs) or inert-gas ion probes, with or without oxygen gas flooding, has achieved extremes in depth resolution or detection limits [1], particularly in semiconductor materials; sub-micron spatial resolution using scanned liquid metal field-ion source, LMFIS, probes of Ga, In, Au [2,3] and Cs [4], and the SIMS application of these ions has been demonstrated in elemental imaging, spectral analysis, line scans and depth profiles of a variety of samples including oxide-insulators of multi-layer or fibre structure [4,5] and polymer fibre [6] as well as semiconductor device structures [4,7]; imaging molecular or "STATIC" SIMS on polymer materials has been demonstrated [8] with coincident ion and electron beams (to maintain charge neutrality) while the introduction of neutral fast-atom beams, "FAB", has made the spectral analysis of all types of insulators as easy as for conductors. The latest addition is the scanned FAB probe [9] which should make molecular imaging of insulators at 5-10 micron resolution a realistic possibility. In pace with these advances in primary probes the collection efficiency of the best quadrupoles has been improved to around the 1% level, so that the potential of SIMS for 3D micro-volume analysis either at the trace-element level or at the major molecular species level is now being realised in UHV. The amount of data generated in multi-ion imaging or multilevel depth profiling is unmanageable without either automatic data-reduction during acquisition as used in depth-profiling (ie, frame integration with raster-gate) which wastes useful information, or photographic superposition, microdensitometry etc, for image analysis. An excellent solution to this data management problem is a FRAMESTORE DATASYSTEM which allows data acquisition with minimum automatic data reduction of the stored raw data but allows unlimited retrospective trials of data reduction and output without altering the raw data. Some simple forms of reduction are : selected area-gate in-depth profiles to avoid edge signal or other artifacts; matrix-signal referencing for minimising surface topographical contrast; multiple image registration with independent sensitivity factors for displaying chemical segregation; smoothing; selected area integration. The scanned probe technique with pulse-counting detection is ideally suited for direct control from, and input to, such a framestore datasystem.

2. THE VG SIMSLAB

The SIMSLAB is a UHV instrument incorporating many of the basic system designs evolved for the popular ESCALAB MkII but with the analysis chamber designed as a multi-probe dedicated SIMS system. It uses the MM12-12S quadrupole mass spectrometer fitted with a new ion-transfer optics, HTO-100,

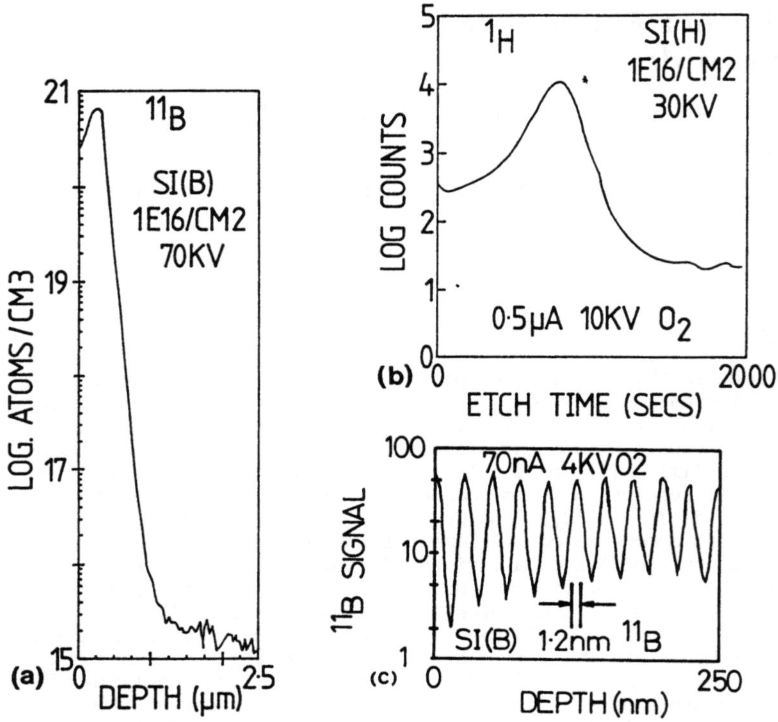

Fig. 1 : DP50 profiles of dopants from 0.5mm² craters in silicon : (a) High dynamic range at almost 3Å/s (collapsed raster method), (b) Mass 1(H) using the 1-800 amu MM12-12S, (c) Good depth resolution at low probe energy on a multi-layer device.

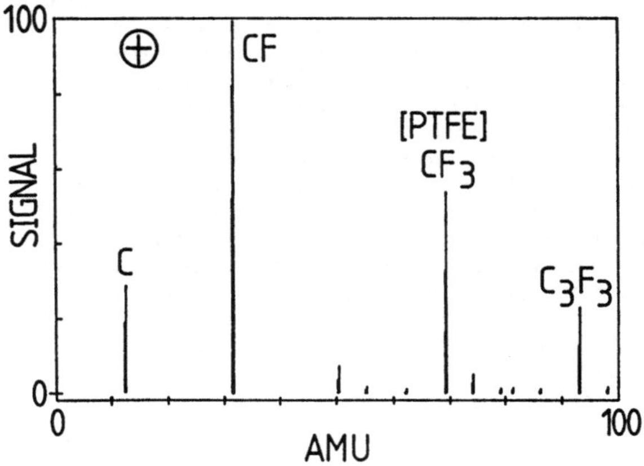

Fig. 2 : A FAB spectrum from PTFE Total dose (estimated from electron emission currents): 5E13/cm² 3kV Ar°

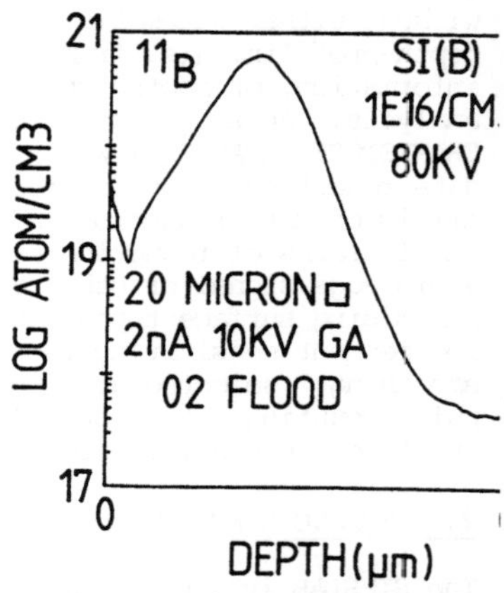

Fig. 3 : A small area profile (gate area 20x20µm²) at 1Å/sec to allow oxidiation and allows the simultaneous fitting of three different primary ion columns: the DP50 for reactive or inert-ion trace analysis depth profiling; the CIG200 LMIS (gallium) microfocussed probe for imaging and small area depth profiling; and the FAB61 neutral atom probe for fast spectral analysis of insulators and STATIC SIMS. The DP50 column is being adapted to accept both the DP5 duoplasmatron and the General Ionex 133 caesium source. For ion probe analysis of insulators two electron guns can be fitted,including a sub micron electron probe, LEG200, suitable for AES. The Auger analyser fits in a position suitable for simultaneous SIMS/AES depth profiling. Alternatively the CIG200 LMFIS (caesium) microprobe can take this position. A higher stability version, SIMSLAB II, offers 50nm resolution with the MIG300 [6]. All the probes (ion, electron and FAB61 in ion mode) can be rastered under control of the VGS5100 framestore. The VGS5100 is based on the DEC LST-11 computer with Winchester disk system (20Mbyte fixed, 10Mbyte removeable) and a 512x512 pixel x 8 bit framestore. Secondary electron and secondary ion images plus 3D images/profiles and spectral scans can be acquired . A fast reduced area scan mode is used for setting probe focus and sample positioning. (An optical microscope monitors initial positioning). The fixed memory can be up to 140 MByte for full 3D capability.

A comparison of probe spot size v. beam current of the various LMFIS probes and the DP50 duoplasmatron probe was published recently [6] and various applications of the LMFIS probes appear in [2,4,5,6,7,8]. Figure 1 shows examples of depth profiles obtained with the DP50 which show high dynamic range or depth resolution. Figure 2 shows a typical application of the FAB61 in the spectral analysis of PTFE which shows peaks characteristic of the molecular structure. Such peaks degrade at low primary dose levels so that fast and easy acquisition is essential. An important new result for the gallium probe, Fig 3, is the small area depth profile of a boron-in-silicon implant. This reveals a detection efficiency of over 2E-4 for boron (ions detected/atoms removed) which compares well with other conventional large-area data [1]. This indicates that there are new applications for the LMFIS probes in semi-conductor device analysis. There is considerable scope for improving on this result with some modification of technique. An example of elemental image overlay by the framestore system is shown in Fig 4 using the gallium probe.

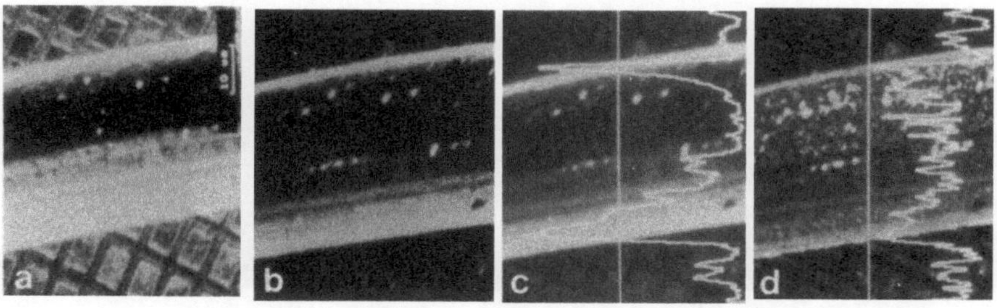

Fig. 4 : Reduction of topographic contrast by matrix signal referencing : A steel wire with copper mesh background : The O^- map (b) is smoothed (c) and ratioed to total ion map (a) to produce a more informative map (d).

Acknowledgements

We thank Drs M. Dowsett and J.B. Clegg for the boron implant samples.

References :

1) C.W. Magee and R.E. Honig : Surf Interface Anal., 4, 35 (1982)
2) A.R. Waugh, A.R. Bayly, and K. Anderson : Proc. 29th Field Emission Symp. (Almquist and Wiksell, Stockholm 1982) p 409
3) A.Wagner : Nucl. Instrum. Method., 218, 355 (1983)
4) A.R. Waugh, A.R. Bayly, and K. Anderson : SIMS IV, Proc. 4th Int. Conf. on SIMS, Osaka, Japan, 1983 (Springer-Verlag, Berlin) p138
5) A.R. Bayly, A.R. Waugh and K. Anderson : Nucl. Instrum. Method. 218, 375 (1983)
6) A.R. Bayly, A.R. Waugh and P. Vohralik : Spectrochim. Acta, 40B, 717(1983)
7) A.R. Bayly, A.R. Waugh and K. Anderson: Scanning Electron Microscopy, 23, (1983)
8) D. Briggs : Polymer, 25, 1379 (1984)
9) A.J. Eccles, J.A. van den Berg, A. Brown and J.C. Vickerman to be published

Chemical Characterisation of Insulating Materials Using High Spatial Resolution SSIMS – An Analysis of the Problems and Possible Solutions

A. Brown, A.J. Eccles, J.A. van den Berg, and J.C. Vickerman

Department of Chemistry, University of Manchester, Institute of Science and Technology, Manchester M60 1QD, U.K.

1. Introduction

The surface chemical characterisation of materials requires not only an analysis of the elemental composition but also a description of the chemical structure. Static SIMS is extremely successful in providing both of these requirements, [1,2]. In combination with the recently developed liquid metal ion sources, with their very high spatial resolution, in principle SSIMS provides the possibility of a true chemical microscopy. Although the potential of this development is great, at the present time the realisation is limited by a number of problems. To illustrate the potential and to analyse the limitations of SSIMS microscopy we consider the application of the technique to the study of oxide and polymer materials.

2. Experimental

A VG SIMSLAB was used which incorporated a MIG 100 liquid Ga beam line with minimum beam diameter of 0.2μm at 10 keV; a non-scanning Ar atom/ion beam source of our own design yielding beam fluxes equivalent to 0.1-10 nA cm^{-2} at 2 keV (flux measurements for atom beams described in [3]); a focussed scanning atom beam line also designed and built at UMIST having a minimum beam diameter of 8 μm at 10 keV and atom currents of 0.05-20 nA cm^{-2} (full specification described in ref.[4]); and a VG MM12-12 quadrupole analyser with an energy analyser after the design of Wittmaack [5]. Electron or ion images are acquired photographically or into a DEC Professional computer.

3. Potential and problems

Good spatially resolved chemical analysis is of particular value in the investigation of polymer surface coatings and oxide gate structures on semiconductor devices. In principle the distribution of the oxide or polymer can be monitored by chemically characteristic cluster ions [1]. With present equipment based on quadrupole mass analysers and liquid metal ion beams, these studies show that there are three serious problems in satisfactorily carrying out the analyses. (a) Since these materials are generally poor conductors electron beam neutralisation is usually necessary. Many organic materials are very susceptible to electron - stimulated desorption with the resulting loss of spatial resolution for these components, [2]. (b) Different parts of the surface being analysed may take up different surface potentials due to varying conductivities, resulting in spurious peak intensity variations. This is vividly demonstrated in figure 1. These are $SiOH^+$ images of the gate oxide structures on a Si substrate obtained using a Ga^+ ion beam. The only difference is the sample bias. At -30 V the ions from the thinner surface oxide are collected. At -45 V those from the thicker gate oxide are able to pass through the energy filter. Almost exact image reversal is observed depending on the surface potential.

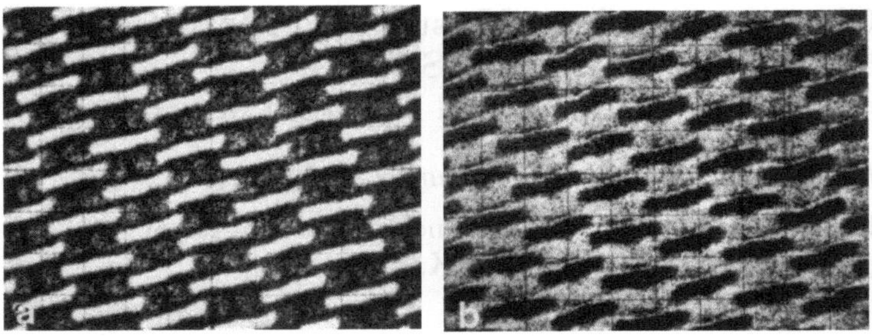

Fig 1 $SiOH^+$ images of an oxide device - target bias (a) -30 V; (b) -45 V. Screen divisions correspond to 10 μm.

(c) Ion bombardment damages organic materials about 10^3 more rapidly than inorganics [3], and even for these, at magnifications above 1000 it will be impossible to stay in the static regime and obtain sufficient sensitivity, Table 1. This is most severe for the chemically interesting cluster ions.

Table 1. Monolayers removed as a function of ion current and magnification

MAG	I_p 1 pA	0.05 nA	0.1 nA	1 nA	10 nA	Assumptions
100	0.0001	0.005	0.01	0.1	1	Sputter yield
500	0.003	0.15	0.3	3	30	=2
1000	0.01	0.5	1	10	100	Frame time =
5000	0.25	12.5	25	250	2500	100s
10000	1	50	100	1000	10000	Nos. frames = 1

4. Possible Solutions

Work in this laboratory has shown that in 'broad beam' SIMS the use of neutral beams (FAB) can obviate many of the problems associated with surface potential and electron beam neutralisation. This is illustrated by the ease with which positive ion polymer spectra can be obtained without electron neutralisation, and furthermore chemically informative negative ion spectra can be generated with a very small flux of electrons where ESD is not a problem, fig 2 [6]. To date all the negative ion spectra of which the authors are aware are low mass 'C' and 'O' species. The small surface charge produced even under neutral flux inhibits the release of the negative ions. Using accurately calibrated ion and atom beams reveals another advantage of atom bombardment in polymer analysis in that it leads to 4 or 5 times lower rate of damage as compared with ion bombardment [3]. There is good evidence that reduction of oxides is also considerably less under neutral beams. The requirement to dissipate implanted positive charge must result in structural disprution in insulating materials. Our knowledge of the mechanisms of fragmentation in organic mass spectrometry would suggest that primary particle charge is a significant factor in generating damage in insulators although it has been neglected to date. Clearly therefore, it would be of great benefit for the imaging of delicate insulating surfaces if a high spatial resolution scanning neutral beam system could be constructed. Although the problem of beam steering was thought to rule out such a device it has been constructed in our laboratories and patents are being sought,

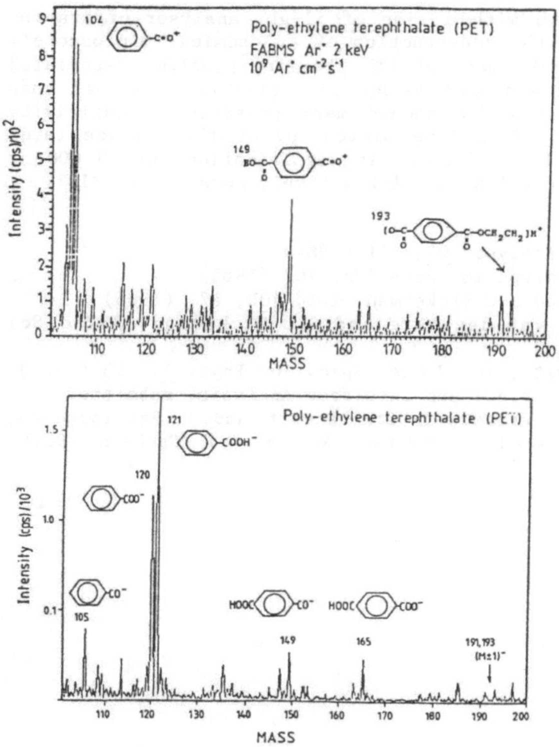

Fig 2 Positive (Ar^o) and negative ($Ar^o + e^-$) ion spectra of PET

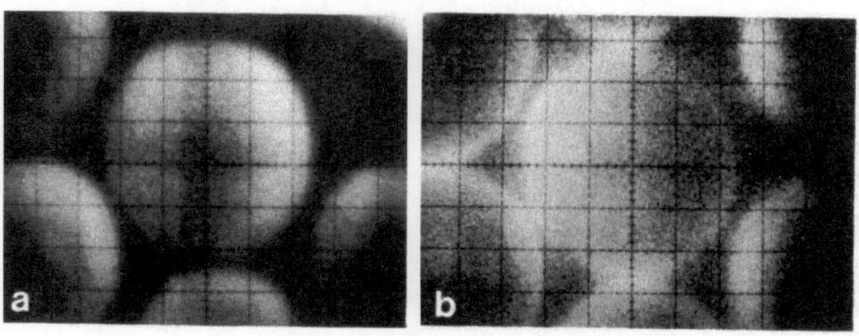

(a) Screen divisions = 100 μm (b)

Fig 3 SEM image (a) and an $SiOH^+$ image (b) of Si spheres under an atom beam

[4]. It utilises a scanning ion beam with a charge neutralisation after the focussing and scanning stage. This results in steerable neutral beam of diameter 8 μm with a maximum flux density of 10^{11} particles. Fig 3 shows an SEM and $SiO(H)^+$ image of some glass spheres. Under ion bombardment SEM imaging was impossible. This beam line is being further improved.

Despite these improvements the low transmission of quadrupole systems (around 10^{-2}) will continue to limit the potential of chemical microscopy on delicate materials due to the rapid damage rate. The combination of high resolution liquid metal sources with a time of flight analyser offers the most exciting possibility for the construction of a 'chemical microscope'. A pulsed ion beam should obviate many of the charging problems associated with ion beams, however charge-induced damage may still be higher than with atom beams. Nevertheless TOF should make possible a sensitivity increase of 10^3 to 10^5, [7]. Thus true microscopy of the surface layer would be posssible under static conditions at magnifications of 20 000 or 50 000. This is now a prospect which is being actively pursued at UMIST.

References.

1 A Brown and J C Vickerman, Analyst 109, 851 (1984)
2 D Briggs and M J Hearn, Spectrochim. Acta 40B, 707 (1985)
3 A Brown, J A van den Berg and J C Vickerman, ibid 40B, 871 (1985)
4 A J Eccles, A Brown, J A van den Berg and J C Vickerman, J Vac Sci Technol., submitted.
5 K Wittmaack, J Maul, F Schultz, Int J Mass Spec. Ion Phys. 11, 23 (1973)
6 A Brown and J C Vickerman, Surface and Interface Analysis, submitted
7 P Steffens, E Niehuis, T. Friese, D Greifendorf and A Benninghoven, SIMS IV, Springer Series in Chemical Physics, Vol 36, 404 (Springer 1984)

Application of Digital SIMS Imaging to Light Element and Trace Element Mapping

D. Newbury and D. Bright

National Bureau of Standards, Gaithersburg, MD 20899, USA

D. Williams and C.M. Sung

Lehigh University, Bethlehem, PA 18015,USA

T. Page and J. Ness

University of Cambridge, Cambridge, England

1. Introduction

The determination of the relationship of compositional microstructure to morphological microstructure is often critical for elucidating structure-property relationships in materials. Traditionally, the study of the lateral distribution of elemental constituents has been carried out in the electron probe x-ray microanalyzer. Because of the inherent properties of x-rays, electron probe compositional maps suffer several significant deficiencies: (1) Light elements with atomic number < 11 (sodium) are detected with poor sensitivity; below carbon detection in an imaging mode effectively ceases. (2) While detection limits for heavier elements can be as low as 100 ppm for analysis at a single location, the practical detection limit in images with 256x256 pixels is about 0.1 weight percent. Imaging by secondary ion mass spectrometry (SIMS) offers two special capabilities which extend substantially beyond the limitations of electron probe microanalysis. These capabilities, light element and trace element imaging, have been employed in this work in attacking two materials science problems: (1) the distribution of lithium in Al-Li alloys and (2) the origin of differences in the secondary electron coefficient observed in crystals in reaction-bonded SiC.

2. Experimental Conditions

A direct imaging ion microscope was used to prepare the secondary ion images. The conventional film camera recording system was replaced by a television camera controlled by a digital image acquisition and processing system. The conventional single stage channel plate was also replaced with a two stage channel plate for additional gain. All ion images used in this work were recorded for accumulation times of 4 seconds or less. Scanning electron microscope (SEM) images were obtained with an Everhart-Thornley detector collecting backscattered and secondary electrons. Subsequent image manipulation was carried out with a combination of conventional image-processing algorithms and special purpose algorithms developed in our laboratory [1]. Conventional operations typically used included superposition of multiple images through primary color addition of single-band images, simple arithmetic operations and manipulation of intensity relationships by selection of the slope and intercept of linear input-output functions. More sophisticated image processing included warping of images to establish superposition when simple translation/rotation operations proved to be inadequate. Reference tie points between SIMS images and other images were selected manually by observation. Generally, tie points were established at prominent features in the microstructure, such as grain boundary triple points.

3. Observations on Aluminum - Lithium Alloys

Aluminum-lithium alloys form a class of high strength-to-weight materials of particular interest in the aircraft industry [2]. The lithium constituent is virtually inaccessible by any other microanalysis technique, particularly at the concentration levels of interest. In this study, the microstructure was observed in a series of Al-Li alloys, with concentrations of Li in the range 1 to 4 wt%. Samples were prepared by conventional metallographic grinding and polishing. A striking change was observed in the lithium distribution in these samples as a function of depth as shown in Figure 1. An image taken at the outer surface is characterized by segregation of the lithium over a lateral scale of tens of micrometers. The true compositional structure of the material, as confirmed by structural observations in the electron microscope, is only seen after the removal of approximately 500 nm of material. This structure consists of a lithium-rich second phase distributed nearly uniformly through the aluminum matrix, with the notable exception of denuded zones near grain boundaries, where the second phase precipitate appears to have been lost and segregated to the nearby boundaries. The surface structure possibly results from leaching in the aqueous environment during grinding and polishing, since it is known that Al-Li alloys will undergo chemical changes in the presence of water. However, the SIMS images suggest that conventional optical metallography images of the surfaces of Al-Li samples prepared in this manner may be highly misleading.

4. Observations on Reaction-Bonded Silicon Carbide

Reaction-bonded silicon carbide is a two-phase material consisting of SiC grains cemented by Si[3]. The material is prepared by heating SiC in the presence of carbon and excess silicon. As a result,

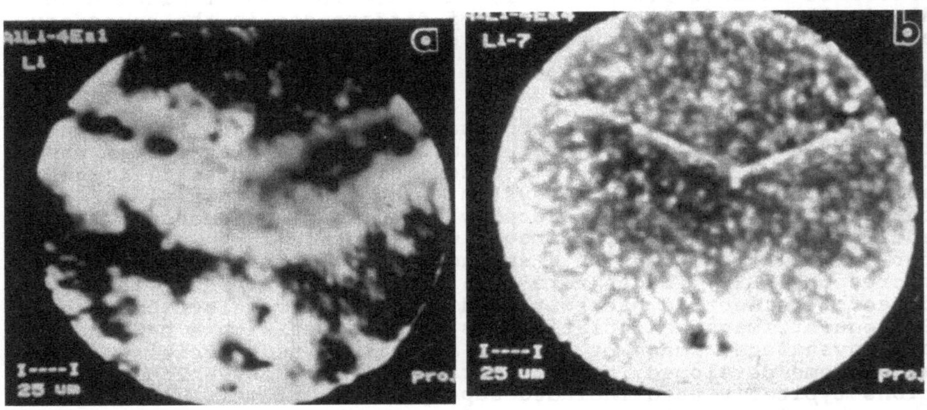

Figure 1 Aluminum - 3.6 w/o lithium alloy; SIMS Li-7 images: (a) surface, as prepared by conventional metallographic grinding and polishing; (b) interior at same location as (a) but following removal of approximately 500 nm by sputtering.

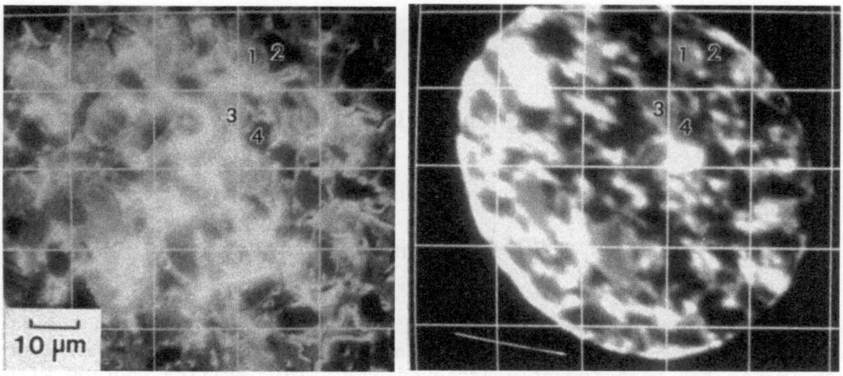

Figure 2 Reaction-bonded silicon carbide: (a) scanning electron microscope image (positively-biassed Everhart-Thornley detector, secondary electron plus backscattered electron mode); (b) SIMS Al-27 image after warping. A grid is superimposed to aid in establishing corresponding points. The numbered pairs show regions in the two images with similar contrast relative to adjacent grains.

the original SiC grains are partially melted and serve as nuclei for subsequent growth from the melt. Although the original and newly grown SiC have the same orientation in a given grain, there may be a difference in the impurities between the two regions. When viewed in the SEM, the reaction-bonded SiC displays anomalous secondary electron contrast within the grains, as shown in Figure 2. This unusual effect arises despite the lack of apparent compositional differences between the regions which show the strong contrast, as measured with the EPMA. Examination of the trace constituents by SIMS reveals the presence of aluminum at an approximate level of 100 ppm. Secondary ion images of the Al distribution show a structure similar in form to the SEM images. The general shapes of silicon and SiC regions can be distinguished, as well as a fine-scale precipitate with a high aluminum content. A significant degree of correlation between the aluminum concentration and the secondary electron emission pattern seen in the SEM is observed, as shown in Figures 2 a and b. Based on these observations, the origin of the difference in secondary electron emission is thought to be due to alteration of the electronic band structure of the semiconductor SiC by the impurity atoms, creating occupied energy states in the band gap which lead to the observed differences in secondary electron emission. Bright areas in the SEM image of the SiC grains are generally found to be correlated with higher levels of aluminum in the corresponding SIMS image.

5. References

1. D.S. Bright: Microbeam Analysis-1984 (San Francisco Press) 173.
2. D.B. Williams:"Microstructural Characteristics of Al-Li Alloys" in Aluminum-Lithium Alloys, ed. T.H. Sanders, Jr. and E.A. Starke, Jr. (Metallurgical Society of AIME, 1980) 89.
3. G.R. Sawyer and T.F. Page: J. Mats. Sci. 13, 885 (1978).

SIMS Imaging of Silicon Defects

R.S. Hockett

Monsanto Electronic Materials Company, 800 North Lindbergh Boulevard, St. Louis, MO 63167, USA

D.A. Reed and D.H. Wayne

Charles Evans & Associates, 1670 South Amphlett Boulevard, San Mateo, CA 94402, USA

The ability to simultaneously measure quantitative SIMS profiles and ion microscope images has been recently achieved [1]. This ability, coupled with digital image storage and processing, provides a powerful tool for silicon defect characterization. Examples presented here include oxygen precipitation and the coaggregation of O and C in thermally processed, carbon-doped silicon.

Secondary ion image acquisition is accomplished using a Resistive Anode Encoder on a CAMECA IMS-3f Ion Microanalyzer. A secondary ion is collected in an image plane by the Ion Microanalyzer and strikes the front side of a microchannel plate assembly, producing a pulse of 10^6 electrons out the reverse side. This charge pulse impacts a layer of resistive material and the charge is distributed to four electrodes at the corners of the anode with the charge pulses differing in magnitude depending on the impact position of the original pulse. The outputs from the position computing electronics of this imaging detector include both the X and Y position of the incident secondary ion and a strobe pulse for each processed signal. An imaging detector based on the RAE concept represents, therefore, a truly quantitative image-acquisition system which requires no calibration of the input signal intensity and output image intensity. The digitized X, Y position signals can be accumulated in an IBM-AT for subsequent processing and display, while a conventional SIMS depth profile can be acquired by recording the total ion arrival rate versus time at the RAE as would be done with the electron multiplier.

The first example of SIMS imaging of silicon defects relates to oxygen precipitates. A recent report [2] has shown SIMS profiles of outdiffused O in thermally processed silicon where the O profiles had "spikes" or intensity fluctuations. The presence of these spikes correlated with defects delineated by Wright-etching. Our first interest was to correlate the intensity spikes with the presence of microscopic SiO_2 precipitates observed previously in the SIMS imaging mode and in TEM studies. As described above, the RAE system permits simultaneous acquisition of conventional depth profile data and the recording of ion image depth profiles. The conventional depth profile for a test sample is shown in Fig. 1 with an oxygen profile exhibiting the spikes and a carbon profile to serve as a check for instrumental artifacts. The sample had been thermally processed using a CMOS type thermal simulation. FTIR showed 21.4 ppma (ASTM F 121-76) interstitial O out of 32.2 ppma had been precipitated. TEM on a nearby area of the sample showed the material had $1 \times 10^{10}/cm^3$ oxygen precipitates of size 0.2 µm and $1 \times 10^{10}/cm^3$ precipitate dislocation complexes of size 0.7 µm [3].

The oxygen profile is typical of the many oxygen profiles we have observed with a near-surface zone depleted in oxygen with respect to the bulk level (generally called "oxygen denudation") and the presence of the oxygen ion intensity "spikes" several μm into the material as discussed above. The smoothness of the lower intensity carbon background signal shows that the fluctuations in the oxygen signal are not due to instrumental artifacts. Oxygen ion images were simultaneously recorded with the total ion profiles of Fig. 1. Figure 2 shows the lateral distribution of oxygen taken at Point A in the expanded profile segment of Fig. 1. The image shows the presence of oxygen segregation over the 150 μm diameter imaged area. Images taken at Point B show a uniform oxygen distribution devoid of any lateral segregation. The one second image acquisition time represents a disk of material sputtered which is 23.4 nm thick with the images of Point A and B separated by 83.7 nm. Point A on the top of a spike has an image showing segregated oxygen with an oxygen background, similar to the background at B which is 6.7×10^{17} oxygen atoms/cm^3. The four segregated oxygen spots at Point A have about 7×10^{18} oxygen atoms/cm^3, as found by integrating the local areas in the image. Each SIMS oxygen point, therefore, corresponds to the oxygen in a disc of silicon with a volume of about 4×10^{-10} cm^3. Within that disc we are detecting about 1-4 segregated oxygen groups at 14.5 μm deep and 8 to 12 at 49 μm deep. This equates to an oxygen aggregate density of 0.25-1.0×10^{10}/cm^3 at 14.5 μm deep or 2-3×10^{10}/cm^2 at 49 μm deep. This compares well with 1-2×10^{10}/cm^3 from the TEM data. Furthermore, if the average oxygen spot at Point A represents SiO_2 for a 0.2 μm diameter disc of the depth 23.4 nm, then the amount of oxygen (3×10^7 atoms) would be close to the number of oxygen atoms (5×10^7) actually detected by SIMS for this spot, but shown as spread over 20 μm.

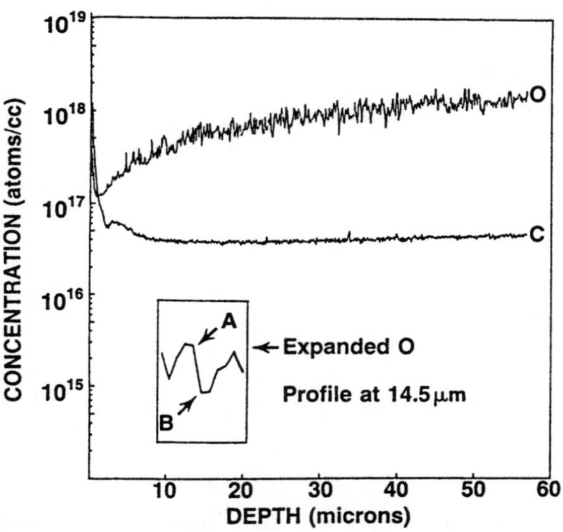

Fig. 1
SIMS O and C profiles in thermally processed silicon

The agglomeration of carbon atoms in silicon has been proposed to enhance the precipitation of the SiO_2 [4]. In order to study the cosegregation of carbon and oxygen in thermally processed silicon, SiO_x precipitation was induced in intentionally carbon-doped silicon. Figure 3 shows the carbon and oxygen profiles from the sample prepared as above. Although the fluctuations in the carbon profile are less pronounced than those in the oxygen profile, a definite correlation in the carbon and oxygen fluctuations can be seen. Sequential ion images of carbon and oxygen show a correlation of the oxygen

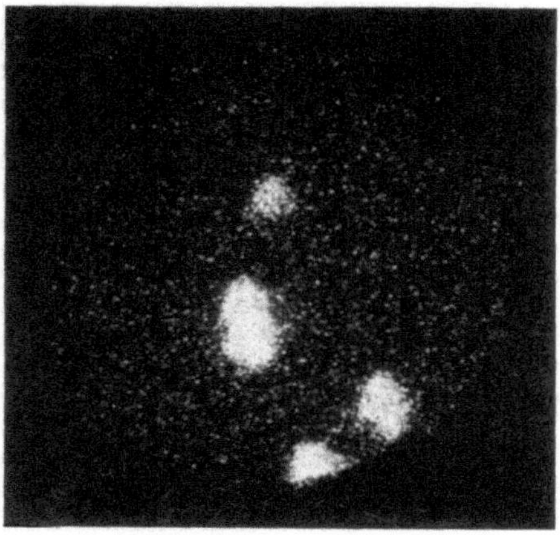

<u>Fig. 2</u> SIMS-RAE O images (150μm field of view) at point A of Fig. 1.

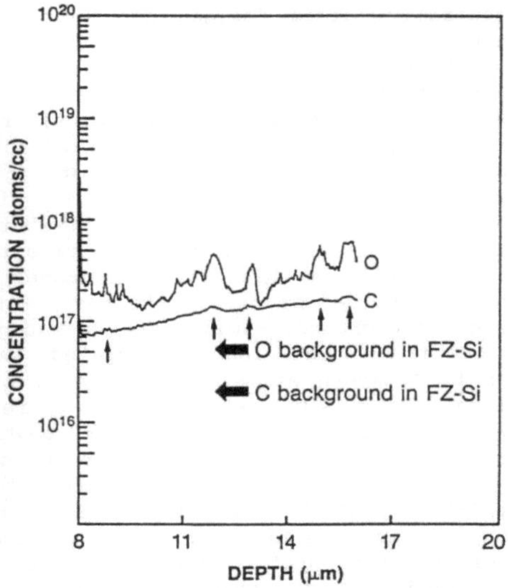

<u>Fig. 3</u> SIMS O and C profile in thermally processed, C-doped silicon.

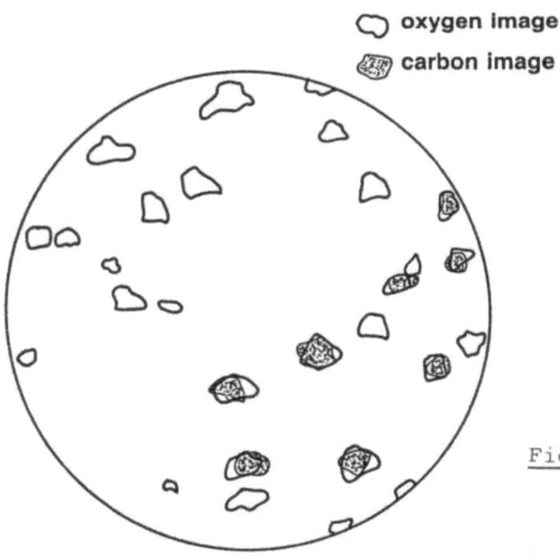

<u>Fig. 4</u> Schematic map of superimposed SIMS-RAE O and C images (150μm field of view) in thermally processed, C-doped silicon.

and carbon segregate even though the two images are from analytical volumes 100 Angstroms apart in depth, as shown in the composite ion image of Fig. 4. The localized carbon signals cannot be due to an ion yield enhancement from the SiO_x precipitates, because the enhancement would then be expedted to show localized carbon for all of the localized oxygen, which is not the case as shown in Figure 4. In order to further test for such an ion yield enhancement, a sample containing oxide precipitates, but no carbon, was analyzed under conditions which enhanced the incorporation of carbon from the vacuum. The carbon image under these conditions showed no significant lateral heterogeneties. This is the first direct evidence for the coaggregation of carbon and oxygen in thermally processed Czochralski-silicon.

In summary, we have employed ion image depth profiling to elucidate abnormal features in a conventional SIMS depth profiles and to obtain quantitative insight into an important materials phenomenon, SiO_x precipitation. In this case, the quantitation resulted from the use of a pulse counting detector, which permits direct correlation of ion image intensity with the intensity of an ion implant standard.

References

1. R. W. Odom, D. H. Wayne and C. A. Evans, Jr.: Secondary Ion Mass Spectrometry SIMS IV, Ed. A. Benninghoven, et al., Springer-Verlag, New York 1984, pp. 186-188.

2. F. Shimura, W. Dyson, J. W. Moody and R. S. Hockett: VLSI Science and Technology/1985, Ed. W. M. Bullis and S. Broydo, ECS Vol. 85-5, (1985) p. 507.

3. Private communication, G. K. Fraundorf, Monsanto Electronic Materials Company.

4. F. Shimura, R. S. Hockett, D. A. Reed and D. H. Wayne, Applied Physics Letters, 47, 794 (1985).

Part IX

Depth Profiling and Semiconductor Applications

Quantitative SIMS Depth Profiling of Semiconductor Materials and Devices

P.R. Boudewijn and H.W. Werner

Philips Research Laboratories, P.O. Box 80.000,
NL-5600 JA Eindhoven, The Netherlands

1 Introduction

The combination of high sensitivity with good depth and lateral resolution makes SIMS an analytical technique well suited for application in the semiconductor industry. Areas of interest include materials research, development of new processes, and controlling the quality of existing processes. Although a fraction of the SIMS analytical work only needs a qualitative interpretation of the results (e.g. comparitive studies of 'good' and 'bad' materials or devices) the major part involves quantitative analysis, particularly in the field of process or device development.

The required accuracy and precision of the analytical results necessitates the use of sophisticated quantification procedures.

The purpose of the present paper is to review the currently existing SIMS quantification methods, to indicate which of these method can be best applied to semiconductor materials, and to discuss some of the factors influencing the accuracy of the analytical results.

2 Quantitative analysis

The aim of a quantitative analysis is to determine the concentration c(A) of a given element A. In fact $\bar{c}$(A) is determined: the average value of the concentration across the probed volume. For inhomogeneous samples this concentration may not be representative for the 'true' concentration averaged over the total sample volume. Discrepancies between SIMS data and the results of bulk analytical techniques are often due to this sampling problem.

An estimate of the analytical accuracy for heterogeneous samples can be obtained using the sampling constant concept proposed by SCILLA and MORRISON [1]. By deriving the degree of heterogeneity from secondary ion images [2] this method allows the precision of the analytical results from a given area to be estimated, or it allows to determine the analyzed area required to obtain a certain precision.

The level of performance of any quantitative analysis is determined by the presence of random uncertainties (statistical errors, precision) and systematic uncertainties (systematic errors, accuracy). The influence of random uncertainties on the analytical result is generally given by the relative standard deviation. In many cases, however, the major contributions to the total uncertainty of the result arise from systematic uncertainties. Assessment of quantification procedures therefore should include a discussion about the origin and importance of systematic uncertainties.

3 Methods for quantitative SIMS analysis

The purpose of a SIMS quantification procedure is to relate the measured secondary ion intensity I(A) of element A to the concentration of element A in the sample. This relation may be expressed as:

$$I(A) = I_p . f_p . f_f . f_i . c(A) \quad (1)$$

where I_p is the primary ion current, f_p is a parameter determined by the physical processes leading to the emission of secondary ions, f_f is a fudge factor and f_i is an instrumental factor. According to eq. (1) three different methods of quantification may be distinguished: (a) first principle methods (f_f=1), (b) semi-theoretical methods (f_f=1), (c) empirical methods where the proportionality factor between I(A) and c(A) is derived from measurements on suitable calibration standards.

3.1 First principle methods

Quantification is achieved by a first principle calculation of the relative secondary ion intensities and assuming f_i to be known or determined by experiment. These calculations are based upon theoretical models describing the emission of secondary ions (for an overview see ref. [3]). The results obtained with these models are in general only valid for a given set of sample and experimental conditions, and certainly not universely applicable. The uncertainties in quantitative SIMS analysis range from 10-20% in optimal cases, up to several orders of magnitude. Note that even these methods require well-characterized calibration samples in order to check the theoretical predictions.

3.2 Semi-theoretical methods

With these methods a simplified model of the secondary ion emission process is used for calculating the SIMS intensities [3,4]. The shortcomings of these approaches can be compensated for by introducing parameters that can be adjusted by the experimental results (fudge factor approach).

Among these methods the local thermodynamic equilibrium (LTE) model [4] has been the most widely used. The application of the various versions of this model for the analysis of a number of different materials has been the subject of several previous review papers [3,5]. It can be concluded from these works that semi-theoretical methods using fitting parameters are adequate for quantitative SIMS analysis with uncertainties between 25% and a factor of 2.

3.3 Empirical methods

In these methods the relation between I(A) and c(A) is determined by means of a so-called calibration curve [3]. A calibration curve for element A in matrix T is generated empirically by measuring calibration samples T_i containing element A in various concentrations. The slope of the calibration curve gives the absolute sensitivity factor ρ_A of element A in matrix T. The relative sensitivity factor $\rho_{A,r}$ is defined similarly from a plot of I(A)/I(R) versus the relative concentration c(A)/c(R) where R is an internal reference element, usually the matrix or a main component. For the determination of the concentration in an unknown sample X the calibration curve is used as an analytical curve (c(A) is considered as a function of I(A)). The validity of this procedure requires that the experimental conditions are kept constant and that the matrices of T and X are identical.

Sensitivity factors are strongly dependent on the matrix surface properties, especially the oxygen content; in principle every matrix composition requires its own set of sensitivity factors. Selecting the proper set for the unknown sample would be greatly facilitated if a measure of the matrix surface properties was available. It has been shown [6] that ratios of matrix ions can serve as internal indicators for indexing matrix surface properties. Improvement in quantitative SIMS analysis using this indexed relative sensitivity factors or matrix ion species ratio (MISR) method has been reported for the case of metal samples [7,8]; in favourable cases relative accuracies of 10% were obtained.

The problem of instrument-dependence of sensitivity factors has recently been addressed by RÜDENAUER et al. [9]. By using a metallic glass sample as a calibration standard and by tuning different SIMS instruments in such a way that relative sensitivity factors obtained from this sample coincide closely, these authors were able to show that sensitivity factors of six further elements determined on different types of SIMS instruments agreed within a factor of 1.7. It was suggested that such a cross-calibration procedure of SIMS instruments leads to the transferability of relative sensitivity factors for any class of homogeneous materials.

Summarizing this section on quantification procedures, it must be concluded that only the empirical methods offer sufficient accuracy (10%) to make SIMS a valuable tool for the assessment of semiconductor materials. The main drawback of these methods, i.e. the selection of standards of matching matrix composition, is of minor importance since semiconductor matrices are generally well characterized.

4 Requirements for quantitative analysis

A quantitative SIMS analysis employing an empirical quantification method requires the following: (a) minimization of the contribution of random uncertainties to the analytical results by proper choice of instrument and experimental conditions [3], (b) identification of all possible sources of systematic uncertainties (see ref. [3] and section 5), (c) availability of well-characterized calibration standards, since any quantitative analysis will be as good as the standard used.

4.1 Requirements for a standard

The general requirements for a SIMS standard are [10]: (a) lateral homogeneity on a micron scale (even submicron scale for semiconductor applications), (b) composition stable over an extended period of time, (c) assessment of elemental composition by one or more independent analytical techniques must be possible. A further, but less stringent, requirement is that the concentration of the element of interest in the standard must be variable over a large concentration range. Additional information on the in-depth distribution of elemental composition should be available from standards applied in quantitative SIMS depth profiling.

Well-characterized standards for the bulk analysis of glasses, minerals and metals are the Standard Reference Materials provided by the National Bureau of Standards; for analysis of Fe-based alloys metallic glasses have been proposed as standards [11]. For quantitative depth profiling, especially in semiconductor materials, ion-implanted standards have been used extensively [12-14]; it is planned by the EEC to provide certified standards [15].

4.2 Ion implantation standards

The following characteristics make ion-implanted standards very attractive: (a) almost any element/matrix combination is possible, (b) monitoring the total implanted fluence allows quantitative ion implantation [16,17], (c) lateral homogeneity is assured by adequate rastering of the ion beam, (d) a large range of concentrations can be obtained by varying the fluence, (e) from the projected range of the profile, if known, the sputtering rate can be estimated, (f) information on system performance (dynamic range, background intensity) can easily be derived from the characteristic shape of the implantation profile.

Furthermore, Rutherford Backscattering Spectrometry (RBS) and Neutron Activation Analysis (NAA) [11] can be used as independent analytical techniques for the assessment of ion-implanted standards.

For quantification of SIMS data the integration method [12,13] is generally used. The sensitivity factor ρ_A is derived from the integrated countrate of the implantation profile. Basic assumptions in this method are that the SIMS intensities depend linearly on the concentration (generally valid for concentrations well below 1 atomic percent) and that the sputtering rate is constant throughout the measurement. The accuracy of the method depends on an accurate knowledge ($<5\%$ [16,17]) of the implanted fluence and on the accuracy of the depth measurement (approximately 10% for stylus-type surface profilers).

For practical work two approaches have been developped: (a) implantation of the element of interest (or an isotope of this element) directly into the unknown sample (standard addition method [18]), (b) application of external implanted standards [12].

The first approach offers the advantage of identical matrix composition and measurement conditions for standard and unknown. A disadvantage is the partial masking of the depth profile in the unknown sample by the implant if the element of interest is monoisotopic; further, the availability of the implanter might be a limiting factor when many samples have to be analyzed. The method becomes impractical in those cases when the implanted fluence must be chosen so high (to seperate the implant signal from the analytical signal) that damage to sample can be expected due to sputtering of the surface [16,17], temperature raise during high current implantation or radiation-enhanced diffusion of mobile species in the sample.

The external standard approach, requiring identical matrix composition of standard and unknown and constant measuring conditions, is generally adopted for SIMS depth profiling of dopant elements in semiconductors. As mentioned before, the selection of the proper matrix composition for the standard presents no problems, since the materials are well characterized; furthermore, e.g. at our laboratory, the standard is remeasured several times a day, when analyzing a series of related samples, to check the long-term stability of the measurement conditions. Using this procedure, an accuracy and reproducibility of 10% can be obtained.

4.2.1 Preparation of standards by ion implantation

When analytical problems require the preparation of new implantation standards, the SIMS analyst often is confronted with the problem of correctly choosing the implantation parameters, i.e. substrate material, angle of implantation ('random' or 'aligned'), fluence, energy and ion species (isotope, charge-state). In this section the factors which determine such a choice will be reviewed shortly.

The substrate material is chosen in accordance with the matrix composition of the samples to be analyzed. The crystalline structure of the substrate (monocrystalline, polycrystalline or amorphous) tends to be of minor importance, as indicated by the analysis of various boron-implanted substrates [19]. Preferably the substrate should be free of the element of interest, allowing relevant information on system performance like dynamic range, background intensities and presence of interferences to be extracted from the measured profile.

Although the actual shape of the implantation profile is not important for the accuracy of the analytical results - the relevant quantity being the integrated intensity - it is preferred for monocrystalline substrates to implant in a random direction; the characteristic shape of such an implantation profile is easily recognized (standard addition method) and comparison with theoretical results is often possible (LSS theory [20]).

The maximum concentration of implanted species n_{max} depends on the fluence F and the implantation energy E. It can be estimated from $n_{max}=F/2.5\ \Delta R_p(E)$ where $\Delta R_p(E)$ is the straggling of the distribution function. Values for $\Delta R_p(E)$ can be found in handbook tables [21] or calculated from LSS theory [20]. The maximum concentration must not exceed 1 at. % to assure linearity of the SIMS signals.

In practice,large fluences (1-5.10^{15} at./cm^2) are preferred,as higher signal intensities reduce the contribution of random uncertainties to the integrated countrate. Moreover,the signal due to the implant is easier separated from mass spectral interferences.

In some cases a lower fluence must be chosen. When analyzing samples containing extremely low concentrations of a certain element,the experimentally found concentrations are often systematically too high if a high fluence implantation is used for calibration. This can be attributed to memory effects, i.e. material sputtered in the high concentration part of the profile is deposited in the instrument, and resputtered and detected during following measurements [22].

A serious point of concern is the crystal damage introduced into the sample by the high fluences required for SIMS standards. Previous work has shown that SIMS ion yields were not affected by this damage [18]; it was argued that damage induced by the SIMS primary ion beam overshadows the defects due to implantation. Although this argument might apply to the ion yield, it is not always valid for the practical ion yield,which includes an instrumental factor. This is demonstrated in fig. 1 by the phosphorous profiles of a p-type silicon sample implanted with 1.10^{15} $^{31}P/cm^2$ at 70 keV. The profile obtained with a quadrupole-type instrument (ATOMIKA A-DIDA 2000) differs from that obtained with a magnetic sector-type instrument (Cameca IMS 3F). The anomaly at a depth of 0.13 μm, also apparent in the $^{28}Si^{++}$ matrix reference signal, is attributed to a change of conductivity in this region caused by damage introduced during ion implantation. The shift in secondary ion energy distribution due to a small charging of the sample surface leads to a decrease in signal intensities for the quadrupole instrument,whereas for the sector-type instrument signal intensities remain unchanged; this can be explained by the difference in energy acceptance between the two mass spectrometers (a few eV and 25 eV, respectively). Although the origin of the observed phenomena is not completely understood [23], their relation with implantation damage was confirmed by depth profiling a sample implanted with a ten times lower fluence; no deviations

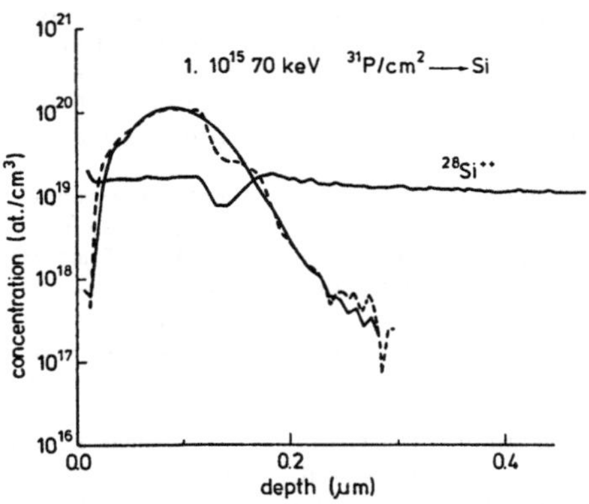

Figure 1 Phosphorous depth profiles obtained with a Cameca IMS 3F instrument (——) and an Atomika A-DIDA 2000 instrument (---). The $^{28}Si^{++}$ signal obtained with the latter instrument is also given.

from the expected profile were observed. A further point apparent from Fig. 1 is the change in magnitude of the $^{28}Si^{++}$ signal going from the implanted to the non-implanted region of the sample. Presumably a small shift in the secondary ion energy distributions caused by a damage-related change in surface potential explains this effect; the consequences for quantitative analysis will be discussed in the next section.

The results mentioned above show that damage effects must be taken into account in selecting the proper implantation fluence. Although the inaccuracy in the integrated countrate is relatively small for the case shown in fig. 1, it was found that it becomes unacceptably high at a fluence of 10^{16} at./cm^2.

The energy of the implant must be chosen in accordance with the requirements of constant sputtering rate and linear dependence of signal intensities on concentration (see section 4.2).

Both conditions are generally not met during the initial phases of a SIMS measurement due to non-steady state sputtering phenomena [24], while also surface contaminants can give ion yield enhancement (oxide) or produce mass spectral interferences. Consequently the implantation energy must be chosen sufficiently high to assure that the sputtered depth over which the aforementioned effects play a role (3-30 nm, dependent on primary ion energy) only forms a small fraction (<10%) of the total implanted depth.

On the other hand, very deep implantations must be avoided, since they require an excessively long measuring time. In view of the above considerations it is advisable for practical work to have available a series of standards implanted with various fluences and energies; dependent on the experimental conditions (primary ion energy, sputtering rate) chosen for the analysis of the unknown sample the proper standard can be selected.

A final point concerns the implanted ion species. Standards implanted with the same isotope as monitored during the SIMS analysis are to be preferred; this helps reproducible mass tuning and reliable background signal determination. Implantation of doubly-charged species should be avoided, as charge-exchange processes during the acceleration stage in the implanter may lead to inaccuracies of the implanted fluence. Similar inaccuracies may come from mass spectral interferences or insufficient mass resolution of the implanter analyzing magnet. By using SIMS we were able to show that for a $^{28}Si^+$ implantation 6% of the implanted fluence was due to $^{14}N_2^+$, while in the case of a $^{123}Sb^+$ implantation (on a 'bad' machine) more than 10% of the implanted fluence was due to $^{121}Sb^+$.

5 Systematic uncertainties

The result of every SIMS analysis may be influenced by systematic uncertainties. Many of these uncertainties have been identified [3] and proper measures can be taken to avoid them. In this section, emphasis will be placed upon two types of uncertainties particularly relevant for semiconductor depth profiling.

The first type of artefact may occur when depth profiling across a p-n junction present in the semiconductor material. Under conditions of perpendicular oxygen bombardment, and using a mass spectrometer with a small (few eV) energy acceptance, variations in signal intensities across the p-n junction can be observed which are not related to concentrational changes. This is illustrated in fig. 2, for the case of a phosphorous diffusion into a sample consisting 0.5 μm poly-Si grown on a p-type silicon substrate. At the depth of the p-n junction (after 6500 sec. sputtering) a change in the $^{28}Si^{++}$ matrix signal is observed. Such behaviour was reported earlier [25], and it was shown that it could be ascribed to a change in surface potential, shifting the secondary ion energy distributions. Note that this effect allows identification of the exact position of a p-n junction.

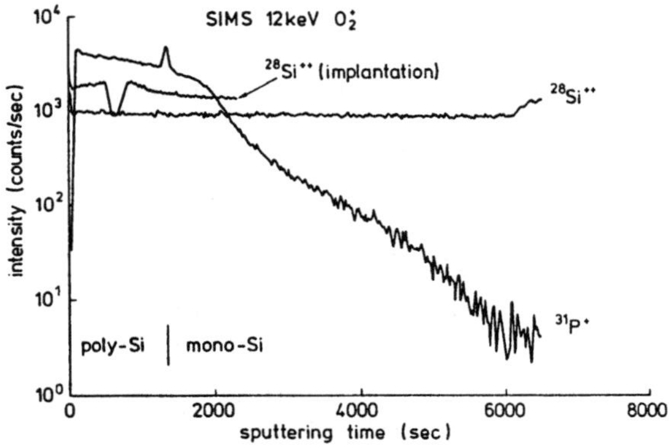

Figure 2 Depth profiles of $^{31}P^+$ snf $^{28}Si^{++}$ obtained from a phosphorous diffused sample consisting of 0.5 μm polycrystalline silicon grown on a p-type monocrystalline substrate. The $^{28}Si^{++}$ signal from the implantation shown in fig. 1 is also given.

As ratioing to matrix signals is common practice to correct for small changes in instrumental conditions, the analytical results depend on whether the ($^{31}P^+/^{28}Si^{++}$) ratio is determined in the n-type or p-type part of the profile. The situation is further complicated when the implantation standard of Fig. 1 is used for calibration. The $^{28}Si^{++}$ matrix signal of this implantation is also given in fig. 2. It can be seen that only the unimplanted region of the standard and p-type region of the phosphorous-diffused sample give matrix signals of equal magnitude, indicating equal sensitivity factors. A comparison of the matrix signals in the implanted region and the n-type region (fig. 2) shows that the sensitivity factors differ by a factor of 2. Thus by assuming constant sensitivity factors, i.e. neglecting the influence of conductivity type and implantation damage (section 4.2.1) on the practical ion yields, the phosphorous concentration determined for the profile of fig. 2 would be too low by a factor 2. In fact,this factor 2 was obtained by comparison of results on the same sample obtained with a different type of instrument; this triggered us to study these effects in more detail. It is obvious that such inaccuracies are unacceptable in quantitative SIMS analysis.

Another type of uncertaintly was found during a SIMS investigation of Zn-doped InP. Generally Cs^+ bombardment, using $CsZn^+$ as the analytical ion, is applied for obtaining Zn depth profiles. Signal intensities of $CsZn^+$ obtained from different samples containing equal concentrations of Zn were found to show large variations; ratioing to the matrix P^+ signal did not improve the reproducibility. We found that this could be attributed to different positioning of the sample holder (in a given instrumental set-up). This is illustrated in fig. 3 which shows the relative $CsZn^+/P^+$ ratio measured on a InP sample homogeneously doped with Zn, as a function of sample holder y position. For comparison,the results for the CsP^+/P^+ ratio are also included. Variations up to a factor 3 were observed for the $CsZn^+/P^+$ ratio, primarily caused by variations in the $CsZn^+$ intensity; the P^+ signal showed a less strong dependence on sample holder position.

Reproducible sample holder positioning for every sample is therefore absolutely mandatory for accurate quantitative analysis of these specific set of samples (at least with the Cameca IMS-3F instrument used in the present study). Multiple sample holders cannot be used in this case.

The origin of these effects is not completely understood, but is probably related to the difference in energy distribution between the $CsZn^+$

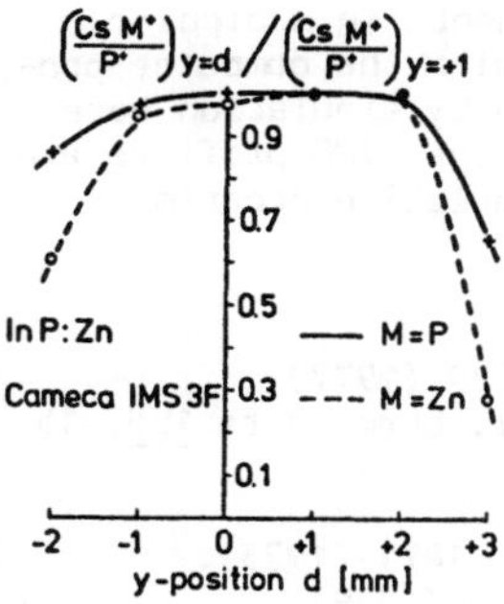

Figure 3 Relative $CsZn^+/P^+$(o) and CsP^+/P^+ (+) ratios obtained from a InP sample homogeneously doped with zinc, as a function of sample holder y-position. The drawn lines only serve to guide the eye.

cluster ion and the P^+ atomic ion; the energy distribution of the cluster ion $CsZn^+$ was found to be narrow compared to that of the atomic P^+ ion. Variation in sample holder position possibly changes slightly the extraction field lines in front of the target which influences the trajectories of the $CsZn^+$ and P^+ in a different way. Ratios of secondary ions having a better match in energy distribution are thus expected to show a much less position-dependent behaviour. This was confirmed by the results from a silicon sample; displacement of the sample holder over several mm's gave only a slight variation in the Si^+/Si^{++} ratio (for the same instrumental set-up).

6 Conclusions

It can be concluded that for application to semiconductor materials SIMS can be regarded as a quantitative analysis technique. Although lack of understanding of the secondary ion emission process prohibits the use of (semi) theoretical quantification methods, the empirical approaches give analytical results generally with a total relative uncertainty of $\leq 10\%$, provided suitable standards are selected and systematic uncertainties are avoided; in this respect SIMS is nearly as good as any other quantitative analytical technique employing standards.

The importance and reliability of present quantitative SIMS depth profiling is probably best illustrated by the results given in fig. 4. This

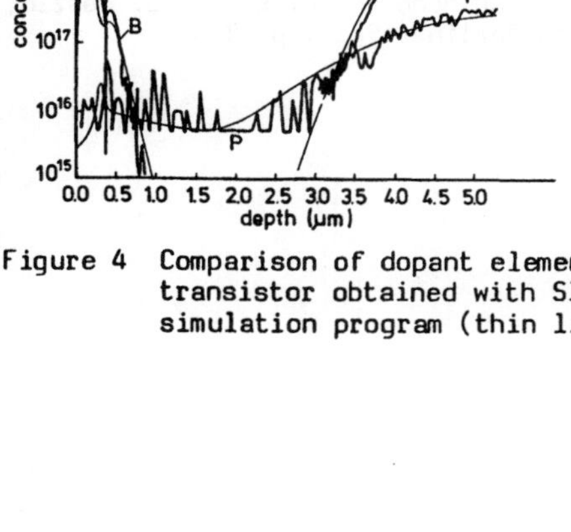

Figure 4 Comparison of dopant element depth profiles in a BIMOS bipolar transistor obtained with SIMS (thick lines) and with the SUPREM3 simulation program (thin lines).

figure shows the depth profiles of various dopant elements in a bipolar transistor in comparison with the results calculated with the computer program SUPREM3. Whereas in the past the results obtained by simulation were considered more reliable than the SIMS data, nowadays the SIMS profiles are used for calibrating the parameters in the process simulation program.

References

1 G.J. Scilla and G.H. Morrison, Anal. Chem. 49, 1529 (1977).
2 D.M. Drummer, J.D. Fasset and G.H. Morrison, Anal. Chem. Acta 100, 15 (1978).
3 H.W. Werner, Surf. Interf. Anal. 2, 56 (1980).
4 C.A. Andersen and J.R. Hinthorne, Anal. Chem. 45, 1421 (1973).
5 M.A. Rudat and G.H. Morrison, Anal. Chem. 51, 1179 (1979).
6 J.A. McHugh in Methods of Surface Analysis, edited by A.W. Czanderna, Elsevier, Amsterdam (1975), p. 223.
7 J.D. Ganjei, D.P. Leta and G.H. Morrison, Anal. Chem. 50, 285 (1978).
8 W.H. Christie, R.E. Eby, R.L. Anderson and T.G. Kolle, Appl. Surf. Sci. 3, 329 (1979).
9 F. Rüdenauer, W. Steiger, M. Riedel, H.E. Beske, H. Holzbrecher, H. Düsterhöft, M. Gericke, C.E. Richter, M. Rieth, M. Trapp, J. Giber, A. Solyom, H. Mai and G. Stingeder, Anal. Chem. 57, 1636 (1985).
10 D.E. Newbury and D.S. Simons in Secondary Ion Mass Spectrometry SIMS IV, edited by A. Benninghoven, J. Okano, R. Shimizu and H.W. Werner, Springer, Berlin (1984), p. 101.
11 M. Riedel, H. Gnaser and F.G. Rüdenauer, Anal. Chem. 54, 290 (1982).
12 H.W. Werner, Acta Electronica 19, 53 (1976).
13 D.P. Leta and G.H. Morrison, Anal. Chem. 52, 514 (1980).
14 J. Giber, A. Solydom, D. Marton and I. Barsony in Secondary Ion Mass Spectrometry SIMS IV, edited by A. Benninghoven, J. Okano, R. Shimizu and H.W. Werner, Springer, Berlin (1984), p. 89.
15 P.H. Davies (private communication).
16 W.H. Gries, Int. J. Mass. Spectr. Ion Physics 30, 97 (1979).
17 W.H. Gries, Ing. J. Mass. Spectr. Ion Physics 30, 113 (1979).
18 D.P. Leta and G.H. Morrison, Anal. Chem. 52, 277 (1980).
19 W.K. Hofker, H.W. Werner, D.P. Oosthoek and H.A.M. de Grefte, Appl. Phys. 2, 265 (1973).
20 J. Lindhard, M. Scharff and H.E. Schiott, Mat. Fys. Medd. Dan. Vid. Selsk. 33 (14) (1963).
21 J.F. Gibbons, W.S. Johnson and S.W. Mylroie, Projected Range Statistics, Dowden, Hutchinson and Ross, Stroudsburg (1975).
22 H.W. Werner in Electron and Ion Spectroscopy of Solids, edited by L. Fiermans, J. Vennik and W. Dekeyser, Plenum, New York (1978), p.324.
23 P.K. Chu, D. Zhu and G.H. Morrison, Rad. Eff. 61, 201 (1982).
24 P.R. Boudewijn, H.W.P. Akerboom and M.N.C. Kempeners, Spectrochim. Acta 39B, 1567 (1984).
25 M. Maier, D. Bimberg, H. Baumgart and F. Philipp in Secondary Ion Mass Spectrometry SIMS III, edited by A. Benninghoven, J. Giber, J. László, M. Riedel and H.W. Werner, Springer, Berlin (1982), p. 336.

High-Accuracy Depth Profiling in Silicon to Refine SUPREM-III Coefficients for B, P, and As

C.W. Magee and K.G. Amberiadis

RCA David Sarnoff Research Center, Princeton, NJ 08540, USA

Over the past years, models have been developed which make it possible to simulate IC fabrication process steps. Programs like SUPREM-III[1] (Stanford University PRocess Engineering Model) were developed to help the designers of semiconductor devices to make their work more effective. For these programs to accurately simulate the results observed in a particular processing sequence, some default coefficient values of the physical models have to be refined to match those corresponding to measured characteristics, i.e. the programs have to be calibrated. For the purpose of calibrating SUPREM-III, extensive experiments were made at RCA Laboratories using Secondary Ion Mass Spectrometry (SIMS) to measure dopant distributions.

The object of this study was to obtain SIMS measured profiles of ion implant distributions after furnace annealing, which we would then match with the SUPREM-III process simulation program. Before we could address these problems, however, we had to establish the reproducibility and accuracy of the SIMS technique. To do this, we used ion implantation and furnace annealing to produce five samples as nearly identical as could be made. We then analyzed the samples to compare the profiles for concentrations, depths, and areal densities. Boron, P, and As were used because they are the major dopants in Si. Sensitivity calibration was achieved using an unannealed sample of each dopant and normalizing the total number of detected secondary ions to the implant dose of 4×10^{15} atom/cm^2.

In performing an instrument calibration in this manner there are effects which must be corrected for if one is to obtain the accuracy needed for this study. These corrections are needed to compensate for instrument drift during analysis and to accommodate differences in instrument set-up between the standard and unknown samples. The simplest correction to make is that for sputtering rates. The average sputtering rate of the analysis can be obtained by measuring the depth of the sputtered crater, and assuming that the flux of primary ions to the sample was constant throughout the analysis. For the analyses in study, this was a valid assumption because the surface-ionization cesium ion source used for primary ion generation is inherently stable to better than 2-3% per hour, and each analysis lasted only 1/2 hour. We estimate that the depth scales on these profiles are accurate to $\pm5\%$.

Inclusion of a matrix ion in the depth profile is important for other reasons as well. Since it is difficult to obtain precisely the same operating conditions for the standard and unknown samples, it is necessary to have some reference in the data that allows one to compensate for these inevitable differences in overall sensitivity between samples. In this study, a silicon-containing peak in the secondary ion mass spectrum was used as a matrix reference. When analyzing for phosphorus as the atomic ion $^{31}P^-$, we used the matrix ion $^{30}Si^-$. Boron and arsenic, on the other hand, are more sensitive as silicide cluster ions $^{11}B^{28}Si^-$ and $^{75}As^{28}Si^-$. These polyatomic species are known to have lower average kinetic energies and more sharply peaked energy distributions than atomic negative ions. Thus, polyatomic species and atomic species can be expected to be affected differently by such things as small changes in lens voltages, irreproducible sample placement in front of the secondary ion extraction lens, and small differences in sample surface charging under the primary ion beam. All of these factors make it necessary to choose a polyatomic silicon-containing matrix reference peak when depth profiling dopants as polyatomic species, as is the case with boron and arsenic. The matrix reference chosen when profiling these elements was the silicon dimer negative ion $^{29}Si^{30}Si^-$. Thus, by having a matrix reference depth profile which responds in a similar way as the dopant depth profile to instrument instabilities, one can correct for them by always referencing to the instantaneous intensity of the matrix reference signal. This procedure is very important when trying to obtain better than ±25% accuracy in SIMS analysis. As one can see from Table I, we have achieved accuracies of the order of ±2-5%.

Table I - Retained areal density of dopants implanted to $4x10^{15}/cm^2$; annealed in N_2 @ 950°C for 30 min.
Total Retained Fluence - (10^{15} at/cm^2)

						Average
Boron	3.85	3.86	4.02	4.01	3.91	3.93±0.08
Phosphorus	4.08	4.10	4.20	4.09	4.04	4.10±0.06
Arsenic	3.38	3.72	3.71	3.56	3.66	3.59±0.13

SIMS depth profiles were then used to adjust the default parameters used in SUPREM such that the simulation matched the measured profile. The samples were B, P, and As implanted into separate <100> silicon wafers to a dose of $4 \times 10^{15}/cm^2$. After implantation, a 350Å oxide was grown in steam at 800°C followed by a deposition of 2000Å of SiO_2. This structure was annealed at 950°C in dry nitrogen for drive-in times of 30 and 60 minutes. The samples were then profiled by SIMS after stripping off the SiO_2.

The resultant SIMS depth profiles were then compared with the SUPREM simulations. Only for phosphorus did the default parameters yield a simulated profile that matched the SIMS profile. In order for SUPREM to give a good simulation of the boron profiles, the segregation coefficient had to be changed from $1.126x10^3$ to $1.126x10^4$. The pre-exponential coefficient for the diffusivity due to neutral vacancy charge states had to be changed from $2.22x10^8$ u^2/min to $5.6x10^9$ u^2/min;

and the pre-exponential coefficient for the diffusivity due to positive vacancy charge states had to be changed from 4.32×10^9 u^2/min to 6.5×10^9 u^2/min. These changes were needed to bring up the simulated surface concentration as well as to match the tail region of the profile as shown in Figure 1. SUPREM has no way of simulating the boron precipitation and/or damage decoration which is shown in the SIMS profiles at the original Rp of the B implantation. For arsenic, the concentration dependent diffusivity had to be changed from 7.2×10^{10} u^2/min to 2.2×10^{11} u^2/min. This was necessary in order to correct for the fast diffusion of As at high concentration. The adjusted coefficients also resulted in accurate simulations for 60-minute anneals, but other temperature comparisons have not yet been made.

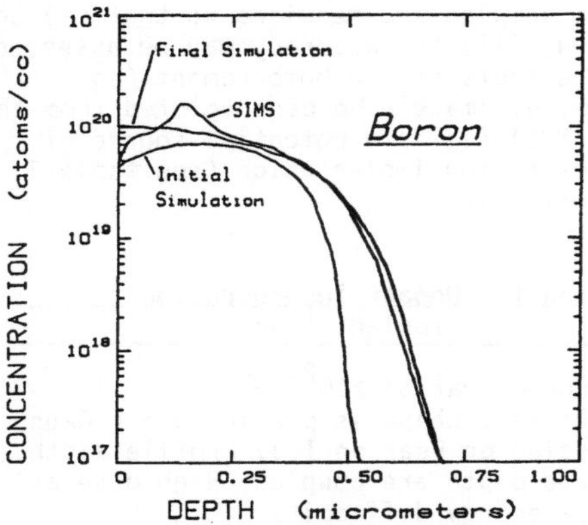

Figure 1. Depth profiles for B implanted into Si at 40 keV, $4 \times 10^{15}/cm^2$, and diffused at 950°C, in N_2 for 30 min. Shown are the initial SUPREM simulation, the SIMS profile, and the final SUPREM simulation using refined coefficients.

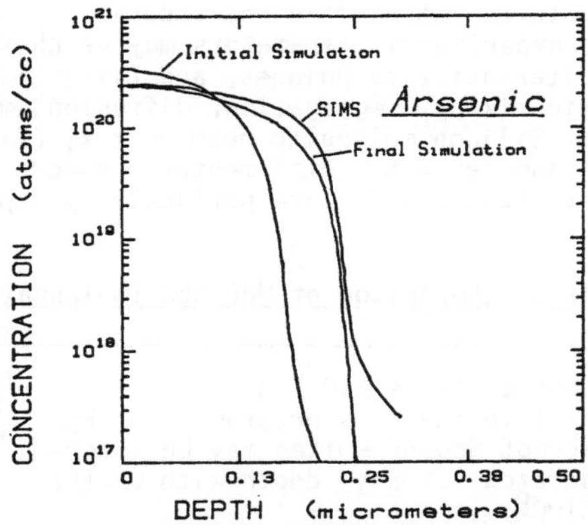

Figure 2. Depth profiles for As implanted into Si at 190 keV, $4 \times 10^{15}/cm^2$, and diffused at 950°C, in N_2 for 30 min. Shown are the initial SUPREM simulation, the SIMS profile, and the final SUPREM simulation using refined coefficients.

References:

1) C. P. Ho and S. E. Hansen, Stanford Electronics Laboratory, Stanford University, Stanford CA., Technical Report SEL-83-001, July 1983.

The Use of Silicon Structures with Rapid Doping Level Transitions to Explore the Limitations of SIMS Depth Profiling

M.G. Dowsett, D.S. McPhail, R.A.A. Kubiak, and E.H.C. Parker

Solid State Research Group, Sir John Cass Faculty of Physical Sciences and Technology, 31, Jewry St., London EC3N 2EY, U.K.

Semiconductor devices continue to shrink, with an immediate prospect of spatial dimensions ~1μm laterally and <100nm in depth. Silicon requires a SIMS depth resolution of ~1nm (2 atomic planes) in order to delineate fine dopant structures with sufficient accuracy. Test samples are required so that (i) the best experimental parameters may be chosen, (ii) the accuracy may be assessed by alternative techniques, and (iii) the effects of ion bombardment (eg. atomic mixing, segregation, diffusion) may ultimately be deconvoluted from the data. Silicon molecular beam epitaxy (Si-MBE) [1] is a potential source of test samples with complimentary properties to ion implantation (see table 1 where items 3 and 4 are particularly important).

Table 1 Comparison of MBE and Implantation for Dopant Incorporation

	MBE	Implantation
1	Doping levels $>10^{20}cm^{-3}$	Concentration $>10^{21}cm^{-3}$
2	Profile shape is programmable. Eg. abrupt dopant spikes may be incorporated at any depth with widths <100Å.	Profile shape is predictable (Gaussian or Pearson IV). Profile width and depth are coupled. High dose at low energy difficult.
3	Defects distributed and dependent on growth conditions. Eg. Si particles: $10^{3}-10^{4}cm^{-2}$, dislocations: $10^{2}-10^{3}$ cm^{-2}.	Damage close to implant peak and dose dependent. Complex effects on dopant diffusion, including precipitation.
4	Dopant is incorporated 100% active at time of growth, no further processing required, so electrical profiling (eg. [1]SRP, [2]CVP) is useable on as-grown structure.	Need thermal processing for dopant activation. Diffusion and precipitation alter and broaden shape. Activation level unpredictable for doses in excess of $10^{14}cm^{-2}$.
5	Atomic concentration quantified from carrier profile or standard.	Dose gives atomic concentration.
6	Concentration change 3 decades in 100Å in a repeated structure.	

1 SRP=spreading resistance profiling [3]. 2 CVP=capacitance-voltage profiling.

In Si-MBE single crystal Si epilayers are deposited by thermal evaporation under UHV conditions onto silicon or other substrates. Programmed distributions of B and Ga may be produced by source flux or shutter control. Similar distributions for As and Sb may be incorporated by the technique of potential enhanced doping (PED) pioneered in our laboratories [2]. Provided the substrate temperature is kept sufficiently low (to reduce diffusion), abrupt features may be incorporated at any depth.

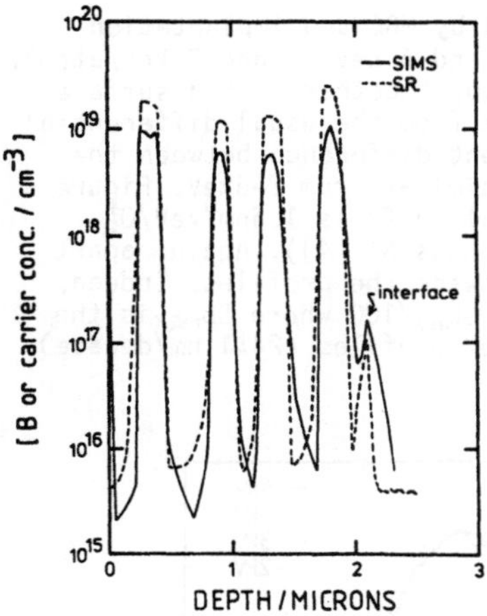

Fig 1 Atomic (SIMS) and carrier (SRP) profiles of Si-MBE B-doped epilayer. Growth: 2um/hr, 850°C. Deepest spikes have diffused during growth

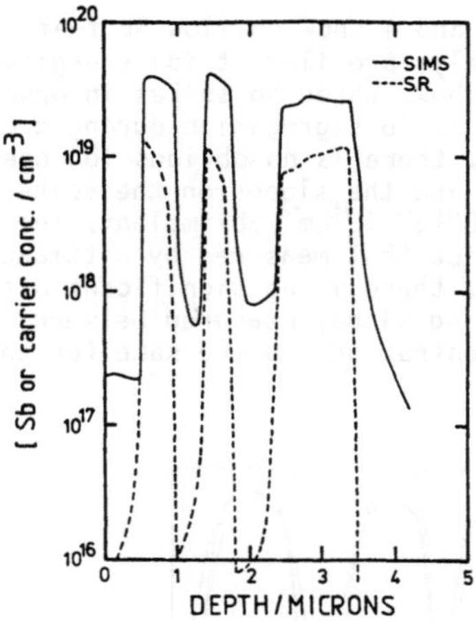

Fig 2 Similar profiles of Sb-doped epilayer. Growth: 4µm/hr, 850 C. Primary ions: 4keV $^{16}O_2^+$, normal inc. No diffusion apparent

The epilayers were grown in our V80 Si-MBE system (VG Scientific Ltd.) [1] and SIMS analysis was performed in the EVA2000 system (self-built). Figures 1 and 2 show SIMS (solid lines) and SRP [3] (dotted lines) profiles from B and Sb doping structures several µm in depth. The SIMS analyses were carried out using 4keV $^{16}O_2^+$ ions at normal incidence. Considering that the two techniques were individually depth-calibrated and quantified, the agreement is excellent, particularly for the positions and widths of the spikes. The vertical displacement of the SRP and SIMS profiles is due to analysis of different areas of sample and the combined error in quantification. In the B profile (Fig. 1) both techniques achieve a similar dynamic range. It is evident from the rounding of the deeper (older in growth time) spikes that some diffusion has taken place at the growth temperature of 850°C. Differences between SIMS and SRP show up when the slopes of the shallow and deep edges on each spike are examined. Measuring the slopes between the 10^{17} and 10^{18} levels, for the shallow edges of each spike, SIMS gives slopes of 10, 16, 20 and 20 nm/decade (a depth resolution of ≃1.6% after the first edge, had the features NOT been broadened by diffusion), whilst SRP gives 23, 30, 28, and 25 nm/decade (an absolute depth resolution ≃27nm/decade). For the deep slopes, SIMS gives 16, 28, 43, and 61 nm/decade (ie. depth resolution ≃3%), whearas SRP gives 9, 11, 16, and 32 nm/decade (depth resolution ≃1.5%). Thus the two techniques give contrary information about the spike shape-SIMS says that the shallow edges are steepest, SRP says the deep edges are. For fig. 2 the Sb diffusion coefficient was about 1/10th that of B, and the growth rate was twice that in fig. 1, and no rounding of the spikes is apparent. The dynamic range in the carrier profile is greater than that of the SIMS. SIMS results indicate that there is a level of $8x10^{17}$ Sb cm^{-3} between the spikes which is electrically inactive, and possibly trapped in extended defects. For the shallow and deep edges of the spikes SIMS gives depth resolutions of 1.2% and 4% respectively. SRP gives 23 and 46 nm/decade. On the first shallow edge SIMS gives a width of 5nm/decade (SRP gives 25nm). If this width is due to atomic mixing, then such effects will be masked once the departure from flatness of the crater exceeds 5nm (eg. at 1um if the primary beam density is 0.5% non-uniform.)

Figures 3 and 4 show shallow Sb features produced by MBE and implantation respectively, profiled at ion energies between 2 and 6 keV (1 and 3 keV/atom). Figure 3 shows three Sb spikes incorporated by PED, together with a surface spike due to Sb segregation during growth. Apart from the usual differential shift [4], there is no obvious ion energy-dependent difference between the profiles, and the slopes on the spike edges are similar from 2-6keV. Figure 4 shows a 100keV $10^{15}cm^{-2}$ Sb implant. The differential shift is 1.6nm/keV/O_2^+ (About twice that measured by Wittmaack for amorphous Si [4]). Again, apart from this, there is no significant difference between the profiles. Indeed, the trailing slope, measured between $Sb_{max}/10$ and $Sb_{max}/100$ where Sb_{max} is the peak concentration, is the same for the 2 and 6keV profiles (24±1 nm/decade).

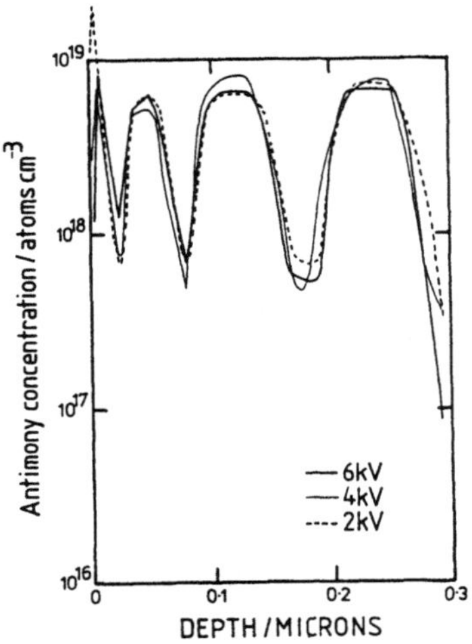

Fig 3 Shallow Sb doped epilayer profiled at different O_2^+ energies. Surface spike is due to Sb segregation during growth

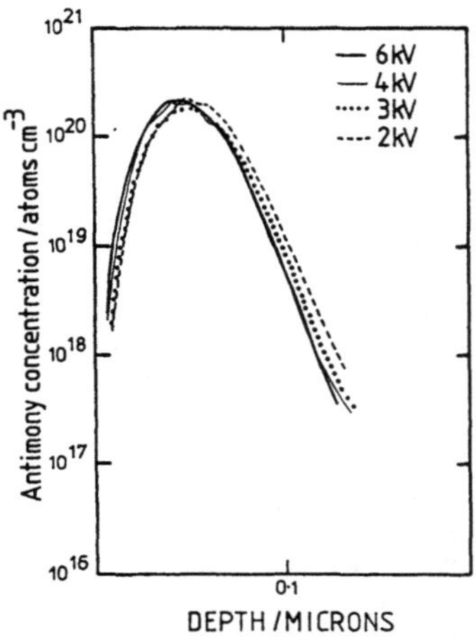

Fig 4 100keV $10^{15}cm^{-2}$ $^{121}Sb \rightarrow$ Si (100) profiled under same conds. as fig. 3. Only difference between profiles is shift towards surface of 1.6nm/keV/^{16}O

Despite the lower diffusion for Sb, the data in figs 1 and 2 show roughly the same percentage SIMS depth resolution. The lack of energy dependence in figs 3 and 4 shows that atomic mixing is not the limitation and that SIMS is measuring the intrinsic slopes. It is likely that defects and flakes currently characteristic of MBE growth (see table 1) are contributing to surface roughness, and limiting the depth resolution/atomic distribution. It is anticipated that improvements in Si-MBE will result in samples more suitable for fundamental SIMS investigations.

Our thanks to M Pawlik (GEC Hirst Research Centre) for SRP measurements.

References

1 R A A Kubiak, W Y Leong and E H C Parker, J. Vac. Sci. Technol, B3(2)(1985)592-595

2 R A A Kubiak, W Y Leong and E H C Parker, Appl. Phys Lett. 46(6)(1985)565-567.

3 M. Pawlik, ASTM Special technical testing Publication 850 (1984)390-408

4 K Wittmaack and W Wach, Nucl. Inst. and Meth. 191(1981)327-334.

Temperature Dependent Broadening Effects in Oxygen SIMS Depth Profiling

F. Schulte and M. Maier

Institute of Semiconductor Electronics/SFB 202,
Aachen University of Technology, D-5100 Aachen, F. R. G.

In sputter depth profiling, the ultimate depth resolution is limited by ion bombardment-induced broadening effects [1]. The combined effect of atomic mixing and selective sputtering was recently shown to be responsible for the usually observed exponential tail in SIMS sputter depth profiling [2,3]. Segregation induced by surface oxidation was suggested to explain the unusually long tails observed for Ag [4] and other elements [5] in Si using Ar primary ions and in-situ oxygen flooding. Recently, radiation enhanced diffusion was substantiated by SIMS profiling with Ar primary ions at elevated temperature in the systems Mo/Si [6] and Ni/Cu, In/Cu [7]. In this study, we investigated the broadening of SIMS depth profiles of the dopants As and Ga implanted into Si using O_2 primary ions and varying the sample temperature. The main purpose was to obtain more information concerning the role of oxidation-induced impurity segregation in sputtering with oxygen ions.

(100) Si substrates were preamorphized by Ar bombardment at an energy of 40 keV with a dose of 10^{15} cm^{-2}. As and, for comparison, Ga were implanted at an energy of 20 keV with a dose of 10^{16} cm^{-2} and $5x10^{14}$ cm^{-2}, respectively. The sample temperature monitored by means of a thermocouple (accuracy $\pm 10^oC$) was varied between -100^oC and 290^oC. SIMS depth profiles of ^{75}As, ^{69}Ga and of ^{30}Si for reference purpose were measured using an Atomika instrument. The probing O_2 beam hit the sample at an angle of $\vartheta_x=2^o$ away from surface normal direction in the plane of secondary ion deflection ($\vartheta_y=0$). The sample was also tilted by an angle ϑ_y around an axis being parallel to the plane of secondary ion deflection and almost ($\vartheta_x=2^o$) perpendicular to the ion beam. Energy and current density applied were 5 keV and 50 $\mu A/cm^2$. The total pressure in the sample chamber during SIMS analysis was 10^{-6} Pa. The sputtered crater depth was determined by a calibrated profilometer (accuracy ±5%).

SIMS depth profiles of As determined at different sample temperatures are shown in fig.1. The room temperature profile is seen to be in agreement with literature data [8,9]. A pronounced exponential tail is found due to the rather high O_2 beam energy of 5 keV. At 120^oC and 290^oC, we monitor a drastic enhancement both of the pre-peak and of the decay length (drop to 1/e) of the exponential tail of the As depth profile. The -100^oC profile is seen not to be significantly different from the room temperature profile. It is important to note that the room temperature profile was reproduced upon returning to room temperature after a temperature cycle. The area under the profiles tends to decrease. The Ga profile determined at 290^oC was very similar to the corresponding As profile.

Average values of the decay length obtained from several temperature cycles are depicted in fig.2 as a function of temperature for different angles of ion incidence. At vertical ion incidence, a distinct increase of

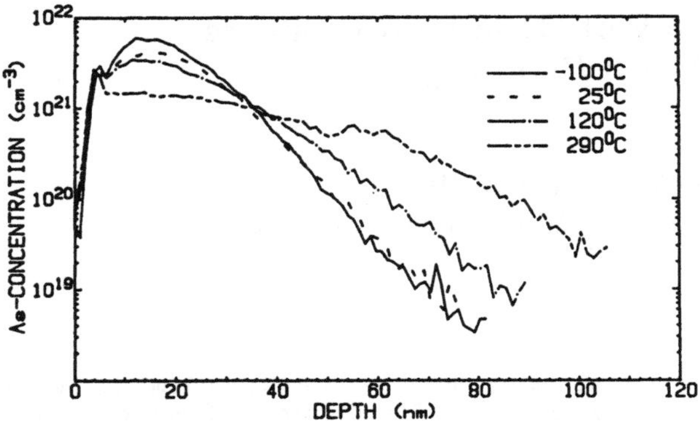

Fig.1 Depth profiles of As implanted in Si observed at different sample temperatures

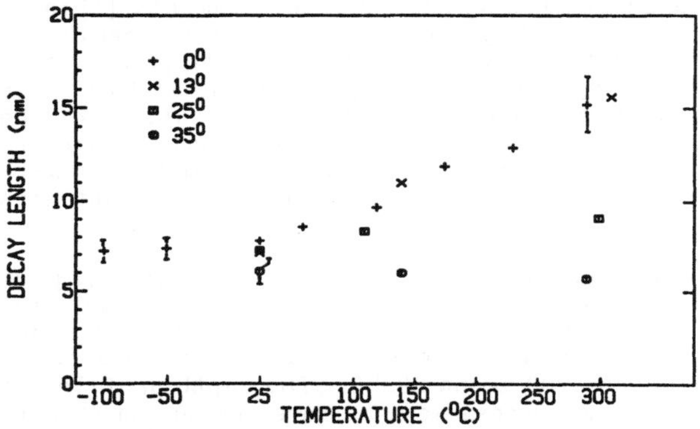

Fig.2 Temperature dependence of decay length of As profiles at different angle of ion incidence

the decay length with increasing temperature can be seen. The variation of the decay length with temperature is reduced at oblique ion incidence for $0^\circ < \vartheta_y < 35^\circ$ and is independent of temperature for ion incidence at 35°.

The probing (at room temperature) of the altered layer built-up in the profiling experiment at 290°C exhibited a pronounced accumulation of As[10]. This finding is positive evidence for dopant segregation to occur. A recent RBS study of the altered layer resulting from (room temperature) depth profiling of As in Si showed up an As distribution which could be interpreted in terms of segregation, too [11]. In accordance with the segregation behaviour of As in Si technology [12], we assume that As is enriched on the Si side of the interface between the SiO_2 film being due to oxygen ion bombardment [13] and the underlying Si substrate. Correspondingly, the SiO_2 film is depleted of As. We attribute the observed profile broadening to the reduced emission rate of As ions escaping from the topmost monolayers of the depleted SiO_2 film. From the slope of the Arrhenius plot of the temperature-dependent part of the decay length, an activation energy of 130 meV is derived [10]. This rather low activation energy [13] indicates a high diffusivity of As in the altered layer, as to be expected from the disorder due to ion bombardment. The low activation energy explains the occurring of dopant segregation already at room temperature and the quickly growing influence of this effect above room temperature. Ga also showed strong evidence of segregation that was expected due to the the segregation coefficient (equilibrium concentration of impurity in the silicon / equilibrium concentration of impurity in the oxide) of As and Ga being both larger than unity [12].

The depth of the pronounced depletion behind the pre-peak (see fig.1) was found to coincide with the depth where the ^{30}Si signal intensity reaches its stationary value. The built-up of the SiO_2 film is completed in that depth. Accordingly, the depletion can be explained by the strong onset of segregation already before the completion of the built-up of the SiO_2 film. The tilting of the sample out of the position of normal ion incidence was recently reported to drastically change the stoichiometry of the silicon oxide film by the generation of suboxides [15]. Assuming that a coherent film of SiO_2 stoichiometry is mandatory for segregation, the decreasing temperature dependence of the exponential tail (see fig.2) can be ascribed to the reduction of the influence of segregation.

In conclusion, we have shown that oxidation induced impurity segregation plays a major role in SIMS sputter depth profiling in Si with oxygen ions at near normal incidence. A distinct variation of the decay length with sample temperature is observed above about room temperature. However, for As there is no increase in depth resolution with sample cooling down to -100 ^{o}C.

We are grateful to D. Dunkmann for preparing the implantations.

1. K. Wittmaack: Vacuum 34, 119 (1983)
2. P. Blank and K. Wittmaack: Rad. Eff. Lett. 43, 105 (1979)
3. K. Wittmaack: J. Appl. Phys. 53, 4817 (1982)
4. P. Williams and J. Baker: Nucl. Instr. and Meth. 182/183, 15 (1981)
5. S.M. Hues and P. Williams: Nucl. Instr. and Meth.(B), to be published
6. B.V. King, D.G. Tonn and I.S.T. Tsong: Nucl. Instr. and Meth. B7/8, 607 (1985)
7. P. Macht and V. Naundorf: Nucl. Instr. and Meth.(B), to be published
8. W. Vandervorst, H.E. Maes and R.F. De Keersmaecker: J. Appl. Phys. 56, 1425 (1984)
9. W. Vandervorst and F.R. Shepherd: Appl. Surface Sci., 21, 230 (1985)
10. F. Schulte and M. Maier: Nucl. Instr. and Meth.(B), to be published
11. W. Vandervorst, M.L. Swanson, H.H. Plattner, O.M. Westcott, I.V. Mitchell and F.R. Shepherd: Nucl. Instr. and Meth.(B),to be published
12. A.S. Grove, O. Leistiko, Jr.,and C.T. Sah: J. Appl. Phys. 35, 2695 (1964)
13. W. Reuter and K. Wittmaack: Appl. Surface Sci. 5, 221 (1980)
14. B.J. Masters and E.F. Gorey: J. Appl. Phys. 49, 2717 (1978)
15. W. Reuter: Nucl. Instr. and Meth.(B), to be published

Sputter-Induced Segregation of As in Si During SIMS Depth Profiling

W. Vandervorst and J. Remmerie IMEC, Belgium

F.R. Shepherd BNR, Ottawa, Canada

M.L. Swanson AECL, Chalk River, Canada

1. Introduction and Experimental

The analysis of very abrupt profiles with SIMS is limited by sputter-induced redistribution processes such as collisional mixing [1-4] and radiation-enhanced diffusion [5,6]. The enhancement of the ionisation yield for a silicon substrate subject to normal incidence oxygen bombardment, or to oxygen flooding, leads to the formation of a silicon dioxide layer at the surface [7]. The presence of this oxide leads to another profile redistribution mechanism i.e. a chemically enhanced directional diffusion or segregation [8]. In a previous investigation, the redistribution of As in Si, under oxygen bombardment, was monitored by grazing exit RBS ; unusually large decay lengths were observed for samples sputtered with 1700eV O_2^+ in a reactive ion etcher (6). In this paper we present further analysis and develop a model to explain these observations.

A (<100>) silicon substrate was implanted with 12 keV As^+($5x10^{14}$ cm^{-2}). SIMS analysis was performed with an Atomika Ion Microprobe using an O_2^+ beam at normal incidence (θ=2°). A stationary electron beam (0.5mm, $2<I<8\mu A$, $150<E<350eV$) was positioned in the crater of the rastered area during the electron flooding experiments.

2. Results and Discussion

Figure 1 shows the profile measured with a 2 keV O_2^+-beam, with and without electron flooding (5µA, 350eV). The decay length of the profile without flooding (8.7nm) increases to 15.5nm with electron flooding. Note that not only the decay length has increased but the profile shape at the leading edge has also changed. This enhanced segregation of As (due to electron beam bombardment) is comparable to recent results of Schulte and Maier who observed temperature-enhanced segregation [9].

In figure 2, the electron beam was turned off, after the initial part of the profile. The remaining profile closely agrees with the profile measured without electron flooding (fig 1) indicating the absence of any redistribution of the deeper lying part of the profile by electron beam heating and outdiffusion. The observed effects can be explained in terms similar to those describing segregation during thermal oxidation [10]. As the oxide front is swept through the profile, (during sputtering or oxide growth) As-atoms are rejected from the oxide and segregate below the oxide at the interface. This rejection from the moving oxide front requires a certain mobility of the As-atoms, which at room temperature must be governed by radiation-enhanced diffusion. The electron beam appears to increase the active defect density and hence the mobility of the As-atoms.

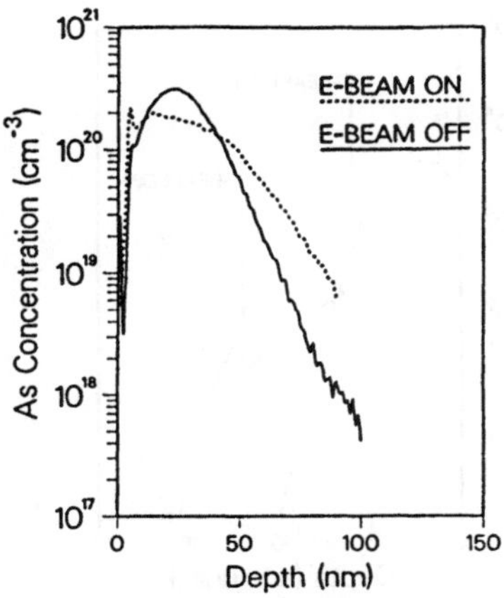

Fig.1. Depth profile recorded without electron bombardment (solid curve) and with electron bombardment (dashed curve).

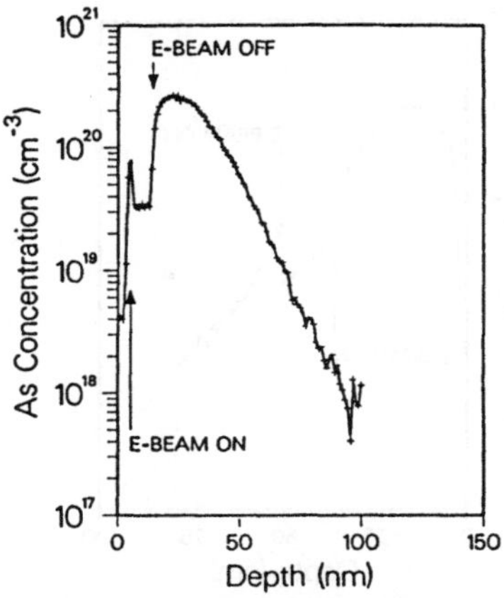

Fig. 2. Depth profile showing the effect of turning the electron beam on and off near the front of the profile

Since the As-mobility is only required at the oxide-silicon interface, there must be a clear dependence on the depth of defect generation. In fact, if the electrons did not penetrate the oxide interface, no enhanced segregation effect was observed. When the oxygen energy (and hence the oxide thickness) was increased to 4 and 6 keV, with the electron beam energy fixed at 350eV, or vice versa, when the oxygen energy was fixed at 2 keV but the electron beam energy was reduced to 150eV, the measured profiles were unchanged from those recorded without electron bombardment.

Numerical simulations have also been performed, where the segregated fraction during depth profiling was determined by the thermodynamic segregation ratio and the As-mobility [11]. The effect of the electron beam can be modelled with an increased mobility. The model has been used to simulate a depth profile in which the electron beam was first turned on at a given depth, and later turned off again. Fig 3a gives the experimental profile measured with 2 keV O_2^+ and a 350 eV electron beam. Fig 3b gives the calculated profile where the change in slope observed is similar to that found experimentally. The good qualitative agreement between the simulated profile and the experimental data supports our model.

3. Conclusions

Simultaneous electron flooding during oxygen bombardment of silicon, leads to larger decay lengths than under standard profiling conditions. The effect can be explained as an enhanced segregation of As below the moving oxide front due to the increased mobility. This effect explains the unusually large decay lengths observed in reactive ion-etched samples, analysed by RBS [6]. A similar segregation is present during standard profiling (but to a lesser extent) where the defects generated by the primary beam play a similar role. The observation of the electron beam effect is important, since it identifies a potential source of error when an electron gun is used for charge compensation during profiling though multilayer structures.

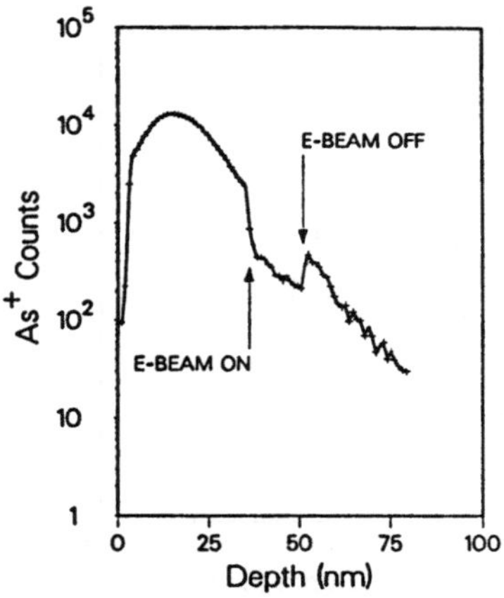

Fig. 3a. SIMS depth profile showing the effect of the electron beam on the back side of the profile.

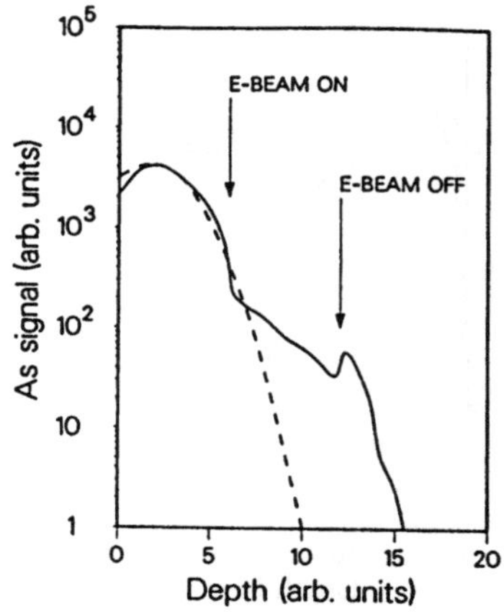

Fig. 3b. Simulated SIMS profile (solid curve) showing the effect of O_2^+ and electron bombardment on a Gaussian As distribution (dashed curve).

4. References

1. U. Littmark and W. Hofer: Nucl. Instr. and Meth 168(1980)
2. P. Sigmund and A. Gras-Marti: Nucl. and Meth 182/183(1981) 25
3. J. Remmerie, W. Vandervorst and H.E. Maes: Proc. 11th Int. Conf. Atomic Collisons in Solids, Nucl. Instr. and Meth (in press).
4. W. Wach and K. Wittmaack: Surf. Interface Anal. 4(1982)230
5. W. Vandervorst, H.E. Maes and R.F. DeKeersmacker: J. Appl. Phys. 56(1984)1425
6. W. Vandervorst, F.R. Shepherd, M.L. Swanson, H.H. Plattner, O.L. Westcott and I.V. Mitchell: Ion Beam Analysis, Berlin 1985, Nucl. Instr. and Meth (in press)
7. W. Reuter: Ion Beam Analysis, Berlin 1985, Nucl. Instr. and Meth (in press)
8. V.R. Deline, W. Reuter and R. Kelly: ibid.
9. F. Schulte and Maier, ibid.
10. R.B. Fair and J.C.C. Tsai: J. Electrochem. Soc. 122(1975)1689
11. W. Vandervorst, D. Avau, J. Remmerie : (in preparation)

SIMS Measurements of As at the SiO_2/Si Interface

C.M. Loxton and J.E. Baker

Materials Research Lab., University of Illinois, Urbana, IL 61801, USA

A.E. Morgan and T.-Y. J. Chen

Philips Research Labs. Sunnyvale, Signetics Corp., Sunnyvale, CA 94086, US

1. Introduction and Experimental

SIMS measurements of concentrations at, or near, an interface are complicated by changes in the degree of ionization of sputtered atoms, in the sputter rate and in any selective sputtering effects [1]. For practical applications, bombardment conditions must be chosen to minimize these matrix effects, and calibration procedures must be established for the interface region. Measurements of dopant concentrations near the SiO_2/Si interface have been accomplished by determining concentration factors to quantify the secondary ion yields using implant standards and using O_2^+ bombardment with, and without, O_2 backfill [2,3]. Appropriate choices of conditions are not always the same; to study B segregation, Morgan et al. [2] have used O_2^+ bombardment with near grazing incidence (using a Cameca instrument), whereas to study As near the interface, Frenzel et al. [3] have used O_2^+ bombardment and O_2 backfill conditions at normal incidence (using an Atomika instrument). In this study, we have investigated the techniques necessary to study As segregation at the SiO_2/Si interface using a Cameca IMS 3f instrument and have used Auger Electron Spectroscopy to confirm the SIMS results.

Arsenic standards in Si and SiO_2 were prepared by implanting $1x10^{15}$ As/cm^2 at 180 keV slightly off normal into a Si wafer half covered by 340nm of oxide. Interface As standards were prepared by implanting $1x10^{16}$ As/cm^2 at 180 keV and $1x10^{15}$ As/cm^2 at 120 keV into 112nm and 90nm oxide on Si wafers, respectively. The samples were kept below 100[C] during implantation. Segregation coefficient measurements were made on a (100) Si wafer, implanted with 160 keV $2x10^{15}$ As/cm^2 and oxidized at 950[C] in 1 atmosphere steam for 40 minutes, thereby growing 182nm of SiO_2.

SIMS depth profiles were measured using a 5.3 keV O_2^+ beam incident at 43°. To reject ions from the crater edges, secondary ions were collected from only the central 80µm of the 500x500µm^2 rastered area. Measurements were made both with ($1x10^{-5}$ Torr) and without O_2 backfill. The Auger measurements were performed with a Physical Electronics 590 SAM using a 5 keV electron beam and peak to peak modulations of 1, 3 and 6V for O, Si and As, respectively. Sputter depth profiling was accomplished using a 2 keV Ar^+ beam incident at 60°.

2. Results

Shown in Fig. 1 are the As concentration profiles derived from the SIMS profiles of the implant standards measured with and without O_2 backfill. The depth scale was assigned using the known oxide thickness (in agreement with stylus measurements) and the measured sputter rates in Si and the oxide. Quantification in the interface region was obtained by interpolating

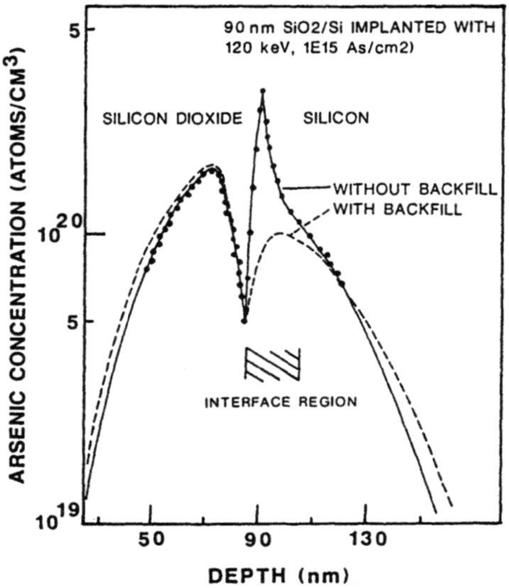

Fig. 1. As concentration profile for As implanted SiO_2 on Si determined using SIMS (see text)

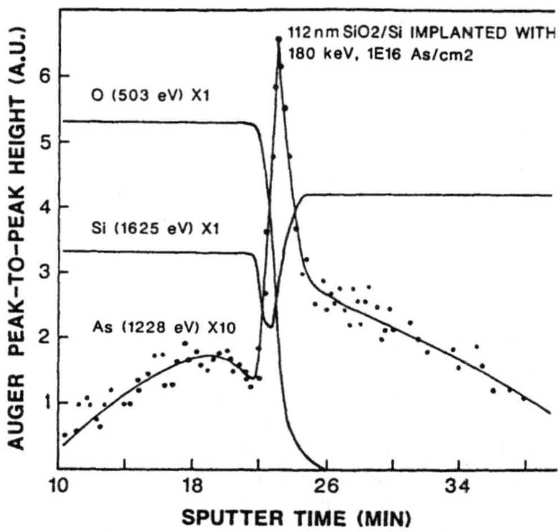

Fig. 2. Auger depth profile of the As implanted SiO_2 on Si (see text)

the concentration factors and sputter rates from the oxide and the substrate. The profile without backfill shows an unexpected pile-up of As which occurs exactly at the depth where the Si^+ yield reaches a mean value between the oxide and the substrate, suggesting that it is situated at the very interface. Because the $^{30}Si^+$ yield does not change through the interface when using backfill conditions, the interface is difficult to locate, and therefore has been assumed to be situated at the peak of the As pile-up as in the previous case. The peak concentration determined with backfill is about three times lower than that obtained without backfill. SIMS analysis of the high dose implant shows similar behavior with a pile-up peak about four times lower with backfill.

Auger results for the high dose implant, shown in Fig. 2, also indicate an As pile-up at the interface in good agreement with the SIMS results obtained without O_2 backfill. Correction of the raw data to account for sensitivity variations, determined from implant standards to be about 20%, has not been performed.

In order to investigate possible artifacts arising from As segregation to the SiO_2/Si interface during analysis by electron-assisted [4] or radiation-induced [4,5] mechanisms, the interface has been analyzed after the oxide layer has been stripped and an amorphous Si layer deposited. In this manner, equilibrium sputtering conditions have been established before profiling through the interface. The results obtained by both AES and SIMS support the results of Figs. 1 and 2. SIMS analysis of the bottom of the sputter craters also indicates a redistribution of As in front of the sputtered surface for oxygen bombardment with O_2 backfill conditions.

The As concentrations in the thermally grown oxide determined from the SIMS measurements with and without O_2 backfill are shown in Fig. 3,

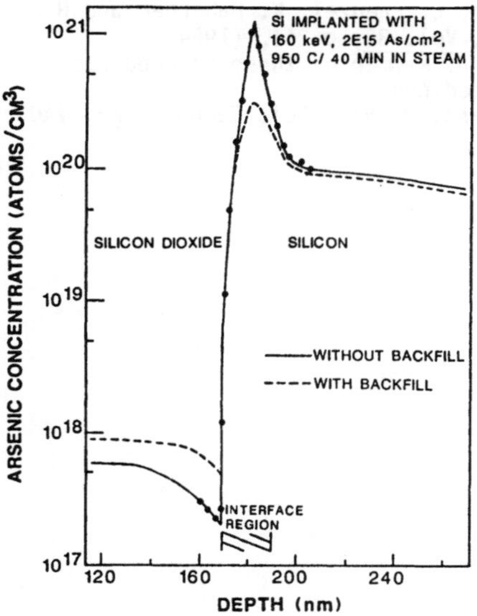

Fig. 3. Arsenic concentration profile for As^+ implanted Si with subsequent thermal oxidation, as described in the text.

clearly demonstrating rejection of As by the growing oxide and its subsequent pile-up at the interface [3]. Quantification was performed in the same manner as described previously, and is expected to have an error of about 50%. The segregation coefficient (ratio of the As concentration in the oxide to that in the Si) determined without backfill is $2x10^{-4}$ but is reduced to $2x10^{-3}$ when using O_2 backfill in saturation.

3. Conclusions

Both SIMS and Auger data show a measured discontinuity at the interface for As implanted SiO_2 on Si which is suggested, from the stripped oxide results, not to be a consequence of the profiling processes. A possible mechanism for this peak to arise during the implantation process has been suggested [6] to support a similar discontinuity observed using SNMS. Alternatively, it may be proposed that the As segregates to the internal interface by radiation-enhanced processes within the collision cascades initiated by the As implantation, similar to the radiation-enhanced As segregation demonstrated during SIMS analysis [4,5].

Although oxygen backfill to saturate the SIMS yields minimizes matrix effects, it causes significant redistribution of the As and a distortion of the SIMS profiles. Comparison with AES data suggest that SIMS measurements to determine As profiles at the interface itself are best performed without O_2 backfill when using the Cameca instrument.

Two of the authors (CML and JB) wish to acknowledge support by the National Science Foundation under MRL grant No. NSF DMR-83-16981.

References

[1] K. Wittmaack, Nucl. Instr. and Meth. B7/8, 750 (1985).
[2] A. E. Morgan, T. Y. Chen, D. A. Reed and J. E. Baker, J. Vac. Sci. Technol. A2, 1266 (1984).

[3] H. Frenzel, J. L. Maul, P. Eichinger, E. Frenzel, K. Haberger and H. Ryssel, in SIMS IV, Springer-Verlag, Vol. 36, p 241, (1984).
[4] W. Vandervorst, F. R. Shepherd, and J. Remmerie, these proceedings.
[5] F. Schulte and M. Maier, these proceedings.
[6] H. Oechsner, H. Paulus and P. Beckmann, J. Vac. Sci. Technol. A3, 1403 (1985).

Elimination of Ion-Bombardment Induced Artefacts in Compound Identification During Thin Film Analysis: Detection of Interface Carbides in "Diamond-Like" Carbon Films on Silicon

P. Sander, U. Kaiser[+]*, O. Ganschow*[+]*, and A. Benninghoven*

Universität Münster, Physikalisches Institut, Domagkstr. 75,
D-4400 Münster, F.R.G.
[+] now with Leybold Heraeus, Cologne

1. Introduction

Amorphous hydrogenated carbon layers on Si (a-C:H) produced by plasma deposition from hydrocarbon vapour are known to be extremely hard and having good adhesion, the reason being high fractions of carbon in a diamond-like (sp3) hybrid state /1/. The hardness depends critically on the substrate bias during deposition. Low substrate bias leads to more graphite-like (sp2) layers, which are soft and crumbly. The good adhesion, in particular on substrates which are known to form carbides, is usually assumed to be caused by interfacial carbide produced by ion-induced reactions during deposition.

In order to prove the a priori presence of carbide by sputter depth profiling, the influence of ion bombardment on the chemical state of carbon has to be clarified.

2. Signature data for SiC and ion-induced carbide formation

In order to obtain reliable standards of thin film analysis, we examined bulk SiC using AES, XPS, and SIMS. SiC is characterized by a chemical shift of Si 2p in XPS (fig.1), a peak broadening of Si LVV in AES (fig.2), and the emission of $SiC^{\pm}$ in SIMS. The

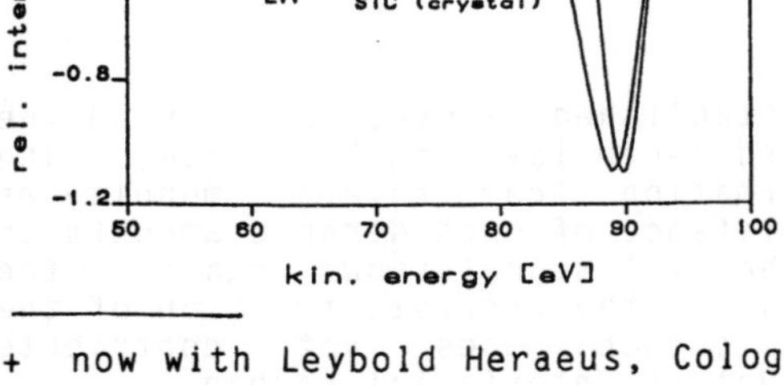

Fig.1: Peak shape of AES Si LVV transition for Si and SiC.

+ now with Leybold Heraeus, Cologne

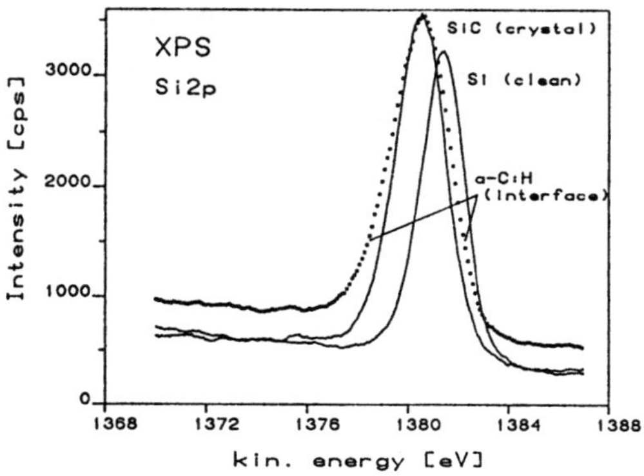

Fig.2: XPS Si 2p signal for Si, SiC and a-C:H/Si interface.

static SIMS signals are, however, not conclusive, as the same species is emitted by adsorbed hydrocarbons, e.g. C_2H_4, where there is no indication of carbide formation.

In order to investigate the influence of ion bombardment, we performed simultaneous C_2H_4 adsorption and ion bombardment with various ion species. The surface-sensitive Si LVV Auger transition indicates a small fraction of SiC (fig.3), whereas the XPS Si 2p line, with its greater information depth, shows no changes in binding energy and peak shape. The $SiC^{\pm}$ emission in SIMS shows some increase parallel to the peak broadening in AES. We conclude that the ion-induced reactions lead to a thin SiC layer at the very surface, which is small enough not to disturb depth profiling applications.

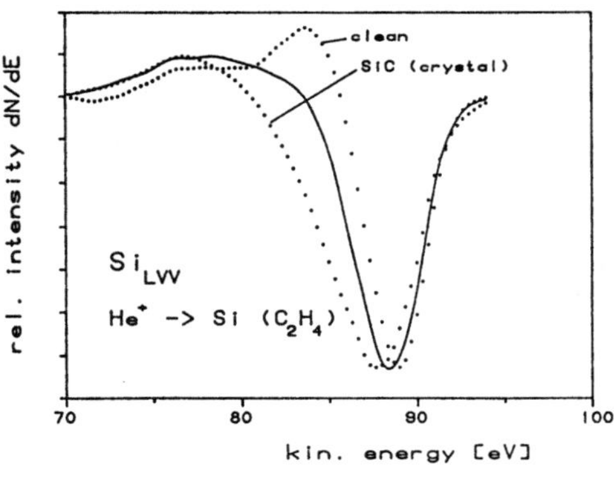

Fig.3: AES Si LVV transition of Si during simultaneous C_2H_4 adsorption and He^+ ion bombardment. The dotted curves indicate the Si and SiC reference spectra.

3. SiC detection

Using the signature data established above, we measured the XPS/SIMS depth profile of a hard a-C:H layer on Si (fig.4). The XPS data, representing information from several monolayers depth, prove the a priori existence of considerable amounts of SiC. It should be noted that the SiC^- signal shows exactly the same variations in intensity as the carbidic fraction of the Si 2p. Thus, non-carbidic carbon does not contribute significantly to the total amount of interfacial carbon.

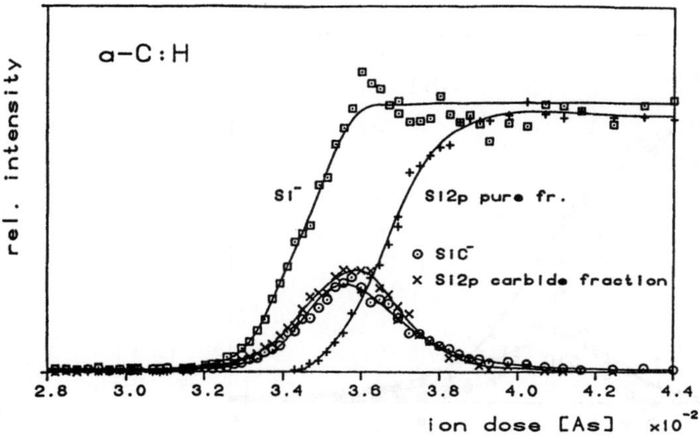

Fig.4: Depth profile of a hard a-C:H layer on Si.

4. Comparison of hard and soft layers

Soft layers deposited at low substrate bias were depth profiled under identical experimental conditions. The comparison of these layers reveals no differences in sputter yield and chemical state and concentrations of the constituents, except for the ratio of C_2^- and C_3^- secondary ions (fig.5). This difference seems to reflect the different amounts of sp3 carbon in hard and soft layers. When heating hard layers, which is known to transfer the diamond-like carbon into the graphitic state, a corresponding decrease of the C_2^-/C_3^- ratio is observed, which supports this interpretation.

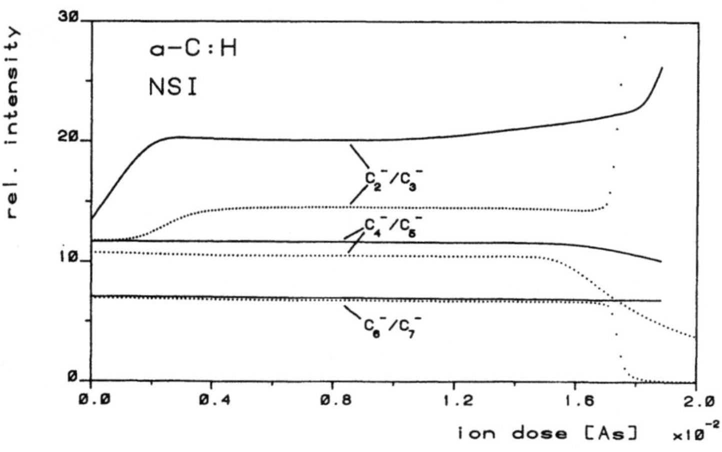

Fig.5: SIMS signal ratio C_{2n}^- / C_{2n+1}^- for hard (solid lines) and soft (dotted lines) carbon layers.

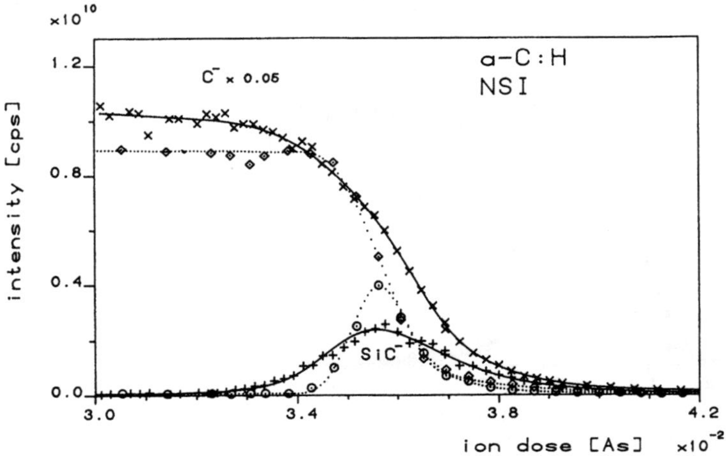

Fig.6: Interface region of hard (solid lines) and soft (dotted lines) a-C:H layers on Si.

Comparison of the interface regions (fig.6) shows that the total amount of interfacial carbide is the same for both types of layers, whereas the interface width is much smaller in soft layers. The narrower interface may be caused by smaller mixing effects at low deposition bias. The extent of this interface, rather than the existence of carbide as such, seems to be responsible for the adhesion properties of these layers.

References

/1/ A.Bubenzer, B.Dischler, G.Brandt and P.Koidl, J.Appl.Phys. 54 (1983), 4590.

Gibbsian Segregation During the Depth Profiling of Copper in Silicon

V.R. Deline, W. Reuter, and R. Kelly*

IBM T.J. Watson Research Center, Yorktown Heights, NY 10598, USA

The segregation of Si to the surface of Cu as inferred by secondary ion mass spectrometry, SIMS, is shown to set in dramatically when O is present at the sputtered surface. Copper depth profiles obtained from Cu implanted silicon wafers or Cu-Si sandwiches using O_2^+ or Ar^+ bombardment at angles of incidence > 30° show no unusual perturbation. However, with O bombardment at ≤15° or with O incorporation from the ambient with either O_2^+ or Ar^+ bombardment, the measured Cu profiles become deeply penetrating. The results can be understood in terms of an interplay between bombardment-induced Gibbsian segregation (Cu is excluded fron the surface of Si in the presence of O) and room-temperature diffusional mobility (Cu is highly mobile in amorphous SiO_2).

1. Introduction

The present work is concerned with composition-depth profiles for the system Cu-Si, where the profiles variously showed enormous extents of Cu persistance, moderate Cu persistance, or no important perturbations at all. The key to understanding the extent of persistance will be shown to lie in the interplay between bombardment-induced Gibbsian segregation and room-temperature diffusional mobility. It may be expected that when the system has a high diffusional mobility, then a non-segregating solute would tend to be excluded from the surface. Owing to cascade mixing, however, the exclusion would not be total, so the non-segregating species would sputter slowly. Similar arguments suggest that a solute which segregates would show enhanced sputter removal. If, on the other hand, the system has a sufficiently low diffusional mobility, then segregation would not occur and the profile would show no unusual perturbatipon.

Bombardment-induced Gibbsian segregation is equivalent to what was in the past termed preferential sputtering. The latter is an operational term for the situation in which one element of a multicomponent system is sensed to be depleted at the bombarded surface, and had long been considered to be the result of mass and chemical binding differences. It is now realized to be in nearly all cases identical with bombardment- induced Gibbsian segregation [1].

It is worth clarifying that what we have termed "Gibbsian segregation" is distinct from the segregation discussed, for example, by WIEDERSICH [2] . Gibbsian segregation, whether thermally activated [3] or bombardment-induced

---

* Present address: IBM Research, San Jose, CA 95193

[1], is driven by themodynamics, whereas the segregation of [2] is driven by point-defect fluxes. With the system Cu-Si, for example, the heat of the reaction $Cu_2O(s) = 2Cu(s) + O$ is 4.4 eV per O atom, whereas that for $SiO_2(s) = Si(s) + 2O$ is 7.3 eV per O atom. The thermodynamic tendency is thus for Si to segregate.

Credit for the first observation of what will be discussed here appears to be due to WILLIAMS and BAKER [4] in a study of Ag films deposited on Si.

2. Experimental

A Cameca IMS-3f equipped with a duoplasmatron was used to obtain depth profiles of Cu at a predetermined angle of incidence, ~40° for the 8 keV O_2^+ or Ar^+ beams. Gaseous O_2 could be introduced into the target chamber through a variable flow jet directed toward the bombarded surface to increase the O incorporation at the sputtered surface. A second SIMS system [5] was used to measure depth distributions as the angle of incidence of the bombarding 8 keV O_2^+ ion beam was changed. This system is equipped with a vacuum transfer into a Hewlett-Packard XPS system for analysis of the extent of Si-O bonding after O_2^+ bombardment at selected angles of incidence. Using the above two systems, Cu depth profiles were obtained from Si wafers implanted with Cu^+ (1×10^{15} cm^{-2} at 200 keV), and from 22.5 nm layers of sputter deposited Cu on Si <100>. Depth calibration was achieved by measuring the crater depth.

Oxygen ion bombardment and O_2 adsorption from the gas phase are used to increase positive secondary ion yields, thus improving detection sensitivities for SIMS analysis. However, in this work the incorporation of the O at the bombarded surface of the Si is shown to lead to strong Si segregation, so that Cu is excluded from the surface.

The sputtering yield of Si under O_2^+ ion bombardment increases considerably from normal incidence, 0°, to more grazing incidence [6]. With O_2^+ at 60°, the XPS results from the Si2p signal show only a small shift in intensity toward higher binding energies over that of Ar^+ sputter-cleaned Si (-98.4 eV), indicating little oxide formation. However, as the angle of incidence is reduced, the shift shows the presence first of Si which is partially coordinated with O (at 45 and 30°) and then of fully coordinated Si, i.e. SiO_2 at (15 and 0°) [7].

In a series of Cu depth profiles obtained from a Cu-implanted Si wafer (1×10^{15} cm^{-2} at 200 keV) bombarded with 8 keV O_2^+ at selected angles of incidence, the measured Cu distribution in some cases showed marked persistence to great depths. Figure 1 shows the measured Cu distributions. For ≥30°, where the XPS showed only partial coordination with O, the decay lengths were 20-30 nm. For 0 and 15°, where the XPS indicated that the Si at the surface was fully coordinated (i.e. SiO_2), the decay lengths were larger by a factor of ~14, namely 290-400 nm.

We here use the term "decay length" to refer to the quantity L when the tails of the depth profiles are fitted to an exponential, exp(-x/L).

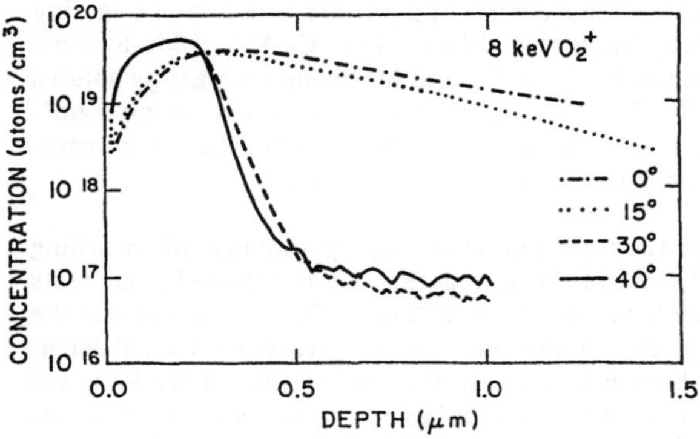

Figure 1. SIMS depth profile from a Cu implant (1×10^{15} cm^{-2} at 200 keV) in Si using 8 keV O_2^+ bombardment at 0, 15, 30, and 40°

Depth profiles were also obtained for thin films of Cu deposited on Si, using the Cameca IMS-3f with 8 KEv O_2^+ or Ar^+ bombardment at a fixed angle of incidence of ~40°. The decay lengths in the absence of gaseous O_2 were 6-10 nm and thus a bit less than those obtained at ≥30° in the angular study of Fig. 1. When O_2 was introduced, SiO_2 formed and the Cu profiles showed decay lengths of 50-75 nm, corresponding to a factor of ~8 increase. Similar Cu retention (or lack of retention) was observed when implanted Cu was profiled with (or without) O_2 adsorption.

3. Conclusions

The basic observation in this work is that, either for Cu implanted into Si or for Cu-Si sandwiches, three types of sputter profiles are obtained. (i) For incident Ar^+ at ~40° or O_2^+ at ≥30°, in both cases in the absence of O_2, the profiles are largely "normal," showing only such perturbations as might arise from recoil effects, cascade mixing, radiation-enhanced diffusion, and surface morphology (decay lengths of 6-30 nm). (ii) For incident O_2^+ or Ar^+ at ~40° with ambient O_2 adsorption, the profiles show a moderate perturbation in which Cu persists with decay lengths of 50-75 nm. Such bombardments lead to surface oxidation appropriate to recoil implantation, thence atomically thin [8]. We suggest that the perturbation arises due to Cu rejection from this thin oxide in response to bombardment-induced Gibbsian segregation of Si. (iii) Finally, for O_2^+ bombardment at ≤15°, the perturbation is very great indeed, characterized by decay lengths of 290-400 nm. Such bombardments lead to surface oxidation appropriate to direct implantation, thence atomically thick [7] . The Cu will now tend to be localized at a relatively deep oxide-Si interface and would be completely resistent to sputtering were it not, owing to cascade mixing, present to some extent thoughout the oxide.

The pattern found here is consistent with other work. Thus Ag-Si, Au-Si, Pb-Si, and Zn-Si resemble Cu-Si in showing moderate persistence of a non-segregating solute when recoil-implanted O is present [4,9] . Cr-Si at 300°C,

Pd-Si, and Cu-Si show the extreme persistence appropriate to a non-segregating solute in the presence of directly implanted O [10 – 11]. Ca-Si and Mg-Si show enhanced removal such as is appropriate to Ca and Mg being segregating solutes [9]. Finally, the systems Al-Si, Cr-Si at $\leq 25°C$, In-Si and Ti-Si show no significant effect at all [9,11], as is to be expected when, whether the species is segregating or non-segregating, the diffusional mobility is low [12].

The principles lying back of the role of Gibbsian segregation in depth profiling are thus three. The potential sense of the segregation can be inferred from heats of reaction for processes of the type $Cu_2O(s) = 2Cu(s) + O$. Whether or not the expected segregation takes place, on the other hand, presupposes a sufficient diffusional mobility. This is known to exist, in the case of amorphous SiO_2, for small ions with charges of 1+ and 2+ but not for large or more highly charged ions [12]. Finally, the extent to which profiles are perturbed depends on whether the surface chemical changes are due to recoil or direct implantation.

A fuller report of this work is in preparation [12]

1. Roger Kelly: Surf. and Interface Anal. 7, 1 (1985)
2. H. Wiedersich: Nucl. Instr. Meth. B7/8, 1 (1985)
3. R. A. van Santen and M. A. M. Boersma: J. Catal. 34, 13 (1974)
4. P. Williams and J. E. Baker: Nucl. Instr. Meth. 182/183, 15 (1981)
5. M. A. Frisch, W. Reuter, and K. Wittmaack: Rev. Sci. Instr. 51, 695 (1980)
6. N. Warmoltz, H. W. Werner, and A. E. Morgan: Surf. and Interface Anal. 2, 46 (1980)
7. W. Reuter: Nucl. Instr. Meth. B (in press)
8. R. Kelly and J. B. Sanders: Surf. Sci. 57, 143 (1976)
9. S. M. Hues and P. Williams: these proceedings
10. P. R. Boudewijn, H. W. P. Akerboom, and M. N. C. Kempeners: Spectrochim. Acta 39B, 1567 (1984)
11. S. D. Littlewood and J. A. Kilner: these proceedings
12. V. R. Deline, W. Reuter, and R. Kelly (in preparation)

Species-Specific Modification of Depth Resolution in Sputtering Depth Profiles by Oxygen Adsorption

S.M. Hues and P. Williams

Department of Chemistry, Arizona State University, Tempe, AZ 85287, USA

1. Introduction

Ion-beam mixing can significantly degrade depth resolution in sputtering depth profiles. Given a composition profile with an initially abrupt interface between an overlayer and the substrate, mixing produces a measured profile with a characteristic exponential decay of the overlayer signal. Mixing redistributes the overlayer species over a depth comparable to the primary ion range. This redistribution leaves only a fraction of the residual overlayer atoms exposed to sputtering in the outer atomic layer, leading to a removal rate much lower than would be expected if all the overlayer material remained at the surface [1]. Mixing effects have been modelled for isotopic markers (i.e. no chemical effects) by several groups [2-5] offering hope of deconvolution procedures to improve depth resolution. However, WILLIAMS [1] showed that mixing decay lengths are element-specific, varying over a factor of ~5 for overlayer elements ranging in mass from H to Pt, sputtered under identical conditions from silicon.

One of the mechanisms suggested [1] to account for the observed elemental differences was surface segregation. If surface segregation could occur in the sputtering collision cascade, it could maintain a surface layer composition different from that which would otherwise be imposed by mixing, leading to a different removal rate of the element enriched in the surface. Surface segregation as a cause of preferential sputtering was discussed at this time by KELLY [6] and by ANDERSEN [7], but no good evidence existed that segregation could occur during the short lifetime of the sputtering collision cascade. In 1981, WILLIAMS and BAKER [8] showed that the removal rate of silver from silicon under Ar^+ bombardment was reduced by a factor of 5 when sputtered in an oxygen ambient. This result was interpreted as due to the antisegregation of silver away from the surface in the steep oxygen gradient arising from adsorption and recoil implantation, and therefore indicated that segregation effects could play a significant role in ion-beam mixing.

Because oxygen adsorption is commonly used to enhance ion yields in secondary ion mass spectrometry, we have investigated oxygen effects on depth resolution in the sputtering of a wide variety of overlayer elements from silicon substrates. The results show that oxygen adsorption can severely alter depth profiles.

2. Experimental

The samples used in this study were single crystal Si substrates onto which 50A of carbon was sputtered, and then several hundred angstroms of metal (Ba,Al,C,Ge,Bi,Zn,Ni, and Pt) were vapor deposited. Samples were depth profiled with a Cameca IMS 3F ion microanalyser using Ar^+ at 8 keV impact energy. Primary current densities were 30 - 50 µA/cm² at an incidence angle nominally 30° to the sample normal. The base pressure in the sample region was typically ~2 x 10^{-8} torr prior to the admission of oxygen. Positive secondary ion signals for both carbon and the deposited metal were monitored. Depth profiles were obtained at base pressure and also with an oxygen jet directed at the sample from a capillary. The pressure measured at a distant ion gauge under these circumstances was ~3 x 10^{-5} torr. The effective oxygen pressure at the surface was at least an order of magnitude higher. In all cases, the oxygen surface concentration was such as to saturate the Si^+ sputtered ion signal.

Decay lengths were determined in the exponential decay region of the depth profile, where the metal and carbon signals had fallen to ~1% of their peak values. The distances over which the signals for the metals fell by a factor of 1/e were determined initially relative to the corresponding distance for carbon. Subsequent depth profiles, using a sample with a thin carbon film only, allowed absolute values for the carbon decay length with and without oxygen to be determined by establishing a depth scale by crater depth measurements. These procedures avoided complications arising from variations in the sputtering yield of the metal overlayers, and the carbon internal standard also offered a degree of immunity against surface roughing effects which can occur when sputtering polycrystalline films.

3. Results and Discussion

The effect of oxygen on profiles for electronegative species was to increase the decay length, as noted in the earlier study of silver. Consistent with the idea that these changes result from segregation in the surface oxygen gradient, the effect of oxygen on depth profiles for electropositive metals (Ca, Mg) was to decrease the decay length. The decay length ratios (ratios of decay lengths with and without the oxygen ambient), are shown in Fig. 1 plotted as a function of the Pauling electronegativity of each element.

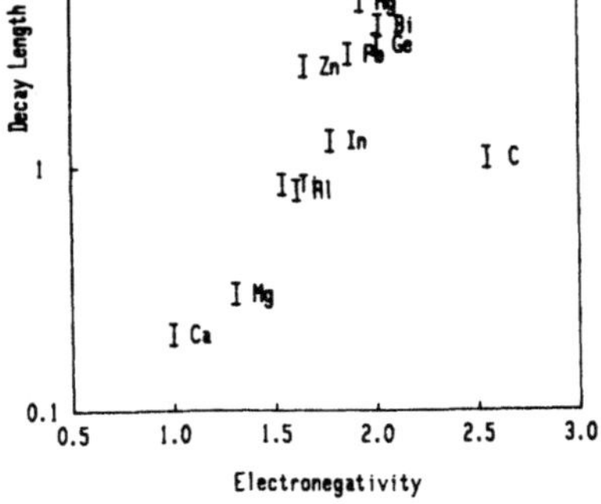

Fig. 1 Decay length ratios (decay length for oxygen ambient profile/decay length at base pressure) for species studied vs. Pauling electronegativity of the overlayer element. The silicon electronegativity is 1.9.

The lack of a significant effect of oxygen on the carbon decay length despite the high electronegativity of carbon may result from an additional chemical sputtering mechanism which removes carbon as CO and CO_2 making the surface a sink for diffusing carbon.

Although the segregation model provides a simple rationalization of the oxygen effects, it seems probable that the process is more complicated. Solubilities of the various elements in the oxygen-rich silicon region may vary. In addition, oxygen may well alter diffusivities in the collision cascade by an amount which depends on the strength of the oxide bonding. To evaluate the latter effect, it would be useful to determine the effect of oxygen on the decay length of a silicon isotopic marker.

4. Conclusion

The data presented above illustrate that, except for very electropositive impurities, the use of an oxygen ambient to increase secondary ion yields in sputter depth profiling can degrade depth resolution considerably under noble gas ion bombardment. The extent of this degradation in a silicon substrate can be related to the Pauling electronegativity of the overlayer species. A model in which segregation or antisegregation of the overlayer species occurs in the near-surface oxygen gradient is consistent with the trend in the experimental data.

Acknowledgements

This work was supported by the National Science Foundation under grant DMR 8206028. We thank John Wheatley and Jim Clark (ASU) for assistance with sample preparation.

References

1. P. Williams: Appl. Phys. Lett. 36, 758 (1980)
2. H. H. Andersen: Appl. Phys. 18, 131 (1979)
3. W. Hofer, U. Littmark: Nucl. Instrum. and Methods 168, 329 (1980)
4. B. V. King, I. S. T. Tsong: J. Vac. Sci. Technol. A2, 1443 (1984)
5. R. Kelly: Surf. Science 100, 85 (1980)
6. H. H. Andersen, in SPIG 1980, ed. M. Matic, Boris Kidric Inst. Nucl. Sci., Beograd (1980) p. 421
7. P. Williams, J. E. Baker: Nucl. Instr. and Meth. 182/183, 15 (1981)

Selective Sputtering and Ion Beam Mixing Effects on SIMS Depth Profiles

E. Lidzbarski and J.D. Brown

Faculty of Engineering Science and the Centre for Interdisciplinary Studies in Chemical Physics, The University of Western Ontario, London, Canada N6A 5B9

1. Introduction

The shape of depth profiles, measured by Secondary Ion Mass Spectrometry, is usually distorted by basic sputtering processes as well as by instrumental factors. Atomic mixing, selective sputtering, radiation-enhanced diffusion and microroughness of the crater bottom limit achievable depth resolution. Uniformity of the primary beam and crater edge effects are also important factors which have to be considered.

Figure 1 shows a typical depth profile of a 37.5 nm Ag film on Si bombarded with 8 keV O_2+. The experimental profile lacks the sharp Ag interface representative of the true distribution. As the sputter erosion proceeds through the interface, the Ag becomes intermixed with the Si substrate and beyond a certain depth secondary ion intensity decays exponentially. In this dilute limit, the intensity can be characterized by

$$I_{(x_2)} = I_{(x_1)} \exp - \left(\frac{X_2 - X_1}{\lambda}\right) \qquad (1)$$

where $I_{(x)}$ is the measured intensity at depth x and λ the decay length parameter. If instrumental factors are eliminated and no microroughness develops, then λ reflects the combined effect of atomic mixing, selective sputtering and radiation-enhanced diffusion.

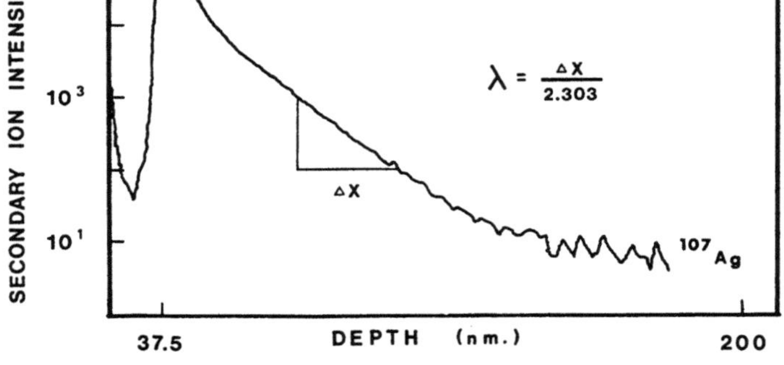

Fig. 1. Depth profile of a Ag film on Si using 8 keV O_2^+

A study by Williams [1] suggests that under 20 keV Cs^+ bombardment, the decay lengths in Si are dependent on the mass of the impurity. However, Wittmaack [2] found no mass dependence when profiling low atomic mass impurities (H→Cl) in Si using Cs^+ primary ions. The study also revealed a linear relationship between λ and the primary beam energy in which the slope was almost independent of the mass of the measured ion. The present study was performed to investigate the energy dependence of the decay length λ for higher mass impurities (Al→Au) in Si and to determine optimum conditions for depth profiling.

2. Experimental

Thin films of Al, Cu, Ag, Sb, Pt and Au deposited on cleaned Si substrates were used in this study. The films ranged in thickness from 30 to 100 nm. Depth profiles were measured on a Cameca IMS 3f using mass analyzed O_2^+, O^- and Cs^+ primary ions of varying energies. The beam was rastered over an area of 250 x 250 m and secondary ions were extracted from a region 60 m in diameter. A scanning ion-imaging system [3] was used to ensure that secondary ions were extracted from the centre of the crater. In the Cameca IMS 3f, the angle of incidence is dependent on the primary beam energy and the polarity of both primary and secondary ions. Table 1 lists the calculated angle of incidence and the energy perpendicular to the surface used in the measurements.

Table 1. Primary ion incidence angles and energy components

Primary Ion	Secondary Ion Polarity	Primary Ion Energy (keV)	Incidence Angle	Perpendicular Ion Energy (keV)
Cs^+ with	-ve	9.5	21.3	8.9
or		12.0	23.3	11.0
O^- with	+ve	14.5	24.5	13.2
		15.9	25.0	14.4
O_2^+	+ve	5.5	42.4	4.1
		9.7	37.2	7.7
		13.4	35.3	10.9

3. Results

Careful SEM investigations of the crater bottoms revealed no evidence of sputtering cones, faceting or other topographical features which would enhance the tail on the profiles. For films ranging in thickness from 30 to 100 nm, no significant differences in the measured decay lengths were found in agreement with the previous work of Wittmaack [2] in which he found the decay lengths of O were identical for adsorbed or surface oxide films.

The dependence of the decay lengths of ions sputtered from Si on the perpendicular component of ion beam energy is shown in Figure 2. The measured values increase linearly with increasing primary beam energy. The rate at which λ changes with energy dλ/dE varies for the different ions. Au and Pt have unusually large decay lengths with practically no energy dependence under cesium bombardment. Extrapolation of the decay lengths to zero energy is indicative of the minimum decay length achievable, λ_o. The extrapolated values are plotted in Figure 3. No clear mass dependence is discernible. Because of the unusual behaviour of Pt and Au under Cs bombardment, λ_o values for these elements are very large compared to the oxygen values. For the other elements, the behaviour is similar, and if total bombarding energy is used for the O_2^+ data, the λ_o values for O are almost identical to the Cs values.

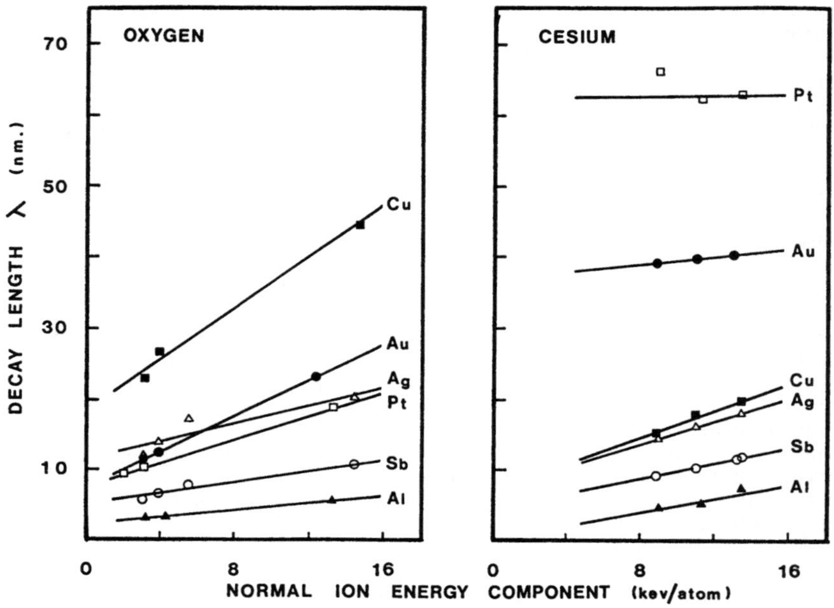

Fig. 2. Decay parameters, in the dilute limit, for Cs and O primary ion beams

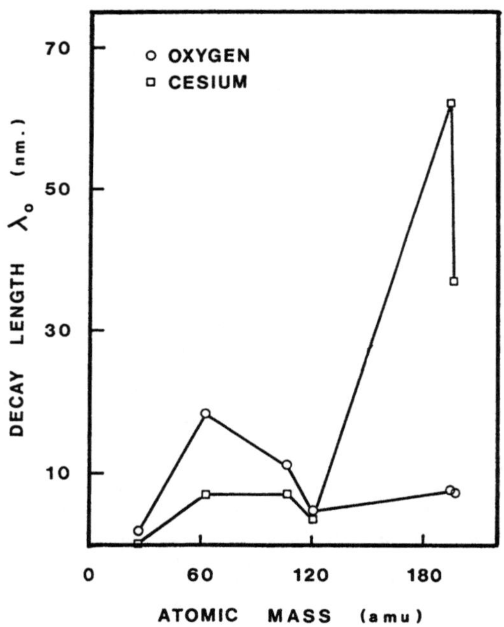

Fig. 3. Decay parameters extrapolatec to zero primary ion energy

4. Conclusions

If there is no limitation on the minimum primary ion energy, then Cs or O will give comparable depth resolutions in depth profiling,and the primary ion which gives highest secondary ion intensities should be used. If instrumental limitations dictate a minimum primary ion energy then the slope of the decay length versus ion energy becomes an important criterion in determining optimum primary ion species. For Pt and Au, an oxygen primary ion beam gives the better depth resolution for all practical primary beam energies.

5. References

[1] P. Williams, Appl. Phys. Lett. 36, 758 (1980).
[2] K. Wittmaack, Nucl. Instrum. Meth., 209/210, 191 (1983).
[3] J.D. Brown, W. Vandervorst, Surf. and Interface Anal. 7, 74 (1985).

The Effect of Temperature on Beam-Induced Broadening in SIMS Depth Profiling

S.D. Littlewood and J.A. Kilner

Department of Metallurgy and Materials Science, Imperial College, London, SW7 2BP, U.K.

In order to explain the observed discrepancies between experimental profile broadening and ballistic mixing theory, the effects of radiation-enhanced diffusion (RED) and segregation must be assumed [1,2]. Under normal incidence O_2^+ profiling of Si, where formation of an SiO_2 surface layer occurs [3], segregation of impurities at the SiO_2/Si interface can cause severe profile distortions (see for e.g. BOUDWIJN et al [4]). SIMS depth profiling over a wide range of temperatures may, therefore, be important for understanding the processes involved, although such effects may not necessarily be thermally activated. In this paper we report the effect of temperature on the profile broadening of the transition metals Pd, Cr, Ti, and the impurities C, O, F and H in Si. The concentration distributions to be analysed contained an infinite slope or step, by which we hoped to produce the expected exponential decay of the impurity signal [1,5]. The observed impurity tails are shown to be useful in studying the effect of temperature on beam-induced broadening in SIMS depth profiling.

Thin layers of Cr and Ti (∿ 3nm), and Pd (∿ 1 nm) were obtained by sputtering onto clean Si substrates. The Pd sample had a further 75 nm of sputtered Si (with high O and C content) and the Cr a further 100 nm of amorphous Si deposited at 580°C by LPCVD. Markers of the interfacial impurities C, O, F and H were made by cleaning a Si substrate with H_2SO_4/H_2O_2 followed by a CF_4 clean, prior to deposition of a 44.2 nm layer of a-Si (deposition as above). Depth profiling was carried out on an Atomika DIDA-II SIMS using 10 KeV O_2^+ primary ions for the transition metal samples and 10 KeV Cs^+ primary ions for the interfacial impurities sample, all at normal incidence. Average background pressure was 10^{-10} Torr. Variation of the sample temperature was achieved using a combined resistance heating and liquid nitrogen cooling stage with a working temperature range - 150°C to + 400°C. To check for any thermal diffusion effects on heating, all samples were profiled again at RT. There was no observable change in the before and after RT profiles, in all cases.

Selected results for Cr/Si are shown in Fig. 1. A slight increase in the length of the Cr tail occurs between - 150°C and RT. From RT to 300°C a distinct shoulder appears after the interface peak with a more rapid increase in the length of the tail. Interrupting the profile at 300°C, just after the shoulder and cooling to RT produces a peak followed by a return to the steeper RT decay. SCHULTE and MAIER [6] observed a similar effect for As in Si and we interpret our result in a similar manner i.e. as being caused by segregation of Cr at the SiO_2/Si interface. The RBS results of MAYER et al [7] lend further support to this argument. They found that ion mixing of the Cr/Si interface is enhanced above RT, where radiation-enhanced diffusion (RED) and chemical driving forces become dominant. In our case, under normal incidence O_2^+ bombardment above RT, segregation (via RED) of the Cr at the SiO_2/Si interface becomes dominant. It should be noted here that, although the observed mixing is temperature independent be-

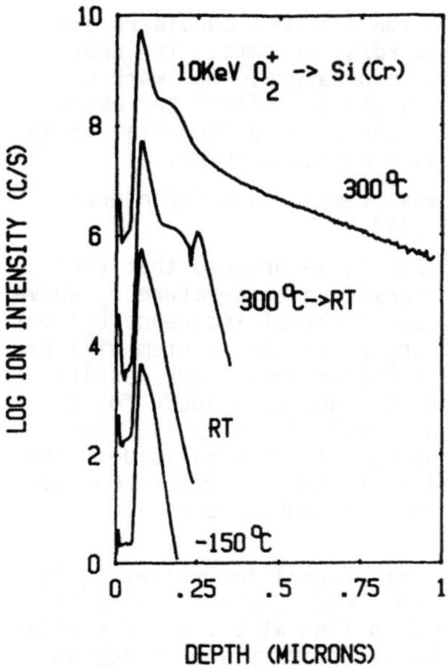

Fig. 1. Depth profiles at various temperatures for Cr/Si. (Note vertical displacement of profiles for clarity)

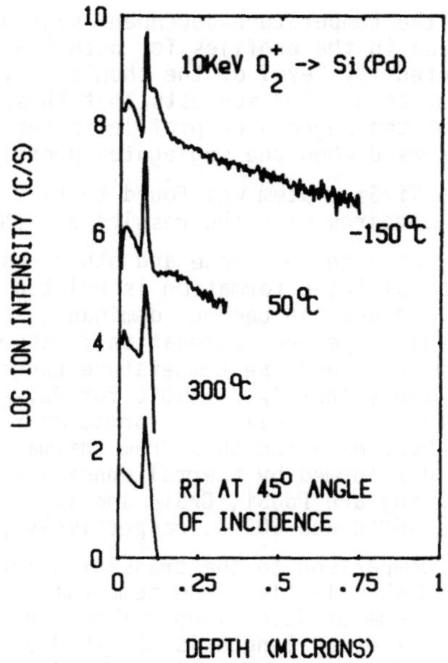

Fig. 2. Depth profiles at -150°C, 50°C, 300°C and RT (45°) for Pd/Si (Note vertical displacement of profiles for clarity)

low RT, ballistic mixing theory underestimates the extent of the broadening [2]. In this temperature regime a contribution from athermal segregation might be expected, on account of the gross segregation effects observed at higher temperatures.

Figure 2 shows the low-temperature results for Pd/Si. The high Pd concentration in the surface layer is most likely due to re-deposition and mixing of the Pd during sputtering of the Si overlayer. As for Cr/Si above RT the profiles at -150°C and 50°C imply that segregation of Pd is occurring at the SiO_2/Si interface, even at low temperatures. The importance of the SiO_2 layer, formed at normal incidence, can be seen by comparing the 50°C profile and the RT profile taken at 45° angle of bombardment. At temperatures above ∿ 200°C the profiles unexpectedly sharpen up (see for e.g. 300°C profile). A plateau is observed in the $^{44}SiO^+$ matrix signal, not shown, indicating the formation of a new layer. MATTESON et al [8] also found a similar transition in the Pd/Si system at a temperature of ∿ 250°C, using low fluence 220 KeV Kr^+ for mixing and RBS for analysis. In view of these results, and noting that ion-assisted silicide formation is known to occur in transition metal/Si systems under low fluence high energy noble gas bombardment [9], we attribute our results for Pd/Si at high temperatures to silicide formation. The silicides in this system form easily by ion mixing due to high atomic mobility and have a high radiation resistance [9]. Sufficient atomic mobility is also a pre-requisite for temperature-dependent segregation. Hence, the low-temperature segregation behaviour observed here, for Pd/Si system, may be related to the ease of silicide formation. The important point to note is that there may be at least two competing processes that affect the temperature dependence. (1) Segregation and (2) Silicide formation.

In the temperature-dependent segregation regime distinct shoulders are observed in the profiles for both the Cr/Si and Pd/Si systems. The depth and intensity level of the shoulders vary with temperature. The work of MARWICK et al [10] suggests that these shoulders merely reflect the evolution of the segregated profile in the surface layers. An exponential decay is observed when the segregated profile attains a constant shape.

The Ti/Si system was found to have a very weak temperature dependence which compares with the results of KING et al [11].

In the light of these and other results [6,7,8,9],we propose that the ease of silicide formation is related to the transition temperature T_c above which segregation becomes dominant, for the case of normal incidence O_2^+ bombardment. Whether segregation is observed in any particular system will depend upon T_c and the temperature range used for measurement. Our results would imply that $T_c < -150°C$ for Pd, $T_c \sim$ RT for Cr, and $T_c > 400°C$ for Ti i.e. the ease of silicide formation follows the order Pd > Cr > Ti. One might have expected this from thermal annealing data for the silicides. The silicides formed by thermal annealing, equivalent to those first formed by ion mixing are Pd_2Si, $CrSi_2$ and $TiSi_2$ with formation temperatures 100 - 300°C, 450°C and 600°C, respectively [9].

In comparison to the transition metals the interfacial impurities C, O, F and H show little or no temperature dependence. Only C shows a significant change of decay length from 13nm at -150°C to 16nm at 250°C. A similar change of decay length is observed at RT if the Cs^+ primary ion energy is changed from $\sim$ 3 KeV to 15 KeV [1].

From a SIMS practical point of view low-temperature profiling does not improve depth resolution for these systems. A possible exception is Pd if a temperature-independent regime exists below -150°C, where profiling at liquid helium temperatures could be important. However, athermal segregation in the temperature-independent regime can be a problem under normal incidence O_2^+ depth profiling of Si systems. Non-normal incidence should be used in these cases, if possible.

References

1. K. Wittmaack: Vacuum, 34, 119 (1984).
2. B.M. Paine and R.S. Averback: Nucl. Instrum. Meth. in Phys. Res. B7/8, 666 (1985).
3. W. Reuter and K. Wittmaack: Appl. Surface Sci. 5, 221 (1980).
4. P.R. Boudewijn et al: Spectrochimica Acta. 39B, No. 12, 1567 (1984).
5. U. Littmark and W.O. Hofer: Nucl. Instrum. Meth. 168, 329 (1980).
6. F. Schulte and M. Maier: Proc. 7th Int. Conf. Ion Beam Analysis, Berlin (1985). to be published in Nucl. Instrum. Meth. B (1986).
7. J.W. Mayer et al: Nucl. Instrum. Meth. 182/183, 1 (1981).
8. S. Matteson et al: ibid. p.43 (1981).
9. M.A. Nicolet et al: 'Formation and Characterisation of Transition Metal Silicides', in VLSI Electronics: Microstructure Sci., Vol. 6, p.330, eds. N. Einspruch and G. Larrabee (Academic Press, New York 1983).
10. A.D. Marwick et al: J. Nucl. Mats., 83, 35 (1979).
11. B.V. King et al: Nucl. Instrum. Meth. in Phys. Res. B7/8, 607 (1985).

Variable Angle SIMS

J.G. Newman

Perkin-Elmer Corporation, Physical Electronics Division,
6509 Flying Cloud Drive, Eden Prairie, MN 55344, USA

1. Introduction

The procedure for performing optimized SIMS analysis on any given sample varies according to the nature of the sample and the type of information that is to be extracted from the analysis. One of the ways in which the data acquisition can be optimized is through the control of the primary ion beam angle of incidence. In this work a quadrupole based SIMS instrument operating with an oxygen primary ion beam was used to demonstrate the effect of altering the angle of incidence on:

A) The ease of charge neutralization of borophosphosilicate glass films.

B) The depth resolution obtained from an AlGaAs/GaAs superlattice structure.

C) The boron sensitivity obtained from a boron-implanted silicon sample.

2. Experimental

The data was obtained on a Perkin-Elmer Physical Electronics Model 6300 SIMS instrument using an oxygen primary ion beam produced by a duoplasmatron source. A beam voltage of 7 KeV was used,except for the analysis of the superlattice structure which was performed at 2 KeV. The size of the beam varied from approximately 30μm to 100μm depending on the beam voltage and current used. The angle of incidence of the primary ion beam (measured as the angle between the primary beam and the sample normal) was varied in-situ by altering the tilt of an externally driven sample stage. Because sputter yield and raster area change with sputter angle, the total primary ion current was adjusted at each angle so as to obtain similiar profile times in each set of analyses. The angle between the ion beam and the analyzer is approximately 60 degrees. This fixed geometry creates a situation where alteration of the sample tilt changes not only the angle of incidence but also the angle at which the secondary ions are extracted from the sample. Charge neutralization was accomplished with the aid of an electron gun appended to the analysis chamber at an angle approximately 90 degrees from the ion gun. The electron gun was operated at 4 KeV and a maximum current setting of approximately 20 μamps. The magnification and size of the rastered electron beam were adjusted at each angle to obtain the maximum secondary ion signal intensity.

3. Data and Discussion

The ease of charge neutralization of borophosphosilicate glass (BPSG) insulating samples was studied as a function of angle of incidence. Figure 1 shows the depth profiles of a 1.0μm BPSG film on silicon at angles of incidence of 30, 45 and 60 degrees. The data indicates a definite trend in the ability to charge-neutralize the BPSG samples as the angle is altered. The signals obtained from the sample analyzed at 60 degrees are stable throughout the film indicating that full charge-neutralization has been achieved. However, the intensities of the B, P and Si secondary ions from the samples profiled at 30 and 45 degrees are all low at some point in the BPSG layer compared to the signals at the "semiconducting" BPSG/Si interface. This indicates that the samples profiled at 30 and 45 degrees were not fully charge-neutralized. Evidently more electron current is required at these angles to achieve full neutralization. There may be several reasons for the increased ease of neutralization at larger angles. First, the sputter yield increases as the angle of incidence is increased from 30 to 60 degrees. Thus, the current density used to obtain a given sputter rate at 60 degrees is much less than that required to obtain the same sputter rate at 30 degrees. Second, increasing the angle of incidence reduces the penetration of the primary beam into the sample. This may allow the sample to more readily dissipate the positive charge introduced by the oxygen ions. A third explanation is that different amounts of sputter-enhanced secondary electron emission occur at different angles of incidence, thereby changing the number of auxiliary electrons which must be added to fully charge-neutralize the sample. Whatever the reason, this experiment shows that, within the fixed charge-neutralization capabilities of the electron gun involved, larger angles of incidence are preferred when depth profiling insulating samples.

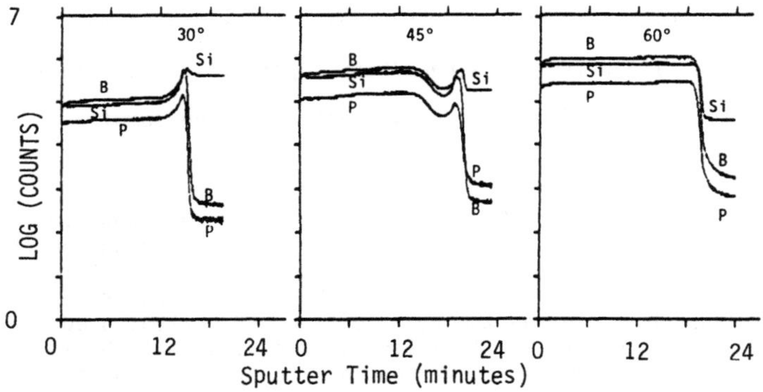

Fig. 1 SIMS depth profiles of 1.0μm BPSG on silicon at angles of incidence of 30, 45 and 60 degrees

Depth resolution is commonly increased in profiling analyses by decreasing the primary ion beam energy and/or by increasing the angle of incidence. This study shows the importance of using fairly large angles of incidence for obtaining good SIMS depth resolution. An MBE grown AlGaAs/GaAs superlattice structure was profiled using a 2 KeV oxygen beam at sputter angles of 30, 45, 60 and 75 degrees. The superlattice structure consisted of a 200Å AlGaAs overlayer on repeating layers of 300Å AlGaAs and 100Å GaAs. Figure 2 depicts the change in depth resolution between adjacent AlGaAs and GaAs layers as the angle of incidence is altered. The resolution is defined in this case as the ratio of the Al ion signal (27 amu) observed in an AlGaAs layer to the minimum Al signal measured in the adjacent GaAs layer. Measurements were taken from the depth profiles at depths of approximately 550Å (first AlGaAs/GaAs interface) and 1700Å (fourth interface) into the superlattice. The increase i resolution with angle is clearly apparent at the 550Å depth, while a much more subtle increase is noted at 1700Å. The reason for the difference in resolutio vs. angle between these two depths is not completely understood, however, it i suspected that changes in surface topography due to preferential sputtering (o

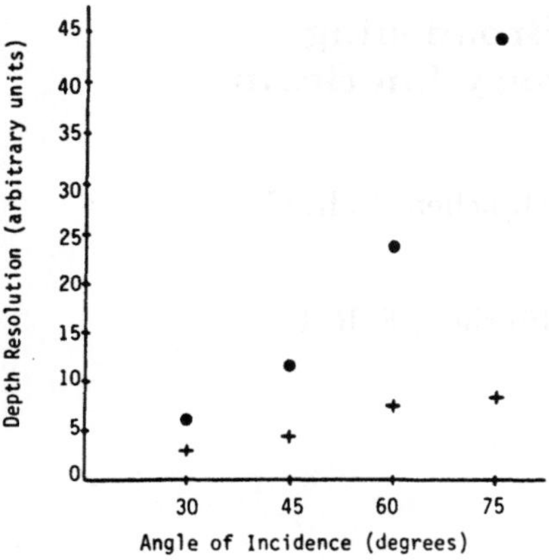

Fig. 2 Depth resolution vs. angle of incidence at depths of 550 Å (●) and 1700 Å (+) into an AlGaAs/GaAs superlattice

other related phenomena) begin limiting the depth resolution as the crater depth is increased.

The need for optimization of the angle of incidence is most apparent in the application of SIMS to the analysis of dopant profiles in semiconductor materials. At a fixed sputter rate and dopant level, the maximum dopant signal intensity obtained at different angles of incidence is governed by two factors: the amount of oxygen incorporated into the sample from the oxygen ion beam and the extraction efficiency of the secondary ion optics. The former of these is optimized at normal incidence, while the latter is optimized at approximately 60 degrees. Figure 3 shows the peak boron secondary ion signal obtained from depth profiles of a 10^{14} atom/cm^2 40 KeV B implant into single crystal silicon at 0, 15, 30, 45 and 60 degrees. The data shows that the maximum boron signal was obtained at an angle of incidence of 30 degrees. This angle is a compromise between the angles at which maximum oxygen incorporation and maximum collection efficiency occur. It should be stressed that the optimum angle of 30 degrees found in this experiment is not necessarily the optimum angle for other SIMS instruments; however, other fixed geometry systems should be characterized at various angles of incidence to make sure that the maximum possible signal is being obtained.

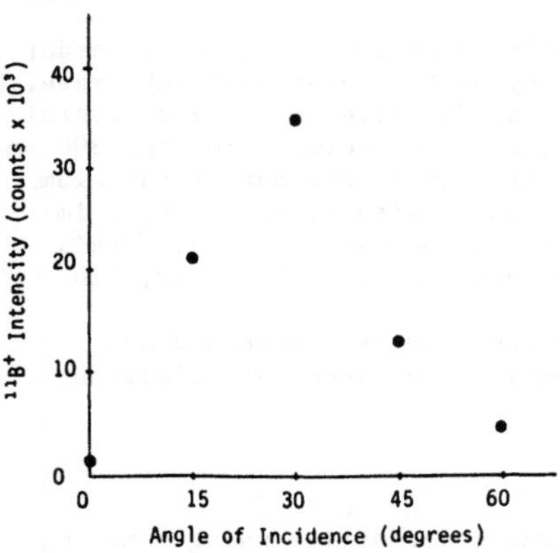

Fig. 3 Peak $^{11}B^+$ secondary ion intensity obtained from the depth profiles of a 10^{14}/cm^2 40 keV boron implant into silicon vs. angle of incidence

Dependence of Sputter Induced Broadening on the Incident Angle of the Primary Ion Beam

H. Frenzel

Atomika Technische Physik GmbH, D-8000 München, F. R. G.

E. Frenzel

Institut für Festkörpertechnologie, D-8000 München, F. R. G.

P. Davies

Intel Corporation, Santa Clara, USA

1. Introduction

A normal incident ion beam is advantageous to obtain highest possible ionisation. It has been previously [1] demonstrated that in case of a SiO_2 layer on Si the Si is completely saturated with oxygen at incident O_2 ion beam to form SiO_2. This leads to the same chemical matrix and differences in ionisation yield need not be corrected.

However, in many analyses different angles may be required. Incomplete oxygen saturation can lead to faster sputtering, which saves analysis time. In addition, the depth resolution is enhanced with 60 or 70 degrees of normal angle because the sputtering cascade develops nearer to the surface [2].

In this paper incidence angle behavior of depth resolution is discussed for silicon and silicides on silicon.

2. Experimental

An A-DIDA 3000 ion microprobe was used in this study. Standard target holders were available to allow the effective incidence angle at the sample surface to be varied between normal (0°) and 60° in 15° increments.

In order to establish the obtainable depth resolution, a MBE grown superlattice structure was analysed that consisted of 19 layers of Si doped with B at $10^{20}/cm^3$ alternating with layers of undoped Si. The periodicity of the layers was 25 nm. The width of the doped regions was expected, according to the MBE growth conditions, to be 1 nm.

The Ta and W-silicides were deposited on polycrystalline silicon. In order to understand the influence of the crystallinity of the underlying substrate, analysis was also performed on tantalum silicides deposited on single crystal Si(100). The thickness of the silicide layer was varied between 100 and 300 nm

To investigate the influence of ion implantation, a second set of tantalum silicides was prepared. These samples were implanted with B, As and P at implantation energies of 50 keV, 80 keV and 80 keV at a dose of $5 \times 10^{15}/cm^2$. A separate set was annealed under Ar and N_2 atmosphere at 900° for 40, 120 and 180 minutes.

SIMS analyses were performed using primary ion energies between 3 and 12 keV. Surface roughness was determined by optical and secondary electron microscopy.

3. Results and Discussion

Figure 1 shows the SIMS depth profile of the MBE superlattice using 3 KeV O_2 primary ions at 60° off normal angle. The B peaks appear broader than the 1 nm expected from the MBE growth. The half-width of the peaks is 6.5 nm. The

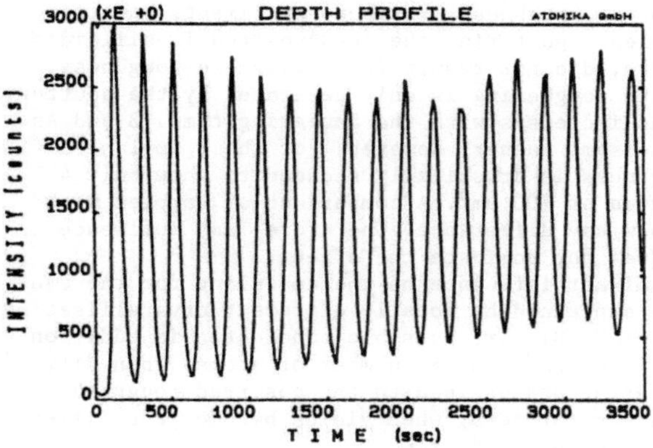

Fig.1 B doped MBE multilayer measured with 3 KeV O_2 at 60° angle.

corresponding widths for 45° and 0° are 7.5 and 10 nm respectively. The decrease of the measured width with angle demonstrates the influence of cascade mixing. However, no improvement was made in reducing the ion energy to 1.5 KeV or using 3 KeV Cs at 60°. It is known from literature [3] that the depth resolution in Si using 2 KeV Cs at 60° is 0.9 nm. It must therefore be concluded that the width expected from MBE growth condition is too small and the true structure is about 6 nm. Therefore the width of 6.5 nm must be interpreted as upper limit for the obtainable depth resolution with 3 KeV O_2 and 60° angle.

In case of silicides the observed interface widths are much broader than expected from cascade mixing. A change of ion energy from 12 KeV to 5 KeV did not lead to any improvement.

With normal incidence ion beam, after sputtering through W-silicides the transition region (90 % to 10 % of the W signal) was 10 nm; for 45 degrees it was 30 nm. This trend of broadening is opposite to the angle behavior that one expects from cascade mixing.

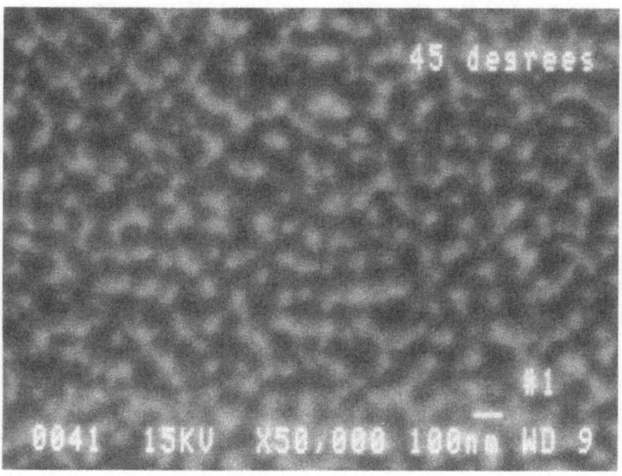

Fig.2 Roughness generated by sputtering of W-silicides at 45 degrees angle.

An SEM micrograph of the surface of the sputter crater, obtained with 45 degrees, shows a considerable roughness (Fig.2). In contrast, similar micrographs obtained after sputtering at normal incidence did not show a detectable roughness at magnifications as high as 50 000. It is concluded that roughening by sputtering is responsible for the interface broadening.

Sputtering of Ta-silicides also exhibited the generation of sputter-induced roughness. However, it was found that the roughness was maximum at normal incidence and invisible by measuring at 45 degrees.

To understand the generation of roughness, further experiments were performed on the Ta-silicide samples. Sputtering the as-deposited Ta-silicides, which were known to be amorphous, did not result in any visible roughness, independent of the donation. The roughening is only generated by the sputtering of the annealed samples and increases with the annealing time. B and As implanted samples behave in the same manner, however, for the P implanted samples the generation of roughness is drastically reduced to show only 4 holes of 1 µm diameter in an area of 100 μm^2. A comparison of samples with poly Si and Si(100) did not show any difference. Therefore, any influence of the silicide-Si interface does not dominate the effect.

The experiments show the dominant influence of the annealing for the evolution of roughness. It may be explained by locally different crystallisation to different phases (Ta_n; Si_n or Ta_2Si_5 are possible candidates in addition to $TaSi_2$) because of fluctuations in Ta and Si number of atoms. Then different sputter yields of the different phases lead to the observed roughness. The generated roughness is smaller (or zero) when yields become less different at smaller oxygen up-take by using 45 degrees.

The abnormal behavior of P cannot be explained without further experiments but apparently P protects the layers against inhomogeneous crystallisation.

For the W-silicides the different angle behavior can be explained by assuming that the enhanced oxygen up-take at normal incidence equalizes the different yields of the different phases present. Different phases are likely for these films because Rutherford backscattering measurements result in a stoichiometry of $WSi_{2.2}$, that means an enrichment of silicon.

However, the different mechanisms are still unclear. More experimental work is necessary to gain a complete understanding of the chemistry of the silicides in connection with the mechanism for the generation of roughness. Currently it is not possible to predict the optimum ion beam angle of incidence for a given silicide. At the present understanding of roughening the unwanted effect can only be minimized by trying experimentally different angles.

4. Conclusion

It was shown that a depth resolution better than 6 nm can be obtained for B in a Si matrix with 3 keV at glancing incidence. However, for silicides on Si the limitation for depth resolution is the generation of sputter-induced roughness. This can be minimized by varying the angle between primary ion beam direction and surface normal.

References

1 H. Frenzel et al, Proceedings of SIMS IV, Springer Verlag, Springer Series in Physics, 36, 241 (1984)
2 H. Frenzel, PhD thesis, Technical University Aachen, FRG (1980)
3 C. McGee and R.E. Honig, Surf. Interf. Anal. 4, 35 (1982)

SIMS Depth Profiling with Oblique Primary Beam Incidence

R. v. Criegern and I. Weitzel

Siemens AG, Research Laboratories, Otto-Hahn-Ring 6,
D-8000 München 83, F. R. G.

1. Introduction

Initiated by the demand for improved depth resolution, we have carried out a series of depth profiling experiments at various angles of incidence between $\beta = 0^{o}$ and $\beta = 85^{o}$, while maintaining a high primary beam energy, namely 12 keV. In order to reduce the experimental difficulties involved with high angles of incidence, we took advantage of our "mini-chip" sample preparation technique and our "checkerboard" data evaluation technique [1].

2. Experimental

2.1 SIMS - Instrument

Quadrupole-type (ATOMIKA GmbH); primary ion beams used: 12 keV O_2^+ and 12 keV Xe^+ (natural mixture of isotopes).

2.2 Sample Preparation and Positioning ("Mini-Chip"-Technique, [1])

The samples were sawn into small square-shaped or rectangular pieces and glued onto the tip of steel needles. A specially designed sample holder wheel would allow the insertion of up to 15 such needles, each of these indented corresponding to one fixed angle of incidence between $\beta = 0^{o}$ and 85°. The axis of inclination was chosen perpendicular to the primary beam and perpendicular to the quadrupole axis (**Fig. 1**), so that the inclined samples were facing towards the entrance aperture of the energy filter. For depth profiling, the size of the raster scan of the primary ion beam was set to be larger than the apparent size of the inclined sample surface.

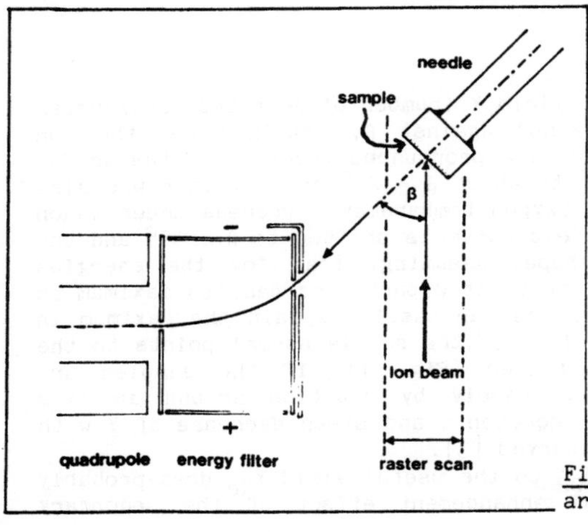

Fig. 1 Schematic of the experimental arrangement (not drawn to scale)

2.3 Test Samples

For the determination of the depth resolution, samples with a steep concentration step of some dopant (^{11}B or ^{75}As) were prepared, making use of well-established processes of the semiconductor manufacturing, e.g. the polysilicon diffusion source technique [2]. A highly doped polysilicon layer was deposited on top of an undoped silicon substrate with a thin (3 nm) SiO_2-layer between.

3. Results

3.1 Sputtering Yield

As shown in **Fig.** 2, the sputtering yield of Si for both, O_2^+ bombardment and Xe^+ bombardment increases rapidly with increasing angle of incidence, β. Accordingly, the time required for depth profiling through a certain silicon layer thickness is clearly reduced, e.g. for O_2^+ at $\beta = 60^o$ by a factor of about five, as compared to normal beam incidence. For β larger than 70°, the sputtering time increases again, because of the ion current density reduction following cos β. Nevertheless, even for $\beta = 85^o$ the sputter time required for a certain layer thickness is shorter than it is for $\beta = 0^o$.

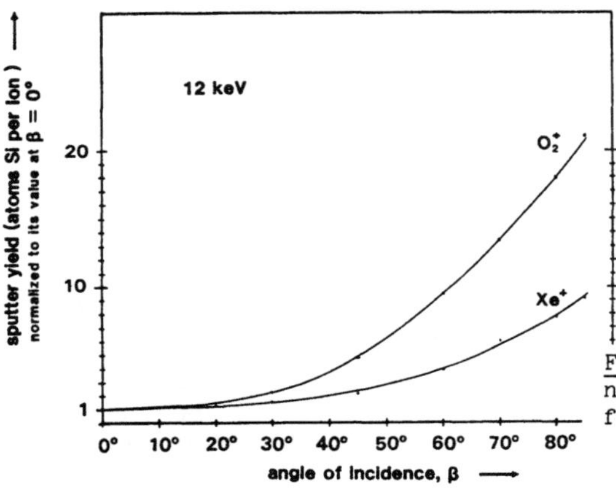

Fig. 2 The sputter yield for silicon normalized to its value at $\beta = 0^o$, for 12 keV O_2+ and 12 keV Xe^+

3.2 Useful Yield

Fig. 3 shows a plot of the useful yield Y (number of detected ions/number of sputtered atoms of the same kind) against β, obtained for the ion species Si^+, SiO^+, and Si_2O^+. There is a pronounced increase of the useful yield Y for the regime from $\beta = 0^o$ to about $\beta = 25^o$ or 30°. This was also found for Boron B^+ (**Fig.** 4) under oxygen bombardment, whereas under Xenon bombardment the maximum of the Y yield curve is shifted to $\beta = 45^o$ and the curve has a broad, symmetrical shape. Assuming, that for the energies involved here, the angular density of emitted particles has its maximum in the direction of the sample normal, one can easily explain the maximum in the yield curves, since for about $\beta = 45^o$, the sample normal points to the entrance aperture of the energy filter (**Fig.** 1). If the samples are inclined in a different direction, namely by rotation around an axis parallel to the quadrupole axis, a monotonic and steep decrease of Y with increasing β right off $\beta = 0^o$ is observed [3].

The ratio of the useful yield Y_{O2} to the useful yield Y_{Xe} does probably essentially represent the chemical enhancement effect of the secondary

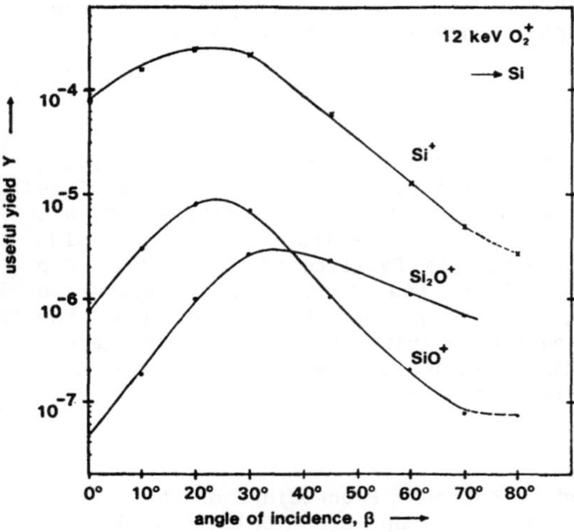

Fig. 3 The useful yield for Si^+, SiO^+, and Si_2O^+ (12 keV O_2^+ onto Si)

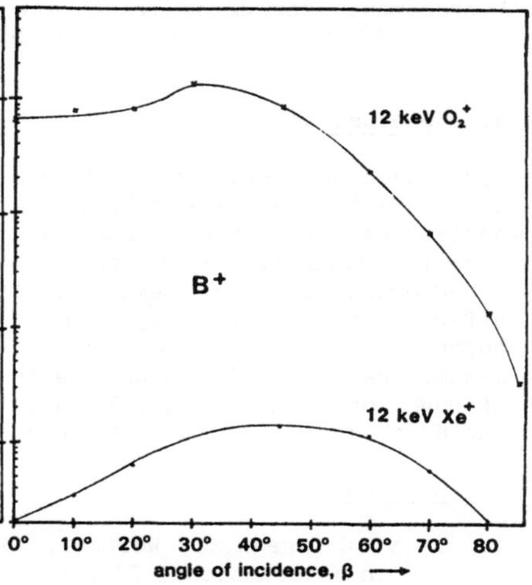

Fig. 4 The useful yield for B^+ (for 12 keV O_2^+ and Xe^+ onto Si)

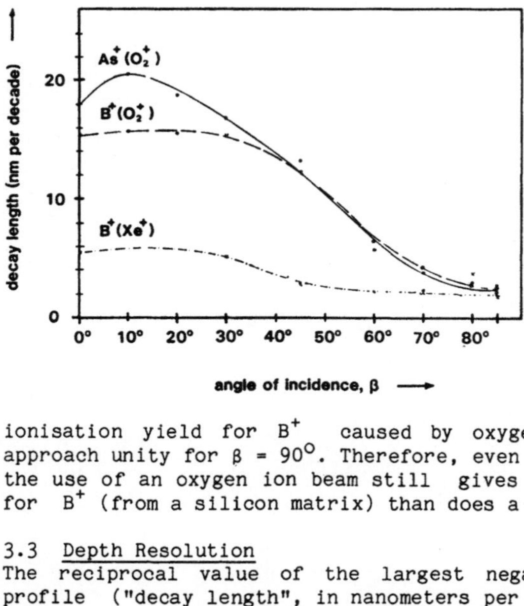

Fig. 5 The decay length for Boron in Silicon (for 12 keV O_2^+ and Xe^+) and for Arsenic in Silicon (12 keV O_2^+), determined at a sharp concentration step with an SiO_2 interlayer

ionisation yield for B^+ caused by oxygen. Remarkably, it does not approach unity for $\beta = 90^o$. Therefore, even for high angles of incidence, the use of an oxygen ion beam still gives a much higher useful yield for B^+ (from a silicon matrix) than does a Xenon beam.

3.3 Depth Resolution

The reciprocal value of the largest negative signal gradient in each profile ("decay length", in nanometers per decade signal drop) is given in **Fig. 5** for the dopants B^+ and As^+. For oxygen bombardment, both curves are similar, only the decay length for As being larger than for B in the regime between $\beta = 0^o$ and $\beta = 30^o$. The curves follow only roughly the cos β law, that one would expect from the reduction of the normal component of the implantation range of the primary ions. A reduction of the decay length by at least a factor of 7 is obtained for $\beta = 80^o$ and 85° as compared to $\beta = 0^o$. At about 2 nm per decade a limitation is probably set by tolerances in the uniformity of the layer thickness of the test samples. For Xenon bombardment, the decay length for Boron is as low as 6 nm per decade even for $\beta = 0^{o}$ and it drops below 3 nm per decade already at $\beta = 45^{o}$.

4. Conclusions

SIMS depth profiling under oblique primary beam incidence while maintaining a high primary ion energy may be a solution for the analyst in a surface analytical application laboratory, who cannot afford to sacrifice analysis time for depth resolution or who does not have available low-energy SIMS or plasma-SNMS [4]. Sample positioning is facilitated and crater edge effects are reduced by application of the "mini-chip" preparation technique. The observed decrease of the mass resolving power and of the dynamic range could probably be in part compensated for, if the arrangement of the ion gun, the sample and the quadrupole were optimized for high angles of incidence instead for normal incidence.

5. References

1. R. v. Criegern, I. Weitzel, and J. Fottner, Proceedings of the Fourth International SIMS Conference, 308, Springer Verlag (1984)
2. A. W. Wieder, Siemens Forsch. u. Entwickl.-Ber., Vol.13, 5 (1984)
3. K. Wittmaack, Nuclear Instr. and Methods, 218, 307, (1983)
4. H. Oechsner et al., J. Vac. Sci. Technol. A3, 3, 1403 (1985)

Deterioration of Depth Resolution in Sputter Depth Profiling by Raster Scanning an Ion Beam at Oblique Incidence and Constant Slew Rate

U. Kaiser[1], *R. Jede*[1], *P. Sander, H.J. Schmidt, O. Ganschow*[1], *and A. Benninghoven*

Physikalisches Institut der Universität Münster, Domagkstr. 75, D-4400 Münster, F. R. G.

1. Introduction

One of the most important parameters describing the performance of thin film analytical instruments is the depth resolution. It is determined by a number of physical factors like atomic mixing [1], surface roughening by ion bombardment [2] or natural substrate topography. In addition, artefacts like cratering and memory effects deteriorate depth resolution. In this paper we deal with a geometrical distortion of depth resolution which occurs at oblique incidence of the primary ion beam. This effect is in principle well known, but its influence on depth resolution has never been discussed, although it is present for instance in most thin film analysis instruments.

2. Ray tracing

Consider a scanning ion beam geometry as depicted in fig.1.

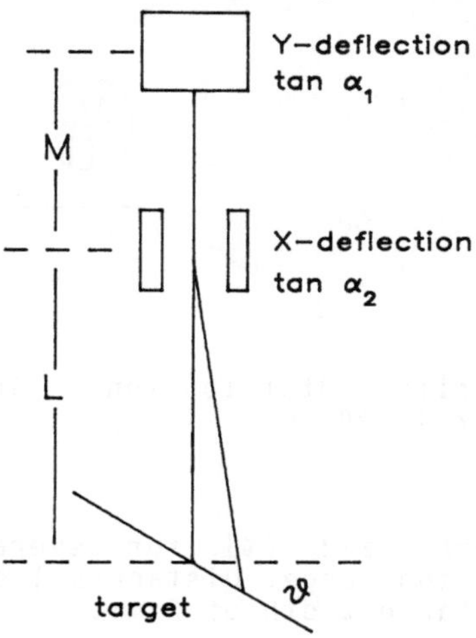

Fig.1: Scanning ion beam geometry

(1) **representation of ion beam:**

$$\begin{pmatrix} 0 \\ M \tan\alpha_1 \\ 0 \end{pmatrix} + w \begin{pmatrix} \tan\alpha_2 \\ \tan\alpha_1 \\ -1 \end{pmatrix}, \quad w \in \mathbb{R}$$

(2) **representation of target surface:**

$$\begin{pmatrix} 0 \\ 0 \\ -L \end{pmatrix} + u \begin{pmatrix} \cos\vartheta \\ 0 \\ -\sin\vartheta \end{pmatrix} + v \begin{pmatrix} 0 \\ 1 \\ 0 \end{pmatrix}$$

Moving the beam from its center position in both directions by the same $\tan\alpha_i$ leads to new spot positions with different distances from the center position. Using the matrix representations of the ion beam (1) and the target plane (2), we obtain the beam position in the target plane as a function of the deflection angles:

[1] now with Leybold Heraeus, Cologne

$$(3)\ u = \frac{L}{\cos\vartheta}\ \frac{\tan\alpha_2}{1-\tan\vartheta\tan\alpha_2} \qquad v = \frac{L\tan\alpha_1}{1-\tan\vartheta\tan\alpha_2} + M\tan\alpha_1$$

With $\tan\alpha_i \sim t$, i.e. constant slew rate scanning, the current distribution on the target plane will be:

$$(4)\qquad j(u,v) \sim \frac{L\cos\vartheta}{(L+u\sin\vartheta)^2\,(L+M+u\sin\vartheta)}$$

This kind of inhomogeneity leads to a linear variation of depth resolution Δz with the eroded depth z. Fig.2a shows the result of a ray trace calculation for ϑ =70°, L=M=20mm and a raster scan area of 8x8 mm, which is typical for combined method machines using SIMS together with XPS. The current density varies more than a factor of 3 in the oblique incidence distribution. In addition, the trapezoidal shape of the crater is seen from this calculation, its contribution to inhomogeneity is, however, less than 5%. The inspection of a number of depth

a)

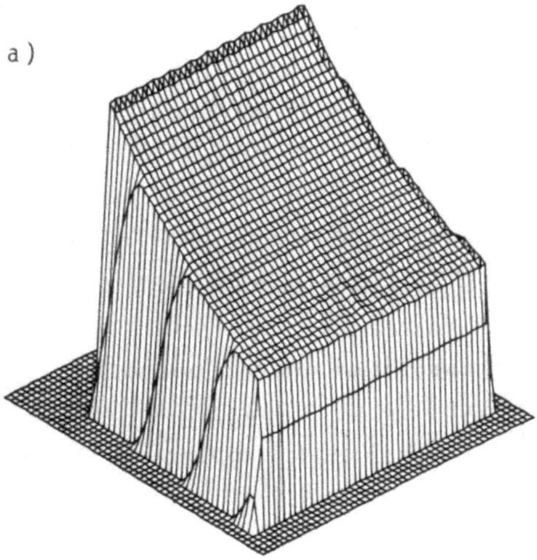

b)

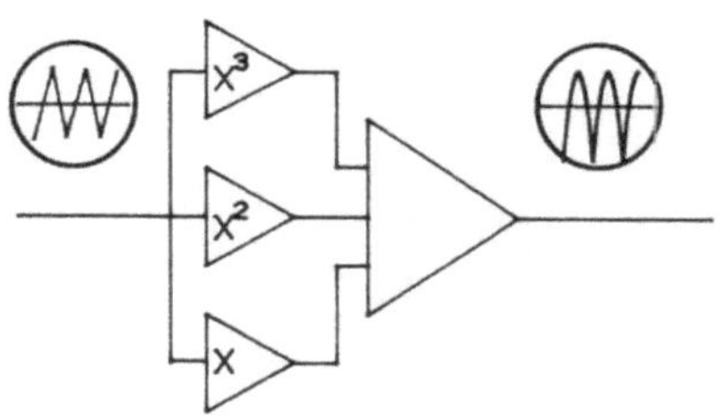

Fig.2: (a) Calculated current density distribution. The displayed area extends over 10x10 mm.
(b) Correction circuit.

Table 1: Calculated depth resolutions $\Delta z/z$ [%] for several incident angles ϑ and deflection target distances l of a typical spectrometer acceptance width of 2 mm.

l[mm] \ ϑ[°]	10	20	30	40	50	60	70	80
10	7	14	20	26	31	35	38	39
20	3	7	10	13	15	17	18	20
30	2	3	5	7	8	9	9	10
40	1	2	3	3	4	4	5	5
50	0	1	1	2	2	2	2	3

profiles taken under those conditions in a combined SIMS-AES-XPS machine [3] told us that the observed depth resolutions were completely reproduced by the calculated data (adjusted for the different acceptance areas) without taking into account other effects like atomic mixing and roughening. Table 1 shows corresponding data for a number of other geometries.

3. Experimental

In order to assess the effect experimentally, we designed a correction by the shaping circuit depicted in fig.2b and implemented it into our raster scan generators. It provides a third-order polynomial correction to flatten j(u,v). For practical purposes, instead of calculating optimized coefficients, an experimental procedure can be used for adjusting the parameters. This is advisable, in particular in situations where the exact geometry is not known.

(a) The deflection sensitivity in the oblique incidence plane is determined in terms of volts/mm.
(b) The beam position on the target plane is recorded as a function of the deflection voltage using a Faraday cup.
(c) After multiplying the ordinates of the curve (b) by the deflection sensitivities (a), a polynomial fit of the data set is taken using a least square method.
(d) Finally, the coefficients from the fit procedure (c) are fixed in the electrical circuit.
(e) To verify the correct procedure, a repetition of (b) must lead to a linear dependence of the beam position on deflection voltage.

4. Application

Fig.3 shows the Ni depth profile based on XPS of a Cr-Ni sandwich structure, fig.4 the XPS and SIMS profiles in the

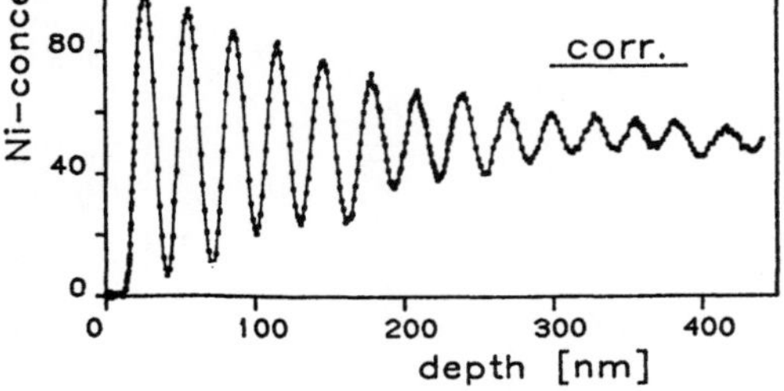

Fig.3: Ni depth profile of a Cr-Ni sandwich structure.

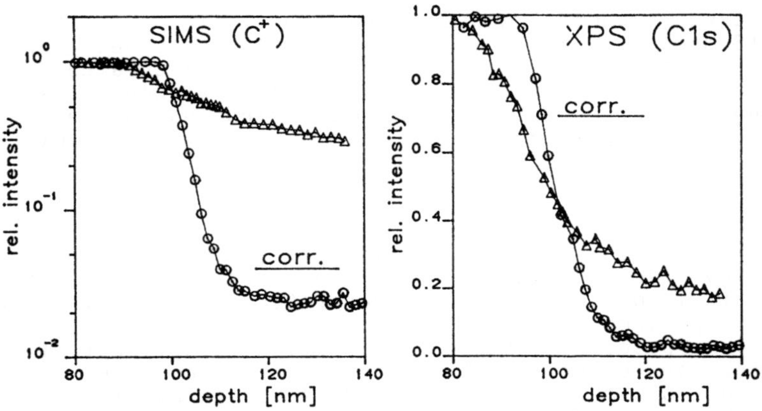

Fig.4: SIMS and XPS profiles in the interface region of a hard carbon layer in silicon.

interface region of a hard carbon layer on silicon [4] with and without correction. The scan width of the primary ion beam is large enough to avoid any crater edge effects. Although this correction method is only an approximate one and does not consider finite ion beam width, all profiles exhibit an obvious improvement in depth resolution by this simple and easily implementable method.

References

1. U.Littmarck, W.O.Hofer, in: Thin Film and Depth Profile Analysis (ed.: H.Oechsner), Topics in Current Physics Vol.37 (Springer, Berlin, Heidelberg, New York, Tokyo 1984).
2. H.J.Mathieu, D.E. McClure, D.Landolt, Thin Solid Films 38, 281 (1976).
3. U.Kaiser, P.Sander, O.Ganschow, A.Benninghoven, Fresenius Z. Anal. Chem. 319, 877, (1984).
4. P.Sander, U.Kaiser, O.Ganschow, A.Benninghoven, these proceedings.

SIMS Analysis of Contamination Due to Ion Implantation

F.A. Stevie[1], *M.J. Kelly*[1], *J.M. Andrews*[2], *and W. Rumble*[3]

[1] AT & T Bell Laboratories, 555 Union Boulevard, Allentown, PA 18103, USA
[2] AT & T Bell Laboratories, Murray Hill, NJ 07974, USA
[3] AT & T Bell Technology Systems, 555 Union Boulevard, Allentown, PA 18103, USA

1. Introduction

The introduction of metallic impurities into a VLSI process can adversely affect device yield and performance [1-4]. Ion implantation would be expected to be relatively free of contamination because of the mass filtering of the ion beam. However, several studies [5-7] indicate that an ion implantation machine can introduce contamination due to sputtering of apertures and other components by the ion beam. Many ion implantation parts are manufactured from commercial grades of stainless steel or Al alloys. Therefore, Fe, Ni, Cr, Al, and Mg are possible contaminants. Other elements commonly present in parts exposed to the beam and susceptible to sputtering are C, Cu, and Ta.

2. Experimental

Si wafers were implanted in PR-30, Extrion 80-10, Extrion DF-5, and AIM-210 ion implantation systems with 10^{16} $^{75}As^+$ ions/cm^2 accelerated to 30 keV.

SIMS analyses were made with a CAMECA IMS-3F ion microanalyzer with a 500 namp O_2^+ primary beam at 6.1 keV net energy and a sputter rate of 1 Å/sec. Some analyses were repeated at 2.5 Å/sec to verify that any redeposition of background gas species did not affect the results. Samples were analyzed with and without the voltage offset technique to determine the effect of possible interferences from polymeric ions. No significant contribution to m/e 27 was detected from $C_2H_3^+$ or BO^+. Similarly, interferences at m/e 24 from C_2^+ and at m/e 52 from SiC_2^+ were discounted. Profiles taken on $^{12}C^+$ showed above background levels at the surface only and did not match the shape of profiles on the metallic species. $^{52}Cr^+$ was used as an indicator for stainless steel because of Si interferences at the primary isotopes of Ni and Fe. Absolute concentrations for Al, Mg, and Cr were determined from calibration implants with respective detection limits of 1×10^{16}, 5×10^{14}, and 7×10^{14} atoms/cm^3.

3. Results

Al, Mg, and Cr were detected in implants from all four implantation systems. Figures 1 and 2 show the contaminant profiles for the PR-30 and Extrion DF-5 implants. The AIM-210 and PR-30 results were similar, as were the Extrion 80-10 and Extrion DF-5. The first 50 Å was not plotted because of the effects of native oxide and primary beam implant range on secondary ion yields. Data points were taken about every 30 Å. The Al/Mg ratio is approximately 100 which correlates with the ratio usually present in construction grade Al alloys [4]. The Al profile represents a flux of about 4×10^{13} ions/cm^2. The detection of Cr implies the presence of Fe and Ni which are the other major components of stainless steel. A

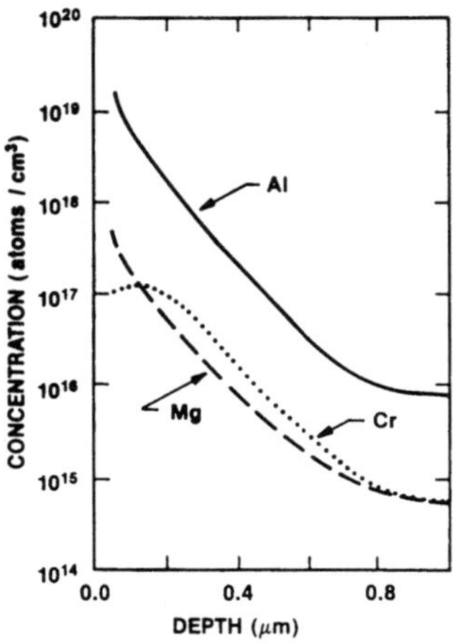

Fig. 1 SIMS depth profile of contaminants from a PR-30 implant

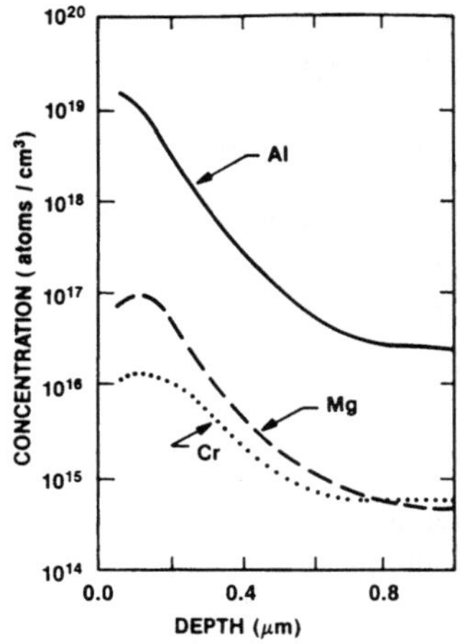

Fig. 2 SIMS depth profile of contaminants from an Extrion DF-5 implant

typical composition would be Fe:Cr:Ni at 4:1:0.5. C, Cu, and Ta were also profiled, but no consistent level above background was detected. Auger analysis of these implants showed no detectable metallic impurity.

Wafers subjected to an H_2SO_4 and H_2O_2 etch which would remove the first 5 to 10 Å of material showed identical profiles as unetched wafers. Therefore the profiles are not the result of knock-on during analysis of a species present only in the first few monolayers.

To further understand the contamination sources, unrastered beam profile implants were made for the PR-30 and Extrion machines. For these samples, the equivalent

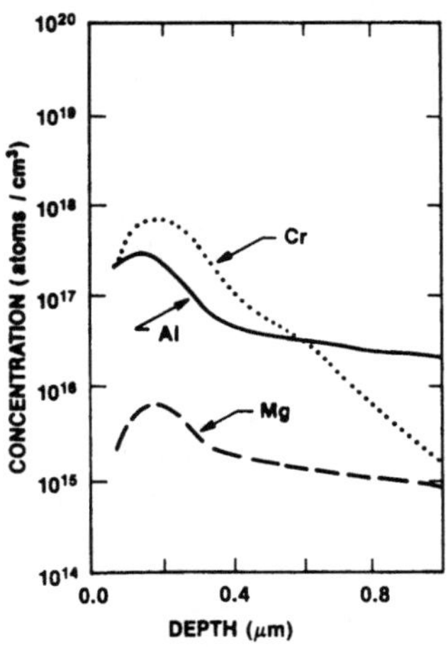

Fig. 3 SIMS depth profile of contaminants from a PR-30 unrastered beam implant

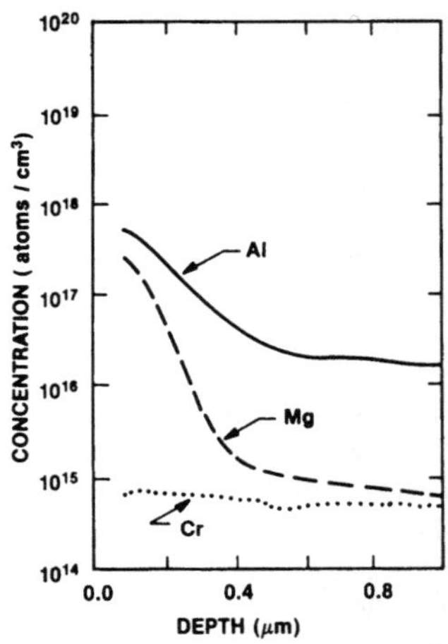

Fig. 4 SIMS depth profile of contaminants from an Extrion DF-5 unrastered beam implant

of a flux of 10^{16} ions/ cm^2 over the entire wafer was concentrated into an area of about 1 cm^2. Figures 3 and 4 show that the Al and Mg levels drop sharply for the unrastered PR-30 and Extrion DF-5 implants when compared to the rastered implants in Figures 1 and 2. The Extrion DF-5 Cr signal disappears entirely in the unrastered mode but the PR-30 Cr signal increases. The PR-30 Cr is due to stainless steel apertures and is increased because the implant was concentrated into a smaller area.

The source of the Al and Mg contamination appears to be deposition at low energy followed by knock-on from the implant species. The shape of a knock-on profile is an exponential decay, which is identical to the Al and Mg profiles observed. Figure 5 shows the effect of a higher energy As implant on the range of the Al impurity. The Al from the 50keV As implant has a deeper profile than the Al from the 30keV As implant. The Cr profiles have a larger high-energy component which would be expected to originate from beam line components.

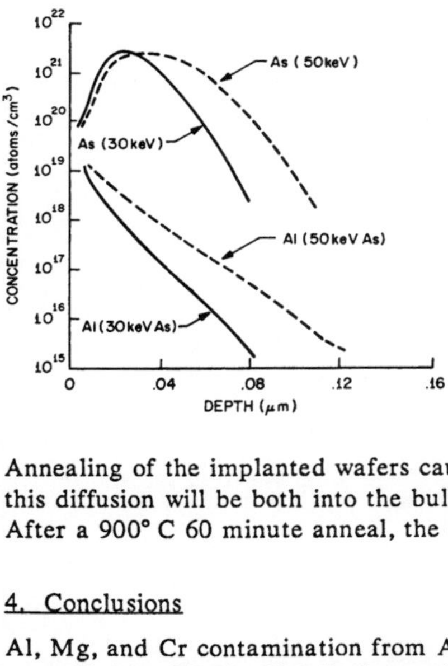

Fig. 5 Dependence of Al implant contamination on energy of implant species

Annealing of the implanted wafers causes diffusion of the contaminants. For Al, this diffusion will be both into the bulk and toward the surface of the wafer [8]. After a 900° C 60 minute anneal, the contaminants were no longer readily detected.

4. Conclusions

Al, Mg, and Cr contamination from As ion implants has been detected and quantified by SIMS. The Cr detected is evidence for the existence of the other stainless steel components, Fe and Ni. Much of the contamination is caused by rastering the ion beam in the implantation system and may vary with erosion of beam line components.

5. Acknowledgements

The authors acknowledge J. B. Bindell and M. J. Vasile for helpful comments and L. E. Plew and T. C. Shane for Auger analysis.

1. Helmut F. Wolf: Silicon Semiconductor Data (Pergamon, Oxford 1969) p. 133-159
2. J. M. Andrews: J. Vac. Sci. Technol. 11, 972 (1974)

3. J. M. Andrews, L. E. Katz, and J. W. Colby: Thin Solid Films 67, 325 (1980)
4. J. M. Andrews and J. M. Morabito: Thin Solid Films 37, 357 (1976)
5. E. W. Haas et al.: Jour. Elect. Mat. 7, 525 (1978)
6. M. Y. Tsai et al.: J. Electrochem. Soc. 126, 98 (1979)
7. A. C. Yen, A. F. Puttlitz, and W. A. Rausch: Secondary Ion Mass Spectrometry, SIMS IV (Springer, Berlin 1984) p. 270
8. H. B. Dietrich, W. H. Weisenberger, and J. Comas: App. Phys. Lett. 28, 182 (1976)

Analysis of Surface Contamination from Chemical Cleaning and Ion Implantation

G.J. Slusser

IBM General Technology Division, Essex Junction, VT 05452, USA

Introduction

Secondary ion mass spectrometry (SIMS) has played an increasingly larger role in the development cycle of semiconductor products. This technique has become deeply involved in the determination of dopant distributions throughout the semiconductor films. At the same time, surface cleanliness requirements have in many cases fallen below the detectability limits of Auger and ESCA into the realm of SIMS.

The requirements for surface contamination studies are usually quite different from those for in-depth profiles. Surface contamination studies require near-static ion beam currents. Quantification of surface levels can be complicated by ion yield changes as a result of surface impurity levels. Standards generation is near impossible.

WILLIAMS et al [1] have suggested deposition of another film above the surface layer, in effect burying the surface and transforming it into an interface. While interfaces themselves present problems, they are typically easier to handle, particularly when the deposited layer is tailored to the analysis. In this report, surface contamination layers on silicon substrates are described in which the buried surface technique is used. Using aluminum and carbon impurities as examples, this technique is found to have several advantages over direct surface analysis methods. Aluminum poses a threat to semiconductor devices when a dual dielectric is used [2], while carbon is a concern for contact resistance.

Experimental

Bare silicon wafers were submitted to an arsenic ion implant at a dose of 1E16 atoms/cm^2 and an energy of 120 keV. One wafer from a lot was held as a control with no implant. Immediately after ion implant, the wafers were coated with a thin (30-100 nm) film of either polysilicon deposited by low pressure chemical vapor deposition (LPCVD) or pyrolytic silicon dioxide. Both films were found to be suitable for the detection of aluminum, while pyrolytic oxide was found to be superior for carbon due to a carbon contamination (2E19 atoms/cm^3) in the LPCVD polysilicon film. The control wafer was used to monitor nonimplant-induced contamination.

SIMS analysis was performed utilizing a Cameca IMS-300 SIMS unit. An O_2^+ ion beam at an apparent energy of 5.5 keV was rastered across a 1E6-µm^2 area of the samples. Secondary ions were collected from a centered 70-µm diameter circle. Current density was in the vicinity of 1 to 5 µamps/cm^2. An oxygen bleed was supplied at the sample surface for maximum sensitivity. ^{27}Al and

^{12}C were ratioed to ^{30}Si. Concentrations and doses were determined by comparison to aluminum and carbon ion implants into silicon.

Results and Discussions

Figure 1 displays an example of SIMS profiles of aluminum contamination found on a silicon wafer with and without the use of the buried surface technique. Without the buried surface technique, it is very difficult to determine the actual amount of aluminum present on the surface. A drop in current density is necessary to reproducibly determine the aluminum coverage. This impacts the detection limits of aluminum severely. In addition, memory effects present in SIMS instrumentation such as used here can disguise the real aluminum contamination level. The first point on Fig. 1A is due to memory effects.

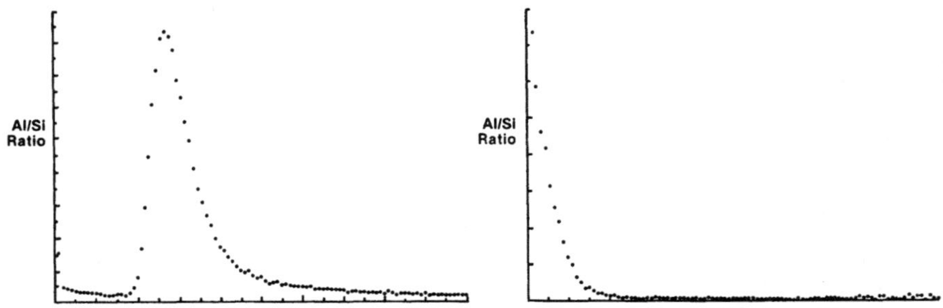

(A) With film deposition. (B) Without film deposition.

Figure 1. SIMS aluminum signal from silicon wafer.

Using the buried surface technique, the aluminum contamination originally on the surface is easily separated from the memory effects. A quantitative determination of the aluminum coverage is possible using either a separate ion implant standard of aluminum or the solid-state standard addition method [3].

Table I shows SIMS measurements for carbon and aluminum on implanted and unimplanted silicon wafers using the buried surface technique. Carbon is particularly susceptible to instrumental conditions. Without a layer of pyrolytic oxide over the as-implanted wafer, carbon detection limits at the surface are 10 to 50 times worse due to masking from vacuum contamination.

The aluminum data in Table I show the value of using a control. The calculated aluminum doses are very high, even on the control wafer. In reality, the majority of the aluminum contamination has been deposited by the precleans used prior to ion implant. Approximately 2E12 atoms/cm^2 have been sputtered away during the ion implant, as noted by the difference in aluminum contamination between the control wafer and the wafer ion implanted using a stainless steel tube. The aluminum contamination coverage actually deposited during ion implantation is only 7E12 atoms/cm^2.

The buried surface technique has been used extensively in this laboratory for the analysis of other contaminants deposited by both ion implantation and chemical cleans. For instance, the aluminum contamination found in the above study was traced to the hydrogen peroxide used in the cleaning steps prior to ion implantation. Metallic contamination from stainless steel is

Table I. SIMS Measurements for Carbon and Aluminum

Wafer ID	Carbon Coverage (atoms/cm^2)	Al Coverage (atoms/cm^2)
Control	< 4E11	4.7E13
Ion implanted with Al flight tube	1.6E13	5.2E13
Ion implanted with stainless steel flight tube	1.1E13	4.5E13

especially suited for this method of analysis, since many of the components of stainless steel (i.e. iron and nickel) have severe interferences with silicon cluster ions. Because high mass resolution is required, sensitivities are decreased and higher current densities are desired.

The choice of the coating is very important. If the matrix is not identical to the material over which the film is deposited, it is necessary to assure that ion yields are identical. For instance, it is not advisable to deposit a copper film over a chromium layer, since ion yields with oxygen are orders of magnitude different. Contamination levels in the film to be deposited must be lower than the contamination levels of concern at the surface. Finally, the physical and chemical conditions upon which the film is deposited must not be severe enough to cause the contaminant of interest to be affected. In the above study, the high temperatures necessary to deposit the pyrolytic oxide or LPCVD polysilicon may have caused removal of some of the carbon. However, these temperatures were appropriate, since the concern was the quantities of material left for the next processing step and the conditions were similiar to the next processing step. In all cases, the control wafer is of extreme importance in monitoring these effects.

Conclusions

Aluminum and carbon contamination from ion implantations and chemical cleans are monitored using the buried surface technique. The choice of the covering layer is dictated by the underlying matrix, the element to be analyzed and the analysis conditions. The importance of a control wafer undergoing identical processing is emphasized.

References

1. P. Williams, J. E. Baker, J. A. Davies, J. E. Jackman, Appl. Phys. Lett., 36(10), 842 (1980).
2. C. Falcony, D.J. DiMaria and C.R. Guarnieri, J. Appl. Phys., 53(7), 5347 (1982).
3. P. K. Chu and G. H. Morrison, Analytical Chemistry, 54, 2113-4 (1982).

Quantification of SIMS Dopant Profiles in High-Dose Oxygen-Implanted Silicon, Using a Simple Two-Matrix Model

G.D.T. Spiller and J.R. Davis

British Telecom Research Laboratories, Martlesham Heath, Ipswich IP5 7RE, UK

1. Introduction

Silicon implanted with high doses of oxygen (in excess of 1×10^{18} atoms cm^{-2}) is a strong contender for the fabrication of silicon on insulator (SOI) devices and very large-scale integrated circuits. With these massive oxygen doses, a buried stoichiometric oxide layer is formed beneath the surface, the thickness and depth of which depend on the dose and energy respectively. After subsequent annealing to remove implantation damage, some redistribution of the oxygen outside the oxide layer occurs, forming a complicated profile which has been investigated extensively by Auger electron spectroscopy (AES) and transmission electron microscopy (TEM) [1]. Implants of B and As are then used to create p- and n-type regions respectively in the material.

SIMS analysis of this material is complicated by the varying oxygen concentration, which results in a change in the degrees of ionisation of dopant and matrix atoms alike. Oxygen profiles recorded by SIMS have been reported by Kilner [2], while Homma [3] has quantified boron implant profiles in this material using an empirical relationship between the B ion yield and the oxygen signal measured in a range of oxygen-doped silicon layers. In the present work, a two-matrix model is developed to quantify boron profiles, incorporating the B ion yields in Si and SiO_2.

2. Experimental and Results

Depth profiles were recorded in a Cameca IMS 3F, using 10.5 keV (net) O_2^+ ions, incident at ~40 deg to the normal, and recording B(11), O(16) and Si (29). An electron flood gun (left running throughout the profile) was used for charge neutralisation in the buried oxide.

Figure 1 shows the profile recorded from an unannealed 2×10^{15} cm^{-2}, 35 keV B implant in Si previously implanted with 1.8×10^{18} cm^{-2}, 200 keV O^+ ions and annealed at 1150°C to remove implantation damage. Unlike oxygen profiling at normal incidence [4], saturation of the ion yields does not occur in this case, and increases in the B and Si signals are observed in regions of high oxygen concentration. However, two fortuitous features arise. Firstly, the sputter rate is approximately constant throughout the material: independent measurements on Si and SiO_2 gave a ratio of 1.07, the rate being slightly higher in the oxide. Secondly, the oxygen profile measured by SIMS is similar to the true O concentration profile, measured by AES (Fig. 2). These two features can be used to advantage to quantify the boron profile.

3. The Quantification Model

The boron ion signal can be quantified if the variations in the ion yield, due to the presence of oxygen in the material, can be accounted for. The

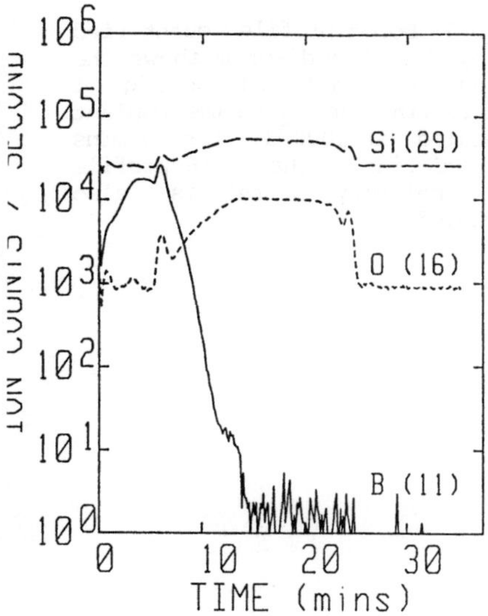

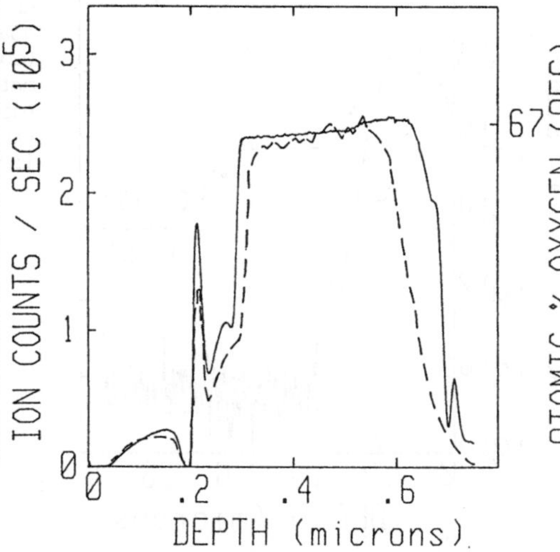

Fig. 1 (left) SIMS depth profile from a 2 x 10^{15} cm^{-2}, 35 keV boron implant in high dose oxygen-implanted silicon, recorded with an oxygen primary beam. (A 25nm gate oxide has been removed before SIMS analysis).

Fig. 2 (right) Comparison of oxygen profiles (from a similar specimen) recorded by SIMS (full line) and by AES (dashed line), plotted on a linear scale. The SIMS analysis conditions are the same as in Fig. 1.

model presented here is based on the assumption that the oxygen in the material, outside the stoichiometric SiO_2 layer, is present mainly as small oxide precipitates. This is supported by TEM studies [5], particularly in the regions of high O content. The total boron signal, I(B), will therefore be the sum of contributions from B in Si and B in SiO_2 weighted according to the local volume fraction, V, of silica precipitates. The B concentration, C(B) is then given by the expression:

$$C(B) = I(B) \{ V.I(Si/SiO_2) / RSF (B/SiO_2) + (1-V).I(Si/Si)/ RSF (B/Si) \}^{-1} \quad (1)$$

where the notation $I(Si/SiO_2)$ denotes the Si signal measured from an SiO_2 matrix, and RSF (B/SiO_2) is the relative sensitivity factor for B in SiO_2 (with Si as reference mass). The RSFs can be measured independently from standards of B in Si and B in SiO_2. It is assumed in (1) that the sputter rates of the two matrices are identical (as supported by the measurements, above), and that the boron is equally likely to be found in the Si and SiO_2 regions.

Figure 3 shows the boron profile of Fig. 1 quantified point-by-point using (1). The reference Si signals are obtained from the buried oxide layer $(I(Si/SiO_2))$ and from the silicon signal on the deep side of the oxide (I(Si/Si)). The local volume fraction of silica, V is calculated to a first approximation from the measured oxygen signal in the same measurement cycle (as a fraction of the increase in the oxygen signal in the buried oxide).

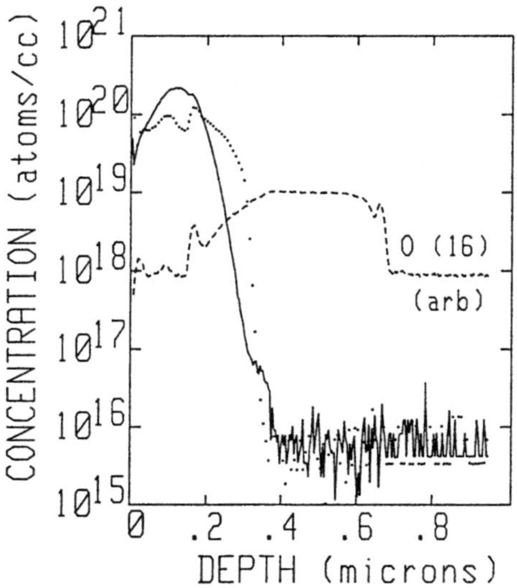

Fig. 3 Boron profiles quantified using (1). The diagram shows the unannealed implant from Fig. 1 (solid line) and the same implant annealed at 1000°C for 15 mins (dotted line). The O-16 profile (in arbitrary units) is also included.

In this unannealed boron implant, the spike in the measured ion signal is almost completely removed by the quantification model, to yield a fairly smooth profile. The general features (peak concentration and depth) are in good agreement with SUPREM III [6] predictions. Fig. 3 also shows the same B implant after annealing at 1000°C for 15 mins. In this case, the profile retains some detailed structure, which may indicate a true localisation of the B in regions of high O concentration. Preferential segregation of B into thermally grown silica layers has been reported by Morgan [4]. Profiles similar to Fig. 3 were also obtained using the empirical relationship developed by Homma [3]; this may indicate that the present model provides a basis for that relationship.

Acknowledgement is made to the Director of Research of British Telecom for permission to publish this paper. We would also like to thank Dr. C.G. Tuppen and Dr. M.R. Taylor of BTRL and Dr. P.L.F.Hemment (Surrey University). Part of this work has been carried out with the support of Procurement Executive, UK Ministry of Defence, sponsored by DCVD.

1. C.G. Tuppen, M.R. Taylor, P.L.F. Hemment and R.P. Arrowsmith: Appl. Phys. Lett. 45(1), 57 (1984)
2. J.A. Kilner, S.D. Littlewood, P.L.F. Hemment, E. Maydell-Ondrusz and K.G. Stephens: Nucl. Instrum. and Methods 218, 573 (1983)
3. Y. Homma, H. Tanaka and Y. Ishii: SIMS IV, 98 (1983)
4. A.E. Morgan, H.A.M. de Grafte, N. Warmoltz and H.W. Werner: Appl. of Surf. Sci. 7, 372 (1981)
5. C. Jaussaud, J. Stoemenos, J. Margail, M. Dupuy, B. Blanchard and M. Bruel: Appl. Phys. Lett: 46, 1064 (1985)
6. J.A. Greenfield and R.W. Dutton: IEEE Trans. Electron. Devices ED-27, 1530 (1980)

Isotope Tracer Studies Using SIMS[1]

J.A. Kilner and R.J. Chater

Dept. of Metallurgy and Materials Science, Imperial College, London SW7 2BP, U.K.

P.L.F. Hemment

Dept. of Electronic and Electrical Engineering, University of Surrey, Guildford, Surrey, GU2 5XH, U.K.

Accurate mass transport data are essential for the interpretation of many important phenomena in materials science. Normally such data can be obtained by radiotracer or other nuclear-based techniques, but these experiments can be difficult to perform. Stable isotope experiments followed by SIMS analysis provide an excellent alternative to the above methods for a large range of elements. This alternative is most commonly used to determine self-diffusion coefficients. Our aim in this paper is to show how the same experimental concepts can be used to generate mass transport information under different conditions, where chemical diffusion effects will be present. To illustrate this we shall describe experiments to study the growth of thin insulating layers of SiO_2 and Si_3N_4 in which we have used SIMS and stable tracers to gain information about the growth processes.

1. The Growth of Thin Oxide Films on Silicon

The formation of SiO_2 films on silicon under most conditions is well understood and can be described by the Deal-Grove linear-parabolic growth law [1]. In this model the oxidant is assumed to diffuse towards the Si-SiO_2 interface, as a molecule, without interacting with the SiO_2 network through which it passes. The oxygen then reacts with the silicon at the Si-SiO_2 interface to form a new layer of oxide. Thin oxides (d < 30 nm), grown under dry conditions, are an exception in that the linear-parabolic growth law is not obeyed. However, there is much technological interest in precisely this growth regime, mainly because such thin layers are required for MOS gate oxides in VLSI circuits.

Attempts at an explanation of this anomalously high growth-rate regime have led to the proposal of a modified zone, approximately 30 nm thick, adjacent to the Si-SiO_2 interface in which the SiO_2 differs from the bulk because of large compressive stresses [2,3]. SIMS analyses may provide evidence for the existence of this zone. Earlier [4], we reported the analysis of thin thermal oxides using negative secondary ions. Near to the Si-SiO_2 interface one negative secondary ion SiO_2^-, shows a sharp peak, 3 to 4 orders of magnitude in intensity above background. The width of this peak is < 20 nm. This strong signal suggests that the chemical nature of the oxide adjacent to the Si-SiO_2 interface differs from that of the bulk material and is the subject of further investigation. The results presented here are intended to show how this ion can be utilised in the interpretation of isotope experiments of the type described below.

A set of thin oxides were grown in dry natural oxygen at 900°C. The samples were then similarly treated in dry $^{18}O_2$ for an incremental period. The samples were analysed using an Atomika SIMS (10 keV Ar^+ primary ions,

---

[1]Work partially supported by U.K. SERC/ALVEY Directorate

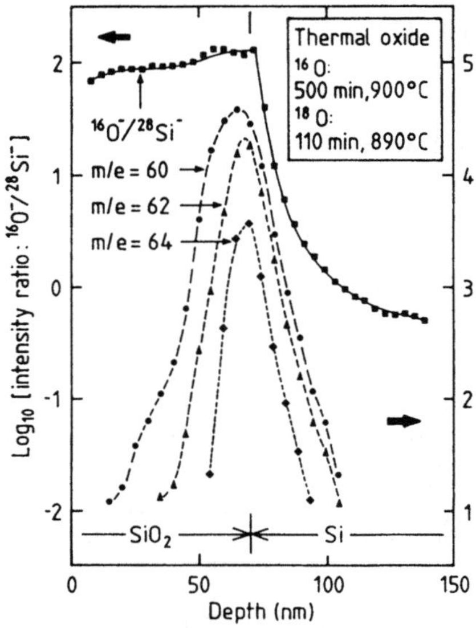

Fig. 1. SIMS depth profile of a thermal oxide (d=70nm) on Si grown first in dry $^{16}O_2$ followed by a shorter oxidation in dry $^{18}O_2$ both at 900°C. Overall oxygen profile shown as determined by the oxygen/silicon ratio with the cluster ion intensities. Analysis conditions (10 KeV Ar^+ primary beam, 500 eV e^- flood, -ve sec ions)

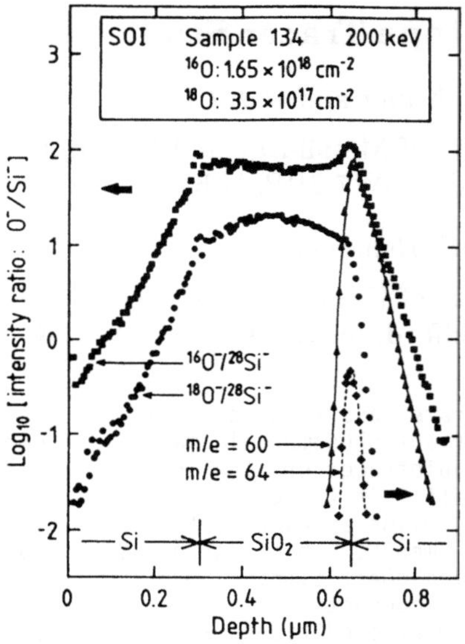

Fig. 2. SIMS depth profile of a buried SiO_2 layer formed by high dose implantation of $^{16}O^+$ followed by a smaller dose of $^{18}O^+$. The four traces show the distribution of the ^{16}O and ^{18}O (presented as intensity ratios because of sample charging) and the cluster ion intensities. Analysis conditions as in Fig. 1.

500 eV e^- flood). The ^{18}O distributions showed good agreement with previous results [5,6], i.e. ^{18}O localised at the SiO_2 surface and the Si-SiO_2 interface with very little in the bulk of the SiO_2 layer. The only major difference was in the isotopic concentration found at the Si-SiO_2 interface (∿20%), this is much lower than the isotopic concentration of the gas (∿80% ^{18}O). To check the isotopic concentration of the interface region we profiled the oxide and followed some of the possible SiO_2^- cluster ions. We chose to follow masses 60, 62, 64 for maximum sensitivity(despite the possible mass interferences : none, $^{30}Si^{16}O^{16}O$, $^{30}Si^{16}O^{18}O$ respectively) since the ions containing ^{28}Si ($Si^{16}O^{16}O$, $Si^{16}O^{18}O$, $Si^{18}O^{18}O$ respectively) are the most intense species. Figure 1 shows the result of this experiment. The SiO_2^- ions show the following intensity relationship $I_{60}>I_{62}>I_{64}$. This is to be expected for a low isotopic concentration of ^{18}O, i.e. the majority of the ^{18}O content should reside in the mixed ion at mass 62. The actual isotopic concentration in the interfacial zone, evaluated from the peak SiO_2^- signal, gives an isotopic concentration, at the interface, of ∿ 24% ^{18}O, confirming the measurement made using atomic signals only. The explanation of this low isotopic concentration could be due to three possible causes: (1) The rate of ^{18}O transport to the interface is lower than expected at 900°C and an incomplete new oxide layer has been formed, (2) some mixing of the interface region has occurred by a process of exchange and diffusion with the overlying Si $^{16}O_2$ network, (3) some ion beam mixing has occurred during analysis, smearing out the thin $Si^{18}O_2$ layer. Experiments are in hand to resolve this problem.

2. Buried Dielectric Layers of Silicon Oxide and Silicon Nitride

Buried SiO_2 layers are formed by high energy, typically 200 keV, high dose implantation of O_2^+ into silicon. The mechanisms of mass transport during layer formation (implantation and annealing) are much less understood than in the thermal case. To investigate the formation of buried SiO_2 layers we have undertaken experiments, similar to those described above for the thermal oxides, except the ^{18}O was introduced into the sample by ion implantation as an incremental dose; full experimental details are given in [4].

A typical result is shown in Fig. 2 which shows the $^{16}O^-$, $^{18}O^-$ and SiO_2^- traces at m/e = 60 and m/e = 64 for an as-implanted sample. The bulk of the ^{18}O is found within the SiO_2 in contrast to the thermal case. Thus much diffusion and exchange of the oxygen occurs within the layer during implantation. The mass 60 trace shows virtually no intensity at the upper Si-SiO_2 interface, again in contrast to the thermal case, whilst at the lower SiO_2-Si interface there is a very intense and broad peak. The mass 64 trace is less intense and is at the leading edge of the 60 trace, also a sharp drop in the atomic ^{18}O signal is found at the same depth. Together this suggests that ^{18}O is not penetrating the lower interface and the growth of new oxide most probably occurs at the upper Si-SiO_2 interface.

Implantation of high doses of energetic N_2^+ into Si leads to the formation of buried Si_3N_4 layers. To analyse these layers cluster ions such as SiN^- and Si_2N^+ must be used, since atomic nitrogen signals are virtually absent. Fortunately these ions show a continuous intensity throughout a nitride layer. In our ^{15}N tracer experiments [7], performed in a similar manner to those with the oxide layer, the distribution of the two nitrogen isotopes can be clearly distinguished from the $^{28}Si^{14}N^-$ and $^{28}Si^{15}N^-$ signals and show rather different distributions. Material implanted first is located at the deeper interface of the buried layer, later incremental doses are situated nearer the surface. As with the buried SiO_2 layers, these results suggest that the growth of the nitride layer occurs at the upper Si/Si_3N_4 interface.

3. Conclusions

(1) SIMS analyses of stable tracer distributions can provide excellent qualitative and quantitative mass transport data as demonstrated for ^{18}O and ^{15}N experiments in SiO_2 and Si_3N_4 layers.

(2) Cluster ions can be used in these experiments to either provide isotopic concentrations of specific regimes or to provide isotopic distributions where no atomic signals are available.

References

1. B.E. Deal and A.S. Grove: J. Appl. Phys. 36, 3770 (1965).
2. A. Fargeix, G. Ghibaudo and G. Karamannos: J. Appl. Phys. 54, 2878 (1983)
3. E.A. Irene: J. Appl. Phys. 54, 5416 (1983).
4. J.A. Kilner, et al: Nucl. Inst. and Meths. B7/8, 293 (1985).
5. F. Rochet, B. Agius and S. Rigo: J. Electrochem. Soc. 131, 914 (1984).
6. J.A. Costello and R.E. Tressler: J. Electrochem. Soc. 131, 1944 (1984).
7. J.A. Kilner, R.J. Chater, P.L.F. Hemment, R.F. Peart, K.J. Reesen, R.P. Arrowsmith and J.R. Davis: Proc. IBA Conf. Berlin 1985 to be publ. in Nucl. Inst. and Meths.

High Dynamic Range SIMS Depth Profiles for Aluminium in Silicon-on-Sapphire

M.G. Dowsett, E.H.C. Parker, and D.S. McPhail

Solid State Research Group, Sir John Cass Faculty of Physical Sciences and Technology, 31, Jewry St., London EC 3N 2EY

Silicon-on-Sapphire (SOS) is the most advanced silicon-on-insulator technology for CMOS-VLSI. We have described techniques for extracting accurate boron profiles from SOS [1] and, elsewhere in these proceedings, methods for obtaining accurate depth profiles from resistive multi-layer structures like SOS [2]. The substrates used for SOS form a potential source of aluminium doping which could lead to poor mobilities and unpredictable device behaviour. Many of the processing techniques for SOS (especially self-implantation and regrowth [3]) are very likely to remove Al atoms from the substrate and redistribute them in the silicon. This SIMS study of the behaviour of Al in SOS is part of a programme to define acceptable processing schedules. Three problems may be anticipated: Firstly, we wish to obtain accurate epi-layer and interface profiles from a resistive structure on an insulator. Secondly, we require quantification from impurity (i.e. $<1\%$) to matrix concentrations, across two different matrices, so calibration standards must be chosen with care, and the interpretation at the interface may be semi-quantitative. Thirdly, the Al^+ signal is only 1 amu away from the intense Si^+ matrix emission. Thus, the dynamic range will be dependent on the **peak-shape** of the mass spectrometer, and the relative ion yields of Al^+ and Si^+.

1 Experimental and Sample Details

Data from the samples in Table 1 are reported here. The samples were mounted so that the Si-epilayers were electrically isolated. The EVA 2000 quadrupole mass spectrometer (QMS) based SIMS instrument built by the authors' research group [4,5,6] was used as follows:- Primary ions: 4keV $^{16}O_2^+$, normal incidence, 200nA, 50um spot, raster 0.3mm, gated area 10%, 10 or 20 s/frame, preclean 10 frames 1mm square. Charge compensation: 500eV electrons, 10uA, 1mm spot co-incident with the crater [1]. Secondary ions: $^{27}Al^+$, $^{12}C^+$, $^{44}(SiO)^+$, $^{43}(AlO)^+$. The latter two channels indicate the accuracy of the charge compensation [2]. The $^{12}C^+$ channel was monitored to ensure that behaviour attributed to Al was not due to a hydrocarbon interference. Ion images of the Al^+ channel were examined throughout the profiles to determine whether or not the lateral distribution was uniform. It was necessary to reduce the transmission in the Al channel at the interface, since the Al intensity rises from a minimum of 1-2 c/s to $>10^8$. Putting a negative offset on the sample potential was found to upset the charge compensation. Instead, the mass in the Al channel was offset 0.5amu to the low mass edge of the peak when the controlling software [6] detected a threshold of 5×10^5 c/s at the interface.

The low mass tail on the peak-shape of the (QMS) was minimised by adjustment of the entrance conditions and axial energy. The DC and RF potentials on the rods were carefully balanced and, by operating at a mass resolution of 500-800 (10% valley), it was possible to achieve a signal level of only 1-2 c/s at mass 27 in the presence of 10^8 - 10^9 at mass 28, i.e. a peak contribution of approx. $1:10^8$.

Table 1 **Sample Details**

Sample No.	Type	Implant/Dopant/Process	Purpose
1	bulk-Si	10^{15} Al cm^{-2}, 100keV	standard
2	0.3um SOS[1]	"	"
3	0.6um SOS	none	control
4	0.3um SOS	SPEG[2]	for analysis
5	"	electron beam anneal[3]	"

1 All SOS grown by chemical vapour deposition (CVD).
2 SPEG = Solid phase epitaxial regrowth = deep Si implant (190keV, 10^{16} cm^{-2}) close to interface + furnace anneal.
3 As 2 with e^- beam anneal of 20 W cm^{-2} replacing furnace anneal.

2 Depth Calibration and Quantification

The depth scale in each profile was established by Tallystep and a previously determined relative sputter yield $Al_2O_3/Si = 0.845$ which was applied after the aluminium concentration had risen above $2x10^{19}$ cm^{-3}.

The change in instrumental transmission caused by the mass offset used to suppress the intense Al^+ signal at the interface complicates the Al^+ quantification. Furthermore, it cannot be assumed a priori that the Al^+ yield is equal in the different media and constant at the interface. Calibration constants were therefore determined as follows: The Al^+ channel in sample 1 was quantified from the dose giving a calibration of $1.9x10^{14}$ for Al^+ from Si. This enables the Al^+ channel in the Si epilayer in sample 2 (Fig. 1) to be quantified. (Implant peak concentration $4.4x10^{19}$ cm^{-3}.) The Al concentration in the Al_2O_3 is $4.69x10^{22}$. The AlO^+ signal (Fig 1) is due to O-loading from the primary beam in the Si, and additionally to matrix fragmentation in the Al_2O_3. Comparing the AlO^+ signal levels from the implant peak and alumina with the known concentrations, it can be seen that the AlO^+ yield from Si is 2.2 times that from alumina. Thus the Al^+ channel may be quantified after the mass offset point by forcing the condition:

$$Al^+(alumina)/Al^+(silicon)=AlO^+(alumina)/AlO^+(silicon)$$

and applying an extra factor of 2.2 to correct the matrix level. At worst, this process will slightly "stretch" some part of the interface between 10^{20} and $5x10^{22}$. The Al^+ calibration constant so derived for alumina was $4.4x10^{18}$.

3 Results and discussion

Figure 2 shows the data from sample 3. Apart from a 50nm wide surface peak, common to most profiles from this material, the Al level in the epilayer is $<$ the instrumental background until the interface. Here, the level rises 7 decades in a depth of about 10nm, and a total of 7.5 decades in 20nm. The dotted line in Fig. 2 shows the effect of surface alumina particles whose presence was revealed by imaging. These cause erroneous and unreproducible Al levels in the epilayer, and distort the interfacial profile. Embedded alumina surface debris are found on most SOS material examined. The particles are probably due to contact with the backs of other wafers, and the residue from cutting and polishing operations. A partial solution is to coat the Si-epilayer with spin-on oxide at the earliest possible stage in the processing and then remove it with HF immediately before loading.

Figure 3 shows a reproducible profile from sample 4 (the SPEG sample). The Si implant has caused a small amount of Al to recoil from the substrate and

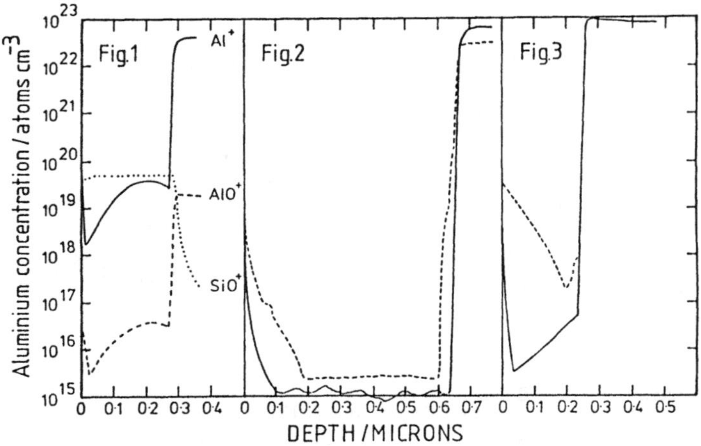

Fig. 1 Al profile from standard SOS sample no. 2, quantified by the process described in section 2. The AlO^+ channel is unquantified and shows the relative magnitudes of the Al and AlO signals

Fig. 2 High dynamic range profile from control (sample 3, solid lines), and sample with surface Al_2O_3 debris (dotted lines). Note distortion at surface and interface in the latter.

Fig. 3 Aluminium profile from thermally annealed (solid line) and electron beam annealed (dotted line) SPEG samples 4 and 5. The electron beam profile is to impurities introduced during anneal

diffuse, producing a level varying between $2x10^{15}$ at the surface an $3x10^{16}$ cm^{-3} at the interface. The artificially high Al level in the alumin shows the maximum error caused by the calibration process out of about 2 profiles. (The level in the Si is accurate, however.) Figure 3 also shows a Al profile from sample 5 (dotted line), similar to other profiles fro electron beam annealed samples. The high levels may well be due to th particulate contamination already mentioned being driven in during the anneal Reproducible profiles could not be obtained from different areas on the sam sample, and imaging revealed non-uniform distributions.

4 Conclusions

It has been shown that a two-matrix model can be applied to obtai quantitative Al depth profiles from SOS epilayers and interfaces. Electro beam flooding gives successful charge compensation, provided the epilayer i left isolated. A quadrupole mass spectrometer can achieve a peak contributio figure of at least $1:10^8$ under the least favorable conditions (i.e. contribution of m+1 to m).

We thank GEC Hirst Research Centre for samples and useful discussions.

1. M G Dowsett, E H C Parker, R M King and P J Mole, J. Appl. Phys. **54**(1983)6340-6345
2. D S McPhail, M G Dowsett and E H C Parker, these proceedings.
3. D J Smith et. al., J. Appl. Phys. **56**(1984)2207-2212
4. M G Dowsett and E H C Parker, Int. J. Mass Spectrom. and Ion Phys. **52**(1983)299-309
5. K Wittmaack, M G Dowsett and J B Clegg, Int. J. Mass Spectrom. and Ion Phys., **43**(1982)31-39
6. M G Dowsett, J W Heal, H Fox and E H C Parker, These proceedings.

Charge Compensation During SIMS Depth Profiling of Multilayer Structures Containing Resistive and Insulating Layers

D.S. McPhail, M.D. Dowsett, and E.H.C. Parker

Solid State Research Group, Sir John Cass Faculty of Physical Sciences and Technology, 31 Jewry Street, London EC3N 2EY

To conduct a SIMS depth profile at constant sensitivity, a fixed fraction of the secondary ion energy distribution should fall within the bandpass of the instrument. Thus, the crater surface potential, V(x,y,t), should not vary by more than a few volts. This condition is unlikely to be met when profiling resistive multilayer structures, and may sometimes be impossible to achieve using fixed instrumental conditions. Thus the use of charge compensation techniques (electron beam flooding, evaporation of conducting films) must be accompanied by sensitive procedures for monitoring the surface potential throughout the profile and for quantifying the change in instrumental sensitivity associated with any charging. It may then be possible to restore the surface potential to its required value by applying a DC offset to the sample holder, provided the charging is uniform (V—▷V+c). This article describes such monitoring procedures with reference to two systems, BF_2 and As implants in silicon-on-sapphire (SOS) epilayers, in which small changes in surface potential can lead to profile distortions.

1 Experimental

Experiments were conducted in EVA 2000, a dedicated SIMS instrument designed and built in-house [1,2]. The instrument is of the electronically gated type and the secondary ion energy filter and quadrupole mass spectrometer define a band pass of ∿10eV. The primary beam ($^{16}O_2^+$, Ip = 50-300nA at 4kV, 50μm spot at normal incidence) is rastered digitally to form a square crater of side 200-400μm. The computer controlling software sets up one channel (mass-target potential V_o) per frame and many combinations can be cycled. Dopant and matrix species can thus be profiled at several target potentials.

Shallow BF_2 and As implants (5x10^{15} cm^{-2}, 10/40 kV) in 0.4μm SOS epilayers were analysed using two alternative charge compensation techniques, electron beam flooding and gold coating, (Fig.1). The electron gun (Ie = 1-10uA at 1kV) produced a stationary 1mm diameter spot that overlapped the crater area. Uniform gold films (500-800 Å thick), deposited in a technical vacuum system, provided a conduction path from sample holder to crater edge, thus short-circuiting the sapphire substrate.

2 Results and Discussion

Initial results for a BF_2 implant in SOS (5x$10^{15}$$cm^{-2}$, 40kV), (Fig 2), profiled using electron beam flooding, showed that the fluorine profile was limited to one concentration decade by an intense (1000 counts/sec) background at mass 19. This was due to electron-stimulated desorption (ESD) of fluorine ions [3], for the signal persisted when the ion beam was switched off. The ESD fluorine was desorbed from all the surfaces in the system and is due to a gaseous adsorbate. The ESD fluorine and ion-sputtered fluorine have identical narrow gaussian energy spectra (fwhm ∿ 10V) and cannot be separated by energy selection [3,4], so an alternative charge compensation technique, gold coating, was tried. Initial results were promising, (Fig.2), but examination of

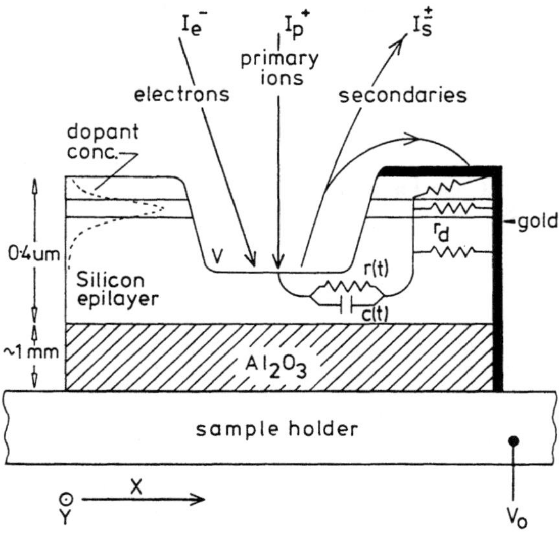

Fig. 1
Sample mounting and charge compensation techniques for SOS epilayers. Incomplete charge compensation can lead to a potential difference between points on the crater surface (x,y) and the target holder, i.e. $V(x,y,t)-V_o$

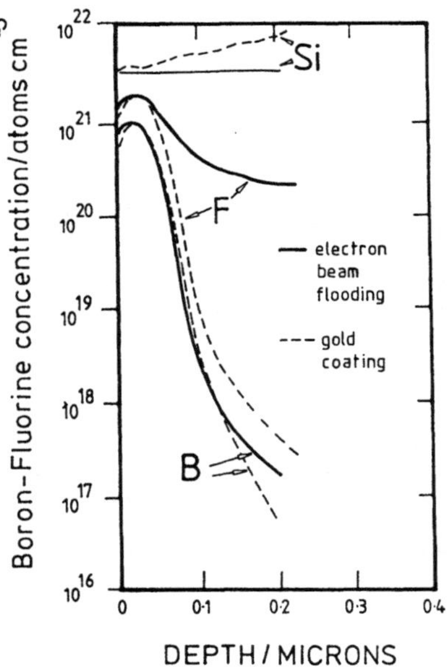

Fig.2 Quantified data for a BF_2 ⟶ SOS implant

the matrix channel revealed that some charging had occurred and so further experiments were set up to quantify this. The charging was monitored by setting up three channels for each of the dopants (B/F). In one dopant channel the target potential V_o was set to the peak of the secondary ion energy distribution, and in the two remaining channels it was set either side of the peak. [ie B=(+25V);(+10V);(-10V) : F= (+27V);(+10V);(+5V)]. A matrix channel was also included, (Si, V_o=-50V). The result of such a multichannel experiment for a BF_2 implant in **bulk silicon** is shown below, (Fig.3i). The matrix channel is constant and the three channels for each dopant rise and fall concurrently. Raw data for the first SOS profile (300nA ⟶ 400µm x 400µm), (Fig.3ii),

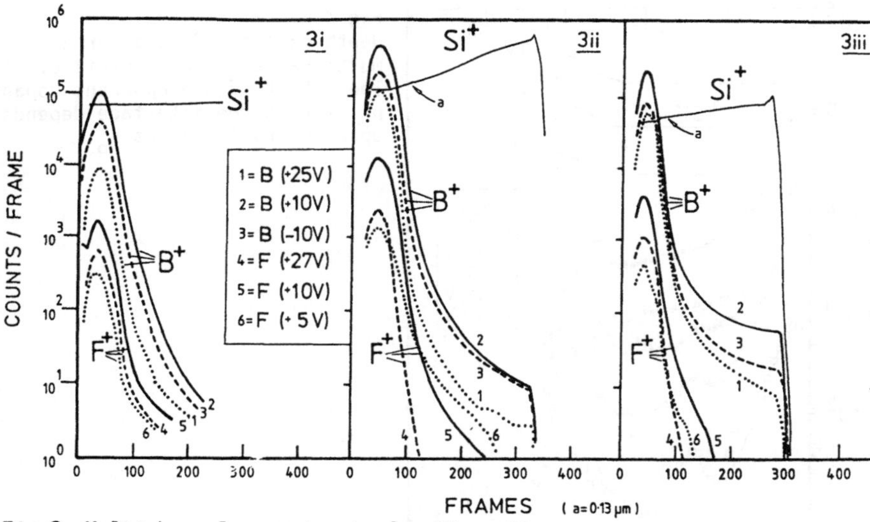

Fig.3 Multichannel experiments for BF_2-->Si and BF_2--> SOS implants (counts vs.time)

indicate some charging. The matrix channel has increased by 42% at frame 100 (0.13μm), equivalent to a **uniform** charging of 10V [from the Si energy spectrum Io-->I as V-->V+dV; d(I/Io)/dV = 4.2%]. Matrix channel chemical imaging, however, revealed that the charging was **non-uniform** with lateral potential gradients of tens of volts, due to the poor conduction path from crater centre to edge. The dopant channels do not vary concurrently and the computed slopes for the trailing edges range from 200-250 Å/decade for boron and 233-410 Å/decade for fluorine. The artificial steepening in the low-energy fluorine channel, V_o=27V, arises because fluorine has a narrow energy spectrum and is particularly sensitive to charging, [d(I/Io)/dV = 50%]. To reduce these effects the crater size was reduced to 250μm and the beam current to 100nA. The result, (Fig.3iii), was a reduction in charging to 4V (at 0.13μm). The trailing edges for the boron are now identical (261 Å/dec compared to 267 for bulk Si) and for fluorine are in the range 305-362 Å/dec (308 A/dec for bulk Si). The profile is accurate to within 200Å of the interface.

Electron beam flooding must be used if SOS profiles are to proceed through the resistive epilayer and into the insulating substrate, as no conduction path can then exist. Inadequate charge compensation at this transition can lead to artifacts such as dopant accumulation or depletion. During recent arsenic profiles in SOS (with electron beam flooding and a -10V tuning point to exclude the mass interference due to the molecular ion Si_2O) we observed an apparent accumulation of the dopant at the interface. However, a multichannel experiment, (Fig.4), showed the dopant profile at the interface was entirely dependent on the tuning point. As the interface is approached the As (20V) channel drops whilst the 10V, 0V, -10V and -40V channels pass through maxima of increasing relative intensity in that order, as the molecular ion Si_2O is swept into the band pass of the instrument. Chemical imaging of the sensitive matrix species SiO at several target potentials, revealed progressive charging towards the centre of the crater, and equipotential mapping was possible. This type of lateral - non-uniform charging, V=V(x,y,t), cannot be rectified by sensing a change in matrix channel and applying a DC offset to the sample

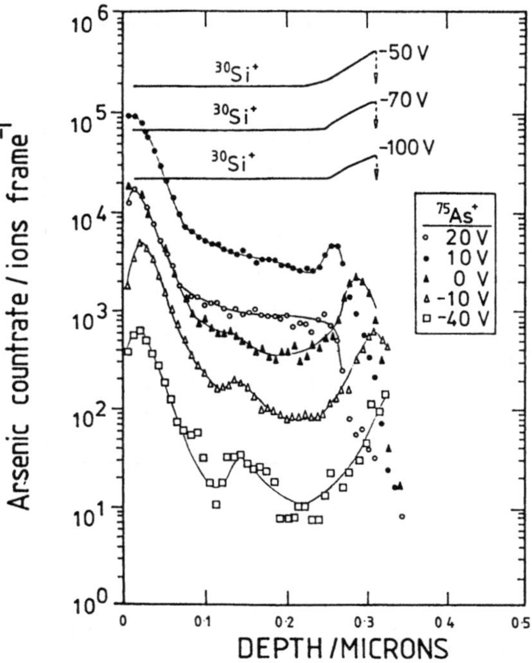

Fig 4
Depth-calibrated data for a multichannel arsenic profile. The extent of the apparent dopant pile-up at the interface depends upon the tuning point V_o.

holder to correct for it, as only part of the crater will be re-optimised for emission. In the above instance the artifact disappeared on increasing the electron beam current density.

In more recent work we use two Si_2O (or SiO) channels, tuned on either side of the narrow energy distribution, as matrix channels (25V, 5V). This is up to ten times more sensitive than Si, [d(I/Io)/dV$\sim \pm$30%], and one can distinguish between charging of a few volts (where one channel will increase whilst the other decreases) and instability in the primary beam current.

3 Conclusions

It is difficult to stabilise the surface potential to better than a few volts when profiling certain resistive structures, and this may lead to profile distortions. Methods for monitoring the surface potential and quantifying these distortions include setting up several channels for the dopant(s) at different tuning points (multichannel experiments), the inclusion of matrix channels and the chemical imaging of matrix channels. The effect of any charging can then be quantified by measuring the energy spectrum of the species and the band pass of the instrument. Changes in instrumental sensitivity associated with lateral potential gradients can only be corrected by better charge compensation, not auto-voltage control.

References

1. M G Dowsett and E H C Parker, Int J.Mass Spec.Ion Phys, 52(1983) 299-309
2. K Wittmaack, M G Dowsett, J B Clegg, Int J.Mass Spec.Ion Phys, 43(1982) 31-39
3. D S McPhail, M G Dowsett, E H C Parker to appear in J.Appl.Phys.
4. P Williams, Phys.Rev.B, Vol.23, 11, p 6187-6190

Ion Deposition Effects in and Around Sputter Craters Formed by Cesium Primary Ions

K. Miethe, W.H. Gries, and A. Pöcker

Forschungsinstitut der Deutschen Bundespost, D-6100 Darmstadt, F. R. G.

The use of a cesium source [1] in an ion microprobe mass analyzer (ARL-IMMA) has led to the development of characteristic topography inside and outside sputtered areas in GaAs.

After depth profiling of an ion implanted non-annealed sample of GaAs a field of "spots" appears outside the area exposed to the cesium primary ions (Fig. 1). In optical microscopy the crater itself looks homogeneous and smooth. This phenomenon is reversed completely when a crater is sputtered in a virgin GaAs material. Now the spots are located inside the crater and few are seen outside (Fig. 2).

It is important to note that these ion deposition effects do not become visible instantaneously, but only after some time (sometimes days) after irradiation, and then only if certain Cs doses are exceeded. In all instances the location of spots relative to the crater depends on whether the GaAs has been pre-irradiated or not.

Using SEM, the spots were found to be asperities (often crystalline). Other structures were observed as well, but are not discussed here. With scanning Auger microscopy (SAM) the spots were found to be significantly enriched

Fig. 1 Cs-sputtered IMMA crater in Si-implanted ($5\cdot10^{15}$ cm^{-2}, 120keV) non-annealed GaAs, crater 65µm x 65µm, Cs dose $3\cdot10^{18}$ cm^{-2}

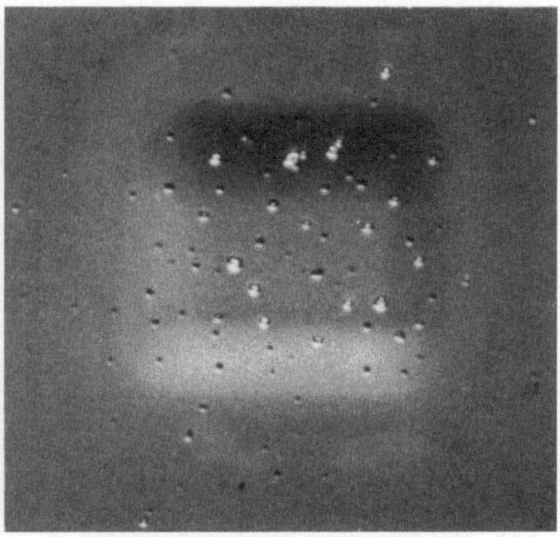

Fig. 2 Cs-sputtered IMMA crater in virgin GaAs n^+ material, crater 58µm x 58µm, Cs dose $2\cdot10^{17}$ cm^{-2}

with cesium. Similar spots enriched with primary beam material have been observed by BROWN and RUEDENAUER [2] on GaAs sputtered with an indium beam from a liquid metal ion source. In contrast to these authors, however, who suppose indium droplets to play a significant role in the spot formation, we can exclude this possibility in the IMMA because of magnetic deflection of the primary beam.

The presence of cesium outside a crater area can be due to three causes, viz. adsorption of cesium "vapour" (from the ion source), deposition of fast neutral cesium atoms (from charge exchange of primary ions in collisions with residual gas) and diffusion of cesium out of the crater. We can not exclude a contribution from the first two causes, although it would be difficult to explain the contrary effects of Figs. 1 and 2 by a major contribution of each or any of the two.

The dose dependence of the formation of spots on virgin conductive GaAs material was investigated with cesium doses ranging from $5\cdot10^{12}cm^{-2}$ to $2\cdot10^{18}cm^{-2}$. For doses higher than $5\cdot10^{15}cm^{-2}$ spots were observed to form inside bombarded areas with a time delay of several days after irradiation.

The distribution of cesium in and around the craters was investigated with SAM and IMMA-SIMS (using O_2^+ primary ions). With SIMS, step scan analysis was used with a 5µm beam, 5µm step width and a sputtertime per step long enough to integrate the total cesium secondary ion signal. For cesium doses up to $5\cdot10^{14}cm^{-2}$ the cesium secondary ion signal from inside the crater bottom was found to rise linearly with the Cs dose. Assuming this fact to reflect quantitative retention, we used this signal/dose relationship to quantify larger signals. This is how the ordinates in Figs. 3 and 4 were calibrated.

Figure 3 is a typical step scan analysis across a crater having spots inside while Fig. 4 is typical for a material with spots outside the crater. When the analyzing oxygen beam hits a spot inside a crater the integrated

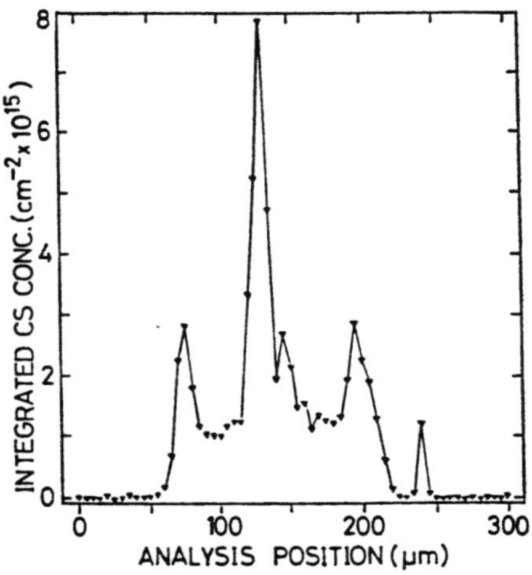

Fig. 3 Step scan analysis across a crater produced by $5\cdot10^{16}$ cesium ions per cm^2 in virgin GaAs

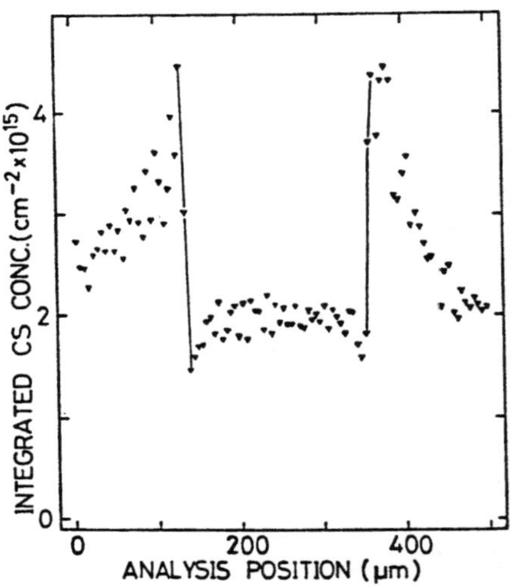

Fig. 4 Step scan analysis across a crater produced by $1.3\cdot10^{18}$ cesium ions per cm^2 in GaAs preimplanted (200keV Si $1\cdot10^{15}cm^{-2}$) and non-annealed

cesium signal peaks, corresponding to a retained cesium dose of up to several $10^{15}cm^{-2}$. In between these peaks the lowest values measured in different craters all correspond to retained doses of about $1 \cdot 10^{15}cm^{-2}$ and are interpreted to be representative of spot-free areas and hence to the retained dose in the bulk. The spots are thus enriched by a factor at least several times the retained dose in the bulk. In pre-implanted non-annealed GaAs the retained cesium dose inside the crater was measured to be $1.8 \cdot 10^{15}cm^{-2}$ independent of the implanted cesium dose which ranged from 0.3 to $1.3 \cdot 10^{18}cm^{-2}$. At the crater edge the cesium secondary ion signal increases sharply to values of two or three times the signal measured inside due to the presence of a high density of cesium-enriched spots. In both cases (cesium-enriched spots inside or outside craters) measurements by SAM and SIMS have shown cesium to be present over large areas outside the crater.

From the observations described above, and others mentioned below, we are led to propose the following model:
As the crater develops, the concentration of cesium at the crater bottom increases in accordance with the well-known model of implantation [3]. At equilibrium the concentration of cesium in the first atomic layer of the sample is estimated to be equivalent to a coverage of less than half a monolayer. For these coverages a dipole character of cesium on GaAs has been suggested [4] to explain observed variations of desorption energy and work function. The valence electron of the cesium is supposed to be attracted towards the bulk, giving rise to a positive cesium "ion" residing on the surface. We could confirm [5] this charge state for at least a significant fraction of the cesium on our samples outside the craters (by observing the movement in an electric field applied to the surface). We assume this to be true also for the cesium inside the craters. Then the mutual repulsion (and possibly also the interaction with the electric field of the positive primary ion beam) should lead to migration of cesium out of the crater during bombardment at a rate sufficient to save a significant fraction from being sputtered.

We propose that the cesium loses its dipole character when it encounters a dislocation on the surface. It can then agglomerate to form phase nuclei and grow with time into the spots which have been observed. Evidence in support of this proposal comes from the observation that spot formation could be induced but not reversed by application of an appropriate electric field. Formation of spots inside a crater proceeds in the same way (at dislocations) after sputtering has ceased. When the cesium inside the crater has been trapped at dislocations, the cesium "ions" outside the crater observe the crater as a field of sinks into which the cesium re-migrates. If the crater is surrounded by dislocations at high density, which is the case in pre-implanted non-annealed material, most of the cesium will be trapped there and does not reach the crater. Hence spots in the crater will not reach sizes which are easily visible.

Further experiments are under way to test the model.

References

1. B.L. Bentz and H. Liebl, SIMS III (Proc. Int. Conf., Eds. A. Benninghoven et al.) Springer Verlag Berlin 1982, p.30
2. J.D. Brown and F.G. Rüdenauer, J.Appl.Phys. 57, 2727 (1985)
3. e.g. J.C.C. Tsai and J.M. Morabito, Surface Sci. 44, 247 (1974)
4. J. Derrien and F.A. d'Avitaya, Surface Sci. 65, 668 (1977)
5. K. Miethe and W.H. Gries, to be published

SIMS Depth Profiling of Si in GaAs

F.R. Shepherd, W. Vandervorst#, W.M. Lau+, W.H. Robinson+ , and A.J. SpringThorpe**

* BNR, Ottawa, Ontario, Canada K1Y 4H7

IMEC, Kard. Mercierlaan 94, Heverlee 3030, Belgium

+ University of Western Ontario, London, Ontario, Canada

Introduction and Experiment

The controllable doping of GaAs layers by silicon, either by ion implantation or by incorporation during molecular beam epitaxial growth, is a critical step in III-V device fabrication. We have therefore undertaken a study to determine the optimum experimental conditions for SIMS depth profiling Si in GaAs, for Si^+ implanted into GaAs, and for thin, abrupt GaAs/AlGaAs MBE layers doped with Si.

SIMS depth profiles were obtained with a Cameca IMS 3-f ion microscope equipped with both duoplasmatron and Cs ion sources, a primary mass filter, and a microprobe imaging system to ensure that the detected signal was extracted from the center of the crater [1]. Ion implantation was carried out with the (100) wafers tilted 7° off axis. The MBE samples consisted of five alternating ~75nm thick layers of GaAs and $Ga_{.5}Al_{.5}As$, the central 30nm wide region of each being doped with Si to ~ $1x10^{18}cm^{-3}$, on top of which were grown 5 $Ga_{.5}Al_{.5}As$ layers, 17nm wide between which were sandwiched GaAs layers of 13nm, 8nm, 4nm and 2nm thickness respectively.

Results and Discussion

Profiling $^{28}Si^+$ using an $O_2{}^+$ primary beam is complicated by the interference at mass 28 from AlH^+, CO^+, $N_2{}^+$ and $C_2H_4{}^+$. Fig 1 shows the mass spectrum, recorded at high mass resolution, from a $^{28}Si^+$-implanted GaAs wafer; the N_2 and C_2H_4 components severely limit the sensitivity for Si. While the situation could be significantly improved by reducing the residual pressure of these contaminants, the contribution of C_2H_4 and CO at mass 28 is essentially reduced to zero, and a higher useful yield is obtained using a Cs^+ primary beam and negative secondary ion detection. All profiles have therefore been measured using Cs^+ primary beams.

An additional complication, even with Cs^+ ions, arises from Si and Al impurities, associated with the primary beam and its neutral component, which are continuously deposited on the specimen surface [2]. The effect of these instrumental impurities on the background or detection limit for Si in GaAs is minimised by using a high erosion rate. In our instrument, which is often used for depth profiling impurities in Si wafers, the detection limit for ^{29}Si in GaAs is typically ~$1x10^{15}cm^{-3}$, while integrating over a depth of ~ 5nm GaAs per point. An order of magnitude improvement in sensitivity can be obtained after lens components have been cleaned.

The profiles measured from wafers implanted with Si were generally found not to be gaussian, but exhibited pronounced tails. To test

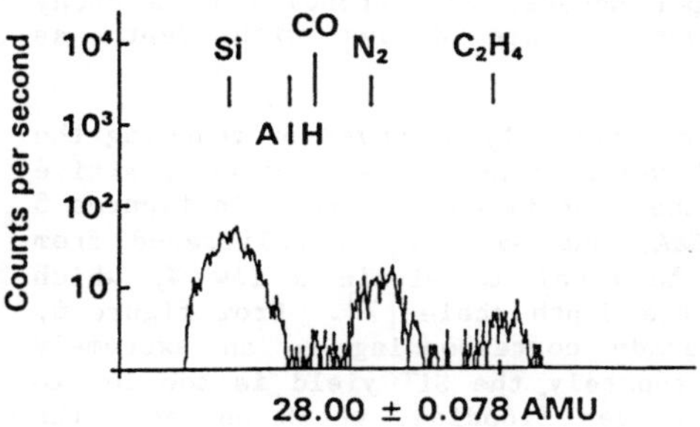

Fig. 1 High mass resolution spectrum of Si in GaAs (O_2^+ bombardment)

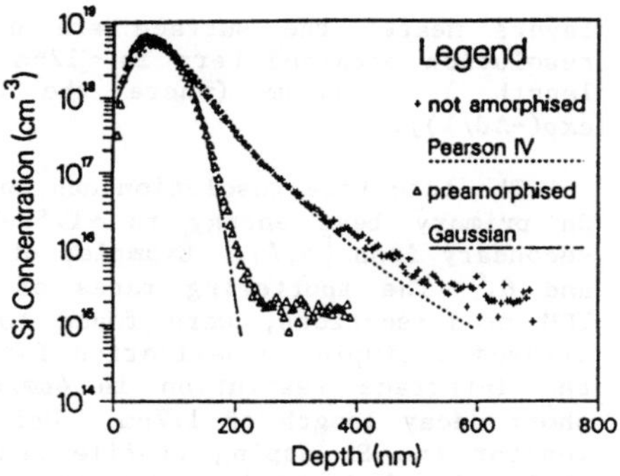

Fig. 2 Profiles of 60 keV$^{29}Si^+$, $5x10^{13} cm^{-3}$ implanted into pre-amorphised and not amorphised GaAs (14.5keV Cs^+ relative to the sample, monitoring $^{104}AsSi^-$)

whether this was an ion beam-induced artifact, the Si implant profile from a wafer preamorphised by As^+ implantation, was compared with the same Si implant into a bare GaAs wafer, under identical measurement conditions. Figure 2 shows that the SIMS profile of the preamorphised wafer is gaussian, whilst implantation into crystalline GaAs results in a pronounced tail, due to channeling of Si^+ during implantation, as previously observed [3,4]. This results in a profile which can be fitted by a Pearson-IV distribution, as for B^+ in Si [5].

MBE device structures, where the layers are often much thinner and dopant profiles are far more abrupt than in implanted layers, provide a more stringent test of depth resolution. Figure 3 shows the depth profile of the MBE structure obtained using 8.5keV Cs^+ bombardment and negative secondary ion detection. At low mass resolution there is an interference from AlH at mass 28 (figure 3), which can be removed by increasing the mass resolution to about 2000 (figure 4). Although the presence of the Si doping spikes is clearly apparent, the thin GaAs

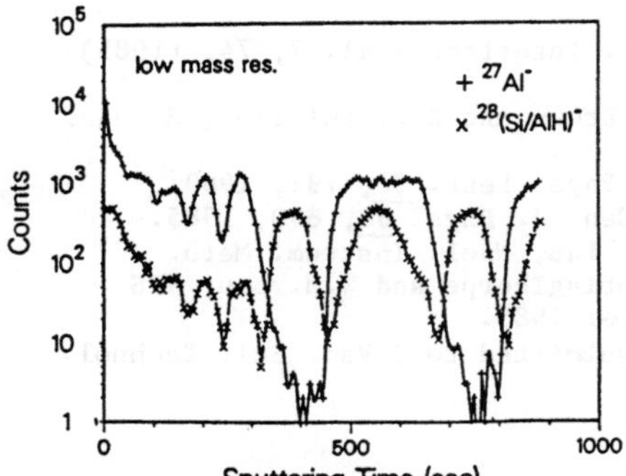

Fig. 3 Depth profile of MBE layers recorded at low mass resolution (8.5keV Cs^+)

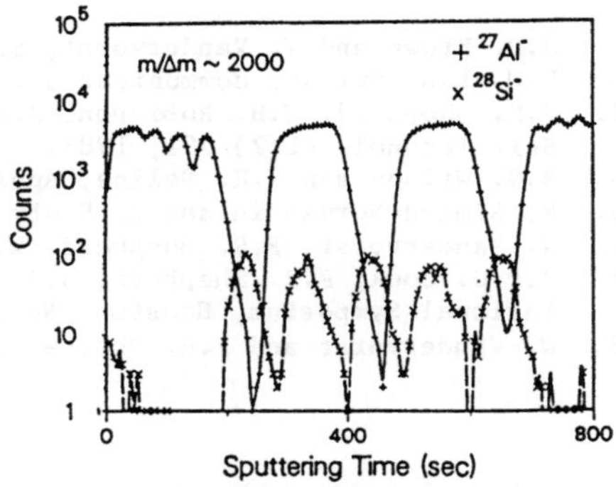

Fig. 4 Depth profile of MBE layers recorded at high mass resolution (14.5keV Cs^+)

layers nearer the surface are not clearly resolved. The interface resolution obtained here is ~12nm per decade, corresponding to a decay length $\lambda \sim 5.3$nm (where the count rate decays with depth as $\exp(-\Delta d/\lambda)$).

The interface resolution can be dramatically improved by reducing the Cs primary beam energy to ~1.5keV which requires monitoring positive secondary ions [6,7]. Examples of these profiles are shown in figures 5 and 6. The sputtering rates of GaAs and $Ga_{.5}Al_{.5}As$, calibrated from TEM cross-sections, were found to be equal to within a few %, which allowed a simple normalization for the depth scale [7]. From figure 6, the interface resolution is 4nm/decade corresponding to an extremely short decay length of 1.7nm. Unfortunately, the Si^+ yield is too low to monitor the Si doping profile and oxygen flooding, which enhances the yield of As^+ in Si [6], is not effective, presumably because of the low sticking coefficient on GaAs [8]. Quadrupole SIMS instruments, where the low extraction field allows the use of 1keV Cs^+ primary beams with negative secondary ion detection, may offer some advantages for depth profiling Si in thin GaAs MBE layers.

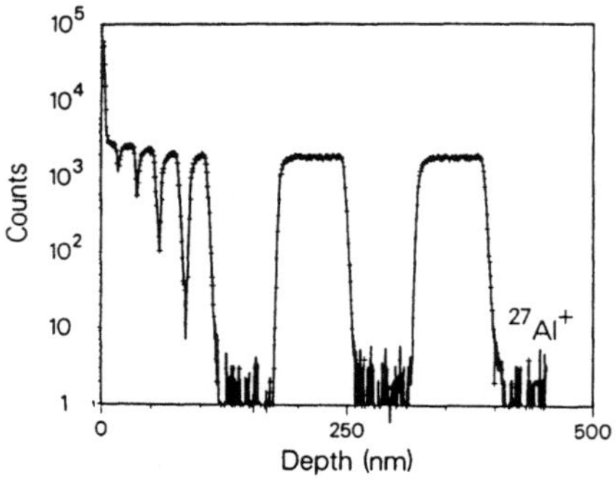

Fig. 5 Depth profile of MBE layers using 1.5keV Cs^+ primary beam, and positive secondary ions.

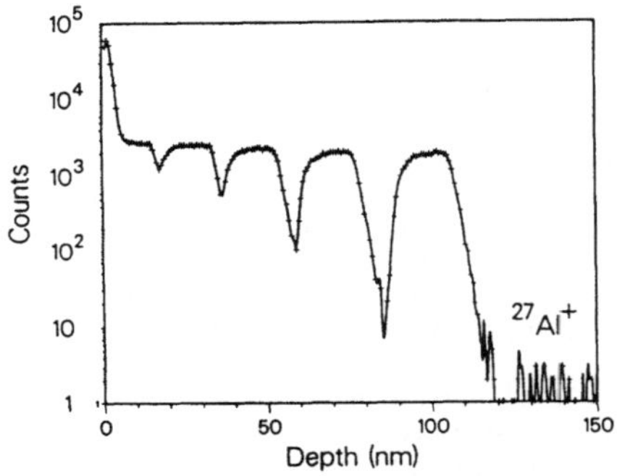

Fig. 6 Enlarged region of fig.5 showing 2nm, 4nm 8nm and 13nm GaAs layers between 17nm $Ga_{.5}Al_{.5}As$ layers.

References

1. J.D. Brown and W. Vandervorst, Surf. Interface Anal. 7, 74, (1985)
2. W.M. Lau, private communication.
3. F.R. Shepherd, W.H. Robinson, J.D. Brown and B.F. Phillips, J. Vac. Sci. Technol. A1(2) 991, 1983.
4. R.G. Wilson and V.R. Deline, Appl. Phys. Lett. 37, 793, 1980.
5. M. Simard-Normandin and C. Slaby, Can. J. Phys. 63, 890, 1985.
6. W. Vandervorst, F.R. Shepherd, W.M. Lau, Nucl. Instrum. Meth.
7. J.P.D. Cook, F.R. Shepherd, A.J. SpringThorpe and W.M. Lau, AVS National Symposium, Houston, November 1985.
8. W. Vandervorst and F.R. Shepherd, submitted to J.Vac. Sci. Technol.

High Purity III-V Compound Analyses by Modified CAMECA IMS 3F

A.M. Huber and G. Morillot

THOMSON-CSF, Laboratoire Central de Recherches, Domaine de Corbeville, B.P. 10, F-91401 Orsay, France

This paper will present the modifications needed to achieve low detection limits (10^{13} at.cm^{-3}) with CAMECA IMS 3F, particularly for Cr and Fe in the III-V compound analysis. Finally, we will give some results relating to the new improved sensitivity of the instrument.

1. Performance in this field

Clegg[1], using an Atomika quadrupole mass analyzer equipped with Ta apertures, reported the following detection limits for Mg, Mn, Cr and Fe respectively : $2x10^{13}$, $4x10^{13}$, $4x10^{13}$ and $1x10^{14}$ at.cm^{-3}. A detection limit of $4x10^{14}$ at.cm^{-3} for Cr in GaAs was published by Kütt et al[2], also by means of Atomika.

With a CAMECA IMS 3F, detection limits for Mg, Mn, Cr and Fe in GaAs of only $3x10^{14}$, $7x10^{14}$, $4x10^{14}$ and $1x10^{15}$ at.cm^{-3} respectively were obtained by authors [3,4]. Using mass filtered primary O_2^+ in IMS 3F, the detection limits for Mg, Cr and Fe were about $1x10^{14}$, $5x10^{14}$, $2x10^{15}$ at.cm^{-3} respectively [5]. This limit was the lowest determined level of elements in very high purity undoped CZ grown bulk GaAs. In the same sample, only $5x10^{13}$ at.cm^{-3} [Cr] and $5x10^{13}$ at.cm^{-3} [Fe] were determined by neutron activation analysis. There is no doubt that Cr and Fe were background level and originated elsewhere in the instrument.

2. Instrument modification

In order to optimize the instrument, the principal requirements are as follows :
i) a primary beam mass filter, ii) all apertures to be made of Pt or Mo, iii) the immersion lens front plate to be Ta. Generally, the IMS 3F meets the last two requirements.

Figure 1 gives a schematic view of a part of the instrument which has undergone modification. The four elements marked with an asterisk were modified as follows :

1) Pt plating of the stainless steel electrostatic lens L3,(used for focusing the beam on the specimen).
2) Replacement of the stainless steel immersion lens by Pt plated Ti material.
3) Removal of deflector 1 from the secondary ion optics.
4) Replacement of the entrance slit with a new slit made of Ta.

In this way the contribution by parasitic sputtering of stainless steel elements to the Cr and Fe background level was suppressed.

3. Results and discussion

Figure 2 shows the dramatically improved Cr level ($5x10^{13}$ [Cr] cm^{-3}) in the tail of the [Cr] profile (dashed line)as compared to the result before modification,($9x10^{14}$ Cr cm^{-3}) background Cr (continous line).The sample is 300 keV

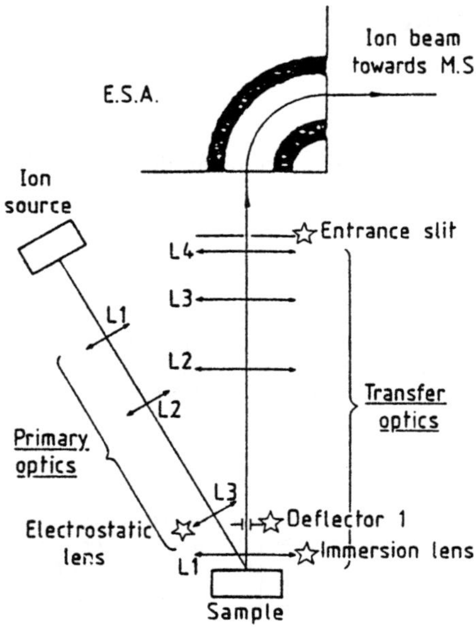

Fig.1 Schematic view of the parts of the instrument in which the modification were carried out.

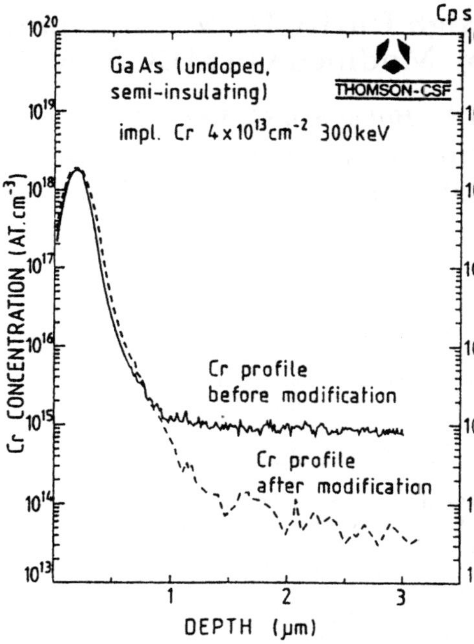

Fig.2 Cr depth profiles in GaAs showing the improved background level in GaAs.

Cr implanted into undoped semi-insulating GaAs with a dose of $4x10^{13}$at.cm^{-2}. The Cr level $5x10^{13}$at.cm^{-3} corresponds to the real Cr concentration determined in the same non-implanted sample by neutron activation.

In all our III-V compound analyses, the surface is scanned with a mass filtered oxygen ion beam (Ip 1,5-3 uA at 10 keV primary ion net energy). The scanned area is 250x250 um and the analyzed region is 150 um in diameter. Offset target voltage -50V is applied to avoid possible interference with ionized hydrocarbons. Ion implanted samples were the standard for quantitative calibration of the instrument.

For a highly sensitive analysis of impurities in GaAs or InP, it is necessary i) to reach, for the $^{150}As^+$ and $^{31}P^+$ internal standards, about $2x10^4$ and $3x10^5$ counts/sec. respectively ; ii) to use 2-5 sec. integration time ; iii) to check that the front plate of the electrostatic lens is clean or has been used only for III-V compound analysis, to avoid memory effect.

The new detection limits of impurities in GaAs and InP measured by the modified IMS 3F are as follows ; ^{24}Mg, ^{52}Cr, ^{55}Mn : $5x10^{12}$at.cm^{-3} ; Fe : $5x10^{13}$at.cm^{-3} ; Si : $7x10^{13}$ at.cm^{-3} (Si in InP).

The background is due to the electron multiplier, amounting to approximately 0,04counts/sec. Under our analytical conditions the ion yield for Cr 1 count/sec is equivalent to about $1x10^{13}$ [Cr] cm ^{3}. This sensitivity allowed us to study the effect of growth parameters on the impurity redistribution in epilayers [6-7]. Fig.3 shows Fe depth profiles in epilayers. By metalorganic chemical vapor deposition (MOCVD), InP layers were grown on Fe doped semi-insulating, Sn and S doped InP substrates, in the same run. Etch pit densities (EPD) of the subtrate were $4x10^4$cm^{-2}, $5x10^4$cm^{-2} and $9x10^3$cm^{-2} respectively. The three (100) off 2° towards (110) oriented wafers were prepared simultaneously side by side. In Fig. 3, one can observe that the Fe accumulation at the interface and the Fe concentration in the epilayer grown on the Fe doped substrate is 10 times lower than in the case of Sn or S doped substrates. The diverse levels of the same element in the layers grown in the same run is most probably due to the substrate quality. The degree of non-stoichiometry on the surface and in the substrates could influence i) the chemisorption,[7] ii) the crystal quality of layers by the formation of vacancies and consequently the impurity redistribution.

Figure 4 presents Mg, Mn, Cr, Fe and Si profiles in the high purity MOCVD InP epilayer. Excellent Hall electron mobilities at liquid nitrogen temperature of 100.000 $cm^2/V\text{-}sec$ are measured for $n = 3x10^{14} cm^{-3}$ doping level. This n level is due to residual Si as determined by SIMS. The layer is very pure. The following results were obtained : Cr $2 \times 10^{13} cm^{-3}$, Fe $8 \times 10^{13} cm^{-3}$, Mg $6x10^{13} cm^{-3}$, Mn $5x10^{12} cm^{-3}$.

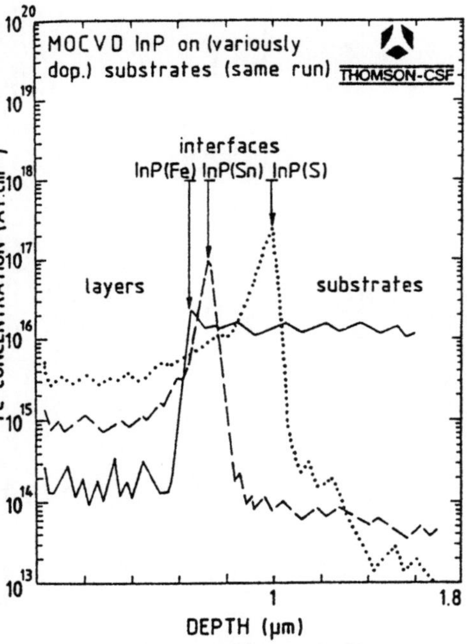

Fig. 3 Fe depth profiles in InP layers grown on InP (Fe) InP (Sn) and InP (S) doped substrates

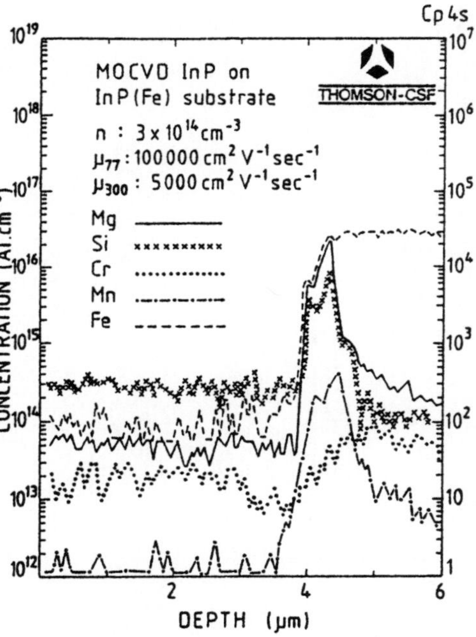

Fig. 4 Mg, Mn, Cr, Fe and Si depth profiles in a high purity MOCVD InP layer

Acknowledgements

We would like to express our thanks to H.N. Migeon,G.Slodzian, C. Grattepain and J.B.Clegg for helpful discussions and to M.Razeghi, M.A.Di-Forté Poisson for providing the InP epilayers.

References

1. J.B. Clegg, J. Appl. Phys. 53 4518 (1982)
2. W. Kütt, D. Bimberg, M.Maier, H. Kräutle, F. Köhl, E. Tomzig, Appl. Phys. Lett. 45 489 (1985)

3. R.N. Thomas, H.M. Hobgood, G.W. Eldridge, D.L. Barett, T.T. Braggins, Solid State Electronics 24 387 (1981)
4. R.G. Wilson, P.K. Vasudev, D.M. Jamda, C.A. Evans, Jr V.R. Deline, Appl. Phys. Lett. 46 215 (1980)
5. A.M. Huber, G. Morillot, A. Friederich, "Secondary Ion Mass Spectrometry" SIMS IV. Osaka 36 278 Springer-Verlag Ed (1983)
6. A.M. Huber, G.Morillot, S.D.Hersee, K. Kazimerski, H. Tews,Proc. of Third Conference on S.I. III-V Materials 466 Edition Shiva (1984)
7. A.M. Huber, M. Razeghi, G. Morillot Proc. of Conference GaAs and Related Compounds Biarritz 223. Conf. Series Number 74 (1984)

Depth Profiling of Dopants in Aluminum Gallium Arsenide

W.H. Robinson[1] *and J.D.Brown*[2]

[1] Surface Science Western, The University of Western Ontario, London, Ontario, Canada

[2] Faculty of Engineering Science, The University of Western Ontario, London, Ontario, Canada N6A 5B9

1. Introduction

SIMS analyses of (Al,Ga)As layered structures are complicated by variations in secondary ion yields as the ratio of Al to Ga or Al to As changes. In addition, molecular species from impurities or major elements in the material, as well as from the primary ion beam and the residual gases in the instrument, can limit sensitivity or require the use of high mass resolution for removal of the interferences. The availability of Cs^+ and O_2^+ primary ion beams offers an additional parameter for optimization of sensitivity for specific dopants or impurities. This study was undertaken to evaluate and optimize the depth profiling of dopants in (Al,Ga)As.

2. Experimental

The specimens used in this investigation were (100) GaAs on which layers of $Al_xGa_{1-x})As$ were grown by liquid phase epitaxy (x = 0.05, 0.20 and 0.40). The specimens were implanted with Si^{28} and Cr^{52} at 25 and 50 keV energy to total doses of $3x10^{13}$ and $1x10^{15}$ atoms/cm^2 at angles of 2° and 7° with respect to [100] for the Si and Cr implants, respectively. Depth profiles were measured in a Cameca IMS 3f ion microscope equipped with a primary beam filter and both Cs^+ and duoplasmatron ion sources. Positive secondary ions were measured when using an O_2^+ primary beam, while negative secondary ions were measured when using a Cs^+ primary beam. The instrument has been modified by installation of a scanning ion-imaging system [1] to ensure that secondary ion extraction is from the centre of the sputtered crater and by changes to the software for depth profiling under high mass resolution conditions [2]. Eighteen secondary ion species, both atomic and molecular, were monitored during low mass resolution depth profiles, while beam currents of 200 to 1000 nA were used to minimize time for data collection. Secondary ions were extracted from the central 60 μm of a 250x250 μm crater. Selected peaks were measured at high mass resolution to evaluate molecular interferences, and beam currents were reduced to approximately 50 nA to improve depth resolution. The depth scales for the profiles were determined on the basis of crater depth measurements using a profilometer.

3. Results

(a) Low Mass Resolution

The dynamic ranges achievable under low mass resolution conditions at high primary beam currents are listed in Table 1 for various secondary ion species and an implant of $1x10^{15}$ atoms/cm^2 at 50 keV. The dynamic range is limited either by the background intensity due to a mass interference or by total count rate where the background falls to the electronic noise of less than 1 count/second. The dynamic ranges achievable in measuring Cr with a

Table 1. Dynamic ranges at low mass resolution and high sputter rates

Cr Implant, O_2^+ Primary Ion Beam $I_p \sim 1000$ nA, 14.5 keV			Si Implant, Cs^+ Primary Ion Beam $I_p \sim 200$ nA, 8.0 keV		
Secondary Ion	Dynamic Range	Limitation	Secondary Ion	Dynamic Range	Limitation
^{52}Cr	10^5	Background	^{28}Si	$10^1 \rightarrow 10^3$	AlH Interference
$^{52}Cr^{16}O$	10^3	Bkgd Interference	$^{28}Si^{27}Al$	50	Count Rate,Bkgd≈0
$^{52}Cr^{27}Al$	10^3	Count Rate,Bkgd≈0	$^{28}Si_2$	50	Count Rate,Bkgd≈0
$^{69}Ga^{52}Cr$	10^4	Count Rate,Bkgd≈0	$^{69}Ga^{28}Si$	50	Count Rate,Bkgd≈0
$^{75}As^{52}Cr$	10^2	Count Rate,Bkgd≈0	$^{28}Si^{75}As$	7×10^2	Background

Cs primary ion beam are poorer because of reduced Cr^- secondary ion intensities and a molecular interference of C_4H_4 from the residual gases in the instrument. Although secondary ion intensities for Si are approximately equivalent with O_2^+ and Cs^+ primary ion beams, the dynamic ranges are poorer when using O_2^+ because of higher backgrounds associated with a CO molecular interference. Thus,under low mass resolution and high primary beam current conditions, Cs^+ is preferred for depth profiling of Si while O_2^+ is preferred for Cr.

An example of the limitation on dynamic range imposed by molecular ion interference and impurity level is shown in the three profiles of $^{28}Si^-$ of Figure 1. (The profiles are offset for clarity, the Si signal in the GaAs substrate is identical for all three.) The background level due to the Si impurity level in GaAs limits the dynamic range for that specimen (top profile). The effect of the higher Al content and hence a higher AlH intensity in $(Al_{0.4},Ga_{0.6})As$ specimen can be seen as a decrease in dynamic range (middle profile) compared to the $(Al_{0.2},Ga_{0.8})As$ specimen (bottom profile). If the rate of sputtering is decreased, to improve depth resolution, then AlH interference becomes increasingly important.

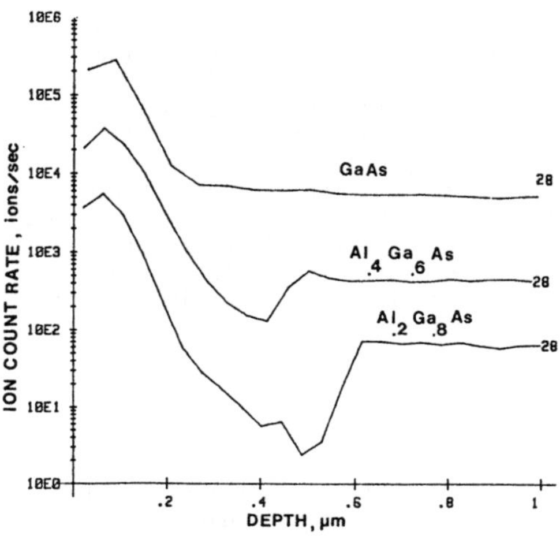

Fig. 1. Low mass resolution depth profiles of Si in GaAs and AlGa As measured with 14.5 keV Cs^+

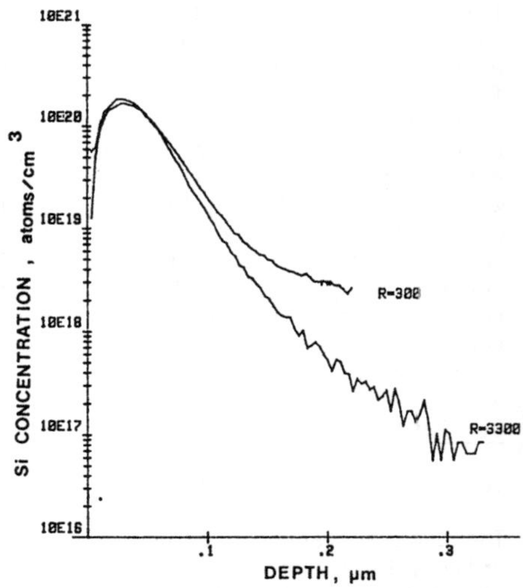

Fig. 2. Low and high mass resolution depth profiles of Si in $Al_{0.4}Ga_{0.6}As$, 14.5 keV Cs^+ primary beam

(b) High Mass Resolution

Figure 2 shows a comparison of depth profiles at low and high mass resolution for $^{28}Si^-$ and a Cs^+ primary beam current of 45 nA. The AlH interference which limits the low resolution profile to less than 2 orders of magnitude, is removed at the higher mass resolution yielding a profile with a dynamic range of over 3 orders of magnitude. The advantage of Cs^+ in dynamic range is removed at high mass resolution and roughly comparable profiles can be achieved with oxygen. Note the improved depth resolution apparent in these profiles compared to Figure 1.

A similar problem exists in the depth profiling of Cr^- using a Cs^+ primary ion beam in which $C_4H_4^-$ originating from the residual gases in the vacuum results in the Cr^- signal being barely discernible. The intensities of the two species are shown in Figure 3 in which the Cr profile shows 2 1/2 orders of dynamic range. In the case of $C_4H_4^-$, the energy distribution of the molecular ion is extremely narrow compared to Cr^-. As a result, voltage offset can be used at low mass resolution to discriminate against low energy ions and improve dynamic range as shown in Figure 4. Voltage offset is not very successful for resolving the AlH interference at mass 28 since AlH ions have a rather broad energy distribution.

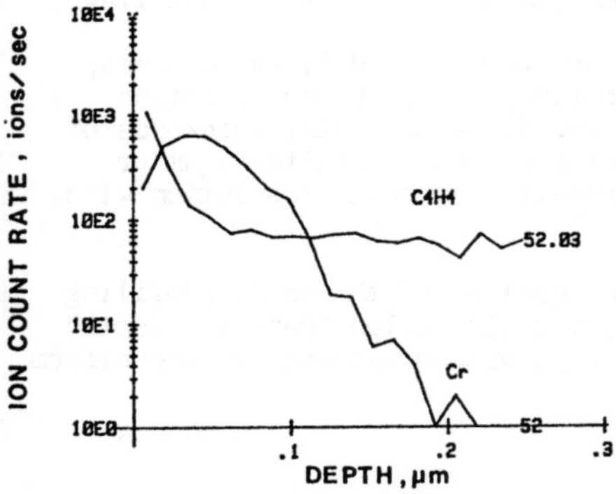

Fig. 3. High mass resolution depth profile at mass 52 of Cr^- and $C_4H_4^-$

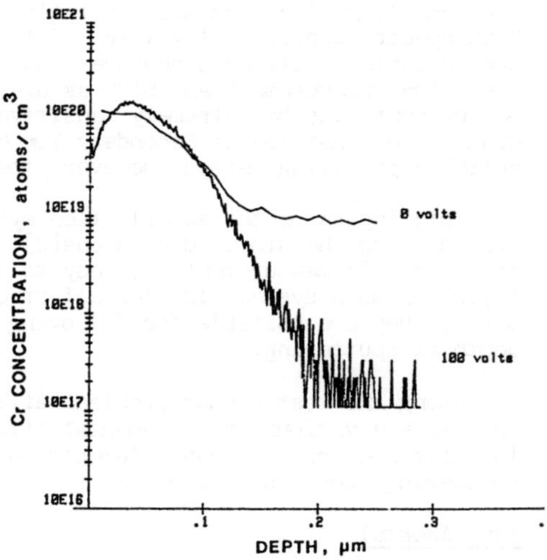

Fig. 4. Effect of voltage offset on Cr depth profiles in $Al_{0.4}Ga_{0.6}As$

4. Conclusions

If depth resolution is not an important consideration in depth profiling of Si and Cr dopants in (Al,Ga)As, then low mass resolution profiles yield maximum dynamic range, and a Cs^+ primary ion beam is to be preferred for Si measurements and O_2^+ for Cr. As requirements for better depth resolution limit sputter rates, then high mass resolution conditions to remove molecular interferences become increasingly important in maintaining good dynamic range.

[1] J.D. Brown, W. Vandervorst, Surf. Ind. Interface Anal. 7, 74 (1985).
[2] W.H. Robinson, J.D. Brown, D.D. Johnston and W.M. Lau, Microbeam Analysis-1984, AD Romig Jr., JI Goldstein, Eds., San Francisco Press, 331(1984).

Depth Resolution in Profiling of Thin GaAs-GaAlAs Layers

J. Gavrilovic

W.C. McCrone Associates, Inc., 2820 S. Michigan Ave., Chicago, IL 60616, USA

Thin gallium-arsenide and gallium-aluminum-arsenide layers (usually called superlattices) are investigated with increasing frequency due to their remarkable electronic and optical properties. These structures, which consist of many (50 to 100 or more) layers of 15 to 500 angstrom thickness, are often doped with elements such as silicon, selenium, zinc, magnesium, etc. to alter their electronic and optical characteristics[1].

Study of the distribution of such dopants in the thin layers presents a real challenge for most microanalytical techniques, including Secondary Ion Mass Spectroscopy. The low level of the dopants virtually eliminates the most common depth profiling techniques (such as Auger, ESCA, RBS), leaving only electrical measurement and SIMS as useful techniques. While electrical measurements may be extremely sensitive (in ppb range) they are not readily quantified. That leaves Secondary Ion Mass Spectrometry (SIMS) as the most suitable profiling method. However, the SIMS technique is not without flaws.

Thin layers of the superlattice system are very well defined in space, even down to the atomic dimensions[2]; therefore, any problems affecting depth profile measurement are easy to observe. Since the total thickness of layers of such systems is, for all practical purposes, infinite (5 μm or more), they are suitable for following degradation of depth resolution with depth of sputtering.

Among the most common problems affecting quality of the depth profiling in SIMS are various interelemental effects, usually called "matrix effects" interference from residual elements in the high vacuum systems and nonuniform sputtering (cone formation).

Experimental

Three SIMS instruments were used to obtain depth profiles of thin layers. An ARL ion microprobe of 1970 vintage was compared to two advanced depth profiling instruments: Physical Electronic 6000 series Secondary Ion Profile $GaAs-Ga_{.8}Al_{.2}As$ layers 170 to 520 angstroms thick. A summary of some results of depth profiling on these samples, using three SIMS instruments, are given in Table 1.

The superlattice samples were analyzed under similar conditions with all three instruments. In addition, with the ARL microprobe, the analyses were performed at two different energies of the primary $^{16}O_2^+$. In this case the depth profile at 10 kV accelerating voltage showed some improvement

*Samples of GaAs-GaAlAs superlattices were received from the Electrical Engineering Laboratory University of Illinois at Urbana-Champaign, Urbana Illinois 61801.

Table 1. Depth Profiling Results for Three Different SIMS Instruments.

	ARL ion microprobe	PHI	RIBER ion microprobe
Depth resolution	86 Å	58 Å	56 Å (2σ)
Dynamic range	1.7×10^{1}	1×10^{3}	0.8×10^{3}
Maximum intensity of $^{27}Al_2^+$	2×10^{4}	1×10^{5}	1×10^{6}
Etching time (min)	23	26	28

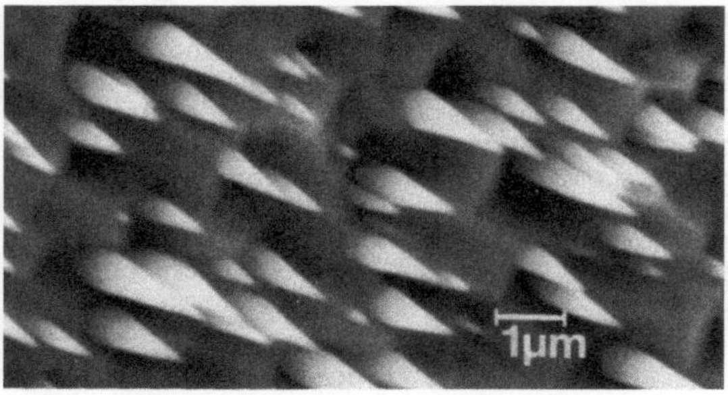

Fig. 1. Cone formation on a superlattice sample after prolonged etching with a $^{16}O_2^+$ ion beam.

in resolution over that at 22 kV. Instrument limitation (the high voltage power supply) prevented use of lower voltages to minimize "cascade mixing effect"[3], which damages depth resolution. The dominant factor that limited depth resolution in the superlattice system, however, was non-uniform etching of the sample. This was revealed by prolonged sputtering of a superlattice sample to a depth of 3.2 µm (Fig. 1).

The resulting cone heights, shown on Figure 1 are larger than 10000 angstroms and are chiefly responsible for resolution degradation from 80 angstroms near the surface to approximately 400 angstroms at 30000 angstrom depth.

The effect of the residual gas contamination in a vacuum system was observed by analyzing a superlattice sample without a silicon dopant in the UHV instrument and a moderate vacuum ARL microprobe. Residual hydrogen (including that from H_2O) in a vacuum system is deleterious to the study of silicon dopant in the presence of larger quantities of aluminum as the sticking coefficient of H_2O for the metal surfaces of the vacuum vessel is extremely high. Consequently the intensity of the aluminum hydride signal is directly related to the quality of the vacuum. The interference of hydride in a SIMS instrument can be minimized by high-resolution mass spectrometry or by improving high vacuum in the analyzing chamber; thus reducing the contribution of hydrides. The first approach eliminates hydrides but, unfortunately, reduces signal intensity of the dopant element. The latter method is efficient and is the main reason that most modern SIMS instruments operate in the UHV range.

A sample of pure Al was etched in aqueous HF, the ion beam cleaned and then analyzed in the ARL microprobe at 7×10^{-7} torr. The mass spectrum signal at mass number 28 corresponded to a hydride contribution of 1% of the $^{27}Al^{+}$ full intensity. In a high purity aluminum this would be nearly 1% hydride contribution. In new high-performance SIMS instruments aluminum dimer $^{27}Al_2^{+}$ is usually plotted instead of $^{27}Al^{+}$ due to the high intensities of the singly-charged ion. This dimer is 1/400 of the intensity. Therefore, even a small contribution of the hydride to $^{28}Si^{+}$ may completely overpower the Si dopant signal.

Conclusion

Depth profiling of GaAs and GaAlAs thin layers is a suitable method for comparing the performance of different SIMS instruments due to the highly regular and repetitive nature of the "superlattices". Such a comparison was performed between a vintage ARL microprobe and two high performance UHV SIMS instruments. Depth resolution is somewhat improved in the two modern instruments. The major improvement in the performance of the new generation ion microprobes is a significant increase in the dynamic range, i.e., sensitivity. The two new instruments have comparable depth resolution in depth profiling of the superlattices with the added advantage of UHV, which eliminates bothersome hydride interference. The most serious limitation in improving the 50-60 angstrom depth resolution of the present instruments seems to be related to the sample nature and associated sputtering artifacts.

References

1. P. Gavrilovic, K. Meehan, N. Holonyak, Jr., and K. Hess. Solid State Communications, Vol. 45 (1983), p. 803.
2. J.M. Brown, N. Holonyak, Jr., M.J. Ludowise, W.D. Dietz, and C.R. Lewis, Electronic Letters, Vol. 20 (1984), p. 204.
3. C.W. Magee, R.E. Honig, and C.A. Evans, Jr., Springer Series in Chemical Physics, Vol. 19 (Springer, Berlin, Heidelberg, NY 1982), p. 172.

Elemental Quantification Through Thin Films and Interfaces

A.A. Galuska and N. Marquez

Materials Sciences Laboratory, The Aerospace Corportion, El Segundo, CA 90245, USA

Due to its high sensitivity for most elements and good depth resolution, SIMS has great potential for characterizing thin films and interfaces. However, quantitative SIMS analysis is extremely difficult at interfaces where ion yields (τ) and sputtering rates ($\dot{z}$) can change dramatically as a function of matrix composition (matrix effects). While these variations have not generally been well characterized, linear variations of τ and $\dot{z}$ with matrix composition have been observed for III-V semiconductors [1,2] and metal silicides [3]. This linearity may be precisely calibrated and used for quantitative thin film analysis [4]. In this paper, methods of applying these calibration lines to elemental quantification through thin films and interfaces are discussed.

$Al_xGa_{1-x}As/GaAs$ (x = 0, .22, .3, .34), Ta_xSi_{1-x}/Si, (x = 0, .33, .43, .56, .61) and Pt_xGe_{1-x}/Ge (x = 0, .33, .40, .50) samples were fabricated by molecular beam epitaxy and sputter deposition. Dopants were introduced by implantation, and matrix composition was verified by Rutherford backscattering spectrometry (RBS). All of the matrices were analyzed using an ARL IMMA (Primary beam: 18 keV O_2^+, 90° incident angle). The $Al_xGa_{1-x}As$ matrices were also analyzed using a CAMECA IMS-3F (Primary beam: 5.5 keV O_2^+, 60° incident angle). Depth measurements were performed using a Dektak IIA profilometer. To minimize the influence of instrumental variations, relative ion yields ($R\tau$) and relative sputtering rates ($R\dot{z}$) were used to determine the calibration lines [1].

Both the $R\tau$ and $R\dot{z}$ calibration had good linearity ($r^2 \hat{=} 0.95$). The slopes of the $R\tau$ calibration lines were positive for $Al_xGa_{1-x}As$ and Pt_xGe_{1-x}, and negative for Ta_xSi_{1-x}. Alternatively, the slopes of the $R\dot{z}$ lines were negative for $Al_xGa_{1-x}As$ and Pt_xGe_{1-x}, and positive for Ta_xSi_{1-x}. The $Al_xGa_{1-x}As$ $R\tau$ calibration lines were much steeper (2-50 times) when obtained with the IMS-3F indicating the presence of more extensive matrix effects when using this instrument. Matrix effects are reduced in the case of the IMMA because the steady-state concentration of oxygen in the sample surface is greater, as a result of a greater primary beam energy and normal primary beam incidence. In contrast to the ion yield lines, the $R\dot{z}$ lines varied little when obtained by the different instruments.

The concentration C_p (atoms/cm^3) of an analyte at any point of a depth profile can be determined using the relationship

$$C_p = I_p/(\tau_p \cdot \dot{z}_p \cdot A) \qquad (1)$$

where I_p, τ_p and $\dot{z}_p$ are the ion intensity (counts/sec), useful ion yield (ions detected/atoms sputtered), and sputtering rate (cm/sec) associated with a given point of a depth profile. A is the analysis area (cm^2).

When a linear relationship between τ, $\dot{z}$, and x exists, τ_p and $\dot{z}_p$ can be determined from the $R\tau$ and $R\dot{z}$ calibration lines as a function of x [4]. Thus, the only factor-limiting quantification through layered multimatrix structures is a knowledge of the matrix composition (x) as a function of depth. The depth distribution of x can be determined using either a complimentary analytical technique, or the SIMS data alone.

In many instances, the matrix composition of the discrete layers is known from growth parameters, or can be determined using a second analytical technique, such as RBS. In those cases, the determination of x through the interface represents the only difficulty. As shown by Galuska et al. [2], a linear x gradient may be effectively used to model certain well-characterized interfaces. For ill defined interfaces, such as that shown in Fig. 1 for 2600 Å of $TaSi_2$ on Si, it is better to let the ion intensities of the matrix elements define the interface. In Fig. 2, the x distribution was determined independently from both the Ta and Si distributions by assuming that the change in ion intensity with matrix was proportional to the change in x ($\Delta I \propto \Delta x$). The average of the two x distributions was then used in the correction procedure. While this method is not rigorous, it has allowed a quantitive correction of the depth scale and the implant distributions. Rather than exhibiting a buildup at the interface, the corrected B and P distributions are roughly Gaussion in the $TaSi_2$ and broaden into the Si. This behavior is expected for implants through a high density layer into a low density substrate.

Providing there is a known relationship between the compositions of the layers, the x distribution can also be determined from the SIMS data alone. For example, the concentration of As in layers of $Al_xGa_{1-x}As$ is always constant. Galuska et al. [4] incorporated this fact and the calibration lines for $R\dot{z}$ and $R\tau_{As}$ into (1), and performed an iterative procedure to obtain x at each data point. A similar procedure may be performed for binary compounds, such as Pt_xGe_{1-x}, where the atomic fractions of the two constituents always equal 1. Two equations relating two unknowns (C_{Ge}/C_{Pt} and x) can be obtained, and solve for

$$C_{Ge}/C_{Pt} = (\tau_{Pt} \cdot I_{Ge})/(\tau_{Ge} \cdot I_{Pt}) \tag{2}$$

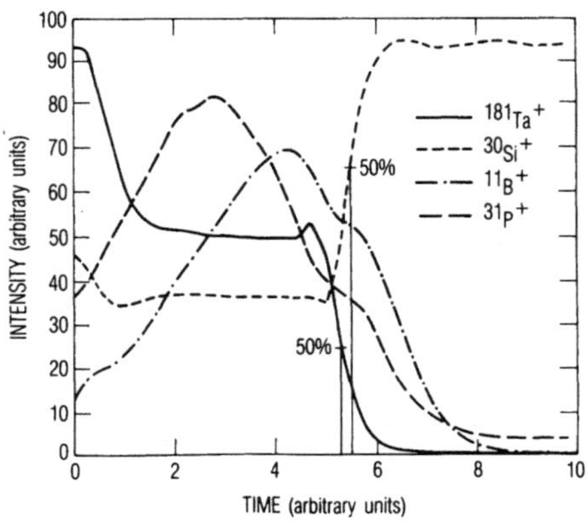

Fig. 1 Uncorrected profile of B and P implants through 2600 Å of $TaSi_2$ on Si.

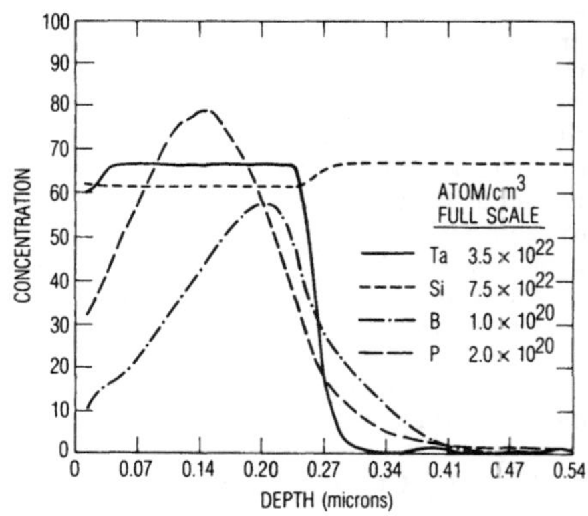

Fig. 2 Concentration profile of B and P implants through 2600 Å of $TaSi_2$ on Si.

where $\tau = F(x)$

$$x = 1/(1 + (C_{Ge}/C_{Pt}) \tag{3}$$

x at each point of a depth profile using an iterative procedure. Figure 3 is the uncorrected SIMS profile of 2100 Å of $PtGe_2$ on Ge, verified by RBS and x-ray diffraction. Due to matrix effects, the intensity distribution of Ge is inversely related to its concentration distribution. Upon performing the iteration procedure, this distorted profile is corrected to the profile in Fig. 4. The atomic fractions and the depth of the interface match the RBS data quite well.

In conclusion, linear relationships between τ, $\dot{z}$ and x can be calibrated for silicides, germanides, and III-V semiconductors using relative ion yields and relative sputtering rates. These calibration lines may then be applied in a variety of ways to quantitative thin film and interface analysis.

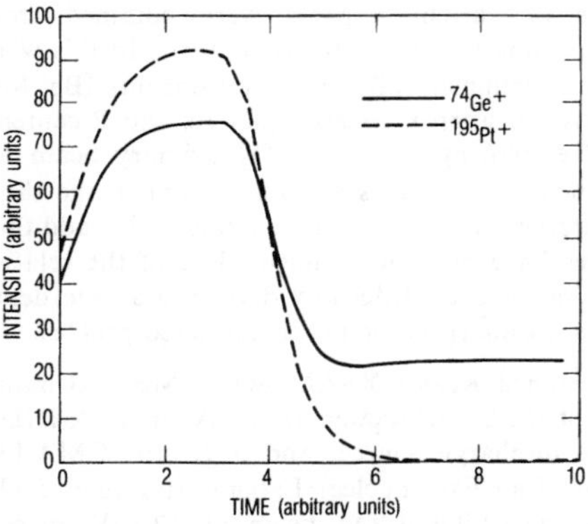

Fig. 3 Uncorrected profile of 2100 Å of $PtGe_2$ on Ge.

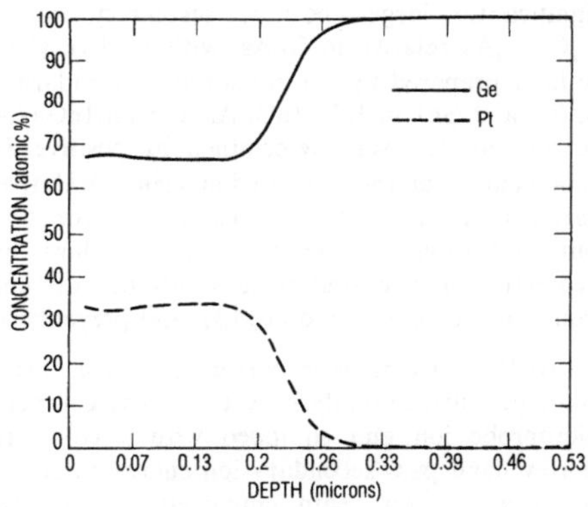

Fig. 4 Corrected profile of atomic percentages of Ge and Pt 2100 Å of $PtGe_2$ on Ge.

References

1. A. A. Galuska and G. H. Morrison: Anal. Chem. 55, 2051 (1983).
2. A. A. Galuska and G. H. Morrison: Int. J. Mass Spectrom. Ion Processes 61, 59 (1984).
3. J. E. Chelgren, W. Katz, V. R. Deline, C. A. Evans, R. J. Blattner and P. Williams: J. Vac. Sci. Technol. 16, 324 (1979).
4. A. A. Galuska and G. H. Morrison: Anal. Chem. 56, 74 (1984).

Matrix Effect Quantification for Positive SIMS Depth Profiling of Dopants in InP/InGaAsP/InGaAs Epitaxial Heterostructures

W.C. Dautremont-Smith

AT & T Bell Laboratories, Murray Hill, NJ 07974, USA

Quantitative depth profiling of a single dopant through an optoelectronic heterostructure wafer requires a reference standard of that dopant in each layer composition encountered, typically InP, $In_{0.53}Ga_{0.47}As$, and one or more lattice-matched quaternary alloy compositions $In_{1-y}Ga_yAs_xP_{1-x}$.[1] This matrix calibration problem has previously been addressed by GALUSKA AND MORRISON [1] and MEYER et al. [2] for the $Al_xGa_{1-x}As/GaAs$ heterostructure system. Significantly larger positive secondary ion yields of various dopants were obtained from $Al_xGa_{1-x}As$ relative to GaAs, with relative yield increasing linearly with Al content. In this work we have prepared by ion implantation standards of five technologically important dopants (Be, Mg, Si, Zn and Sn) in InP, InGaAs and an InGaAsP layer of approximately equal As and P content. Dopant profiles were determined by positive SIMS employing a reactive O_2^+ primary beam for enhancement of the positive ion yield. Relative secondary ion yields of each dopant in the three matrices are presented, in each case for two different secondary ion energies. In addition, sputtering rates and yields of the three host matrices have been determined. Use of the relative secondary ion and sputtering yields permits conversion of the SIMS raw data into a true depth profile in the heterostructure [3], and prevents serious misinterpretation of uncorrected profiles.

SIMS analyses were carried out in a custom-designed Kratos XSAM 800/SIMS 800 multi-technique surface analysis system. Key components of the SIMS section are an Atomika A-DIDA microprobe ion gun equipped with a cold cathode discharge source, and a Kratos CMA-like narrow band-pass secondary ion energy filter in front of an Extranuclear Laboratories quadrupole. The primary ion beam employed was a 1.00 μA mass-filtered O_2^+ beam at 12 keV energy, impinging on the sample within a few degrees of normal incidence. Positive secondary ions emitted at about $(45 \pm 25)^\circ$ to the sample normal were collected by the energy filter, which because it employed a very small field of $\sim 1 Vcm^{-1}$ for ion collection had little effect on secondary ion trajectories. Since the energy filter transmitted ions of a fixed energy, sample biasing was used to vary the as-emitted energy of the detected ion [4,5]. After initial optimization, all secondary ion collection and detection parameters were kept constant throughout this work.

For analyses of a particular dopant from the different matrices, the fraction of time spent counting on the dopant peak remained constant. Since the object of this work is to determine relative secondary ion yields of a given dopant from various matrices, it was imperative to minimize instrumental sensitivity variations between measurements. Accordingly, small pieces of each matrix were mounted side-by-side on a single sample mounting stud, with analyses performed sequentially. Etched crater profiles in both raster directions were measured on a Sloan Dektak IIA stylus instrument, whose microcomputer was used to calculate the crater cross-sectional area along each axis. The product of the two cross-sectional areas divided by the mean crater depth gave the sputtered volume used to determine sputtering yields.

Samples with a given dopant were prepared by simultaneous ion implantation of that dopant into the three matrices, at the energies and doses shown in Table 1. Implant energies were chosen

1 Henceforth referred to in abbreviated form as InGaAs and InGaAsP, respectively.

Table 1 Ion implants performed simultaneously into InP, InGaAsP and InGaAs.

Implanted Ion	Dose (10^{15}cm^{-2})	Energy (keV)	Mean Range (μm)
$^{9}Be^{+}$	1.00	50	0.22
$^{24}Mg^{+}$	1.00	100	0.14
$^{28}Si^{+}$	1.00	120	0.13
$^{64}Zn^{+}$	5.00 5.00	190 300[a]	0.11 0.16
$^{120}Sn^{+}$	1.00	380	0.11

[a]into InP and InGaAs only

to give a mean projected range between 0.11 and 0.22 μm, so that only a small fraction of the dose was within the initial 25 nm required for the sample surface to achieve its dynamic equilibrium oxygen content, and hence its asymptotic level of positive ion yield enhancement. SIMS depth profiles for each dopant/matrix combination were recorded down to the background level for that dopant, with a dynamic range of at least two decades. After background subtraction, the total number of dopant ions detected was obtained by numerical integration over the depth profile. The 1×10^{15}cm^{-2} doses produced peak concentrations of $4-6\times10^{19}$cm^{-3}.

InGaAs and InGaAsP layers, of thickness 0.5 to 0.8 μm, were grown by liquid phase epitaxy on n-InP substrates. All layers were Zn doped, except those used for Zn implantation, which were undoped. The InGaAsP used for the Be, Mg and Si implants was of composition $In_{0.80}Ga_{0.20}As_{0.44}P_{0.56}$, with $In_{0.78}Ga_{0.22}As_{0.48}P_{0.52}$ and $In_{0.74}Ga_{0.26}As_{0.57}P_{0.43}$ used for the Zn and Sn implants, respectively.

Secondary ion energy distributions of selected atomic and molecular ions emitted from InP and InGaAs were measured in order to characterize the energy filter performance and to indicate the actual detected secondary ion energies corresponding to given sample bias settings. The energy distributions of As^{+} ions from InGaAs are similar to those from GaAs, as reported by CLEGG [5]. Firstly, a sharp drop in ion intensity with sample bias in the range +3v to +6v indicates the narrow energy pass of the filter, estimated to be about 1 eV. Secondly, since the ion intensity goes to zero at zero energy, zero sample bias corresponds to sampling ions of 6 eV energy, just above the peak in the distributions at ~4 eV. Similarly, sample bias levels of -20v and -30v correspond to detection of secondary ions of initial energy 26 eV and 36 eV respectively, in each case with 1 eV resolution.

The major points of this work are clearly discernible in Fig. 1, which shows superimposed the SIMS Be^{+} intensity versus time (measured in number of scans) profiles from the Be implanted InP, InGaAsP and InGaAs samples. These spectra were recorded at a secondary ion energy of 26 eV. Firstly, the total number of Be^{+} ions detected increases from InP to InGaAsP to InGaAs. Secondly, since the mean projected range of the implant is about equal in the three matrices, the sputtering rates of InGaAs and InGaAsP are about half and two thirds that of InP respectively. Similar profiles are obtained at a 6 eV ion energy, except that the ion intensities are larger and the ratios of the areas under the InGaAs and InGaAsP profiles relative to the InP profile are somewhat reduced.

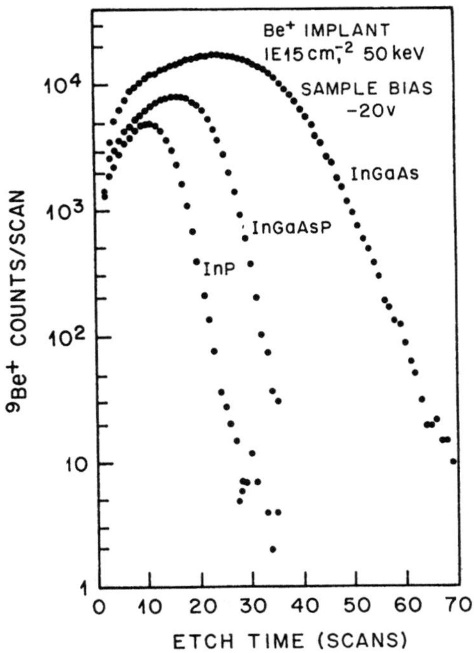

Fig. 1 $^{9}Be^{+}$ ion intensity as a function of ion etch time, i.e., raw SIMS depth profiles of Be, from Be implanted InP, InGaAsP and InGaAs. Be concentration at the peak is 4×10^{19} cm^{-3} in each case.

The secondary ion yields from InGaAs and InGaAsP relative to InP, for Be and for Mg, Si, Z and Sn are shown in Fig. 2, plotted as a function of the As fractional group V content of th matrix. Data for two different secondary ion energies, corresponding to the sample bias value indicated, where one energy is just above the peak in the energy distribution and the other is 20 o 30 eV more energetic, are included. It is evident that for all five dopants the secondary ion yiel increases sublinearly with fractional As content of the matrix, and that this matrix effect is mor pronounced for more energetic secondary ions. The magnitude of the matrix effect increases wit the first ionization potential of the dopant, or in general decreases with increasing yield from InP.

Finally, in Fig. 3 the sputter etch rates of the three matrices are plotted as a function of the A fractional group V content. The fractional errors in the etch rates for the LPE grown epitaxia layers are larger than for the bulk InP due to surface roughness contributions, and due to laye thickness limitations on the etched crater depth. The right hand ordinate in Fig. 3 shows the etc rates converted into absolute sputtering yields, defined as the total number of atoms sputtered pe incident molecular O_2^+ ion. There is a single conversion factor from sputtering rate to yield for th full composition range, since we are only considering lattice-matched epitaxial layers. A typica etched crater profile obtained in this work is shown in the inset of Fig. 3, with the gated regio indicated.

The higher secondary ion yields obtained for increasing As (and Ga) content may be due to higher dynamic equilibrium O surface concentration, causing a greater enhancement of positiv secondary ion yield. The larger magnitude of the transient occurring in the initial ~30nm whil dynamic equilibrium is being achieved supports this model. A higher surface O concentratio during O_2^+ ion etching is to be expected by virtue of the lower sputter removal rate (Fig. 3) an the greater tendency to oxidation (i.e., the higher O sticking coefficient) of InGaAs relative to InP A further minor contribution may come from the slightly smaller implantation depth of 12 keV O_2^+ ions into InGaAs. A similar explanation based on the first two points applies for the increase

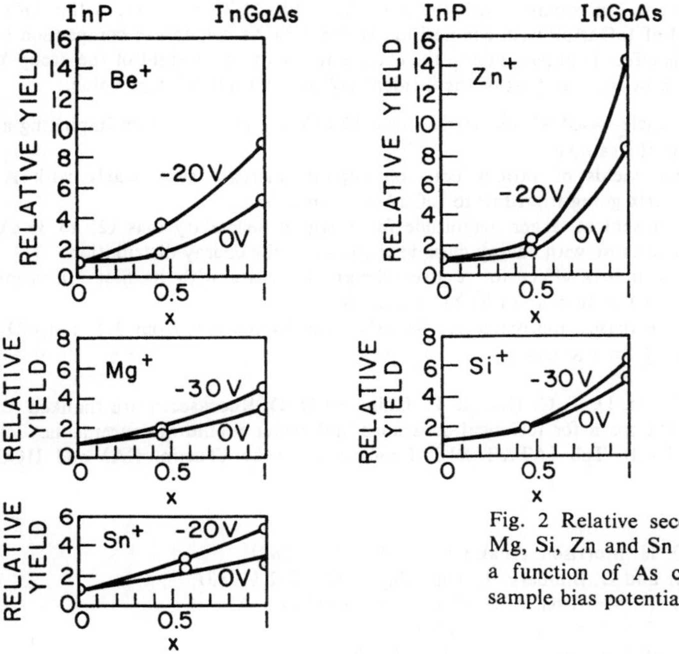

Fig. 2 Relative secondary ion yields of Be, Mg, Si, Zn and Sn from $In_{1-y}Ga_yAs_xP_{1-x}$ as a function of As content, measured at the sample bias potentials indicated.

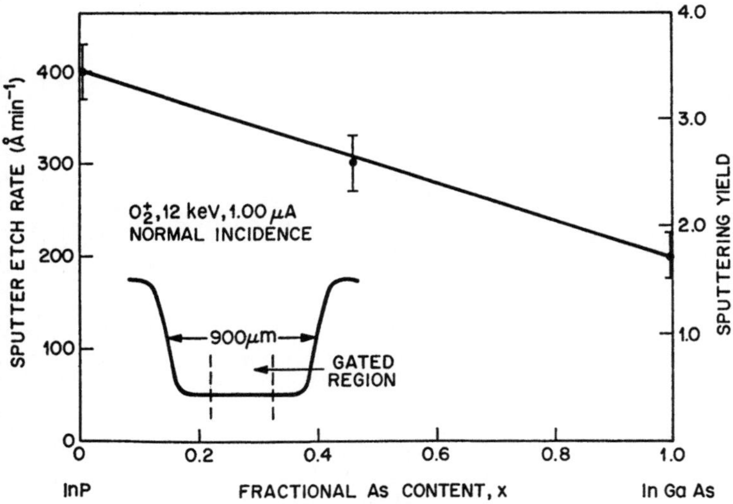

Fig. 3 Sputtering rates and yields (atoms/O_2^+) of $In_{1-y}Ga_yAs_xP_{1-x}$ as a function of composition.

secondary ion yields from $Al_xGa_{1-x}As$ relative to GaAs [2]. A difference between the results here for InGaAsP and those for AlGaAs [1,2] is the linear dependence of the relative secondary ion yield on Al content of the AlGaAs, compared to the sub-linear dependence (Fig. 2) on As content in InGaAsP. Note that this dependence remains sub-linear if the yields are plotted versus Ga content in place of As content. In common with the AlGaAs results [1,2], however, is the linear decrease in sputtering yield of InGaAsP with composition (increasing As content, in comparison to increasing Al content). This effect is in part due to the change in the atomic weight of the group V component from a value almost equal to that of the incident O_2^+ ion to a much larger value.

In summary, in a quadrupole-based SIMS system, for a 12 keV O_2^+ primary beam impinging at close to normal incidence on the sample:

(i) Positive secondary ion yields of various common dopants increase sub-linearly with As content of the host matrix going from InP to InGaAsP to InGaAs.

(ii) The matrix effect is present at larger magnitude for energetic secondary ions (25 to 35 eV range) in comparison to those with energy close to the peak in the energy distribution.

(iii) The magnitude of the matrix effect for a given dopant increases with its first ionization potential, and can be greater than an order of magnitude.

(iv) The sputtering yield decreases approximately linearly with As content, from 3.4 atoms/O_2^+ for InP to 1.7 atoms/O_2^+ for InGaAs.

Acknowledgements J. V. Florio, D. T. C. Huo, B. H. Chin and D. D. Roccasecca are thanked for the epitaxial layers, as is J. Lopata for technical assistance and crater profile measurements. Ion implants were performed by B. Tell (AT&T BL), Leonard J. Kroko (Tustin, CA) and IICO (Santa Clara, CA).

References

1. A. A. Galuska and G. H. Morrison, Anal. Chem. *55*, 2051 (1983).
2. Ch. Meyer, M. Maier and D. Bimberg, J. Appl. Phys. *54*, 2672 (1983).
3. A. A. Galuska and G. H. Morrison, Anal. Chem. *56*, 74 (1984).
4. K. Wittmaack, Appl. Phys. Lett. *29*, 552 (1976).
5. J. B. Clegg, Surface and Interface Analysis *2*, 91 (1980).

A Comparison of Electron-Gas SNMS, RBS and AES for the Quantitative Depth Profiling of Microscopically Modulated Thin Films

H. Oechsner[1], *G. Bachmann*[1], *P. Beckmann*[1], *M. Kopnarski*[1], *D.A. Reed*[2], *S.M. Baumann*[2], *S.D. Wilson*[2], *and C.A. Evans, Jr.*[2]

[1] Fachbereich Physik der Universität Kaiserslautern, D-6750 Kaiserslautern, F. R. G.

[2] Charles Evans & Associates, 1670 South Amphlett Boulevard, Suite 120, San Mateo, CA 94402, USA

A variety of surface analytical techniques are in wide use today for the characterization of thin film systems, particularly in the semiconductor industry. Traditionally, the most popular technique for the characterization of thin film composition and interface impurity analysis levels is Auger electron spectrometry (AES). This technique provides the ability to perform high resolution depth profiling using low energy ion beam sputtering (1 to 3keV). For quantitative stoichiometry and thin film thickness measurements we employ Rutherford backscattering spectrometry (RBS) using MeV helium ions. In an attempt to find a technique which can provide the quantitative capabilities of RBS, depth resolutions as good or better than AES, and improved detection limits, we have begun an investigation of electron gas sputtered neutral mass spectrometry (SNMS). This paper will address the results of this study as they relate to the depth resolution of the SNMS versus AES, with RBS used to characterize the film compositions.

The samples employed for this study were a series of W-Si films deposited by placing a silicon substrate on a rotating platen and sequentially depositing W and Si from a two-gun sputtering system. The nominal total thickness of the W-Si films is 2400 angstroms. According to the RBS measurements the compositions of the deposited films vary from $WSi_{1.8}$ to $WSi_{3.3}$.

A direct comparison of the analyses of films for the same wafer with RBS and SNMS is presented in Figs. 1 and 2. Contrary to the RBS results, the strong modulation of the SNMS signal for W with sputtered depth (Fig. 2) demonstrates a layer structure of the films. By additional SNMS measurements it has been shown that the Si signal is modulated inversely, i.e., that maximum Si signals coincide with minimum W. From the total layer thickness, the partial thickness of each of the 25 W-Si double layers detected by SNMS in Fig. 2 is found to be 105 angstroms. By integrating the modulated W and Si signals over the sputter time, the mean concentration values obtained by SNMS are confirmed by RBS.

The measurement shown in Fig. 2 was performed with the Leybold-Heraeus INA-3 (SNMS) system being operated under the conditions for optimum depth resolution, i.e., extremely high lateral homogeneity of the bombarding ion current and a bombarding energy in the 10^2 eV regime [1,2,3]. The detailed experimental conditions are given in Table 1 together with those used for a comparative investigation of the same sample with AES sputter depth profiling. For AES the primary electron beam (5 keV, raster width 40 mm) was centered in the sputtered area (3.2 x 3.2 mm^2) to avoid crater effects.

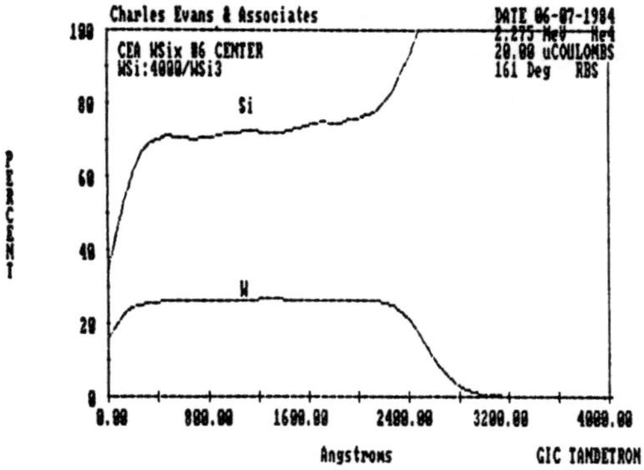

Fig. 1 RBS depth profiles of W and Si for a sputter deposited WSi layer (nominal composition $WSi_{2.9}$) on Si.

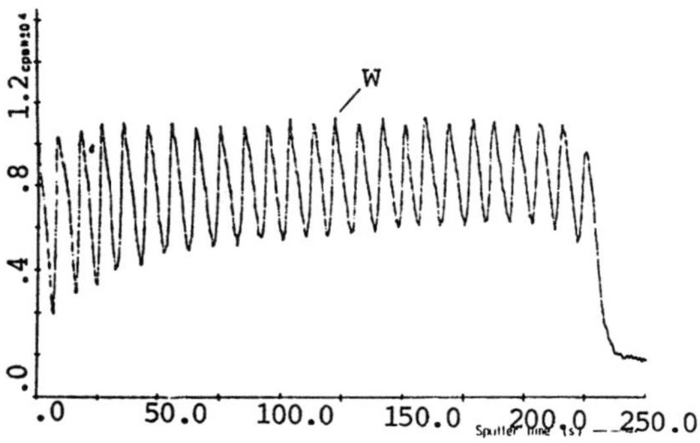

Fig. 2 Low energy SNMS depth profile of W for the same sample as in Fig. 1. For experimental details see Table 1.

The periodic structure detected by SNMS was confirmed by the AES measurements. The modulation of the AES signals, however, was less pronounced. The corresponding values of the modulation ratio R between the maximum and the minimum signals at the beginning and the end of the depth profiles obtained with SNMS and AES are included in Table 1, together with the experimental interface width between the W-Si layer system and the Si substrate. $\Delta z_{90/10}$ corresponds to the width of the 90% to 10% variation of the signals at that interface. Both the smaller modulation ratio and the broader interface measured with AES should presumably be not only attributed to the higher ion bombarding energy but also to the larger information depth of AES. While the mean escape depth for the Auger electrons from the 1736eV MNN transition of W used for the present measurements is about 30 angstroms, the information depth for SNMS with Ar^+ ions of 280eV should not exceed two atomic distances [2]. The complete experimental results and more details will be presented in a forthcoming paper.

While the structure of the multilayer system discussed so far was detectable also by AES, the modulated structure of a different set of sputter deposited W-Si samples with a periodicity of 30-40 angstroms could be resolved solely by low energy SNMS. An example measured with the SNMS equipment in Kaiserslautern is shown in Fig. 3. More than 65 oscillations of the W signal (or of the alternatively varying Si signal) have been determined with a constant periodicity of 32 angstroms, corresponding to the thickness of the individual W-Si layers. As the modulation structure in Figure 3 is resolved down to the film substrate interface, the low modulation ratio in this case is presumably not only due to limitations in depth resolution but

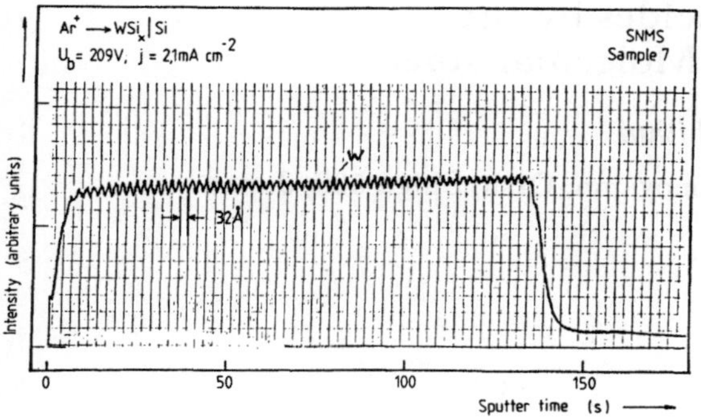

Fig. 3 Low energy SNMS depth profile for W in a highly structured sputter deposited silicide layer (nominal composition also $WSi_{2.9}$).

Table 1

	SNMS		AES	
Bombarding conditions	Ar^+; 280eV; 9.5 x 10^{-4} Acm^{-2} normal incidence		Ar^+; 1 keV; 7.9x10^{-6} Acm^{-2} 150° against normal	
Modulation ratio R				
Begin. of profile	W: 3.4	Si: 2.3	W: 1.7	Si: 2.0
End of profile	1.8	1.6	1.3	1.4
$\Delta z_{90/10}$ (exp)	86 angstroms		280 angstroms	
Profiling time	233 sec		12 480 sec	

also governed by the intrinsic stoichiometry variations in the sample. The composition of this sample corresponds to $WSi_{2.9}$. From this the average atomic distance in the layer is 2.54 angstroms using the mean atomic radii in elemental W and Si. Hence, the thickness for a single sublayer of W would correspond to 8 angstroms or approximately 3 atomic layers in a system consisting of elemental sublayers. However, a complete separation of such thin sublayers is unlikely with respect to atomic roughness effects and residual atomic mixing during sputter deposition. The total sputtering yield for the $WSi_{2.9}$ layer obtained from the SNMS profile in Fig. 3 amounts to 0.92 atoms/ion for normal bombardment with 209 atoms/ion for W and Si, respectively [4]. This indicates again that in agreement with the low modulation ratio of the SNMS signals the sample in Fig. 3 is a compound rather than a sequence of well-separated elemental layers. Nevertheless, SNMS depth profiles as shown in Fig. 3 demonstrate the possibility to resolve chemical variations of only a few atomic distances wide across a depth of several 1000 angstroms.

In conclusion, low energy SNMS has been shown to display unique properties for compositional depth profiling with high depth resolution. The present study demonstrates the advantages being achieved when different analytical techniques with specific figures of merit are applied for the same analytical problem.

References

1. See, e.g., paper by H. Oechsner in the present volume.
2. H. Oechsner, H. Paulus and P. Beckmann, J. Vac. Sci. Technol. A3, 1403 (1985).
3. E. Stumpe, H. Oechsner and H. Schoof, Appl. Phys. 20, 55 (1979).
4. N. Laegreid and G. K. Wehner, J. Appl. Phys. 32, 365 (1961).

Characterization of Silicides by the Energy Distribution of Molecular Ions

K. Okuno, F. Soeda, and A. Ishitani

Toray Research Center, Inc., Sonoyama, Otsu, Shiga 520, Japan

1. Introduction

Dynamic SIMS is not normally considered to give information about chemical states, although it has advantages of high sensitivity, good lateral resolution and depth profiling capability[1]. Although analyses of chemical states by the fragmentation patterns of molecular ions[2,3] or by energy distributions of secondary ions[4-6] have been tried, no attempts have been done to obtain such structural information under dynamic SIMS conditions with the above-mentioned advantages. It is most desirable to combine both aspects of SIMS from the analytical viewpoint.

We propose a new method to study chemical states by SIMS utilizing the energy distributions of molecular ions. This technique has the potential to analyze chemical states of species even at low concentration and to carry out rapid depth profiling.

In this paper, we study chemical states of titanium silicides. Refractory metal silicides are being used as alternatives to polysilicon gates in MOS-FET devices. Detailed knowledge of composition and chemical states of silicides is essential to design the function of the device. We have measured the energy distributions of molecular ions, $^{76}(TiSi)^+$, $^{56}(Si_2)^+$ and $96(Ti_2)^+$ in several Ti-Si systems to study their dependence on chemical states and carried out depth profiling of the chemical state variation in titanium/titanium silicide/silicon(sub.) multilayers.

2. Experimental

An O_2^+ primary beam (10.5kV, 2μA and 500μm square raster) was used both for measurement of energy distributions of molecular ions, $76(TiSi)^+$, $^{56}(Si_2)^+$ and $96(Ti_2)^+$, and for depth profiling, in a CAMECA IMS-3F. Both energy distributions and depth profiles of molecular ions were measured with an energy window of 10eV width. $TiSi_2$(~2500Å) and TiSi(~2500Å) layers were prepared as model systems by plasma CVD on Si substrates and their structures were confirmed by X-ray diffraction. The composition of interest, Ti(~1000Å)/TiSi (~1000Å)/Si(sub.), was fabricated by annealing at ~700°C after evaporating Ti onto a Si substrate.

3. Results and Discussion

Normalized energy distributions of molecular ions, $^{76}(TiSi)^+$, in TiSi, $TiSi_2$, Si coated Ti and Si are shown in Fig. 1. The ralative intensity of high energy component of $^{76}(TiSi)^+$ is different in the four samples. The high-energy component of $^{76}(TiSi)^+$ in TiSi is larger than that of $TiSi_2$. This is due to the difference in their chemical states. The high-energy component of $^{76}(TiSi)^+$ in Si coated Ti is much smaller than that of TiSi or $TiSi_2$.

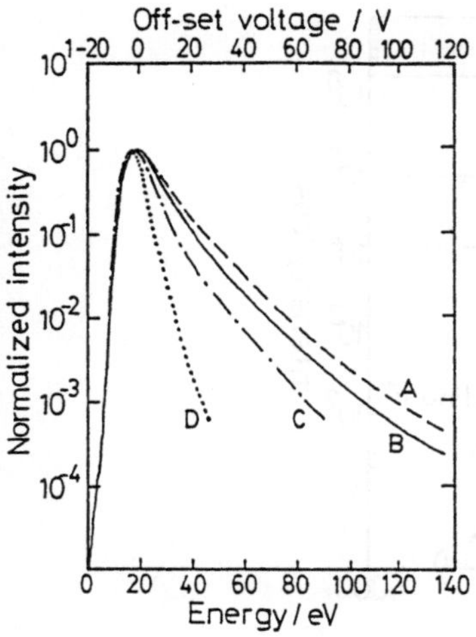

Fig. 1 Energy distribution spectra of $^{76}(TiSi)^+$ in four different Ti-Si systems; A: TiSi, B: $TiSi_2$, C: Si coated Ti substrate, D: Si wafer

This is due to the lack of bonding between Ti and Si. The mass 76 ion from the Si wafer has a drastically different energy distribution because it is a different molecular ion consisting of three atoms, $^{76}(^{30}Si_2{}^{16}O)^+$. A similar difference due to a change in chemical state is observed for the energy distributions of $^{96}(Ti_2)^+$ and $^{56}(Si_2)^+$ in Ti, TiSi and $TiSi_2$.

We have introduced a parameter, A = $^{76}(TiSi)^+$[30V]/$^{76}(TiSi)^+$[0V], to express difference of the energy distribution. [30V] and [0V] are offset voltages. The values of A calculated for four different Ti-Si systems are given in Table 1. This parameter A can be used to monitor the chemical state in depth profiling of multilayers containing different types of titanium silicides.

Figure 2 shows an example of depth profiling with parameter A, together with $^{76}(TiSi)^+$, $^{30}Si^+$ and $^{47}Ti^+$ in Ti(~1000Å)/$TiSi_x$(~1000Å)/Si(sub.). The $TiSi_x$ middle layer in Fig. 2 is identified as $TiSi_2$ from the parameter A value. The value of A is larger in Ti layer than in TiSi or $TiSi_2$, and indicates that in the Ti layer Si bonds with Ti with higher coordination number. Profiles of $^{56}(Si_2)^+$ and $^{96}(Ti_2)^+$ in Ti/$TiSi_x$/Si(sub.) system were also measured. They give similar information for the chemical state of bonding in Ti-Si system.

Table 1. Values of parameter A for various Ti-Si systems with different $^{47}Ti^+/^{30}Si^+$ ratios.

	TiSi	$TiSi_2$	Si coated Ti substrate	Si in Ti layer in Fig. 2*
$^{47}Ti^+/^{30}Si^+$	11	5	37	40
Parameter A $^{76}(TiSi)^+$[30V]/ $^{76}(TiSi)^+$[0V]	7.7×10^{-2}	5.2×10^{-2}	1.9×10^{-2}	1.3×10^{-1}

* This region is pointed with an arrow in Fig. 2.

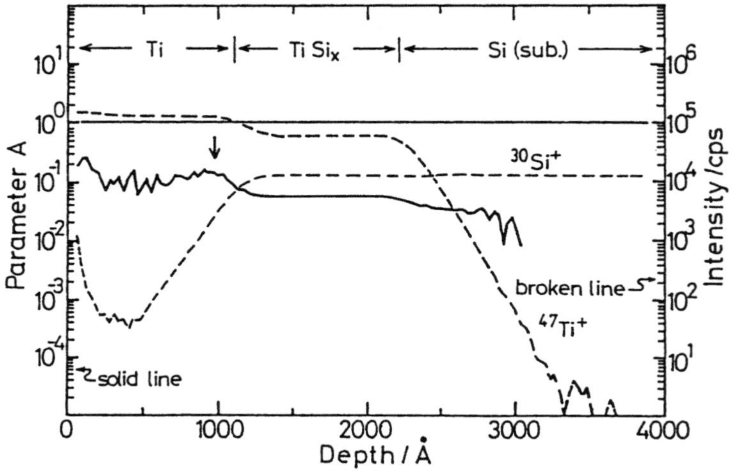

Fig. 2 Depth profiles of $^{76}(TiSi)^+[30V]$ (parameter A), $^{76}(TiSi)^+[0V]$ normalized by $^{76}(TiSi)^+[0V]$, $^{30}Si^+$ and $^{47}Ti^+$ in Ti(~1000Å)/TiSi$_x$ (~1000Å)/Si(sub.)

4. Conclusion

It is confirmed that the energy distribution spectra of molecular ions are different in the chemical variation of titanium silicides. This makes it possible to analyze the chemical states of Ti-Si system with the dynamic SIMS condition. The high sensitivity and depth profiling capability of SIMS can now be fully used to study chemical bonding states of heterogeneous layer composition and also of even low concentration species.

5. References

1. C. A. Evans, Jr. and R. J. Blattner: Ann. Rev. Mater. Sci., 8, 181 (1978)
2. H. W. Werner, H. A. M. de Grefte and J van den Berg, Advances in Mass Spectrometry, edited by A. P. West (Applied Science, Barking, Essex, 1974), vol. VI, p.673
3. H. W. Werner: Mikrochim. Acta. Suppl., 8, 25 (1979)
4. G. Plog, L. Wiedmann and A. Benninghoven: Surf. Sci., 67, 565 (1977)
5. H. Oechsner: Proc. SIMS III, p.106, (1982)
6. M. Riedel and H. Düsterhöft: Proc. SIMS IV, p.73, (1984)

Investigation of Interfaces in a Ni/Cr Multilayer Film with Secondary Ion Mass Spectrometry

M. Moens[1], *F.C. Adams*[1], *D.S. Simons*[2], *and D.E. Newbury*[2]

[1] University of Antwerp (U.I.A.), Dept. of Chemistry, Universiteitsplein 1, B-2610 Wilrijk, Belgium

[2] Center for Analytical Chemistry, National Bureau of Standards, Gaithersburg, MD 20899, USA

1. Introduction

Surface and depth studies of different metal interface systems [1,2] have indicated the potential seriousness of profile distortion phenomena. We used the NBS Ni/Cr multilayered thin film Standard Reference Material (SRM 2135) [3] to study these phenomena and to optimize sputtering conditions so as to achieve maximum interface resolution.

2. Experimental

The periodically modulated metal thin film structure, issued by NBS as Standard Reference Material 2135, consists of nine alternating elemental films (Cr/Ni/Cr...) on a Si-substrate. The Cr and Ni layers have nominal widths of 53 and 66 nm, respectively. Characterization of the thin film structure by Auger sputter depth profiling, Rutherford backscattering spectrometry and neutron activation analysis indicates that the accuracy of the thickness of the metallic layers is better than 5% and that the AES sputter profiles are well defined and reproducible [3,4].

In this work, sputter depth profiles of the thin film structure were obtained with oxygen, argon and cesium ion bombardment in the CAMECA IMS 3F. The primary ions were accelerated to different energies and oxygen bleeding onto the sample was used in some experiments.

3. Results and discussion

3.1. Oxygen ion bombardment

A sputter profile obtained under normal operating conditions used in the CAMECA IMS 3F (8 keV O_2^+ bombardment) is represented in Fig.1a. In the figure, the maximum intensity of each constituent is arbitrarily chosen. The sputtering time is related to the erosion depth assuming that the erosion rates are equal for Cr and Ni. Only the first few layers are well resolved and the resolution becomes increasingly worse with depth. The cause of this poor depth resolution is the formation of surface topography, as was observed with scanning electron microscopy.

Amorphous surface layers generally reduce the amount of surface roughening. Therefore, the system has been studied using oxygen flooding of the sample during ion bombardment. From Fig.1b, it is clear that adsorption of reactive gases on the sample surface leads to better defined layers.

The resolution has also been studied as a function of the primary ion impact energy (Fig.1a and 1c). The measured interface widths at 2 keV O_2^+

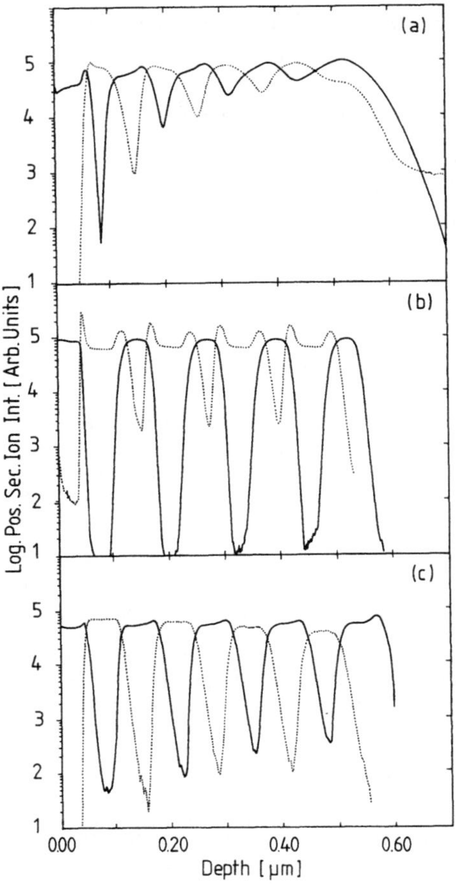

Fig. 1 (left) SIMS depth profiles of Cr (—) and Ni (...) in a Ni/Cr multilayered sample obtained by (a) 8 keV O_2^+ bombardment, (b) 8 keV O_2^+ bombardment and oxygen flooding ($p_{O_2} = 1.10^{-5}$ Torr) and (c) 2 keV O_2^+ bombardment

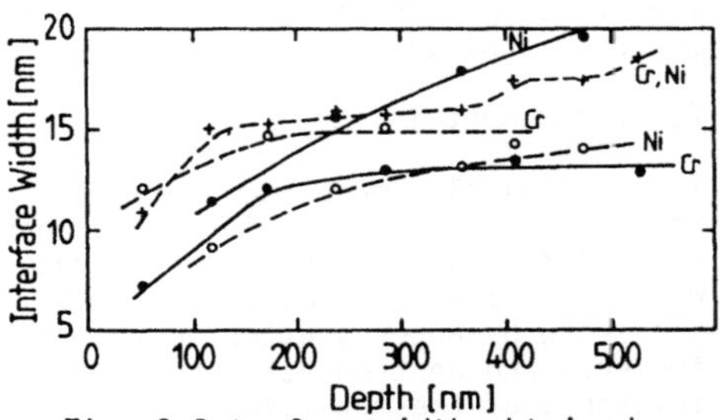

Fig. 2 Interface width obtained as a function of sputtered depth using 2keV O_2^+ bombardment (•), 2 keV Ar^+ bombardment (∘), and 14.5 keV Cs^+ bombardment (+)

bombardment, defined by the drop from 84% to 16% of the plateau secondary ion intensity, are shown in Fig.2. The lower impact energy alters the resolution significantly. Since the penetration depth and the damage-producing effect of the primary ions are strongly energy dependent, the induced roughening is expected to be much smaller at low primary ion impact energies. At the same time, lowering the primary ion energy in the CAMECA IMS 3F results in an increase of the angle of incidence with respect to the surface normal. It is accepted that narrower interfaces can be obtained at lower beam energies and at near-grazing incidence [4], but these effects cannot be treated separately in our instrument.

Looking at the secondary ion intensities in Fig.1, one sees some anomalies, e.g. bombardment at high oxygen backfilling pressures results in an increase of the Ni^+ intensity at each interface. This makes it difficult to determine the exact interface width, since the 100% level is hard to define. The varying ion yield results from matrix effects. PIVIN et al. [5] found that under oxygen saturation Ni^+ ions are about 20 times more effectively produced when sputtered from a NiCr alloy than from pure Ni.

3.2. Argon ion bombardment

The effect of Ar^+ bombardment on the interfacial resolution for 2 keV impact energy is also represented in Fig.2. The figure shows that argon sputtering results in smaller interface widths for the Ni layers,while oxygen bombardment resolves the Cr layers better. These effects are likely to be associated with ion yield variations,depending on the nature of the primary ion and the other experimental conditions used, since they do not occur in the AES sputter profiles [3,4].

3.3. Cesium ion bombardment

Cesium ion bombardment was performed at a rather high impact energy (14.5 keV) but yielded only a small degradation in interface resolution when compared to low energy Ar^+ or O_2^+ bombardment,as can be seen in Fig.2. The heavy mass Cs^+ ion is moving somewhat faster than the light ions and thus produces more damage. Cesium ions at the same low energy would produce a much better resolution, since the extent in depth of the collision cascade resulting from the impact of a heavy mass ion is much smaller than that of a light ion.

4. Conclusion

The best interfacial resolution was obtained using low impact energies, high angles of incidence, oxygen bleeding onto the sample and high mass primary ions. Secondary ion yield variations were observed, in particular at interfaces.

5. References

1. W.O. Hofer and H. Liebl: Appl. Phys. 8, 359 (1975)
2. W.O. Hofer and P.J. Martin: Appl. Phys. 16, 271 (1978)
3. J. Fine and B. Navinsek: J. Vac. Sci. Technol. (in press)
4. J. Fine, P.A. Lindfors, M.E. Gorman, R.L. Gerlach, B. Navinsek, D.F. Mitchell and G.P. Chambers: J. Vac. Sci. Technol. (in press)
5. J.C. Pivin, C. Roques-Carmes and G. Slodzian: Int. J. Mass Spectr. Ion Phys. 26, 219 (1978)

Sputter Depth Profiles in an Al-Cr Multilayer Sample

D. Marton and J. László*

Physical Institute of the Technical University Budapest, POB 112,
XI. Budafoki ut 8., H-1111 Budapest, Hungary

1. Introduction

Interface broadening due to sputter depth profiling can now easily be checked using Ni-Cr multilayers [1]. Other structures could also be studied in order to establish general rules related to interface broadening as well as to distinguish the various factors involved. Two problems often arise in such studies: (1) the quantitative evaluation of interface widths when the observed profiles are not completely resolved (i.e., the measure of concentration does not encompass the full range from 100% to 0%); (2) the interpretation of distorted profiles or those which seem to be the results of various unexplained artifacts.

In the case of SIMS measurements, depth profiles can be seriously affected by sensitivity factor changes. Such results were found for Ni-Cr layer structures using an oxygen ion beam by W. REUTER [2], and also by one of the authors [3]. Similar results, although with moderate artifacts, have been shown at this conference by MOENS et al [4].

2. Evaluation of depth profiles

Some factors contributing to the broadening of an originally sharp interface lead to an error-function-like, symmetric profile shape (e.g. surface roughness, statistical sputtering, spike diffusion). Some others lead to more or less asymmetric profile shapes (e.g. preferential sputtering, matrix effects, etc.). In many cases studied thus far the first group of factors is probably the greater, and thus the error function superposition approximation (EFSA) can be used to evaluate these depth profiles [5,6].

The basic idea of the EFSA method can be understood from Fig.1. The rectangular shape of the "real profile" is changed to an error function-like shape due to sputtering. The "measured profile" is obtained as a superposition of

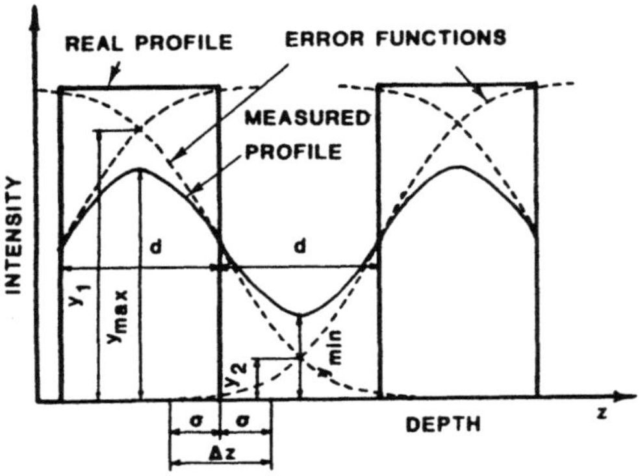

Fig.1 Schematics of the EFSA method. See explanation in the text.

---

*Temporary address: Surface Science Division, National Bureau of Standards, Gaithersburg, MD 20899, U.S.A.

these error functions. Due to the rapid decay of the error function with the depth, the contributions of the underlaying layers can, in general, be neglected. Thus the minimum value of the measured profile (y_{min}) is twice the value of the error function (y_2) at a distance d/2 from its center, where d is the thickness of the layers when uniform layer thicknesses are used. Similarily, the maximal value of the measured profile is

$$y_{max} = 1-2(1-y_1) = 2y_1-1 \tag{1}$$

where the full height of the real profile (Fig. 1) is taken to be equal 1. Using the definition for the depth resolution (Δz) that it is twice the standard deviation of the error function, one obtains

$$\Delta z = d[\sqrt{2}\,\mathrm{erf}^{-1}(\frac{y_{max}}{y_{max}+y_{min}})]^{-1} \tag{2}$$

This result may be extended for the case of non-uniform layer thicknesses. As a first approximation we again take into account only two neighbouring layers with thicknesses d_o and d_e. The relation of the minimal and maximal values of the depth profiles are then expressed as

$$\frac{y_{min}}{y_{max}} = \frac{1-\mathrm{erf}\frac{d_e}{\sqrt{2}\Delta z}}{\mathrm{erf}\frac{d_o}{\sqrt{2}\Delta z}} \tag{3}$$

(Here the minimum is in the layer denoted with subscript e). This equation may be further extended for a second neighbour approximation,which can be used for very precise measurements (better than 1%):

$$\frac{y_{min}}{y_{max}} = \frac{(1-\mathrm{erf}\frac{d_e}{\sqrt{2}\Delta z})+(1-\mathrm{erf}\frac{3d_e+2d_o}{\sqrt{2}\Delta z})}{\mathrm{erf}\frac{d_o}{\sqrt{2}\Delta z}+(\mathrm{erf}\frac{3d_o+2d_e}{\sqrt{2}\Delta z}-1)} \tag{4}$$

While using eq. (2) the evaluation of Δz is straigthforward, the use of eqs. (3) and (4) needs somewhat more computation. Nevertheless, it is advisable to determine Δz as quantitatively as possible.

3. Experimental Results

The depth profiles of a Cr-Al multilayer structure obtained by SIMS and SAM measurements are shown in Figs. 2 and 3, resp. The SIMS measurement was carried out using Balzers quadrupole equipment with Ar^+ ions of 3keV energy and 60° angle of incidence to the surface normal. The static ion beam was of fairly uniform current density ($j=10^{-5}Acm^{-2}$, D=2.5mm). Using the same instrument we have obtained good results for Ni-Cr multilayers [5,6]. The SAM depth profile was obtained in a PHI 545 equipment using 3 keV Ar^+ ions (scanning the beam over 2mm, $j=4x10^{-5}Acm^{-2}$) with an incident angle of $\sim 60^o$. The sample was vacuum evaporated on a smooth glass substrate at a background pressure of 10^{-6}Torr. The layer thicknesses were monitored by a quartz meter, and were nearly uniform (10 ± 2nm, except the 4th Al layer, which was about 14nm thick). The carbon and oxygen content were monitored simultaneously, but no significant O or C was detected except on the surface of the sample.

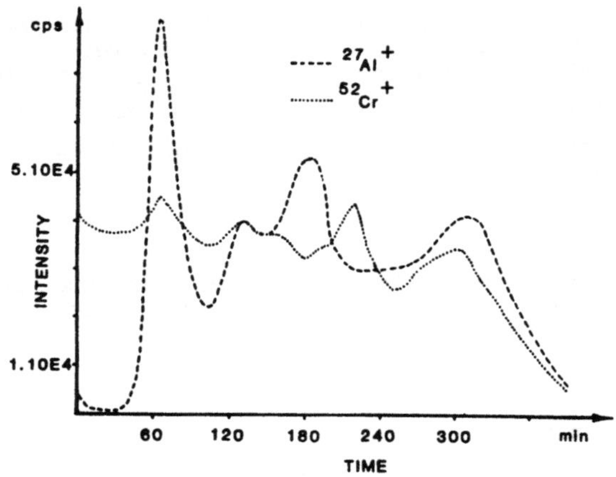

Fig. 2 SIMS depth profile obtained on an Al-Cr multilayer sample containing 6 layers of each type. Ar^+ ions, 3 keV. Such results are typical, but the reproducibility is poor.

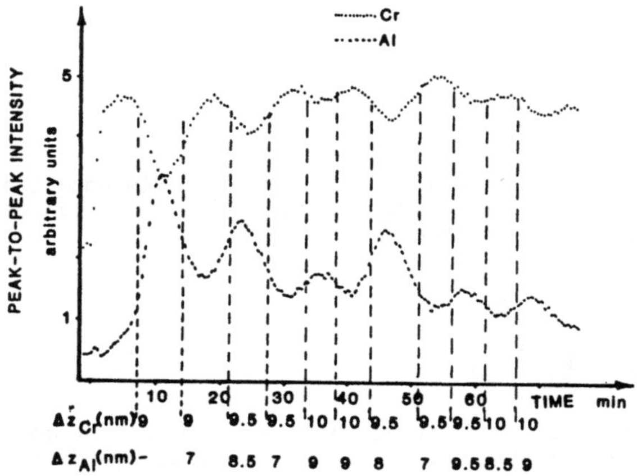

Fig. 3 SAM depth profile obtained on the same sample used in Fig. 2. The layer thicknesses are ~10nm. The interface widths indicated in the bottom of the figure were obtained using the EFSA method.

4. Conclusions

The SIMS depth profiles of Cr-Al films do not result in error-function-like interface profiles and are difficult to interpret. It is most disturbing that such distorted profiles can occur even when sputtering with inert gas ions. It is possible that such profiles result from the significant Al enrichment occurring due to preferential sputtering in the Cr-Al compound. The Al enrichment then may lead to enhanced Al^+ yield in the Cr layer, and the preferential sputtering of Cr may lead to an enhanced Cr^+ yield from the interfaces. Other factors such as diffusion and oxygen-induced processes may play a role, too.

The depth resolutions observed in the SAM measurement are not very good, but they fit in the usual range of interface broadening for polycrystalline metals. It is to be noted, however, that the depth resolution seems to be nearly constant in the depth interval analysed.

References

1. J. Fine, B. Navinsek: J. Vac. Sci.Technol. A 3, 1408 (1985)
2. W. Reuter: private communication, 1982
3. J. Giber, D. Marton, 1981, unpublished
4. M. Moens, F.C. Adams, D. Simons and D.E. Newbury, this conference
5. J. Giber, D. Marton and J. Laszlo: J. de Physique 45, C2-115 (1984)
6. D. Marton, J. László, J. Giber and F.G. Ruedenauer: Vacuum, in print

SIMS Depth Profiling of Multilayer Metal-Oxide Thin Films – Improved Accuracy Using a Xenon Primary Beam

D.D. Johnston[1], *N.S. McIntyre*[1], *W.J. Chauvin*[1], *W.M. Lau*[1], *K. Nietering*[2], *and D. Schuetzle*[2]

[1] Surface Science Western, University of Western Ontario, London, Ontario, Canada N6A 5B7

[2] Ford Motor Company, Dearborn, MI 48121, USA

1. Introduction

The accuracy of SIMS depth profile information, particularly at interfaces, is strongly affected by reactions occurring between the primary ion beam and the surface. Ion beam mixing can result in broadened and distorted profiles [1]. Less well-known are the effects of field-induced migration of mobile species which can accompany the charging of a surface [2]. This paper is concerned with several extreme examples of such chemical reactions near interfaces and with the correction of such effects using alternate primary ions over a range of energies.

2. Experimental

SIMS studies were carried out using a Cameca IMS-3f ion microscope. It was found that intense, stable beams of Ar^+ and Xe^+ could be produced from the duoplasmatron source if a gas mixture of 80% oxygen and 20% either argon or xenon was used. The appropriate primary ion beam was selected using a mass filter. Other primary ions; O^-, Cs^+ and O_2^+ were produced by standard sources in the instrument.

For depth profiling, the primary beam was rastered over a 250 x 250 µm area and a 60 µm analysis area was selected. Beam fluxes of 40-120 µA/cm^2 were typically used, and beam energy at the sample surface could be altered between 2.5 and 10.5 keV.

Rutherford backscattering spectrometry measurements were made using a 2nA beam of 1.6 MeV $^4He^+$ ions produced by a Van de Graaff accelerator. To obtain optimum separation of the backscattered peaks, each specimen surface was mounted at an angle to the beam, typically 30-60°. The detector (resolution fwhm ≤ 20 keV) was mounted at the same angle.

Multiple layers of nominal composition ZnO(50nm)/Ag(10 nm)/ZnO(50nm)/Sn/SiO were prepared at the Glass Technology Center of Ford Motor Company by dc sputtering in an Airco-Temescal ILS-1600 in-line coater with metallic targets and, where necessary, with O_2 gas. Depositions were made onto float glass surfaces containing a thin layer of tin.

3. Results

RBS measurement clearly showed the presence of each separate deposited oxide and metal layer, in addition to silicon from the underlying glass. The absolute thickness of each layer was calculated, assuming the volume density of the metal and oxide layers to be the same as in the bulk materials. The thicknesses, calculated from the Rutherford data are:

ZnO(52.5 nm)/Ag(7.5 nm)/ZnO(30.30 nm)/substrate. Inspection of the symmetry of metal and oxide peaks showed no variance of the peak shape from an apparent Gaussian distribution. No anomalies can be detected at any interface.

Depth profile analysis of the film was undertaken using an O_2^+ primary beam with energy ranging from 5.5 to 8.0 keV. In addition, an O^+ primary beam of 7.5 keV was employed in some studies. In all cases using positive oxygen bombardment the silver distribution was strongly distorted, to the extent that the silver ion signal appeared immediately prior to the signals from the glass interface. A representative profile is shown in Fig. 1(a). Evidence for a relatively sharp ZnO/Ag interface is seen at I_1, but no evidence of the second Ag/ZnO interface (I_2) can be found. At the glass interface (I_3) there is an abrupt suppression of all signals, presumably due to the onset of additional charging. Attempts to reduce the effects of specimen charging by gold coating of the surface and electron beam flooding were unsuccessful, and no change was observed in the silver profile.

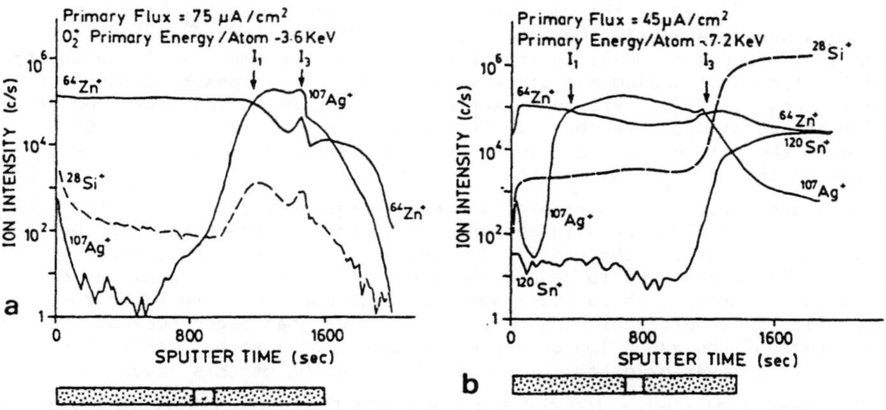

Figure 1. SIMS depth profiles of metal/metal oxide specimen using (a) an O_2^+ primary beam and (b) an O primary beam. Hatched bars show the expected distribution of silver and zinc oxide according to RBS.

Distortion of the silver distribution was also observed in profiles obtained with an O^- primary beam. In this case, however, a significant concentration of silver was detected after only 20% of the ZnO/Ag/ZnO film had been penetrated.

This specimen was also profiled using Ar^+ as the primary beam at energies from 2.5 - 10.5 keV. For energies above 3.5 keV the silver layer could be distinctly separated from the ZnO layers on either side. At the highest beam energies a broad silver peak was observed with a half-width about twice that expected from RBS measurements.

Use of a xenon primary beam resulted in significantly sharper profile shapes and profile widths which more accurately reflect the RBS data. Profiles obtained with primary energies between 4 and 10.5 keV gave silver intensity half-widths of 7.5 nm. By reducing the primary energy to 5.5 keV the relative thickness of the two ZnO layers was determined to an accuracy of better than 10%.

Analysis of the same film with a 5.5 keV Cs^+ primary beam and positive secondary ions gave a sharp profile accurately representing the width of the silver layer, but detecting the layer to be somewhat closer to the interface than its position determined by RBS. Severe distortion of the profile occurred for primary energies outside the range 3.5 - 7.5 keV.

4. Discussion

The effects of O^- and O_2^+ primary ion bombardment are the most severe on the detected distribution of the silver layers. The fact that these two primary ions have the opposite effect on the observed position of the silver secondary ions is believed to be related to the differing charges imparted to the surface by the primary ions. Recent work on insulator charging in this laboratory [3] has indicated that, in the Cameca IMS-3f, Ar^+, Xe^+, O_2^+ and Cs^+ beams lead to positive charging, while O^- results in a negative charge on the surface. Charging of the surface, even by a few tens of eV, is expected to establish an electric field over a 10 nm distance, sufficient to cause migration of ions formed at this distance. The use of O_2^+ primary beams in glasses has been previously found to cause inward migration of sodium [4]. The opposing effects of positive and negative primary beams has also been observed in studies of sodium migration in SiO_2 [5]. The opposing effects of O_2^+ and O^- shown in Fig. 1 are thus ascribed to ionic migration away from or toward surfaces of opposite charge. The enhanced migration caused by O_2^+, compared to the rare gases, is likely due to the enhanced ionization yield of silver in the presence of oxygen.

Using argon greatly improves profile shape presumably because of the lesser degree of ionization with the inert gas ion. As the energy of the primary beam is reduced, the profile distortion increases. While lower energies could be expected to reduce the range of mixing between oxide and metal at the interface, these lower energies also enhance charging on the surface [5]. Xenon produces more accurate profiles than argon because the shorter range of the xenon ion decreases the amount of internal ionization in the metallic layer prior to the onset of sputtering of this layer.

It is clear that charge-induced profile distortions are likely to occur in metal/metal oxide films even if the most judicious choice of primary ion and energy is exercised. Our SIMS studies have demonstrated that beam-induced diffusion of many different elements can occur in oxide matrices. This work attributes much of the diffusion observed to the effects of surface charging. Corrective measures suggested in this work are partly effective in restricting the migration, and additional measures, such as specimen cooling [6] may be of additional benefit.

The authors gratefully acknowledge the assistance of K. Shankar and J. MacDonald, of the University of Guelph with the RBS measurements as well as the financial assistance of the Natural Sciences and Engineering Research Council.

References

1. K. Wittmaack: Vacuum 34, 119 (1984).
2. H.W. Werner and A.E. Morgan: J. Appl. Phys. 47, 1232 (1976).
3. W.M. Lau, N.S. McIntyre, J.B. Metson, D. Cochrane and J.D. Brown: submitted to Surf. Interfac. Anal.
4. H.L. Hughes, R.D. Baxter and B. Phillips: IEEE Trans. Nucl. Sci. NS 19, 256 (1972).
5. D.V. McCaughan, R.A. Kishner and V.T. Murphy: Phys. Rev. Lett. 30, 614 (1973).
6. B.F. Phillips, A.E. Austin and H.L. Hughes: NBS Spec Pub. 400-23, 65 (1974)

Part X

Metallurgical Applications

Metallurgical Applications of SIMS

F. Degrève and J.M. Lang

Cégédur Péchiney, Centre de Recherches BP 27, Voreppe, France

Metallurgical materials are characterized by a complex chemical composition and by heterogeneous microstructure. The surface is different from the bulk and the bulk itself contains many grains, intermetallic phases, inclusions, segregations,... For the investigation of such samples, the intrinsic advantages and the difficulties associated with SIMS will be discussed by examples taken from the literature and from the author's laboratory. Particular attention will be paid to the necessary complementarity with other microscopical techniques.

1. Mass Spectrum

Mass spectra at low resolution ($M/\Delta M = 300$) of the positive and negative ions give a rapid survey of the major elements and of the minor impurities [1] present in an unknown sample. Correct analysis of an element requires high mass resolution ($M/\Delta M = 3{,}000-10{,}000$) [2] to resolve possible interferences.

2. Ion Imaging

With a lateral resolution of 1 µm, direct imaging (Ion Microscopy [3]) is irreplaceable to display instantaneously the lateral distribution of every element, in particular hydrogen [4-6]. With the development of very sensitive channel plate detectors, resistive anode encoders [7] and quantitative image processors [8], ion micrographs of trace elements are now obtained in a few seconds.

At the sub-micron scale, ion microscopy is no longer operative and a scanning ion microprobe [9] must be used. High brightness Liquid Metal Ion Sources (LMIS) [10]providing probes 0.02-0.1 µm in size,are now able to produce very high resolution micrographs.

Both techniques can be used on the same instrument [11].

To gain a maximum of information from ion micrographs, they can be compared with other micrographs of the same area : Optical Microscopy for the pre-location of phases and Scanning Electron Microscopy (SEM) for a semi-quantitative chemical analysis (X ray emission for $Z > 5$ and backscattered electrons for mean Z contrast). For light elements ($Z < 5$), SIMS gives

generally the missing information. This will be illustrated [12] for the characterization of a Al-Li alloy using the following data :
- identification of a new quaternary phase at the interdendritic regions by superposable ion micrographs of Li, Mg, Al and Cu
- visualization of the segregation of Li in the solid solution within the dendrites
- establishment of a calibration curve for quantitative analysis of Li in the matrix by selecting a small area (= 25 μm^2) between phases with a variable field diaphragm (Cf.§3.1.).

The crystallographic contrast [13] observed in the matrix ion micrograph, due to the differential sputtering of adjacent grains (under Ar bombardment mainly), is very useful to relate structural and analytical information. This will be illustrated by the grain boundary enrichment of B [14] and Mn [15] in steel and of Be and Na in aluminium [16].

3. Local Quantitative Analysis

In a multiphase material the quantitative analysis is made difficult [17] since the sputtering vield and the ionization probability change locally ; standard samples of similar structure are needed.

3.1. In the matrix

It is essential to know the solubility limit of an element. Electron Probe Micro Analysis (EPMA) can be used [18], but with the following limitations : inoperative for light elements ($Z < 5$), rather poor detection limit (10-1000 $\mu g.g^{-1}$) and information depth of about 1 μm. On samples with a very fine microstructure, small phases below the surface can contribute to the measured signal, distorting the analysis of the matrix. How SIMS can obviate these difficulties will be illustrated by the determination of the solubility limit of Fe present at a total concentration of $105 \pm 8\ \mu g.g^{-1}$ in an 7X75 aluminium alloy (Al-Zn-Mg-Cu). EPMA gave a histogram ranging between 80 and 170 $\mu g.g^{-1}$. The smallest values probably correspond to the concentration in the solid solution. The higher values correspond to the contribution of Fe rich areas. The SIMS analyses in phase-free areas selected on the Fe^+ ion micrograph gave a well-defined value of $101 \pm 10\ \mu g.g^{-1}$.

In the case of light elements at high concentration, EPMA can be replaced by the Nuclear Microprobe (NM). Line scan analysis of Li across the thickness (800 μm) of a sheet of heat treated Al-Li (2.8 % Li) alloy were carried out by SIMS with a lateral resolution of 1.5 μm and by NM using a 5 μm proton beam (1.2 MeV) ; detection was performed by the $^7Li(p,\alpha)\ ^4He$ reaction [19]. The normalized profiles given by the two methods were superposable and showed a significant depletion of Li extending to about 150 μm from the surface on each side of the sheet.

3.2. Phases, Small Precipitates

The intensity of molecular and atomic ions can constitute a signature for phase identification [20]. When the dimension of the phase exceeds the lateral resolving power of the SIMS instrument, a direct analysis is possible. For instance, PIVIN et al. [21] could establish the stoechiometry of complex oxides of Cr and Ni from the relative intensities of selected ions. Namdar et al. [22] could distinguish between the three iron oxides Fe_2O_3,Fe_3O_4 and FeO by monitoring the significant change in the intensity of Fe_3O^+. When the size of the precipitates is lower than the lateral resolving power, the signal includes a contribution from the matrix and from the precipitates. In this way, NAMDAR et al. could [23] characterize the precipitation rate of NbC and AlN in steel by monitoring the radio $I(Nb^+)/I(Fe^+)$ and $I(Al^+)/I(Fe^+)$.

4. Depth Profiles

For heterogeneous structures in depth, the position of the subsurface interface(s) with respect to the initial sample surface is of prime importance. When naturally parallel (oxide layer, multilayered thin films,...), the sample can be analyzed "as-received". When randomly orientated (precipitates/matrix, grain boundaries, matrix/fiber,...), they have to be preorientated and sample preparation is necessary.

4.1. Thin Oxide Layers

Oxide layers a few nanometers thick covering various metals have been investigated by SIMS [16,24-27] in combination with other surface analytical techniques like AES, XPS and ISS. Important matrix effects due to abrupt variations of sputtering yield and ionization probability arises in the oxide/metal interfacial region. Such analytical artifacts can be minimized by flooding oxygen to saturate the surface, thus imposing a constant oxide stoichiometry. With optimal operating conditions, unique information on the 3D distribution of major and minor elements can be obtained in the oxide, at the interface and in the metal bulk. This will be illustrated by the tremendous surface enrichment (10^3-10^4) of alkaline (Li, Na) and alkaline-earth impurities (Be, Mg) with respect to the bulk of a heat-treated polycrystalline Al sheet [16].

4.2. Multilayered Film

The rapid succession of different matrices makes difficult the calibration of both depth and composition. Complementary data can be given by (Scanning) Transmission Electron Microscopy (TEM/STEM) on cross-sections prepared by ultramicrotomy. Slices of material, typically 50 nm thick and cut perpendicularly to the interface, are investigated in the TEM mode at high magnification (up to x 300,000) to display the size and the morphology. Positioning an e- nanoprobe (several nm in diameter) in the STEM mode gives local X-ray semi-quantitative chemical analysis and electron microdiffraction. Examples on aluminium alloys will be given for a conversion coating layer (50 nm thick oxide) and for 4 deposited metallic layers of different thickness.

4.3. Concentration Profile at the Matrix/Precipitate Interface

Alloys in the as-solidified state usually exhibit a segregated structure, for instance a concentration profile of minor elements in the solid solution between two interdendritic areas, and specially very close to intermetallic phases. High spatial resolution (< 100 nm) and low detection limit (< 10 $\mu g.g^{-1}$) are necessary. Complementary use of SEM and SIMS can resolve the problem as follows [28]. Sub-surface Fe-rich precipitates in the micron range were located in an Al matrix using backscattered electrons at different primary energy (10 - 60 keV) in the SEM. These pictures were then correlated with the surface Fe^+ ion micrograph of the same area. A triangulation procedure allowed the precise location of a precipitate below the surface (0.5 - 1µm). Finally, depth profiles of this small area (8 µm in diameter) showed a significant enrichment of Fe in the solid solution (300 $\mu g.g^{-1}$) near intermetallic Al-Fe phases. The profile shape could be quantitatively interpreted by a theoretical solidification model [29].

4.4. Bulk Diffusion

The distance Z concerned by unidirectional diffusion can be estimated by $Z = 2(Dt)^{1/2}$ where D is the diffusion coefficient at temperature T and t the heat treatment time. The diffusing element has to be deposited as a pure thin layer (10 - 30 nm) on, or within [30] the substrate prior to the heat treatment. For low D (> 10^{-19} $cm^2.s^{-1}$), depth profiling of depth Z up to several µm is suitable [31] . For higher D, depth profiles at different lateral positions along a 500 µm line on a bevelled sample (1°) are recorded [32].

4,5. Diffusion, Segregation at Grain Boundaries and Interfaces

By preparing well-characterized bicrystals, GUST et al. [33] could determine the grain boundary diffusion coefficient of In in Cu and Ni. Depth profiles at different lateral positions on a bevelled section were recorded.

On real industrial samples, the grain boundary plane is not so well defined. In this case, using bevelled sections of rolled material, ion imaging of the grains allows the identification of grain boundaries nearly parallel to the surface : the boundary of the grain of interest moves rapidly with time as the matrix is sputtered away. Depth profiles across the identified grain boundary can then be carried out on a small area to yield a good resolution. An example of the heterogeneous segregation of Na at the grain boundary of an Al-Li alloy will be presented.

The same technique can be used to characterize the interface between the matrix and the fiber in a metal matrix composite. An example of Mg enrichment at the Al matrix/SiC fiber interface will be presented.

5. Use of Isotopic Tracers

^{18}O, D_2, $D_2\,^{16}O$, $H_2\,^{18}O$,... are useful to study chemical reaction mechanisms such as oxidation and corrosion processes. Thermal [16,27, 34-36] and anodic

[37,38] oxidation of different metals have been investigated with respect to the composition of the gaseous atmosphere and of the electrolyte.

6. Conclusion

Due to its intrinsic properties (imaging capability of every element, quantitative analysis, excellent depth resolution), SIMS constitutes an original method and a complementary tool for microscopic and microanalytical techniques for the characterization of heterogeneous metallic materials.

REFERENCES

1. M. Grasserbauer, H.M. Ortner, P. Wilhartitz, M. Pimminger and R. Leuprecht, Fresenius Z Anal.Chem., 317, 539(1984).
2. F. Degrève, R. Figaret and P. Laty, Intern.J.Mass Spectrom.Ion Phys., 29, 351 (1979).
3. G.H. Morrison and G. Slodzian, Anal. Chem., 47,942A(1975).
4. P. Williams, C.A. Evans Jr, M.L. Grossbeck, H..K.Birnbaum, Anal.Chem., 48, 964(1976).
5. M.L. Grossbeck, P. Williams, C.A. Evans Jr and H.K. Birnbaum, Phys.Stat.Sol., 34, k97(1976).
6. T. Ohtsubo and K. Suzuki, Proc. SIMS IV, Springer Verlag, Berlin (1984), p 426.
7. R.W. Odom, B.K. Furman, C.A. Evans Jr, C.E. Bryson, W.A. Petersen, M.A. Kelly and D.A. Wayne, Anal.Chem., 55, 578(1983).
8. B.K. Furman and G.H. Morrison, Anal.Chem., 52, 2305(1980).
9. H. Liebl, Scanning, 3, 79(1980).
10. R. Levi-Setti, G. Crow and Y.L. Wang, this conference.
11. G. Slodzian in "Applied Charged Particle Optics", Adv. in Electronics and Electron Physics, Suppl.13B, Ed. A. Septier, Academic Press, N.Y. (1980), pl.
12. B. Dubost, J.M. Lang and F. Degrève, 3th Intern.Al-Li Conference, Oxford (UK), July 1985.
13. J.D. Fassett and G.H. Morrison, Anal.Chem., 50, 1861(1978).
14. C.A. Evans Jr and D.H. Wayne, Private Communication(1985).
15. N. Namdar and D. Loison, Mém.Et.Sc.Rev.Métal., to be published (1985).
16. F. Degrève and J.M. Lang, MRS Applied Materials Characterization Symposium, San Francisco (Ca), April 1985.
17. G.H. Morrison, Proc. SIMS III, Springer Verlag, Berlin (1982), p 214.
18. D.E. Newbury and D, Simons, Proc. SIMS IV, Springer Verlag, Berlin (1984), p 11.
19. F. Bodart and G. Demortier, LARN, Université Namur (Belgium), Private Communication (1984).
20. H.W. Werner, H.A.M. de Greft and J. Van den Berg, Rad.Eff., 18, 305 (1974).
21. J.C. Pivin, C. Roques-Carmes and G. Slodzian, Appl.Phys., 51(8), 4158(1980).

22. R. Namdar, J.Microsc.Spectrosc.Electron., 2, 293(1977).
23. R. Namdar, D. Loison and R. Tixier, J.Phys.,45,C2-673(1984).
24. L. Habraken, T.D. Shungu and V. Leroy, Le Vide, les Couches minces, 199(1979).
25. J.C. Joud, R. Faure, R. Namdar, C. Pichard and P. Poyet, Mém.Et.Sc.Rev.Métal., 235(1982).
26. M. Texor and R. Grauer, Corr.Sc., 23, 41(1983).
27. J.C. Pivin and D. Loison, J.Phys., 45, C2-647(1984).
28. F. Degrève, S. Dermarkar and A. Dubus, J.Microsc.Spectrosc.Electron., 7, 243(1982).
29. T.F. Bower, H.D. Brody and M.C. Flemings, Trans.Met.Soc.MAIME, 236, 624(1966).
30. M.P. Macht and V. Naundorf, J.Appl.Phys., 53(11), 7551(1982).
31. W. Gust, M.B. Hintz, A. Loding, H. Odelius and B. Predel, Acta metall., 28, 291(1980).
32. W. Gust, C. Ostertag, B. Predel, U. Roll, A. Lodding and H. Odelius, Phil.Mag. A, 47(3), 395(1983).
33. W. Gust, M.B. Hintz, B. Predel and U. Roll, Acta.metall., 28, 1235(1982).
34. J.C. Pivin, D. Loison, C. Roques-Carmes, J. Chaumont, A.M. Huber and G. Morillot, Proc. SIMS III, Springer Verlag, Berlin (1982), p 274.
35. D. Loison, J.C. Pivin, J. Chaumont and C. Roques-Carmes, Nucl.Instrum.Meth., 209/210, 975(1983).
36. J.C. Mikkelsen, Jr., J.Electron.Mat., 11, 541(1983).
37. C.A. Evans, Jr. and P.P. Pemsler, Anal.Chem., 42, 1060(1970).
38. M.J. Graham, Proc.Int.Corr.Congress, Toronto, May 1984, pl.

Sputtering of Au-Cu Thin Film Alloys

T. Koshikawa[1] *and K. Goto*[2]

[1] Dept. of Applied Electronics, Osaka Electro-Communication Univ., Neyagawa, Osaka 572, Japan

[2] Dept. of Fine Mesurements, Nagoya Institute of Technology, Gokiso-cho, Showa-ku, Nagoya 466, Japan

Surface composition changes in alloys sputtered with ions are very important in the field of plasma-wall interaction in the fusion reactor, thin film preparation with sputtering, surface and interface analysis with ion beam techniques and so on. Ion radiation-enhanced segregation(IRES) and ion radiation-enhanced diffusion(IRED) are very important in composition changes [1-6] . The composition change of Au-Cu thin film alloys (made by coevaporation techniques), which were sputtered with Ar, Kr and Ne ions of 2KeV (ion current density:1-5μA/cm^2), were studied so as to know the influence of IRES and IRED on preferential sputtering. Computer simulations of the sputtering process including IRES and IRED were also carried out [5,7] and applied to the thin film alloys.

Compositions of the Au-Cu thin film alloys were monitored with high energy Auger spectra (239 and 920 eV for Au and Cu) during evaporation and sputtering. Au-Cu alloys have a sold solution and are suitable for the study of preferential sputtering. The film thickness is between 100 and 150Å. Five sets of experiments have been carried out. (1) L-L-L condition: Au-Cu thin film alloys were prepared on a W sheet which was cooled with liquid nitrogen and sputtered at low temperature ($<$-120°C). (2) L-R-L condition: Au and Cu were coevaporated at low temperature then the sample holder was warmed up to room temperature and the holder was cooled down again with liquid nitrogen. The sputtering was carried out at low temperature. (3) L-R-R condition: Au and Cu were coevaporated at low temperature and the sputtering was carried out at room temperature. (4) R-R-L: Au and Cu were coevaporated at room temperature then the holder was cooled down with liquid nitrogen and the sputtering was carried at low temperature. (5) R-R-R: Au and Cu were coevaporated at room temperature and sputtered at room temperture.

We can check the influence of temperature on the film composition change after warmed up to room temperature with condition (1) and (2). The effect of IRES and IRED can be observed with condition (1) and (3). Segregation and diffusion of the constituent atoms during evaporation can be observed with condition (4).

The details of the simulation model have been published elsewhere [5,7] so we mention the model briefly here. This model includes three processes sputtering, IRES localized at the surface and IRED. The extent of IRES and IRED were estimated with experimental results from the $Au_{0.56}Cu_{0.44}$ alloy [6].

The results of the condition of L-L-L, L-R-L, L-R-R are shown in Fig.1 (a),(b) and (c). Primary ion is Ne. The composition (except at the top

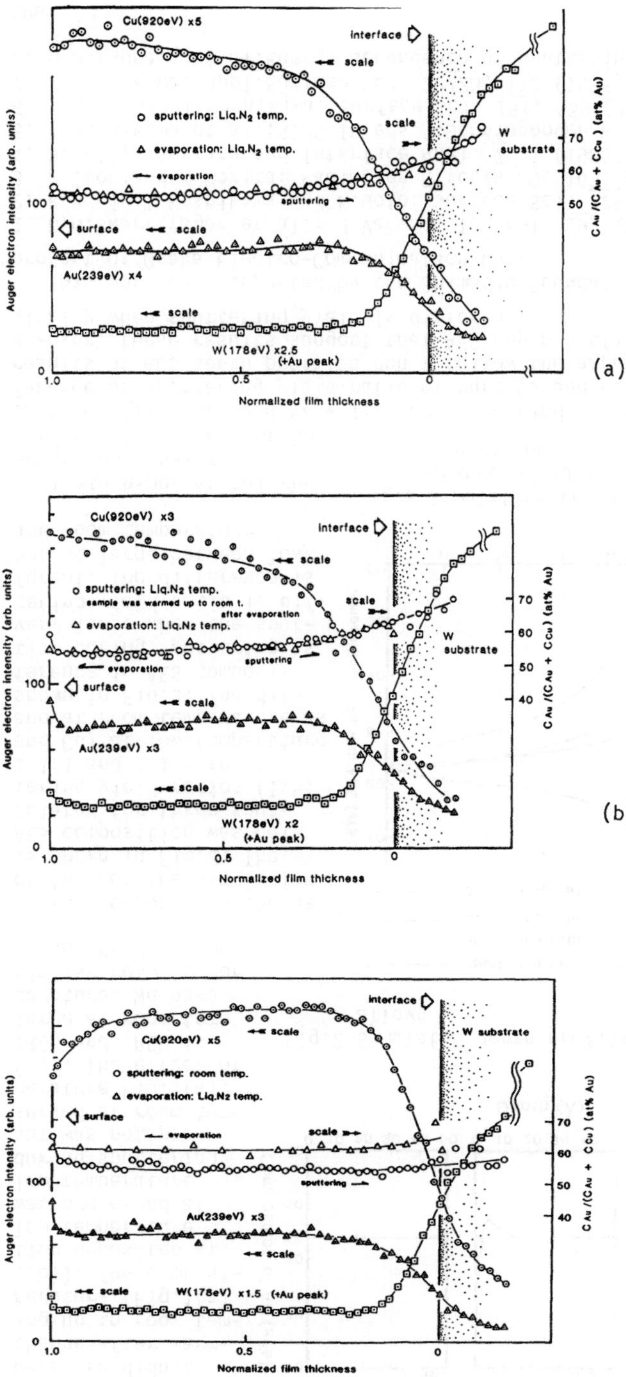

Fig.1
Composition profiles of Au-Cu coevaporated thin film alloys sputtered with Ne ions of 2KeV.
(a) shows L-L-L condition. (b) shows L-R-L condition. (c) shows L-R-R condition.
Triangles and circles show the composition of Au during evaporation and sputtering. Film thickness is 100-150Å.

surface layer) deposited at low temperature didn't change after warming up to room temperature (Fig.1(a),(b)). The composition deposited at low temperature was preserved at low temperature during sputtering but was not preserved at room temperature Fig.1(a),(c)). The effect of IRES and IRED is large at room temperature. We have similar results for Ar and Kr 2KeV ions.

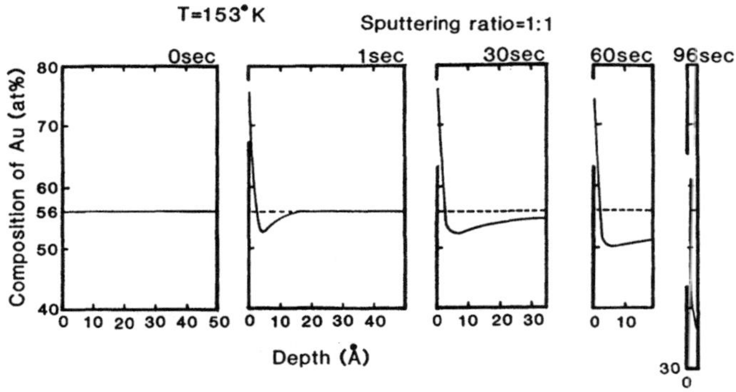

Fig.2 Simulated depth profiles of Au-Cu thin film alloys

The composition change of Au from the simulation is shown in Fig.2. The AES composition was calculated for three sputtering yield ratios (1.5:1,1:1 and 1:1.5 for Au and Cu) at low temperature and at room temperature as shown in Fig.3. The difference in AES composition during sputtering is very large when the sputtering yield ratio is different. The difference is not so large between low and room temperature.

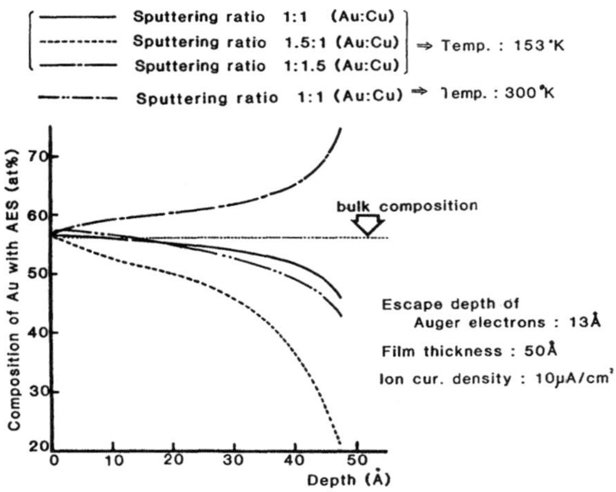

Fig.3 Simulation results of AES composition change of Au in Au-Cu thin film alloys for various sputtering condition.

These experimental results show that the sputtering ratio of Au and Cu in thin films may be nearly 1:1 for three kinds of ions in spite of the difference of sputtering yield ratio of pure Cu and Au [8]. The simulation results of AES depth profiles don't follow the experimental ones exactly. However, these results suggest that AES depth profiles may change drastically when sputtering yield is different.

This work was supported by the Shimadzu Foundation and the special project of Osaka Electro-Communication Univ.

1. D.G.Swartzfager et al.: J.Vac.Sci.Technol. 19, 185 (1981).
2. R.S.Li, T.Koshikawa and K.Goto: Surface Sci. 129, 192 (1983).
3. N.Itoh and K.Morita: Radiation Effects 80, 163 (1984).
4. R.Kelly: Surface and Interface Anal. 7, 1 (1985).
5. T.Koshikawa et al.:SIMS IV eds.A.Benninghoven et al.(Springer,1984)p.31.
6. R.S.Li and T.Koshikawa: Surface Sci. 151, 459 (1985).
7. T.Koshikawa: Appl.Surface Sci. 22/23, 118 (1985).
8. N.Matsunami et al.Energy dependence of sputtering yields (Nagoya Univ.)

Hydrogen Profiling in Titanium

R. Bastasz

Sandia National Laboratories, Livermore, CA 94550, USA

1. Introduction

Measurements of the hydrogen concentration in metals using secondary ion mass spectrometry (SIMS) can be made by monitoring either atomic hydrogen ion emission or the ratio of metal-hydride to metal ion emission. BENNINGHOVEN and co-workers [1,2] have shown that the latter approach is less susceptible to matrix effects and more quantitative. For titanium, overlap between metal-hydride and metal ion emission occurs in typical SIMS measurements because the five naturally occurring Ti isotopes have adjacent masses. This complication has led to the development of several methods to extract the TiH^+/Ti^+ ratio from SIMS data [3]. This paper will describe the various methods and evaluate them in terms of accuracy and data collection efficiency.

2. The Exact Method

For n adjacent-mass isotopes having surface abundances A_i ($i=1,2,\ldots,n$), the observed signals S_i ($i=1,2,\ldots,n+1$) arising from unresolved atomic and monohydride secondary-ion emission are given by

$$S_1 = (1-f)A_1; \tag{1a}$$
$$S_i = (1-f)A_i + fA_{i-1}, \qquad i = 2,3,\ldots,n; \tag{1b}$$
$$S_{n+1} = fA_n; \tag{1c}$$

where f is the formation probablity for monohydride ions.

Given normalized values for S_i, it has been shown [3] that f can be related to S_i by eliminating every A_i from (1a-c), which yields

$$\sum_{j=0}^{n} a_j f^{n-j} = 0 \qquad \text{where} \qquad a_j = (-1)^j \sum_{i=j}^{n} \binom{i}{j} S_{i+1}. \tag{2a, b}$$

Consequently, it is possible to find f without knowing the surface abundances A_i. Once f is known, each A_i can easily be determined with (1a-c).

For the case of Ti, $n = 5$ and the appropriate coefficients in (2) are

$$\begin{aligned} a_0 &= 1, \\ a_1 &= -(S_2 + 2S_3 + 3S_4 + 4S_5 + 5S_6), \\ a_2 &= S_3 + 3S_4 + 6S_5 + 10S_6, \\ a_3 &= -(S_4 + 4S_5 + 10S_6), \\ a_4 &= S_5 + 5S_6, \\ a_5 &= -S_6. \end{aligned} \tag{3a–e}$$

Solving the polynomial in (2) with these coefficients gives f. The quantity $f/(1-f)$ is equal to the TiH^+/Ti^+ intensity ratio.

3. Approximate Methods

In addition to the above analysis, there are several approximations to the f ratio which may be useful for measuring H in Ti. If one has independent knowledge of the actual surface abundances of the various isotopes, it is a simple matter to find f given S_i. For example,

$$\frac{f}{1-f} = \frac{S_6}{S_1}\frac{A_1}{A_5}. \tag{4}$$

Actual surface abundances may differ from known natural isotopic abundances during SIMS analysis due to preferential sputtering effects, but the difference is small in the case of Ti and (4) is a very close approximation.

The ratio of SIMS signals from adjacent masses can also be used to approximate $f/(1-f)$. If the ratio of the Ti isotope abundances at two adjacent masses is known and the hydride ion component in the lower mass signal can be neglected, then

$$\frac{f}{1-f} = \frac{S_{i+1}}{S_i} - \frac{A_{i+1}}{A_i}, \qquad i = 1,\ldots,5. \tag{5}$$

This is a good approximation when the signal ratio from either masses 47 and 46 or masses 49 and 48 is used. Since about 74% of naturally occurring Ti is ^{48}Ti, while only about 8% is ^{47}Ti, a relatively small error is incurred by neglecting $^{47}TiH^+$ emission.

Another approximation is to consider only the heaviest Ti isotope and its monohydride ion, taking the signal ratio of masses 51 to 50 as being equal to f. This is valid when $A_4 = A_5$, which is nearly correct for naturally abundant Ti, and has the advantage that the signal at mass 51 is free from isotopic interference.

4. Discussion

The above methods can be compared using the criteria of accuracy in evaluating the f ratio and data collection efficiency. The exact method given by (2) has no error in determining f, while the approximations have varying degrees of error ranging from nearly zero for the methods given by (4) and by (5) with $i=1$ to about 9% at $f=1$ for method (5) with $i=3$. The signal ratio of masses 51 to 50 approximates f with a maximum error of about 3% at $f=1$. These error estimates assume that the Ti isotopes are present in their natural abundances.

Data collection efficiency can be estimated by considering the total sampling time per data point, the weighting factor given to each signal, and the signal strength of each Ti isotope sampled. The exact method is very inefficient because of the large weighting factors given to the relatively weak signals S_5 and S_6. In contrast, method (5) with $i=3$ is the most efficient because it monitors only the most abundant Ti isotope and its hydride.

For quantitative analysis, the linearity between the f ratio and hydrogen content must be established and the appropriate proportionality factor determined. This can be done using standards such as Ti implanted with hydrogen ions of known energy and fluence.

An example of the application of these methods is shown in Fig. 1, which depicts depth profiles of a standard hydrogen implant in Ti. The four methods produce similar results which compare favorably with an implantation profile calculated using the Monte Carlo TRIM model [4]. Deviations from the expected distribution are attributed to recoil mixing and instrument background levels. From the standpoint of data collection efficiency, the method given by (5) is preferred.

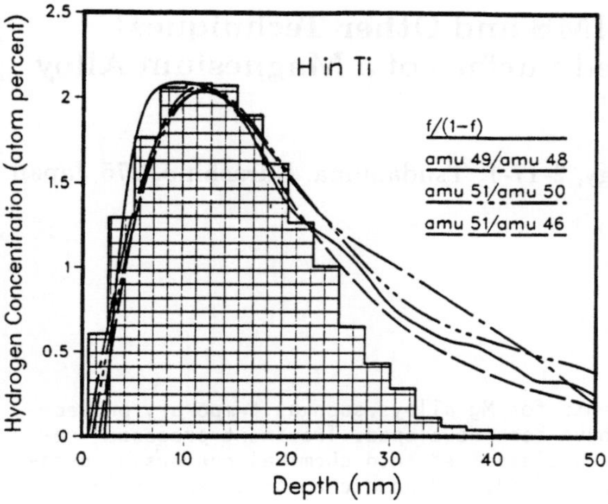

Fig. 1 Hydrogen depth profiles obtained from a SIMS measurement of Ti implanted with 2 keV H_3^+ using the exact method (solid curve), adjacent mass ratios (chain-dotted curves), and the $^{50}TiH^+/^{46}Ti^+$ ratio (dashed curve). The shaded region represents the expected distribution based on a TRIM calculation. The curves are smoothed fits to the data.

5. Conclusion

Several SIMS data collection methods can be used to measure hydrogen levels in Ti. Unlike SIMS measurements of atomic hydrogen ion emission, the TiH^+/Ti^+ intensity ratio is not strongly perturbed by matrix effects and is a good measure of H content. An exact method for obtaining this ratio is available which does not require knowledge of the isotopic abundances of the Ti isotopes. However, the data collection efficiency of this method is poor. An efficient approximation of the intensity ratio can be obtained by monitoring the most abundant Ti isotope and its hydride.

Acknowledgments

It is a pleasure to acknowledge the technical assistance of D. H. Morse and K. L. Black. This work was supported by the U. S. Department of Energy.

References

1. A. Benninghoven, P. Beckmann, D. Greifendorf, K. -H. Müller, and M. Schemmer: Surf. Sci. 107, 148 (1981)
2. M. Schemmer, P. Beckmann, D. Greifendorf, and A. Benninghoven: In Secondary Ion Mass Spectrometry SIMS-III, ed. by A. Benninghoven, J. Giber, J. László, M. Riedel, and H. W. Werner, Springer Series in Chemical Physics, Vol. 19, (Springer, Berlin, Heidelberg, New York 1982) p. 405
3. R. Bastasz: J. Vac. Sci. Technol. A 3, 1363 (1985)
4. J. P. Biersack and L. G. Haggmark: Nucl. Instrum. Meth. 174, 257 (1980)

The Application of SIMS and Other Techniques to Study the Anodized Surface of a Magnesium Alloy

S. Fukushi and M. Takaya

Chiba Institute of Technology, 2-17-1, Tsudanuma, Narashino, 275 Japan

1. Introduction

Various anticorrosive treatments for Mg alloys, such as temporary protective coatings and paintings, have been developed. Treatment processes reported in the literature may be classified into chemical conversion process [1], electroplating process [2-4], and anodic coating process [5]. Above all, anodized treatment is important to increase corrosion resistance of the Mg alloy, but structure and compositions of the coated materials have not been satisfactorily clarified in previous reports.

The present work has been carried out to determine compositions of the surface materials of anticorrosive Mg alloy which was produced by anodizing in alkaline bath, and to pursue a root cause of the anticorrosivity of the anodizing surface. For these purposes, the SIMS method would be most suitable [6-8]. On the other hand, to verify the SIMS's data, other measurements such as ESCA, EPMA and some spectroscopic methods were applied.

2. Experimental

Preparation of sample; Sheet Mg alloy was used as a starting material. Its analytic data is (wt. %): Mn 0.82, Fe 0.005, Si 0.003, Zn 0.014, Cu 0.001, Ni 0.001, Al 0.00017 and the remainder is Mg.

An electrolyte was prepared by mixing KOH (165 g/l), KF (35 g/l), $Al(OH)_3$ (30 g/l) and distilled water. Anodizing conditions are; Anode-Mg alloy, cathode-stainless steel (SUS 304), current density-6 mA/cm^2, anodizing voltage-from 0V up to 120 V(D.C.), increasing stepwise and kept at 120 V for 15 min., temp. of bath-10°, 20°, 30°, 40° and 50° C. Samples were made by anodizing under above conditions.

Instruments used and their operating conditions were as follows:

1) SIMS; Argon pressure: 10^{-4} Torr., Voltage: 10 KV (CAMECA)
2) ESCA; Target: Mg, Voltage: 10 KV, Current: 20 mA
3) EPMA; Voltage: 20 KV, Current: 0.03 μA
4) X-ray diffractometer; Target: Cu, Voltage: 30 KV, Current: 20mA

3. Results and Discussion

3.1 Voltage-time curve

Fig. 1 shows voltage-time curve during anodization of Mg alloy in alkaline bath. From this curve it was shown that the electrolytic voltage increased rapidly in a short time during anodization, and some breakdown phenomena were observed with increasing voltage [9].

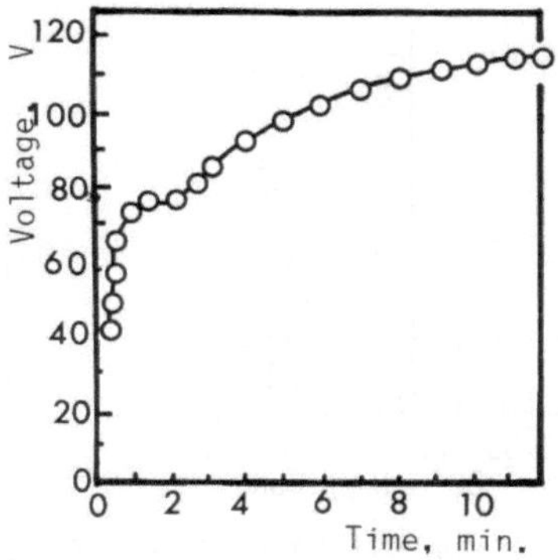

Fig. 1 Voltage-time curve during anodization

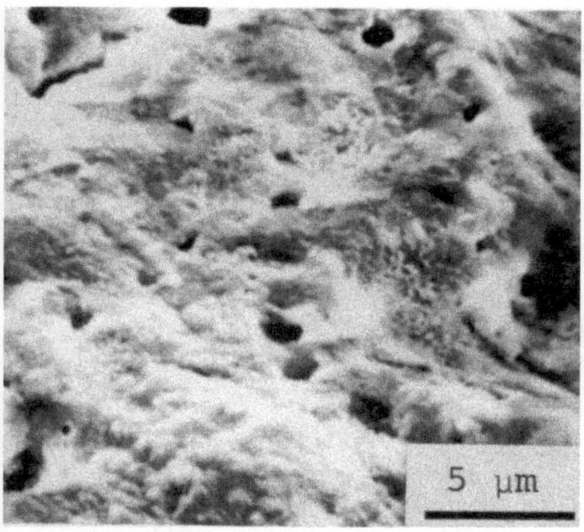

Fig. 2 SEM micrograph of the coating

3.2 Image of sample by SEM photograph

Fig.2 shows SEM photograph of an anodized coating which was created by electrolysis for 15 min. at 120 V. The photograph shows that there are some coarse irregularities on the sample's surface which may be produced during breakdown electrolysis.

3.3 Analysis by SIMS

Fig.3 shows the positive-ion spectra running between 1 and 50 (m/e). From this chart, we could conclude that H^+, O^+, Mg^+, Al^+, MgO^+ and AlO^+ ions will segregate to the surface region of the coating. The depth profile of Al shows that the distribution of Al is large at the surface, but decreases with increasing depth. On the contrary, the distribution of Mg increases with increasing depth, and depth profile shows this tendency (Fig. 4).

We could make a conclusion that Al ion added to bath is oxidized during the electrolysis and finally segregate to surface of the coating as a solid oxidized compound. If the ratio of MgO to AlO in the coating is the same or close to that of highly anticorrosive materials, then the anticorrosivity of the coating will be attributable to the formation of that material. For comparison, the SIMS spectra of spinell was measured (Fig. 5), and the similarity

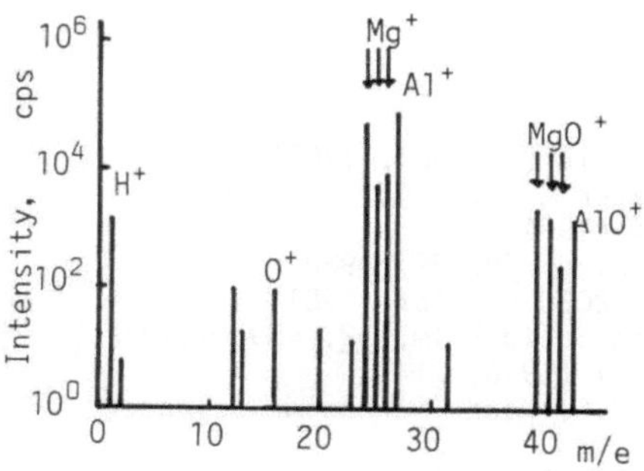

Fig. 3 Positive SIMS spectra of the coating

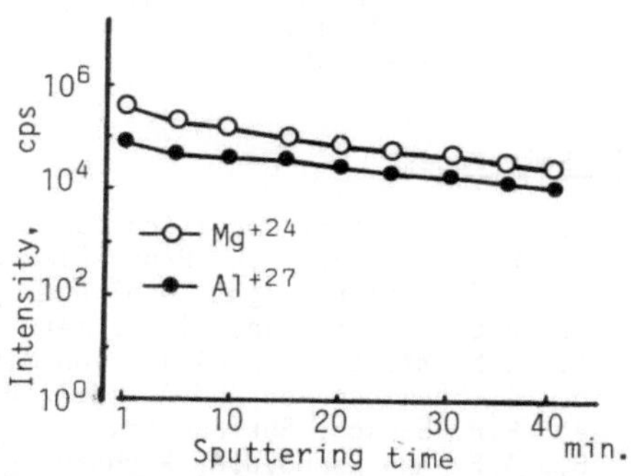

Fig. 4 Depth distribution of Mg^{+24} and Al^{+27} in the coating by SIMS

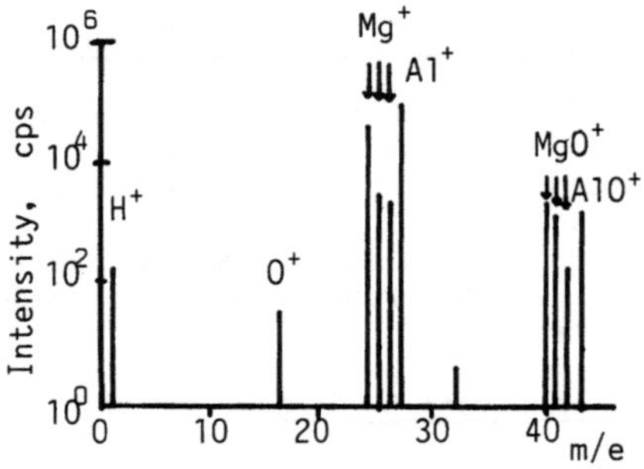

Fig. 5 Positive SIMS spectra of spinell

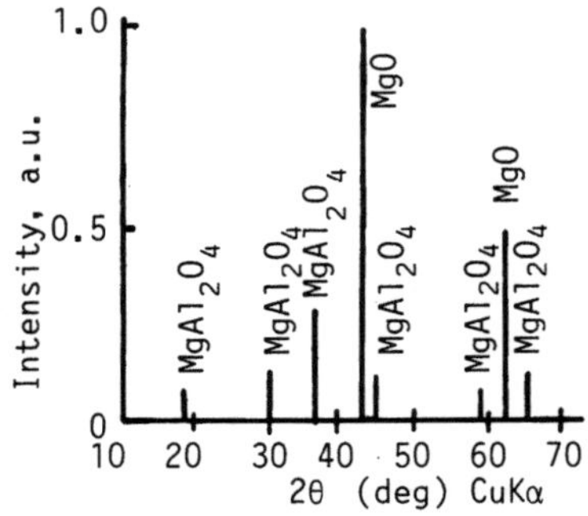

Fig. 6 X-ray diffraction pattern of the coating

of patterns among Fig. (3) and Fig. (5) leads us to a conclusion that spinell-like material has been formed in the coating.

3.4 Results by other analytic techniques

Informations obtained by SIMS have been verified by the data obtained by other analytic techniques such as EPMA, ESCA and X-ray diffraction. Only X-ray diffraction pattern of the coating is shown in Fig. 6 to give an example.

4. Conclusion

Using SIMS, the composition and structure of anodized coatings on a Mg alloy have been inferred. By using other analytic techniques such as EPMA, ESCA and X-ray diffraction, informations obtained from SIMS's data could be expanded further.

In conclusion:

(1) The composition of the anodized coating is a composite material which consists of spinell-like $MgAl_2O_4$ and MgO.

(2) There is an inverse relation between distribution-patterns of Al and that of Mg against depth in the coating.

(3) A high anticorrosivity of the anodized coating on Mg alloy would likely be attributed to the formation of spinell-like material.

5. References

1. U.S. Military Specification MIL-M-3171C (11, Jul. 1966)
2. H.K. Delong: Tech. Proc. Electroplater's Soc., 48, 90 (1961)
3. I.C. Hepfer: Light Metal Age, 32, 4 (1964)
4. A.L. Olsen: Trans. Inst. Met. Finishing, 58, 29 (1980)
5. U.S. Military Specification MIL-M-45202C (1, Apr. 1981)
6. A. Benninghoven and A. Mueller, Thin Solid Film, 12, 439 (1972)
7. P.H. Dawson: Surface Science, 57, 227 (1976)
8. A.F. Abd Rabbo, J.A. Richardson and G.C. Wood: Corrosion Science, 18, 117 (1978)
9. A.E. Yaniv and H. Schick: Plating, 58, 1294 (1968)

Incorporation of Chromium in Sputtered Copper Films and Its Removal During Wet Chemical Etching

R.J. Day, M.S. Waters, and J. Rasile

IBM Corp., Systems Technology Division, 1701 North Street, Endicott, NY 13760, USA

1. Introduction

In modern computing equipment, semiconductor devices are normally mounted on a substrate that acts as an interface between the circuit card or board and the silicon chip. The circuitry deposited on such substrates combined with pins through holes in the substrate material allow communication between I/O sites on a device and the card or board. In some cases the circuitry is generated by screening on a paste that, when fired, will produce lines of a conductive glass. In other cases, the metallurgy is vacuum-deposited using either evaporation or sputtering and is made of metallic components. One such metallic system consists of chromium and copper, deposited to give a Cr/Cu/Cr/ceramic film structure [1]. These layers are deposited on the substrate and circuitry formed by photolithography followed by wet chemical etching.

When Cr/Cu/Cr films deposited onto glass slides were examined using secondary ion mass spectrometry, we found that easily detectable concentrations of chromium were present in the copper layer. The observed chromium signals were not simply experimental artifacts such as might be caused by sputter-induced roughening or atomic mixing. Rather, for films composed of approximately 800 Angstroms Cr, 80,000 Angstroms Cu and 800 Angstroms Cr, and Cr:Cu ratio in the copper layer reached a constant value only a few hundred Angstroms below the Cr/Cu interface during SIMS depth profiles. The Cr:Cu ratio did not decrease during the remainder of the profile, but held a constant value until the bottom chromium layer was reached. These results implied that significant amounts of chromium were being incorporated in the eight-micrometer-thick copper film. Such a conclusion was unexpected in that studies of chromium/copper interdiffusion have shown that no diffusion occurs for single crystal samples, while limited intermixing occurs for polycrystalline films [2]. Interdiffusion in the Cr/Cu system occurred almost completely at grain boundaries [2]. It was, therefore, of interest to examine how chromium was incorporated in the copper layer of Cr/Cu/Cr structures and to assess how easily this chromium was removed by wet chemical etching solutions.

Chromium was etched with alkaline potassium permanganate solution (20 g/l, pH 12.7) at room temperature for times of 10, 30, 45, 60, 90 and 180 seconds. No chromium was visible on the copper film after the 90-second etch, but the 180-second etch was performed to assess the effect of overetching. Films were characterized by secondary ion mass spectrometry using commercial equipment. Experimental parameters are shown in Table I.

2. Results and Discussion

The effect of the KMnO4 etch on the minimum chromium concentration in the films is shown in Figure 1. Cr:Cu ratios were measured at the minimum in

Table I. Experimental conditions

Primary ion:	O_2^+
Ion energy:	10 keV
Ion current:	1-2 µA
Raster size:	600 µm
Gated area:	180 µm

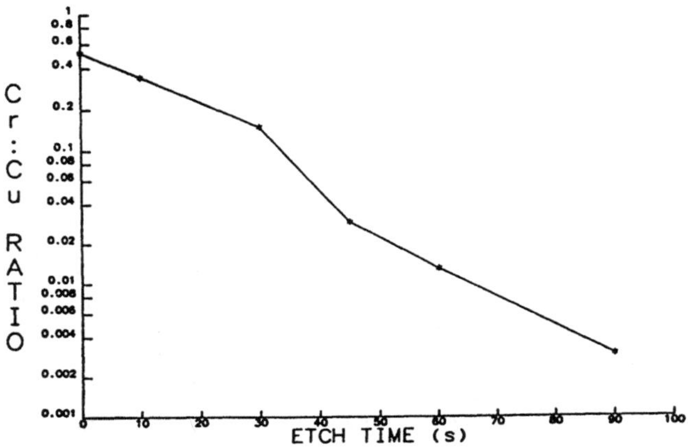

Figure 1. Plot of chromium/copper ratio versus etch time for Cr/Cu/Cr films

the Cr profile just before the crater bottom reached the lower chromium layer. Figure 1 shows that this ratio decreases logarithmically with etch time. A square root of time dependence would be expected if diffusion of etchant into the film structure were rate determining. The exponential relationship of Figure 1 therefore suggests that diffusion is not rate determining. In fact, the speed with which chromium was attacked suggested facile etch penetration that could be explained by film porosity. If the copper film was porous enough to allow significant etch penetration, it could also allow chromium incorporation during deposition of the top Cr layer. This was tested by profiling Cu/Cr films that had never seen top Cr, then depositing a top layer of Cr and profiling again. The results showed that chromium penetrates deep into the copper film during deposition. To assess the speed with which dissolved species could move into or through the Cr/Cu/Cr films, several samples were immersed in NaCl solution for one hour and rinsed for various times. Unrinsed samples showed high levels of sodium throughout the film thickness. Rinsing even for times as short as ten seconds reduced the sodium concentration all the way down to the bottom chromium layer. Such rapid removal of ions is also consistent with the copper films being porous. As a comparison, electroplated copper was exposed to a one-hour immersion in NaCl and profiled. Essentially, no sodium penetration occurred for this sample.

1. D. P. Seraphim and I. Feinberg, IBM J. Res. Dev. 25, 617, 1981.

2. J. E. E. Baglin, V. Brusik, E. Alessandrini and J. F. Ziegler, "Thin Film Interdiffusion of Chromium and Copper," in Applications of Ion Beams to Metals, S. T. Picraux, E. P. EerNisse and F. L. Vook, Eds. (Plenum Press, New York, 1974).

Application of SIMS to the Study of a Corrosion Process – Oxidation of Uranium by Water

S.S. Cristy and J.B. Condon

Oak Ridge Y-12 Plant, Martin Marietta Energy Systems Inc.
Oak Ridge, TN 37831, USA

Corrosion of metals is an extremely important field with great economic and engineering implications. To effectively combat corrosion one must understand the processes occurring. This paper shows the utility of SIMS data for elucidating the processes occurring in one particular corrosion process--the oxidation of uranium by water--and for validating a theoretical model.

Although much has been written about the oxidation of uranium by water [1-2], before these SIMS studies the mode of diffusion, the role of hydrogen, the location of the rate-limiting step, and the locus of inhibition by oxygen were unknown. Some of this work has been reported [3-5].

High purity uranium, γ quenched from 825°C and then annealed at 625° was used. The surfaces were polished with 0.25μm diamond in ethanol. Exposures to isotopically altered reactants or natural abundance reactants were made in selected sequences in an UHV system equipped with a constant temperature bath. Depth profiles were followed using an ARL IMMA. Typically a 5nA, N_2^+ ion beam, focused to 4μm and rastered over an area 31 by 39μm was used for sputtering. The secondary ion current was electronically apertured.

As reported at the SIMS II conference [3] a simple alternation of exposures to $^{18}OH_2$ and $^{16}OH_2$ shows that the water adsorbed last produces the oxide nearest the metal and thus the oxygen (in some form) is diffusing to the oxide/metal interface to react. Multiple exposure experiments with five alternating layers showed that the oxygen species migration left the permanent oxide lattice positions essentially unaffected by this passage (Figure 1). The slope of the $^{16}O^-$ signal between the first and second layers was the same as between the third and fourth layers. Therefore the slopes between layers are due to the final usage or precipitation of the migrating species and not to interdiffusion in the oxide lattice. Thus the diffusion mode is interstitial rather than vacancy.

In earlier studies [1], the rate-controlling step was assumed to be the diffusion of water through fine capillaries to the oxide surface. The profiles in Figure 2 demonstrate that the rate-controlling step for the reaction is not the diffusion of reactant; as in a final 5 minute exposure to $^{16}OD^2$ after two exposures totalling 24 hours, oxide equivalent to 86 minutes of exposure or 6% of the total oxide is exchanged. Thus the diffusion rate of reactant far exceeds the oxidation rate. Rapid changes in oxide isotopic composition as a function of third exposure times from 0 to 10 minutes were found. At 10 minutes essentially the maximum exchange is reached.

OH^- has been postulated as the diffusing species in the water oxidation of uranium [1] and the formation of uranium hydride is suspected as accountable for the higher oxidation rates observed with water compared to pure oxygen [1,6]. Indeed, hydrogen gas is evolved in the reaction and the

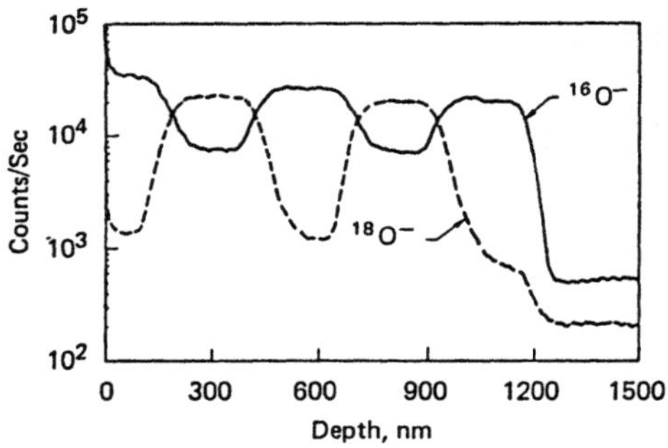

Figure 1. Depth profile on uranium sample exposed to $^{16}OH_2$, $^{18}OH_2$, $^{16}OH_2$, $^{18}OH_2$, $^{16}OH_2$ for 6h each (80°C, 0.6kPa

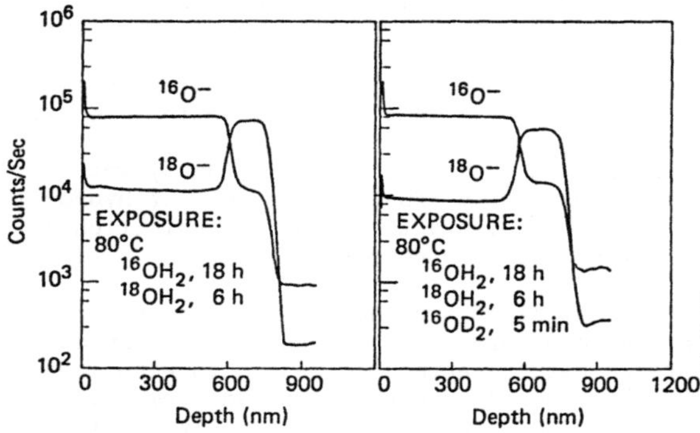

Figure 2. Depth profiles illustrating rapidity of diffusion in water formed uranium oxide

presence of hydride in the oxide has been confirmed [1,7]. To test the role of hydrogen, uranium was exposed to $^{18}OH_2$ followed by $^{16}OD_2$. The deuterium profile was found to mimic the coprecipitant oxide ^{16}O, but no evidence for a hydride intermediate of finite lifetime was found. However, the deuterium substitution did reduce the oxidation rate, implying a role for hydrogen in the mechanism. The deuterium concentration left in the oxide was found to be approximately 2 to 3% atomic and was incorporated as a stable portion since no exchange was observed when additional exposures to OH_2 were made.

The model adopted for the water oxidation of uranium is derived from a model successfully applied to the hydriding of uranium [4,8,9]. In this model, OH in some form is transported rapidly through the oxide to the oxide/metal interface. The rate of the reaction then depends on the rate of hydriding to cause spalling or breaking up of uranium metal which is known to occur when 1.4 to 3% of the uranium is hydrided [9]. The oxidation then occurs rapidly with the microscopically divided uranium. The mathematics of the model provides two equations which can be used to check its validity. The first (1) predicts the rate of reaction:

$$V = \left[-kC^\circ/\ln U\right](D/akU)^{1/2}, \tag{1}$$

where V is the velocity of the moving front, k is a combination of rate constants, C° is the solubility for hydrogen in the metal, U is the metal concentration at the transformation boundary (spall front), D is the diffusion coefficient for hydrogen in the metal, and a is the stoichiometric ratio for the reaction. A comparison of the calculated rates and data compiled by RITCHIE [10], COLMENARES [11], and HOPKINSON [12] is shown in Figure 3. The second (2) predicts the product profile:

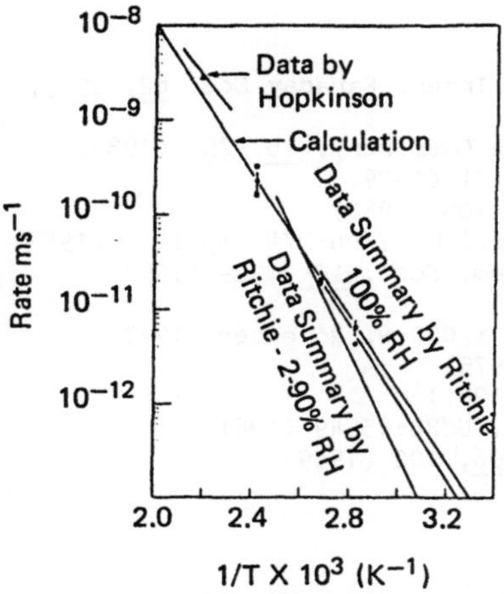

Figure 3. Calculated vs. experimental rates of oxidation of uranium by water

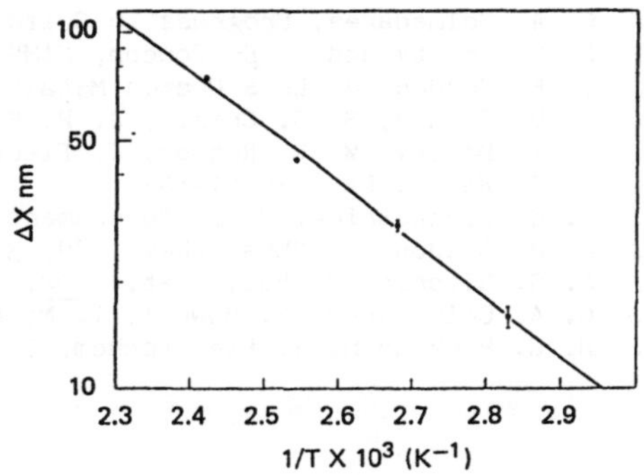

Figure 4. Calculated vs. experimental Δx (Points experimental)

$$d \ln P/dx = U[akU/D]^{1/2}, \quad (2)$$

where P is the product concentration and x is the location of the transformation boundary. The isotopic SIMS profiles show these product profiles clearly between isotopic layers and therefore, when Δx (the distance for the product concentration to drop by a factor of 10) is measured as a function of temperature, a second test of the model can be made. The outcome of this test is illustrated in Figure 4.

It has long been known that the oxidation of uranium by water is retarded by the presence of oxygen gas and the retardation has been assumed to occur by site blocking at the surface [1,2]. However, when alternate isotopic exposures were made followed by exposure to a mixture of $^{16}O_2$ and $^{18}OH_2$, the rapid exchange of ^{16}O and ^{18}O occurred in the oxide layer, but the further oxidation by water in this and subsequent exposures was retarded for up to 21 hours (Figure 5). This shows graphically that OH_2 is not held up at the surface and that the retarding mechanism is effective at the oxide/metal interface rather than at the surface. The effectiveness of the O_2 to retard the further water oxidation was much reduced if no water-formed oxide layer were present. The effectiveness was also crystallite related.

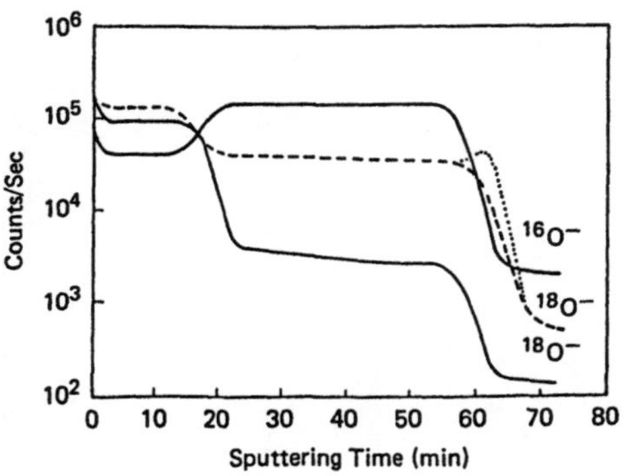

Figure 5. Profiles showing inhibition of water oxidation by oxygen at oxide/metal interface. Exposures are 0.6kPa (4.6 torr) $^{18}OH_2$ for 6.5h, 0.6kPa $^{16}OD_2$ for 18h at 80°C (solid lines). After exposure to 1.3kPa $^{16}O_2$ and 0.6kPa $^{18}OH_2$ for 6h, the ^{18}O profile shifted to the dashed line. Additional exposures to pure $^{18}OH_2$ for 12 hours result in the ^{18}O final profile (dotted line)

References

1. M. McD. Baker, L. N. Less, and S. Orman, Trans. Faraday Soc. 62, 2513, 2515 (1966).
2. C. A. Colmenares, Progress in Solid State Chemistry, 15, 257 (1984)
3. S. S. Cristy and J. B. Condon, SIMS II, 151 (1979)
4. J. B. Condon, J. Less Common Metals, 73, 105 (1980)
5. J. B. Condon, S. S. Cristy, J. R. Kirkpatrick, Y/DU-274, April 5 (1983)
6. J. E. Draley, W. E. Ruther, J. Electrochem. Soc., 104, 329 (1957)
7. J. T. Weber, LA-2035 (1956)
8. J. R. Kirkpatrick, U. S. Government Rep. K/CSD-9, November, 1977
9. J. B. Condon, J. Phys. Chem., 79, 392 (1975)
10. A. G. Ritchie, J. Nucl. Mat., 102, 170 (1981)
11. C. A. Colmenares, R. Howell, T. McCreary, UCRL-85549 (1981)
12. B. E. Hopkinson, J. Electrochem. Soc., 106, 102 (1959)

SIMS Trace Detection of Heavy Elements in High-Speed Rotors

J.C. Keith, M.I. Mora, O. Olea, and J.L. Peña[1]

Instituto Nacional de Investigaciones Nucleares, Agricultura 21, Mexico, D. F. 11800, Mexico

[1] Also at Cinvestav-IPN, Physics Department, Apdo. Postal 14-740, Mexico, D. F. 07000, Mexico

1. Introduction and Background Overview

Encouraged by contacts with J.W. Beams, KEITH [1] in the late 50's initiated quantitative experimental studies to determine residual frequency decays df/dt of small magnetically supported high-speed steel rotors coasting in ultra high vacuum. FREMEREY [2] extended these studies to even higher precision. Beyond eddy current and Coriolis drags observed by both authors, KEITH [3] in the early 60's noted marked anomalous high-frequency decays of form:

$$\frac{df}{dt} = -A(f)\, e^{uf^2} \tag{1}$$

The value of u agrees well with the theoretical form for rapid stress-enhanced [4], grain boundary self-diffusion of Fe in stretched rotors.

In the Ultracentrifuge Lab. at the Mexican Nuclear Center, we have engaged since the start of the 70's in experimental physical and materials studies of trace intermetallic diffusion of heavy elements in similar high-speed rotors spun near rupture. This work has covered elements such as Te, W, Au, Tl, Pb, and U. As our main interest was in rapid defect and internal surface diffusion, we chose ambient temperature operation, where such effects dominate due to their lower activation energy. A suitable rotor design was the cold metallic insert configuration of Fig. 1 below.

The problem was how to analyse trace quantities of heavy elements force diffused under some five million Earth gravities routinely obtained within our 2.54 cm spherical rotors. SIMS offered desired options such as submillimeter spatial resolution, depth and mass resolution, energy discrimination of interfering masses, dynamic range, tracking, and excellent sensitivity into the ppb range, with capabilities of profiling rotor samples at different sputter rates from fractions to hundreds of A°/s [5]. The latter factor provides the flexibility needed to study grain boundary and surface diffusion effects.

2. Rotor Sample Preparation and Analysis

High strength steel rotors of 2.54 cm diameter were drilled as in Fig. 1 and subjected to thermal treatments for desired material hardness and homogeneity. Heavy metal bars were inserted, rotors introduced into high vacuum, and spun to elastoplastic regions under preselected regimes aimed at generating desired diffusion flows. Rotors were coasted for one to several hours while the frequency decay was monitored, the latter correlating with mass flows as later confirmed by SIMS. Rotors were then braked magnetically, removed from vacuum, and serially sectioned with a low-speed diamond saw, polished, and degreased.

Analyses of rotor samples, standards, and blanks were carried out using either NBS Thermal Ion or Leybold-Heraeus QMG-511 Quadrupole Mass Spectrometers,

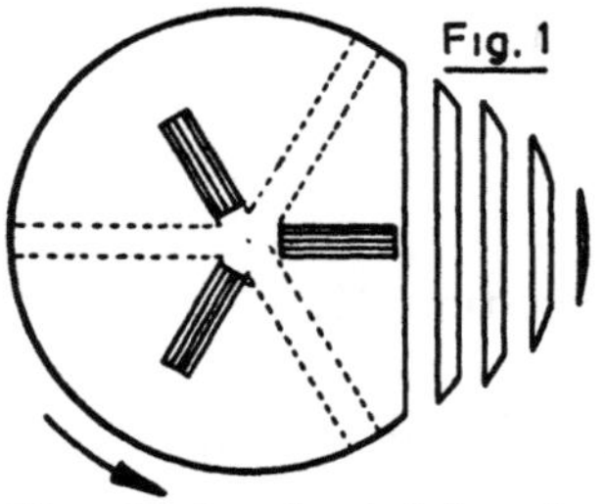

Fig. 1 Equatorial bar insert configuration, also indicating positioning of samples serially sectioned for SIMS analysis

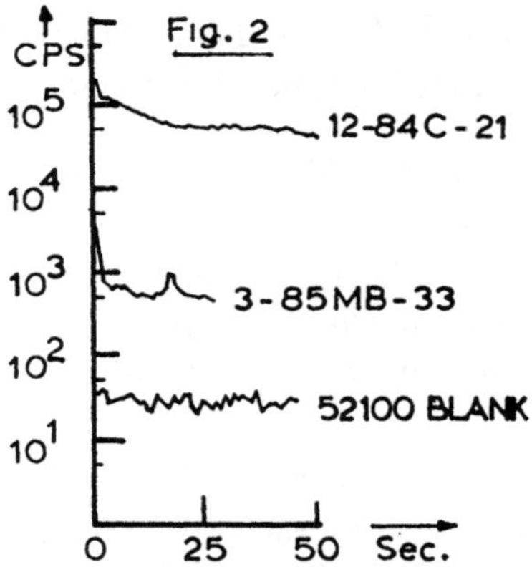

Fig. 2 CAMECA IMS-3f depth profiles of U^+ secondary ion intensity vs time at nominal 100 A°/s sputter rate. Selected positions (250 micrometer raster, 150 image) on different rotor samples are compared with 52100 steel blank. Signal/molecular interference ratio is improved 500x by energy discrimination

and later a double focussing CAMECA IMS-3f SIMS ion microprobe. Primary ion beams of Ar^+ and O_2^+ at 4 to 12 keV were used to bombard samples. Oxygen was leaked into the analysis chamber or used as primary ion source to enhance the signal [6, 7]. Concentration distributions of heavy elements were observed as functions of displacement with respect to bar inserts. Depth profile and controlled raster techniques have located grain boundary segregation/precipitation resulting from forced diffusion migration through rotors.

3. Summary of Results

Recent SIMS diffusion data, cf. Fig. 2 show relative concentration levels more than three orders of magnitude above naturally occurring heavy-metal backgrounds (1/10 ppm), depending on specific material and spin regimes. Observation of large lateral inhomogeneities and secular heightening of count rates at the beginning and often during these analyses, suggest presence of large intergranular contributions of active material in the grain boundaries. A maximum density "accumulation layer" also appears to form at the periphery, backing up into the rotor as the flow proceeds.

Thus, heavy atoms essentially appear to "seep" around grain/subgrain boundaries and over internal surfaces, sustaining strong radial drift under action of enormous centrifugal forces present in rotors. Measurements agree well with global drift-velocity theory predictions with $mv^2 \gg 2KT$ and flow rate densities given by $\vec{J} = - D\,\vec{\nabla}c + c\vec{v}$, where drift velocity $\vec{v} = Bm\,\omega^2\vec{r}$, as illustrated by comparison with the compiled data in Fig. 3. Data curves closely follow theoretical nonequilibrium and equilibrium concentration distributions near bar inserts and rotor periphery respectively:

$$c(r.t) \longrightarrow c(r_0)\, r_0^2/r^2 \quad \text{as } r \longrightarrow r_0 \tag{2}$$

$$c(r,t) \longrightarrow c(b)\, \exp\left\{-\frac{ab^2}{2D}(1 - r^2/b^2)\right\} \quad \text{as } r \longrightarrow b \tag{3}$$

Data of Fig. 3 are in good agreement with the theoretical slope of the equilibrium distribution $ab^2/2D = m\,\omega^2 b^2/2kT$ predicted from Einstein's relation, which for our trials with heavy elements lies between 20 and 32. Also, since diffusing atoms reach the rotor periphery in runs lasting for about

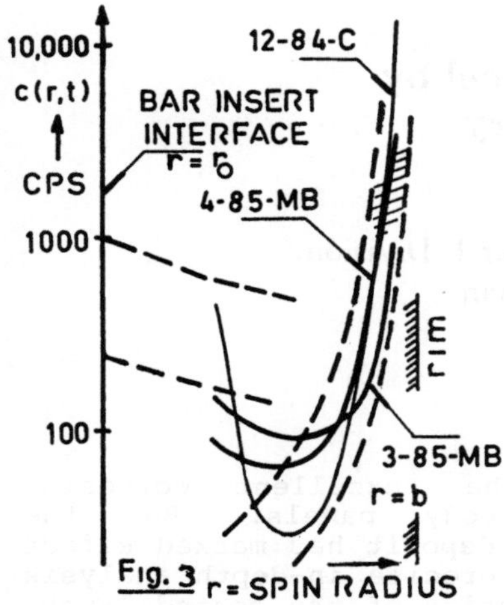

Fig. 3 SIMS diffusion data for three rotors run at different speeds with heavy-metal bar inserts. Slopes of data are in good agreement with theory, cf. dashed lines: c(r,t) runs from 1/10 to 100 ppma

an hour at high speeds, a lower limit is placed on the mobility connected with force-diffused heavy elements of $a = mB\omega^2 > 10^{-4}$ sec^{-1}. The diffusion thus involves activation energies of only small fractions of an eV.

In conclusion, SIMS is able to detect trace volume & surface distributions of force-diffused heavy elements in high-speed rotors, as well as persistent accumulation layers of diffused material on the periphery. SIMS Quadrupole and Ion Microprobe techniques have thus served as unique tools in the solid state ultracentrifuge diffusion field. The authors extend their thanks for support from CONACYT (MEXICO) and Sistema Nacional de Investigadores (MEXICO).

4. References

1. J.C. Keith: TCREC Tech. Rep. 62-54, Proj. ERG, Fig. 6, p. 17, U. of Det. Res. Inst., prep. by D.J. Kenney under contract # DA-44-177-TC 519, 1962
2. J.K. Fremerey: "Significant Deviation of Rotational Decay from Theory at a Reliability in the 10^{-12} sec^{-1} Range", Phys. Rev. Lett. 30, 753 (1973)
3. J.C. Keith: "Magnetic Torques and Coriolis Effects on a Magnetically Suspended Rotating Sphere", J. of Res. NBS-D, Radio Prop., 67D 533 (1963)
4. L.A. Girifalco: Atomic Migration in Crystals (Blaisdell, NY 1964)
5. C.W. Magee, R.E. Honig and C.A. Evans: "Depth Profiling by SIMS", SIMS III, Springer Ser. Chem. Phys. 19 (Springer, Berlin 1982)
6. A.E. Morgan and H.W. Werner: "Quantitative SIMS Studies with a Uranium Matrix", Surface Science, 65, 687 (1977).
7. V.R. Deline, W. Katz, C.A. Evans, Jr. and P. Williams: Appl. Phys. Lett. 33, 832 (1978).

Quantitative Analysis of Zn-Fe Alloy Electrodeposit on Steel by Secondary Ion Mass Spectrometry

T. Suzuki, Y. Ohashi, and K. Tsunoyama

Iron and Steel Laboratories, Technical Research Division,
Kawasaki Steel Corporation, Chiba 260, Japan

1. Introduction

Zn-Fe alloy electroplated steel has excellent corrosion resistance and is used for autobody panels. But the distribution of alloy element in the deposit has marked effect on the corrosion resistance and its precise in-depth analysis is required to determine the construction of the coated layer. Secondary Ion Mass Spectrometry is one of the most useful techniques to accomplish the quantitative in-depth analysis. However the sputter rate of high alloy varies with the concentration of alloy element [1] and it is necessary to anticipate this variation of sputter rate with Fe concentration for Fe-Zn deposit. A new technique based on the calibration curve method [2] was developed to estimate the true quantitative in-depth profile of alloy element in the deposit.

2. Experimental

Analysis was performed with ARL-IMMA. Primary ion species was oxygen accelerated at 18.5keV. The secondary ions were detected from the area of about 25% of primary beam scanning area. Zn-Fe alloys with various compositions were electrodeposited homogeneously on Cu-disk and used as standard samples. Chemical compositions of the standard samples are listed in Table 1. Total coated weights and Fe concentrations were determined by atomic absorption spectrometry.

3. Results and Discussion

Fig. 1 shows the calibration curve obtained by using the standard samples. A linear increment of ion intensity ratio with the concentration was observed in spite of wide variation of Fe concentration from 9 to 81%. This fact indicates that the matrix effect is negligible in Zn-Fe alloy electrodeposit.

It is necessary to convert the sputter time to the sputtered depth in order to obtain accurate in-depth profile. Fig. 2 shows the in-depth profiles of Zn, Fe and Cu of standard sample No. 1 and 3 obtained under the same sputtering conditions. Total coated weights (not the thickness) of these samples were 18 or 17 g/m^2 respectively. In spite of smaller amount of deposit, longer time was required for No.3 sample. This result suggested that the sputter rate decreased as Fe concentration of Zn-Fe alloy increased.

The sputter rate D(x) for Zn-X%Fe alloy is given by

$$D(x) = W(x) \cdot S \ / \ i_p \cdot T \qquad (1)$$

Table 1. Chemical compositions of standard samples

Sample No.	1	2	3	4	5	6
Fe wt%	9.1	18.1	27.3	47.4	67.7	81.1
Coated Weight g/m^2	18.1	17.1	16.9	15.6	14.8	12.3

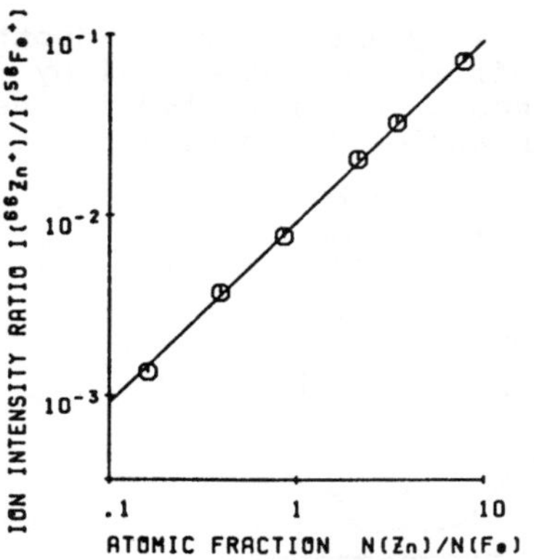

Fig.1 The calibration curve for Zn-Fe alloy electrodeposit (O_2+ 18.5keV)

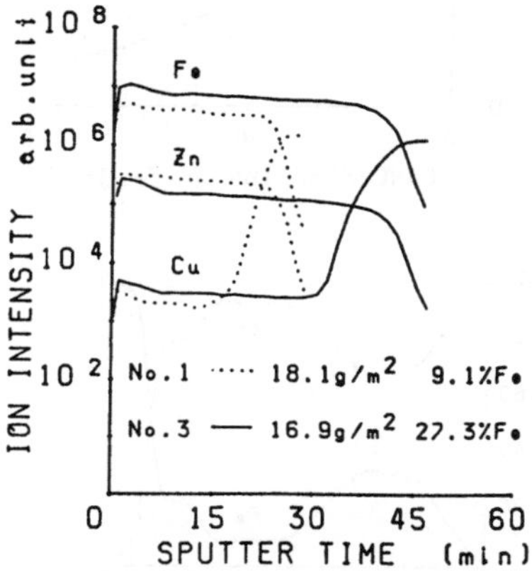

Fig.2 In-depth profiles of Zn,Fe and Cu for sample No.1 and 3 (O_2+ 18.5keV)

where W(x) is coated weight per unit area, S is the scanning area of primary ion with intensity i_p, and T is the time required to sputter off the coated layer and is defined as mean value of T_{Zn} and T_{Fe}. T_x is the time when secondary ion intensity of element X in the deposit is reduced by half.

The relation between the sputter rate D(x) and Fe concentration X obtained by measuring the standard samples is shown in Fig. 3. The sputter rate of Zn-Fe alloy electrodeposit decreased monotonically with Fe concentration, but marked deviation from the straight line connecting sputter rates of pure Zn and pure Fe was observed.

The in-depth analysis of practically produced Zn-Fe deposits was carried out by correcting the sputter rate sequentially. The ratio of ion intensity measured at t_i was converted to the Fe concentration X and then the corresponding sputter rate $D_i(X)$ was estimated by using the sputter rate-Fe concentration curve. The sputtered amount during Δt_i is given as the product $D_i(x)\cdot\Delta t_i\cdot i_p$. Total sputtered weight per unit area, Wg/m^2, is given by integrating the sputtered amount during each period Δt_i from the surface t=0 (i=1) to the time t (i=n) and corrected by scanning area S:

$$W = \sum_{i=1}^{n} D_i(x)\cdot\Delta t_i\cdot i_p/S \qquad (2)$$

This correction method was applied to the analysis of a practically produced Zn-Fe deposit. The results before and

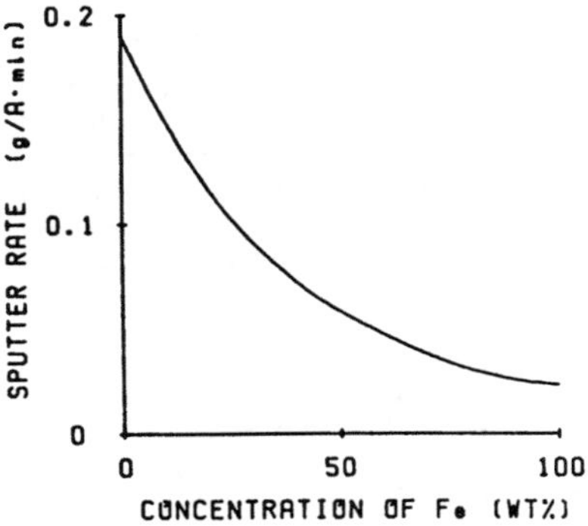

Fig.3 Variation of the sputter rate with Fe concentration

Fig.4 Quantitative in-depth profiles of Zn-Fe alloy deposit electroplated in the conventional condition

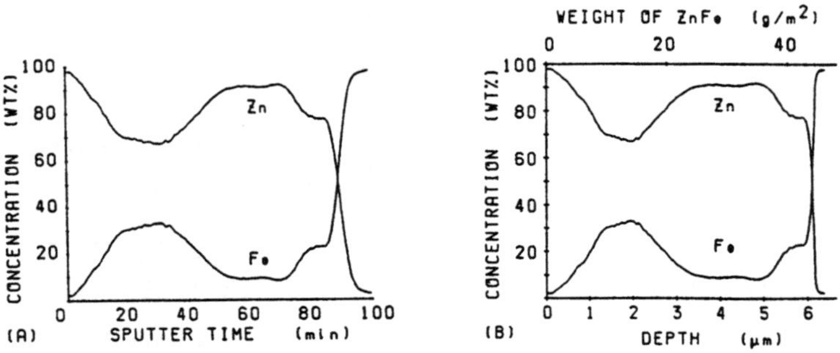

after the correction of the sputter rate variation with Fe concentration are shown in Fig. 4. Clear improvement of in-depth resolution was observed at the deposit-substrate interface. Total coated weight can be obtained by integrating from t=0 (i=1) to t=T (i=N) in eq (2) and this value agreed well with the result by chemical analysis. The correction method in this study gives the total coated weight as well as the in-depth profiles of alloy elements and impurities in the deposit. This method can be applied to other alloy electrodeposits or multi-layered electrodeposits.

4. Conclusions

A new method was developed for quantitative analysis of Zn-Fe alloy electrodeposit on steel by using SIMS.

1) Fairly linear calibration curve was obtained for Zn-Fe alloy electrodeposit provided oxygen ion beam was used as a primary beam.
2) Accurate in-depth profile was obtained by considering the sputter rate variation with alloy concentration.
3) Total coated weight was obtained by integrating the sputtered amount layer by layer, and the results obtained by this method agreed well with the results obtained by chemical analysis.

5. References

1) K.Tsunoyama and T.Suzuki;Proc.SIMS IV (1983),8
2) K.Tsuruoka,K.Tsunoyama,Y.Ohashi and T.Suzuki; Jpn.J.Appl.Phys.,2(1974),391

SIMS Analysis of Zn-Fe Alloy Galvanized Layer

K. Takimoto, K. Suzuki, K. Nishizaka, and T. Ohtsubo*

R & D Labs.-I, Nippon Steel Corp., 1618 Ida, Nakahara-ku, Kawasaki, Japan

* Nittetsu Techno Research Co., Ltd., 1618 Ida, Nakahara-ku, Kawasaki, Japan

1. Introduction

Recently a few kinds of double-layered zinc-alloy plating have been developed in Japanese steel industry for auto-body steel sheet. The conventional chemical analysis or the electrochemical analysis can not provide us with the chemical composition and the plating weight of each layer, because the plating has two electrogalvanized layers, both layers contain zinc and other alloying elements such as Fe or Ni and the concentration of the elements is different in each layer. In this paper we report the quantitative SIMS analysis of Zn-Fe alloy plating based on the experimental examination of the sputtering yield and the secondary ion yield in relation to the alloy composition.

2. Experimental

Experimental conditions are as follows: Primary ion: O_2^+ 12.5 kV, primary current density: 0.8 - 4.8 mA/cm^2, analysis area: 150μm ϕ, vacuum pressure: 5 x 10^{-9}Torr, measured mass: 66 for Zn (natural abundance: 27.8%) and 54 for Fe (natural abundance: 5.8%). The test samples used for basic experiments were prepared by a single electroplating to cover various Zn-Fe compositions. The layer depth was calculated using the layer weight and the chemical composition obtained by the chemical analysis of the chemically dissolved solution of the layer and the estimated specific gravity. According to the X-ray diffraction, the layer was identified as α-phase when Zn content was lower than 55% and it was identified as Γ- or δ_1-phase when Zn content was higher.

3. Results and Discussion

From the SIMS depth profiles of the various single-layered Zn-Fe alloy platings, the sputtering time of the alloy was obtained. The relationship between the Zn content (12-94%) of the alloy and the sputtering yield (S.Y.=(C.W./a.w.)/(S.T.·C.D./E.C.), where C.W. is the coating weight, a.w. is the average atomic weight of the plating composition, S.T. is the sputtering time until the Zn ion intensity decreases to half of the plateau intensity, C.D. is the current density of the primary ion beam and E.C. is the elementary electric charge) is shown in Fig.1 and is presented in a quadratic equation in the figure. The sputtering yield for Fe using O_2 primary ion beam was found to be 0.25 atoms/ion and this value is nearly equal to the reported value[1; 0.31 atoms/ion]. That for Zn was 1.2 atoms/ion and the yield has not been reported before. The ratio of the sputtering yields for Zn and Fe is thus 4.8 for O_2 primary ion beam and is close to the value obtained by Ar ion beam [1; 3.7]. The effect of the primary ion current density (0.8-4.8 mA/cm^2) was not evident.

The possibility of the preferential sputtering can be neglected because the very similar depth profile has been obtained using the glow discharge optical spectrometry [2; 4mm in glow lamp diameter] to the one obtained by SIMS for the same sample, and the GDS depth profile has been well accepted based on the many comparative experiments with chemical analysis [3,4].

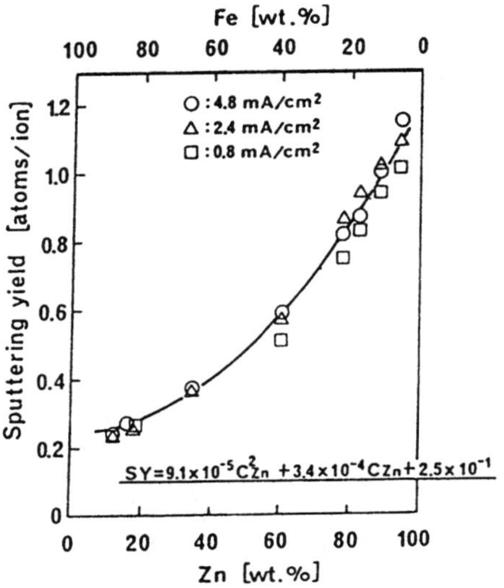

Fig.1 Relationship between the alloy composition and the sputtering yield

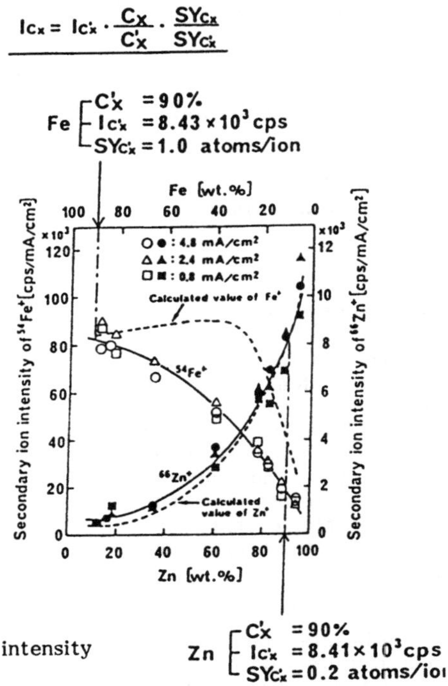

Fig.2 Relationship between the alloy composition and normalized secondary ion intensity

The relationship between the alloy composition and the secondary ion intensity per primary ion current density (the normalized secondary ion intensity) is shown in Fig.2 (solid lines). The secondary ion intensity of Zn abruptly increases according to the Zn concentration, while that of Fe increases almost linearly depending on the Fe content. If it is assumed that the secondary ion yield ($n_{ix}^{+} = K \cdot n_p \cdot Sx \cdot Aix \cdot Cx$ ····(1) where n_{ix}^{+} is the number of positive ion of isotope i of element x, K is a constant which includes the ion collection efficiency and an apparatus constant, n_p is the number of the primary ion, Sx is the secondary ion yield, Aix is the natural abundance of isotope i, and Cx is the concentration of element x in the alloy [5].) is constant and independent of the alloy composition, it is possible to calculate the estimated secondary ion intensity (normalized to 90% Fe alloy for Fe ion, and to 90% Zn alloy for Zn ion) using both the equation given in Fig.2 and the sputtering yields shown in Fig.1. The estimated results are also shown in Fig.2 (broken lines). The solid line for Zn is higher (approximately 2 times) than the broken line for the Zn content range lower than 40%, while the broken line for Fe is higher (2–1.2 times) than the solid line for the Fe content range between 70 and 10%. The descrepancy between the solid line and the broken line for both Zn and Fe suggests that the secondary ion yield does change depending on the alloy composition. (The effect of the primary ion current density is not clear in this figure.) Therefore, the equation (1) should be rewritten using f(Cx), the secondary ion yield as a function of alloy composition;

$$I_{ix}^{+} = Ip \cdot f(Cx) \cdot A_{ix} \cdot Cx \quad \cdots\cdots\cdots \quad (2)$$

where I_{ix}^{+} is the secondary ion intensity, and Ip is the primary ion current density. And f(Cx) can be obtained from equation (2) using the experimental results.
The experimentally obtained relationship between f(Cx) and Cx is shown in Fig.3. It is obvious that $f(C_{Fe})$ and $f(C_{Zn})$ increase greatly when Zn content is decreased lower

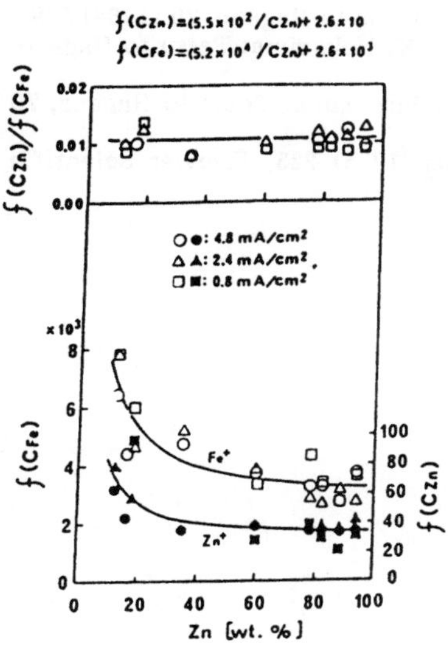

Fig.3 Relationship between the alloy composition and the secondary ion yield as a function of alloy composition (f(Cx))

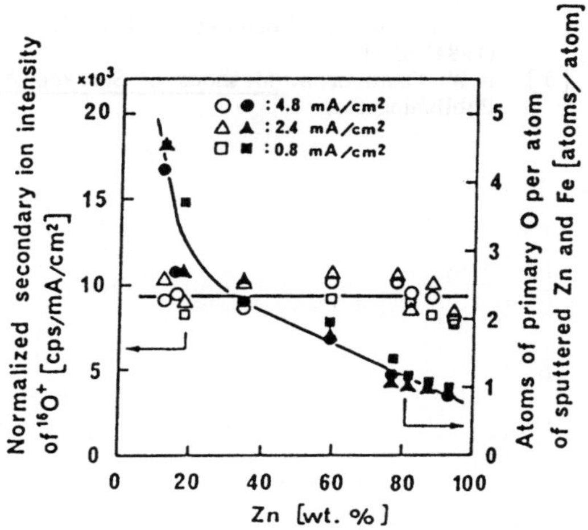

Fig.4 Relationship between the alloy composition and normalized secondary ion intensity of oxygen and the inverse of the sputtering yield.

than 20%. (The relationship is also given in equations as shown in Fig.3). It is evident that the ratio of $f(C_{Zn})$ to $f(C_{Fe})$ is independent of the alloy composition and the primary ion current density.

The relationship between the alloy composition and the inverse of the sputtering yield (the ratio of the number of primary O_2 atoms to the number of sputtered atoms) is shown in Fig.4. It is clear that the ratio increases greatly when the Zn content is decreased to lower than 20% and the Fe content is increased to higher than 80%, while the normalized ion intensity of O_2^+ is constant and independent of the alloy composition. The results shown in Figures 3 and 4 strongly suggest that the secondary ion yield is greatly enhanced with O_2 when the ratio (of the number of O_2 atoms to the sputtered atoms) is higher than about 2.5.

4. Conclusion

1) The sputtering yield varies from 0.2 atoms/ion (for 20% Zn) to 1.2 atoms/ion (for 80% Zn) according to the Zn-Fe alloy composition.
2) The secondary ion yields for Zn and Fe are not constant and vary according to the alloy composition.
3) The change of the secondary ion yield can be explained by the oxygen enhancement effect.
4) The ratio of the secondary ion yield of Zn to that of Fe is constant and independent of the alloy composition.
5) Based on the discussion above, the double-layered Zn-Fe alloy plating can be quantitatively analyzed using SIMS.

Reference

[1] R. Behrisch et al.: Sputtering by particle Bombardment I, 47 (1981) 167, 173 Springer-Verlag

[2] N. Henmi, K. Nakazima, K. Suzuki, T. Ohtsubo: Tetsu to Hagane, **70** (1984) S296
[3] K. Suzuki, K. Nishizaka, T. Ohtsubo, N. Henmi, K. Nakazima: Tetsu to Hagane, **71** (1985) A109
[4] Y. Ishibashi, Y. Yoshioka, T. Ishii, I. Fukui, T. Fukayama: Tetsu to Hagane, **70** (1984) S1044
[5] A.W. Czanderna: Methods of Surface Analysis (1975) 225, Elsevier Scientific Publishing Co.

Part XI

Biological Applications

Biological Microanalysis Using SIMS – A Review

R.W. Linton

Department of Chemistry, The University of North Carolina,
Chapel Hill, NC 27514, USA

1. Introduction

The utility of SIMS as a biological microanalysis technique is a reflection of the analytical merits of high-performance mass spectrometry coupled to recent advances in ion optics. Of special interest in this review is the application of SIMS to element localization in soft biological tissue. Although subcellular spatial resolution has been demonstrated in numerous instances [1-4], the quantitative relationship between secondary ion intensity observed in a fixed specimen and absolute concentration variations in the original tissue is highly complex. Indeed, many of the published ion images are probably influenced significantly by artifact contrast mechanisms.

Much of the interest in SIMS/biological microanalysis of soft tissues continues to center within a small number of research groups who currently utilize the Cameca IMS-3f ion microscope. Principal investigators responsible for many of these studies include Pierre Galle (Creteil), George Morrison (Cornell) and Margaret Burns (U.C.-Davis). My research group has become increasingly involved in this applications area over the past five years. The limited scope of collective research activity probably reflects a number of factors: i) the expense of instrumentation, ii) a continuing lack of substantive exposure of the biological/medical community to the technique, iii) the difficulties encountered in specimen preparation and quantification, iv) the limits of lateral spatial resolution with respect to the dimensions of many subcellular structures of physiologic interest. Thus, there are several current research areas, both instrumental and applied, that are essential to the sustained evolution of SIMS/biological microanalysis as outlined in the following sections.

2. Specimen Preparation

Since soft tissue is composed primarily of water, recent attention has been focused on the application of cryotechniques to specimen preparation. Although such methodologies have been exploited to a much greater extent in electron microscopy [5], the benefits in SIMS should be analogous in that they involve minimization of tissue damage and element translocation during fixation. Many elements of physiologic interest (e.g. Na, K, Mg, Ca) have high secondary ion yields. However, these elements also may be weakly bound and highly subject to diffusion during conventional chemical fixation. The alternatives are to examine freeze-fixed specimens, either freeze-dried or in a frozen hydrated state for high vacuum compatibility. The utility of freeze-drying is limited by surface topographic artifacts, including differential shrinkage observed in either cultured cell monolayers or tissue cryosections [6]. Although ice crystal damage is a major concern in high resolution electron microscopy, it is less important in SIMS since the spatial resolution is usually much more limited. To avoid freeze-drying artifacts, the examination of frozen hydrated materials is of interest. Initial ion microscopy results from such preparations is the subject of a separate paper in this symposium [7].

3. Sputter/Ion Yield Considerations

One of the most troublesome issues is the influence of tissue microheterogeneity, both physical and chemical, on sputter yields and useful secondary ion yields. Experiments to isolate the quantitative contributions of each yield factor above have not been reported for soft biological tissues. However, apparent differences in sputter yields or rates within cells or tissues have been observed using pre- and post-SIMS electron microscopy [6,8]. The use of "apparent" in the above statement is necessary, since variable sputter resistance of cell structures reflects non-uniformity of specimen chemical composition, density and thickness in addition to the sputter yield (atoms sputtered/incident ion) variations. Prior SIMS/SEM studies of dehydrated whole cells or tissue sections suggest that cell membranes, nuclear envelopes and nucleoli are relatively sputter resistant compared to cytoplasm or nuclear heterochromatin [6,8]. "Burn-through" maps [6,9], i.e. image - depth profiles for selected ions, indicate that nuclei disappear faster than the cytoplasm. These observations also may be a function of specimen type and/or preparation. For example, a paper by BURNS in this symposium suggests nuclei sputter less rapidly than surrounding tissue [10]. Normalization of ion images to atomic carbon ion intensity has been proposed as a method to compensate for these apparent preferential sputtering artifacts [11]. This is inherently limited by the assumptions that carbon is uniformly distributed in the specimen and that carbon secondary ion yields are independent of local variations in sputter yield.

Even more difficult are quantitative issues related to evaluation of local variations in useful ion yields. These also can reflect tissue microheterogeneity, including topography and chemical composition. They also may reflect variable implanted primary beam concentrations resulting from apparent variations in the sputter yields of cell microstructures. Use of ion beam implantation to incorporate an internal standard element has some utility in trying to compensate for these effects [10,12]. However, the implant will be subject to non-uniformities in projected range and depth distribution for the same reasons as described above. Thus, variations in the intensity of the internal standard element may not easily be related to variations in sputter ion yields of the physiologic elements of interest. Further discussions of matrix effects, including the influence of heavy metal staining and ion implantation studies, are presented in a contributed paper to this symposium [10]. Future advances in quantitation will include the use of frozen, hydrated tissues (to provide a more uniform, oxygen-rich specimen with perhaps less variation in sputter and useful ion yields), and the use of post-ionization techniques for sputtered neutrals [13] (to reduce the extent of useful ion yield variations). The latter also will be particularly valuable in improving detection limits of trace elements of toxicological interest which have relatively poor ion yields owing to both low electron affinity and high ionization potential (e.g. As, Cd, Hg, Tl).

4. Spatial Resolution

The recent improvements in focused liquid metal ion sources have enhanced the possibilities for lateral spatial resolution in the 100 nm range [14]. For example, LEVI-SETTI and coworkers have demonstrated the high resolution capabilities of ion-induced secondary electron and total secondary ion imaging of biological specimens [15]. However, it must be remembered that the ultimate limiting factor in SIMS is the amplitude of the secondary ion signal from a given analytical microvolume. Without concomitant improvements in useful ion yields, this implies that depth resolution must be traded-off against lateral resolution to retain sufficient analytical volume and corresponding detection sensitivity [16]. This issue is highly relevant to SIMS/biological microanalysis for several reasons. Specimens usually are highly heterogeneous on the micron-scale in all three dimensions, and many applications require the determination of trace elements or elements of low ionization probability. It is clear that improvements in primary optics must be integrated with secondary optics design, such that SIMS depth resolution is not unduly sacrificed because of useful ion yield considerations.

5. Molecular Information

The goal of molecular SIMS microscopy of biological tissues is inherently difficult. A major concern is the seemingly contradictory need to couple low primary ion dose, static SIMS conditions, as needed to optimize molecular information, with high lateral spatial resolution SIMS capability that normally requires high primary ion dose, dynamic conditions. Studies of organic polymer films have demonstrated the dramatic loss of molecular fragment ion intensities as doses exceed 1E14 ions/cm^2 [17]. For a modest resolution primary beam from a liquid metal ion source (1 nA current in a 1 µm spot), the total dose to the specimen is about 1E18 ions/cm^2-sec. Thus, even if the effects of a high instantaneous current density on the specimen are ignored, primary beam pulses of <100 µsec would be required to optimize molecular information. A promising avenue for future study is the combination of pulsed liquid metal ion sources with advanced time-of-flight instrumentation capable of high mass range and high secondary ion transmission [18].

Another possible instrumental approach is the application of the Cameca IMS-3f ion microscope. Although limited to a mass range of 500 amu, it is possible to sputter large areas (500 µm x 500 µm) while retaining about 1 µm lateral resolution with the stigmatic secondary ion optics. Primary oxygen or cesium beam currents as small as 0.1 nA can be achieved corresponding to doses on the order of 1E12 ions/cm^2-sec. For spatially modified organic polymers, mass-selected ion images can be acquired using doses <5E14 ions/cm^2-sec with limited polymer degradation [19]. An analogous experiment has not been reported to our knowledge for a biological specimen. However, it may have considerable promise for cases in which molecular xenobiotic, pharmaceutic or physiologic agents are sequestered locally within tissue. Even if direct molecular fragment ion imaging is too insensitive or subject to irreconcilable mass interferences, other analytical possibilities exist. These might involve various "tagging" schemes to enhance useful secondary ion yields, for example the use of isotopic labels or chemical derivatives as employed in recent SIMS studies of spatially modified organic polymers [18-20].

A third instrument for possible SIMS/biological molecular microanalysis applications is a scanning neutral beam system that has already been used for static SIMS microscopy of non-conducting organic coatings [21]. Not surprisingly, limits on useful ion yield and analytical volume restrictions cause severe problems with sensitivity in preliminary studies [21].

6. Digital Imaging

The general history of digital imaging in SIMS through 1983 is available in a recent review by RUDENAUER [22]. A SIMS-V symposium addresses more recent advances in detail. The two research groups most active in the current application of SIMS/digital imaging to biological microanalysis are those of MORRISON and LINTON [2,23]. Demonstrated advantages of digital manipulation of biological images include the use of various standard image-processing and enhancement algorithms (signal averaging, smoothing, histogram expansion and equalization, image overlays, pseudo-color display, etc). Corrections for artifact contrast contributions such as apparent preferential sputtering and non-uniform detector response also are available [9,12]. The advent of image depth profiling permits the exciting possibility of 3D reconstructed images of tissues, assuming that empirical corrections for sputter and useful ion yield variations can be developed.

To date, little effort has been directed toward the use of advanced image restoration algorithms for resolution enhancement in SIMS [24], certainly in comparison to the more intense activity in electron microscopy. A major difficulty in the SIMS application is that the enhancement of resolution is, by definition, required at the boundary between two dissimilar phases. Thus, matrix effects and useful ion yields may vary dramatically, and in a non-linear fashion, in the interfacial region. Apparent improvements in spatial resolution may be irrelevant to quantitative questions about actual concentration gradients at interfaces (e.g. biological membranes.)

7. Correlative Microscopy

Although ion images are a reflection of tissue morphology, the direct relationship of ion image to cell microstructure is often equivocal. This is a reflection of several factors including: i) many cell structures are on the borderline of ion image spatial resolution, ii) SIMS is sufficiently sensitive that trace contaminants potentially introduced during specimen preparation may contribute to the ion image [25], iii) ion image contrast may vary substantially as a function of sputter time, reflecting various SIMS artifacts and tissue microheterogeneity as described earlier. Thus, the ability to correlate ion images with other morphochemical data is particularly useful. This is often done on serial tissue sections such that specimen preparation can be optimized for each microanalytical technique. Recent examples of correlative chemical microscopy approaches include the comparison of ion imaging and microautoradiography for detection of U introduced into renal tissues by injection of uranyl acetate [26]; correlation of laser and ion microprobe data on trace Pb distributions in atherosclerotic aorta [27]; comparison of ion and X-ray microanalysis for Al localization in brain tissues [25]; comparison of laser, ion, and X-ray microanalysis of Al, Si, Fe in particle-laden human alveolar macrophages [28]; comparison of ion and X-ray microanalysis of an intestinal abscess containing barium sulfate particles [25].

A regimen permitting correlative microscopy of a single histologic section also has been published recently [25]. Conventional light microscopy (LM), secondary electron imaging (SEI), backscattered electron imaging (BEI), and electron probe microanalysis (EPMA) can be performed on the specimen both before and after ion microanalysis (SIMS). Advantages include the selection of tissue areas of interest for subsequent chemical microanalysis, direct comparison of data from various morphochemical techniques, and observation of apparent sputter yield variations or tissue contaminants that may influence ion image contrast. A fundamental limitation in this regard is the variation in depths of observation. Although LM views the entire specimen thickness, EPMA extends only to a maximum depth of about 10 µm. BEI is more surface specific with <2 µm as the usual sampling depth, while SIMS views only the approximate depth of the section sputtered during image acquisition, typically between 0.1 and 100 nm. Thus, a tissue structure visible by LM in a histologic section of 10 µm thickness would be visible by BEI, EPMA or SIMS only if it is located in the appropriate analytical volume for observation.

The ability to digitally process images for multitechnique correlations will facilitate future investigations in this area. We have demonstrated this aspect in a recent study showing digital overlays of ion images on light micrographs in a study of Al localization in osteomalacic bone [23]. Finally, the evolution of liquid metal ion sources is an exciting development for high spatial resolution correlative microscopy within the SIMS analysis chamber. Ion-induced secondary electron images, ion-induced total secondary ion or mass selected ion images potentially can be obtained from the same image field. All three image types have different contrast mechanisms [15] and will help to facilitate the relation of a secondary ion image to tissue microstructures.

References

1. M. Meignan, J.P. Berry, F. Escaig, A. Joczsa, J.P. Delauney, R. Masse, G. Madelaine, P. Galle: Phys. Med. Biol. 29, 927(1984).
2. S. Chandra, W.C. Harris, Jr., G.H. Morrison: J. Histochem. Cytochem. 32, 11(1984).
3. M.S. Burns: J. Microsc. 127(3), 237(1982).
4. R.W. Linton, I.H. Musselman, S.R. Bryan, "Laser and Ion Microprobe Mass Spectrometry - Applications to Human Tissues", in Microprobe Analysis in Medicine, J.D. Shelburne, P. Ingram, eds., Hemisphere Publishing, in press, 1985.
5. Fourth Annual Symposium on Advances in Electron Microscopy, Cyromicroscopy and Electron Beam Sensitivity, Duke University Marine Laboratory, Beaufort, NC, Sept., 1985.
6. R.W. Linton, M.E. Farmer, P. Ingram, S.R. Walker, J.D. Shelburne: Scanning Electron Microsc. III, 1191(1982).
7. S. Chandra, M.T. Bernius, G.H. Morrison: "Ion Microanalysis of Frozen-Hydrated Biological Specimens Using a Cold Stage in the Cameca IMS-3f Ion Microscope", SIMS-V, in press, 1985.
8. R.W. Linton, M.E. Farmer, P. Ingram, J.R. Sommer, J.D. Shelburne: J. Microsc. 134(1), 101(1984).
9. A.J. Patkin, S. Chandra, G.H. Morrison: Anal. Chem. 54, 2507(1982).
10. M.S. Burns: "Observations Concerning the Existence of Matrix Effects in SIMS Analysis of Biological Specimens", SIMS-V, in press, 1985.
11. W.C. Harris, Jr., G.H. Morrison: "Quantitative Ion Microscopy of Soft Biological Tissues", in Microbeam Analysis, K.F.J. Heinrich, ed., 227(1982).

12. W.C. Harris, Jr., S. Chandra, G.H. Morrison: Anal. Chem. 55, 1959(1983).
13. D.L. Donohue, W.H. Christie, D.E. Goeringer, H.S. McKown: Anal. Chem. 57, 1193(1985).
14. N. Anazawa, R. Aihara, M. Okunuki, R. Shimizu: Scanning Electron Microsc. IV, 1143(1982).
15. R. Levi-Setti: Scanning Electron Microsc., 1(1983).
16. P. Williams: "Limits of Quantitative Analysis Using SIMS", Scanning Electron Microsc., in press, 1985.
17. S.J. Simko, D.P. Griffis, R.W. Murray, R.W. Linton: Anal. Chem. 57, 137(1985).
18. P. Steffens, E. Niehuis, T. Friese, D. Greifendorf, A. Benninghoven: SIMS-IV, A. Benninghoven, J. Okano, R. Shimizu, H.W. Werner, eds., Springer-Verlag, New York, 1984, p. 404.
19. S.J. Simko, S.R. Bryan, D.P. Griffis, R.W. Murray, R.W. Linton: Anal. Chem. 57, 1198(1985).
20. S.J. Simko, R.B. Cole, R.W. Linton, R.W. Murray, C.R. Guenat, S.R. Bryan, D.P. Griffis:"Surface and Microanalysis of Modified Organic Polymer Films Using SIMS", SIMS-V, in press, 1985.
21. A. Brown, A.J. Eccles, J.A. van den Berg, J.C. Vickerman: "Chemical Characterization of Insulating Materials Using High Spatial Resolution SIMS", SIMS-V, in press, 1985.
22. F.G. Rudenauer: Surf. Interface Anal. 3, 132(1984).
23. S.R. Bryan, W.S. Woodward, D.P. Griffis, R.W. Linton: J. Micros. 138(1), 15(1985).
24. B.G.M. Vandeginste, B.R. Kowalski: Anal. Chem. 55, 557(1983).
25. K.G. Kupke, J.P. Pickett, P. Ingram, D.P. Griffis, R.W. Linton, P.C. Burger, J.D. Shelburne: J. Electron Microsc. Tech. 1, 299(1984).
26. P. Galle: SIMS-IV, A. Benninghoven, J. Okano, R. Shimizu, H.W. Werner, eds., Springer-Verlag, New York, 1984, p. 495.
27. R.W. Linton, S.R. Bryan, P.F. Schmidt, D.P. Griffis: Anal. Chem. 57, 440(1985).
28. R.W. Linton, J.D. Shelburne, D.S. Simons, P. Ingram, 40th Ann. Proc. Electron Microscopy Soc. Amer., Washington, D.C., 1982, p. 370.

Observations Concerning the Existence of Matrix Effects in SIMS Analysis of Biological Specimens

M.S. Burns

Department of Ophthalmology, School of Medicine, University of California, Davis, CA 95616, USA

1. Introduction

Matrix effects have been documented for analysis of a variety of materials analyzed by SIMS. Matrix effects in analysis of soft biological materials have been documented as being due to differential sputtering effects and attributed to differences in ion yield. GALLE, et al., [1] showed faster sputtering of plasma and slower sputtering of red blood cells in a smear of human blood. FARMER, et al., [2] demonstrated preferential etching of heterochromatin and Z lines due to oxygen sputtering of fixed frog skeletal muscle in both unstained and heavy metal stained tissues, using TEM and SEM criteria. LINTON, et al., [3] showed that the nuclear and plasma membranes were etch-resistant in fixed and osmium - stained kidney tissue and in isolated cells. Frozen and freeze-dried cell monolayers of several types also showed etch-resistant nuclear and plasma membranes and nucleolus. Cone formation was ubiquitous in all tissues and cells examined, as well as in pure epoxy resins. PATKIN, et al., [4] used osmium-fixed plant tissue and sputtered through to a tantalum substrate to assess relative sputtering rates, and reported that cell walls and nuclei sputtered more rapidly than cell cytoplasm. The existence of differences in ion yield due to the nature of the chemical bond was reported by Quettier and Quintana [5] in a comparison of carbon ion emission from tissues of known concentration. The chemical concentration ratio of organic to inorganic carbon was ca. 5, but the SIMS measured ion current ratio from the two areas was 10 for C^+ and 14 for C^-.

During the course of our studies on ion emission from soft biological tissue, we have made observations that indicate the existence of matrix effects, encompassing both differential sputtering and ion yield effects. We attribute these to local variations in physical density and electron density of cells.

2. Observations

2.1. Differential Sputtering Effects

Following positively charged oxygen bombardment of retinal tissues embedded in epoxy resin, there is cone formation that is most prominent in certain structures. These include red blood cells and pigment granules. The extent of cone formation is dependent upon the length of sputtering time. Although the location of cone formation is qualitatively similar in both freeze-dried or osmium-fixed material, our impression is that there are quantitative differences between these two preparative methods. The existence of cone formation from red blood cells and pigment granules is considered to occur because these structures are physically dense, as measured in sedimentation experiments. Moreover, in transmission electron microscopy studies, both structures are inherently electron-dense. Electron density can also be induced in tissues by fixation with osmium, which binds selectively to cell membranes and nuclei. This could be responsible for the clear delineation of plasma membranes and nuclei seen in osmium-fixed tissue as compared to freeze-dried material. Since cone formation also occurs in pure epoxy resin [3], this differential sputtering effect can not be attributed solely to the existence of density differences, but the density differences may be critical in controlling ion emission quantitatively.

2.2. Ion Emission Effects

2.2.1. Treatment with Osmium Vapor

Freeze-dried retinal tissue shows a characteristic emission pattern for Na, K, and Ca. When that same tissue is suspended in osmium vapor and then analyzed by SIMS, the emission is radically changed. In the case of calcium emission, the pattern now resembles that of tissue which has been fixed with aqueous solvents and osmium. There is not, however, a translocation of elements, as can occur with aqueous fixation techniques, since the normal barium emission from the choroid is unchanged. Also, the total current from calcium is increased and shifted to new electron-dense elements, the nuclei. We interpret this as being due to enhanced calcium emission from electron-dense structures.

2.2.2. Boron Implantation

Boron implantation of osmium-fixed material followed by imaging of boron in the implanted tissue shows increased emission from pigment granules, shown to exhibit cone formation in preceding studies. We interpret this as either that boron is implanted at different depths in microareas of the material, or that it is differentially emitted. In order to resolve this, non-averaged depth profiles in implanted tissue need be measured.

2.2.3. Vanadium-Epoxy Infiltrated Tissue

Vanadium-doped epoxy resins were used to infiltrate retinal tissue sections. The assumption was that the vanadium could be used as an internal standard to reference ion counts from the tissue areas of interest [6]. Instead, the emission of vanadium from tissue areas was increased relative to that of the epoxy resin alone. Moreover, images from these areas indicated that the vanadium was more emissive in areas of tissue, indicating enhanced ionization of vanadium due to the chemical environment.

3. Discussion

These observations, in combination with the literature reports of others, document the existence of differential sputtering effects in heterogeneous biological tissue. In the tissues we have studied, this differential sputtering appears to be qualitatively related to local increased physical density and electron density of the structures refractory to ion etching. To what extent these phenomena are quantitatively important remains for future study.

It appears that the formation of sputter-resistant structures is related to relatively high ion emission from some elements, so that enhanced ionization may be associated with these structures, an important variable for developing quantitative analysis of tissues by SIMS. This occurs despite specimen properties of amorphous material and high oxygen content, both of which should optimize and stabilize ion yields [3,7].

These considerations suggest that detailed quantitative study of the physical and chemical properties of cells and subcellular organelles must be accomplished for individual tissues before programs for quantitation can be applied.

4. Literature References

1. P. Galle, G. Blaise and G. Slodzian: Seventh International Congress of Electron Microscopy, Grenoble (1970).
2. M.E. Farmer, R.W. Linton, P. Ingram, J.R. Sommer, J.D. Shelburne: J. Microscopy. 124, RP1.
3. R.W. Linton, M.E. Farmer, P. Ingram, J. Shelburne: Scanning Electron Microscopy, Part III. p. 1191 (1982).
4. A.J. Patkin, S. Chandra, and G.H. Morrison: Anal. Chem. 54, 2507 (1982).
5. A. Quettier and C. Quintana: Compt. Rend. Acad. Sci. (Paris) 289, 433 (1979).
6. M.S. Burns-Bellhorn and D.M. File. Anal. Biochem. 92, 213 (1979).
7. M.S. Burns: "Secondary Ion Mass Spectrometry", in Analysis of Organic and Biological Surfaces (ed. P. Echlin) John Wiley and Sons (1984).

Ion Microanalysis of Frozen-Hydrated Cultured Cells

S. Chandra, M.T. Bernius, and G.H. Morrison

Baker Laboratory of Chemistry, Cornell University, Ithaca, NY 14853, USA

1. Introduction

Inorganic cations (Na^+, K^+, Ca^{++}, etc.) play important roles in cell function, and cultured cells provide an excellent system to understand the role of ions in the state of health and disease. Ion microscopy, based on secondary ion mass spectrometry (SIMS), provides a powerful technique for such studies (1). The unique ion optics of the instrument are capable of producing visual ion images showing cell morphology with a lateral resolution of ~0.5µm. However, reliable means of sample preparation and an understanding of SIMS effects is necessary before such studies can be undertaken.

Analysis of frozen-hydrated biological samples has long been considered as an ideal sample preparation method for diffusible elements. A recently available cryogenic sample stage for the Cameca IMS-3f ion microanalyzer (2) has been used for a preliminary study in analyzing frozen-hydrated cultured cells and tissue sections(3). The present study extends these observations in an attempt to understand the effect of the hydrated matrix on the secondary ion signals from different cellular compartments of cultured cells.

2. Experimental

3T3 mouse fibroblasts and PtK_2 potoroo kidney cells were grown according to standard tissue culture techniques on sterilized silicon wafer pieces. The cells were fast frozen in liquid nitrogen slush and cryofractured using a sandwich technique(4,5,6). The silicon wafers containing cells were stored under liquid nitrogen before analysis. The transfer of these samples to the precooled (-182°C) sample stage of the Cameca IMS-3f ion microscope was carried out using a cryogenic sample holder according to the procedure described previously(2,3).

An 8.0 KeV O_2^+ primary beam was used and positive secondaries were monitored in this study. A 150 nA primary beam of approximately 100µm diameter was rastered over an area of 250 x 250 μm^2. The cold stage was maintained at -182°C during the analysis.

3. Results and Discussion

Image evolution in PtK_2 cells along with water images at mass 18 (H_2O^+) are shown in Fig. 1. The pre-sputtering requirement for a few minutes that allowed recognizable clear morphological details remained the same (3). These images were recorded in succession with continuous ion bombardment to understand the relationship between the water signal and the enhancement of secondary signal observed for the K image. Figure 1a shows the $^{39}K^+$ ion image. A water image (Fig. 1b) recorded immediately after this image reveals the presence of water in these cells. A one-to-one matching of the water signal clearly indicates that hot spots in image 1a representing

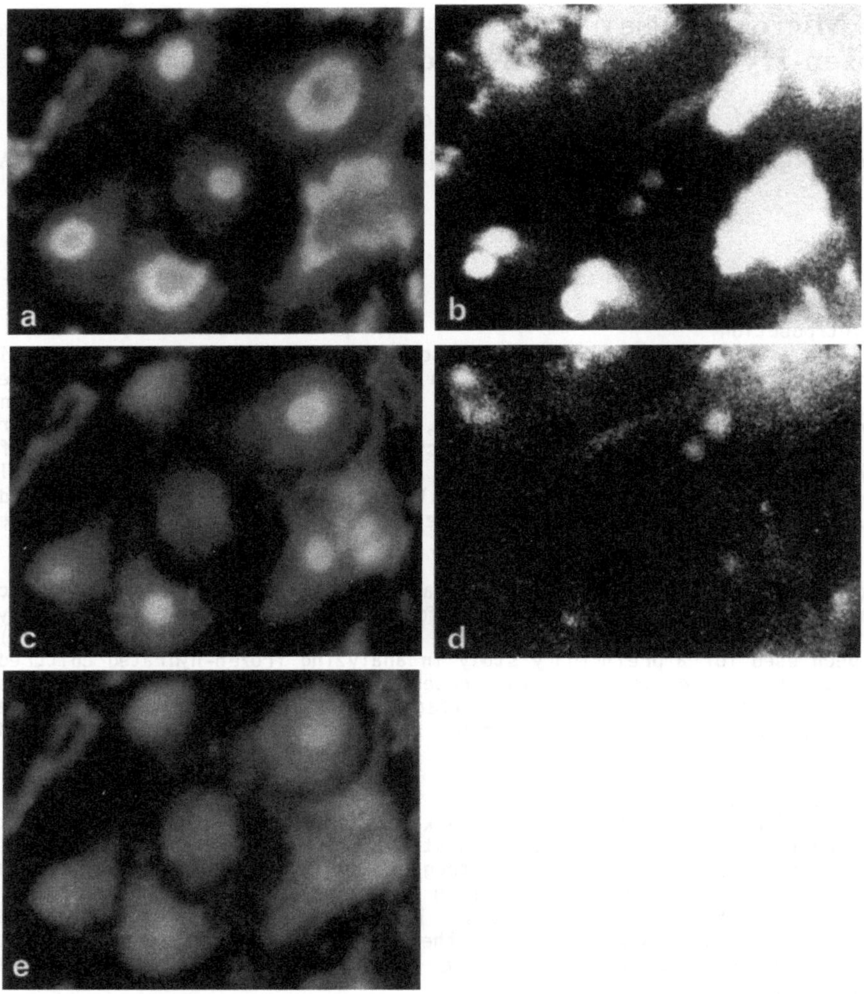

Fig. 1: Image evolution in frozen-hydrated Pt K_2 cells. Images of $^{39}K^+$ and water at mass 18 (H_2O^+) were recorded in succession with continuous ion bombardment. Initial K distribution and corresponding H_2O^+ signals are shown in images "a" and "b", respectively. Images "c" and "d" representing $^{39}K^+$ and H_2O^+ signals were recorded again from the same field of view after an additional 4 min. of continuous ion bombardment. Image "e" showing K distribution from the same cells was recorded after an additional 3 min. of ion bombardment. Image exposure times: K images = 1/4sec.; H_2O^+ images = 60sec.; Bar = 50µm.

nuclei and surrounding areas correspond perfectly to the water signals. A continuous bombardment of this area for an additional 4 min. reveals that the K image has been changed (Fig. 1c). The shrinkage of hot spots in size and even a total disappearance from a few cells is observed now. A water image (Fig. 1d) recorded immediately after image 1c indicates a substantial decrease in the water signal as compared to Fig. 1b. Once again the remaining hot spots in the K image (Fig. 1c) can be matched with the water signals in Fig. 1d. An additional bombardment for 3 minutes of this field of view revealed no enhancement of K signal from any cellular compartments (Fig. 1e), and after this image the water signal was found to be too low for decent imaging. Similar observations were made for 3T3 cells. It should be noted that cryofractured and freeze-dried cells do not show such image change.

The appearance of higher water signals from nuclei and surrounding areas is not surprising due to the physiological spread of these cells. In these cells lines generally, the thicker portion of the cell is towards the nuclei. The water enhancement of the K signal is interesting and is found to be irrespective of the elements studied. It is plausible that there is a signal enhancement due to the water matrix because of its oxygen content; an effect that would be consistent with signal enhancement due to the SIMS oxygen effect.

The preferential removal of water in the Z-direction from the cell matrix is also interesting, since freeze-drying of the specimen at -182°C is least likely. In addition, the localized beam heating is not anticipated to be more than a few degrees Kelvin under the instrumental parameters used in this study. A detailed study to understand these effects is presently underway in our laboratory.

4. Acknowledgment

The authors would like to thank Christine Coulter Wolcott and Lisa Kapustay Turner for technical assistance. Howard Smith is acknowledged for constructive criticism of the cold stage work. The Cornell Biophysics Laboratory is acknowledged for culturing cells. Financial support for this project was provided by the National Institutes of Health.

References

(1) G.H. Morrison and G. Slodzian, Anal. Chem. 47, 932A-943A (1975).
(2) M.T. Bernius, S. Chandra, and G.H. Morrison. Rev. Sci. Instr. 56, 1347-1351 (1985).
(3) S. Chandra, M.T. Bernius, and G.H. Morrison. Anal. Chem. (in press)
(4) S. Chandra and G.H. Morrison. Science 228, 1543-1544 (1985).
(5) S. Chandra, G.H. Morrison, C.W. Coulter, and S.E. Bloom. J. Cell Biol. 99(4) part 2, 424a (1984).
(6) S. Chandra, G.H. Morrison, C.C. Wolcott. J. Cell Biol. (submitted)

Imaging Intracellular Elemental Distribution and Ion Fluxes in Cryofractured, Freeze-Dried Cultured Cells Using Ion Microscopy

S. Chandra and G.H. Morrison

Baker Laboratory of Chemistry, Cornell University, Ithaca, NY 14853, USA

1. Introduction

Elements play an important role in intracellular regulatory events. Cells grown in cultures provide an excellent model for studying the distribution and transport of ions under normal and pathological conditions. The unique ability of the ion microscope to provide ion images with cell morphology is ideally suited to such studies. However, analysis of cultured cells at the subcellular level has been problematic in ion microscopy (1,2,3). Due to the highly diffusible nature of physiologically important elements (Na, K, Ca, etc.), cryo-techniques are the only acceptable means of sample preparation. Using a cryo-fracture technique, we have been able to study subcellular elemental distribution in several cell lines.

2. Experimental

Chinese hamster ovary (CHO), normal rat kidney (NRK), and 3T3 mouse fibroblast cultures were grown directly on the smooth surface of sterilized silicon wafer pieces according to standard tissue culture techniques. The cells were sandwiched using another smooth silicon wafer surface and fast frozen in liquid nitrogen slush in order to immobilize the diffusible elements in their native states. The sandwich was fractured by separating the two halves under liquid nitrogen. This procedure produces large areas containing fractured cells on the silicon substrate where the fracture plane has gone through the apical cell surface. The details of this technique are described elsewhere (4). After freeze-drying at -80°C for 24 hrs., the fractured cells were analyzed in the ion microscope to study the subcellular elemental distribution.

A Cameca IMS-3f ion microscope was used for ths study. A 200 nA primary O_2^+ beam of approximately 100 μm diameter was rastered over an area of 250 x 250 μm^2. High mass resolution analysis confirmed no significant contribution of NaH^+ to $^{24}Mg^+$ secondary signal.

3. Results and Discussion

The subcellular distribution of Na, K, Ca and Mg from cryofractured and freeze-dried NRK cells is shown in Fig. 1. No major differences are observed for Na, K and Mg distribution between the nuclei and the cytoplasm. However, Ca concentrates more in the cytoplasm than the nuclei of cells. Even within the cytoplasm Ca is not homogeneously distributed. These observations are suggestive of calcium binding sites in the cell cytoplasm. The possible nucleoli reveals higher Ca intensities than the rest of the nucleoplasm. Similar observations were made for other cell lines.

The variation of sodium intensities in the individual cells is clearly evident, and perhaps is due to the different physiological states of these unsynchronized cells. One cell in this field of view (arrow) shows a relatively higher Na signal and an excessive loading of Ca. Such abnormal ion fluxes are typical for injured cells, and have been discussed in detail elsewhere (4).

The fractured cells are consistently high potassium-low sodium cells. The dead cells can be easily discriminated due to their high Na-low K signals. This simple cryofracture technique allows fracturing of cells under liquid nitrogen without removing them from the substrate. In addition, it eliminates the need for washing the nutrient media away and

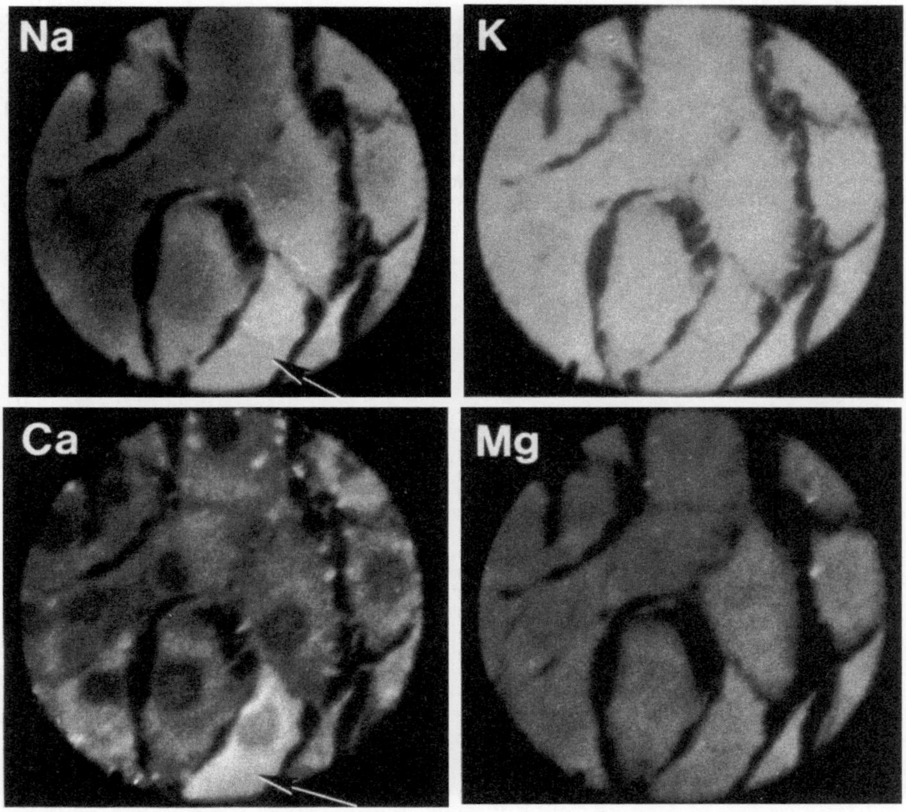

Fig. 1: Intracellular distribution of Na, K, Ca and Mg in cryofractured and freeze-dried NRK cells. Brighter spots in the nuclei of some cells in the Ca image perhaps represent the possible nucleoli. One cell in the field of view (arrow) shows high Na - high Ca signals, a typical symptom of cell injury. Images were recorded in the order of K, Na, Ca and Mg. Individual images were optimized for the image quality. Image exposure times: Na=1sec.; K=1/8sec.;Ca=50sec.; Mg=50sec. Image field of view = 150µm in diameter.

cryo-sectioning before ion microanalysis. This method has been successfully used to image not only intracellular elemental distribution but also ion transport in cultured cells (5). The Standardization of ion microscopic methodology to cell culture systems makes it a powerful tool for studying the role and transport of ions under physiological, pathological and toxicological states.

4. Acknowledgment

C.C. Wolcott and the Cornell biophysics facility are acknowledged for culturing cells. Financial support for this project was provided by the National Institutes of Health.

References

(1) R.W. Linton, S.R. Walker, C.R. DeVries, P. Ingram, J.D. Shelburne: Scanning Electron Microsc. II, 583 (1980).
(2) R.W. Linton, M.E. Farmer, P. Ingram, S.R. Walker, J.D. Shelburne: Scanning Electron Microsc. III, 1191 (1982).
(3) S. Chandra, G.H. Morrison; "Cell Cultures: An Alternative in Biological Ion Mass Spectrometry SIMS IV; Benninghoven, A.; Okana, J.; Shimizu, R.; Werner, H.W., Eds.; Springer-Verlag: New York, 1984; pp. 489-491.
(4) S. Chandra, G.H. Morrison, C.C. Wolcott. J. Cell Biol. (submitted).
(5) S. Chandra and G.H. Morrison. Science 228, 1543-1544 (1985).

Cellular Microlocalization of Mineral Elements by Ion Microscopy in Organisms of the Pacific Ocean

C. Chassard-Bouchaud, P. Galle , and F. Escaig

Laboratoire de Biologie et Physiologie des Organismes Marins, Université P. et M. Curie, 4 pl. Jussieu, F-75230 Paris, Cedex 05, France
and
Centre de Microanalyse Appliquée à la Biologie-Laboratoire de Biophysique de la Faculté de Médecine, 8 rue du Général Sarrail, F-94000 Créteil, France

1. Introduction.

Secondary ion mass microanalysis was shown in previous investigations to be suitable for the detection and localization of elements in marine organisms (1).

The purpose of this paper is to present new data concerning benthic organisms living in the Pacific Ocean in two different biotopes and having different feeding habits and physiology.The following species were investigated: - *Charybdis japonica* : Crabs collected from the coastal waters of Japan (Tanabe,Honshu Island 34°N),with omnivorous feeding habits.

- *Alvinella pompejana* :Polychaetes worms (juvenile forms),collected from the deep sea hydrothermal vents of the East Pacific Rise (21°N) recently discovered and which partly feed on sulfide-oxidizing bacteria living on the Polychaete body.

2. Instrumentation.

Microanalytical data were obtained by using two analytical ion microscopes:
- CAMECA IMS 300 ,equipped with an electrostatic deflector and using O_2^+ as primary ions for the study of positive secondary ions.
- CAMECA IMS 3F ,using Cs^+ as primary ions for the study of negative secondary ions.Experiments were carried out on this instrument,using a method of suppressing molecular ions (2).

3. Preparation of specimens.

Frozen dehydrated tissues and paraffin-embedded tissues after chemical fixation (sectioned at 5 µm) were used,according to the methods previously described (1).

4. Results and discussion.

As results obtained with both cryo and chemical fixations were identical, we choose to present those obtained with the second technique,the ion images being better.

In *C.japonica* the digestive gland was shown to be the major site of the element concentration.In *A.pompejana* the major sites were shown to be the dorsal body epidermis,the parapodia epidermis and the digestive tract lumen.Beside the classical elements existing in every organism (sodium,calcium,phosphorus,potassium,magnesium) together with strontium and barium existing in every marine animal,the following elements were detected within both species:lithium (Li),fluorine (F),aluminum (Al),iron (Fe),copper

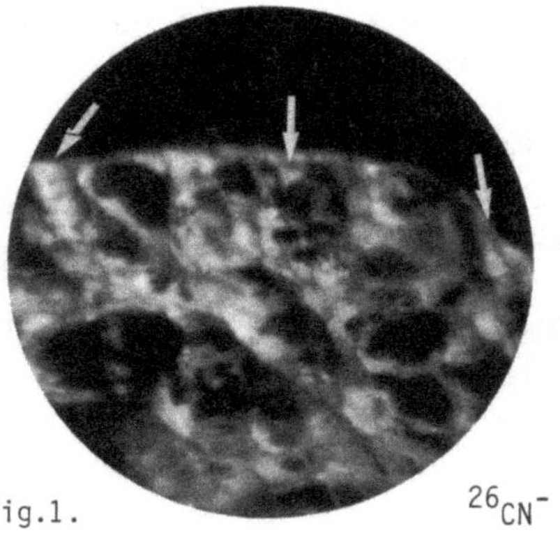

Fig.1. $^{26}CN^-$

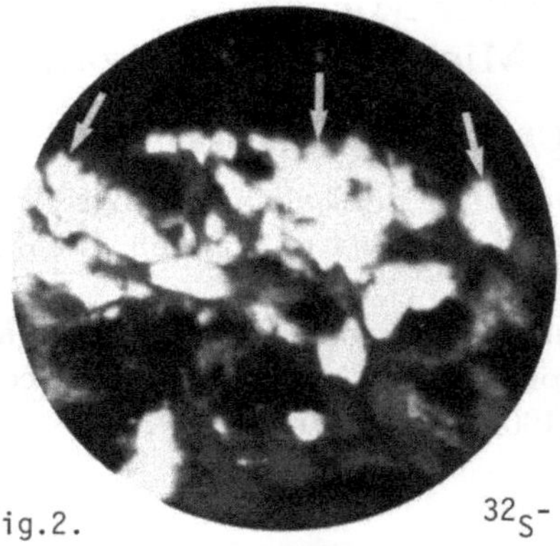

Fig.2. $^{32}S^-$

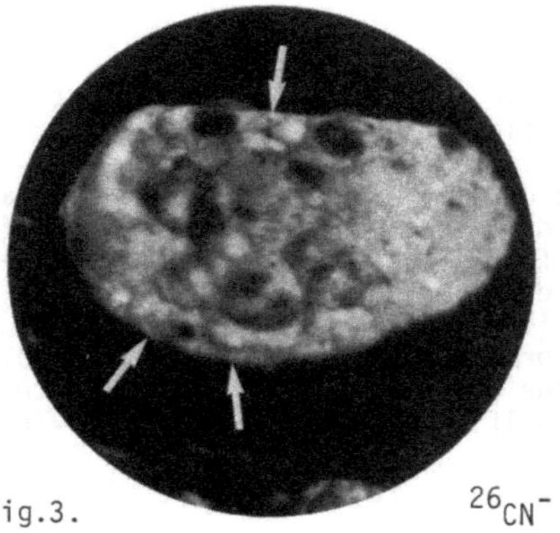

Fig.3. $^{26}CN^-$

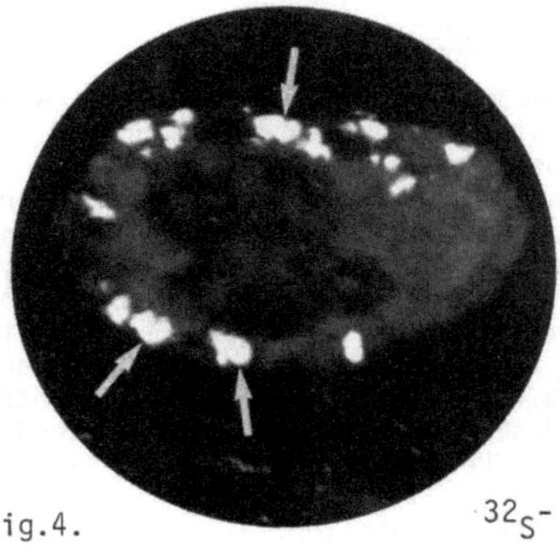

Fig.4. $^{32}S^-$

Alvinella pompejana - Fig.1 & 2:dorsal epidermis. Fig.3 & 4: parapodia.
Fig.1: CN ion micrograph showing the histological structure of the epithelial cells(arrows).
Fig.2. S ion micrograph obtained from the same section as in Fig.1,showing the sulfur concentration within the epithelial cells (arrows).
Fig.3. CN ion micrograph showing the histological structure of the epithelial cells (arrows).
Fig.4. S ion micrograph obtained from the same section as in Fig.3,showing the sulfur concentration within the epithelial cells (arrows).

(Cu),silver(Ag),iodine (I) and rare earth elements (REE) such as lanthanum (La) and thulium (Tm).All of these were previously shown to be usually accumulated by marine Invertebrates collected from various coastal waters (1)(3).

Moreover,some elements were specifically detected in each species (see table 1).Among those detected in *C.japonica*,uranium (U) was previously demonstrated to accumulate in Molluscs and Crustacea from contaminated waters (4).In *A.pompejana* ,sulfur (S) is shown on ion images to occur in an unusual high concentration in the surface epidermis(fig.1,2) and in the parapodia epidermis (fig.3,4).The presence of sulfur in the cells where bacteria occur,suggests that *A.pompejana* could partly feed from sulfide-oxidizing bacteria (5).While arsenic (As),lead (Pb) and zinc (Zn) were detected in this species by X-ray spectrometry (6),chromium (Cr),manganese (Mn) and particularly thallium (Tl)were never reported before.As the geochemical con-

Table 1. Major isotopes of the elements specifically detected within each species by SIMS. Emission intensities of the elements normalized to carbon.

	^{51}V	^{58}Ni	^{79}Br	^{209}Bi	^{238}U	^{239}Pu	
Charybdis japonica digestive gland	2.10^{-1}	3.10^{-1}	3.10^{-1}	5.10^{-1}	2.10^{-2}	4.10^{-2}	
	^{32}S	^{52}Cr	^{55}Mn	^{64}Zn	^{75}As	^{205}Tl	^{208}Pb
Alvinella pompejana dorsal epidermis	9.10^{-1}	1.10^{-3}	7.10^{-1}	1.10^{-1}	2.10^{-1}	2.10^{-1}	3.10^{-1}
digestive tract lumen	3.10^{-3}	2.10^{-1}	1.10^{-2}	3.10^{-1}	3.10^{-1}	9.10^{-1}	4.10^{-1}

ditions existing in the hydrothermal vent areas are known only in parts(7), the significance of unusual and toxic elements such as thallium remains still unknown.

Suppressing of molecular ion interferences in the spectrum offers major advances.The application of SIMS as a practical method allows detection of major and trace elements taken up from the environment and then concentrated, often at high levels,by marine organisms.

5. Acknowledgements.
This work was performed with the financial assistance of C.N.R.S. and I.N. S.E.R.M.,service commun n° 27 and with the scientific collaboration of Pr. Hirota, Koshida, Laubier and Desbruyères.

References.

1. C.Chassard-Bouchaud,F.Escaig and P.Hallégot: J.Microsc.Spectrosc.Electron. 9, pp.481-498 (1984).
2. J.B.Metson,G.M.Bancroft,N.S. McIntyre and W.J.Chauvin: SIMS IV,ed.by Benninghoven,Okano,Shimizu and Werner.Springer Verlag,N.Y.,pp.466-468 (1984).
3. C.Chassard-Bouchaud ,P.Galle, F.Escaig et M.Miyawaki:C.R.Acad.Sc.Paris 299,série III,18,pp.719-724.(1984).
4. C.Chassard-Bouchaud: SIMS IV,ed.by Benninghoven,Okano,Shimizu and Werner.Springer Verlag N.Y.pp.492-494,(1984).
5. L.Laubier,D.Desbruyères and C.Chassard-Bouchaud: Mar.Biology Letters 4 pp.113-116 (1983).
6. F.Gaill,S.Halpern,C.Quintana and D.Desbruyères : C.R.Acad.Sc.Paris,298 série III,12,pp.331-335,(1984).
7. G.Klinkhammer,H.Elderfield and A.Hudson :Nature 305, pp.185-188 (1983).

Quantitative SIMS of Prehistoric Teeth

P. Fischer[1], *J. Norén*[1], *A. Lodding*[2], *and H. Odelius*[2]

[1] Dental School, University of Gothenburg, S-40033 Gothenburg, Sweden

[2] Physics Department, Chalmers University of Technology, S-41296 Gothenburg, Sweden

1. Introduction

Biological mineralized tissues, such as dry substance of teeth, consist mainly of apatite. Although the secondary ion spectra of apatite are complex /1,2/, sensitive and quantitative SIMS routines have been established for most elements in biomineralizations, with extensive application chiefly in the odontological field /2,3/. In the hard tissue of enamel and dentine, essentially hydroxy-apatite, numerous impurities and elemental substitutions can be found, some of considerable clinical and/or environmental significance. Also, the apatite structure may locally transform to other phosphates. Nevertheless SIMS has been shown to give reproducible elemental analysis of dental tissues.

Table 1 lists representative concentrations of elements in teeth, as determined in recent years by other techniques, such as neutron activation, thermionic or spark-source mass spectrometry, or absorption/emission spectrometry /4/. Also shown are the detection limits in SIMS of apatites (recalculated from ref./2/; defined at 2x background level; assuming an energy-pass window of 80 eV, high voltage offset 100 V). It is to be emphasized that while the other techniques yield average concentrations from essentially macroscopic portions of enamel or dentine, the SIMS sensitivities are formulated for probing analysis at topographically well-defined small areas on the specimen, with a lateral resolution of the order of 10 µm and a depth resolution better than ca 0.1 µm. The sensitivity of analytical SIMS is seen to be ample for detecting and measuring more than 20 elements at their "normal" concentrations in dental hard tissue. Concerning many other elements SIMS may still be adequate for tracing concentrations locally elevated due to biological or evironmental factors.

Elemental composition of dental tissues in living humans is influenced endogenously (nourishment, blood metabolism) as well as exogenously (oral fluids). Post-mortem teeth continue to interact with their environment. Particularly from the archaeological point of view, it may be of considerable interest whether the distribution of elements may mirror the lifetime habitat. Also, the systematics of absorption and leaching on teeth in burial storage should relate to questions of dating and site identification.

However, as different regions of tooth topography exhibit widely varying element concentrations (see, e.g., ref./3/), meaningful comparison between different teeth appears possible only as long as the morphological location of each point of analysis is well specified. The task therefore requires a technique combining high detection sensitivity with a sufficient spacial resolution. SIMS satisfies these criteria in a unique way, and has in recent years proved useful in environmental studies /3/. This has motivated the present pilot study, an attempt to explore the applicability of SIMS in the field of archaeometry. The work is to form part of an extensive investigation /5/ of successive cultures on Cyprus, with material from ca 2000 to 800 BC.

Table 1. Concentrations of elements (order-of-magnitude, from ref./4/) in normal human enamel (c_E) and dentine (c_D); compared to approximate detection limits (DL) in SIMS of apatite-base biomineralizations. All in weight-ppm.

	c_E	c_D	DL		c_E	c_D	DL
Li	1		10^{-3}	Co	10^{-3} - 10^{-1}	10^{-3} - 1	50
Na	5×10^{3}	5×10^{3}	10^{-2}	Rh	$< 10^{-2}$		10
K	5×10^{2}		10^{-3}	Ir	$< 5 \times 10^{-2}$		10
Rb	10^{-1} - 10^{2}		10^{-1}	Ni		1	10^{2}
Cs	5×10^{-2}		1	Pd	$< 5 \times 10^{-2}$	$< 10^{-2}$	5
Be	$< 10^{-2}$		10^{-1}	Pt	$< 5 \times 10^{-2}$		50
Mg	$> 10^{3}$	$> 5 \times 10^{3}$	1	Cu	10^{-2} - 50	10^{-1} - 50	10
Sr	10^{2}	10^{2}	10^{-1}	Ag	10^{-3} - 0.5	10^{-3} - 5	10
Ba	1 - 10^{2}	10^{2}	1	Au	10^{-4} - 5×10^{-2}	5×10^{-2}	10^{2}
Sc	$< 10^{-1}$		10^{2}	Zn	10^{2} - 5×10^{2}	10^{2} - 5×10^{2}	10
Y	$< 10^{-1}$		1	Cd	10^{-4} - 5×10^{-1}	10^{-1}	10
La	$< 5 \times 10^{-2}$		10^{-1}	Hg	$< 10^{-1}$ - 5		10^{2}
Ce	5×10^{-2}		5×10^{-1}	B	1 - 10		10^{-1}
Pr	5×10^{-2}		5×10^{-1}	Al	10 - 10^{2}	> 50	1
Nd			1	Ga	$< 5 \times 10^{-2}$		5
Th			1	In			10
U		10^{-3} - 5×10^{-2}	10^{-1}	Tl	$< 5 \times 10^{-2}$	5×10^{-3}	1
Ti	10^{-1}	10	1	C	$> 10^{2}$	$> 5 \times 10^{2}$	50
Zr	10^{-1}		5×10^{-1}	Si	50	10^{2}	10
Hf	$< 10^{-1}$		1	Ge	5×10^{-2}		50
V	$< 5 \times 10^{-2}$		5×10^{-1}	Sn	10^{-1} - 10^{2}	10^{2}	5
Nb	$< 5 \times 10^{-1}$		5	Pb	1 - 50	5 - 50	5
Ta	$< 10^{-1}$		5×10^{-1}	N			5×10^{2}
Cr	5×10^{-3} - 5	5×10^{-3} - 5	1	As	$< 10^{-2}$ - 10^{-1}	10^{-2} - 10^{-1}	10^{2}
Mo	5×10^{-2} - 10	< 5	5	Sb	10^{-1} - 1	10^{-1} - 1	10
W	$< 5 \times 10^{-1}$	< 5	10	Bi	5×10^{-3}	10 - 50	10
Mn	10^{-1} - 1	10^{-1} - 10	50	S	50 - 5×10^{2}	5×10^{2}	10^{2}
Re	$< 5 \times 10^{-2}$		10	Se	10^{-1} - 1	10	5×10^{2}
Fe	5 - 5×10^{2}	10 - 5×10^{2}	10	Te			5×10^{2}
Ru	$< 5 \times 10^{-2}$		50	F	> 10	10 - 10^{3}	1
Os		$< 10^{-1}$	50	Cl	10^{3} - 10^{4}	10^{3}	10
				Br	1 - 50	1 - 10^{2}	5×10^{2}
				I	5×10^{-2}		10^{2}

2. Experimental

Each tooth was cut perpendicular to the surface into two parts. The smooth sections were resin-embedded in the SIMS sample holder and coated with ca 100 nm Au. The profiling was performed with a Cameca IMS-3F instrument mainly according to earlier described routines /3,6/. Step-scan analysis took place along a line from pulpal dentine to outermost surface. The step length was between 10 and 200 µm. The O^- primary ions, ca 0.5 µA, were accelerated through ca 15 kV. The bombarded area was ca 60 µm, analyzed area 10 µm in diameter. At each step 30 seconds were allowed for the beam to penetrate the Au coating, after which the ion currents were cyclically counted for 1 or 10 seconds: 7-Li, 11-B, 12-C, 19-F, 23-Na, 24-Mg, 28-Si, 35-Cl, 38-K, 44-Ca, 54-Fe, 56-CaO, 57-CaOH, 59-CaF, 64 and 66-Zn, 77 CaCl, 88-Sr, 118 and 120-Sn, 138-Ba, 207 and 208-Pb. To reduce spectral background and to minimize differences in sensitivity between different elements /7/ a relatively high offset in sample potential was employed, allowing only ions with kinetic energies between 60 and 140 eV to reach the analyzer. The drift in sample potential due to the build-up of surface charge was automatically compensated in each measuring cycle /7/. For the conversion of the recorded ion currents (all positive and related to 44-Ca) into the the respective concentrations, relative sensitivity factors were used. For all elements except Fe, Si and Sn, these factors were based on external standards. The normalized elemental ion yields (except for the halogens, see e.g. ref./2/) were found to depend on the first ionization potentials, in reasonable agreement with the LTE formalism /8,2/, implying an "ionization temperature" of ca 14000 K. The relative sensitivity factors for the three elements without external standards were therefore calculated according to the systematics of LTE. The main interest of this pilot study was in comparative figures between archaeological and modern tooth material, rather than in absolute concentrations.

3. Results and Discussion

a) Shapes of in-depth profiles

Figure 1, on an arbitrary depth scale, shows representative profiles for F, C Mg, Pb and Ba, stretching from outermost enamel (OE) through inner enamel (IE) dentinal enamel, outer dentine, central dentine (CD) and pulpal dentine (PD). The total length of a profile is generally between ca 3 and 6 mm. The broken line plots are typical of a recently extracted tooth of a young adult, here used as reference. The dashed curves, from a tooth of a young female warrior buried ca 1800 BC, shows several features typical of the archaeological material in comparison with modern teeth: relatively small differences in regard of Mg and C between dentine and enamel leaching-out tendencies for Mg, C, and Pb at pulpal dentine; clearly higher concentrations of Ba and F, especially in inner enamel and central dentine.

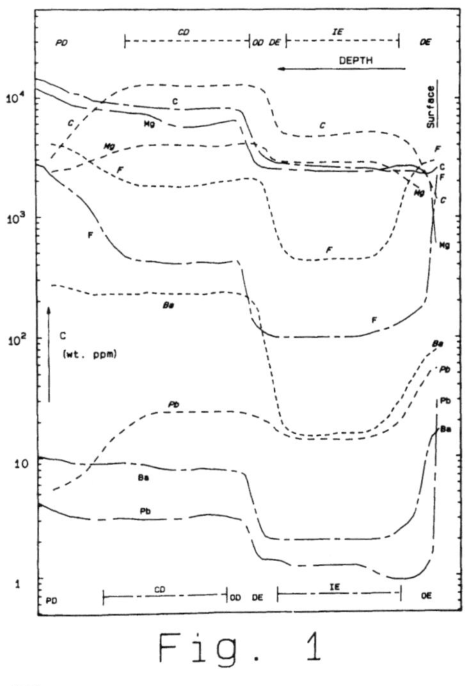

Fig. 1

The "plateau" regions, IE and CD, are obviously optimal for quantitative comparison between different teeth. Table 2 lists the mean concentrations of 15 elements measured at IE and CD in ten specimens. The following tendencies can be seen as consequences of prolonged burial:

1) Strong absorption of F in both regions. A scrutiny of the individual teeth shows (see below) a clear dependence of F in the ID zone on burial time, and only slight differences between individual burial sites /5/.

2) Clear absorption at both levels of Li, Sr, Ba, Si, Zn; in IE also Pb and Sn; in CD also B. These concentrations are found /5/ to differ considerably between different specimens, and at least for Li, Ba and Si appear dependent on the geological environment. For Pb and Zn, some correlations with implements and other objects found at

Table 2. Element concentrations (weight-ppm) at two depth levels (inner enamel and central dentine) in teeth from excavations on Cyprus (averages of 8, aged ca 2900 to 3700 years) and modern reference teeth (averages of 2).

Level	Li	Na	K	Mg	Sr	Ba	Zn	B	C	Si	Sn	Pb	Fe	F	Cl
IE, ref.	0.007	4400	265	2500	160	4	100	55	2250	20	1.0	5.5	44	97	1750
IE, Cyprus	0.22	4400	215	2450	300	33	315	82	3790	108	4.3	35	47	700	2050
CD, ref.	0.007	4950	225	6300	160	11	135	55	6700	47	3.3	15	58	360	1710
CD, Cyprus	0.47	3960	155	2990	745	298	466	165	10200	290	7.0	35	180	2940	1960

the respective grave sites /5/ imply, although so far with insufficient statistics, influences from the cultural habitat.

3) Clear leaching-out tendencies for Mg, especially in dentine, are probably connected with the original presence of Mg mainly as organic phase in pores.

4) The remaining major substitutions in the apatite phase, viz. Na and Cl, remain nearly unaffected by burial.

b) Fluorine in dependence on burial time

It is known that F concentrations (c_F) in the outermost layers of enamel tend to saturate at values (c_o, say) in a restricted range, between ca 1500 and 4000 ppm, only slightly dependent on F presence in ambient environment /6/. It is also often suggested that c_F in inner enamel is lower for young adults than for older persons. This seems well borne out by the present study (see Fig.2): the plateau concentration in IE (c_i, say) appears clearly dependent on the age of the teeth, ranging between ca 300 and 1000 ppm in the archaeological material. In living adults, on the other hand, c_i is seldom found to exceed 200 ppm /6/.

If c_o is accepted as a nearly constant entity (given by the microstructure of enamel, see refs./9,10/), then fluorine must penetrate from the surface into the IE zone with a flux

$$J_F = - D_F(dc_F/dx) \qquad , \qquad (1)$$

where x is the depth coordinate and D_F the diffusion coefficient of F in enamel. The geometric conditions may be simplified /5/ so as to yield. as an approximate solution of the diffusion equation,

$$1 - \Delta c_i/(c_o-c_i)_{t=0} = \exp(-D_F t/a^2) \qquad , \qquad (2)$$

where Δc_i is the fluorine uptake of IE in time t, and 2a the thickness of IE.

In Fig. 2 one sees c_i as measured for 6 teeth of different burial periods from one particular site. With $c_o \cong 2000$ ppm, $a \cong 0.15$ cm, and $c_i \cong 0$ at t=0, the plot of $c_i/(c_o-c_i)_{t=0}$ in Fig. 2 yields $D_F \cong 5\times10^{-14} cm^2/s$. This is of a similar magnitude as measured /10/ diffusion coefficients of, e.g., Sr in enamel, and compatible with F diffusion in enamel in a slightly alkaline environment.

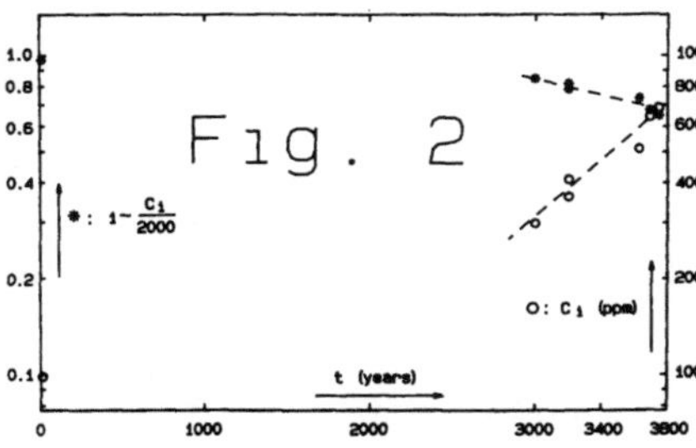

Fig. 2

Regardless of physical interpretation: if the observed dependence of c_i on t is found reproducible, the SIMS study of fluorine in well-defined regions of teeth might provide a new dating technique for archaeological material.

4. Conclusions

The combination of high detection sensitivity and spatial resolution, typical of modern SIMS, appears well suited for the purpose of archaeometry of dental material. A pilot study has been performed, utilizing a quantitative SIMS technique for charting the in-depth distributions of elements in teeth from burial sites on Cyprus, ca 3000 to 4000 years old.

Of fifteen investigated elements, twelve exhibited systematic distribution differences between excavated teeth and recently extracted reference teeth. In several cases environmental factors could be traced. In particular, quantitative measurements of fluorine on morphologically well-defined portions of inner enamel suggest a feasible technique for archaeological dating.

References

1. A. Lodding, S.J. Larsson and H. Odelius: Z. Naturforsch. 33a, 697 (1978)
2. A. Lodding, H. Odelius and L.G. Petersson: in "SIMS IV" (Benninghoven & al, eds.; Springer Ser. Chem. Phys. 36, Berlin - N.Y.), pp 478-484 (1984)
3. A. Lodding: Scanning Elec. Microsc. 83-III, 1229 (1983)
4. G.V. Iyengar, W.E. Kollmer and H.J.M. Bowen: The Elemental Composition of Human Tissues and Body Fluids (Verlag Chemie, New York 1978)
5. P. Fischer: Docent Thesis, Univ. of Gothenburg (P. Åström, Gothenburg 1985)
6. A. Lodding, G. Frostell, G. Koch, J.G. Norén, H. Odelius & L.G. Petersson: J. Microsc. Spectrosc. Electr. 6, 201 (1981)
7. A. Lodding, H. Odelius, D.E. Clark and L.O. Werme: Microchimica Acta, in press (1985).
8. C.A. Andersen and J.R. Hinthorne: Analyt. Chem. 45, 1421 (1973)
9. G.J. Flim, Z. Kolar and J. Arends: J. Bioeng. 1, 209 (1977)
10. A. Lodding, J. Norén and H. Odelius: in "Surface and Colloid Phenomena in the Oral Cavity" (Frank & Leach, eds.; IRL Press, London), pp 11-26 (1982)

Part XII

Geological Applications

Ion Probe Determination of the Abundances of all the Rare Earth Elements in Single Mineral Grains

E. Zinner[1,2] *and G. Crozaz*[1,3]

[1] McDonnell Center for the Space Sciences, [2] Physics Department and [3] Earth and Planetary Sciences Department, Washington University, St. Louis, MO 63130, USA

1. Introduction

Although the measurement of rare earth element (REE) abundances in mg aliquots of geological samples has contributed much to the understanding of petrogenetic processes, probe techniques, which would allow the measurement of compositional variations between grains and within grains (crystal zoning), have so far been of limited usefulness for REE determinations. The electron microprobe detection limit for these elements is ~500 ppm and the concentrations of only a few REE can be determined, usually with large uncertainties.

Previous attempts to measure REE abundances in minerals in the ion microprobe were made by Reed et al. [1] and Metson et al. [2] who respectively used high mass resolution and an extreme level of energy filtering ("isolated specimen technique"), to overcome the problem of molecular interferences.

The present measurements, made with a CAMECA IMS 3F, use moderate energy filtering (100V offset from the low-energy edge of the energy distribution, with an energy window of 32.5V) which eliminates complex molecular ion interferences. The mass spectrum between the atomic masses 133 and 191 is deconvoluted into the elemental components Cs to Hf and the monoxide components CsO to LuO. Concentrations are determined from sensitivity factors, relative to a major element, obtained from a standard. The detection limit of all REE is less than 1 ppm and the analyzed areas are 5 to 20 μm in diameter.

2. Sensitivity Factors in Phosphates and Silicates

Our method was first developed for calcium phosphate [3,4]. Calcium was used as a reference element and the REE concentrations were obtained from the equation

$$\left[\mathrm{REE_i}\right] = \mathrm{F_i^{(Ca)}} \frac{\mathrm{REE_i^+}}{\mathrm{Ca^+}} \left[\mathrm{CaO}\right]$$

where REE_i^+ and Ca^+ are the measured ion signals for each REE and Ca, [CaO] is the Ca oxide concentration and $F_i^{(Ca)}$ are the sensitivity factors for the REE. The $F_i^{(Ca)}$ determined from a phosphate standard [4] are given in column 2, table 1. Column 3 contains the REE ion yields (REE^+/Ca^+) relative to Ca^+ as measured in the phosphate standard.

In order to extend the REE determinations to silicates, Ca-Al-silicate glasses containing different combinations of REE in known concentrations, at the level of several percent [5], were analyzed. The sensitivity factors $F_i^{(Ca)}$ and (REE^+/Ca^+) obtained from these measurements are given in columns 4 and 5 of Table 1. For most REE, the $F_i^{(Ca)}$ determined in the silicate glasses differ by less than 10% from those determined in phosphate matrix. For the elements Gd, Tb, Yb and Lu, it is likely that the determination of the $F_i^{(Ca)}$ in the silicate glasses represents an improvement over the measurement in the phosphate standard. In phosphates, not only are there fluoride interferences but the determination of Yb and Lu concentrations also require the subtraction of Gd and Tb oxides, calculated from interpolated GdO^+/Gd^+ and TbO^+/Tb^+ ratios [4]. Since the glass containing Gd and Tb does not contain Yb and Lu, these oxide to element ratios (MO^+/M^+) could be

Table 1

Element	$F_i^{(Ca)}$ in phosphate	Ion yields REE^+/Ca^+ in phosphate	$F_i^{(Ca)}$ in silicate glass	Ion yields REE^+/Ca^+ in silicate glass	MO^+/M^+ in silicate glass	$F_i^{(Si)}$ in silicate glass	Ion yields REE^+/Si^+ in silicate glass	Angra dos Reis ion probe ppm	Angra dos Reis INAA [6] ppm (*)
La	3.37	0.74	3.35	0.75	0.158	0.92	3.46	5.7±0.1	5.4 (40)
Ce	3.73	0.67	3.80	0.66	0.195	1.05	3.04	18.7±0.1	
Pr	3.05	0.82	3.36	0.75	0.141	0.93	3.46	3.0±0.1	
Nd	3.02	0.85	3.12	0.82	0.121	0.86	3.78	16.1±0.1	20.0 (25)
Sm	2.56	1.05	2.86	0.91	0.0648	0.79	4.19	5.4±0.1	5.9 (10)
Eu	2.20	1.23	2.61	1.03	0.0476	0.72	4.75	1.6±0.1	1.66 (5)
Gd	2.80	1.00	3.21	0.87	0.1033	0.89	4.01	7.3±0.1	
Tb	4.18	0.68	3.59	0.79	0.0962	0.99	3.64	1.36±0.02	1.48 (40)
Dy	3.64	0.79	3.74	0.77	0.0654	1.03	3.55	9.4±0.1	8.5 (10)
Ho	4.12	0.71	3.82	0.77	0.0637	1.05	3.55	1.95±0.03	
Er	4.31	0.69	4.16	0.72	0.0657	1.15	3.32	6.21±0.15	
Tm	4.37	0.69	4.02	0.75	0.0523	1.10	3.46	0.89±0.04	
Yb	5.21	0.59	4.15	0.74	0.0465	1.15	3.41	5.27±0.06	4.2 (5)
Lu	5.81	0.53	5.41	0.57	0.067	1.49	2.63	0.79±0.03	0.65 (10)

* Maximum error (in %) due to counting statistics.

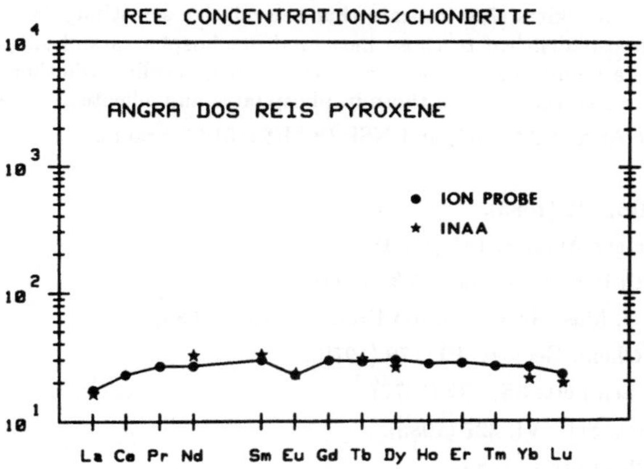

Fig. 1: REE abundances (normalized to chondrites) measured in the achondrite Angra dos Reis by ion probe analysis and instrumental neutron activation analysis (INAA).

measured directly for the first time (see column 6). Columns 7 and 8 give $F_i^{(Si)}$ and (REE^+/Si^+) with Si as the reference element according to

$$[REE_i] = F_i^{(Si)} \frac{REE_i^+}{Si^+} [SiO_2].$$

The sensitivity factors $F_i^{(Ca)}$ (from silicate glass) were used to calculate the REE concentrations in pyroxenes of the meteorite (achondrite) Angra dos Reis. Figure 1 shows the chondrite-normalized REE concentrations measured in the ion probe. As expected [6], the REE pattern is smooth with a slight negative Eu anomaly. The ion probe results, in column 9, are compared with the INAA data of Ma et al. [6] given in column 10. With the exception of Yb and Lu, there is agreement within the maximum percent errors quoted by [6], added in parenthesis in column 10.

The $F_i^{(Si)}$ factors were used for the determination of REE in a terrestrial zircon. The measured concentrations normalized to chondrites are shown in Figure 2.

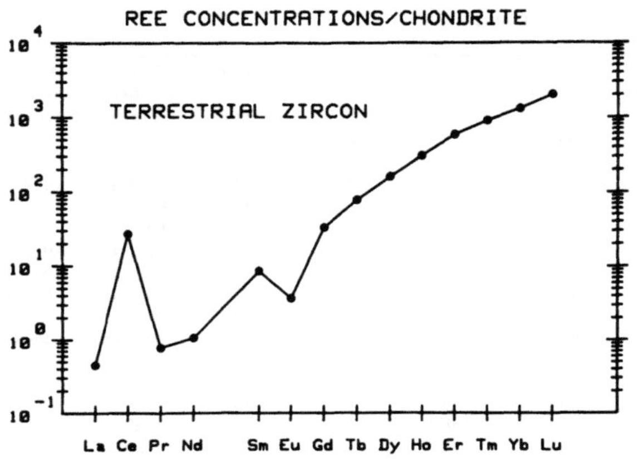

Fig. 2: REE abundances in a terrestrial zircon.

3. Other mineral phases

We are presently trying to determine relative REE sensitivity factors in oxide minerals. Previous analyses of hibonites and perovskites [7,8] indicate that the relative sensitivity factors between REE in oxides are almost identical to those measured in phosphates and silicates. INAA measurements of oxide standards are now in progress to determine whether absolute ion yields relative to Ca are also the same as those in phosphates and silicates.

This work was supported by NASA (NAG9-55) and NSF (EAR8415168) grants.

4. References:

1. S. J. B. Reed et al., Nature **306**, 172 (1983).
2. J. B. Metson et al., Surf. Interface Anal. **5**, 181 (1983).
3. G. Crozaz and E. Zinner, Earth Planet. Sci. Lett. **73**, 41 (1985).
4. E. Zinner and G. Crozaz, Int. J. Mass Spectrom. Ion Proc., in press (1985).
5. M. J. Drake and D. F. Weill, Chem. Geology **10**, 179 (1972).
6. M. S. Ma et al., Earth Planet Sci. Lett. **35**, 331 (1977)
7. A. J. Fahey et al., Lunar Planet. Sci XVI, 227 (1985).
8. A. Fahey et al., Meteoritics **20**, in press (1985).

Ion Microprobe Studies of the Magnesium Isotopic Abundance in Allende and Antarctic Meteorites

J. Okano[1] *and H. Nishimura*[2]

[1] Institute of Geological Sciences, College of General Education, Osaka University, Machikaneyama, Toyonaka, Osaka 560, Japan

[2] Naruto University of Teacher Education, Takashima, Naruto, Tokushima 772, Japan

1 Introduction

In recent years, it has been generally accepted that the early solar nebula consisted of two or more components with different nucleosynthetic histories and that the nebula was not completely homogenized at the time of accretion of meteorite parent bodies. This concept has arisen mainly from the evidence of isotopic anomalies for primitive meteorites. The most striking evidence was found by CLAYTON et al.[1]. They found an excess of ^{16}O in Ca-Al-rich refractory inclusions of Allende, suggesting mixing of normal oxygen with extra ^{16}O. Another piece of evidence was the excess of ^{26}Mg found by GRAY and COMPSTON[2], and LEE and PAPANASTASSIOU[3]. This excess was found in Ca-Al-rich inclusions with high Al/Mg ratios in Allende and explained to be the decay product of extinct ^{26}Al.

Recently, NISHIMURA and OKANO[4-6] reported that an excess of ^{24}Mg has been found in the primitive meteorites Allende(C3), Y-74191(L3), ALH-77278(LL3) and ALH-77304(LL3). This excess has not yet been found in Y-74190(L5-6), Bruderheim(L6), Leedy(L6), Potter(L6), Y-74159(achond, Euc.), Bondoc Peninsula(stony-iron, Mes.) and Y-75028 (L3 or H3). In this paper, the results of detailed isotopic ratio measurements of magnesium for four Antarctic meteorites and the results of remeasurements for Allende will be described. The correlation of the excess of ^{24}Mg with Si/Mg ratio and with Fe/Mg ratio will also be shown.

2 Experimental

A modified Hitachi IMA-2A ion microprobe mass spectrometer used for this work. 9keV O_2^+ ions were used to sputter the sample. The beam current was about 1μA and the spot diameter was 70-100 μm. The base pressure of the sample chamber was about 6×10^{-6} Pa and the pressure under the operating conditions was $1\text{-}2\times10^{-5}$ Pa.

The meteorites studied in this work were Allende (C3), ALH-77003(C3), ALH-77216(L3), ALH-77304(LL3) and ALH-764(LL3). A piece of each meteorite sample was put in a crucible-like stainless steel container together with molten pure indium. After cooling, the top surface was polished with No. 600-1200 SiC emery cloths. The polished sample and the container were ultrasonically cleaned in acetone and mounted in the sample holder of the ion microprobe. A terrestrial forsterite-rich olivine from Ehime,

Japan was used as a laboratory standard. The method of preparation and mounting was the same as that for the meteorite sampl

Measurements of magnesium isotopic ratios $^{24}Mg/^{25}Mg$ and $^{26}Mg/^{25}Mg$ were made on Mg-rich portions of the specimens. Mg-rich portions were selected because we found the excess of ^{24}Mg in such portions in the previous work and because contributions of interference species to the magnesium isotopic peaks would be minimized. Although interferences had been throughly examined in the previous work, the examination was repeated by taking correlations of the magnesium isotopic ratios with the secondary ion intensity ratios of Na^+, Al^+, Ca^+, Ti^+ and Cr^+ to Mg^+. The interference of MgH^+ was sensitively examined by plotting the measured isotopic ratios in a three-isotope plot. The overall contributions of interferences to the magnesium isotopic peaks were estimated to be less than 1 permil.

3 Results and Discussion

The isotopic ratios of $^{24}Mg^+/^{25}Mg^+$ and $^{26}Mg^+/^{25}Mg^+$ were measured for twenty to sixty Mg-rich portions of each meteorite sample an compared with those for the terrestrial forsterite-rich olivine measured under the same experimental conditions. The isotopic ratios were then plotted in the figures of three-isotope plot of magnesium taking the data of the terrestrial olivine in each run as the origin. The results for Allende and ALH-77003 are shown in Fig. 1a and 1b. In this paper, the excess of ^{24}Mg, Δ_{24}, was taken as the displacement of the data point from the normal mass fractionation line (NMF) in the direction of $\delta(24/25)$ axis.

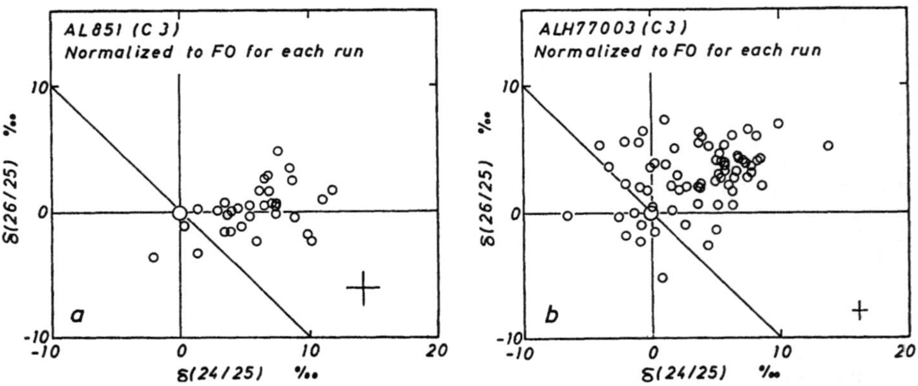

Fig.1 Three isotope plot of magnesium for Allende(a) and ALH-77003(b) The error bars are typical $2\sigma_m$ for each data point.

Figure 2a shows the correlation between $^{28}Si^+/^{24}Mg^+$ and Δ_{24} for Allende. The results reveal that there is a strong correlation between them. Similar but weaker correlations were found for ALH-77003(Fig. 2b), ALH-77216 and ALH-77304. The correlatior are taken to indicate that minerals with high Mg/Si ratios are accompanied with the ^{24}Mg-enriched Mg.

It was found that there were weak correlations between $^{56}Fe^+/^{24}Mg^+$ and Δ_{24} for Allende and ALH-77003, as shown in Fig. 3a and

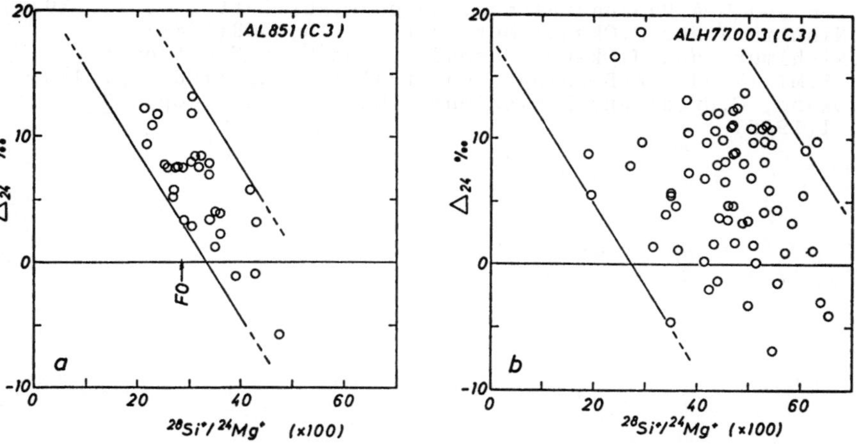

Fig.2 Correlations of the excess of ^{24}Mg with $^{28}Si^{+}/^{24}Mg^{+}$ for Allende(a) and ALH-77003(b). In (a), FO corresponds to the data for the forsterite-rich olivine

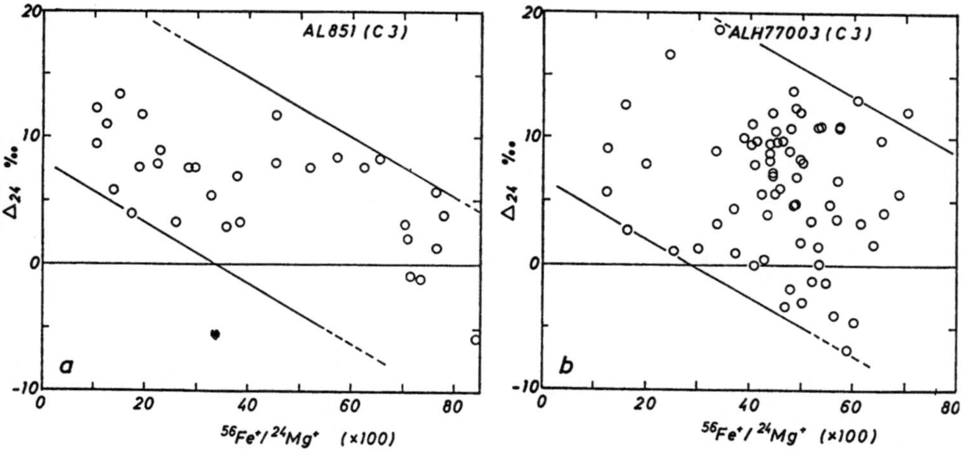

Fig.3 Correlations of the excess of ^{24}Mg with $^{56}Fe^{+}/^{24}Mg^{+}$ for Allende(a) and ALH-77003(b)

3b. The results reveal a tendency that the smaller the $^{56}Fe^{+}/^{24}Mg^{+}$ is, the larger Δ_{24}. As for the other meteorites studied in this work, the correlations were not clear.

From the results obtained so far, it seems that some primitive meteorites contain ^{24}Mg-rich materials and that the isotopically anomalous materials exist in minerals with high Mg/Si ratios.

References

1. R.N.Clayton, L.Grossman and T.K.Mayeda, Science, 182, 485(1973)
2. C.M.Gray and W.Compston, Nature, 251, 495(1974)

3. T.Lee and D.A.Papanastassiou, Geophys.Res.Lett., 1, 225(1974)
4. H.Nishimura and J.Okano, Meteoritics, 16, 368(1981)
5. H.Nishimura and J.Okano, Secondary Ion Mass Spectrometry SIMI IV eds. A.Benninghoven et al.(Springer1984) p. 475
6. J.Okano, C.Uyeda and H.Nishimura, Mass Spectroscopy, 33, 1(1985)

Rock and Mineral Analysis by Accelerator Mass Spectrometry

J.C. Rucklidge, G.C. Wilson, L.R. Kilius, and A.E. Litherland

IsoTrace Laboratory, University of Toronto, Toronto, Ontario M5S 1A7, Canada

1. Introduction

Accelerator Mass Spectrometry (AMS) has mainly been applied to the analysis of rare radioactive isotopes whose natural abundances fall below 1 part in 10^{12}. Examples include ^{14}C, ^{10}Be and ^{36}Cl, which are difficult to measure by decay counting. AMS differs from conventional mass spectrometry in that negative ions to be counted are accelerated to MeV energies, and in the process molecules of the same nominal mass are destroyed, resulting in the total elimination of background. Potentially interfering isobars may be discriminated against by various techniques. For example, the ^{14}N interference with ^{14}C is avoided by the fortunate fact that the negative nitrogen ion is too unstable to be accelerated.

The same apparatus can be used to look at stable isotopes present in very low concentrations, and an accelerator coupled to an ion microprobe would seem to be a natural combination. Conventional ion microprobes suffer badly from molecular interferences, and various strategies have been adopted to deal with these [1,2]. In all cases, while significant improvements may be made, it is done at the expense of transmission efficiency. In AMS, high transmission may be preserved while interferences are eliminated. This paper describes some attempts to measure directly elements in the sub-ppm concentration range in minerals, prepared as a flat surface exposed to the primary Cs^+ ion beam. Another geochemical area where AMS may be applied is in the measurement in situ of the natural isotope ratios of major elements in minerals, such as those of C, O and S. While these determinations remain elusive at the present time, a strategy by which they may be obtained by AMS will be outlined.

2. Experimental Technique

The equipment at IsoTrace is shown in Fig. 1. A sample stage, steerable in X, Y and Z directions, holds a section of the material to be analysed. A beam of Cs^+ ions bombards the sample at 45°, and sputtered negative ions are extracted normal to the sample surface. The secondary ions are passed through a 45° spherical electric analyser which defines the energy (E/q) of the ions to be injected into the 90° analysing magnet. A single mass thus selected is accelerated to 2MeV at the terminal of the accelerator where it passes through an argon gas canal. Electrons are stripped from the -ve ions which are transformed into a variety of +ve charge states, and are subsequently accelerated back to ground potential. Any molecules of charge state +3 or greater disintegrate in the terminal, so by defining the E/q of ions passing through the high energy 15° electrostatic analyser to be 8/3 only atomic ions will be seen in the high-energy faraday cup. For ultra-sensitive analysis further magnetic filtering is needed, but signals gathered directly after the high-energy ESA can span a dynamic range of 12

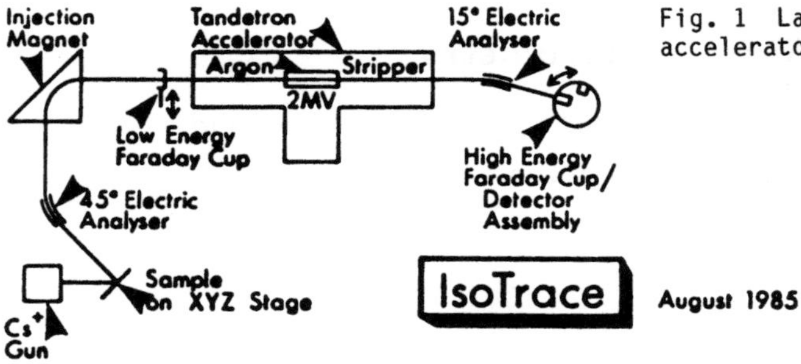

Fig. 1 Layout of IsoTrace accelerator mass spectrometer

decades, and sub-ppm sensitivity is thus accessible for many elements. A faraday cup measures currents in the range from 1uA down to 100fA, at which point a silicon surface barrier detector, under computer control, is moved into the faraday cup position and another 5 decades of sensitivity are acquired. This arrangement lends itself to broad band mass scanning, so complete mass spectra can be taken and minor elements may be expected to reveal themselves. In other applications, where attention is on a single element or a narrow band of isotopes, e.g. platinum group elements, further charge changing and magnetic analysis may be employed [3].

3. Geological Applications

As part of a survey of carbonaceous materials from precious metal ore deposits, several samples of graphite have been studied by the method described above, with the aim of discovering trace element patterns. Figure 2 illustrates a typical mass spectrum of a natural graphite, from which it can be seen that there are a remarkable number of peaks showing up at intensities less by factors exceeding a million compared with the main peak of carbon. While concentrations cannot be inferred directly from these peaks, it is clear that a great many minor elements are present. At these levels it becomes apparent that the simple model of background elimination described above is failing to suppress the background in certain parts of the spectrum. Scattering of the intense negative carbon ions in various parts of the beam line, especially in the 90° magnet box, gives rise to a continuum in the mass range 13 to 21 which limits sensitivity in this region. Similar bands are seen around peaks ascribed to Al, Cs and Ta, which are elements introduced from the primary beam or

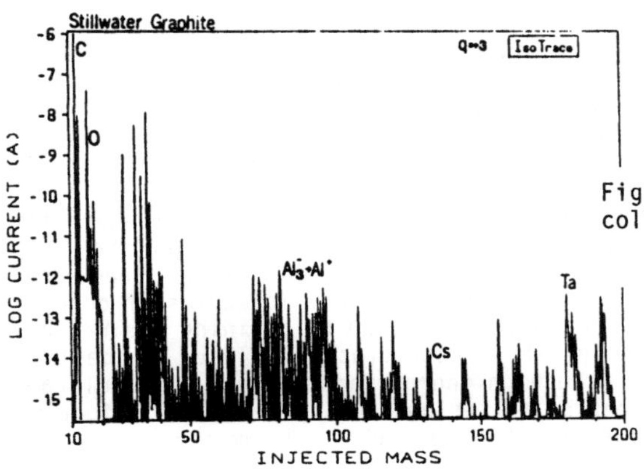

Fig. 2 Mass spectrum of graphite collected by HEFC/detector assembly

from the sample holder. In the case of a heterogeneous sample one may expect such scattering to degrade sensitivity in many parts of the spectrum, but with the problem exposed it is possible to correct the design of the beam line to minimize the trouble.

The difficulty of studying natural, unprepared materials, such as minerals, was shown by Kilius et al. [3] who used AMS to detect ^{194}Pt in a nickel sulphide mineral. In this case charge changing from +3 to +10 was employed at the high-energy end, together with subsequent magnetic filtering. Peaks other than that of Pt occur, arising from low probability charge-changing processes on other elements in the beam. Despite these, it is estimated that Pt can be detected below levels of 1ppb in this type of sample, and therefore the technique compares well with neutron activation, and has the advantage of spatial resolution and small sample size.

4. Isotope Ratios

In natural materials isotope ratio variations can be produced by different processes. Non-linear processes result in the enrichment of one isotope of an element because of the decay of a neighbouring element e.g. ^{26}Mg forming from ^{26}Al. This effect and other similar ones have been described by ion microprobe measurements [2]. Large deviations, over 100%, may be found.

Equilibrium processes prevailing at the time of formation of a crystal, fractionate all isotopes of an element in a linear manner. To determine the absolute ratio, the slope of the fractionation line must be determined and compared with a standard. To do this by sputtering ions from the material of interest,two difficulties arise. The first is that hydrides of light isotopes will invariably interfere with the heavier isotopes of the same nominal mass, e.g. ^{12}CH will contribute significantly to the signal of ^{13}C. High-mass resolution on modern ion microprobes may be able to resolve the small-mass differences, but a more general and certain way of solving this problem is to remove interfering molecules entirely, as can be done by AMS, which at the same time preserves high transmission. However, even when this is done, and isotope ratios in the right range are measured, there is still the problem that sputtering introduces more matrix-dependent fractionation. Because the natural fractionation and the sputtering fractionation processes are both linear, only the sum of the two can be measured. Hence the natural fractionation remains elusive if one is unable to eliminate the sputtering fractionation, as shown by Shimizu and Hart [4]. To resolve this problem,one must examine exactly what it is that brings about the fractionation. Sputtering fractionation of negative ions can be derived from the expression:

$$\text{Ion current} \propto \exp(-(\phi - A)/cv)$$

where ϕ = work function of surface, A = electron affinity of the element, v = velocity of the sputtered ion and c is constant [5]. If v is large the exponential approaches unity, i.e. there is no fractionation due to sputtering, so if one measures only ions which have high initial velocity they will be observed in the true natural ratios. The intensity of the high-energy tail of an ion, several keV above the peak, will be very low, but AMS reduces backgrounds to such low levels that it should be possible to find signals of the order of a few hundred counts per second in the tail of the isotope of interest,even at energies 2 to 3keV above the main peak. If measurements are made at several positions along the tail, a plot of log(ratio) against 1/v should produce a line which will extrapolate to the true ratio as 1/v approaches zero.

References

1. J.B.Metson, G.M.Bancroft, N.S.McIntyre and W.J.Chauvin: Surf. Interface. Anal. 5, 181 (1985)
2. J.C.Huneke, J.T.Armstrong and G.J.Wasserburg: Geochim. Cosmochim. Acta 47, 185 (1983)
3. L.R.Kilius, J.C.Rucklidge, G.C.Wilson, H.W.Lee, K.H.Chang, A.E.Litherland, W.E.Kieser, R.P.Beukens and M.P.Gorton: Nucl. Instr. and Meth. B5, 185 (1984)
4. N.Shimizu and S.R.Hart: J. Appl. Phys. 53, 1303 (1982)
5. J.K.Nørskov and B.I.Lundquist: Phys. Rev. B19, 5661 (1979)

Part XIII

Symposium: Particle-Induced Emission from Organics

Organic Secondary Ion Mass Spectrometry: Theory, Technique, and Application

R.J. Colton

Chemistry Division, Naval Research Laboratory, Washington, DC 20375, US.

1. Introduction

SIMS has become a diverse and valuable analytical tool for the study of many substances such as metals, semiconductors, inorganic compounds, and organic compounds including polymers and biomolecules [1]. This paper presents a fundamental overview of recent developments in organic SIMS analysis -theory, technique, and application.

2. Beam-Surface Interactions and Ionization Processes

2.1 Sputtering

In organic SIMS, molecules that have been deposited onto a solid or liquid surface are desorbed from the surface by the impact of an energetic particle. When the molecules are deposited onto metal surfaces, the sputtering process can be described by the treatment of a collision cascade [2] where the momentum of the primary particle is transferred to the lattice atoms which, through multiple collisions, ultimately deposit some energy at the surface. Harrison, Garrison, Winograd and coworkers [3-5] have made considerable progress using molecular dynamics calculations to understand the formation and emission of polyatomic species from elemental solids and from simple atomic or molecular adsorbates on metals. Molecular adsorbates such as CO and C_6H_6 are found to eject as intact molecules without rearrangement, and to form metal-adsorbate clusters over the surface via reactions between sputtered metal atoms and the molecules.

Molecular and polyatomic species are produced by atomic collisions that transfer enough energy into the surface species to desorb and ionize it. The amount of energy transferred to the "nascent ion molecule" depends on the mass ratio and homogeneity of the atoms involved in the collision cascade [6] and on the momentum of the primary particle. Figure 1 shows the effect of the momentum of the particle on the emission of ethyldimethylcetylammonium ions from silver.

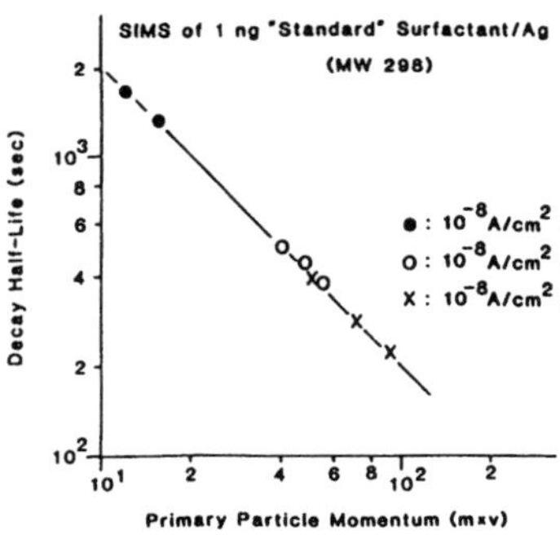

Fig. 1. Effect of the momentum of the primary particle on the secondary ion emission process.

Here, the decay half-life of the secondary ion abundance (or the amount of time required to remove half of the molecules) is plotted against the momentum of the primary particle. The momentum of the primary particle is proportional to the amount of energy deposited at the surface. Particles with higher momenta are able to desorb a larger number of surface species from the area around the impact site. This area may also increase with increasing particle momenta provided that the initial energy is not deposited too deeply within the solid.

2.2 Secondary Ion Emission

Researchers in organic SIMS have defined phenomenologically three distinct ionization processes based on the type of ions created and their relative ease of ionization [7]. The first process involves electron ionization to form radical molecular ions, $M^{+\cdot}$. This process has been observed primarily for nonpolar molecules. The second process involves the formation of protonated or cationized molecules, i.e. $[M+H]^+$ or $[M+C]^+$, where the cationizing species C is usually a metal ion from the substrate, matrix, or an impurity. These ions fragment by the loss of neutral molecules, which is common in chemical ionization mass spectrometry. The third process involves the direct emission of intact (or "preformed") charged species from the solid state as $[M\text{-anion}]^+$ or $[M\text{-cation}]^-$ ions. SIMS studies of organic salts give intense cationic and anionic species with little fragmentation [8]. The relative ionization efficiency (number of secondary ions detected/number of molecules deposited) for these processes in organic SIMS is: direct emission > cationization > electron ionization. The direct emission process lowers the detection limits for organic salts such that picogram quantities can be detected [9].

Figure 2 compares the relative ionization efficiency between cationization and direct emission for cortisone [11]. When 1000 ng of cortisone are deposited from solution onto the surface of silver, the protonated molecule $[M+H]^+$ (m/z 361), the silver cationized molecule $[M+Ag]^+$ (m/z 467 and 469), and the fragment ion at m/z 407 and 409 are observed (Fig. 2a). With 100 ng of cortisone, only $[M+Ag]^+$ is observed.

The SIMS mass spectrum of derivatized cortisone shows a single peak corresponding to the molecular ion of the derivatized species and an ion of low abundance corresponding to the same fragmentation observed for the underivatized species (Fig. 2b). The derivatized cortisone was detected at 10 ng with the same signal-to-noise ratio as that for 100 ng of underivatized cortisone.

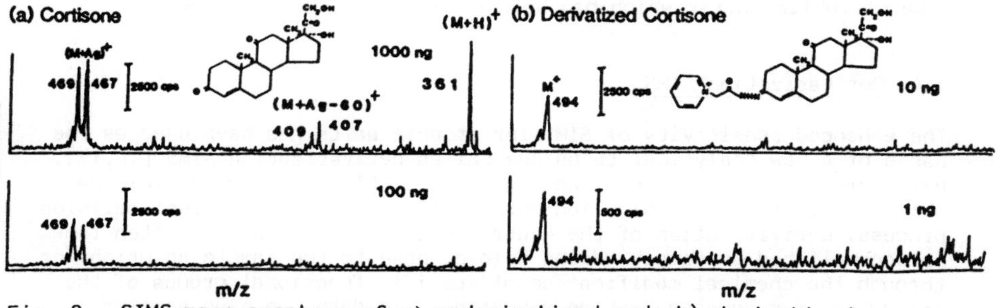

Fig. 2. SIMS mass spectrum of a) underivatized and b) derivatized cortisone.

2.3 Cationization

Cationization of organic molecules in SIMS occurs with alkali, transition, or noble metals. The cationization efficiency of many metals is directly related to the strength of the interaction between the metal atom and the molecule. There is, in fact, a direct relation between the binding state of a molecule on the surface and the charge state of the species that is emitted. The enhanced emission of organic salts or "preformed" ions also supports the idea of a precursor charge state [11].

3. Polyatomic and Molecular Ion Formation/Emission

3.1 Sample Preparation

The methods of sample preparation affect the chemical and physical properties of the sample molecules, and hence the secondary ion formation/emission process. In earlier organic SIMS studies the samples were prepared by placing a dilute solution of the sample onto an acid-etched silver foil, or by burnishing (rubbing) the sample onto metal foils or mixing it with metal salts. Benninghoven and coworkers [11] developed a molecular beam deposition technique to deposit molecules onto atomically clean metal surfaces under ultra-high vacuum conditions. These well-defined surfaces were used to study the influence of the substrate material and its chemical environment on the molecular secondary ion emission process.

3.2 Matrix-Assisted SIMS

Molecular ion emission from a number of solid-state and liquid matrices been investigated recently. There are two types of solid-state matrices used in organic SIMS, namely, low-temperature matrices (such as the rare-gas solids [12] and molecular solids [13]) and room-temperature matrices (such as ammonium chloride, NH_4Cl [14] and carbon [15]). One important property of these matrices is their ability to matrix isolate or dilute the sample molecules. In addition, NH_4Cl matrices can also enhance the ionization efficiency of organic salts by forming large clusters of solvated ions that subsequently dissociate,leaving the intact analyte ions [14]. This model is similar to one discussed by Vestal [16] for ion emission from liquids.

The recent use of liquid matrices such as glycerol in organic SIMS has led to several significant accomplishments, particularly in the analysis of biomolecules by fast-atom bombardment (FAB) mass spectrometry [17]. The mobility of the sample molecules in the liquid matrix is an important property of this matrix. The surface of a liquid metal is also another type of mobile matrix which has been used in organic SIMS [18].

3.4 Derivatization SIMS

The enhanced sensitivity of SIMS for organic salts has been used as the basis of a new analytical technique called derivatization SIMS [10,19]. Here, organic molecules are chemically derivatized to form organic salts which are emitted with high ionization efficiencies by a direct emission process. Derivatization of the neutral organic molecules can often times be accomplished by simply adding acid or base to the sample solution, or through the chemical modification of specific functional groups of the molecules, e.g. quaternization reactions. To date, we have applied derivatization SIMS to analyze drug compounds from urine [20], to detect aldehydes and ketones [10] in air or adsorbed on activated charcoals [21], and to sequence peptides [22] and other biomolecules. In addition, in order to determine (and improve) the sensitivity and applicability of this SIMS method, we measured the relative secondary ion abundances of several quaternary ammonium salts as a function of their ionic sizes [23]. As the size of the cation increases, i.e. $Et_4N^+ < Pr_4N^+ < Bu_4N^+ < Pe_4N^+$, the ion abundance of the sputtered cation increases. Therefore, forming large-size quaternary salts during derivatization increases the ion abundance of the analyte.

3.5 Fragment and Impurity Ions in Organic SIMS

During the formation and emission of molecular ions from organic molecules adsorbed on a metal surface, the energy transferred to the molecule by the collision cascade must be partitioned in several ways: the molecules must be ionized and desorbed from the surface. The order of the two events is subject to debate,and probably depends on the species of interest.

The molecule may also be desorbed with excess energy that may appear as molecular excitation or fragmentation. The fragmentation reactions in organic SIMS are usually associated with unimolecular processes,such as rearrangements that typically proceed along low-energy pathways. Preformed ions usually show less fragmentation. Some of the lower mass ions observed in SIMS mass spectra are not due to fragmentation, however. For example, the SIMS mass spectra of diquaternary salts contain many singly-charged ions [24]. Dications are known (or assumed) to suffer many fates which form singly-charged ions. These include charge separation, one-electron reduction by electron capture, anion attachment, and fragmentation via low-energy pathways. In addition, we find that low level impurities of monocation species in the diquat salt give intense secondary ion signals, presumed to be due to the dication. Figure 3 shows the SIMS mass spectrum of dimethyl-triethylenediamine (TEDA) diiodide that contains a 1% impurity of tetrabutyl-ammonium bromide. Note the intense secondary ion abundance for Bu_4N^+ compared to the ions from the TEDA salt.

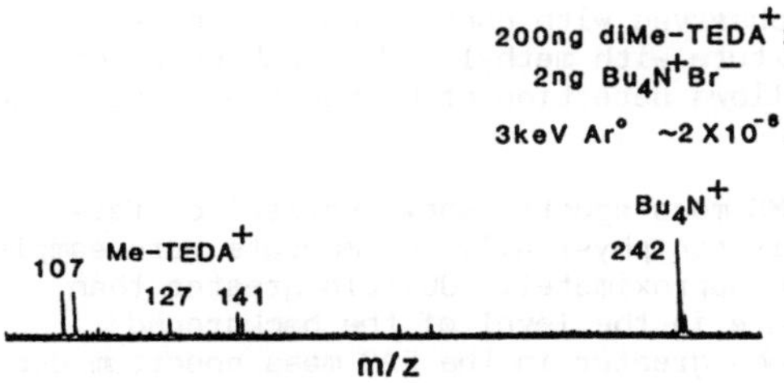

Fig. 3. SIMS mass spectrum of dimethyl triethylenediamine diiodide and 1% tetrabutylammonium bromide.

Monocation impurities can be formed during synthesis. For example, the ion species at m/z 141 in Figure 3 (interpretable as $[M-H]^+$ from dimethyl-TEDA) increases in intensity when the dimethyl-TEDA salt is heated with K_2CO_3 in EtOH. The diquaternary salt reacts via a Hofmann elimination reaction to form a monocation impurity with m/z 141. The SIMS mass spectrum therefore becomes more indicative of the monocation impurity than that of the dication salt.

3.6 Comparison of FABMS and SIMS Mass Spectra

Figure 4 is a comparison of the FAB and SIMS mass spectra of a mixture of five basic drugs. All of the spectra were obtained with a VG ZAB-2F mass

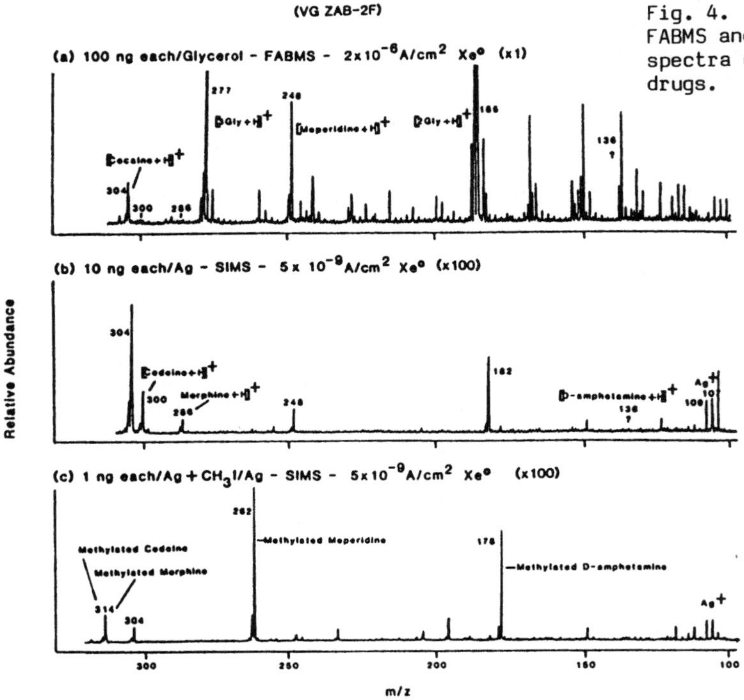

Fig. 4. Comparison of FABMS and SIMS mass spectra of five basic drugs.

spectrometer. Figure 4a is the FAB mass spectrum of 100 ng each of the drugs in glycerol employing the usual FAB conditions. Protonated meperidine and cocaine are observed while the other three drugs are not observed above the background. Figure 4b is the static SIMS mass spectrum of 10 ng each of the drugs sputtered from silver. All of the protonated drug species except D-amphetamine are observed with good signal-to-noise ratios. Derivatization of the mixture with methyl iodide enhances the sensitivity of the analysis and allows detection of D-amphetamine, as shown in the spectrum in Figure 4c.

Comparison of the FABMS and SIMS mass spectra shows several differences. For example, the FAB matrix can physically accommodate more sample and the secondary ion currents are approximately 100-fold greater than those from SIMS. Another difference is the level of the background secondary ion current, which is much greater in the FAB mass spectrum due to the liquid matrix and the high primary beam current density. Although FABMS gives higher secondary ion abundances, SIMS allows analyses of smaller quantities of these drugs with superior signal-to-noise ratios, or in other words, with higher sensitivity.

References

1. R.J. Colton, J.E. Campana, D.A. Kidwell, M.M. Ross, and J.R. Wyatt, Appl. Surf. Sci. 21, 168(1985).
2. P. Sigmund, in:" Sputtering by Particle Bombardment I, R. Behrisch, Ed., Topics in Applied Physics, Vol. 47 (Springer-Verlag, Berlin, 1981) p.9.

. D.E. Harrison, Jr., Radiat. Effects 70, 1(1983).
. B.J. Garrison and N. Winograd, Science 216, 805(1982).
. N. Winograd, Prog. Solid St. Chem. 13, 285(1982).
. C. Plog and W. Gerhard, Surf. Sci. 152/153, 127(1985).
. R.J. Day, S.E. Unger, and R.G. Cooks, Anal. Chem. 52, 557A(1980).
. L.K. Liu, S.E. Unger, and R.G. Cooks, Tetrahedron 37, 1067(1981).
. S.E. Unger, T.M. Ryan, and R.G. Cooks, Anal. Chem. Acta. 118, 169(1980).
0. M.M. Ross, D.A. Kidwell, and R.J. Colton, Int. J. Mass Spectrom. Ion Proc. 63, 141(1985).
1. A. Benninghoven, Springer Ser. Chem. Phys. 36, 342(1984).
2. J. Michl, Int. J. Mass Spectrom. Ion Phys. 53, 255(1983).
3. G.M. Lancaster, I. Honda, Y. Fukuda, and J.W. Rabalais, J. Amer. Chem. Soc. 101, 1951(1979).
4. B.H. Hsu, Y.-X. Xie, K.L. Busch, and R.G. Cooks, Int. J. Mass Spectrom. Ion PHys. 51, 225(1983).
5. M.M. Ross and R.J. Colton, Anal. Chem. 55, 50(1983).
6. M.L. Vestal, Mass Spectrom. Rev. 2, 1(1983).
7. M. Barber, R.S. Bordoli, G.J. Eliott, R.D. Sedgwick, and A.N. Tyler, Anal. Chem. 54, 645A(1982).
8. M.M. Ross and R.J. Colton, Anal. Chem. 55, 1170(1983).
9. K.L. Busch and R.G. Cooks, Science 218, 805(1982).
0. D.A. Kidwell, M.M. Ross, and R.J. Colton, Biomed. Mass Spectrom. 12, 254(1985).
1. M.M. Ross, D.A. Kidwell and J.E. Campana, Anal. Chem. 56, 2142(1984).
2. D.A. Kidwell, M.M. Ross, and R.J. Colton, J. Am. Chem. Soc. 106, 2219(1984).
3. M.M. Ross, D.A. Kidwell, R.J. Colton and J.E. Campana, these proceedings.
4. T.M. Ryan, R.J. Day, and R.G. Cooks, Anal. Chem. 52, 2054(1980).

Mechanisms of Organic Molecule Ejection in SIMS and FABMS

D.W. Brenner and B.J. Garrison

Department of Chemistry, The Pennsylvania State University, University Park, PA 16802, USA

1. Introduction

Why is the study of organics on surfaces important? First, metal surfaces have long been known to promote certain organic reactions. These reactions can produce long-lived surface intermediates which can be studied by various techniques, including SIMS and FABMS. Another reason for studying organics on surfaces is based on the pioneering work of BENNINGHOVEN, where it was shown that large involatile organic molecules could be adsorbed and subsequently sputtered nearly intact [1]. This allowed the easy study of these large molecules by mass spectrometry. In the first case mentioned above, the adsorbed species undergo reactions to form different surface compounds. In the second case, one would like the initially adsorbed species and the surface species to remain essentially identical, so that the surface (and bombarding particle) during SIMS and FABMS act simply as a vehicle for obtaining gas phase ions. In both cases, however, one would like the detected ions (both parent and fragments) to be indicative of what was on the surface. It is this requirement that makes the study of what happens to the surface plus adsorbed species during the bombardment process very important.

In a recent review of the development of SIMS, HONIG [2] states that "at this time it seems that there are as many models (of organic ejection) as there are theoreticians studying the problem". Unfortunately, at first glance this statement appears true. However, along with the differences, certain underlying themes seem to be present in most models. The purpose of this review then, is to point out the differences and similarities between these models and how each model explains experimental results.

The models of organic ejection can be separated into two categories. The first approach is based on empirical interpretations of experimental results. Intuitive models are used here with very few actual calculations being performed. While important thoughts on the sputtering mechanism have come from these models, the results of several of the approaches mentioned next show that the interpretation of experimental results is not as simple as these ideas indicate. The other category of models is based more on the results of actual calculations, which are then interpreted in terms of experimental observables. Examples include classical molecular dynamics [3], Monte Carlo calculations [4], non-equilibrium thermodynamics [5] and kinetic rate theories [6]. Of this category, the classical dynamics, while still containing inherent weaknesses, provides the best feel for what physically occurs during the bombardment process, and so can act as a bridge between the two categories. Therefore, this review will focus on the similarities and differences between the intuitive ideas and the results of classical dynamics simulations. Finally, only low-energy primary particles are examined. This means that the nuclear collisions are important in transfering momentum and energy and that high-energy mechanisms (electronic stopping, electron-hole pairs, repulsive electronic states, etc.) for the most part can be ignored.

2. Experimental Results

Probably the most striking general experimental result is the fact that organic molecules, with bond strengths on the order of 4-5 eV, can be sputtered intact by primary particles with energies of a few keV. Another almost equally unexpected result is that the fragmentation patterns of the sputtered material are very similiar for seemingly different sputtering techniques. This seems to indicate that the sputtering process may be indifferent to the mechanism of energy transfer and chiefly dependent on the initial system. Both of these results are encouraging in view of the original goal of interpreting sputtering data in terms of what is on the surface.

Other general results have been seen which seem to yield more information on the actual sputtering process. First, the kinetic energies of sputtered fragments are higher than those of intact parent ions [7]. The angular distributions of parent ions are also different from those of the fragment ions [8]. These two results can be interpreted in several ways, as is discussed below.

3. Ejection Mechanisms

In order to follow the ejection mechanism in detail, we have broken it up into several steps. Each step is then examined in terms of what events can influence the experimental results.

3.1 The Initial System

As the incoming primary particle approaches the surface, it exchanges energy and momentum with the initial system. As this energy is dissipated through the crystal, surface species can begin to move and be ejected. This is an obvious start for any model of sputtering. However, the influence which this energy dissipation has on the ejected species is not agreed upon from model to model.

One school of thought is that ejected fragments are a product of direct impacts in the vicinity of the initial bombardment [9]. The intact parent ions, it is envisioned, then come from a region further away from the impact point where the energy has dissipated sufficiently, so that no fragmentation takes place, yet the intact surface species can still eject. This scenario accounts for the fact that the energy of the fragments is higher than that of the parents. It also accounts for the angular distributions, where analytic theories of sputtering predict broader cosine-like distributions for species sputtered by random collision cascades [10] similar to that seen for the intact ions.

While Monte Carlo calculations have indirectly supported these ideas [4], molecular dynamics calculations predict a somewhat different conclusion [11]. For several types of organic molecules on clean surfaces, it has been shown that organic species can eject intact from areas both near and away from the initial impact. The dynamics calculations give three contributing explanations for intact ejection. First, it has been shown that the energy of the primary particle can be dissipated very quickly by the substrate.

Also, the molecular species contain many internal vibrational modes which are efficient in absorbing energy from an energetic collision. Finally, multiple atoms in the molecular species may be struck by a larger substrate atom,causing them to move away from the surface in a concerted fashion. Also in contrast to the qualitative model discussed above, the dynamics calculations show that the fragmented species can also arise from points away from the impact area. Not all direct fragments are from collisions with the primary particle.

Energy distributions from computer simulations also have shown higher energy fragments as compared to parent ions. This is due to the simple fact that the more energetic parent species fragment, thus depleting the distribution in the higher energy regions and increasing the number of high energy fragments. Calculated angular distributions also show a more cosine like distribution for ejected parent ions than for fragments [11]. This is because the smaller fragments formed at the surface are more susceptible to the influence of the matrix, which can change these distributions.

Another popular concept is that of a precursor or preformed ion pair [9,12]. In this model it is assumed that an initially charged surface species will retain its charge (and composition) during the bombardment process. While classical dynamics in its pure form does not include charge transfer, these calculations have indicated that the composition (and hence charge in some cases) can change during sputtering. This is discussed more below. Furthermore, other experiments and calculations involving charge transfer with inorganic systems clearly show that there is charge transfer between the surface and the ejected species as they leave the surface [13]. These ideas have also been generalized to organic systems [14]. While experimental results indicate that the idea of a precursor is somewhat valid [15], clearly the composition and charge of detected species in SIMS and FABMS is somewhat more involved.

3.2 The Near Surface Region

On the way to the detector, the ejected parent molecules, fragments and single atoms pass through the near surface region as highly reactive species. If more than one of these species are ejected in a single bombardment, there is a probability that reactions can take place, especially for fragments and single atoms. Here the nearby surface can act as an energy sink or source. There is also a high probability of electron transfer with the surface in this region. Both of these will clearly effect what is seen at the detector.

While computer simulations do show the ejection of intact molecules and clusters from the surface, they also show that ejected species can recombine in the near-surface region,so that their composition as seen at the detector is not what was on the surface. This has been seen in simulations of benzene on nickel [11], where ejected single Ni atoms combined with ejected intact benzene molecules above the surface. The single nickel atom can supply the charge necessary to be detected. Clearly no charged benzene molecule had to exist on the surface for the cationized species to be detected. This type of near-surface reaction is applicable to the transfer of other atomic species, including the gain or loss of hydrogens, the attachment of other cations or anions and combinations of each. Because of these types of reactions, both the charge and the composition of detected ions does not always give a direct indication of the surface species.

3.3 Away from the Surface

If a sputtered molecule leaves the near-surface region with sufficient internal energy, it may unimolecularly decay before it reaches the detector [16]. While this is generally accepted in all models of sputtering, the amount of fragmentation in this region, as compared to that in the previous two steps, has been debated. MOON [17] has shown by way of a simple classical description of the fragmentation of an ejected molecule away from the surface that the polar angle distributions of daughter ions from a single parent ion must peak at the same position. This principle was then applied to the sputtering of several organics on Ag(111) to show that gas phase unimolecular decay was not the chief mechanism for the formation of fragments. Experiments by CHAIT [16] and ENS et al. [16] clearly show, on the other hand, fragmentation on the way to the detector. The primary mechanism of fragmentation is still an open question.

4. Conclusion

Are the number of approaches and interpretations as different as the number of researchers in the field? As this review has shown, there are still some very fundamental questions which still need to be answered before the ejection mechanism is understood. It should be emphasized here that only charged species are detected experimentally, so a detailed comparison between theory and experiment is dependent on a thorough understanding of charge transfer during sputtering as well as the experimental detection of neutral species. Progress is currently being made in both of these areas [18]. As more experimental results become available and more sophisticated calculations are performed, the effect of the properties of the systems being sputtered on the experimental results are being better understood.

Acknowledgement

The financial support of the National Science Foundation, the Office of Naval Research, the Camille and Henry Dreyfus Foundation and the IBM Corporation is gratefully acknowledged.

References

1. A. Benninghoven, Surf. Sci. 53,596(1975); The desorption of involatile organics had also been studied previous to SIMS by ^{252}Cf PDMS. D.F. Torgerson and R.D. Macfarlane, Science 191,920(1976).
2. R.E. Honig, Int. J. Mass. Spect. Ion Proc. 66,31(1985).
3. B.J. Garrison, Int. J. Mass. Spect. Ion Phys. 53,243(1983).
4. T. Ishitani and R. Shimizu, Phys. Lett. 46A,487(1974); C.W. Magee, Int. J. Mass. Spect. Ion Phys. 49,211(1983).
5. F.R. Krueger, Z. Naturforsch. 38a,385(1983).
6. B.V. King, A.R. Ziv, S.H. Lin and I.S.T. Tsong, J.Chem.Phys. 82,3641 (1985).
7. Ion Formation from Organic Solids, A. Benninghoven, ed.,(Springer 1983) p. 69.
8. Secondary Ion Mass Spectrometry-SIMS III, A Benninghoven, et al., ed., (Springer 1982) p. 438.
9. A. Benninghoven, Int. J. Mass. Spect. Ion Phys. 53,85(1983).
10. M.W. Thompson, Phil. Mag. 18,377(1968)
11. B.J. Garrison, J. Am. Chem. Soc. 104,6211(1982); N. Winograd, B.J. Garrison and D.E. Harrison, Jr., J. Chem. Phys. 73,3476(1980); M.R. Liberatore, Master's Thesis, The Pennsylvania State University (1982).

12. R.G. Cooks and K.L. Busch, Int. J. Mass Spect. Ion Phys. 53,111(1983).
13. M.L.Yu and N.D. Lang, Phys. Rev. Lett. 50, 127(1983); J.K. Norskov, B.I. Lundqvist, Phys. Rev. B. 19,5661(1979); J.-H. Lin and B.J. Garrison, J. Vac. Sci. Tech. A1(2),1205(1983).
14. R.D. MacFarlane, Acc. Chem. Res. 15,268(1982).
15. Secondary Ion Mass Spectrometry-SIMS IV, A. Benninghoven, et al.,ed., (Spring 1979), p. 116; K.L. Busch, S.E. Unger, A. Vincze and R.G. Cooks, J. Am. Chem Soc. 104,1507(1982).
16. B.T. Chait, Int. J. Mass Spect. Ion Phys. 53,227(1983); W. Ens, R. Beavis and K.G. Standing, Phys. Rev. Lett. 50, 27(1983).
17. D.W Moon and N. Winograd, Int. J. Mass Spect. Ion Phys. 51,217(1983).
18. J.A. Olson and B.J. Garrison, J. Chem. Phys. 83,1392(1985); F.M. Kimock, J.P. Baxter, D.L. Pappas, P.H. Kobrin and N. Winograd, Anal. Chem. 56 ,2782(1984).

A Thermodynamic Description of 252-Cf-Plasma Desorption

R.D. Macfarlane

Texas A & M University, Department of Chemisty,
College Station, TX 77843, USA

1. Introduction

The objective of this paper is to discuss progress that has been made in elucidating the fundamentals of 252-Cf-plasma desorption mass spectrometry (252-Cf-PDMS). This method is based on the electronic excitation of a condensed matrix by fast (1 MeV/u) heavy ions. While the primary interaction is different from that of the keV ions and atoms of SIMS and FAB which involve an initial nuclear excitation, the mass spectrum of secondary ions is remarkably similar. There are important differences. There is no high-energy sputtering component in the 252-Cf-PDMS spectrum. The molecular ion yield appears to be considerably higher in 252-Cf-PDMS as well as the ion multiplicity (number of ions emitted per incident ion) [1]. The difference in molecular ion yield increases with mass.

ENS has recently used incident Li ions in an energy regime where there is a significant electronic and nuclear component [1]. That work has demonstrated clearly that the underlying processes that lead to molecular ion emission of organic molecules are sensitive to the energy derived from the incident ion. It does not matter whether the primary interaction is electronic or nuclear in origin.

What is the nature of this dynamic environment is the fundamental question that we are presently confronting. In the approach to be discussed in this paper, we shall assume that the similarities of 252-Cf-PDMS and SIMS mass spectra are evidence that the form of the primary energy deposited has been randomized prior to molecular ion emission to an extent that details of the primary excitation have been lost. The continued time evolution of the excitation region can be described by statistical thermodynamics. The limit of this evolution is a "hot spot" in the substrate close to the primary impact site with a thermodynamic temperature, and that it rapidly cools. During the approach to the statistical thermodynamic limit and in the cooling stage, it is possible for molecular ion emission to occur. We shall examine the experimental and theoretical evidence that has brought us to this thermodynamic model for 252-Cf-PDMS.

2. Electronic Excitation and Energy Relaxation

When a fast heavy ion passes through a condensed matrix it transfers energy to the matrix by electronic excitation. For a matrix of organic molecules, the amount of energy transferred is dependent on two properties of the incident ion: the ionic charge and its velocity. Both of these properties have been verified to be directly linked to the probability for molecular ion emission using accelerator beams of heavy ions with variable ionic charge and velocity [2]. The peak of the energy-transfer curve occurs when the ion velocity is 0.2 cm/ns which is close to the velocity of 252-Cf fission fragments and corresponds to a linear energy-transfer rate of 10 keV/nm. Several mechanisms operate to convert electronic excitation to vibrational excitation of molecules and molecular aggregates in the region surrounding the ion track. This includes exciton-exciton annihilation at high ionization density and phonon excitation by "hot" electrons which has recently been measured by PFLUGER [3]. The dynamic properties of this vibrationally hot surface region lead to the population of vibrationally-unstable states resulting in fragmentation and molecular ion emission.

3. The Rapid-Heating Model of LUCCHESE and TULLY [4]

Rapid heating was first proposed by FRIEDMAN [5] as a means for desorbing molecular ions of involatile biomolecules without inducing thermal decomposition. Recent desorption studies using ns to ps pulsed lasers have extended the validity of the rapid heating concept to heating rates several orders of magnitude faster than what FRIEDMAN used in his studies. The time scale for energy deposition by a fast heavy ion is on the order of 0.1 fs. A detailed statistical treatment of desorption by a "nonselective deposition of energy into the surface region" has recently been published by LUCCHESE and TULLY [4]. Although the system studied was comparatively simple (desorption of NO from LiF), generally applicable features were revealed in the computer simulation of the rapid heating-desorption process that may be used to interpret some of the kinetic energy and angular distribution data obtained by WIEN [6] for ions desorbed from fast heavy ion tracks.

If one were to describe the desorption of molecular ions in 252-Cf-PDMS within the framework of the LUCCHESE-TULLY model using the available kinetic energy and angular distribution data, it would go as follows. The fission fragment (fast heavy ion) is an agent for rapid heating of the surface of a solid with a peak heating rate on the order of 10^{15} K/s. The molecule to be desorbed (as a molecular ion) is the adsorbate and the ensemble of molecules remaining at the site of the adsorbate is the substrate. Vibrational energy flows into the substrate from the ion track and is statistically distributed to give a surface temperature. The time constant for rapid heating is so short that it is controlled by the dielectric response of the substrate. Energy is transferred to the adsorbate in thermal contact with the substrate and desorption occurs prior to equipartition of energy within the desorbed species. The translational and internal degrees of freedom of the desorbed species serve as a thermometer for the relaxation of energy between the adsorbate and substrate in the desorption process. The very short heating time-constant reduces the amount of energy going into vibrational excitation of the adsorbate; fragmentation intensities are sensitive to this effect. Also, the component of translational energy of the desorbed species normal to the surface of the substrate is considerably higher than the parallel components. This can be measured from an angular distribution pattern using the mean value of cos θ to parameterize the distribution. The limited data available for 252-Cf-PDMS indicates that this value is on the order of 0.95 [6,7] for molecular ions of valine and for alkali metal ion emission. (The statistical limit for equipartition of translational energy is 0.667.) Such a high value is indicative of a very short heating time constant and high surface temperature. The degree of freedom with the shortest relaxation time is the component of translational energy normal to the surface, which has been found to be 1.5 to 2.5 eV for atomic and molecular ion emission from fast ion tracks [6,7] which corresponds to a temperature of 9000-15000 K. The surface temperature at the desorption site will be hotter. A variable that is part of the LUCCHESE-TULLY model is the mean residence time of the adsorbate, the time between excitation of the matrix by the fast ion and desorption. For polyatomic species, this can be greater than 1 ns [8]. While no measurements of residence time have been made for 252-Cf-PDMS, it is possible, in principle, to obtain this information from time-of-flight (TOF) mass spectra to times of 0.1 ns using the electron TOF to define a "zero" residence time.

In summary, the LUCCHESE-TULLY model clearly identifies the experimental measurements that most likely will reveal details of the dynamics of the surface processes that are responsible for the mass spectrum of an organic molecule in a condensed medium: the kinetic energy and angular distributions of all secondary ions and their residence times.

4. The TSONG Model of Non-Cascade Sputtering [9]

In our development of a thermodynamic description for 252-Cf-PDMS, we have adopted the LUCCHESE-TULLY formalism of rapid heating of the substrate followed by energy relaxation to the adsorbate system. Because of the large cross-section for electronic excitation by the incident ion, it is likely that some electronic energy is directly transfered to the adsorbate by the fast heavy ion as it penetrates the surface layer and recedes from the surface region. This is highly probable for large molecules like insulin, where a fission fragment trajectory through the molecule would deposit on the average 15 keV of electronic energy. Thus, it is possible for the desorbed molecule to be excited to a very high state. This is clearly demonstrated in the complex pattern of fragment ions for large molecules such as chlorophyll, which shows multiple bond cleavage when excited by fast heavy ions [10]. The TSONG model uses RRKM theory to interpret fragmentation pattern intensities in terms of the amount of energy deposited into a molecule, how many degrees of freedom are excited and what the activation energies are for final product pathways. This calculation has recently been carried out by TSONG et al [9] using data from the fast ion-induced desorption of insulin [11]. They found that in order to account for the intensities of the a- and b-chain relative to that for the molecular ion, the insulin molecule contained 780 eV of vibrational excitation involving 450 of the 2358 normal modes. Energy of activation data are also obtained in the fit of theoretical intensities to experiment.

This is the first attempt to apply a theoretical model to the desorption of a complex molecule by a rapid heating mechanism, and it shows how details of the desorption process can be deduced from the mass spectrum. This study also brings out some of the features of the desorption process that will be part of the "final" description. This includes energy transfer between molecules on the surface and vibrational energy flow within a polyatomic molecule, both topics of current interest and which are being studied using ps and fs laser spectroscopy [12].

5. Influence of Matrix in the Desorption Process

In both of the theoretical studies discussed above, the role of the matrix was implicitly involved. In the LUCCHESE-TULLY study, the matrix was the medium for rapid heating and in the TSONG study, relaxation of energy between molecules at the surface was necessary to reduce the vibrational energy of the excited insulin molecules to a level where survival of an intact desorbed insulin molecule was possible. In most of the 252-Cf-PDMS studies carried out so far, multilayer deposits were used,which means that the substrate and adsorbate are the same species and they are coupled through van der Waals interactions. While this arrangement gives good mass spectra, it is difficult to understand the events responsible for molecular ion emission, because it is not easy to separate substrate effects from adsorbate interactions. In an attempt to simplify the problem, we have been using Langmuir adsorption at a solid/liquid interface to prepare submonolayer to multilayer deposits of organic molecules on a common substrate [13]. This makes it possible to study the influence of surface interactions on molecular ion emission for a common substrate. The following observations have been made. Mylar and polypropylene surfaces have cation exchange properties; the molecular ion intensity of Rhodamine 6-G (which forms a stable organic cation) increases linearly with surface concentration reaching a saturation value at monolayer coverage. Rhodamine B emits a molecular ion and fragment ion (loss of HCOOH) when the surface concentration is low. The fragment ion intensity is quenched at high surface concentration and at low surface concentration in the presence of a submonolayer of Au. We interpret this as evidence for quenching of dissociative electronic excited states in Rhodamine B.

Acknowlegements This research was supported by the National Institutes of Health (GM-26096), the National Science Foundation (CHE-82-06030) and the Welch Foundation (A-258).

References

1. W. Ens, PhD Thesis, University of Manitoba, Winnipeg, 1984.
2. P. Hakansson, E. Jayasinghe, A. Johansson, I. Kamensky and B. Sundqvist: Phys. Rev. Lett. 47, 1227 (1981); A. Albers, K. Wien, P. Duck, W. Treu and H. Voit: Nucl. Inst. Meth. 198, 69 (1982).
3. P. Pfluger, H.R. Zeller, and J. Bernasconi: Phys. Rev. Lett. 53, 94 (1984).
4. R.R. Lucchese and J.C. Tully: J. Chem. Phys. 81, 6313 (1984).
5. R.J. Beuhler, E. Flanigan, L.J. Greene, and L. Friedman: J. Am. Chem. Soc. 96, 3990, (1974).
6. N. Furstenau, W. Knippelberg, F.R. Krueger, G. Weiss and K. Wien: Z. Naturforsch 32a, 1084 (1977).
7. R.D. Macfarlane, unpublished results.
8. J. Heidberg, H. Stein, E. Riehl, and I. Hussla in Surface Studies with Lasers; p. 226 (Springer, Berlin, 1983).
9. B.V. King, A.R. Ziv, S.H. Lin, and I.S.T. Tsong: J. Chem. Phys. 82, 3641 (1985).
10. B.T. Chait and F.H. Field: J. Am. Chem. Soc. 106, 1931 (1984).
11. P. Hakansson, I. Kamensky, B. Sundqvist, J. Fohlman, P. Peterson, C.J. McNeal, and R.D. Macfarlane: J. Am. Chem. Soc. 194, 2948 (1982).
12. A.J. Taylor, D.J. Erskine C.L. Tang: Chem. Phys. Lett. 103, 430 (1984)
13. R.D. Macfarlane, C.J. McNeal, and C.R. Martin: Anal. Chem. (in press)

Laser Desorption Mass Spectrometry. A Review

F. Hillenkamp

Institut für Biophysik der Universität Frankfurt,
Theodor-Stern-Kai 7, D-6000 Frankfurt/M. 70, F. R. G.

1. Introduction

Lasers can be used in a variety of different ways to generate atomic or molecular ions, suitable for mass spectrometric analysis. Most obvious is the photoionization of neutrals in the gas phase. Because the ionization energy of almost all species is larger than the photon energy of the commonly used lasers, photoionization is almost exclusively done by resonant or nonresonant two-photon excitation; the resulting removal of an electron necessarily leads to radical ions. For a few selected references of this technique see e.g. [1-6]. Gas phase irradiation of ions with lasers can also be used for photodissociation in structural analysis [7, 8]. For elemental analysis of solid samples spark plasma ionization mass spectrometry is a well-established technique. Replacement of the spark- by a laser plasma has obvious advantages e.g. for nonconducting samples, and in cases, in which spacial resolution is of interest. Generation of suitable laser plasmas requires short pulses and irradiances on the sample of 10^9 W/cm^2 and above. For recent development in this field see e.g. [9-11].

The topic of this review is laser desorption. Desorption is the release of neutrals or ions from the condensed, usually solid phase into the gas phase. Though desorption, particularly of neutrals can in many cases be induced thermally, i.e. by direct or indirect sample heating, in which case the laser is used as a heat source only, desorption is of most interest in the analysis of samples of very limited volatility, e.g. of large organic molecules. Such desorptions are considered "soft", because molecules or ions are released with very little internal excitation, leading to limited fragmentation, but, as in other desorption techniques, the initial event certainly constitutes a strong perturbation of the system and still high irradiances in the range of 10^6-10^8 W/cm^2 are required. Differentiation between the three methods is not always easy and in practice at least two of them may contribute to ion formation. At high total laser energies, large spot sizes and closed sources, the density of ions, electrons and neutrals directly above the sample surface may well lead to a plasma of low density and temperature with ensuing collisional reactions [12, 13]. Conversely, dense plasmas created above a surface may interact with this surface resulting in desorption-type ions [14]. At wavelengths in the visible and ultraviolet and the high irradiance used, photoionization or photofragmentation of desorbed species above the sample surface also may gain some probability.

2. Techniques of Laser Desorption

The very wide range of techniques and parameters tested and used in Laser Desorption Mass Spectrometry (LDMS) has been discussed in detail in a former review [15]. Since the publication of that review in 1983 a few trends in LDMS have evolved, though there may well be merits to one or the other technique, which has not since been pursued extensively. The use of time-of-flight mass spectrometers (TOF) has increased substantially along with the general trend in the field[3, 16-29] . The combination of LD-ion sources with Cyclotron Resonance Fourier Transform spectrometers has more recently gained high interest[16, 30-32]. Among others, the laser in this combination offers the great advantage of ease of technical implementation and of very low disturbance of the vacuum. Several other papers in this volume will discuss LD-FTMS (CASTRO and RUSSELL, MCIVER et.al. and CODY et.al.). Sector instruments[9, 11, 13, 33, 34] and quadrupoles[35-38] have been used less frequently and mostly for special investigations.

All lasers, currently used for LD-MS, are pulsed in the range of 5 ns-50 ns Such short pulses are ideal in conjunction with TOF-instruments with immediate ion extraction, but offer advantages for FT-instruments as well, because of the very short disturbance of the vacuum. One desorption study with ps-laser pulses has been reported[39] . Pulses of CO_2-lasers used have a nominal pulsewidth of app. 50 ns, but have a tail of app. 1 µs in duration, which may contain about 50% of the total energy. This pulse structure may be important for the interpretation of some results, particularly the ones obtained with TOF instruments and delayed drawout pulse. About half of the published work has been done with pulsed CO_2-lasers at 10.6 µm wavelength[13, 16, 17, 18, 30, 31, 38], the other half with short UV-laser pulses (quadrupled Nd-YAG at 265 nm[20-29, 39], excimer lasers at 193 nm or 248 nm[1, 3, 7, 28] , dye-laser [28, 35]). Both wavelength ranges in the far infrared and far ultraviolet offer the possibility of resonant excitation of respectively vibrational or electronic transitions. At the fundamental wavelength of the Nd-YAG laser at 1.06 µm, also used by some groups, energy transfer takes place via the substrate, because of lacking absorption of almost any sample at this wavelength [9, 11, 32-43, 36, 37] . This transfer process may cause side effects of delayed and/or prolonged heating, usually an undesirable effect. For wavelengths in the UV, nonlinear optical processes such as two-photon absorption may become important[27]. Ion source- as well as irradiation/extraction geometries still vary in a wide range, making comparison of results difficult, particularly in cases of delayed ion extraction such as in the FT-MS-instruments and the TOF-MS of COTTER et.al.[16-19]. No standard, optimal procedure for sample preparation has as yet evolved, another obstacle for intercomparison of results.

It seems to be well established that the ion/neutral ratio is in the range of about 10^{-5}-10^{-3}, depending on the sample and the laser parameters. Post-ionization of the emitted neutrals therefore is an attractive scheme both for a better understanding of the desorption process and for increased sensitivity in practical applications. Chemical ionization has been used by several investigators[16, 34, 35, 38]. In addition, EI-postionization has been applied [16] and several reports have appeared reporting a combination of laser-desorption and laser-post-photoionization[3, 39].

3. Mechanisms of Laser Desorption

The basic question, whether or not LD is a thermal process, i.e. whether thermal equilibrium is obtained during desorption, seems to be still open. Even considered that the terms "thermal" and "equilibrium process" are used with somewhat different meanings by different authors, the fundamental question remains. Observation of "non volatile", large organic parent molecular ions with relative molecular masses above 1000[18, 21, 22, 29, 31, 36] and often very limited fragmentation supports the notion of a nonequilibrium process. Very little is on the other hand known about the true volatility of most substances investigated and some of the improvements in the range of detected large ions also result from increased sensitivities of the instruments used. The prolonged ion emission over tens of microseconds as reported by COTTER[17, 18], on the other hand, seems to rather support a thermal model. Some doubt, however, remains in that case, whether the observed phenomena really result from prolonged ion emission, or rather delayed gas-phase reactions in the rather closed source. Initial kinetic energies of desorbed ions, measured by several investigators [13, 17, 42] all certainly do not reflect realistic sample temperatures during desorption, but it is difficult to assess the contributions of adiabatic expansion, Coulomb repulsion, charge separation and varying loci of ion formation to these energies. One publication [32] reports slightly higher internal excitation of LD- vs. FAB-ions of phosphonium salts, as inferred by CAD studies (Nd-YAG-laser, 1.06 µm, 9 ns), but this may not be representative for the other, more frequently used lasers. The thermal nature of CW-CO_2-laser desorption has been well established some years ago [15]. Experimental evidence now seems to indicate that pulsed CO_2-laser desorption is closer to a thermal process, if not truly thermal, than UV-laser desorption. Among the parameters that have been demonstrated experimentally to strongly influence the desorption process are the wavelength, if vibrational or electronic resonance absorption can be achieved [26, 27, 40, 41] and the sample matrix [29]. There is good reason to believe that both affect not only the primary energy-transfer process, but also the ensuing chemistry [26]. Substrates, if heated appreciably by the laser, may strongly influence results, as does the sample preparation. At UV-wavelengths non-linear processes such as two-photon absorption has been shown to be of importance for nonabsorbing samples and, under certain conditions, even for absorbing ones [27]. A few reports on the metastable decay of LD-ions have appeared in the literature [19, 25]. Such studies can also help to eventually clarify the desorption process or, more likely, the processes. Besides the direct ion formation by laser radiation, SEYDEL et.al. [22] have postulated that at very high irradiances (10^{11} W/cm^2) in a LAMMA 500 geometry with thick sample layers (d= 20 µm), laser-generated shock waves can also lead to the desorption of large ions of nonvolatile substances.

4. Status and outlook

The highest mass ions of nonvolatile, organic ions, detectable or detected with the different desorption techniques, has over the last few years dominated the discussion and intercomparison, even though other criteria may be at least as important from the practical point of view. Triggered by the advent of liquid matrix FABMS and improved systems in plasma desorption, new world records have been reported almost every month or so. LD had a rather slow start, but considering the limited number of investigators, progress has been quite remarkable nonetheless. A variety of bioorganic molecules with relative molecular masses in the range of 1000-2000 have been detected with all the different techniques [18, 21, 22, 29, 31, 36]. Fig. 1 shows the positive ion spectrum of the unprotected linear polypeptide mellitin (26 aminoacids, one absorbing tryptophane) of relative molecular mass 2844, desorbed out of a nicotinic acid matrix with a LAMMA 1000 instrument. For organic polymers ions up to mass 7000 have been detected (pulsed CO_2-laser, FTMR) [31]. Clusters of $(CsI)_nCs^+$ up to n= 57 at mass 15000, desorbed out of a CsI/sucrose mixture,

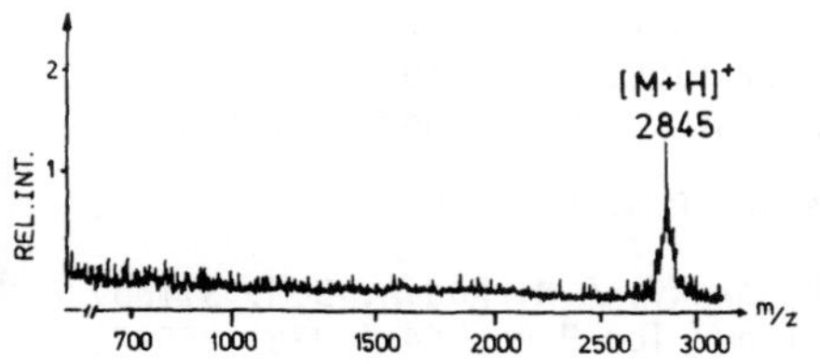

Fig. 1 Positive Ion Spectrum of the Polypeptide Mellitin, desorbed out of a nicotinic acid matrix with LAMMA 1000 (265 nm, 10 ns, threshold irradiance, sum of 12 shots).

have also been observed in the authors group. The intensity of the clusters decreases in these spectra by only a factor of about 10 from n= 1 to n= 57, in contrast to SIMS and plasma desorption experiments. This is another indication of the strong influence of the matrix in LD. Possibly more important than the highest detectable mass is the attainable signal-to-noise ratio for high mass ions. As a general observation, LD-spectra show significantly less background than the chemical noise in FAB spectra. This is also evident from Fig.1. Systematic fragmentation for structural analysis is another important factor. Generally, LD-spectra show less, often also less systematic fragmentation as compared to FAB-spectra. This difference may actually be of minor importance, as real structural analysis seems to require additional techniques such as CAD MS/MS or, very attractive, resonant laser photodissociation. In selected cases the fragmentation of LD-and FAB-ions may yield complementary information. GROSS has observed e.g. that LD-spectra of digitonine show preferentially fragments of the sugar moiety, whereas in FAB-spectra of the same substance fragments originate preferentially from the steroid moiety. Similar observations have been made by VESTAL [37].

In the authors mind, all the different techniques are capable of generating the same type and size ions as far as the desorption itself goes. Differences observed in practice are most probably the result of different boundary conditions or experimental parameters such as matrix, time scale of excitation and metastable decay, laser wavelength, detector etc. Laser desorption may offer the advantage of more free variables for adjustment in a given analytical problem. At the same time, this very fact makes experimentation and systematic interpretation of results difficult, and it is the reason for the somewhat slow start of the technique. In the end, the different techniques will in all likelihood evolve as mostly complimentary rather than competitive, but LD, no doubt, can accept the challenge.

References

1. M.P. Irion, W.D. Bowers, R.L. Hunter, F.S. Rowland and R.T. McIver Jr.: Chem. Phys. Lett. 93 (1982) 375-379.
2. T.J. Carlin and B.S. Freiser: Anal. Chem. 55 (1983) 955-958.
3. S.E. Egorov, V.S. Letokhov and A.N. Shibanov: Chemical Physics 85 (1984) 349-353.
4. H. Kühlewind, H.J. Neusser and E.W. Schlag: J. Phys. Chem. 88 (1984) 6104-6106.
5. H.-J. Neusser, U. Boesl, R. Weinkauf and E.W. Schlag: Int. J. Mass Spectrom. Ion Proc. 60 (1984) 147-156.
6. M. Stuke: Appl. Phys. Lett. 45 (1984) 1175-1178.
7. W.D. Bowers, S.-S. Delbert, R.L. Hunter and R.T. McIver Jr.: J. Am. CHem. Soc. 106 (1984) 7288-7289.
8. E.K. Fukuda and J.E. Campana: Anal. Chem. 57 (1985) 949-952.
9. R.J. Conzemius, S. Zhao, R.S. Houk and H.J. Svec: Int. J. Mass Spectrom. Ion Proc. 61 (1984) 277-292.
10. G.I. Ramendik, O.I. Kryuchkova, D.A. Tyurin, T.R. Mchedlidze and M.Sh. Kaviladze: Int. J. Mass Spectrom. Ion Proc. 63 (1985) 1-15.
11. T.K. Sanderson: Analytical Proceedings: 22 (1985) 118-119.
12. P. Feigl, B. Schueler and F. Hillenkamp: Int. J. Mass Spectrom. Ion Phys. 47 (1983) 15-18.
13. G.J.Q. van der Peyl, W.J. van der Zande and P.G. Kistemaker: Int. J. Mass Spectrom. Ion Proc. 62 (1984) 51-71.
14. G.M. Arzumanian, D.D. Bogdanov, Yu.A. Bykowsky, A.M. Rodin, S.M. Silnov and G.M. Ter-Akopian: Int. J. Mass Spectrom. Ion Proc. 64 (1985) 255-264.

15. F. Hillenkamp: Laser Induced Ion Formation from Organic Solids (Review), in Ion Formation from Organic Solids, A. Benninghoven ed., Springer Series in Chem. Phys. 25, Springer Verlag, 1983.
16. R.B. van Bremen, M. Snow and R.J. Cotter: Int. J. Mass Spectrom. Ion Phys. 49 (1983) 35-50.
17. J.C. Tabet and R.J. Cotter: Int. J. Mass Spectrom. Ion Phys. 54 (1983) 151-158.
18. J.C. Tabet and R.J. Cotter: Anal. Chem. 56 (1984) 1662-1667.
19. J.C. Tabet, M. Jablonski and R.J. Cotter: Int. J. Mass Spectrom. Ion Proc. 65 (1985) 105-117.
20. D.E. Mattern and D.M. Hercules: Anal. Chem. 57 (1985) 2041-2046.
21. U. Seydel, B. Lindner, U. Zähringer, E.Th. Rietschel, S. Kusumoto and T. Shiba: Biomed. Mass Spectrom. 11 (1984) 132-141.
22. B. Lindner and U. Seydel: Anal. Chem. 57 (1985) 895-899.
23. J.F. Muller, A. Ricard and M. Ricard: Int. J. Mass Spectrom. Ion Proc. 62 (1984) 125-136.
24. J.F. Muller, G. Krier, F. Verdun, M. Laboule and D. Muller: Int. J. Mass Spectrom. Ion Proc. 64 (1985) 127-138.
25. J. Rosmarinowsky, M. Karas and F. Hillenkamp: Int. J. Mass Spectrom. Ion Proc., in print.
26. M. Karas, D. Bachmann and F. Hillenkamp: Anal. Chem., in print.
27. F. Hillenkamp, D. Holtkamp, M. Karas and P. Klüsener: Proceedings of the 3. Int. Conf. on Ion Formation from Organic Solids, Springer Series in Chem. Phys., in print.
28. U. Bahr and F. Hillenkamp: Proceedings of the 10. Int. Mass Spectrom. Conf., J.F.J. Todd, ed., John Wiley & Sons Ltd., (1986) in print.
29. M. Karas, D. Bachmann and F. Hillenkamp: Proceedings of the 10. Int. Mass Spectrom. Conf., J.F.J. Todd, ed., John Wiley & Sons Ltd., (1986) in print.
30. D.A. Mc Crery, E.B. Ledford Jr. and M.L. Gross: Anal. Chem. 54 (1982) 1435-1437.
31. Ch.L. Wilkins, D.A. Weil, C.L.C. Yang and C.F. Ijames: Anal. Chem. 57 (1985) 520-524.
32. D.A. Mc Crery, D.A. Peake and M.L. Gross: Anal. Chem. 57 (1985) 1181-1186.
33. J.F.K. Huber, T. Dzido and F. Heresch: J. Chromat. 271 (1983) 27-33.
34. D.V. Davis, R.G. Cooks, B.N. Meyer and J.L. Mc Laughlin: Anal. Chem. 55 (1983) 1302-1305.
35. R.J. Perchalski, R.A. Yost and B.J. Wilder: Anal. Chem. 55 (1983) 2002-2005.
36. E.D. Hardin, T.P. Fan, C.R. Blakley and M.L. Vestal: Anal. Chem. 56 (1984) 2-7.
37. T.P. Fan, E.D. Hardin and M.L. Vestal: Anal. Chem. 56 (1984) 1870-1876.
38. L. Ramaley, M.-A. Vaughan and W.D. Jamieson: Anal. Chem. 57 (1985) 353-358.
39. V.S. Antonov, V.S. Letokhov, Yu.A. Matveyets and A.N. Shibanov: Laser Chemistry 1 (1982) 37-43.
40. R. Stoll and F.W. Röllgen: Z. Naturforsch. 37a (1982) 9.
41. M. Mashni and P. Hess: Chem. Phys. Lett. 77 (1981) 541.
42. E. Michiels, T. Mauney, F. Adams and R. Gijbels: Int. J. Mass Spectrom. Ion Proc. 61 (1984) 231-246.

Time-of-Flight Measurements in Secondary Ion Mass Spectrometry

K.G. Standing, R. Beavis, G. Bolbach[1]*, W. Ens, D.E. Main, and B. Schueler*

Physics Department, University of Manitoba, Winnipeg, Canada, R3T 2N2

1. Introduction

In recent years a number of mass spectrometers have been constructed for measuring the mass spectra of secondary ions by time-of-flight methods [1]. The secondary ions may be produced by bombardment of the sample with high energy (MeV) primary ions [2] or with low-energy (keV) ions. Instruments using both low-energy ion bombardment (SIMS) and time-of-flight (TOF) are listed in table 1.

Table 1 TOF - SIMS Instruments

	Type	Primary Ions	References
Manitoba I	Linear TOF	$Cs^+(K^+, Na^+, Li^+)$	[3,4]
Münster I	Toroidal Electrostatic Deflector	Ar^+	[5-8]
Rockefeller	Linear TOF	In^+	[9]
Uppsala	Linear TOF		[10]
Johns Hopkins	CVC(Bendix) linear TOF	Ar^+	[11]
Münster II	Linear TOF + Mirror	Ar^+	[12]
Manitoba II	Linear TOF + Mirror	Cs^+	[13]

Their principle of operation is illustrated most simply in the linear TOF spectrometer [3,4]. There the primary beam consists of a regular series of ~3 ns Cs^+ ion pulses spaced 250 μs apart. The pulses are produced by sweeping a focused beam of ions across a narrow slit [3]. Secondary ions liberated from the target are accelerated across a potential difference ~10 kV between the target and a grid a few mm away. The time interval between the arrival of a pulse of ions at the target and of a secondary ion at the detector determines the mass of the secondary ion.

This instrument is similar to the spectrometer of Macfarlane and Torgerson [2,14] except for the use of low-energy primary ions instead of fission fragments to produce the secondaries. The measuring technique is also very similar [15]. Because of these similarities it has been possible to make fairly direct comparisons of the mass spectra observed with the two methods of ion production [16].

On the other hand, the method of ion production used here is similar to the one used in other types of SIMS instruments but the spectrometer and the measuring technique are different. For comparison of results it is important to understand the distinctive features of TOF measurements. Some of these are described below.

[1]Present address: Institut Curie, Paris, France.

2. Efficiency

An outstanding feature of TOF spectrometers is their high efficiency, where efficiency is defined here as the ratio of the number of ions recorded in the spectrum to the number ejected from the target. The high efficiency is a result of two factors. First, the transmission is high since no slits are needed, at least for the simple geometry described above. All ions leaving the target assembly within the angle subtended by the detector will strike it; e.g. all ions with $\frac{1}{2} m v^2 < 1.5$ eV, where v is the ion ejection velocity perpendicular to the axis, for detector radius 2 cm, flight path 160 cm and accelerating voltage 10 kV. Thus a substantial fraction of the ejected ions will strike the detector [17]. The problem of then obtaining a measurable signal from the detector is the same problem as in other types of spectrometer, but the relatively high energy of the ions here is an advantage. However, it is still desirable to supply post-acceleration, particularly for ions of large mass [7,18].

Second, the whole mass spectrum can be recorded for each primary ion pulse; all ions giving pulses in the detector are integrated to form the final spectrum. Early TOF instruments were not able to use this feature because of electronic difficulties, but a number of suitable data systems have now been developed, [e.g. 2-7,10,11,15,19,20]. Most of these have been custom-built, but recently it has become possible to assemble a satisfactory system from commercially available components [20].

The latter feature may be contrasted with the usual procedure followed when using sector-field or quadrupole spectrometers. There the mass spectrum is scanned; when ions of a given mass are being measured ions of all other masses are discarded. Our procedure is equivalent to the classical mode of operation of a sector-field instrument as a mass spectrograph with photographic plate recording, or to other forms of "array detection" [21]. Clearly this factor also gives a large increase in efficiency. Benninghoven has estimated that the combined factors increase the efficiency by a factor $\sim 10^5$ compared to a typical quadrupole mass spectrometer [5].

3. Sensitivity

Sensitivity is determined partly by the spectrometer efficiency, but also by the method of ion production. The most popular production technique at present is the bombardment of a sample dissolved in a liquid matrix by low energy ions or neutrals (FAB) [22]. It has the advantage that the products of radiation damage are continuously removed by sputtering, exposing fresh sample in the bulk target material. Cotter has used this type of target for TOF [11]. His instrument is well suited to the technique, since its electric field at the target is transient and small, but the large vapor pressure of the liquid matrix is likely to cause difficulty in other TOF spectrometers.

Although the use of a liquid matrix has been very successful, it has some limitations. Considerable "chemical noise" is present; typically a background signal is observed at every mass number. Moreover the self-renewing surface is only useful in cases where considerable sample material is available. A sample of 1 pmole, for example, is sufficient to provide only a single monolayer of molecules over a target area ~ 1 mm^2 (assuming ~ 100Å^2 per molecule). Once the surface molecules are destroyed the signal disappears.

The highest sensitivities reported to date have in fact been obtained on TOF spectrometers using monolayer or sub-monolayer targets deposited on a metallic substrate, preferably a noble metal [5-7,9,23].

Benninghoven has reported that ion yields are the same for metallic and liquid substrates [23], at least for low masses. However, the solid targets are more compatible with the high electric fields normally used in TOF instruments. Clear molecular ion signals have been observed from less than 10^{-13} moles of bradykinin [24] and from less than 10^{-16} moles of compounds such as crystal violet [9,24].

As yet it has not been possible to extend this high sensitivity to compounds of mass ≥ 5000 u. Indeed, molecular ions from insulin have not yet been observed in TOF-SIMS, while they are produced fairly readily by fission fragments [2]; MeV primaries are apparently much more efficient at producing such ions. Perhaps the use of different matrices [2,27] may improve the situation. [Insulin ions have also been measured in sector-field spectrometers with FAB, probably because of the large primary currents usable in such instruments.]

4. Time Scale

For both organic and inorganic targets, many of the secondary ions produced by particle bombardment are metastable [3,25,26]. When these decay in the flight path of a linear TOF spectrometer, the centre of mass of the fragments retains the parent ion velocity. Thus the fragments appear in the mass spectrum at the mass number of the parent; the peak is broadened because of the energy released in the decay. The mass spectrum in a linear TOF instrument is therefore measured at an effective time immediately after acceleration, typically ~100 after emission. By contrast, ions in a sector-field or quadrupole spectrometer must survive the whole transit time through the instrument (tens or hundreds of microseconds) to be detected at their original mass. This difference in time scale can produce very large effects in the measured spectrum [26]. Linear TOF spectrometers are thus much better suited to measure the mass spectrum of ions actually ejected from the target.

5. Resolution

TOF mass spectrometers have traditionally been regarded as low-resolution instruments with $M/\Delta M_{FWHM}$ = 500 to 1000 [1]. However, in TOF-SIMS the production of ions on a plane equipotential surface makes it possible to do much better with proper measuring technique; $M/\Delta M > 3000$ has recently been reported [18,20]. This value appears to be determined by the velocity spread of the ejected ions. To do better requires velocity focusing [28]. Perhaps the simplest method of doing this is the use of an ion mirror, as pointed out some years ago by Mamyrin [29]. This method has been used successfully in a fission fragment spectrometer [30], and more recently in TOF-SIMS [12,13] where resolution of 13000 was reported for crystal violet [12]. This appears to be competitive with sector-field instruments for routine organic measurements.

6. Acknowledgements

This work was supported by grants from the U.S. National Institutes of Health and from the Natural Sciences and Engineering Research Council of Canada. G. Bolbach received support from CNRS (France).

References

1. For a review of earlier work on TOF mass spectrometers see D. Price and G.J. Milnes, Int. J. Mass Spectrom. Ion Processes 60, 61 (1984)
2. See papers by R.D. Macfarlane and by B. Sundqvist; these proceedings
3. B.T.Chait & K.G.Standing, Int. J. Mass Spectrom. IonPhys 40,185(1981)
4. K.G. Standing, R. Beavis, W. Ens and B. Schueler, Int. J. Mass Spectrom. Ion Phys 53, 125 (1983)
5. P. Steffens, E. Niehuis, T. Friese and A. Benninghoven, in Ion Formation from Organic Solids, Ed. A. Benninghoven, Springer Series in Chemical Physics (Springer, Berlin, 1983) Vol. 25, p. 111; A. Benninghoven, ibid. p. 64
6. P. Steffens, E. Niehuis, T. Friese, D. Greifendorf and A. Benninghoven, in Secondary Ion Mass Spectrometry SIMS IV, Ed. A. Benninghoven et al, Springer Series in Chemical Physics (Springer, Berlin, 1984) Vol 36, p. 404; A. Benninghoven, ibid. p. 342
7. P. Steffens, E. Niehuis, T. Friese, D. Greigendorf and A. Benninghoven J. Vac. Sci. Technol. A3, 1322 (1985)
8. W. P. Poschenrieder, Int. J. Mass Spectrom. Ion Phys 9, 357 (1972); G.H. Oetjen and W.P. Poschenrieder, ibid. 16, 353 (1975)
9. B.T. Chait and F.H. Field, 32nd Annual Conference on Mass Spectrometry and Allied Topics (Am. Soc. for Mass Spectrometry 1984) p. 237
10. B. Sundqvist, private communication
11. R.J. Cotter, Anal. Chem. 56, 2594 (1984)
12. E. Niehuis, T. Friese, H. Feld and A. Benninghoven, these proceedings
13. K.G. Standing et al, to be submitted
14. R.D. Macfarlane & D. Torgerson, Int. J. Mass Spectrom. Ion Phys. 21, 81 (1976)
15. R.D. Macfarlane, Anal. Chem. 55, 1247A (1983)
16. W.Ens, K.G.Standing, B.T.Chait & F.H.Field, Anal. Chem. 53,1241(1981)
17. K.G. Standing, B.T. Chait, W. Ens, G. McIntosh, and R. Beavis, Nucl. Instr. and Meth. 198, 33 (1982)
18. B. Schueler, R. Beavis, W. Ens, D.E. Main and K.G. Standing, Surface Sci. 160, 571 (1985)
19. R.F. Bonner, D.V. Bowen, B.T. Chait, A. Lipton, F.H. Field, and W.F. Sippach, Anal. Chem. 52, 1923 (1980)
20. W. Ens, R. Beavis, G. Bolbach, D. Main, B. Schueler and K.G. Standing, submitted to Nucl. Instr. and Meth.; also these proceedings
21. J.F. Holland, C.G. Enke, J. Allison, J.T. Stults, J.D. Pinkston, B. Newcome and J.T. Watson, Anal. Chem. 55, 997A (1983)
22. See the paper by K.L. Rinehart, Jr. in these proceedings
23. A. Benninghoven, M. Junach and W. Sichtermann, 32nd Annual Conference on Mass Spectrometry and Allied Topics (American Society for Mass Spectrometry 1984) p. 251
24. A. Benninghoven, E. Niehuis, T. Friese, D. Greifendorf, and P. Steffens, Org. Mass Spectrom. 19, 346 (1984)
25. B.T.Chait and F.H.Field, Int. J. Mass Spectrom. Ion Phys. 41,17(1981)
26. W. Ens, R. Beavis and K.G. Standing, Phys. Rev. Lett. 50, 27 (1983)
27. P. Demirev, M. Alai, R. van Breeman, R.J. Cotter and C. Fenselau, these proceedings
28. H. Wollnik in Ion Formation from Organic Solids, (IFOS III) Münster 1985, to be published by Springer
29. B.A. Mamyrin, V.I. Karataev, D.V. Schmikk and V.A. Zagulin, Sov. Phys. JETP 37, 45 (1973)
30. S. Della Negra and Y. Le Beyec, Int. J. Mass Spectrom. Ion Processes 61, 21 (1984)

Fast Atom Bombardment Mass Spectrometry of Biomolecules

K.L. Rinehart, Jr., [1], *J.C. Cook*[1], *J.G. Stroh*[1], *M.E. Hemling*[1], *G. Gäde*[2], *M.H. Schaffer*[3], *and M. Suzuki*[4]

[1] School of Chemical Sciences, University of Illinois at Urbana-Champaign, Urbana, IL 61801, USA

[2] Institut für Zoologie IV, Universität Düsseldorf, D-4000 Düsseldorf, F.R.G.

[3] Department of Psychiatry, University of Chicago, Chicago, IL 60637, USA

[4] Faculty of Pharmacy, Meijo University, Nagoya 468, Japan

Since its introduction in early 1981 [1], fast atom bombardment mass spectrometry (FABMS, liquid SIMS) has become a routine tool in many mass spectrometry laboratories [2]. In our Urbana laboratory, which serves scientists at other institutions as well as those on our campus, four double-focusing spectrometers (ZAB-SE, 7070E, 731, 311A) have FAB ion sources, and about as many FAB spectra as electron ionization spectra are run. The technique is still developing, however, and it is appropriate to describe here some recent advances as well as associated problems.

The matrix used to obtain a FAB spectrum is critical to the success of the experiment, and every laboratory has its favorite. We most often employ a liquid mixture of dithioerythritol and dithiothreitol [3] as the matrix for positive ion FAB, especially of peptides, since the mixture combines the acidity of thioglycerol with the non-volatility of glycerol. Other workers regularly use acids [4] or bases such as diethanolamine to promote ionization of peptides or sugars [5].

Routine high-precision measurement of the masses of all ions remains an elusive goal of FABMS. For single ions, we obtain high-resolution data by using a split stage with the mass marker occupying one-half the stage and the sample the other half [6]. The halves are shifted mechanically so that each is located alternately in precisely the same spot, thereby eliminating differences in ion-source focusing. Employing the split stage, we routinely obtain data accurate within 4 ppm to *m/z* 2000. Our preferred mass marker for these positive ion measurements is poly(propylene glycol) (PPG).

For high-precision studies of negative ions, we have shown recently that combinations of poly(ethylene glycol) (PEG) or PPG with triethylenetetramine provide excellent mass marking systems to *m/z* 1900 or 2300, respectively [7]; measurements accurate to 1-4 ppm are made routinely to *m/z* 1400.

Attempts to assign partial structures on the basis of FABMS cleavages or fragmentations encounter a recurring problem. Although many peptides have been studied most successfully by FABMS, the fragmentation of others is limited or even non-existent, as in our recently completed study of the AKH II insect neuropeptides, adipokinetic hormones from the locusts *Schistocerca nitans* and *Locusta migratoria*, and of HGF I, a hyperglycemic factor from the Indian stick insect *Carausius morosus*. Protonated molecular ions $(M + H)^+$ were abundant for all the peptides, even for 1-μg samples, and the observation of identical $(M + H)^+$ ions for extracts of *S. gregaria* and

S. nitans led to the presumption that the two species contained identical peptides. Those ions confirmed the reported amino acid analyses [8], indicated the presence of glutamic acid in a >Glu unit and of aspartic acid as asparagine, and showed the C-terminus to be an NH_2 group. Even for the more available AKH II from S. nitans, no fragmentations were seen in the normal FAB spectrum other than loss of the C-terminal NH_2 group.

Lack of fragmentation can be addressed by a number of chemical and mass spectrometric methods, but, in view of the small sample size and our instrumental capabilities, we elected a B/E linked scan, which revealed fragmentations between the five C-terminal amino acids (Scheme I) as well as loss of the >Glu group from the N-terminus. By analogy to other known insect neuropeptides, a sequence was proposed and the corresponding synthetic polypeptide [9] was examined. It showed the same fragmentations in the B/E linked scan, and one additional amino acid, Phe, was cleaved from both the N- and C-directions with this larger sample. In addition to showing essentially identical behavior by FABMS, the synthetic peptide was identical by HPLC and bioactivity to the natural insect hormone derived from both S. nitans and S. gregaria.

Scheme I. FABMS, B/E linked scan[a]

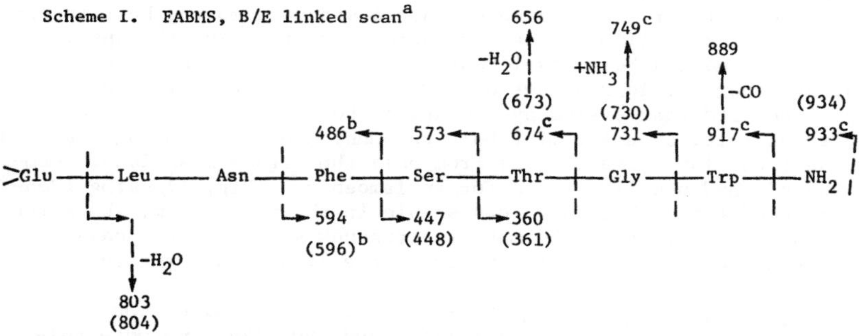

[a]Observed ions in ( ). [b]Observed only in synthetic peptide's spectra. [c]Observed in S. gregaria spectra.

The corresponding AKH II from L. migratoria [10] behaved in the same fashion, showing no fragmentation in the ordinary FAB spectrum [$(M+H)^+$ = 904], which did, however, assign a molecular formula, confirm the previously identified [10] amino acids, and indicate the presence of >Glu, Asn, and terminal NH_2 units. Here, the B/E linked scan sequenced the four amino acids at the C-terminus and the N-terminal >Glu unit, and cleavage was observed between Asn and Phe. Although neither the Leu-Asn nor the Phe-Ser sequence was demonstrated, the sequence could be inferred by homology to the compound from S. gregaria, with Ala^6 in place of the Thr^6 of Scheme I.

The normal mass spectrum of the hyperglycemic factor from C. morosus [11] showed more cleavages (Scheme II). This can be attributed in large measure to the Pro unit, which provides a focus for fragmentation as we have previously shown for insect neuropeptides M I and M II [3]; here, this cleavage gave the major peaks at m/z 574 and 573. In this case, the B/E linked scan assigned a complete structure, if one assumes preferential cleavage left of Pro and the >Glu N-terminus. It is apparent from these peptide studies that a B/E linked scan usually provides more fragmentation information than does a simple FAB spectrum. Because only small samples were available, we were unable to carry out collisional activation experiments, which sometimes provide additional information.

Scheme II. FABMS, B/E linked scan[a]

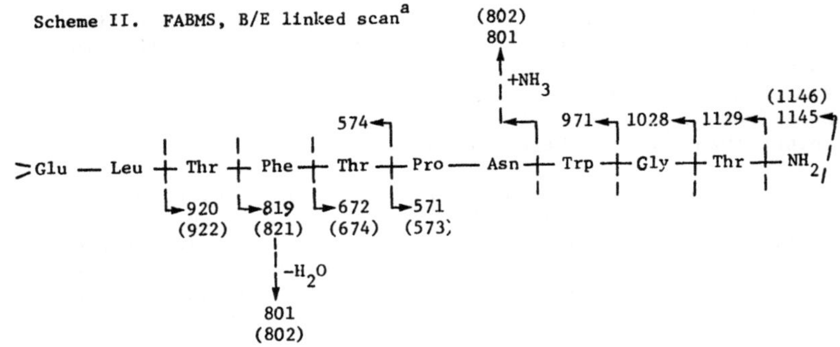

[a]Observed ions in ( ).

We have attempted in the past two years to develop a successful combination of FABMS, a powerful method of ionizing polar compounds, and reversed-phase high-performance liquid chromatography (RP-HPLC), the method of choice for separating polar compounds. The LC/FAB system employed for this effort has been described earlier [12]; we noted that a standard moving belt interface had been modified by eliminating the infrared heater used to vaporize compounds, that the moving belt itself was bombarded in the ion source by xenon atoms, and that sample residues were removed by a methanol wash. We have recently modified that system by replacing the earlier spray device by a stainless steel frit, by using microbore columns, and by including a program to compute the total ion current (TIC) from only the higher mass ions. Comparison of the old and new TIC traces for antiamoebin I (Fig. 1), using identical scanning conditions but 10 x less sample in the new system, shows the improved peak shape. Moreover, the mass spectra now show isotope peaks, previously lacking because of low intensity in the higher mass region.

Using this system to study the antibiotic CC-1014 [13], also reported under the names leucinostatin [14], P168 [15], and 1907 [16], we were able to demonstrate that CC-1014A and leucinostatin A have identical LC chromatographic behavior and give identical FAB mass spectra. We also observed two unreported leucinostatins, C and D, and assigned their structures on the basis of MS, as shown in Scheme III. The complex C_{11} amino acid in leucinostatins A and B is replaced in C and D by a leucine unit.

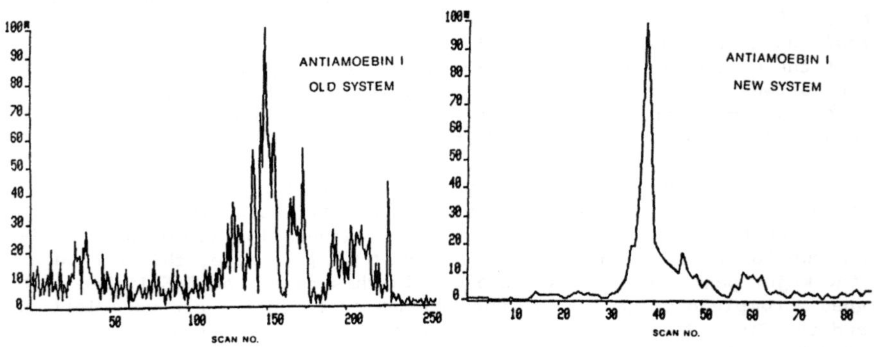

Fig. 1. TIC for antiamoebin I (Ac-Phe-Aib-Aib-Aib-Iva-Gly-Leu-Aib-Aib-Hyp-Gln-Iva-Hyp-Aib-Pro-Phol)

Scheme III

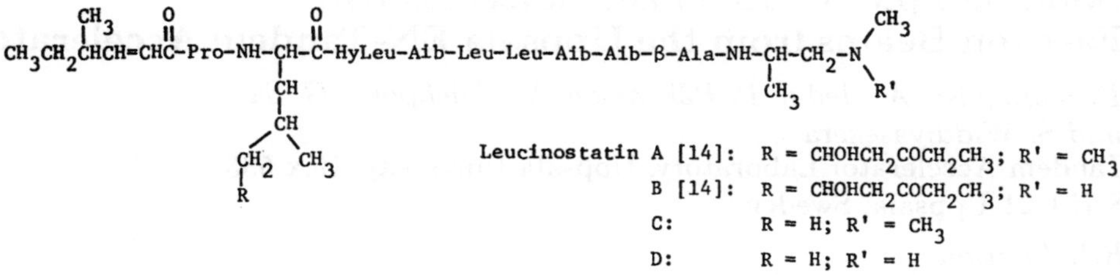

In conclusion, FABMS is a valuable tool for routine use in both the positive and negative ion modes and at both low and high resolution. Fragmentations, while sometimes missing in ordinary FAB spectra, can be studied by alternative means such as B/E linked scans. FABMS is also amenable to combination with LC employing a moving belt interface.

Acknowledgement. This work was supported in part by grants from the National Institutes of Health (GM27029, RR01575) and the National Science Foundation (PCM-8121494).

References

1. M. Barber, R. S. Bordoli, R. D. Sedgwick, A. N. Tyler: J. Chem. Soc., Chem. Commun. 1981, 325-327 (1981); D. J. Surman, J. C. Vickerman: ibid., 324-325.
2. K. L. Rinehart, Jr.: Science (Washington, D.C.) 218, 254-260 (1982).
3. J. L. Witten, M. H. Schaffer, M. O'Shea, J. C. Cook, M. E. Hemling, K. L. Rinehart, Jr.: Biochem. Biophys. Res. Commun. 124, 350-358 (1984).
4. J. L. Gower: Biomed. Mass Spectrom. 12, 191-196 (1985).
5. K.-I. Harada, M. Suzuki, H. Kambara: Org. Mass Spectrom. 17, 386-391 (1982).
6. J. C. Cook, B. N. Green: Abstracts, 32nd Annual Conference on Mass Spectrometry and Allied Topics, San Antonio, TX, May 27-June 1, 1984, p. 835.
7. M. E. Hemling, J. C. Cook, K. L. Rinehart, Jr.: Anal. Chem., submitted.
8. J. Carlsen, W. S. Herman, M. Christensen, L. Josefsson: Insect Biochem. 9, 497-501 (1979).
9. D. Yamashiro, S. W. Applebaum, C. H. Li: Int. J. Peptide Protein Res. 23, 39-41 (1984).
10. G. Gäde, G. J. Goldsworthy, G. Kegel, R. Keller: Hoppe-Seyler's Z. Physiol. Chem. 365, 393-398 (1984).
11. G. Gäde: Biol. Chem. Hoppe-Seyler 366, 195-199 (1985).
12. J. G. Stroh, J. C. Cook, R. M. Milberg, L. Brayton, T. Kihara, Z. Huang, K. L. Rinehart, Jr., I. A. S. Lewis: Anal. Chem. 57, 985-991 (1985).
13. K. L. Rinehart, Jr., R. C. Pandey, M. L. Moore, S. R. Tarbox, C. R. Snelling, J. C. Cook, Jr., R. H. Milberg: In "Peptides: Structure and Biological Function," Proc. 6th Am. Peptide Symp., E. Gross, J. Meienhofer, Eds., Pierce Chemical Co., Rockford, IL, 1979, pp 59-71; K. L. Rinehart, Jr.: Trends Anal. Chem. 2, 10-14 (1983); P. F. Wiley, J. M. Koert, L. J. Hanka: J. Antibiot. 35, 1231-1233 (1982).
14. K. Fukushima, T. Arai, Y. Mori, M. Tsuboi, M. Suzuki: J. Antibiot. 36, 1613-1630 (1983).
15. A. Isogai, A. Suzuki, S. Tamura, S. Higashikawa, S. Kuyama: J. Chem. Soc. Perkin Trans. I, 1405-1411 (1984).
16. M. Sato, T. Beppu, K. Arima: Agric. Biol. Chem. 44, 3037-3040 (1980).

Solid Sample-SIMS on Biomolecules with Fast Ion Beams from the Uppsala EN-Tandem Accelerat(

B. Sundqvist, A. Hedin, P. Håkansson, M. Salehpour, G. Säve, and S. Widdiyasekera
Tandem Accelerator Laboratory, Uppsala University, Box 533, S-751 21 Uppsala, Sweden

R.E.Johnson
Department of Nuclear Engineering and Engineering Physics, University of Virginia, Charlottesville, USA

1. Introduction

Secondary Ion Mass Spectrometry (SIMS) on bioorganic solids was first performed by R.D. MACFARLANE and coworkers [1] in 1974 usir fission fragments from a Cf-252 source as primary particles. Fission fragments are fast particles i.e. with a velocity larger than the Bohr velocity (v_0=2.0·10^8 cm/s) interacting mainly with the electrons in a medium. In most SIMS-studies performed so far slow particles are used as primaries. Slow particles interact pri marily via elastic collisions with the atoms in a solid. Macfarla called the mass spectrometric method based on the use of fission fragments from Cf-252 and time-of-flight (TOF) analysis, ^{252}Cf-Plasma Desorption Mass Spectrometry (PDMS), and proposed a new formation process different from that assumed for the secondary ion formation by keV-ions. 1976 BENNINGHOVEN et al. applied static SIMS to bioorganic solids. Very similar spectra were found and therefore the same ion formation process for PDMS and SIMS was postulated [2]. A major contribution to the field was made in 1981 by BARBER et.al. [3] with the introduction of slow ion-liquid sample-SIMS with a magnetic sector instrument (FAB).Already at an early stage Macfarlane et.al. showed that fission fragments were very effective to desorb and ionize large and labile biomolecules [4] and a joint study by the Texas and Uppsal groups showed that the fast primaries are more efficient than slo primaries and more so the larger the molecule is [5].

During the last few years our research group at the Tandem Accelerator Laboratory at Uppsala has been working with a project on fast ion-SIMS on biomolecules. The project has two main aims, namely to study mechanisms for fast ion interactions with bio-organic solids and to develop analytical methods in biomolecular mass spectrometry. This paper is a short summary of the earlier achievements in, and a presentation of some recent results from this project.

2. Experimental Arrangements

Fission fragments from a ^{252}Cf-source are a mixture of fast ions with a distribution in energy, mass and charge state [6]. In orde to study the influence of primary ion parameters on the desorptio ionization process, an accelerator is therefore preferable, since i can produce a well-defined beam of fast ions with a specific velo city, charge state and angle of incidence. This was independently realized and used at Erlangen and Uppsala in the late seventies

[7,8]. Similar projects using a "high" energy accelerator were later started at GSI, Darmstadt [9] and ALICE, Orsay [10]. By now a fairly large data set on the influence on secondary biomolecular ion yields of primary ion parameters like angle of incidence [11,12], velocity [7,8,10] and charge state [9,13,14] have been published. So far, mainly secondary ions have been studied, with a recent exception, namely a study of SALEHPOUR et.al. [15] indicating that the yields of intact neutral molecules ejected is about four order of magnitude larger than the yield of secondary ions.

The experimental setup is shown schematically in Fig. 1.

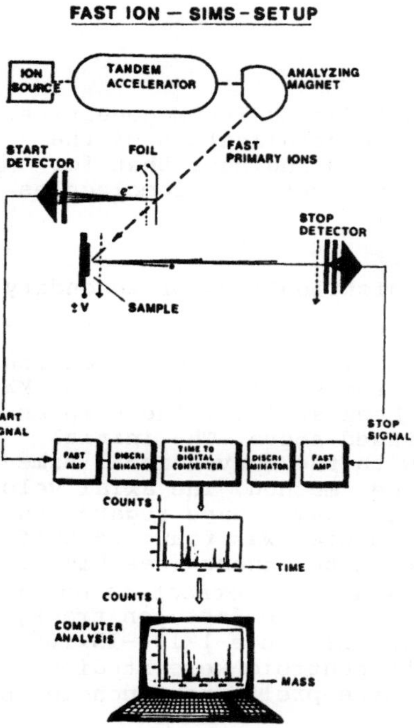

Fig. 1 The principle of the on-line SIMS-setup at Uppsala used in the accelerator experiments. The accelerator is equipped with a Cs^+-sputter negative ion source which means that fast ions of almost any element in the periodic table can be acceleratred except for the noble gases, as negative ions are accelerated. In the studies described here, an acceleration voltage of 20 kV was used in the TOF system. In the setup shown, also mass spectrometric studies are performed and molecular ions of bovine insulin (MW 5733) were observed for the first time [16]. The main technical advantages of this system as compared to a conventional PDMS spectrometer [1] are that the fast ions bombard the sample from the front side so that thick backings can be used, and in principle microbeams can also be considered.

3. Neutralization of secondary ions

Samples of the peptide LHRH (Releasing Hormone for Luteinizing Hormone) (MW 1182) of different thickness, measured with ellipsometry, were spincasted on Si-backings [17]. The samples were bombarded with 90 MeV $^{127}I^{14+}$ ions and mass spectra of secondary ions studied with the time-of-flight spectrometer shown in Fig. 1. The yields of secondary ions were studied as a function of sample thickness (see Fig. 2). The solid lines are parameter fits of the same functional form to the four sets of experimental points. The ellipsometry method is reliable down to 100-200 Å thicknesses. Therefore the thickness dependence found can be questioned. The thickness dependence measured might be due to inhomogenous layers. However, the fact that a specific molecular fragment gives a different thickness dependence than the molecular ion may indicate that

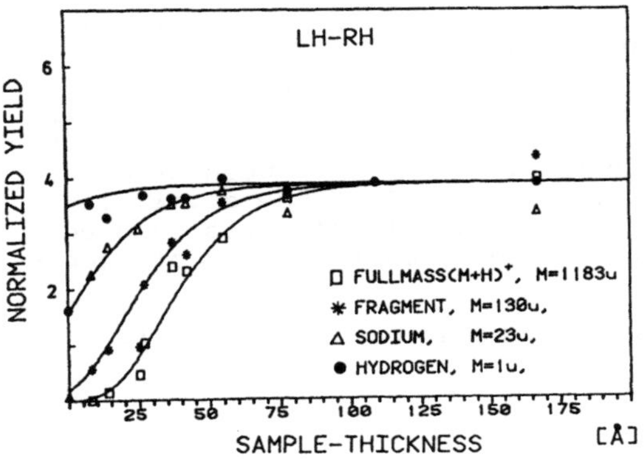

Fig. 2 Yields of secondary ions as a function of sample thickness.

the thickness dependence measured is not due to inhomogeneities. The mass dependence found suggests that neutralization of the secondary ions is taking place. The results indicate that for effective fast ions-SIMS analysis of monolayers of biomolecules insulating backings should be used [18].

4. The dependence of initial velocity distributions of secondary ions on sample thickness

The sample thickness dependence of initial velocity distributions of H^+ desorbed by 90 MeV $^{127}I^{14+}$ ions from samples of valine (MW 117) and bovine insulin (MW 5733) has been studied. The samples were prepared in the same way as described above. The initial velocity distributions were measured with the conventional time-of-flight technique calibrated with a new method. The axial velocity distributions were found to have the same shape regardless of the sample thickness. The centroid, however, was found to shift to higher energies with increasing sample thickness (see Fig. 3) namely by 0.005 eV/Å. In order to check if this effect is due to macroscopic charging or "local" charging in the fast ion track, the primary ion intensity was varied in the range $1 \cdot 10^3$-$15 \cdot 10^3$ particles per second and shifts in the centroid were studied. As no significant shifts were observed, the preliminary conclusion

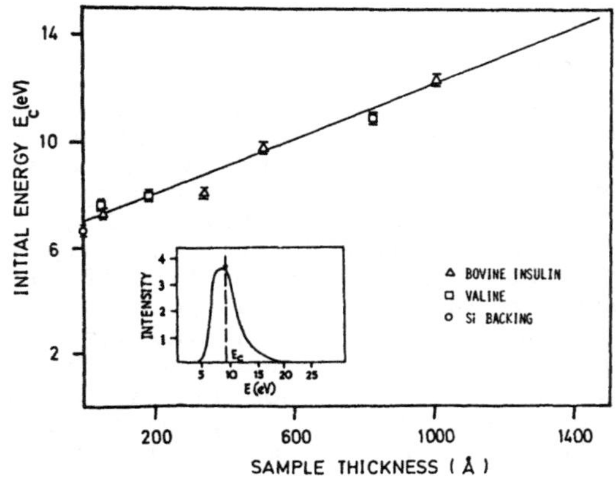

Fig. 3 The centroid of H^+ axial velocity distributions as a function of sample thickness. The inset shows an example of an axial H± velocity distribution.

is that the H^+ ions experience an additional repulsion from the the charged center of the track [19].

References:

1. D.F. Torgerson, R.P. Skowronski and R.D. Macfarlane: Biochem. Biophys. Res Commun. 60, 616 (1974)
2. A. Benninghoven, D. Jaspers and W. Sichtermann: Appl. Phys. 11, 35 (1976)
3. M. Barber, R.S. Bordoli, R.D. Sedgewick and A.N. Tyler: I. Chem. Soc. Chem. Commun. 325, (1981)
4. R.D. Macfarlane and D.F. Torgerson: Science 191, 920 (1976)
5. I. Kamensky, P. Håkansson, B. Sundqvist, C.J. McNeal and R.D. Macfarlane: Nucl. Instr. Meth. 198, 65 (1982)
6. B. Sundqvist: Nucl. Instr. Meth. 218, 267 (1983)
7. P. Dück, W. Treu, H. Fröhlich, W. Galster and H. Voit: Surf. Sci. 95, 603 (1980)
8. P. Håkansson, A. Johansson, I. Kamensky, B. Sundqvist, J. Fohlman and P. Peterson: IEEE Trans. Nucl. Sci. NS-28-2, 1776 (1981)
9. W. Guthier, O. Becker, W. Knippelberg, K. Weikert, K. Wien, S. DellaNegra, Y. LeBeyec, P. Wieser and R. Wurster: Int. J. Mass. Spectrom. Ion Phys. 53, 185 (1983)
10. S. DellaNegra, D. Jacquet, I. Lorthiois, Y. LeBeyec, O. Becker and K. Wien: Int. J. Mass. Spectrom. Ion Phys. 53, 215 (1983)
11. P. Håkansson, I. Kamensky and B. Sundqvist: Surf. Sci. 116, 302 (1982)
12. E. Nieschter, B. Nees, N. Bischof, H. Fröhlich, W. Tiereth and H. Voit: Surf. Sci. 145, 294 (1984)
13. P. Håkansson, E. Jayasinghe, A. Johansson, I. Kamensky and B. Sundqvist: Phys. Rev. Lett. 47, 1227 (1981)
14. E. Nieschter, B. Nees, N. Bischof, H. Fröhlich, W. Tiereth and H. Voit: Rad. Eff. 83, 121 (1984)
15. M. Salehpour, P. Håkansson, B. Sundqvist and S. Widdiyasekera: TLU 123/85, Tandem Laboratory Report, Uppsala, Sweden, 1985. Presented at the Int. Conf. Atomic Coll. in Solids, Washington D.C., 1985 (To appear in Nucl. Instr. Meth.)
16. P. Håkansson, I. Kamensky, B. Sundqvist, J. Fohlman, P. Peterson, C.J. McNeal and R.D. Macfarlane: J. Am. Chem. Soc. 104, 2948 (1982)
17. U. Jönsson, G. Olofson, M. Malmquist, G. Säve, P. Håkansson, J. Fohlman and B. Sundqvist: TLU 129/85, Tandem Laboratory Report, Uppsala, Sweden, 1985
18. G. Säve, P. Håkansson, B. Sundqvist and U. Jönsson: TLU 127/85 Tandem Laboratory Report, Uppsala, Sweden, 1985
19. S. Widdiyasekera, P. Håkansson, B. Sundqvist and G. Säve: TLU 124/85, Tandem Laboratory Report, Uppsala, Sweden, 1985

Fourier Transform Mass Spectrometry for High Mass Applications

M.E. Castro, L.M. Mallis, and D.H. Russell

Department of Chemistry, Texas A & M University, College Station, TX 77843, USA

1. Introduction

Fourier transform mass spectrometry (FTMS) has enormous potential as an analytical mass spectrometry, especially for the analysis of large molecules [1,2,3]. The potential advantages of this instrument over magnetic sector and time-of-flight instruments are the available mass range and ultra-high mass resolution. In addition, as with the time-of-flight system, the FTMS experiment utilizes a pulsed, low duty cycle ionization source which should result in greater sensitivity than can be obtained with CW ionization sources.

The fundamental advantages of FTMS for the analysis of large biomolecules can be illustrated by examining the equations describing the mass range (equation 1) and mass resolution (equation 2) of the ion cyclotron resonance (ICR) principle [4].

$$M = 1.536 \times 10^7 \, B/f \quad \text{(f in hertz)} \qquad (1)$$

$$M/\Delta M = 1.733 \times 10^4 \, eBt/M \qquad (2)$$

B is the magnetic field strength (in tesla); f is the cyclotron frequency of the ion, e is the charge on the ion, and t is the observation time of the transient signal.

Thus the mass range for FTMS is directly proportional to B and mass resolution is directly proportional to B and t and inversely proportional to M. Assuming a lower limit of f of 10 kHz (arbitrarily selected, at some point the detection of low frequencies is limited by electronic/background noise), a mass limit of ca. 4,000 daltons is obtained for an FTMS system with a 3 tesla magnet (the standard commercial magnet system). These same considerations give a value for mass resolution of 13,000 at m/z 4,000 for t = 1 s. Extending t to 10 s increases $M/\Delta M$ to 130,000 at m/z 4,000. The two principle considerations for high-mass FTMS are f and t. Previous work has shown that low-frequency detection (2-3 kHz) and t in excess of 10 s are feasible [5]. These considerations, combined with the available magnet technology (in excess of 11 tesla), suggest that the mass range of FTMS should extend beyond 30,000-40,000 daltons, and that better than unit resolution at these masses is feasible.

2. Experimental

An important consideration in developing high-mass FTMS is the signal observation time, t. The observed signal in FTMS is generated by forming a coherent packet of ions which produces an image current in the radiofrequency receive plates. The image current decays if the coherence of the ion packet is lost by collisions with residual neutrals in the vacuum system. Thus, the mass resolution is very sensitive to the pressure of the vacuum analyzer. For this reason, the ionization source used should not contribute neutral species to the system, i.e., gaseous discharge guns and liquid matrices are not suitable for FTMS. To overcome this fundamental limitation, several approaches for performing desorption ionization have been tried, viz. laser desorption [6,7] and Cs^+ ion SIMS [8]. An alternative approach is to produce the ions external to the vacuum system [9,10].

3. Results and Discussion

To date, the emphasis of our work has dealt with system design and evaluating the general capabilities for high-mass FTMS. The system consists of a Cs^+ ion gun, a lens assembly for directional control of the Cs^+ ion beam, the ICR cell, and direct sample introduction assembly. Typical operating conditions for the Cs^+ ion gun are beam energy of 3-6 keV and beam density of $1\text{-}100 \times 10^{-9}$ A/cm^2. The Cs^+ ion beam is pulsed on for 3-5 milliseconds. Using these conditions, ample secondary ions are produced; however, in most cases the dominant molecular ions observed correspond to $[M+Na]^+$ and/or $[M+Na-(x-1)H]^+$. We have studied this effect in some detail and proposed a mechanism for desorption ionization involving a thermal mechanism [11]. In a separate paper in this volume results pertinent to this mechanism are discussed [12].

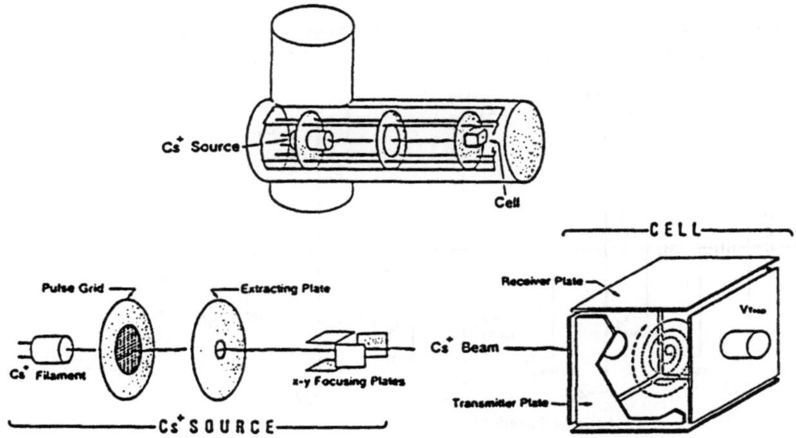

Fig. 1 Schematic of the ion cell and ion gun configuration used for the Cs^+ ion secondary ion-FTMS studies.

More recent work with peptides demonstrates that secondary ion-FTMS spectra are very similar to those obtained with laser desorption FTMS and that the observed fragment ions contain significant structural information. Figure 2 contains the Cs^+ SI-FTMS spectrum of leucine enkephalin. In the region of the molecular ion, intense signals at m/z 594, 632 and 670 are observed, which correspond to $[M+K]^+$ and $[M+xK-(x-1)H]^+$ ions where x = 2,3. Since potassium was admixed (as KBr) with the sample, the presence of $[M+K]^+$ molecule ions is not surprising. The fragment ions observed, however, arise almost exclusively from an unobserved $[M+Na]^+$ complex. The most intense fragment ions in the spectrum are observed at m/z 325 and 327 which correspond to cleavages A_3 and X_2 (all containing Na) as shown in Figure 3. Other less intense ions are also observed,

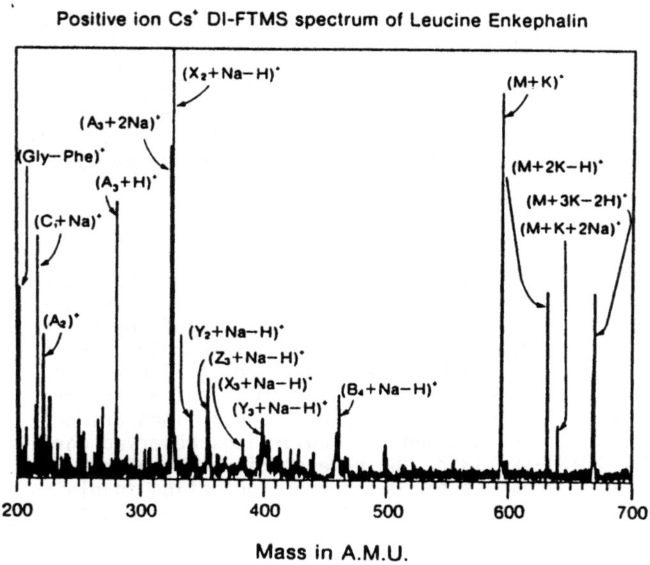

Fig. 2 The Cs^+ SI-FTMS spectrum of Leucine Enkephalin obtained with an incident beam energy of 5 KeV and beam current of 10×10^{-9} A cm^{-1}.

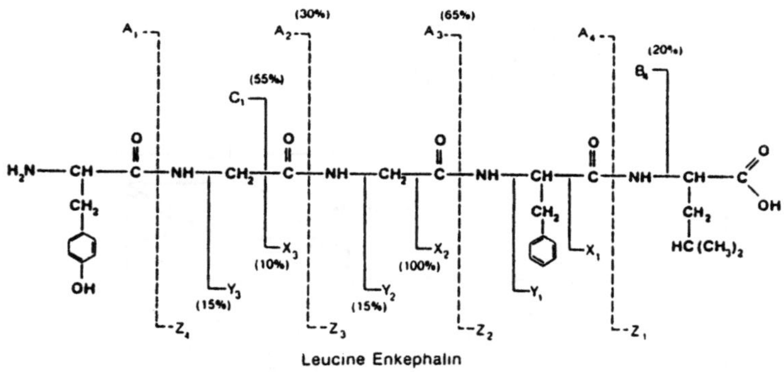

Fig. 3 Structure and proposed fragment ions of Leucine Enkephalin.

however, in all cases the fragment ions contain sodium vis-a-vis potassium. Other work, on the hippuryl-histidyl-leucine (HHL) molecule, has shown that the reasons for this are related to the relative stability of the $[M+Na]^+$ ions vis-a-vis the $[M+K]^+$ ions. That is, the binding energy of Na^+ to the peptide is greater than the binding energy of K^+. The enhancement of fragmentation reaction product ions formed by CID (using fast atom bombardment ionization) increases from Cs to Li. This suggests that as the size of the alkali metal attached to the molecule decreases, the more stable is the complex. Therefore, when equimolar or greater amounts of alkali metal cations (e.g., sodium and potassium) are present with the molecule of interest, in most cases, the molecule ions will contain the alkali metal in greatest abundance, while the fragment ions will contain the smaller alkali metal with the smallest size or greater binding energy.

In order to evaluate the high mass range (low-frequency range) of FTMS, studies were performed on the Cs^+ SI-FTMS of CsI. This system is well suited to such objectives, since cluster ions of the type $(Cs(CsI)_n^+$ are readily formed by particle bombardment of CsI [13]. In the Cs^+SI-FTMS spectra of CsI, ions are observed up to m/z 16,241, e.g., n = 62 [5]. These studies demonstrate several important points. (1) The detection of ions in the low-frequency range (less than 5 kHz) is feasible. In fact, comparing the relative abundance of the high-mass cluster ions with the data of Campana [14] and Standing [15] suggest that mass discrimination of the FTMS detection scheme is less than that for the charged particle detector. (2) In spite of the limited dynamic range of FTMS (10^3-10^4) ample signal-to-noise is obtained for the high-mass cluster ions. (3) The long time-scale of the FTMS experiment does not influence the sensitivity of the method. However, it should be noted that the relative abundances of some of the cluster ions is attentuated in the SI-FTMS spectrum. This observation is consistent with Standing's explanation [15] of the "anomalous cluster ion intensities" reported by Campana [14].

4. References

1. R.T. McIver, Presented at the workshop on newer aspects of Ion Cyclotron Resonance (Fourier Transform Mass Spectrometry), 29th Annual Conference on Mass Spectrometry and Allied Topics, Minneapolis, MN, p. 791(1981).
2. C.L. Wilkins and M.L. Gross, Anal. Chem. 53, 1661A (1981).
3. E. Onyiriuka, R.L. White, D.A. McCrery, M.L. Gross and C.L. Wilkins, Int. J. Mass Spectrom, Ion Phys. 46, 135 (1983).
4. M.B. Comisarow Transform Techniques in Chemistry, P.R. Griffiths, Ed; Plenum Press, New York, N.Y., Chapter 10 (1978).
5. M.E. Castro, D.H. Russell, S. Ghaderi, R.B. Cody, I.J. Amster and F.W. McLafferty, Anal. Chem. (in press).
6. D.A. McGrery, E.G. Ledford and M.L. Gross, Anal. Chem. 54, 1437 (1982).
7. C.L. Wilkins, D.A. Weil, C.L.C. Yang and C.F. Ijames, Anal. Chem 57 520 (1985).
8. M.E. Castro and D.H. Russell, Anal. Chem. 56, 578 (1984).
9. D.F. Hunt, J. Shabanowitz, R.T. McIver, Jr., R.L. Hunter and J.E.P. Syka, Anal. Chem. 57, 765-768 (1985).
10. D.F. Hunt, J. Shabanowitz, J.R. Yates, III, R.T. McIver, J.E.P. Syka and J. Amy, Anal. Chem. (in press).
11. M.E. Castro, L.M. Mallis and D.H. Russell, J. Am. Chem. Soc., 107, 5652 (1985).
12. L.M. Mallis, M.E. Castro and D.H. Russell, *Some Aspects of the Chemistry of Ionic Organo-Alkali Metal Halide Clusters Formed by Desorption Ionization*, This Volume.
13. J.E. Campana, T.M. Barlak, R.J. Colton, J.J. DeCorpo, J.R. Wyatt and B.I. Dunlap, Phys. Rev. Lett. 47, 1046 (1981).
14. T.M. Barlak, J.R. Wyatt, R.J. Colton, J.J. DeCorpo and J.E. Campana, J. Am. Chem. Soc. 104 1212 (1982).
15. K.G. Standing, B.T. Chait, W. Ens, G. McIntosh and R. Beavis, Nucl. Instrum. Methods 198, 33 (1982).

Part XIV

Organic Applications Including Fast Atom Bombardment Mass Spectrometry

Energy Distribution of Secondary Organic Ions Emitted by Amino Acids

L. Kelner and T.C. Patel

National Institutes of Health, Biomedical Engineering and Instrumentation Branch, 9000 Wisconsin Ave., Bethesda, MD 20892, USA

1. Introduction

Knowledge of the energy distribution (ED) of secondary organic ions sputtered from the surface of the target is helpful in understanding the mechanism of organic sputtering and in optimization of solid and liquid SIMS experiments.Based on our previous experiments [1,2] we proposed that the primary mechanism responsible for the emission of secondary organic ions is linear cascade sputtering, since this process results in the production of ions with a lower kinetic energy than in the case of inorganics. The energy distribution of various amino acids as well as effects of liquid matrix on this parameter has been studied. Experimental data were obtained on the instrument described previously by us in detail [3]. A computer system based upon an IBM PC-XT was used to scan the main cylinder of the electrostatic energy analyser (Bessel Box) and to control the quadrupole mass spectrometer. The target bias remained fixed during the energy scan, and the energy distribution was determined as the function of ion abundance versus the voltage difference between the main cylinder and the target. Complete mass spectra were stored for each step of the voltage change. These voltage steps could be varied depending on the number of data points required to obtain a reasonable energy distribution curve.

2. Experimental

In our previous report [2] we examined the effect of various matrices on the ED of molecular and fragment ions of arginine. We have extended our experiments to include amino acids of different structures, such as phenylalanine, 2-aminobenzoic acid, 3-aminobenzoic acid as well as arginine. Trifluoromethane sulfonic acid (TFMSA) was used to determine the effect of a protonating agent on the energy distribution of organic ions. Measured amounts of the acid in glycerol and addition of TFMSA in concentrations ranging from 0.1 to 10M were prepared in advance before placing the specimen on the target.

3. Results and Discussion

The most significant effect on the ED of molecular ions of amino acids has been the addition of the protonating agent of TFMSA at concentrations ranging from 1M to 10M. When TFMSA is present on the target, the maximum of ED shifts a few eV's to lower energies. This shift to lower energies is dependent on the type of amino acid. It is somewhat more pronounced for phenylalanine in glycerol than for 2- and 3-aminobenzoic acid and

Table I. Shift of ED Maximum for(M+H)+ Ions of Amino Acids in Glycerol Matrix

	ED Maximum Shift for , eV			
	Arginine	Phenylalanine	2-aminobenzoic acid	3-aminobenzoic acid
1M TFMSA	1.40 ±0.1	3.65 ±0.1	3.20 ±0.1	2.70 ±0.1
10M TFMSA	1.80 ±0.1	3.90 ±0.1	3.70 ±0.1	3.50 ±0.1

and considerably more than for arginine. Table I represents shift of the maximum of ED as a function of TFMSA concentration for various amino acids relative to the sample without TFMSA.

Experiments with glycerol and glycerol with TFMSA (without sample) revealed that there is a comparable shift in the molecular, fragment and cluster ions of glycerol. This result suggests that the ED shift observed is due to the change in electrical properties of the liquid target which, in turn, depends on the concentration of pre-formed protonated ions in the liquid phase when TFMSA is added. The practical implication of these results might be in the ability to regulate the abundance of peaks in the secondary ion mass spectra of the compound of interest by changing the target bias. This may affect the desorption process of secondary organic ions when a protonating agent such as TFMSA is added to the liquid phase. It also suggests that the target bias should be used in liquid SIMS experiments in order to obtain optimal results.

As previously noted [2], Eds of sample as well as matrix ions very often exhibit a complex structure (two peaks). The double-peak structure of ED observed in this case can be explained by the assumption that sputtering of secondary organic ions is not just a pure surface phenomenon, but it also involves considerable disturbance of the adjacent volume (in the tens of angstroms). So-called recoil ions can produce lower-energy secondary ions, while ions sputtered from the surface are of higher energy. The two-peak structure of the ED changes with the type of matrix, concentration of the sample in the matrix, and with "doping" by protonating agents such as TFMSA.

Dependence of the ED structure on the energy of primary particles reinforces our hypothesis that this "splitting" of the ED into two peaks is a result of superposition of two processes, one of which involves the monolayer at the surface and the second of which involves the volume of the target. The "splitting" of the ED is less pronounced for phenylalanine and for aminobenzoic acids, than it is for arginine.

The energy of the primary beam affects the ED of secondary ions by increasing the width of the ED curve for higher energies of primary particles, and by increasing the number of secondary ions in the lower-energy range. This effect can be understood if it is assumed that some of these ions are produced as a result of secondary processes occurring in the volume of the target. To minimize this effect, one may want to lower th energy of the primary beam to around 1-2 keV, which is sufficien to reduce this effect to a negligible level, and to be able t produce mass spectra resulting mainly from the surface desor ption.

References

1. L. Kelner, S.P. Markey, J. Mass Spectrom. Ion Proc., 5 157 (1984).
2. L. Kelner, S.P. Markey, T.C. Patel, 33rd Conf. Mass Spec trom. Allied Topics, San-Diego (1985).
3. L.Kelner et al., Int. J. Mass Spectrom. Ion Proc., 62 2 (1984).

Analytical Application of a High Performance TOF-SIMS

A. Benninghoven, E. Niehuis, D. Greifendorf, D. van Leyen, and W. Lange

Physikalisches Institut der Universität Münster, Domagkstr. 75, D-4400 Münster, F. R. G.

1. Introduction

Until 1982 our SIMS investigations were carried out using quadrupol mass spectrometers. Due to the limited mass range and the low and mass dependent transmission of this type of instruments, applications were restricted to relatively small molecules like mononucleotides and nucleoside molecules /1/. This situation greatly improved when in 1982 our first time of flight instrument became working /2/. The high sensitivity and large mass range of this instrument allowed SIMS to be used for identification and structure analysis of large molecules with molecular weight in the range of > 5000. The new instrument was applied for extended investigations of polymers (e.g. polyethylenes, polystyrenes, nylons /3/) and biopolymers (e.g. oligonucleotides, peptides /4/, oligosachcharides).

Oligomeric desoxyribonucleotides (DNA) are of major importance in gentechnology, where protected oligonucleotides play a keyrole in DNA syntheses. Classical mass spectrometry of these molecules is almost impossible because of their thermal lability. This is different for secondary ion mass spectrometry, where sputtering results in the formation and emission of parent like and structural informative secondary ions.

In this paper some results from investigations of a series of unprotected, partially protected, and fully protected oligonucleotides are presented.

2. Experimental

2.1 Instrumentation

The TOF instrument is composed of a primary ion beam system, a Poschenrieder-type multiple (energy and angle) focusing time-of-flight analyzer, and a single ion counting registration system /2/. The primary ion source generates a pulsed and mass separated ion beam (typical: 10 ns, 12 keV, Ar^+) hitting the sample in an area of about 1 mm^2. Average ion current densities in the range of 100 pA/cm^2 can be obtained. A number of samples are mounted on a sample holder which is introduced into the mass spectrometer through a vacuum lock system allowing several target exchanges per hour. Secondary ions extracted from the sample surface are accelerated to about 3 keV before entering into a first order energy and angular focusing bended flight path. They are postaccelerated and hit the ion detection system with an energy of 20 keV, which guarantees a high counting efficiency even for large organic molecules.

2.2 Sample preparation

A micro syringe technique was used to deposit some µl of a 10^{-3} to 10^{-4} molar solution of the corresponding nucleotide on the surface of etched silver or gold foils. For the unprotected oligonucleotides water was used as solvent. The protected species were dissolved in chloroform. After evaporation of the solvent an area of several mm^2 was covered by roughly a monolayer of sample material, which is the optimum coverage range.

3. Results

From the group of fully protected nucleotides we investigated the dimers CT, TC, CC, GA, and TG, the trimer GAC, and the tetramer AGGA. As an example the upper mass range (> 800) of a negative SIMS spectrum of fully protected AGGA is shown in fig. 1. The main parent ions emitted are $(M-2H+Ag)^-$, $(M-TBE)^-$, and $(M-TBE+Ag)^-$. In addition high intensities of the 3'P-sequence ions, e.g. $(MMTrA_{bz}Pocp G_{ib}Pocp O)^-$ and $(MMTrA_{bz}Pocp G_{ib}Pocp O + Ag)^-$ are emitted. In the positive spectrum the parent ions $(M+Ag)^+$ and $(2M+Ag)^+$ are found. Protonated and deprotonated bases, secondary ions originating from protection groups, and phosphate fragment ions are emitted. The secondary ion emission behaviour of the dimers and trimers is very similar to that of the tetramer. In all cases $(M+H)^+$ and $(M-H)^-$ ions could not be detected.

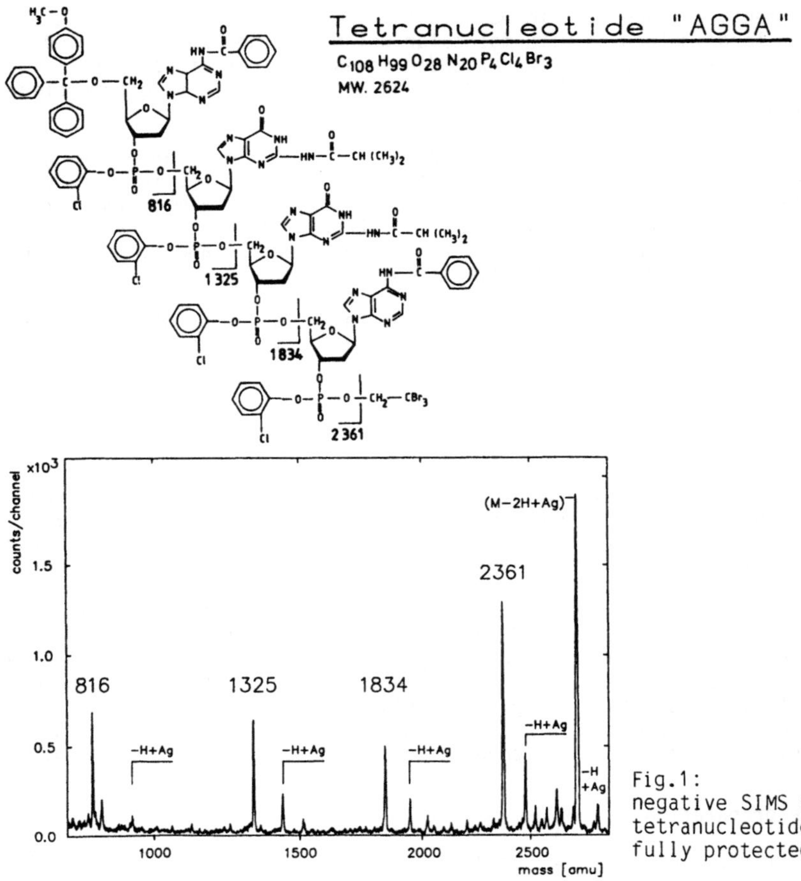

Fig.1:
negative SIMS spectrum of tetranucleotide AGGA, fully protected

The partially protected dimers AG, GA, CT, and TC, and the hexamer GTCGAC were investigated. In all these molecules the protection group TBE was missing. The dominating parent ion is $(M-H)^-$. In addition 3'P and 5'P sequence ions and their cationized analogues are detected allowing bidirectional sequence determination. In the positive spectra only weak parent ion intensities are observed. The fragment ion emission in the low mass range is similar to that observed for fully protected molecules.

Three of the unprotected nucleotides were investigated, i.e. GTAA, CTGCC, and TCACTT. Figure 2 presents the high mass range of a negative TCACTT spectrum. It is dominated by the parent ions $(M-H)^-$ and $(M+(-H+Ag)_n)^-$ clusters ($N<=8$). In the positive spectrum $(M+H)^+$ and $(M+Ag)^+$ are observed. Again, in the low mass range protonated and deprotonated bases are emitted. The main difference to the spectra of protected nucleotides is the absence of noticeable sequence ions.

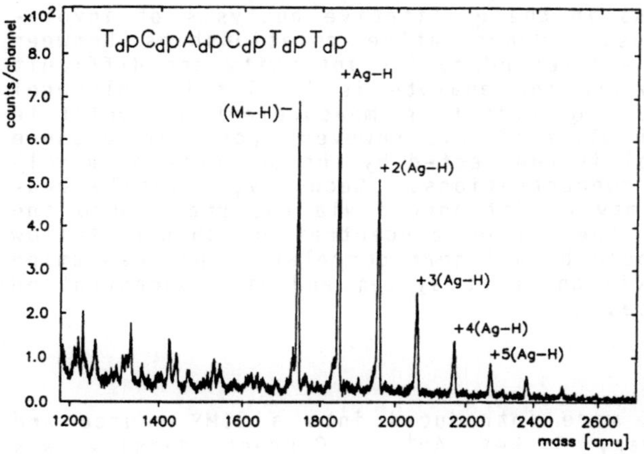

Fig.2:
negative SIMS spectrum of nucleotide TCACTT, unprotected

4. Summary

- protonated molecular ions $(M+H)^+$ are found for unprotected monomers only.
- cationized SI $(M+Ag)^+$, $(M+Na)^+$ are found for all except unprotected oligomers.
- deprotonated molecular ions $(M-H)^-$ are found for all except fully protected oligomers.
- deprotonated molecules with a number of H - Ag exchanges are found for all unprotected oligomers. In partially or fully protected molecules at most one hydrogen is exchanged.
- sequence information is contained in the spectra of partially or fully protected nucleotides but not in the spectra of the unprotected species.

1. A.Eicke, W.Sichtermann, A.Benninghoven, Org.Mass Spectrom. 15, 6 (1980)
2. P.Steffens, E.Niehuis, T.Friese, D.Greifendorf, A.Benninghoven, J.Vac.Sci.Technol. A3 (3), 1985
3. I.V.Bletsos, D.M.Hercules, D.van Leyen, A.Benninghoven, in: IFOS III, Springer Series in Chemical Physics, in press
4. W.Lange, D.Greifendorf, D.van Leyen, A.Benninghoven, in: IFOS III, Springer Series in Chemical Physics, in press

Organic Secondary Ion Intensity and Analyte Concentration

P.J. Todd and C.P. Leibman

Analytical Chemistry Division, Oak Ridge National Laboratory, Oak Ridge, TN 37831, USA

1. Introduction

Secondary ion emission from organic analyte-glycerol solutions is an important process in the qualitative analysis of involatile organic compounds. Quantitative relationships between analyte concentration and secondary ion intensity are difficult to evaluate because while the analyte is involatile, glycerol evaporates in vacuum. Quantitative measurement of volatile analytes dissolved in glycerol is, however, possible because evaporation of glycerol is unaffected by the presence of a volatile analyte at low concentrations. Secondly, volatile analytes such as amines may be introduced via gas phase onto the glycerol surface and the amine concentration should follow Henry's Law; there should be a linear correlation between amine pressure, controlled via an inlet system and its concentration on the glycerol surface.

2. Experimental

Neat glycerol samples were introduced into a SIMS source and irradiated with primary 5 keV Ar^+. Current density was varied from 1-10 $\mu A/cm^2$ as measured by a Faraday cup. Various amines were introduced into the ion source to establish partial pressures between 10^{-7}-10^{-8} Torr via an all-glass heated inlet system (200°C) which had been previously calibrated. Secondary ion mass spectra were obtained by using a double-focusing mass spectrometer, with the secondary ions accelerated 8.0 keV.

3. Results and Discussion

Upon initial introduction of the amine, the intensity of secondary protonated amine ions $[A+H]^+$ rises as shown in Fig. 1. The rate of this increase varies with the molecular weight of the amine. Secondly, the shape of the plot $[A+H]^+$ vs time is characteristic of diffusion of the amine into the matrix. The reference plots shown in Fig. 1 were calculated for diffusion of volatile compounds with varying diffusion coefficients into a liquid [1]. Diffusion coefficients estimated for the amines from these plots are similar to those obtained for nitroalkanes of similar chain length [2]. This indicates that a requirement for enhanced secondary emission is solvation. This result is consistent with other results which showed that volatile phosphonates do not yield characteristic secondary ions when adsorbed onto a glycerol surface.

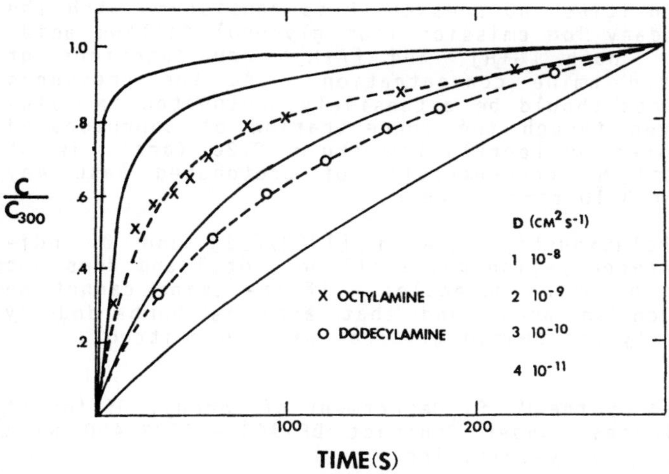

Fig. 1. Plot of analyte concentration in the penultimate surface layer with respect to the concentration at t = 300 s, when the analyte is introduced at t = o (solid line) and diffuses into the liquid with diffusion constants as indicated. Secondary amine $[A+H]^+$ intensity vs time since introduction (dashed line)

The steady-state behavior of secondary emission of these amines is demonstrated in Fig. 2, which shows a plot of the secondary octylamine $[A+H]^+$ intensity relative to the secondary glycerol $[A+H]^+$ intensity as a function of primary ion flux, ϕ, and octyamine pressure P. The relationship between

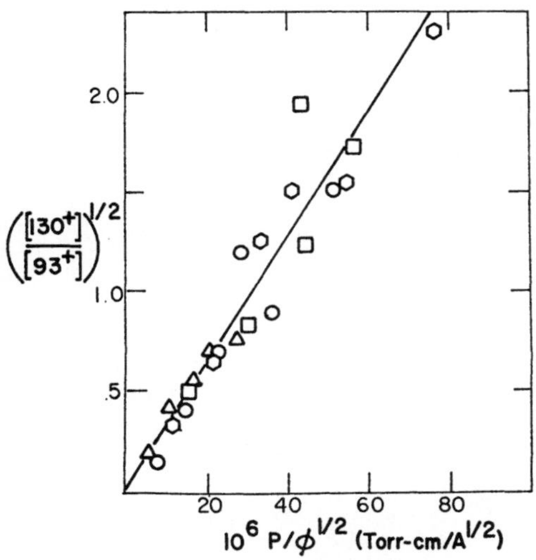

Fig. 2. Plot of $\sqrt{([130^+]/[93])}$ vs the ratio of octylamine pressure (Torr) and $\emptyset$. Primary current densities of ($\mu A/cm^2$) are as indicated: △, 1.3; 0, 2.4; □, 5.3; ⬡, 10

the ratio of intensities and pressure P is consistent with the notion that secondary ion emission from glycerol follows acid-base equilibria: Both $[G+H]^+$ and $[A+H]^+$ are functions of solvated protonated amine concentration. At the pressures employed, the amines should be extensively protonated in solution. Hence, even though the concentration of unprotonated amine, as determined by Henry's Law (k_H = 2.26 Torr), is of the order of 10^{-6} M, concentration of protonated base may be of the order of 3-10 times larger.

The inverse relationship between [130]/[93] and Ø indicates that a sputtered region may still be "hot," and does not permit adsorption of amine molecules. If the amine cannot be adsorbed from such an area, and that area is bombarded by subsequent Ar^+, only protonated glycerol will be emitted.

Research sponsored by the U. S. Department of Energy, Office of Basic Energy Sciences, under Contract DE-AC05-84OR21400 with Martin Marietta Energy Systems, Inc.

4. References

1. J. Crank: The Mathematics of Diffusion (Clarendon Press, Oxford, 1979).
2. J. E. Evanoff and W. E. Harris: Can. J. Chem. 56, 574 (1978).
3. G. S. Groenewold and P. J. Todd: Anal. Chem. 57, 886 (1985).

Characteristics of Ion Emission in Desorption Ionization Mass Spectrometry

M.M. Ross, J.E. Campana, R.J. Colton, and D.A. Kidwell

Naval Research Laboratory, Chemistry Division, Washington, DC 20375, USA

1. Introduction

Desorption ionization mass spectrometric methods, secondary ion mass spectrometry (SIMS) and fast-atom bombardment mass spectrometry (FABMS), are being used in numerous applications for the analysis of biomolecules [1,2]. Our work using derivatization/SIMS [3,4], in which target compounds are derivatized to make quaternary ammonium salts (preformed ions) to enhance the sensitivity of SIMS analysis, has led us to study some fundamental aspects of preformed ion emission. The objective of these studies is to gain a better understanding of the secondary ion emission process and how different factors affect the process in order to use SIMS more efficiently. This paper reports fundamental studies on the secondary ion emission of "model" tetraalkyl ammonium bromides and applications of derivatization/SIMS to the analysis of a selected drug compound.

2. Tetraalkyl Ammonium Salts

Figure 1 shows the secondary ion mass spectrum of an equimolar mixture of four tetraalkyl ammonium bromides. These salts were chosen to study the effect of analyte molecular size on secondary ion abundances. It is evident from Figure 1 that as the size of the cation increases, Et_4N^+ < Pr_4N^+ < Bu_4N^+ < Pe_4N^+, the ion abundance of that sputtered cation increases. As the size of the alkyl group increases, the energy required to separate the cation/anion pair decreases, because that energy is inversely proportional to the ionic radius. The lower energy requirement results in a greater secondary ion abundance.

The observed increase in secondary ion abundance with an increase in cation size may be related to an increase in sputtering rate, as well. We investigated this possibility by monitoring the secondary ion abundance of each tetraalkyl ammonium cation versus time at a constant primary beam current density. These secondary ion abundance decay curves are shown in

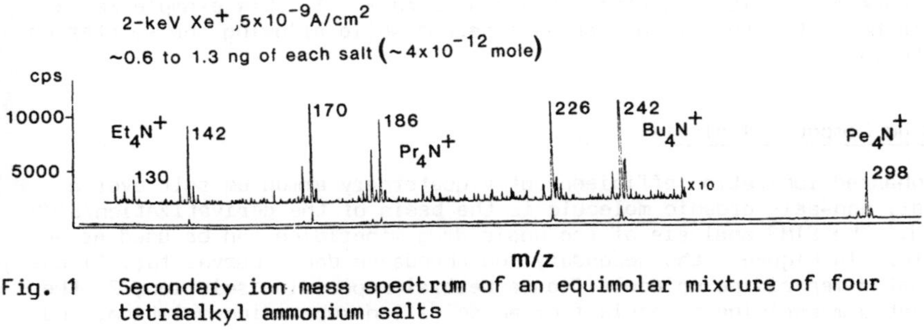

Fig. 1 Secondary ion mass spectrum of an equimolar mixture of four tetraalkyl ammonium salts

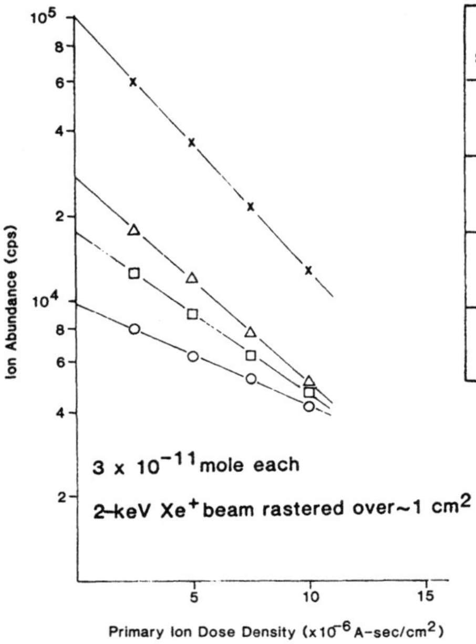

Graph Symbol	R_4N^+	Decay Half-Life		Ioniz. Effic.[2]
		Seconds[1]	A-sec/ cm^2	
○	Et_4N	1650	8.2 (x10-6)	~3%
□	Pr_4N	1100	5.5	~6%
△	Bu_4N	850	4.3	~9%
x	Pe_4N	650	3.2	~15%

[1] (using 5×10^{-9} A/cm^2)

[2] ± 15%

Fig. 2 Secondary ion abundance decay curves for the four tetraalkyl ammonium salts

Figure 2. The slope of a decay curve is a measure of the sputtering rate of the particular species or, the effect of the primary beam on the analyte molecules. For the tetraalkyl ammonium salts, as the size of the cation increases, the sputtering rate increases as well. The area under each curve is a measure of the total number of ions observed, and the ionization efficiency (number of secondary ions detected/number of sample molecules analyzed) can be calculated using this area. The ionization efficiency increases with increasing cation size.

In order to separate the effect of the cation/anion distance from the size of the molecule, we sputtered an equilmolar mixture of tetraethylammonium bromide and triethylcetylammonium bromide ($(C_2H_5)_3(C_{14}H_{29})NBr$). The slopes of the secondary ion abundance decay curves for the cations from the mixture were significantly different, but the ionization efficiencies of the the species were nearly identical. The ionization efficiency is governed by the cation/anion distance, which is approximately the same for both salts. However, the triethylcetylammonium cation is physically much larger than the tetraethylammonium cation, and this difference causes it to sputter at a faster rate. In this example we were able to hold the ionization process constant while allowing the sputtering rate to vary.

3. Drug Compound Analysis

The enhanced ionization efficiency of a quaternary ammonium salt over a neutral, non-salt organic molecule is the basis of the derivatization/SIMS method. The SIMS analysis of the basic drug meperidine can be used as an example. In Figure 3 the secondary ion abundance decay curves for: 1) the protonated meperidine molecule from a neutral meperidine solution, 2) the protonated meperidine molecule from an acidified meperidine solution, and

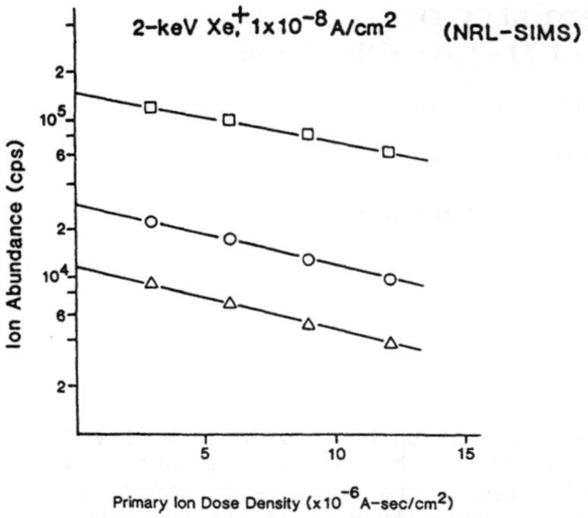

Graph Symbol	Sample	Ion	Ioniz. Effic.
△	1 ng Meperidine/Ag	$[M+H]^+$	~1%
○	1 ng Meperidine + HOAc/Ag	$[M+H]^+$	~4%
□	1 ng Meperidine + CH_3I/Ag	$[M+CH_3]^+$	~15%

Fig. 3 Secondary ion abundance decay curves for meperidine and derivatized meperidine

3) the methylated meperidine molecule from a mixture of meperidine and methyl iodide, are compared. Addition of acid to the meperidine solution prior to analysis results in an enhancement of the $[M+H]^+$ ion abundance. Methylation of meperidine results in a further secondary ion abundance enhancement; $[M+CH_3]^+ > [M+H]^+$. The derivatization produced a greater than 10-fold increase in the ionization efficiency of the analyte species. It is interesting to note that the slopes of the three curves are nearly identical, because the size of the analyte species is not altered by derivatization.

4. Summary

Fundamental studies of the characteristics of the emission of preformed ions by desorption ionization indicate that the sputtering rate is determined by the cation size, and the ionization efficiency is determined by the cation/anion intercharge distance. These results have consequences for the analytical use of SIMS. Neutral organic molecules can be derivatized to form quaternary ammonium salts. This derivatization results in a significant enhancement in the ionization efficiency of the analyte and, therefore, in the sensitivity of organic SIMS as well.

5. References

1. K.L. Rinehart, Jr., Science 218 254 (1982).
2. K.L. Busch and R.G. Cooks, Science 218 247 (1982).
3. M.M. Ross, D.A. Kidwell, and R.J. Colton, Int. J. Mass Spectrom Ion Proc. 63, 141 (1985).
4. D.A. Kidwell, M.M. Ross, and R.J. Colton, Biomed. Mass Spectrom. 12, 254 (1985).

Some Aspects of the Chemistry of Ionic Organo-Alkali Metal Halide Clusters Formed by Desorption Ionization

L.M. Mallis, M.E. Castro, and D.H. Russell

Department of Chemistry, Texas A & M University,
College Station, TX 77843, USA

1. Introduction

Although fast-atom bombardment (FAB) ionization has proven to be a versatile and sensitive ionization method for many biological compounds, the FAB-MS spectra are frequently complicated by the presence of impurities and/or adduct ions of the sample and liquid matrix. For these reasons several workers have proposed the use of tandem mass spectrometry (TMS) in combination with FAB ionization for structural characterization of large molecules [1,2]. During the development of mass spectrometry, several methods for enhancing specific dissociation reactions have been proposed [3,4,5]. In this paper a potentially useful and simple method is described for enhancing the structural information obtained by FAB-TMS. The method relies upon the comparison of the FAB-TMS spectra of $[M+H]^+$ ions and organo-alkali metal ions of the form $[M+A]^+$ or $[M+xA-(x-1)H]^+$, where A is the alkali metal and x = 1-3.

Owing to the fact that molecules such as peptides contain highly polar functional groups, and these functional groups have different H^+ and A^+ ion affinities, it follows that the binding sites of H^+ and A^+ may differ. Thus, the site of interaction of the H^+ or A^+ ion with the molecule, as well as the nature of the ionic complex, may give rise to significant differences in the FAB-TMS spectrum. Although such changes in SIMS spectra have been reported [6], the data reported in this paper for the FAB-TMS of the hippuryl-L-histidyl-L-leucine (HHL) $[M+H]^+$ and $[M+A]^+$ ions is a striking example of this behavior.

2. Experimental

The studies reported here were performed with a Kratos MS 50 triple analyzer (TA) and fast-atom bombardment (FAB) ionization. Xenon was used for the bombarding atom beam; typical operating conditions were beam energies of 6-8 keV and neutral beam currents equivalent to 20-30 µA. Collision-induced-dissociation (CID) studies were performed using mass-analyzed ion kinetic energy (MIKE) scans. Helium was used as target gas on an incident ion potential of 8 kV. All CID spectra were recorded with a collision gas pressure corresponding to a 50% attenuation of the molecular ion beam. A signal-to-noise ratio of at least 5:1 was obtained by signal averaging 16 scans (20 seconds per scan), using a Nicolet Inst. Corp. 1170 signal averager. Spectra were plotted on a standard X-Y recorder. A solution of HHL was prepared by dissolving ca. 1.5 mg of sample in 500 µL of HPLC grade methanol. Typically 2 µL of this solution was placed on a gold plated copper probe tip and air dried. To this sample ca. 2 µL of a 1:1 mixture of thioglycerol/glycerol (Liquid Matrix) and ca. 1 µg of the alkali metal chloride salt were admixed on the probe tip.

3. Results and Discussion

In the CID spectrum of the $[M+H]^+$ of HHL (Figure 1A), two dominant ions are observed at m/z 105 and m/z 110. The m/z 105 ion is indicative of a benzoyl group (e.g., $C_7H_5O^+$) and the m/z 110 ion is characteristic of peptides containing histidine residues (e.g., $C_5H_8N_3^+$) [1]. These two ions correspond to fragments *a* and *b*, respectively (Figure 2A). Two additional structurally significant fragment ions are also observed at m/z 269 and m/z 299. The m/z 269 is assigned to loss of the benzoyl glycine (hippuryl) portion of the molecule, while the m/z 299 ion is formed by loss of the leucine moiety, e.g., cleavage of the LEU-HIS peptide bond. These two ions correspond to fragments *b(1)* and *d*, respectively (Figure 2A and 2B).

The CID spectrum of the $[M+Na]^+$ ion of HHL is markedly different from the $[M+H]^+$ ion. The most important ions, in terms of structural significance, are observed at m/z 363, 338, 321, and 294. These ions

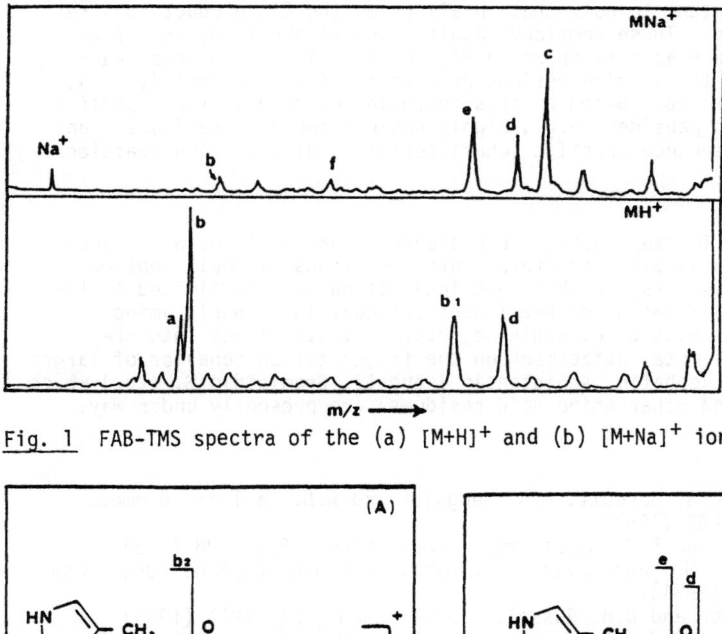

Fig. 1 FAB-TMS spectra of the (a) $[M+H]^+$ and (b) $[M+Na]^+$ ions of HHL

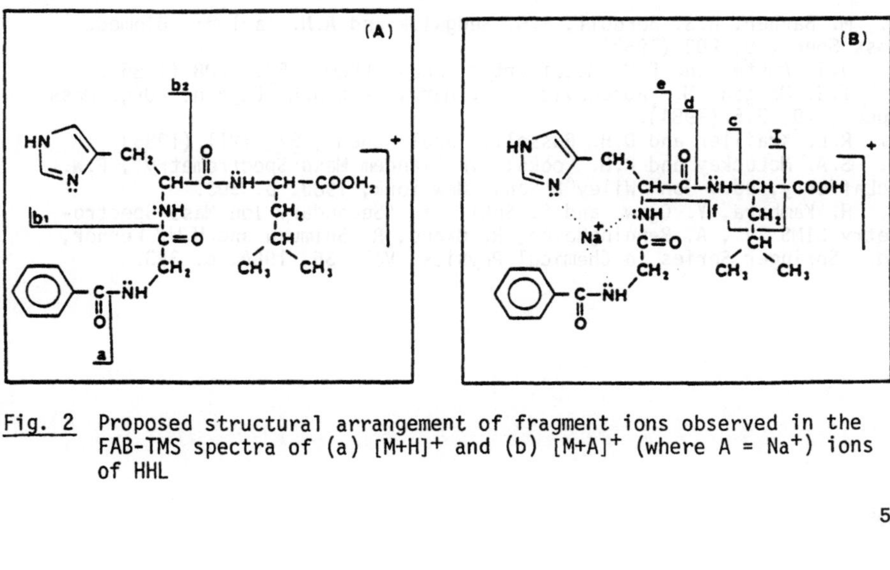

Fig. 2 Proposed structural arrangement of fragment ions observed in the FAB-TMS spectra of (a) $[M+H]^+$ and (b) $[M+A]^+$ (where A = Na^+) ions of HHL

account for ca. 70% of the total CID product ions. Conversely, the two most structurally significant CID product ions (i.e., loss of hippuryl and leucine residues) in the FAB-TMS spectrum of the $[M+H]^+$ ion accounts for less than 25% of the total product ions. The m/z 363 ion in the FAB-TMS spectrum of the $[M+Na]^+$ ion is assigned to the simultaneous loss of $CH(CH_3)_2$ and CO_2H from the leucine portion of the molecule (fragment *I*, Figure 2B). The m/z 338 ion (fragment *c*, Figure 2B) is explained as a hydrogen transfer from the leucine residue followed by cleavage of the NH-CH bond of leucine. The two remaining ions at m/z 321 and m/z 294 correspond to loss of the leucine residue and loss of leucine plus the adjacent (histidyl) carbonyl group (fragments *d* and *e*, respectively, Figure 2B).

Attention is also drawn to the m/z 132 ion, which is assigned as the histidyl moiety plus the Na^+ ion (i.e., fragment *b*, Figure 2A, plus Na^+). Owing to the highly basic nature of the imidizole nitrogen, it is not surprising that the Na^+ ion is preferentially attached to this position, viz., the Na^+ ion is not bound to the C-terminus of the HHL molecule. It is also important to note that in all cases the CID product ions retain the Na^+ ion. These combined results suggest the formation of an ionic complex such as that shown in Figure 2B. Direct cleavage reactions involving the leucine residue produce the fragment ions *I*, *c*, *d*, *e* shown in Figure 2B. Based on this reasoning we suggest that specific binding sites in peptides, e.g., highly basic amino acid residues, can be employed to enhance specific (characteristic) dissociation reactions.

4. Conclusion

One finds that, in many cases, the presence of an alkali metal attachment can lead to simpler structural interpretations of small peptide molecules. It is possible that this interaction will be limited to the structural identification of small peptides containing basic amino residues such as histidine, arginine, etc. Investigations into the effect of alkali metal attachment on the fragmentation behavior of larger peptides (leu-enkephalin, Angiotensin I and II, Gramicidin S, etc.) which contain basic and other amino acid residues, are presently under way.

5. References

1. M. Barber, R.S. Bordoli, R.D. Sedgwick and A.N. Taylor: Biomed. Mass Spec., 9, 208 (1982).
2. J.I. Amster and F.W. McLafferty: Anal. Chem., 57, 1208 (1985).
3. T.G. Morgan, M. Rabrenovic, F.M. Harris and J.H. Beynon: Org. Mass Spec., 19, 315 (1984).
4. R.E. Krailler and D.H. Russell: Anal. Chem., 57, 1211 (1985).
5. S.A. McLuckey and R.G. Cooks: in "Tandem Mass Spectrometry", F.W. McLafferty, ed., John Wiley & Sons, New York, 1983, p. 303.
6. H. Kambara, Y. Ogawa and S. Seki: in "Secondary Ion Mass Spectrometry SIMS IV", A. Benninghoven, R. Okano, R. Shimuzu and H.W. Werner, ed., Springer Series in Chemical Physics, Vol. 36, 1984, p. 383.

Detection of Biomolecules by Derivatization/SIMS

D.A. Kidwell, M.M. Ross, and R.J. Colton

Chemistry Division, Naval Research Laboratory,
Washington, DC 20375, USA

The sensitivity and specificity of static SIMS can be increased dramatically by using chemical derivatization to produce "preformed" ions. This sensitivity can extend to the parts-per-billion (ppb) range for selected components in complex mixtures such as body fluids. We have applied derivatization/SIMS to several classes of compounds, including peptides [1,2], saccharides [3], ketones and ketosteroids [4] and amine-containing drugs [5].

Figure 1a shows a SIMS spectrum of four underivatized drugs. Methamphetamine is not observed in the underivatized spectrum. Derivatizing the mixture with methyl iodide in a basic medium converts all the amines to quaternary ammonium salts. Figure 1b shows a SIMS spectrum of this derivatized mixture. The ion abundance of the molecular ions are enhanced by at least a factor of ten over the protonated species. Also, methamphetamine is now observed.

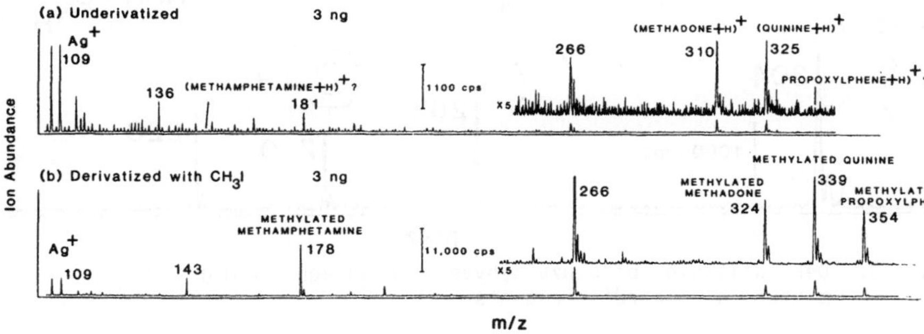

Figure 1. Human urine spiked with four drug standards

The increase in ion abundance may be due to one of two factors: increased secondary ion yield or increased sputtering yield. Figure 2 compares the sputter rates for the drug meperidine as its protonated and derivatized forms. Since the slopes of the lines are identical, the sputtering rates for these species are identical. The areas under the curves are proportional to the ion yields and show that ion production from the derivatized meperidine is greatly increased. These two observations indicate that increased ionization efficiency of the derivatized species is responsible for the observed increase in the ion abundance.

Figure 3 shows an equal molar mixture of quaternary ammonium salts. These salts were produced by derivatizing excess benzaldehyde with different Girard's reagents [3]. As the size of the cation increases, the

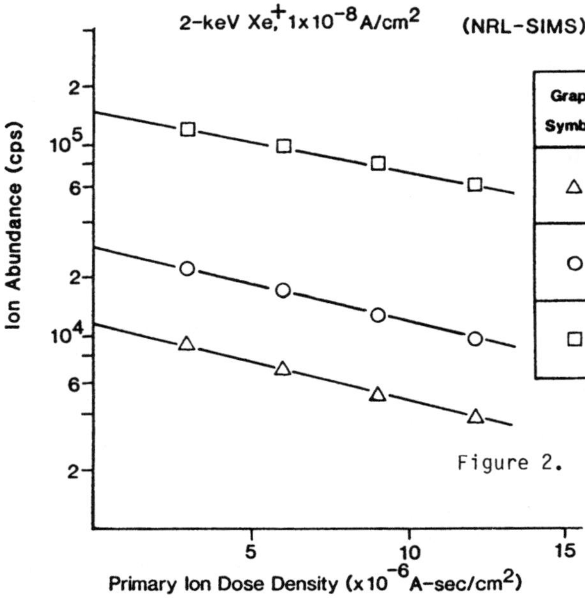

Graph Symbol	Sample	Ion	Ioniz. Effic.
△	1 ng Meperidine/Ag	$[M+H]^+$	~1%
○	1 ng Meperidine + HOAc/Ag	$[M+H]^+$	~4%
□	1 ng Meperidine + CH_3I/Ag	$[M+CH_3]^+$	~15%

Figure 2. Secondary Ion Abundance Decay Curves for Meperidine

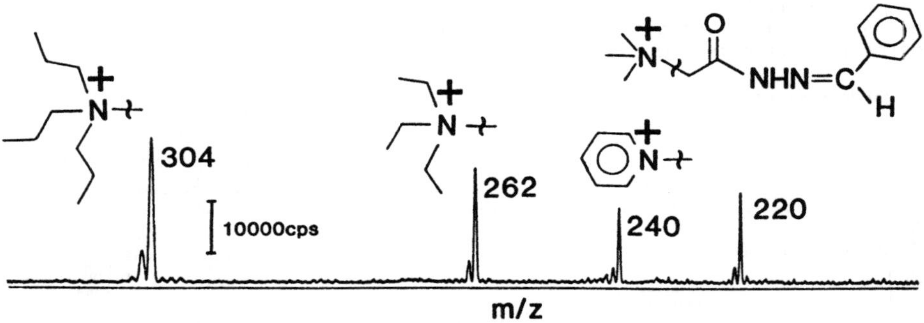

Figure 3. Derivatization of benzaldehyde with an equal molar mixture of Girard's reagents

ion yield also increases because less energy is necessary to separate the cation from the anion the larger the distance between them.

Similar size quaternary ammonium salts have similar ion yields. Mixtures of ketosteroids and aliphatic ketones derivatized with the same Girard's reagent give relative ion abundances proportional to their relative concentrations. This facilitates the quantitation of mixtures of materials.

To achieve maximium sensitivity one should employ the largest quaternary ammonium salt possible. By examining simple alkyl ammonium salts, we have found that the increase in ion abundance levels-off after the tetrapentyl ammonium salt. However, other factors must be considered in selecting a derivative. For example, the synthetic yield in forming larger quaternary ammonium salts decreases with longer alkyl chains.

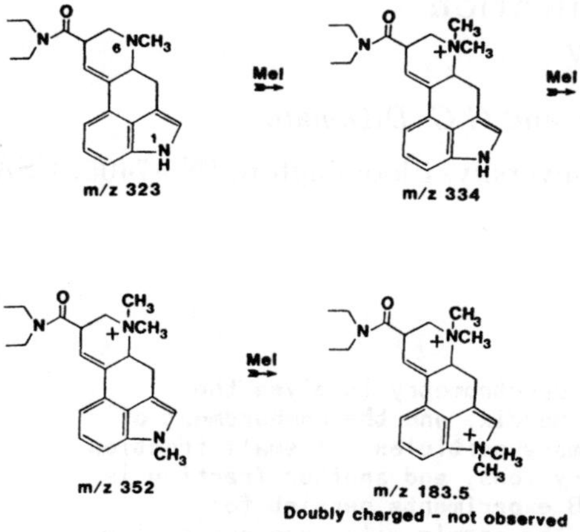

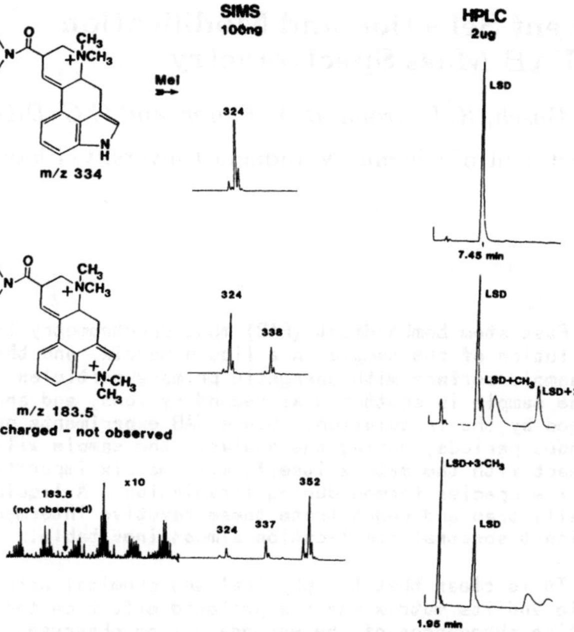

Figure 4. Derivatization of LSD

Figure 5. SIMS and HPLC analysis of methylated LSD derivatives

The detection and quantitation of other drugs such as LSD in human body fluids is a non-trivial problem due to the low dosage needed to produce a hallucinogenic response and LSD's extensive and unknown metabolism. Since LSD contains two potential sites for derivatization (Figure 4), most alkylating agents produce doubly-charged molecules which are only weakly observed by SIMS. The reaction with methyl iodide was followed by HPLC and SIMS (Figure 5). A decrease in signal was observed with time as the doubly-charged species formed. Successful detection of LSD was accomplished using protonation with weak acids.

In summary, quaternary ammonium salts make excellent SIMS labels, with higher sensitivity compared to protonated and non-ionic compounds. The larger size quaternary salts have higher sensitivity. As in the example of LSD, indiscriminant use of derivatization can greatly decrease sensitivity, if doubly charged derivatives are formed.

References

1. D.A. Kidwell, M.M. Ross, R.J. Colton; J. Amer. Chem. Soc., 106 2219(1984).
2. D.A. Kidwell, M.M. Ross and R.J. Colton; in Springer Ser. Chem. Phys. 36 412(1984).
3. D.A. Kidwell, M.M. Ross and R.J. Colton; 33rd Annual Conference on Mass Spectrometry and Allied Topics, San Diego, CA, May 1984.
4. M.M. Ross, D.A. Kidwell, R.J. Colton; Int. J. Mass Spectrom. Ion Proc., 63 141(1985).
5. D.A. Kidwell, M.M. Ross and R.J. Colton; Biomed. Mass Spectrom., 12 254(1985).

Solvent Selection and Modification for FAB Mass Spectrometry

K.L. Busch, K.J. Kroha, R.A. Flurer, and G.C. DiDonato

Department of Chemistry, Indiana University, Bloomington, IN 47405, USA

Fast atom bombardment (FAB) mass spectrometry involves the dissolution of the sample in a liquid matrix, and the bombardment of the sample surface with energetic primary particles. A small fraction of the sample is sputtered as secondary ions, and another fraction is damaged by the irradiation. Since FAB experiments persist for extended periods, during the analysis the sample will have ample time to react with the matrix itself, with matrix impurities, and with reactive species formed during irradiation. A liquid matrix can actually trap and concentrate these reactive intermediates, making the resultant spectral complication almost inevitable.

It is clear that the physical and chemical nature of both the sample and its matrix exert a profound effect on the nature and the relative abundances of the secondary ions observed. Proper matrix selection increases the sensitivity of the analysis by decreasing spectral and chemical interferences. Conversion of the sample into ionic forms which congregate at the surface of the liquid increases sensitivity. Functional group specific reactions have been designed to accomplish this derivatization [1]. Finally, the segregation of cations and anions within the matrix decreases losses from ion neutralization. This paper describes work on the first two aspects of matrix selection and modification. Our work constructs a quantitative basis on which the solvent can be selected for ability to enhance or suppress any of the various processes of secondary ion formation [2]. A summary of derivatization reactions which create preformed ions from the sample is also described [3,4]. Since ion addition reactions can be chemically selective, they provide a tool for the targeted analysis of functional groups in complex mixtures by both FAB mass spectrometry and FAB mass spectrometry/mass spectrometry (FAB MS/MS).

The ideal FAB solvent is chemically inert, spectroscopically transparent, and able to dissolve a wide variety of compounds. Without surprise, most of the commonly used FAB solvents fail to fully meet all of these criteria. Glycerol is an excellent general solvent, but produces a significant spectral background containing a series of ions at m/z 93, 185, 277,..., and also a low-level signal at almost every integer mass. Glycerol, or other solvents, may react with the sample in the irradiated sample/solvent mixture. The acidity of glycerol and thioglycerol, while often promoting the formation of the protonated molecular ions used for molecular weight determination, interferes with the FAB spectra of inorganic compounds and of charge transfer complexes [3]. Compilations of solvents, each found to be well-suited for the analysis of a particular class of compounds, are empirical rather than quantitative. Sensitivity for a give compound in the various solvents varies widely. For standard compounds, this variation can be used to generate a quantitative description of FAB solvents. A series of compounds have been investigated for use as standards. Each compound is chosen to represent one of the parameters which control the abundances of secondary ions in FAB: 1) direct sputtering of preformed ions, 2) surfactancy, 3) acid/base properties, 4) charge-transfer complexation, 5) charge solvation, 6) cluster formation, and 7) intermolecular reactivity. The standards initially chosen are, respectively, tetraphenylphosphonium bromide, cetyltrimethylammonium bromide, methyl red, the charge-transfer complex between hexanitrosobenzene and naphthalene, heimcholinium-3-hydrate, cesium perfluorobutanesulfonate, and carnitine hydrochloride. Solvents investigated are glycerol, diethanolamine, Fomblin (a vacuum-compatible perfluorinated ether), and 1,2,4-butanetriol. Under optimum conditions, reproducibility in determination of the sensitivity factor for this solvent series is ±10% on a day-to-day and sample-to-sample basis. The exact protocol of sample introduction into the source of the mass spectrometer accounts for part of this variation.

For glycerol and diethanolamine, secondary ions are formed most efficiently from surfactants. Sensitivity for the surfactant standard was a factor of ten higher than that for the direct sputtering of a preformed ion. This conclusion, demonstrated quantitatively for the first time, underscores the importance of derivatization reactions which not only create precursor ions in the matrix, but also ions which possess surfactant properties in the FAB matrix used. The direct sputtering of preformed ions has been found to be a favored process, but again exhibits a solvent dependence. The intact cation from the standard tetraphenylphosphonium bromide is produced about 20 times as readily from glycerol as from diethanolamine. Diethanolamine seems to better stabilize the creation of doubly-charged ions in the FAB mass spectrum. Neither glycerol nor diethanolamine could be used to form the expected radical cation from the charge-transfer complex, but this ion could be observed in the absence of any matrix at all. The detailed nature of the spectra has been discussed elsewhere [2]. This work does not imply that all of the standards represent independent processes of secondary ion formation. Those which accurately describe the performance of solvents in a range of FAB analyses can only be established with further study.

Derivatization by ion addition [1] is a straightforward means of increasing both the sensitivity and selectivity of a FAB analysis. We have described a variety of reactions specific for various functional groups (Table 1). Most useful have been the reactions designed to form ionic derivatives from hydroxyl and amino groups, which are common functional groups in biomolecules. The reagent 2,4,6-trimethylpyrylium tetrafluoroborae reacts with primary amino groups to form N-pyridinium derivatives, which then produce abundant ions corresponding to the intact cation in the positive ion FAB spectrum. This reaction has been used to derivatize lysine, amino sugars, and even enzymes containing multiple lysine residues. The reagent 2-fluoro-N-methylpyridinium p-toulenesulfonate reacts with hydroxyl groups to form N-pyridinium derivatives bound through an ether linkage to R of the original sample ROH. This reaction has been used to derivatize steroids, as an alternative to the use of Girard's reagents for the same purpose [4]. Figure 1 is the positive ion FAB spectrum of corticosterone so derivatized. The ions at m/z 150 and below correspond to the solvent used, as does the ion at m/z 299. The intact cation at m/z 438 is observed with a high relative abundance in a region of low chemical noise from the matrix.

Table 1: Summary of ion addition derivatization reactions developed for functional group analysis

Compound Type	Reagent or Product
Primary amines	Pyrylium salts
Sugars	Dopamine hydrochloride Aminobenzoic ethyl ester
Aldehydes and ketones	Pyrrolodinium perchlorate Bisulfite
Secondary amines	Aryl-1,2-dithiolium salts
Steroids	Gerard's Reagent's P and T
Alcohols	Fluoromethylpyridinium salts

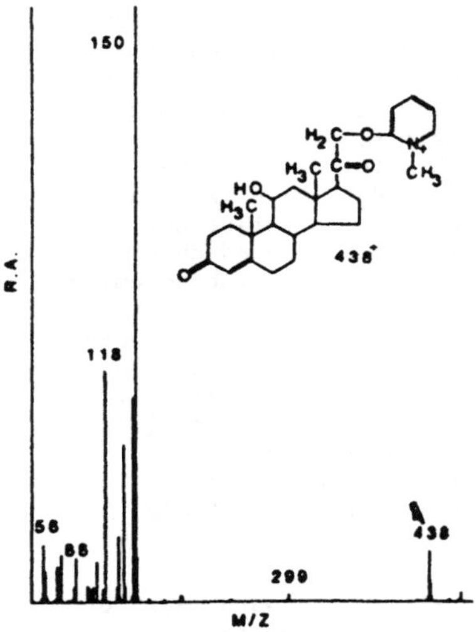

Fig. 1. Corticosterone can be derivatized to an N-pyridinium salt which appears as an abundant intact cation in the positive ion FAB spectrum.

Acknowledgments

Acknowledgment is made to the Donors of the Petroleum Research Fund, administered by the American Chemical Society, for partial support of this research. The mass spectrometer was purchased with support from the National Science Foundation (NSF CHE-81-11957).

References

1. K.L. Busch, S.E. Unger, A. Vincze, R.G. Cooks, and T. Keough, J. Am. Chem. Soc., 104, 1507 (1982)..
2. K.L. Busch and K.J. Kroha: 33rd Annual Conference on Mass Spectrometry and Allied Topics, San Diego, CA, May 1985.
3. G.C. DiDonato and K.L. Busch: Anal. Chim. Acta, 171, 233 (1985).
4. K.L. Busch, G.C. DiDonato, K.J. Kroha, and L.R. Hittle, 1985 Pittsburgh Conference, February 1985, page 371.

The Use of Negative Ion Fast Atom Bombardment Mass Spectrometry for the Detection of Esterfield Fatty Acids in Biomolecules

W.M. Bone[1] *and C. Seid*[2]

[1] Div. of Biochemistry and [2] Div. of Communicable Diseases and Immunology, Walter Reed Army Institute of Research, Washington, DC 20307, USA

1. Introduction

Endotoxins are lipopolysaccharides (LPS) which occur on the outer cell membrane of gram-negative bacteria such as E. coli, P. aeruginosa, N. gonorrhea, N. meningococci and S. minnesota. These endotoxins are responsible to a great extent for the pathogenicity exhibited when a host is infected with any of these bacteria. The endotoxins are composed of three main parts one of which is the Lipid A region. The Lipid A consists of a diglucosamine, one or more phosphate units attached to the sugar along with fatty acids. The fatty acids are necessary for anchoring the entire LPS to the bacterial cell membrane and are essential for the LPS to exhibit toxicity. Both the Lipid A and the entire LPS are being investigated by a number of laboratories for the development of vaccines against the organisms that produce these endotoxins [1].

Other mass spectral results have focussed on the high-mass region (>1000 amu) to obtain structural information regarding endotoxins [2]. While much useful information is available from high-mass studies, few laboratories have the instrumentation necessary for this type of analysis. Structural information can be obtained from the low-mass region (below mass 400) using FAB conditions and this type of data should be available to more investigators on a routine basis with virtually any type of mass spectrometer. Examination of the low-mass region clearly shows that under NI-FAB conditions free acids are readily cleaved from both the Lipid A and LPS. Comparison with fatty acid standards and previously obtained GC results provides confirmation that the ester-linked fatty acids are indeed the ones detected.

2. Materials and Methods

The LPS products used in this study had previously been prepared as described [3]. They consisted of both the LPS and Lipid A obtained from various strains of E. coli, N. gonnorrhea, S. minnesota and P. aeruginosa. Analysis of the fatty acids, both ester- and amide-linked, had been accomplished and the data were available for comparison with the NI-FAB/MS results.

Initial mass spectrometric results using only glycerol as the FAB matrix and sample solvent were unsuccessful in that no useful ions were produced at high or low mass. The insolubility of both the LPS and Lipid A samples in the glycerol was apparently the problem. Through experimentation we determined that both LPS and Lipid A could be dissolved using 10 uL each of water, methanol, acetonitrile with 3 uL di-iso-propylamine, followed by sonication and/or vortex mixing. Normally 3-5ul of the resulting solution/suspension was applied to 1-2 uL of glycerol on the FAB probe tip. Mass spectra were then recorded using a Kratos MS-50 and DS-55 data system. A 4-6 keV "neutral" xenon primary beam was used to sputter all samples [4].

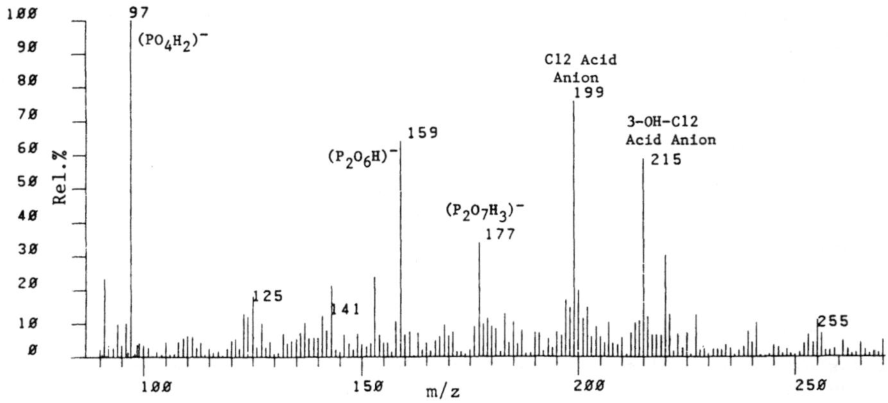

Fig. 1. NI-FAB Mass Spectrum of a N. gonorrhea lipid A sample.

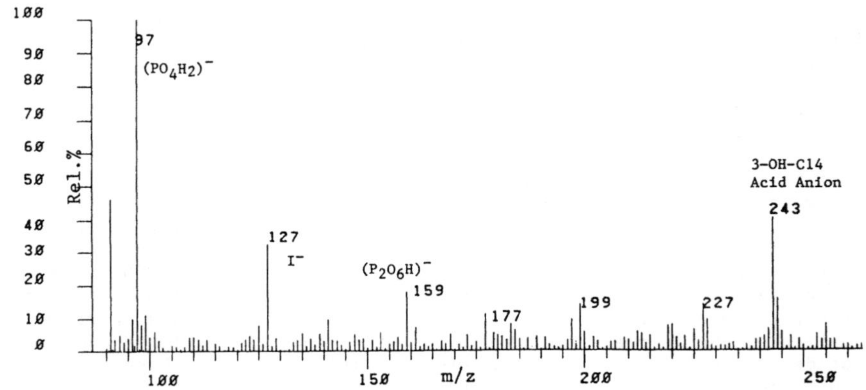

Fig. 2 NI-FAB Mass Spectrum of a S. minnesota LPS sample.

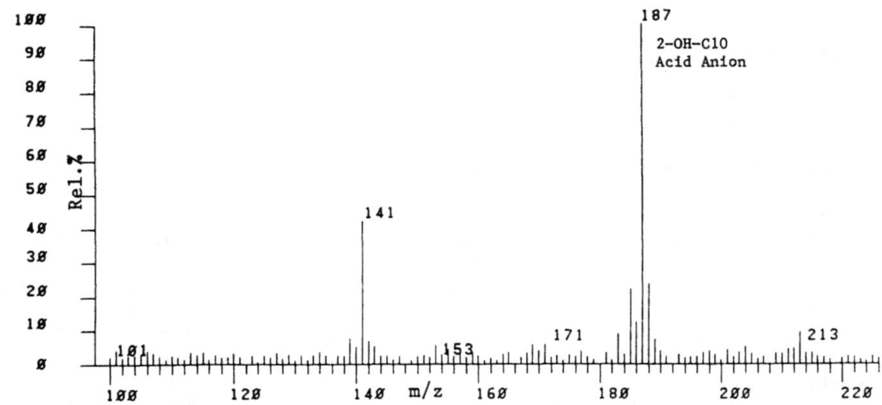

Fig. 3 NI-FAB Mass Spectrum of 2-Hydroxy-decanoic acid methyl ester.

3. RESULTS AND DISCUSSION

Shown in Figs. 1 & 2 are typical mass spectra of a N. gonorrhea lipid A and a salmonella LPS, respectively. The base peak in both cases at m/z 97 corresponds to phosphate anion and m/z 177 a diphosphate anion, with a fragment derived from loss of water m/z 159. The other major peaks in these spectra are due to saturated as well as hydroxy-fatty acid anions. The fatty acids anions occur at masses 199, 227 and 255 indicative of C12 C14 and C16 chains respectively. Beta-hydroxy fatty acids are also present at masses 215 (C12) and 243 (C14) in at least one case or the other. Alpha-hydroxy acids are known to occur in biological systems, however these have been assigned as beta-hydroxy acids based on GC results and the fragment ions which occur in the mass spectra.

A typical hydroxy-fatty acid (alpha-C10) methyl-ester NI-FAB mass spectrum is shown in Fig. 3. The ester-linkage is readily cleaved under FAB conditions to give the acid anion as the base peak. A collection of fatty acids and their methyl-ester derivatives all produced the free acid anion as the base peak as did the corresponding alpha- and beta-hydroxy-fatty acids.

4. CONCLUSIONS

Under the mass spectral conditions employed, both the LPS and the Lipid A produced readily detectable mass fragment ions characteristic of both the saturated and the hydroxy-fatty acids ester-linked to the Lipid A portion of various bacterial endotoxins. The method appears applicable to other biological matrices as well, and offers the convenience of rapid analysis and elimination of sample workup procedures.

5. REFERENCES

1. R. Germanier, Ed. : Bacterial Vaccines (Academic Press, Inc., New York 1984) and references therein.

2. E.Th. Rietschel, Ed. : Chemistry of Endotoxin, Vol.1 (Elsevier, New York 1984) and references therein.

3. R.C. Seid, Jr. et. al. : Anal. Biochem. 124 , 320 (1982) ; J. Biol. Chem. 259 , 9028 (1984) ; J. Biol. Chem. 256 , 7305 (1981).

4. W.V. Lignon: Int. J. Mass Spectrom. and Ion Phys. 41 , 205 (1982).

Fast-Atom Bombardment Mass Spectra of O-Isopropyl Oligodeoxyribonucleotide Triesters

L.R. Phillips[1], *K.A. Gallo*[1], *G. Zon*[1], *W.J. Stec*[2], *and B. Uznanski*[2]

[1] Division of Biochemistry and Biophysics, Office of Biologics Research and Review, Food and Drug Administration, 8800 Rockville Pike, Bethesda, MD 20892, USA

[2] Polish Academy of Sciences, Centre of Molecular and Macromolecular Studies Department of Bioorganic Chemistry, 90-362 Lodz, Poland

1. Introduction

Oligodeoxyribonucleotides with alkylated internucleotide phosphate moieties are not only resistant to degradation by nucleases, but are also incorporated by cells, thus targeting this class of compounds for potential use as antiviral drugs and drug-carriers [1]. Subsequent to synthesis, these compounds must first be analyzed to verify the correctness of chemical structure. Fast-atom bombardment (FAB) mass spectrometry [2] has been used to examine some unprotected [3-7] and protected [8-10] oligonucleotides. Herein is reported the application of FAB mass spectrometry to the determination of several features characteristic of the primary structure of synthetic O-isopropyl oligodeoxyribonucleotide triesters, which are now accessible by means of recently reported methodology [11].

2. Experimental

The solid-phase synthesis of the O-isopropyl oligodeoxyribonucleotide triesters was accomplished [11] using base-protected, 5'-dimethoxytrityl nucleosides which have an O-isopropyl phosphomorpholodite functionality attached to the 3-oxygen atom. Positive and negative fast-atom bombardment mass spectra were recorded with a Kratos MS-50 mass spectrometer equipped with a Kratos DS-55 data system (Data General Nova 4X minicomputer, employing the RDOS operating system). Samples were first dissolved in a mixture of acetonitrile and water (30:70) to give a final concentration of 5 nanomoles of sample per uL of solution. Glycerol (0.3 uL) was deposited on a copper sample probe-tip, and to this was added an aliquot (1 uL, 5 nanomoles) of the previously prepared sample solution. Bombardment was achieved with a 25-uA beam of 7.5 keV fast xenon atoms generated in a Saddle Field neutral-beam gun (Ion Tech, Ltd., Teddington, England).

3. Results and Discussion

There are several features found common to the FAB mass spectra of the O-isopropyl oligonucleotide triesters of d(AA), d(AG), d(AC), d(AT), d(TA), d(TG), d(TC), and d(TT) (see Figs. 1 and 2). First, all compounds gave prominent molecular species in both positive and negative FAB. The ability to observe both $[M+H]^+$ and $[M-H]^-$ ions for these compounds permits easy verification of the molecular species. The second common feature is the strong presence of an ion fragment corresponding to free phosphate (*m/z* 99 in positive FAB, and *m/z* 97 in negative FAB). In addition, the negative FAB mass spectra displayed ions of moderate abundance at *m/z* 139, indicative of the presence of phosphate with an attached O-isopropyl group. The corresponding positive ions at *m/z* 141 were also observed in the positive FAB mass spectra, although the ion abundance was low. Collectively, these ions may prove to be a valuable tool in the identification of the alkylated moiety present

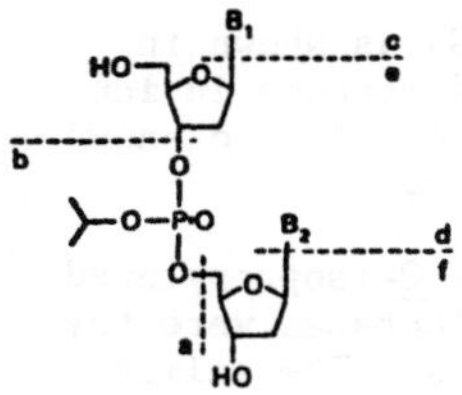

Fig. 1 Generalized fragmentation of O-isopropyl oligodeoxyribonucleotide triesters by fast-atom bombardment (FAB) mass spectrometry

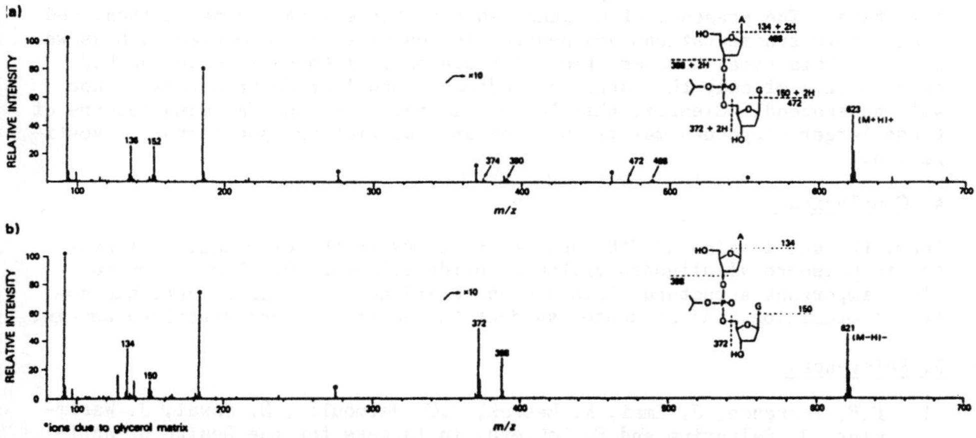

Fig. 2 Positive (a) and negative (b) fast-atom bombardment (FAB) mass spectra of O-isopropyl triester of d(AG)

in other synthetic and mutagen-derived oligodeoxyribonucleotides with alkylated internucleotide phosphate groups. The third common feature of the FAB mass spectra of these compounds is the presence of ions due to the nucleobases (fragments c and d in Fig. 1). While useful in ascertaining the nucleobase composition, additional information is required for more complete characterization of the structure.

Loss of a single nucleobase leading to fragment e or f can only be observed in positive FAB mass spectra of these compounds, although these ions are of very low abundance. There appears to be no discernible preference for the formation of either fragment.

Of particular interest are the sequence ions, fragments a or b, which are always present with moderate abundance in the negative FAB mass spectra of the O-isopropyl oligodeoxyribonucleotide triesters, but are of low and highly variable intensity, and are occasionally altogether absent, in the positive FAB mass spectra of these compounds. It quickly becomes a non-trivial matter to make the distinction between the 3' and the 5' termini. However, two generalizations could be derived from the data of the negative FAB mass spectra,which usually allows distinction between the 3' and the 5' termini of these compounds. First, the abundance of fragment b ions (3') tended to be greater than that of the abundance of fragment a ions (5'). Secondly, the abundance of fragment c ions (nucleobase of the 5' termini) tended to be greater than that of the abundance of fragment d ions (nucleobase of the 3' termini). However, this was not always observed (e.g. d(AG) as shown in Fig. 2). Despite this apparent deficiency, significant differences in ion abundances allow the intercomparison of "unknown" FAB mass spectra to known compounds to permit unambiguous identification of the former.

The positive and negative FAB mass spectra of completely O-isopropylated trimeric and tetrameric oligodeoxyribonucleotides (all nucleobases were thymine) were also recorded under the aforementioned conditions. The salient features of the FAB mass spectra of these "polyester DNAs" are the same as those noted above for the FAB mass spectra of the dimeric triesters. In addition, however, there are other ions present in the mass region from m/z 500 and below which are not present in the FAB mass spectra of the dimeric triesters. The presence of at least some of these ions can be rationalized from simple fragmentations and neutral losses of small molecules such as water. At this time, however, the added presence of these ions in the FAB mass spectra of only the larger oligodeoxyribonucleotide triesters is not well understood. Clearly, this lower mass region of the FAB mass spectra of these larger compounds may yield important information upon further investigation.

4. Conclusions

Thus, the application of FAB mass spectrometry to the examination of synthetic O-isopropyl oligodeoxyribonucleotide triesters has been shown to yield important structural information regarding molecular weight, nucleobase composition, and sequence, subject to the limitations described herein.

5. References

1. P.F. Torrence, J. Imai, K. Lesiak, J.C. Jarmoulle, H. Sawai, J. Warimmier, J. Balzarium and E. DeClerq, in Targets for the Design of Antiviral Agents, E. DeClerq and R.T. Walker, editors. Plenum Press, New York (1983), pp. 259-285.
2. M. Barber, R.S. Bordoli, R.D. Sedgwick and A.N. Tyler, J. Chem. Soc., Chem. Commun. 325 (1981).
3. L. Grotjahn, R. Frank and H. Blocker, Nucleic Acids Res. 10, 4671 (1982).
4. L. Grotjahn, R. Frank and H. Blocker, Int. J. Mass Spectrom. Ion Phys. 46, 439 (1983).
5. G. Sindona, N. Uccella and K. Weclawek, J. Chem. Res. (S) 184 (1982).
6. J. Eagles, C. Javanaud and R. Self, Biomed. Mass Spectrom. 11, 41 (1984).
7. B.A. Connolly, B.V.L. Potter, F. Eckstein, A. Pingoud and L. Grotjahn, Biochemistry 23, 3443 (1984).
8. W.J. Stec, G. Zon, W. Egan, R.A. Byrd, L.R. Phillips and K.A. Gallo, J. Org. Chem. 50, 3908 (1985).
9. J. Ulrich, A. Guy, D. Molko and R. Teoule, Org. Mass Spectrom. 19, 585 (1984).
10. D.L. Slowikowski and K.H. Schram, Nucleosides and Nucleotides 4, 309 (1985). Review of FAB/MS of oligonucleotides.
11. W.J. Stec, G. Zon, K.A. Gallo, R.A. Byrd, B. Uznanski and P. Guga, Tetrahedron Lett. 26, 2191 (1985).

Characterization of Biomolecules by Fast Atom Bombardment Mass Spectrometry

S.A. Martin[1], *C.E. Costello*[1], *K. Biemann*[1], *and E. Kubota*[2]

[1] Department of Chemistry, Massachusetts Institute of Technology, Cambridge, MA 02139, USA

[2] JEOL Ltd., Akishima, Tokyo 196, Japan

During the past several years there have been a number of advances in the field of particle-induced desorption which have increased the number of techniques and the type of compounds amenable to mass spectrometric analysis. Currently, plasma desorption (PD), secondary ion mass spectrometry (SIMS), fast atom bombardment (FAB) and laser desorption (LD) are employed for the analysis of large, thermally labile compounds such as peptides and glycoconjugates. These soft ionization techniques primarily produce abundant ions associated with the molecular weight of the sample, and depending on compound structure they may also provide structural information. The type and extent of fragmentation observed in the FAB mass spectra of peptides is dependent on the number of components, pH, matrix, type of ions detected (pos. or neg.) and amino acid sequence. Furthermore, when several components are present in a single FAB sample, often the only strictly reliable information obtained is the molecular weights of the components. In order to characterize the type and extent of fragmentation of peptides and glycoconjugates in FABMS, several closely related sets of compounds of varying compositions have been recorded, in the normal positive ion and negative ion modes and in the positive ion MSMS mode. In the MSMS mode both single and multiple component samples are examined using a JEOL JMS HX110/HX110 four sector mass spectrometer of EBEB geometry with high-mass range and good resolution in MS2.

Results of the study of the effect of amino acid composition and position on the fragmentation of peptides indicate that the presence and position of specific amino acids influence the type, distribution and frequency of occurrence of structure-specific fragment ion series. These parameters differ between the positive ion and negative ion detection modes. The spectra of bradykinin (Arg-Pro-Pro-Gly-Phe-Ser-Pro-Phe-Arg, M_r 1059), des-Arg^1-bradykinin, M_r 903, and des-Arg^9-bradykinin, M_r 903, acquired in the normal positive ion, negative ion and MSMS modes of FABMS are illustrative of this work. In the positive ion mode the spectrum of bradykinin exhibits ion series associated with both the N and C-termini, while that of des-Arg^1-bradykinin exhibits predominantly ions associated with the C-terminus and des-Arg^9-bradykinin with the N-terminus. In the negative ion mode all three spectra are dominated by C-terminal ions with the majority of the fragments containing serine. Another example are the spectra of fibrinopeptide A (Ala-Asp-Ser-Gly-Glu-Gly-Asp-Phe-Leu-Ala-Glu-Gly-Gly-Gly-Val-Arg, M_r 1535) and des-Arg^{16}-fibrinopeptide A, M_r 1379, in which the number of N-terminal ions increase in the positive ion mode with the removal of the C-terminal arginine, whereas in the negative ion mode the presence of four acidic residues controls the fragmentation in both peptides. A comparison of the fragment ion series in the spectra of bradykinin and several other peptides in the normal and MSMS mode (Table 1) indicates that amino acid composition similarly affects the MSMS spectra. The nomenclature for the

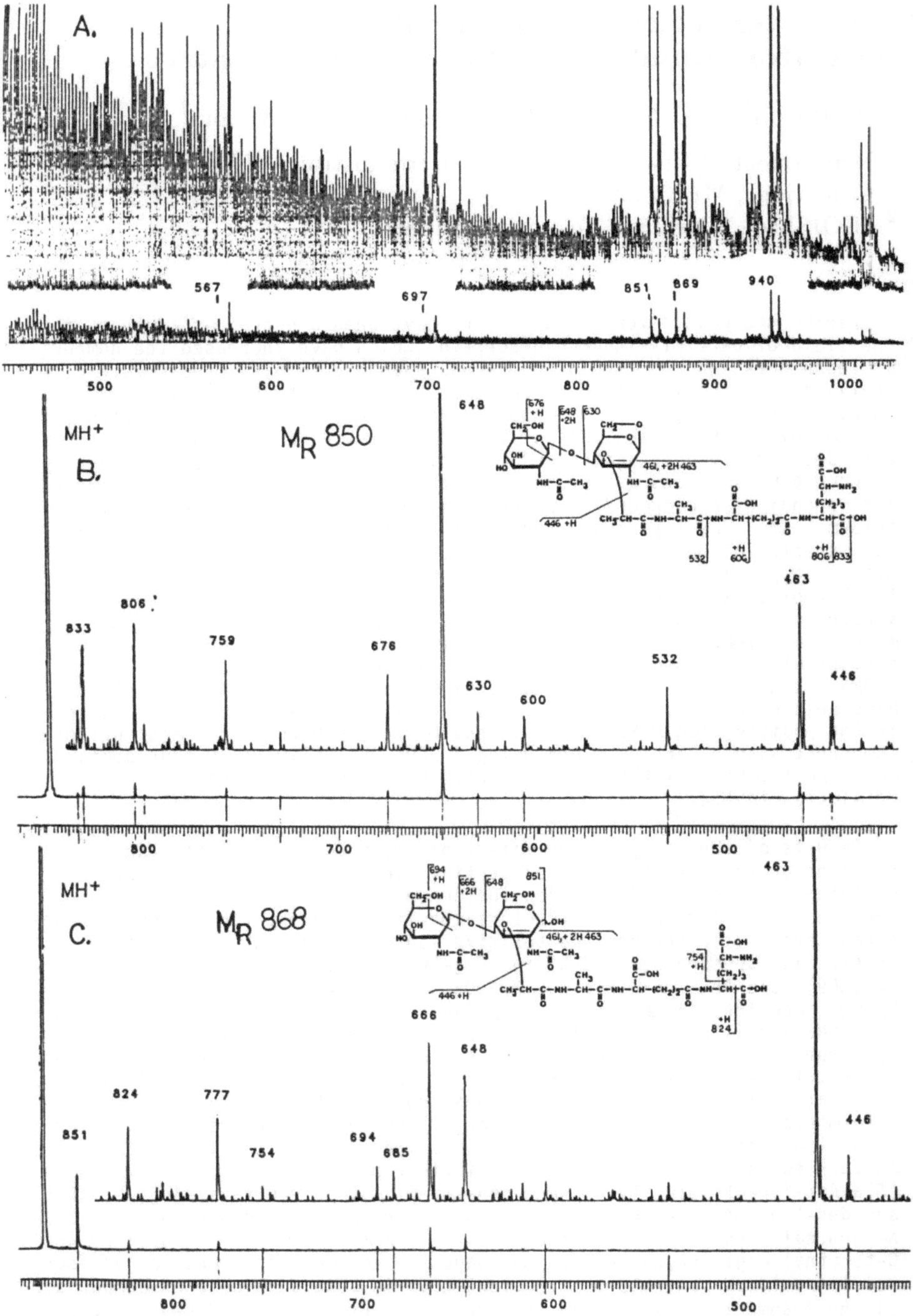

Fig. 1. Normal FABMS and CID MSMS spectra of peptidoglycans (see text).

TABLE 1 Comparison of peptide fragment ion series observed in the normal and MSMS modes of FABMS.

PEPTIDE	SEQUENCE	M_r	FRAGMENT SERIES OBSERVED NORMAL	MSMS
des Arg^1-Bradykinin	P-P-G-F-S-P-F-R	903	Y~Z>A	(Y>A>X~Z)[a]
des Arg^9-Bradykinin	R-P-P-G-F-S-P-F	903	A~C (B,Y)	A>Y
Bradykinin	R-P-P-G-F-S-P-F-R	1059	A~C~Y>Z (B,X)	Y>A (X)
Peptide 1	Ac-G-E-Q-H-H-L-I-G-G-A-K-G-NH_2	1243	C~Y	A~B
Peptide 2	Ac-G-E-Q-H-H-L-I-G-G-A-K(Ac)-G-NH_2	1285	C~Y	A~B
Substance P	R-P-K-P-Q-Q-F-F-G-L-M-NH_2	1346	C (A,Y,Z)	A (B,Y)
Renin[5-14]K	K-I-H-P-F-H-L-L-V-Y-S	1352	A~C (B,Y)	A~B (C,Y)
Fibrinopeptide A	A-D-S-G-E-G-D-F-L-A-E-G-G-G-V-R	1535	Y~Z (A,B,C,X)	Y (X,Z)
α endorphin	Y-G-G-F-M-T-S-E-K-S-Q-T-P-L-V-T	1744	Y~C (B,Z)	A~B~Y~Z
Renin Substrate	D-R-V-Y-I-H-P-F-H-L-L-V-Y-S	1757	(A~B~C~Y~Z)	A
Cross-linked Peptide 1-Peptide 2	Ac-G-E-Q-H-H-L-I-G-G-A-K-G-NH_2 / Ac-G-E-Q-H-H-L-I-G-G-A-K(Ac)-G-NH_2	2510	C~Y	A>B (C,X,Y)
Glucagon	H-S-E-G-T-F-T-S-D-Y-S-K-Y-L-D-S-R-R- -A-Q-D-F-V-Q-W-L-M-N-T	3480	Y (B,C,X)	Z>Y (B,C)

a. Fragment ion series enclosed in brackets are of low abundance and their series are incomplete.

fragment ion series listed in Table 1 are based on the suggestion of Roepstorff et al. (Biomed. Mass Spectrom 1984,11,601).

Without MSMS it is almost impossible to acquire structural information from small amounts of individual components of a mixture. A bacterial cell wall preparation containing peptidoglycan monomers, which otherwise would have to be separated by being subjected to repeated HPLC fractionation prior to FABMS, illustrates the utility of MSMS for this application. A normal FAB scan without HPLC separation (Fig. 1A) indicates at least six components and their lithium adducts but no sequence information. The $(M+H)^+$ parent ions at m/z 851 and m/z 869 were sequentially selected in MS1 and their daughter ion spectra recorded in MS2 (Figs. 1B and C). These well-resolved MSMS spectra define the carbohydrate and peptide portions and identify the structural difference as the formation of a 1-6 anhydro linkage.

Thus FABMS can provide information about structural details, even for mixture components, in addition to its more obvious utility for the determination of the molecular weights of pure compounds. Rules for the interpretation of fragmentation patterns are being developed by correlation of FABMS data acquired in the positive and negative ion modes for normal scans and comparison with MSMS spectra. Results of these studies should contribute toward the elucidation of the structures of unknown compounds, and to an understanding of the fragmentation pathways of specific compound classes.

Acknowledgement. These studies are supported by Grants No. RR00317 and No. GM05472 from the National Insitutes of Health.

Application of Fast Atom Bombardment and Tandem Mass Spectrometry to the Differentiation of Isomeric Molecules of Biological Interest

C. Guenat and S. Gaskell

Laboratory of Molecular Biophysics, National Institute of Environmental Health Sciences, P.O. Box 12233, Research Triangle Park, NC 27709, USA

The advent of fast atom bombardment (FAB) and related techniques of secondary ion mass spectrometry has brought new possibilities to the analysis of a variety of polar and involatile compounds not previously amenable to direct mass spectrometric analysis. While molecular weight information is often readily available, conventional FAB mass spectra are frequently devoid of structurally-diagnostic fragment ions. Information may also be obscured by ions arising from the matrix and/or additives used to enhance $[M+H]^+$. The use of tandem mass spectrometry (MS/MS) may assist in these respects by producing fragment ion spectra derived from selected parents via monitoring of unimolecular decompositions or collision-activated dissociation (CAD) occurring in a field-free region.

We have applied this technique to the differentiation of isomeric compounds of biological interest using a tandem double-focussing mass spectrometer (VG ZAB4F with a BE-EB configuration). The fragmentations of a parent ion selected with B_1E_1 were observed in the third field-free region by performing a linked scan at constant B_2/E_2. The collision cell situated in this region may be floated between 0 V and the accelerating voltage of the ion source, allowing the study of CAD at collision energies ranging from 0 eV to several keV (in the laboratory frame of reference). When the collision cell is grounded, unimolecular fragmentation and CAD are superimposed on each other in the spectrum. Floating the cell leads to a discrimination against the former.

The steroid glucuronides studied here yielded abundant $[M-H]^-$ and $[M+Na]^+$ ions, but their CAD spectra were of very low intensity. $[M+H]^+$ ions however, yielded intense daughter ions in both unimolecular and collision-activated dissociations, with the abundance of major daughters corresponding to 1 to >10% of the abundance of the transmitted parent. Figure 1 shows an example where two isomeric pregnan-triol-one-3-gluronides can be differentiated unambiguously by the unimolecular fragmentation of the $[M+H]^+$ ions (m/z 527). In the case of THS-3-glucuronide, the spectrum presents an intense ion at m/z 351 corresponding to the intact aglycone. This fragment is absent from the spectrum of THB-3-glucuronide. These compounds can also be easily distinguished by their 8 keV CAD spectra (collision cell grounded). The relative abundances of the ions corresponding to the intact aglycone and those resulting from subsequent water losses (m/z 333, 315 and 297) are quite different.

The unimolecular fragmentation spectra of the $[M+H]^+$ ions (m/z 467) from the 5α- and 5β-androsterone-3-glucuronide do not permit one to distinguish between the two isomers. But the 8 keV collision energy spectra (figure 2) show clear and reproducible differences between the relative abundances of the aglycone ion (m/z 291), and those derived from additional water losses (m/z 273 and 255). When the collision energy is

5β-pregnan-3α,17α,21-triol-20-one (THS) 3-glucuronide	5β-pregnan-3α,11β,21-triol-20-one (THB) 3-glucuronide

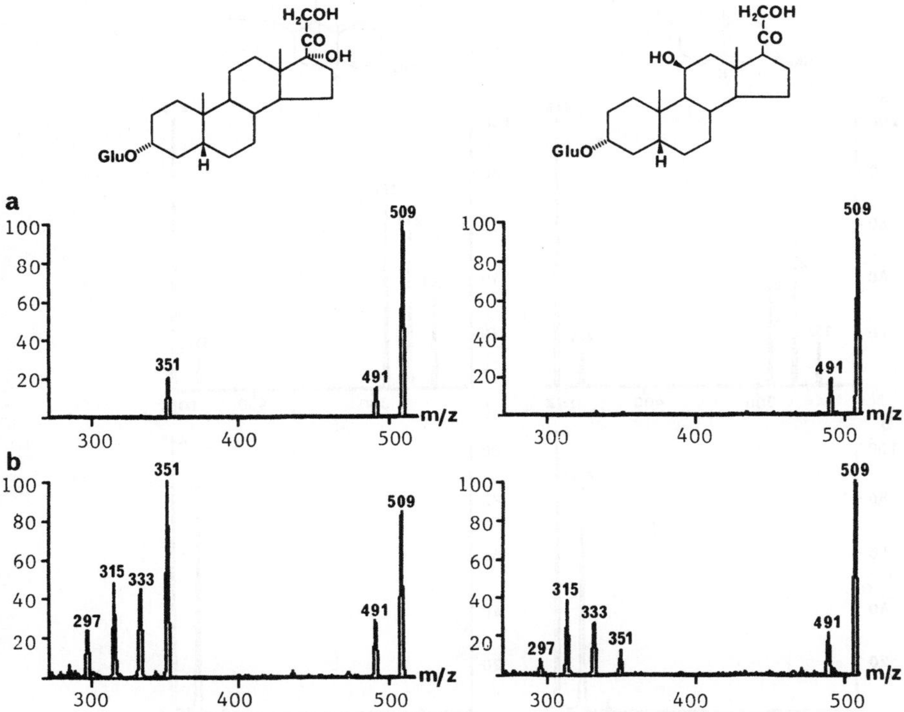

Figure 1. Comparison of the unimolecular fragmentation (a) and the 8 keV CAD (b) of two pregnan-triol-one-3-glucuronide isomers

decreased to ca. 200 eV however, these fragments are absent from the spectra, and distinction between these isomers can no longer be made. This emphasizes the fact that, in this case, high-energy CAD is required to get detailed structural information.

Isomers coming from other classes of compounds such as carnitines [1] or branched peptides have also been differentiated using tandem mass spectrometry. In the latter case, no information was obtained from the $[M+H]^+$ ions, but the technique has been successfully applied to smaller fragment ions produced in the ion source. The examples presented here show the usefulness of FAB MS/MS in differentiating between isomeric biomolecules and that the appropriate conditions (unimolecular dissociations or high-energy CAD) may vary with analyte compounds.

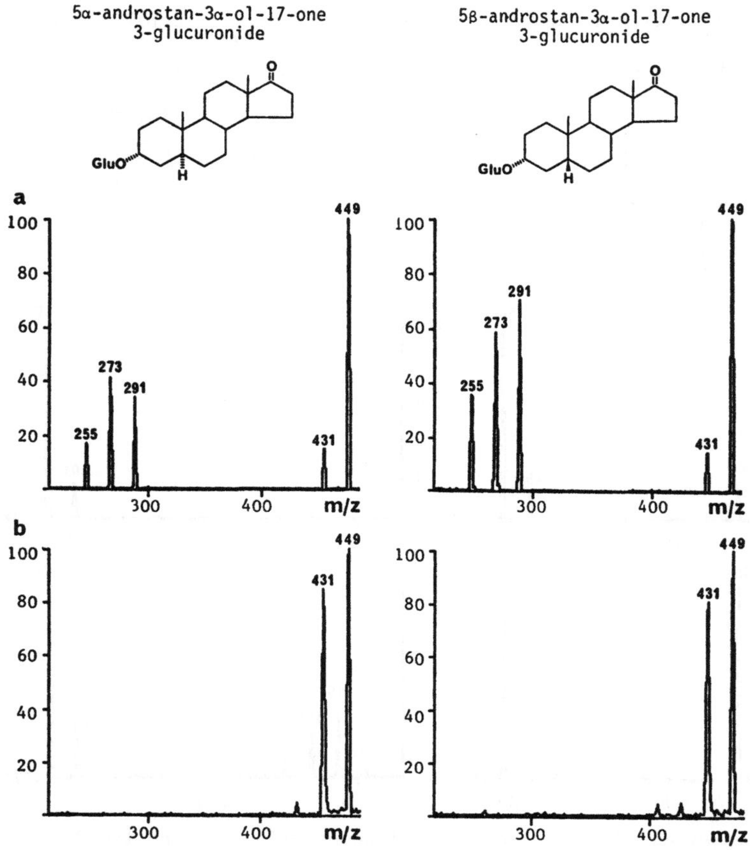

Figure 2. Comparison of the CAD spectra of the α- and β-androsterone-3-glucuronide isomers at 8 keV (a) and 200 eV (b)

[1] S.J. Gaskell, C. Guenat, D.J. Harvan, D.S. Millington and D.A. Maltby: in "Advances in Mass Spectrometry", vol. 10 (J.F.J. Todd, Ed.) John Wiley, Chichester (UK), 1985.

Fragmentation of Heavy Ions (5000 to 7000 Daltons) Generated by PD and FAB

P. Demirev, M. Alai, R. van Breemen, R. Cotter, and C. Fenselau

Johns Hopkins Univ., Pharmacology Dept., 725 N. Wolfe, Baltimore, MD 21205, USA

1. Introduction

The phenomenon of desorbing secondary ions from insulating organic materials in the condensed phase following energy deposition by primary particles in a broad kinetic energy range (1 keV-100MeV) has attracted considerable attention in recent years. The interest stems largely from the fact that the phenomenon is successfully exploited in a number of new techniques capable of generating gas-phase ions from involatile thermally labile biomolecules with weights up to 20 kD [1,2]. The most widely used techniques include fast atom bombardment (FAB) and plasma desorption (PD). These two techniques differ markedly in several points:

- energy of the impinging particles and primary beam flux
- mechanism of energy deposition
- sample preparation and exposure
- method of mass analysis and different time frame of ion detection

Despite these differences, mass spectra obtained by the two techniques are generally similar, for example in the case of smaller peptides [3]. We have studied different peptides in the range 5000 - 7000 D. These include insulins, relaxins, epidermal growth factors (EGF), ovomucoid components and basic pancreatic trypsin inhibitor (BPTI).

Spectra of these heavy compounds run on a variety of analysers have several features which are different from familiar spectra of lighter ions. Few sequence fragments are visible above m/z 2000. Peaks representing molecular ion species have extended envelopes on the low-mass side. Compounds comprising peptide chains held together only by disulfide bonds usually contain peaks corresponding to these discrete chains. Our objectives in this study are to examine the origins of the A and B chain ions and the contribution of prompt fragmentation to the extended envelope on the lighter mass side of the molecular ion peaks.

2. Experimental

The ^{252}Cf PD spectra were acquired on a commercial "Bio-Ion" PD time-of-flight mass spectrometer (Uppsala, Sweden). Data were acquired and subsequently processed by Apple II-E microcomputer. Calibration was carried out using H^+. and Na^+ as reference peaks.

Positive ion FAB spectra were obtained on a Kratos MS-50 (Manchester, UK) double focussing mass spectrometer equipped with a 2.3 T magnet and a 8kV post-accelerator detector. Data were acquired in two modes, utilizing a Kratos DS-55 data system and software written in-house.

I) Wide-range magnet scans at a lower resolving power--2,000, and data collection utilizing the time-binning technique [4].

II) Narrow-range voltage scans of the molecular ion region at a resolving power of 5,000. Mass assignment was achieved by acquiring the CsI spectrum under the analytical conditions in the first case and peak matching with the corresponding CsI clusters in the second one.

High-performance liquid chromatography was done on a Beckman model 114 M liquid chromatograph with a reversed-phase Vydec C_4 protein column, UV detection at 210 nm. The isocratic solvent was 0.05 M KH_2PO_4, 1mM ammonium formate, 5mM(+) Tartaric Acid pH 4.2: Acetonitrile; 10:4.

3. Results

All the compounds examined show sufficiently intense peaks corresponding to the singly and doubly-charged molecular ions (Fig. 1). In the case of peptides consisting of two amino acid chains linked by disulfide bridges (insulins, relaxins), the ions corresponding to the different chains were present in the spectra.

The lower mass range of the spectra consists of exponentially increasing continuous "tailing".

No sequence ions were observed above m/z 2000. Broad "skirts" were present on the light mass side of the molecular ion peak, their width depending on the method of mass analysis. The "closer" examination of the molecular ion region of several of these peptides by narrow voltage scanning (Fig. 2) shows peaks corresponding to losses of approximately 15, 30 and 44 amu and M+15.

Bovine insulin samples were shown by HPLC to be free of A and B chain fragments. Samples dissolved in acetic acid and electrosprayed for PD, as well as samples dissolved in thioglycerol (used as a matrix in FAB) and recovered were found by HPLC to contain substantial amounts of A chain and reduced amounts of intact insulin as a function of time.

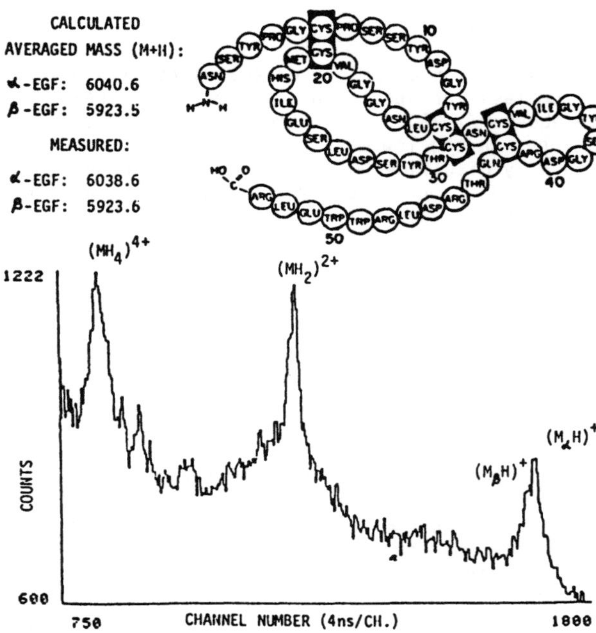

Fig. 1 PD Spectrum of EGF

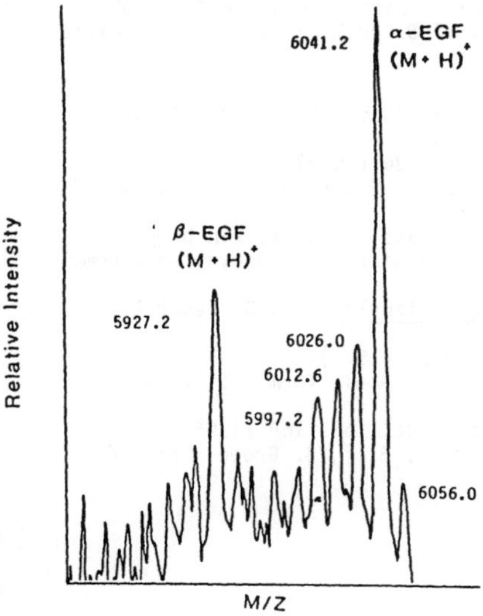

Fig. 2 Narrow Voltage Scan of EGF

4. Discussion

The averaged molecular weights obtained by both techniques allow the differentiation between structurally related peptides, for example natural homologs or chemically generated analogs in the EGF [4], insulin, relaxin, ovomucoid families. No doubt unit resolution mass spectra could provide additional information about the presence of isotope labels,or estimate of the redox state of the disulfide bonds in the peptide [5].

The markedly decreased prompt fragmentation above 2000 D for high-mass peptides has already been noted [7]. A factor that probably leads to the stabilization of the molecular ion in the gas phase despite its high internal energy content is the presence of intramolecular non-covalent interactions--Van-der Waals forces, H-bonds and electrostatic interactions--responsible for the tertiary structure of the peptide. The lack of sequence ion fragmentation has also been predicted by theoretical considerations of unimolecular behavior of large linear polymers [8]. It is predicted that the reaction rate for a fragmentation decreases with the increase of number of atoms in the fragment. Hence the loss of fragments containing fewer atoms would be preferable. It has also been suggested [4] that successive hydrogen loss ("H-stripping") and other small neutral ("peripheral group") losses contribute to the tailing in the wide-scan spectra. The nature of the envelope extending below the molecular ion of insulin has been considered by Chait and Field [8]. They suggest that considerable fragmentation--both prompt and metastable--contributes to this envelope in their system. Further evidence for H-stripping is provided by comparing the spectra of nonhydrogen containing compounds e.g. cesium perfluorosulphonate salts [9].

The presence of structurally significant ions corresponding to the A and B chains in the case of insulin has been discussed in [10] for PD and [11] for FAB. Our HPLC studies suggest that,at least partially,these ions are due to the presence of the neutral fragments prior to ionization.

Acknowledgements: This research was supported in part by a grant from NSF PCM82-09954. We thank C. Schwabe, Med. Univ. of S.Carolina, and M. Laskowski, Purdue Univ. for some of the samples studied.

References

1. J. Cottrell, B. Frank: Biochem. Biophys. Res. Comm. 127, 1032 (1985).
2. B. Sundqvist et al: Science, 222, 696 (1984).
3. J. Fohlman et al: Biomed. Mass Spectrom. 12, 380 (1985).
4. R. Cotter, B. Larsen, D. Heller, J. Campana, C. Fenselau: Anal. Chem. 57, 1473 (1985).
5. B. Larsen, J.A. Yergey, R. Cotter: Biomed. Mass Spectrom. (in press).
6. K. Hyver, J. Campana, R. Cotter, C. Fenselau: Biochem. Biophys. Res. Comm. 130, 287 (1985).
7. D. Williams: in Mass Spectrometry of Large Molecules (Ed. S. Facchetti), Elsevier (1985), p. 93.
8. D. Bunker, F. Wang: J.Am. Chem. Soc., 99, 7457 (1977).
9. D. Heller, C. Fenselau, J. Yergey, R. Cotter: Anal. Chem. 56, 2274 (1984).
10. B. Chait, F. Field: Int. J. Mass Spec. Ion Proc., 65, 169 (1985).
11. M. Barber, R. Bordoli, G. Elliot, A. Tyler, J. Bill, B. Green: Biomed. Mass Spectrom. 11, 182 (1984).

Amino Acid Sequencing of Norwegian Fresh Water Blue-Green Algal (Microcystis Aeruginosa) Peptide by FAB-MS/MS Technique

T. Krishnamurthy[1], *L. Szafraniec*[1], *E.W. Sarver*[1], *D.F. Hunt*[2], *S. Missler*[3], *and W.W. Carmichael*[4]

[1] Chemical Research and Development Center, Aberdeen Proving Ground, MD 21010, USA
[2] University of Virginia, Charlottesville, VA 22901, USA
[3] USAMRIID, Ft. Detrick, MD 21701, USA
[4] Wright State University, Dayton, OH 45435, USA

1. INTRODUCTION

Fresh water blue-green algal (cyanobacteria) blooms have been found to be toxic, and have led to the loss of livestock in several countries. [1] Human health hazards have also been linked to these toxic blooms in several instances.[1] Most of these reported instances have been more due to a single species of cyanobacteria, known as Microcystis Aeruginosa, than any other species of blue-green algae. The majority of the blue-green algal toxins (Microcystins) were observed to be hepatotoxic and the others neurotoxic[1] Eutrophication has been observed to facilitate the growth of the toxic blooms in agricultural, urban, and recreational water sources. The heavy algal mass production leads to an increase in water treatment costs in addition to other economic losses and/or any human health hazards. As a result, several investigations have been initiated in order to elucidate the structures of these toxins and to study their physico-chemical properties.[1] Such information is vital for the detection of the toxins in fresh water supplies and for subsequent decontamination of the infested sources.

The available information on the chemical structures and properties of blue-green algal toxins is very limited. In most cases the toxins have been found to be small peptides with molecular weights in the range of 1,000 Daltons.[1] We have recently investigated the hepatotoxin, termed as Akerstox, isolated from the natural blooms of microcystis aeruginosa in Akersvatn (Akers Lake), Norway. The toxin was also found to be a cyclic peptide with five commonly observed amino acid moieties along with two uncommon ones. The identity of the latter and the amino acid sequence in the peptide were determined from the fast atom bombardment daughter spectral data of the intact peptide, its derivatives, and degradation products. The proton and C-13 nmr spectral data were also in support of the deduced sequence.

2. EXPERIMENTAL

A standard Finnigan-MAT TSQ triple quadrupole mass spectrometer with Ion-Tech FAB gun and Incos data system; and VG model 7070 EQ high-field magnetic sector mass spectrometer with extended geometry and 11-250 data system were used for all measurements. The samples in glycerol or magic bullet matrix [dithioerythratol(3) and dithiothreotol(1)] were bombarded either with 8kV xenon or 8kV krypton atoms. The collisionally activated dissociation (CAD) spectra were obtained using argon maintaining the collision gas pressure and collision energy at 2.2 mTorr and 20 eV respectively.

3. RESULTS AND DISCUSSION

A hepatotoxic peptide (Akerstox) was isolated from blue-green algal (Microcystis Aeruginosa) blooms found in the fresh water lake (Akersvatn) in Norway. It was purified by HPLC using a reverse-phase C_{18} column. Amino acid analysis of the peptide indicated the presence of alanine, arginine, glutamic acid, leucine, and β-methylaspartic acid. The fast atom bombardment (FAB) mass spectrum in glycerol or the magic bullet matrix showed the protonated molecular ion at m/z 995. Absence of the sequence ions in the FAB spectrum showed the peptide to be cyclic. However, the FAB daughter spectrum (Figure 1) of the intact peptide provided the amino acid sequence as well as the identity of two more amino acid moieties present in the molecule. They were found to be n-methyldehydroalanine and 3-amino, 9-methoxy,2,6,8-tri-methyl,10-phenyldeca, 4,6-dienoic acid. The proton and the ^{13}C nuclear magnetic resonance (nmr) spectral data were also in agreement with the structures assigned for these two uncommon amino

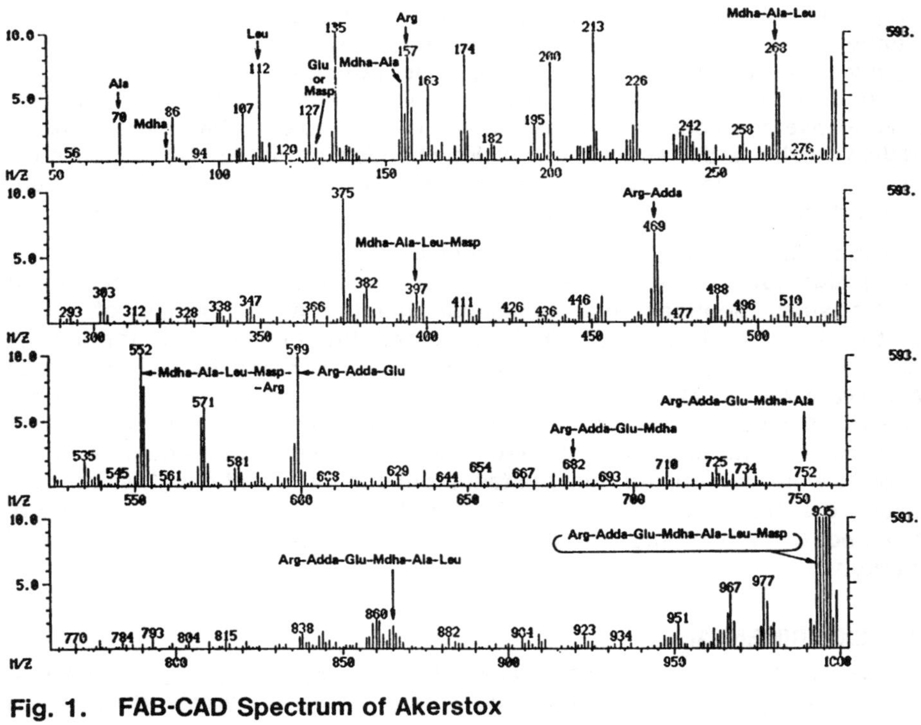

Fig. 1. FAB-CAD Spectrum of Akerstox

acids. Earlier results observed by Botes, et. al., on the hepatotoxin from a South African species of Microcystis Aeruginosa [2] also supported the postulated identity of these two moieties. The observed amino acid sequence in the Akerstox peptide are shown in Figure 1. The fragment ions observed in the daughter spectrum of acetyl-permethylated peptide were also consistent with this deduced sequence. It was further confirmed by the results obtained from the CAD spectra of the hydrolysates of Akerstox and their corresponding methyl esters. The sequence ions observed in the CAD spectrum of a deuterium labelled methyl ester indicated the glutamic acid and β-methylaspartic acid in the peptide to be iso-linked. The structure of the cyclic peptide thus determined is shown in Figure 2.

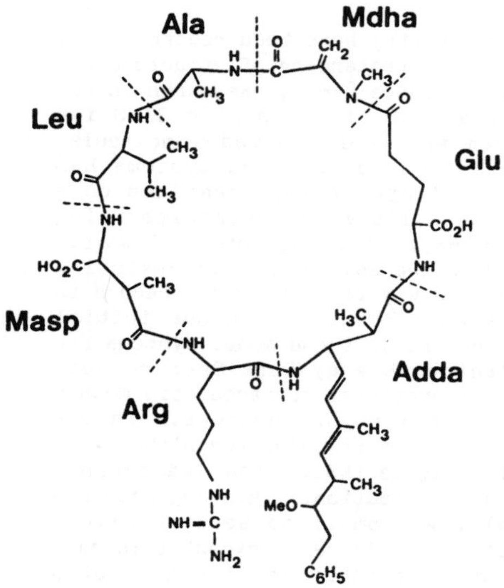

Fig. 2. Norwegian blue-green algal hepatotoxin (Akerstox)

REFERENCES

1. W.W. Carmichael and N.A. Mahmood: "Toxins from Fresh Water Cyanobateria" in Seafood Toxins, Ragelis Ed., American Chemical Society Symposium Series 262, Washington, D.C., p377, (1984).

2. Botes et. al., J. Chem. Soc. Commun. 12, 652, (1983); J. Chem. Soc. Perkin. Trans. 1, 2311 (1984).

Static SIMS Studies of Molecular and Macromolecular Surfaces: Ion Formation Studies

J.A. Gardella, Jr., J.H. Wandass, P.A. Cornelio, and R.L. Schmitt

Chemistry Department and the Surface Science Center, University at Buffalo State University of New York, Buffalo, NY 14214, USA

1. Introduction

Static methods of Secondary Ion Mass Spectrometry have been recognized as a primary method in the development of the understanding of structure and bonding at molecular and macromolecular surfaces [1-3]. Research in our laboratory is directed toward the understanding of the static SIMS ion ejection and formation mechanisms from well-defined model molecular systems. Our approach to design and construction of such systems has utilized fatty acid monolayer and multilayer films organized on a purified water subphase and transferred to solid metal substrate using classical Langmuir-Blodgett methods [3-5]. This approach allows for control of the structure and orientation of the so-called "pre-existing state" of large molecules, while providing a way to develop approaches to the characterization of polymeric surfaces. Results from our initial SIMS work [3] have been combined with the use of these model systems for the investigation of High-Resolution Electron Energy-Loss Spectroscopy (HREELS) [6,7] as a surface-sensitive vibrational spectroscopic method for the examination of molecular and macromolecular surfaces. In our laboratory, the goal is to integrate the results of these techniques with other surface-sensitive spectroscopic techniques (ESCA, Ion Scattering, FT-IR) [8-10] for polymer surface characterization. This approach can also be thought as the analogue of solving a small molecule chemical structure using vibrational and mass spectrometric data, except this same information is desired from the near surface/interfacial region of a large molecular assembly, a crystalline molecular network or a macromolecular solid.

In this paper, results from the static SIMS analysis of Langmuir Blodgett monolayers of fatty acids transferred to plasma-glow-oxidized silver surfaces will be compared to the analysis of fatty acids deposited on chemically oxidized/cleaned silver surfaces by microsyringe techniques [1]. The results will be used to evaluate the effects of 1) orientation of the monolayer 2) chemical structure of the fatty acid and 3) chemical state of the substrate on silver cationization, the limits of detection of molecular ions from organic monolayers and the analysis of multilayers of fatty acids.

HREELS analyses of fatty acid monolayers and multilayers illustrate the extreme sensitivity to the outermost functional groups, showing selective enhancement of relative intensities of vibrational bands associated with methyl groups. Sensitivity to unsaturation is observed for monolayer fatty acids. In addition, multilayers of up to fifteen stearic acids were successfully analyzed, with shifts of vibrational bands demonstrating changes due to intermolecular forces in the multilayer assembly. Assignments are related to both infrared and Raman data for carboxylic fatty acids.

Details of the SIMS and HREELS instrumentation can be found elsewhere [3,7], with a complete description of the Langmuir-Adam trough used in these studies. Briefly, instrumental conditions for the SIMS experiments involved Argon primary ion beam at 5 kV and < 5 $nA\text{-}cm^{-2}$. Conditions for HREELS involved 6.5 eV primary electron energy with electron currents < 10^{-11} Amps.

2. Results and Discussion

Tables 1 and 2 show results of the SIMS and HREELS analyses of the Langmuir-Blodgett mono and multilayer systems.

Table 1 Results of Static SIMS Analysis of Fatty Acids on Silver Substrate Comparison of Ion Formation from LB and Microsyringe Preparations

Acid	Ion	Langmuir Blodgett Monolayer Preparation[a]	Microsyringe from Benzene[b]: Low Current Presputter[c] (mol/cm^2)	Microsyringe from Benzene[b]: High Current Presputter[d] (mol/cm^2)
Stearic	$[M-H]^-$	strong	3.5×10^{-17}	3.5×10^{-17}
$CH_3(CH_2)_{16}COOH$	$[M-H]^+$	not obs	3.5×10^{-17}	3.5×10^{-13}
	$[M+Na]^+$	weak	not obs	not obs
	$[M+Ag]^+$	not obs	3.5×10^{-9}	3.5×10^{-5}
Oleic	$[M-H]^-$	strong	Analysis not completed	
$CH_3(CH_2)_7CH{=}CH(CH_2)_7CO_2H$	$[M+H]^+$	weak	---	---
	$[M+Na]^+$	not observed	---	---
	$[M+Ag]^+$	weak	---	---

Notes: a) 7.5×10^{10} mol/cm Ag b) Limit of Detection for ion of interest c) Ag washed with HNO_3 (50%) followed by DI H_2O rinse, Ar^+ sputter $6nA/cm^2$, 1 hr. d) same as c) + 420 nA/cm^2 1 hr.

From the foregoing data in Table 1, it is apparent that static SIMS possesses adequate sensitivity to detect a quantified organized monomolecular layer and to observe molecular and molecular-like ions in both positive and negative ion spectra for fatty acids. Also, differences in the degree of cationization are seen between "gentle" plasma oxidation utilized for LB preparation and the traditional thick oxide, promoting wet chemical methods [1] used in the microsyringe preparation. Typically, these wet methods produce a high intensity of substrate-cationized parent ions. For example, for microsyringed stearic acid on chemically oxidized silver, signal from the quasi-molecular ion $[M-H]^-$ (and possibly $[M+H]^+$ as well) is detectable at the level 10^{-17} moles. Benninghoven [1] has reported detection levels of 10^{-13} to 10^{-17} moles for various drugs and biomolecules studied by time-of-flight SIMS. In comparison, the LB results show detection of $[M-H]^-$ but not $[M+H]^+$ at monolayer coverage of stearic acid on Ag. The effective sample size of LB monolayers was approximately 7.5×10^{-10} moles of stearic acid, given that monolayers are constructed as 21 Å^2 molecule. This suggests that protonation occurs only where a source of hydrogen is available, either the Ag surface oxide or intermolecular H transfer. The latter is not possible in the LB layer, as heads are all aligned down. We also observe weaker emission of $[M+Ag]^+$ detectable at 10^{-5} moles of stearic

Table 2 HREELS Results (meV) and Assignments

System		$\nu(CH_3),\nu(CH_2)$	$\delta(CH_2)$	$\nu(C-C)$	$\nu(CH_2)$	$\delta_s(CH)_3)$	$t-r(CH_2)$	$\nu(C=C-H)$	$\nu(CO)$	$-CH_2$ wags and stretches
Approximate expected value (meV)		360	180	130	93	170	112	373	160	137-48
stearic/Ag	1	359	N.O.	130	92	173	N.O.	-	159	N.O.
	3	360	178	135	95	169	110	-	162	149
	5	361	180	135	94	170	109	-	157	N.O.
	15	364	170	N.O.	N.O.	N.O.	116	-	164	144
Oleic monolayer	Ag[a]	360	176	133	92	169	114	375	158	142-49
	Au	359	N.O.	134	93	170	110	377	160	146
	Al	360	180	130	93	172	112[b]	373	161	150
	Ge[a]	359	176	133	95	165	113[c]	370	158	N.O.
mixed monolayer		359	176	133	92	170	115	372	163	146

NO Not observed. [a] $\nu(C=O)$ at 207 cm^{-1} observed only in these systems.
[b] Al surface optical phonon
[c] Ge-O valance band vibration

acid on Ag substrates sputter-cleaned under higher conditions. These latter data support the idea [3] of cationization in these systems as coordination of silver ions to the double bonds in unsaturated acids in the selvedge during desorption, the lack of cationization in the saturated fatty acid systems prepared by LB techniques would suggest that silver coordination does not occur at the acid head. Additionally, clear differences are observed in relative abundances of quasi-molecular ions between wet-oxidized syringe preparations from solution and LB preparations of monolayers. This points out the need to control orientation to understand chemical effects on ion production in SIMS. Because of the high abundances of characteristically even-numbered fragment ions, it is possible to distinguish unsaturation in the monolayer.

The results of these HREELS data in Table II allow us to conclude that succesful HREELS analysis may be conducted on organic overlayers approaching 400 Å thickness with adequate resolution to discern and assign vibrational features. Our relative intensity results lend some authority to the opinion that an impact-scattering mechanism might account for these HREELS results, however experiments involving control of surface topology must be implemented to definitely prove this. The possibility of resonance scattering interactions with the multilayer assemblies is an additional factor that will be investigated by the variation of primary electron energy, with a knowledge of valence band structures.

The effect of substrate identity on the character of the HREELS sampling process on polycrystalline materials is still vague, although some evidence has been gathered to show that ease of surface oxidation may increase relative HREELS intensities (neglecting surface roughness) because of a decrease in the elastic peak intensity. If it is shown that collecting such HREELS spectra is independent of the nature of the substrate, then the organic overlayer might possibly be described as a two-dimensional crystal situated on a supporting non-interacting substrate. It is also possible to observe unsaturated vibrational loss features both in homogeneous and mixed systems where dilution of the target molecule in the L-B monolayer has been done.

References

1. A. Benninghoven: Int. J. Mass Spectrom. Ion Phys. 53, 85 (1983).
2. R. J. Day, S. E. Unger, R. G. Cooks: Anal. Chem. 52, 557A (1980).
3. J. H. Wandass, J. A. Gardella, Jr.: J. Am. Chem. Soc., 1985, in press.
4. I. Langmuir, K. B. Blodgett: Phys. Rev. 51, 317 (1937).
5. G. L. Gaines: Insoluble Monolayers at Liquid Gas Interfaces (Wiley Interscience, New York 1966).
6. J. H. Wandass, J. A. Gardella, Jr.: Surface Sci. 150, L107 (1985).
7. J. H. Wandass, J. A. Gardella, Jr.: Langmuir, submitted.
8. J. A. Gardella, Jr.: Trends in Analytical Chemistry 3(5), 129 (1984).
9. J. A. Gardella, Jr., J. S. Chen, J. H. Magill, D. M. Hercules: J. Am. Chem. Soc. 105, 4536 (1983).
10. R. L. Schmitt, J. A. Gardella, Jr., J. H. Magill, L. Salvati, Jr., R. L. Chin: Macromolecules, 1986, in press.

TOF-SIMS of Polymers in the Range of $M/Z = 500$ to 5000

I.V. Bletsos[1], *D.M. Hercules*[1], *A. Benninghoven*[2], *and D. Greifendorf*[2]

[1] Department of Chemistry, University of Pittsburgh, Pittsburgh, PA 15260, USA

[2] University of Münster, Physikalisches Institut, D-4400 Münster, F.R.G.

1. Introduction

Early volatilization methods coupled to mass spectrometry (MS) for characterization of polymers involved mainly pyrolysis (1). Invention of softer volatilization and ionization sources like field desorption MS, (FDMS), (2, 3,4,5), electrohydrodynamic MS, (EHDMS), (6), 252 Cf-plasma desorption (252 Cf-PD), (7), laser desorption (LDMS), (8), and secondary ion MS, (SIMS), (9, 10,11), has expanded MS to include polymer characterization in higher mass ranges, oligomer distributions, and surface analysis. Inherent limitations in each technique, however, impose restrictions due to elaborate sample preparation and control of ionization processes, long data-acquisition times, and low-mass ranges due to mass analyzers used. As part of our effort to characterize polymers in the high-mass range, we reported detection of high-mass fragments for nylons using a time-of-flight secondary ion mass spectrometer (TOF-SIMS) [12].

In this report we present mass spectra of polydimethylsiloxane (PDMSO) and polystyrenes (PS) obtained by a TOF-SIMS, equipped with a mass-selected pulsed primary ion source, an angle and time focusing time-of-flight analyzer, and a single ion counting detector (13). Fragmentation in the low-mass range provided structural information about the repeat unit. Ag+ cationization of polymer fragments containing large numbers of repeat units at high m/z allowed identification of the polymer. Fragmentation patterns were unique for polymers having different repeat units but of equal mass, and identification of such polymers was possible. Oligomer distributions obtained from mass spectra compared well with distributions determined by other techniques (e.g. GPC) for the same polymers.

2. Results and Discussion

A mass spectrum of (PDMSO) is shown in Fig. 1. At $m/z < 500$, extensive fragmentation is observed producing structurally significant fragments. At high mass, from $m/z = 500$ to 10000, intact polymer molecules are detected cationized with Ag+. The most intense peaks at high mass correspond to $(nR+CH_3+Ag)^+$ series, where R = repeat unit. Presence of a terminal $-CH_3$ in ions detected indicates that desorption of intact polymer molecules occurs without fragmentation. The intensity variation of cationized polymer molecules as a function of mass reflects a distribution of the order of polymerization of PDMSO. Loss of terminal $-CH_3$ groups from polymer molecules gives rise to $(nR+Ag)^+$ peaks of lower intensity than $(nR+CH_3+Ag)^+$. The largest ion detected at approximately $m/z = 9600$ is cationized with Ag+, contains a terminal $-CH_3$, and consists of 128 repeat units of PDMSO.

Several polystyrenes with various substituent groups at different positions on the benzene ring or the hydrocarbon backbone were studied. A 1 ul solution of a styrene polymer in toluene (ca. 1×10^{-2} M) was deposited on

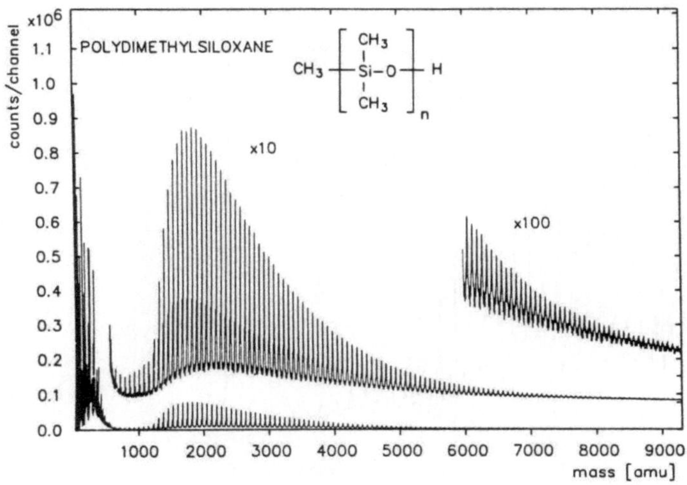

Fig. 1 TOF-SIMS spectrum of polydimethyl siloxane

100 mm^2 of an etched Ag substrate. A target area of about 1 mm^2 was bombarded by 12 KeV Ar+ with an average current of 30 pA, corresponding to a primary ion current density of ~ 10^{-9} A/cm2 (static SIMS). Polymer fragments cationized with Ag+, (nR+Ag)+, produce the most intense peaks; the spacing between them corresponded to one repeat unit. Fragmentation within one repeat unit spacing is consistent throughout the spectra and is characteristic of the PS type studied. For example, the repeat units of P(α-MS) and P(4-MS) have different chemical structures but equal mass (R=118). The most prominent peaks due to Ag^+ cationized fragments for both P(α-MS) and P(4-MS) appear at exactly the same m/z. Fragmentation patterns, however, are different, and P(α-MS) and P(4-MS) can be distinguished. Differences in fragmentation are due to different positions of the $-CH_3$ substituent group. Cleavage occurs statistically at each possible bond, and chemical stability of the fragments produced reflects peak intensities. Detailed discussion on such fragmentation patterns will be published.

Polymer molecular weight distributions obtained from mass spectra were evaluated with polymer standards of having molecular weight distributions. A spectrum of a PS standard with a number average molecular weight Mn = 5100 is shown in Fig. 2. At m/z = 3000 the most intense peaks are due to cationized polymer fragments corresponding to $(nR+Ag)^+$ series. In the range m/z = 3000 to 7000 whole polymer molecules are detected intact with their terminal groups giving rise to a $(nR+C_4H_9+Ag)^+$ series. These peaks are the most intense in this range, and their intensity distribution reflects the number average molecular weight distribution for the PS standard. Loss of terminal $-C_4H_9$ groups results in (nR+Ag)+ series of lower intensity than $(nR+C_4H_9+Ag)^+$. This indicates that the dominant process is the desorption of intact polymer molecules cationized with Ag+. At the low m/z end of the distribution region,polymer fragments containing a $-C_4H_9$, and whole polymer molecules give rise to peaks which are not resolved. Consequently, the starting point of the distribution cannot be exactly determined. The value Mn = 4500 was calculated from peak intensities within 11% of Mn = 5100. A low Mn value is expected because the TOF-SIMS detection efficiency decreases with increasing m/z [13]. An approximate correction was applied to the distribution to compensate for post acceleration and detection efficiencies, and

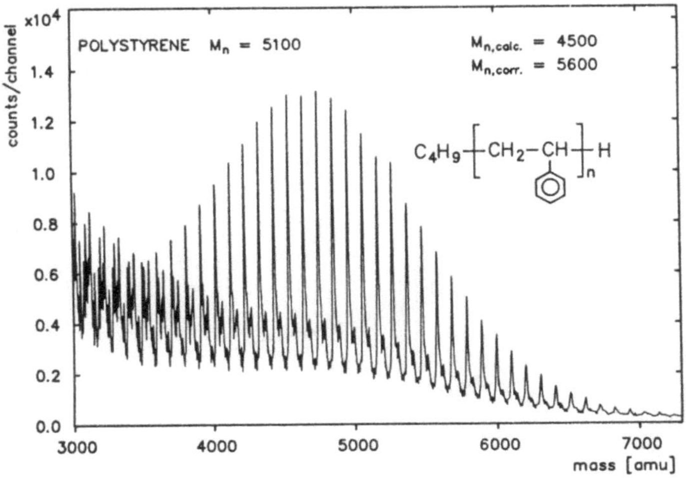

Fig. 2 TOF-SIMS spectrum of polystyrene of molecular weight 5100

a corrected Mn value of 5600 was obtained. Although approximate, the correction shifts the distribution in the right direction. For determination of more accurate molecular weight distributions, appropriate corrections for detection efficiencies are necessary.

In conclusion, it has been shown that TOF-SIMS can be used effectively to characterize polymers in the low and high range and obtain oligomer distributions.

This work was supported by the National Science Foundation under Grant CHE8411835 and by the Deutsche Forschungsgemeinschaft. We are particularly grateful to the Alexander von Humboldt Foundation for providing a senior fellowship for D.M.H., which stimulated this work.

References

1. Y. Sugimura, T. Nagaya, S. Tsunge, Macromolecules, 14, 520, 1981.

2. T. Matsuo, H. Matsuda, I. Katakuse, Anal. Chem., 51, 1329, 1979.

3. R. P. Lattimer, D. J. Harmon, G. E. Hansen, Anal. Chem., 52, 1808, 1980.

4. R. P. Lattimer, H.-R. Schulten, Int. J. Spectrom. Ion Phys., 52, 105, 1983.

5. U. Bhr, I. Luderwald, R. Muller, H.-R. Schulten, Angew. Makromol. Chem., 120, 163, 1984.

6. K. W. S. Chan, K. D. Cook, Org. Mass Spectrom., 18, 423, 1983.

7. B. T. Chait, J. Shpungin, F. H. Field, Int. J. Mass Spectrom. Ion Process, 58, 121, 1984.

8. D. Mattern, D. M. Hercules, Anal. Chem., 57, 2041, 1985.

9. J. Gardella, F. Novak, D. M. Hercules, Anal. Chem., 56, 1371, 1984.

10. D. Briggs, Surf. Interface Anal., 4, 151, 1982.

11. J. E. Campana, M. M. Ross, S. L. Rose, J. R. Wyatt, R. J. Colton in "Ion Formation from Organic Solids", A. Benninghoven (Ed.), Springer-Verlag, Berlin 1983, Vol. 25, pp. 144-155.

12. I. V. Bletsos, D. M. Hercules, D. Greifendorf, A. Benninghoven, Anal. Chem., in press.

13. P. Steffens, E. Niehuis, T. Friese, D. Greifendorf, A. Benninghoven, J. Vac. Sci. Technol., 3, 1322, 1985.

Secondary Ion Mass Spectrometry of Modified Polymer Films

S.J. Simko[1], *R.W. Linton*[1], *R.W. Murray*[1], *S.R. Bryan*[1], and *D.P. Griffis*[2]

[1] Department of Chemistry, University of North Carolina, Chapel Hill, NC 27514, USA

[2] Engineering Research Services, North Carolina State University, Raleigh, NC 27650, USA

1. Solvent-Induced Surface Functional Group Mobility in Chemically-Modified Polyethylene Determined by Static Fast Atom Bombardment Quadrupole Mass Spectrometry

The surface chemical modification of polymers influences a variety of properties important in commercial applications, e.g. adhesion, wettability, solubility, and biocompatibility. Chemical tagging reactions have been helpful in characterizing polymer surface functional groups on modified polymers using X-ray photoelectron spectroscopy (XPS) [1,2]. This approach will be demonstrated to have similar analytical advantages in studies using static fast atom bombardment mass spectrometry (FABMS).

Girards reagent P (GRP) ($NH_2NHCOCH_2N^+C_5H_5$) forms hydrazones with carbonyl groups on the surface of chromic acid-oxidized polyethylene (PE). The FAB mass spectrum of Girards reagent P reacted, acid-etched polyethylene (GRP-PE) is presented in Fig. 1A. The increase of intensity at mass 93 ($CH_2N^+C_5H_5$) is characteristic of the GRP tagging of carbonyl groups introduced on the PE during acid oxidation. Tag intensity variations for different samples are compared using ratios to hydrocarbon peaks (e.g. mass 55-$C_4H_7^+$) to compensate for small differences in primary beam flux and useful ion yields. The mass 93/55 ratio is 0.01 in PE and 0.27 in GRP-PE.

It has been shown that polymer-permeable solvents capable of strong H-bonding cause migration of surface functional groups to subsurface regions [2]. Extraction of GRP-PE with methanol causes a reduction in the mass 93/55 ratio to 0.07. This reflects migration of many GRP tagged carbonyl groups away from the outermost surface monolayers comprising the static FABMS sampling depth (Fig. 1B). The surface GRP concentration can be repopulated somewhat by soaking the methanol-treated film in phosphoric acid as shown in Fig. 1C. Reaction of phosphoric acid with amine groups on the GRP tag probably forms a polymer-insoluble ammonium salt which migrates to the polymer surface. The result is an increase in the mass 93/55 ratio from 0.07 to 0.13. Reaction of the phosphoric acid treated film with the GRP reagent results in a still greater apparent surface GRP concentration (mass 93/55 ratio of 0.46). This value is greater than the 93/55 ratio of the original GRP-PE. This may reflect an increase of unreacted surface carbonyl functionalities during the surface repopulation step with phosphoric acid. The SIMS results were substantiated by quantitative XPS data involving N,C,P atomic ratios.

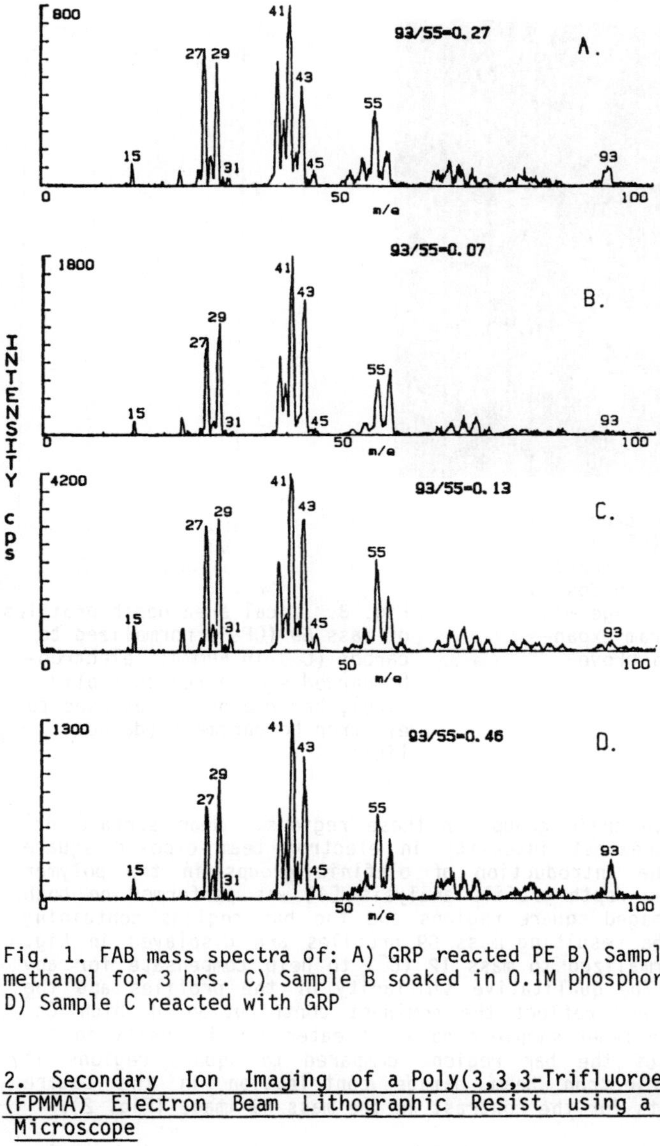

Fig. 1. FAB mass spectra of: A) GRP reacted PE B) Sample A extracted with methanol for 3 hrs. C) Sample B soaked in 0.1M phosphoric acid for 1.5 hrs. D) Sample C reacted with GRP

2. Secondary Ion Imaging of a Poly(3,3,3-Trifluoroethyl methacrylate) (FPMMA) Electron Beam Lithographic Resist using a Cameca IMS-3f Ion Microscope

Radiation sensitivity of FPMMA was characterized by Cs^+ bombardment/negative secondary ion imaging of a selectively electron irradiated film. Sample preparation was similar to that used previously for poly(methyl methacrylate) [3]. In essence, thin polymer films on optically flat, electrically conductive Au substrates were exposed to a 1 x 10^{-3} C/cm^2 electron dose through a minigrid mask. Exposed films were reacted in bromine vapor to chemically tag olefinic groups introduced in the film by electron bombardment. The mass 69 (CF_3^-) ion image is presented in Fig. 2. Intensity is strongest in the bar regions not exposed to electron bombardment, which reflects a greater concentration

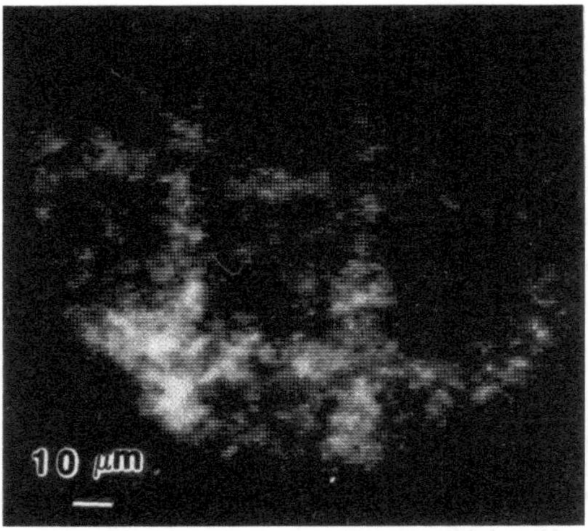

Fig. 2. Cs^+ bombardment/negative ion image of mass 69 (CF_3^-) acquired from electron bombarded FPMMA after a primary ion dose of $3x10^{16}$ ions/cm^2. The image was enhanced using histogram expansion and smoothed to improve contrast

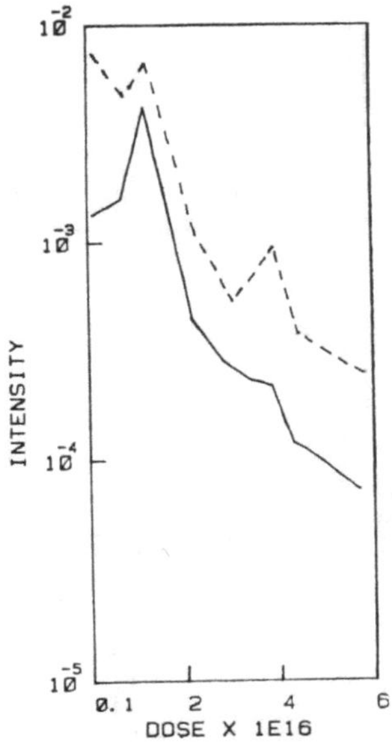

Fig. 3. Local area depth profiles of mass 69 (CF_3^-) normalized to carbon (C^-) in FPMMA: electron-bombarded square region (solid line), bar region not exposed to electron bombardment (dashed line)

of native polymer side chain groups in these regions. Near surface Br^- ion images exhibit greatest intensity in electron beam-exposed square regions, suggesting the introduction of olefinic groups in the polymer matrix. Selected area depth profile analysis [4] was performed on both the electron beam damaged square regions and the bar regions containing the native FPMMA. The resulting mass 69 profiles are displayed in Fig. 3. Profiles were normalized to mass 12 (C^-) to help compensate for any charging artifacts. The qualitative similarity of the profiles and the general decreasing trend reflect the dominant contribution of high Cs^+ primary ion doses to polymer sample damage. Greater ion intensity in the near-surface region of the bar regions compared to square regions is indicative of the electron beam damage contribution in the square regions. This accounts for the contrast in the mass 69 image (Fig. 2).

The above studies demonstrate that SIMS is a promising technique for the study of chemically modified polymer films having a high degree of spatial heterogeneity.

1. D.S. Everhart, C.N. Reilley: Anal. Chem. 53, 665(1980).
2. D.S. Everhart, C.N. Reilley: Surf. Interface Anal. 3. 126(1981).
3. S.J. Simko, S.R. Bryan, D.P. Griffis, R.W. Murray, R.W. Linton: Anal. Chem. 57, 1198(1985).
4. S.R. Bryan, D.P. Griffis, W.S. Woodward, R.W. Linton: "SIMS/Digital Imaging for the Three Dimensional Characterization of Solid State Devices", J. Vac. Sci. Technol. A, in press, 1985.

Application of SIMS Technique to Industrially Used Organic Materials

S. Tomita, K. Okuno, F. Soeda, and A. Ishitani

Toray Research Center, Inc., Sonoyama, Otsu, Shiga 520, Japan

1. Introduction

We reported in the SIMS IV conference establishment of a measurement condition for the analysis of low concentration elements within an organic polymer matrix [1]. Prevention of charge-up by gold coating, dynamic SIMS conditions with O_2^+ and Cs^+ primary ions to suppress molecular ions background and also to attain sensitivity and good depth resolution, were undertaken.
Thin gold film deposition can easily prevent charging up of a polymer matrix because of formation of a conductive thin glassy carbon film on the sputtered surface, which was confirmed by a Raman spectrum as indicated in Fig. 1.

This technique is finding rapidly expanding application area in wide variety of organic materials. Typical examples are illustrated in the paper.

2. Experimental

A CAMECA IMS-3F was used for measurement with Cs^+(14.5kV) and O_2^+(10.5kV). Primary ions were used for negative and positive secondary ion generation respectively. Measurement conditions were primary ion beam current: 0.8 to 7μA, rastered area: 500μm^2, analyzed area: 1.4 to 60μmϕ, vacuum: 10^{-8} to 10^{-9} Torr.

3. Results

Quality control: Quantitative comparison and spatial distribution of low concentration elements from reagents, catalysts, additives and contamination

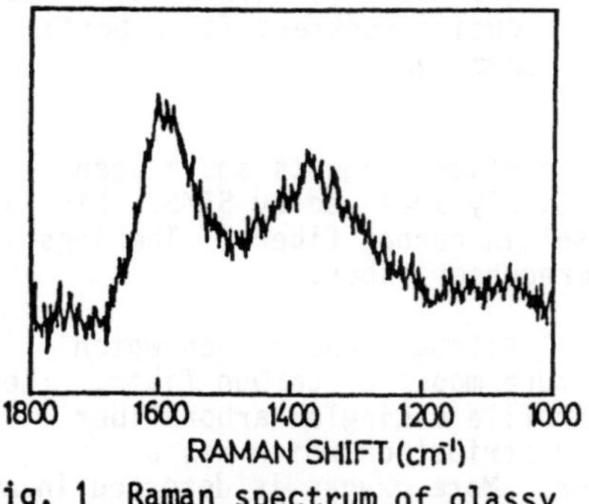

Fig. 1 Raman spectrum of glassy carbon generated on a sputtered surface of cellulose acetate by an O_2^+ primary ion beam

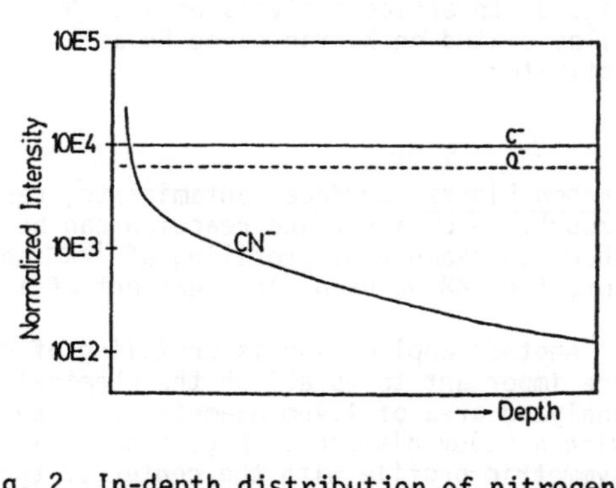

Fig. 2 In-depth distribution of nitrogen in an epoxy resin cross-linked by contact with a secondary amine aqueous solution

in polymers are possible by this technique. An example of depth profiling of an amine used for the cross-linking reaction of an epoxy resin is indicated in Fig. 2. An amine in aqueous solution penetrates into epoxy resin and participates in the reaction.

Metal-polymer interface: Analysis of the interface between polymers and metals plated, evaporated or laminated on them and depth profiling of migrating metal ions within polymer substrates are possible by SIMS. An example of Cu-Ni plated Kapton is given in Fig. 3. A thin layer of Pd nuclei is observed to migrate into the metal layer. A fair amount of copper diffuses deep into the polymer matrix.

Fine structure of blended polymers: The spatial distribution of a polymer with hetero-atoms like silicone and fluorinated polymers within blended system can be studied by SIMS. Polymers without label elements can also be examined by creating contrast with osmic acid dyeing.

An example of depth profiling of a silicone rubber printing plate which contains many kinds of additive polymers is illustrated in Fig. 4. Osmic acid dyeing was tried to detect hydrophilic components. The profile of $^{192}Os^+$ fits that of $^{48}Ti^+$. This reveals that the hydrophilic component is a titanic acid ester polymer.

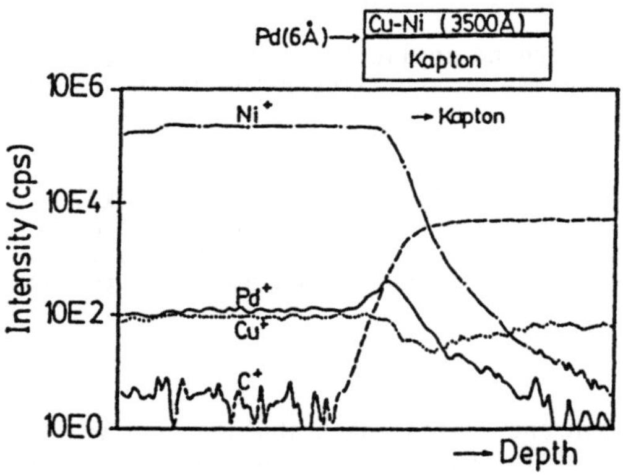

Fig. 3 Interface analysis of a Cu-Ni alloy plated on Kapton using Pd as an activator

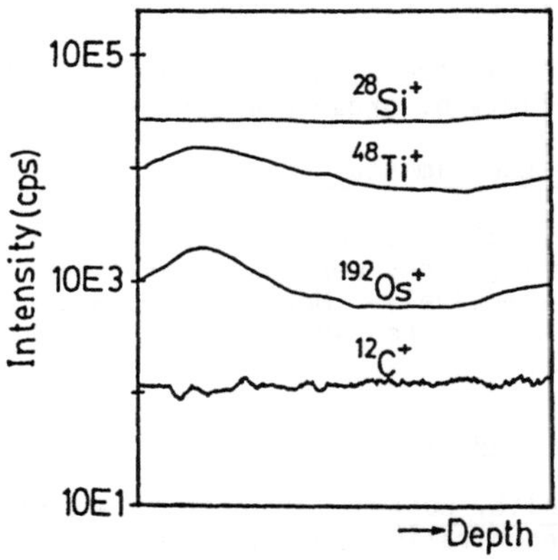

Fig. 4 Depth profiling of a silicone printing plate after dyeing Osmic acic for producing contrast for a particular component

Carbon Fibers: Surface contaminants, residual surface reagents and oxygen brought in by a surface reaction can be effectively analyzed by SIMS. Fig. 5 shows an example of profiling of $^{39}K^+$ and $^{55}Mn^+$ in carbon fibers. The ions come from $KMnO_4$ used for treatment of a PAN precursor fiber.

Another application is profiling of residual nitrogen and oxygen which are important to establish the chemical structure model of carbon fiber. The analyzed area of 1.4µm diameter was used to profile a single carbon fiber with a 5.5µm diameter. Fig. 6 shows an $^{16}O^-$ distribution which has a symmetric profile with the center of the fiber. More oxygen is detected in the skin area. Nitrogen was found to be homogeneously distributed.

Medical Applications: Low concentration metal ions play important roles in human tissues. High sensitivity and good lateral resolution of SIMS has high potentiality for medical application. An example of a study of turbidity of a human eye crystalline lens is given here. Multicharged metal ions are expected to generate cross-linking between proteins, causing turbidity. Fig. 7 shows depth profiling of major metal ions in a turbid lens. Aluminum and silicon indicate an interesting spatial correlation which may be related to the disease.

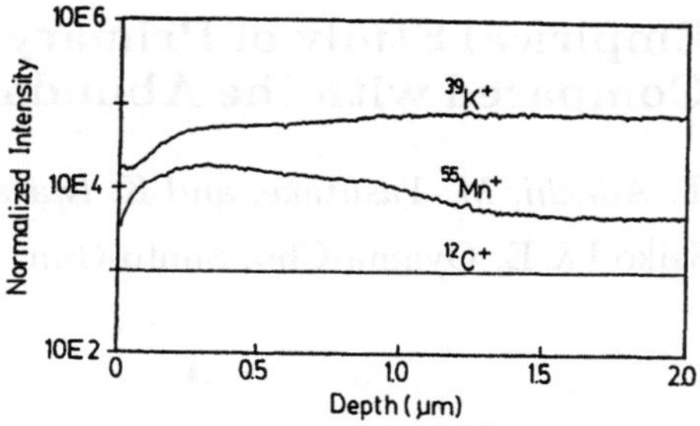

Fig. 5 Depth profiling of residual K and Mn on carbon fibers prepared from $KMnO_4$ treated PAN precursor

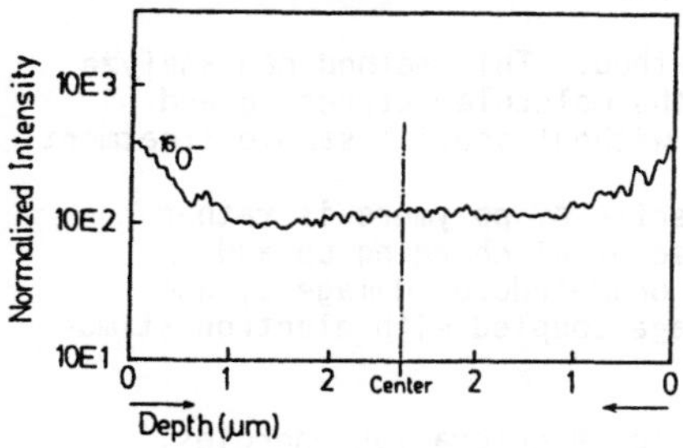

Fig. 6 Profiling of oxygen in a single carbon fiber with a well-focused analyzed area

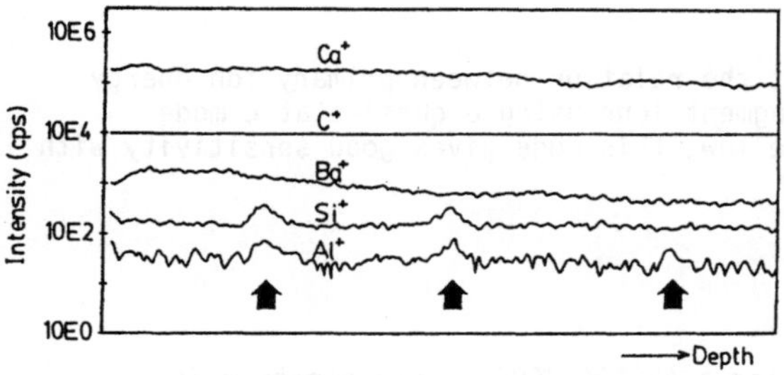

Fig. 7 Depth profiles of major metal ions detected in a turbid human eye crystalline lens

4. Reference

1. K. Okuno, S. Tomita and A. Ishitani: Proceeding 4th SIMS Conference, Osaka 392 (1984) Springer Verlag Ed.

Empirical Study of Primary Ion Energy Compared with the Abundance of Polymer Fragment Ion

T. Adachi, M. Yasutake, and K. Iwasaki

Seiko I & E, Oyama-Cho, Sunto-Gun, Shizuoka 410-13, Japan

1. Introduction

Recent trends show polymeric materials being used in more electronic parts. In these applications, surface, interface, and small area analysis as well as bulk analysis, are of increasing importance.

Static SIMS is one potential analysis method. This method can analyze not only impurity distributions, but also the molecular structure and spacial distribution of molecular ions[1], without special sample treatment.

Getting stable and reproducible SIMS spectra of polymers is rather difficult. These difficulties arise because of a) charging up and uncertain sample surface potential, b) ion beam-induced damage c) and charge neutralization electron-induced damage coupled with electron stimulated ion emisson (ESIE).
Other difficulties are as follows:

d) getting a stable ion beam current using low acceleration energies.
e) damage rates are different for every polymer species.
f) long measurement times are necessary to obtain sufficient secondary ion count rates.

In this paper, we will report the relation between primary ion energy and the abundance of polymer fragment ions using a quasi-static mode (1μA/cm^{-2}). If damage rates are low, this mode gives good sensitivity with short measurement times.

2. Experimental Conditions

Experiments were carried out using a A-DIDA 3000 quadrupole-type ion-microprobe. (ATOMIKA INC.). With this instrument, the acceleration voltages can be set anywhere from 0.3 keV to 15keV with a stable primary current, even in the 0.3keV acceleration voltage range. Primary ion conditions were Ar^+ with 10nA current, using a 200μm diameter beam, raster scanned over a 1mm square. For charge neutralization, we chose the electron flood method [2]. The electron gun used could be adjusted between 0.1keV and 5keV acceleration voltages.

The base pressure of the measurement chamber was 8 X 10^{-10} Torr and the working pressure was 5 X 10^{-9} Torr (10nA Ar^+). Measurement samples were polyethyleneterephthalate (PET $C_{10}H_8O_4$) using 100μm thick, 1cm squares cut from a sheet. Na, K, and Ca were added to the samples. These samples were mounted on a silicon wafer to maintain a flat surface.

3. Results and Discussion

First we investigated the electron-induced damage and ESIE. In consequence of the experiments adjusted the energy between 5keV and 0.3keV of the electron beam bombarded on the PET, the electron beam of 0.3keV, 300nA confirmed free from damage and ESIE.

Next, we checked for ion-induced damage and whether the sample surface potential was constant. To check the damage rate, we measured the main PET fragments, m/z 104 $[C_7 H_4 O]^+$ and m/z 76 $[C_6 H_4]^+$, abundances compared to the time. To check the surface potential stability, the electron gun was adjusted using the X, Y deflectors and focus controller to maximize secondary ions. Measuring the Ca^+ m/z 40 ion, we monitored the stability of the surface potential. As seen in Fig.1, the m/z 104 ion count decreased by about half after 100 seconds. At the same time, the Ca^+ ion kept constant during the entire measurement. This means that the greater part of the decay was derived from ion-induced damage.

We checked the relation between primary energy and fragment ion abundance using the same ion dose ($1\mu A/cm^{-2}$) bombardment conditions.
During measurement of mass spectra, the sample was damaged to the extent that we were not able to get true abundance mass spectra.
To reduce this effect, we had the following choices: (1) reduce the primary ion dose (2) rapidly scan the mass spectrum (3) constantly supply undamaged surface area.

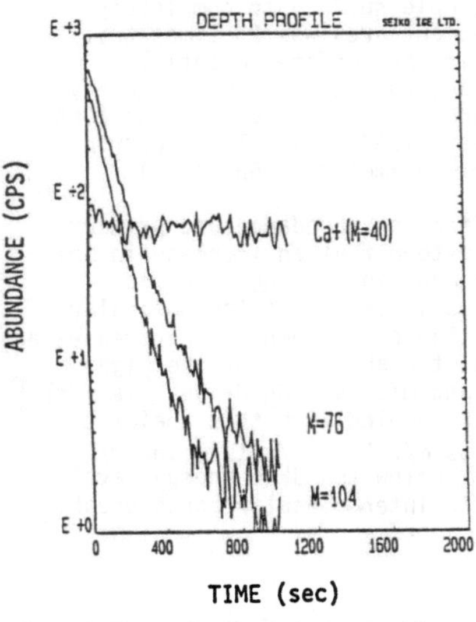

Fig. 1 The relations between time and secondary ion abundance
primary ion : Ar^+, 0.3 keV, 2.5nA, 1 mmsq. ($0.25\ \mu Acm^{-2}$)
electron : 0.2 keV, 60 nA ($0.3\ \mu Acm^{-2}$)

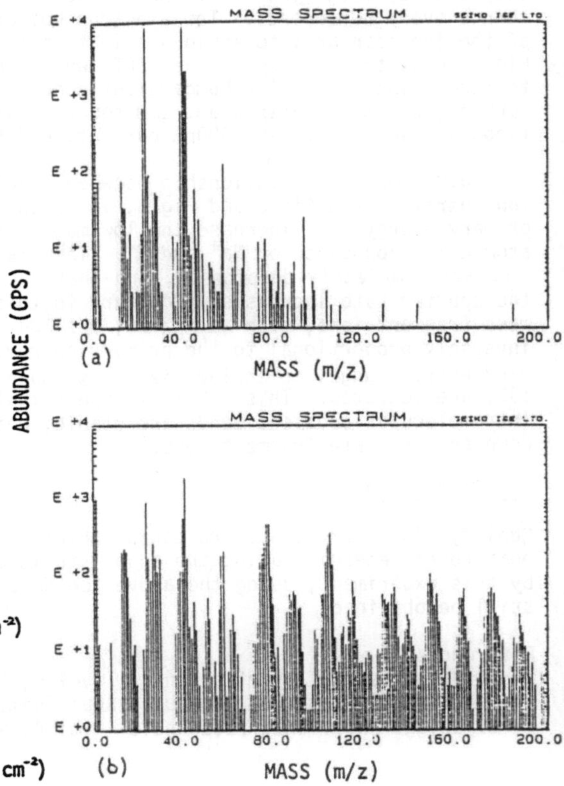

Fig. 2 Secondary ion mass spectra
primary ion : Ar^+, 10nA, 1 mmsq. ($1\ \mu A\ cm^{-2}$) primary ion energy :
(a) 12 keV, (b) 1 keV
electron : 0.3 keV, 300 nA ($1.5\ \mu A\ cm^{-2}$)

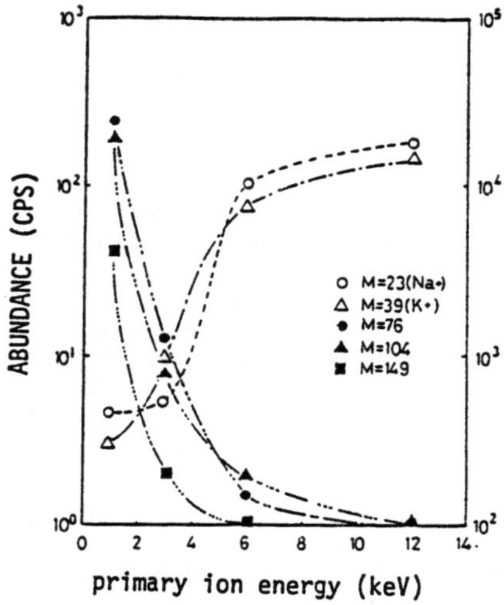

Fig. 3 The relations between the primary ion energy and secondary ion abundance
primary ion : Ar^+, 10 nA, 1 mmsq. (1 $uAcm^{-2}$)
electron : 0.3 keV, 300nA (1.5 $uAcm^{-2}$)

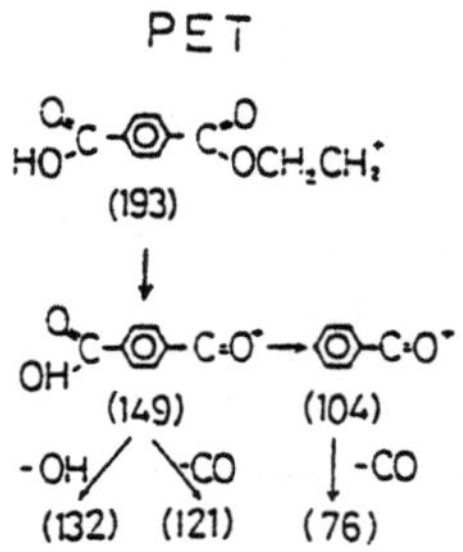

The first method had been tried in papers [3] and it requires long measurement times. The second method can not be adopted because of scan speed limitations. Therefore, we chose the third method. In this method, the sample stage moved just fast enough for the sample surface to completely change every 50 seconds. The electron bombardment area was 25 times that of the ion scan area to avoid a slight shift in the surface potential. Fig.2 shows the mass spectra of PET, while changing the primary ion energy to 12keV, and 1keV. Ion bombardment conditions were Ar^+ at a constant 10nA current, 200μm diameter beam, 1mm square raster (1μA/cm^{-2}). The electron flood was set at 0.3keV, 300nA current, and 5mm diameter (1.5μA/cm^{-2}).

Fig.3 shows the relationship between specific ion abundance and primary ion energy. From Fig.2 and Fig.3, it is understood that an increase in the primary energy will increase the low-mass fragment ions along with the atomic ion abundance of Na^+ and K^+. This makes it seem that the molecular ions are completely decomposed, and that the atomic ion abundance increases as the sputter rate increases. Contrary to this, the abundance of the high-mass fragment ions, such as m/z 76, m/z 104, and m/z 149 $[C_8 H_5 O_3]^+$ is inversely proportional to the primary energy. Coupled with this, below the 3keV energy range, molecular-like ions, such as m/z 191, m/z 192, and m/z 193, are detected. This makes it appear that, below the 3keV energy level, intramolecular covalent bonds are stable, while intermolecular bonds break down and generate fragment ions.

4. Conclusion

Quality finger-print spectra can be obtained from PET samples below 3keV bombardment energies using the quasi-static mode (1μA/cm^{-2}). As determined by this experiment, using the above ion dose conditions, molecular ions can still be obtained.

References

1) D. Briggs surf. Interface anal. vol 5, No.3, 1983.
2) J.A. Gardella and D.M. Hercules Anal Chem. 52, 226 1980.
3) D. Briggs surf. Interface anal. vol 4, No.3, 1982.

Comparison of Laser Mass Spectra Obtained at Ambient Conditions vs. Sample Freezing

J.J. Morelli, F.P. Novak, and D.M. Hercules

Department of Chemistry, University of Pittsburgh,
Pittsburgh, PA 15260, USA

1. Introduction

A significant advantage of laser mass spectrometry (LMS) is its capability to obtain spatial distributions of analytes within organic matrices [1-2]. Our interest in this area has focused on developing LMS for studying organic components within organic matrices. Many systems of interest contain volatile materials not readily suited to LMS analysis [3]. We have utilized two approaches to obtain LM spectra of volatile materials. The first approach uses a cold probe to freeze volatile organics, significantly reducing their vapor pressure. The second employs a thin formvar [poly(vinyl formal)] film as a seal between the sample and the instrument vacuum [3]. Formvar films employed in this manner allow laser mass analysis to be performed at atmospheric pressure; furthermore, no change in the physical state of the sample occurs when using these films. The potential advantage of using formvar lies in the ability to obtain LMS of liquids and solutions as in FAB, whereas, LMS analysis using the cold probe is performed with the samples in the solid state.

Comparing LM spectra of organic compounds in the liquid state with those of identical compounds in the solid state should be useful. This comparison can be realized by employing formvar films and the cold probe separately; therefore, the LMS of seven organic compounds were obtained in both liquid and solid states. The compounds studied represent a variety of organic functionalities. Here we present the preliminary results of this work.

2. Experimental

Atmospheric pressure laser mass spectrometry was accomplished by using a thin formvar film (< 1 μm) as a seal between the vacuum (2×10^{-5} Torr) in a LAMMA-1000 sample chamber and the analyte [3]. This film was formed by submerging a standard glass slide used in optical microscopy in a 0.4% formvar solution and allowing it to dry. The resulting film was peeled off the slide by the surface tension of water, where it remained floating. Analytes of interest were then placed on the film, which in turn was placed on a piece of Zn foil having slightly smaller dimensions than the film. The formvar adhered to the Zn foil, sealing the sample. Details of the technique for obtaining LMS at atmospheric pressure will be published [3].

Sample freezing was accomplished by using a cold probe, designed to allow liquid nitrogen to flow through the probe and vent to atmosphere, thus freezing volatile analytes mounted to the front of the probe and reducing their vapor pressure. Details of the cold probe will be published [4].

Laser mass spectra of the following compounds were obtained using both formvar films and the cold probe: methanol <u>1</u> (m.w. 32), ethanol <u>2</u>

(m.w. 46), acetic acid 3 (m.w. 60), benzaldehyde 4 (m.w. 106), benzyl alcohol 5 (m.w. 108), n-amyl acetate 6 (m.w. 130), and di-n-butylpiperidylphosphoramidate 7 (m.w. 277). The numbers underlined will be used to refer to these compounds for the purpose of subsequent discussion. All compounds were of analytical grade and used without further purification. Compound 7 was obtained from Alfa products; its structure is indicated below.

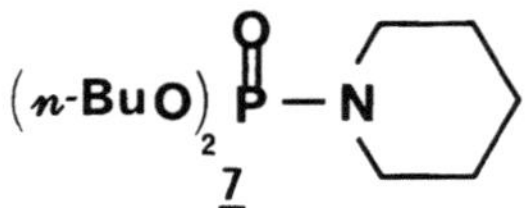

7

All LM spectra were obtained with a LAMMA-1000 laser microprobe mass spectrometer manufactured by Leybold-Heraeus, Koln, FRG; it is described elsewhere [5].

3. Results and Discussion

The types of ions observed in the LM spectra of the compounds analyzed (see Experimental) were primarily molecular-like ions, cluster ions, and structurally relevant fragments. Table I summarizes the molecular-like and fragment ions seen in LMS using both formvar films and the cold probe. The abundance of molecular-like ions can often be enhanced dramatically with the addition of an acid or salt prior to analysis. In this study, LiBr was used to evaluate the effect that such promoters have on the positive ion LM spectra of 7 obtained with the cold probe and at atmospheric pressure. The presence of $[M+Na]^+$ m/z 300 and $[M+K]^+$ m/z 316 in the spectra (Table I) were due to sodium and potassium contamination. The $[M+Zn]^+$ m/z 341 was due to Zn cationization from the substrate employed (see Experimental). One significant difference between the positive LMS obtained using the cold probe compared to that obtained using formvar films was the

Table 1 Quasimolecular and fragment ions observed in positive ion LMS[a]

ambient conditions										
compound #	[3]		[4]		[5]		[6]		[7]	
	m/z	int	m/z	int	m/z	int	m/z	int	m/z	int
quasimolecular ions										
M	x	x	106	w	b	b	x	x	x	x
M+H	61	m	107	w	b	b	131	w	278	s
M-H	x	x	105	w	b	b	x	x	x	x
M+Na	x	x	x	x	x	x	x	x	300	m
M+Li[c]	N.A.		x	x	N.A.		N.A.		283	m
M+K	x	x	x	x	x	x	x	x	316	m
M+Zn	x	x	x	x	x	x	x	x	341	?
fragment ions										
	43	w	91	s			115	s	86	s
			77	m			215	s	222	s
			78	w			71	m		
							61	m		
							43	m		

cold probe										
compound #	[3]		[4]		[5]		[6]		[7]	
	m/z	int	m/z	int	m/z	int	m/z	int	m/z	int
quasimolecular ions										
M	x	x	106	?	108	?	x	x	x	x
M+H	61	m	107	?	109	w	131	s	278	m
M-H	x	x	105	?	107	m	x	x	x	x
M+Na	x	x	x	x	x	x	x	x	300	?
M+Li[c]	N.A.		N.A.		N.A.		N.A.		283	?
M+K	x	x	x	x	x	x	x	x	316	?
M+Zn	x	x	x	x	x	x	x	x	341	?
fragment ions										
	43	w	91	m	91	s	115	m	222	m
			77	w	79	s	71	m	86	s
					77	w	61	m		
							43	m		

a: intesities are given as (s) strong = signal > 50mv, (m) medium = signal >10mv, (w) weak = signal <10mv.
b: chemical noise of backround prevented unequivocal observation of ions
c: applicable only to samples predoped with LiBr otherwise indicated as N.A. (not aplicable)
?: the appearance of this ion was not seen in >10% of the LM spectra obtained
x: this ion was not observed

inconsistency in cationized quasimolecular ions observed with the cold probe. An explanation for this inconsistency is that during the freezing process lithium, sodium, and potassium salts are segregated spatially from the analyte. During analysis, the sample is relatively heterogeneous; under ambient conditions, the sample remains liquid, providing a more homogeneous matrix. The presence of $[M+Zn]^+$ m/z 341 is inconsistent at both ambient and freezing conditions. The reason for this lies in the fact that the presence of $[M+Zn]^+$ is a function of the laser power density and sample thickness. The laser power density was controlled somewhat by using the same laser energy and keeping the focus on the sample constant; however, no measures were taken to insure constant sample thickness.

The cluster ions observed involved either water solvation or solvation of an analyte ion by an analyte molecule. Water solvation resulted in $[M+C+nH_2O]^+$ ions, where **n** varied from 1 to 4, and **C** was a positively charged species, H^+, Na^+, or Zn^+. Solvation of an analyte ion by an analyte molecule resulted in $[nM]^+$ or $[nM+C]$ ions. Again, **C** is H^+ or Na^+; however, **n** varied between 1 and 3. Occasionally $[nM+C+mH_2O]^+$ ions were observed for 6 when analyses were performed with the cold probe; however, the abundances of these ions were extremely weak making assignment dubious. Solvation of ions by analyte molecules was the primary process involved in cluster formation for the non-polar species (i.e., compounds 4, 5, and 7). Figure 1 shows a typical example of solvation within an analyte, in this case the positive ion LMS of 5 obtained using the cold probe. The group of peaks at m/z 108 corresponds to the molecular and molecular-like species of 5 (m.w. = 108) : $[M-H]^+$, $[M]^{\cdot+}$, and $[M+H]^+$ at m/z 107, 108, and 109, respectively. There is also a peak at m/z 105; this fragment corresponds to production of the acylium species $[C_6H_5CO]^+$ identical to the molecular-like ion observed for 4 at m/z 105 (see Table I). Further examination of Fig. 1 indicates the presence of a group of peaks at m/z 217; these correspond to the dimer of 5. Between m/z 109 and 215 there is a series of fragment ions which correspond to combinations of fragments m/z 77, 79, 91, and monomer ions. The ions seen between m/z 109 and m/z 215 in Fig. 1 will be referred to as combination fragments for the purpose of further discussion.

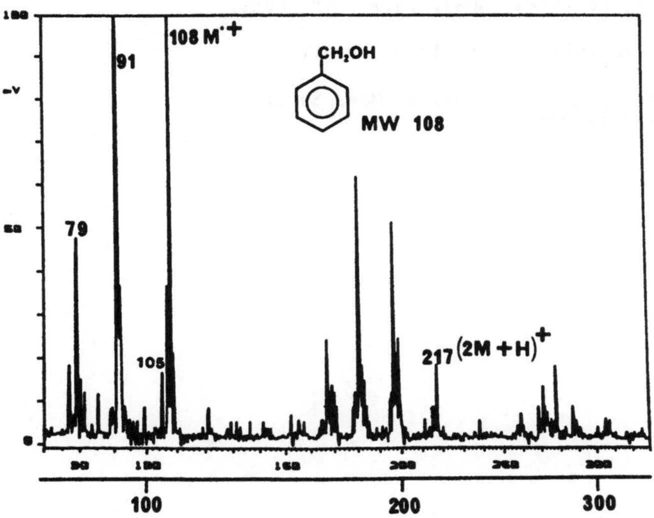

Fig. 1 Positive ion LMS of benzyl alcohol obtained with cold probe. Laser energy - 12 μJ. Defocused 300 μm.

These combination fragments can result from two possible pathways. The first mechanism would require an analyte cluster to be desorbed with subsequent fragmentation producing combination ions. An alternate process would involve ion/molecule reactions, making the term "recombination" a more accurate description. We cannot distinguish between these two mechanisms at this time. Due to combination fragment ions, the spectrum of 5 becomes increasingly complex with the production of higher-order clusters (i.e., 3M in Fig. 1).

Compounds 1 and 2 could not be analyzed successfully. This was because ions expected during LMS analyses of these alcohols could not be separated from ions resulting from Zn, formvar, or contamination present on the cold probe or formvar film. The interference encountered due to matrix fragmentation or contamination is referred to as chemical noise. Chemical noise was observed at laser energies above 2.5 μJ with the laser focused on either alcohol; no signal could be obtained from these alcohols at laser energies below 2.5 μJ. No improvement could be observed by defocusing the laser beam from the sample and increasing the laser energy.

4. Conclusions

Both the cold probe and atmospheric pressure laser mass analysis techniques were successfully employed to analyze a variety of volatile organic compounds. The advantage obtained from leaving the analyte in its liquid state was the increased homogeneity of the liquid analyte and whatever was dissolved or associated with it. The successful use of formvar films for analyzing liquid organics prompts further investigation in liquid-LMS.

References

1. W. H. Schroder, G. C. Fain, Nature, 309, 268-270 (1984).

2. F. P. Novak, Z. A. Wilk, D. M. Hercules, J. Trace Microprobe Tech., 3, 149-163 (1985).

3. J. J. Morelli, D. M. Hercules, Anal. Chem., in press.

4. F. P. Novak, D. M. Hercules, unpublished results.

5. H. J. Heinen, S. Meir, H. Vogt, Int. J. Mass Spec. Ion Phys., 47, 19-22 (2983).

Surface Analysis by Laser Desorption of Neutral Molecules with Fourier Transform Mass Spectrometry Detection

R.T. McIver, Jr., M.G. Sherman, D.P. Land, J.R. Kingsley, and J.C. Hemminger

Department of Chemistry, University of California, Irvine, CA 92717, USA

1. Introduction

We have developed a new instrument that utilizes a pulsed laser to desorb neutral molecules from a surface and a Fourier transform mass spectrometer (FTMS) to detect ions that are produced by electron impact or laser ionization of the desorbed species [1,2]. The Fourier transform detection method can acquire a complete mass spectrum after each laser pulse, and the mass resolution is 10 to 100 times greater than can be achieved with time-of-flight (TOF) detection. With FTMS the signal due to laser-desorbed ethylene (m/z 28.031) can be mass-resolved from that due to background nitrogen (m/z 28.006) or possible silicon (m/z 27.976) impurities in the crystal.

In this paper we report results for laser desorption of several organics from a platinum single crystal, and these are compared with SIMS spectra. For several aromatic molecules, efficient resonance-enhanced multiphoton ionization of the desorbed species has been observed.

2. Experimental

The Fourier transform mass spectrometer and UHV chamber have been described previously [1,2]. Basically, a platinum single-crystal is positioned in front of a hole in one of the plates of a FTMS analyzer cell, and a laser beam (4 mJ, 20 ns pulse at 249 nm) enters the vacuum chamber through a window directly opposite the crystal. The FTMS analyzer cell is centered between the pole caps of a large electromagnet (1.1 T). There is a filament mounted on one of the plates of the analyzer cell, and an electron beam is pulsed through the analyzer cell, parallel to the direction of the magnetic field and 3.1 cm in front of the crystal. When the laser beam strikes the crystal, a small area (0.5 mm^2) is rapidly heated and molecules bound to the surface are desorbed. Some of these are ionized as they traverse the path of the electron beam. The magnetic field traps the ions and causes them to move in small cyclotron orbits. The ions are then accelerated by a radiofrequency pulse, and the coherent image current signals that are produced by their cyclotron motion are detected. Fourier transform analysis of the image current signals yields the mass spectrum.

Prior to each experiment,the platinum crystal was cleaned by heating it for 5 minutes in oxygen at 1100 K and then annealed at 1300 K to remove residual oxygen. The crystal was then cooled and an adsorbate of interest was admitted to the vacuum chamber through a variable leak valve. After a given exposure, the leak valve was closed and the remaining gas was pumped away. Once the pressure in the vacuum chamber had stabilized, the laser-desorption experiments were begun.

3. Results and Discussion

Figure 1 shows typical FTMS data for laser desorption of cyanogen off a platinum crystal at a laser power of 40 MW/cm^2. The narrowband detection mode of FTMS was used in order to maximize the detection sensitivity in the region around the molecular ions of interest. In the narrowband mode the mass resolution is routinely greater than 30,000. Similar results were obtained for ethylene, methanol, carbon monoxide, carbon dioxide, acetone, acetic anhydride, and isobutane adsorbed on platinum.

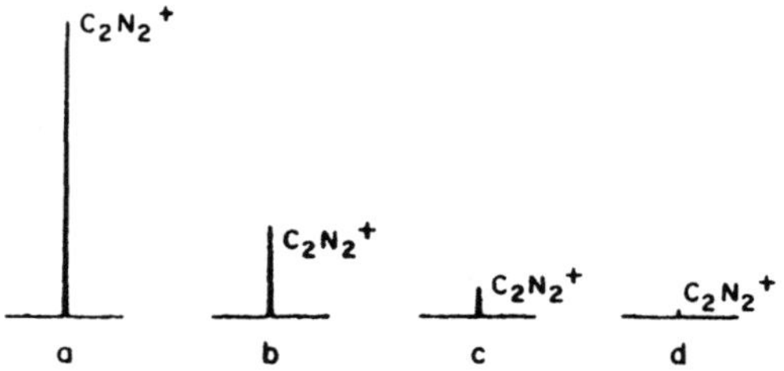

Fig. 1 Laser desorption FTMS mass spectra of C_2N_2 (12.6 L, 150K) on platinum: (a) after first laser pulse, (b) after second laser pulse at the same spot, (c) after a third laser pulse at the same spot, and (d) backgroun spectrum with laser off

In all the cases mentioned, most of the adsorbed molecules are removed by the first laser pulse. Typically, a second laser pulse at the same spot produces a signal that is only 20 percent as large as the first, and the third laser pulse produces an even smaller signal. Also, laser desorption of the neutrals followed by electron ionization produces remarkably simple mass spectra. Molecular ions of the adsorbates were the predominant peaks in the FTMS mass spectra. Figure 1 shows only the molecular ion region of cyanogen, but other experiments performed using broadband FTMS detection showed only a few additional peaks and all were readily assigned as EI fragments of the adsorbates.

Ions are produced at the surface during a SIMS experiment, and many secondary ions are produced by rapid ion-molecule reactions (clustering, charge exchange, and displacement) that occur directly above the surface between adsorbates and the substrate ions. In contrast, the FTMS spectra showed no evidence of reactions between the adsorbates and platinum surface. For the cases mentioned above, ions are not produced by the laser alone at the low-power levels used. This may account for the differences between our results and the SIMS results. It appears that cationization ions do not form in the laser desorption FTMS experiments because the ions are formed after the desorbed molecules have moved away from the surface and have expanded into the vacuum to a much lower density.

Multiphoton Ionization of Adsorbates: We found recently that even a low-power laser pulse at a wavelength of 249 nm produces intense ionization of benzene, naphthalene, phenol, toluene, trimethylamine, and methyl benzoate adsorbed on platinum. It appears that the ions are made by resonance-enhanced two-photon ionization of the adsorbates as they depart from the surface. Our evidence is as follows:

(1) A thermal ionization mechanism is ruled out because ions are not formed directly when 350 nm radiation is used. Since the temperature jumps caused by the 249 nm and 350 nm laser pulses are similar, the formation of ions does not appear to be related to the surface temperature.

(2) Multiphoton ionization of background molecules is ruled out because ions are not observed when the focal point of the laser beam is moved several centimeters away from the surface of the crystal. Also, after three laser pulses at the same spot on the surface, the ion yields are very small.

(3) The ionization potentials of benzene (9.25 eV) and naphthalene (8.12 eV) are such that two photons at 249 nm are sufficient for ionization but two photons at 350 nm are not.

Multiphoton ionization of laser-desorbed molecules has been previously reported [3-6]. Our results provide further support for an electronic process rather than a thermal process for the formation of ions.

Sensitivity of FTMS: Some experiments have been done to test the sensitivity of the laser desorption FTMS method. For carbon monoxide adsorbed on platinum, the detection limit with electron ionization is approximately 0.002 monolayer for an irradiated area of 0.5 mm^2. Multiphoton ionization of naphthalene at 249 nm can detect approximately 3×10^{-4} monolayer for the same irradiated area in a single laser pulse .

4. Conclusions

Laser desorption of neutral molecules followed by electron ionization or multiphoton ionization is a promising method for detecting molecular adsorbates. By separating the desorption and ionization processes, rapid ionic reactions directly above the surface are avoided, and easily interpretable mass spectra are obtained. Fourier transform mass spectrometry provides high-mass resolution and the ability to get a complete mass spectrum for each laser pulse. In addition, high ion-detection sensitivity results, because all ions formed by the electron beam are trapped in the analyzer cell and detected.

Acknowledgements

This work was supported by the National Science Foundation (CHE8511999) and by the Petroleum Research Fund administered by the American Chemical Society.

References

1. M.G. Sherman, J.R. Kingsley, D.A. Dahgren, J.C. Hemminger, and R. T. McIver, Jr.: Surf. Sci. 148, L25 (1985)
2. M.G. Sherman, J.R. Kingsley, J.C. Hemminger, and R.T. McIver, Jr.: Analytica Chimica Acta, in press
3. D.M. Lubman and R. Naaman: Chem. Phys. Lett. 95, 325 (1983)
4. R.B. Opsal and J.P. Reilly: Chem. Phys. Lett. 99, 461 (1983)
5. J.T. Meek, S.R. Long, R.B. Opsal and J.P. Reilly: Laser Chem. 3, 19 (1983)
6. S.E. Egorov, V.S. Letokhov and A.N. Shibanov: Chem. Phys. 85 349 (1984)

Index of Contributors

The individuals whose names are marked with an asterisk in this index are acknowledged for their assistance in reviewing manuscripts for this volume.

MIX
Papier aus verantwortungsvollen Quellen
Paper from responsible sources
FSC® C105338

If you have any concerns about our products,
you can contact us on
ProductSafety@springernature.com

In case Publisher is established outside the EU,
the EU authorized representative is:
Springer Nature Customer Service Center GmbH
Europaplatz 3, 69115 Heidelberg, Germany

Printed by Libri Plureos GmbH
in Hamburg, Germany